Psychrophiles: From Biodiversity to Biotechnology

Rosa Margesin
Editor

Psychrophiles: From Biodiversity to Biotechnology

Second Edition

Springer

Editor
Rosa Margesin
Institute of Microbiology
University of Innsbruck
Innsbruck, Austria

ISBN 978-3-319-86067-1 ISBN 978-3-319-57057-0 (eBook)
DOI 10.1007/978-3-319-57057-0

The 1st edition © Springer-Verlag Berlin Heidelberg 2008 was edited by Rosa Margesin, Franz Schinner, Jean-Claude Marx and Charles Gerday.

Printed on acid-free paper

This Springer imprint is published by Springer Nature
The registered company is Springer International Publishing AG
The registered company address is: Gewerbestrasse 11, 6330 Cham, Switzerland

Preface to the Second Edition

The Earth is a cold planet. The majority of the Earth's biosphere is exposed to temperatures below 5 °C throughout the year. Cold ecosystems are very diverse and range from high mountains to deep oceans. Even the most extreme cold and frozen environments harbour enormously diverse and metabolically active microbial populations. The key feature of microorganisms living in these habitats is the successful surmounting of the negative effects of low temperatures. Cold adaptation includes a complex range of structural and functional adaptations at the level of all cellular constituents. Cold-adapted microorganisms play a key ecological role in their natural habitats, and their mechanisms and strategies to cope with the cold render them particularly useful for biotechnological applications.

Why a second edition of the book "Psychrophiles: from Biodiversity to Biotechnology"? The first edition of this book was published in 2008. Since then, there was a continuously increasing interest in the microbiology of the Earth's cryobiosphere. This can be attributed to several factors, such as the awareness of the consequences of the global climate change on the response of microorganisms in cold ecosystems, the development of the emerging fields of omics approaches enabling the study of the versatility and functional roles of life in the cold biosphere, the interest in the existence of extraterrestrial life and the need for environmentally friendly and sustainable perspectives in biotechnology.

The second edition of the book presents, at the leading edge of knowledge, the newest insights into the psychrophilic lifestyle. Both fundamental and applied aspects are considered. All chapters were written by leading scientists and authorities in the respective field. Some of the chapters are updated contributions to the first edition of the book, while several newly introduced chapters outline topics of interest emerging in the last decade. The book is organized into four thematic Parts:

Part I, "Boundary Conditions for Microbial Life in the Cold", addresses the conditions that influence the low-temperature limit of microbial life. Climatic factors and biological constraints in the cold are discussed in Chaps. 1 and 2.

Part II, "Microbial Diversity and Activity in Cold Ecosystems", presents the current knowledge of microbial communities in the atmosphere and in glacial environments, deep oceans and permafrost. The diversity and activity of bacteria, archaea and eukaryotes as well as of viruses in these habitats are described in Chaps. 3–9.

Part III, "Molecular and Physiological Adaptations to Cold Habitats", is focused on the characteristics and roles of the key features of microbial cold adaptation, such as cold-active enzymes, cryoprotectants, exopolysaccharides and cell membrane components, and reports the latest developments in genomics, metagenomics, proteomics, transcriptomics and metatranscriptomics of psychrophiles. These topics are outlined in Chaps. 10–18.

Part IV, "Biotechnological Perspectives", considers the significance of psychrophiles in various biotechnological applications. The role of psychrophiles as a source of novel cold-active enzymes, antimicrobials and other biopolymers, as expression systems for recombinant protein production, in biomedical applications and in the bioremediation of contaminated soil and water is outlined in Chaps. 19–29.

I wish to express my gratitude to all authors for their excellent contributions. I also thank the Springer team, especially Hanna Hensler-Fritton, Isabel Ullmann and Dr. Andrea Schlitzberger, for their valuable and continuous support during the preparation of this book.

Innsbruck, Austria Rosa Margesin
March 2017

Preface to the First Edition

Most scientists in the middle of the twentieth century would probably not have believed that life was possible at extreme values of environmental factors, such as pH values close to 0 (e.g. sulphurous environments) or to 14 (e.g. soda lakes), salinities of 6 M NaCl (e.g. Dead Sea), hydrostatic pressures approaching 0.1 MPa (deep sea) and temperatures exceeding 100 °C (thermal vents or hot springs) or as low as −20 °C (e.g. polar regions). Of the current studies on extremophiles, approximately 30,000 articles by the year 2007, almost two-thirds have been performed on organisms adapted to outstanding temperatures, but much more attention has been paid to thermophiles than to psychrophiles. However, over the past 10 years, scientific publications on cold-adapted microorganisms have increased by a factor of ten.

If one considers the extent of cold habitats, psychrophiles, i.e. cold-loving organisms, should largely lead in this comparison with thermophiles because a great proportion of the Earth's biosphere never reaches temperatures above 5 °C. Nearly three-quarters of the Earth is covered by oceans whose deep water masses, irrespective of latitude, are constantly between 2 and 4 °C. The large continent of Antarctica also provides a permanently cold terrestrial environment as well as an aquatic niche in the surrounding ice that melts during the summer. Other examples of cold habitats in polar and alpine regions are permafrost soils, high alpine soils, cold deserts, cold caves, marine sediments, snow, glacier and sea ice. Cold ecosystems host a wide diversity of psychrophiles, including bacteria, archaea, yeasts, filamentous fungi and algae. These microorganisms have evolved a number of strategies to thrive successfully in cold habitats where they play key roles in nutrient cycling, such as nitrogen fixation, nitrification and denitrification, photosynthesis, sulphur oxidation and reduction, methanogenesis and transformation of organic compounds.

This book is focused on psychrophiles and describes, at the edge of knowledge, representative groups of cold-adapted microorganisms as well as the habitats in which they live and their strategies to cope with the cold. It is subdivided into four main sections:

1. Boundary conditions for microbial life at low temperatures
2. Biodiversity
3. Molecular adaptations
4. Biotechnological aspects

thus covering almost all the fields of knowledge in "cold" microbiological research.

It is certainly not by chance that this book is published during the International Polar Year 2007–2008, which is the fourth polar year following those in 1882–1883, 1932–1933 and 1957–1958 and involving over 200 projects, with thousands of scientists from over 60 nations examining a wide range of physical, biological and social research topics. Therefore, this book perfectly matches the current demands and trends and provides an additional source of information to all those scientists who are interested in "cold" microbiology.

Last but certainly not least, the editors of this book want to thank all the authors, who are the leading scientists in the respective field, for having accepted to write a chapter of this book, even though all these persons are also very busy and highly solicited scientists. We also thank Springer—Life Sciences, especially Dr. Dieter Czeschlik and Dr. Jutta Lindenborn, for their continuous support and trust in our capacity to successfully achieve the editing of this book.

Innsbruck, Austria — Rosa Margesin
Innsbruck, Austria — Franz Schinner
Liege, Belgium — Jean-Claude Marx
Liege, Belgium — Charles Gerday
July 2007

Contents

List of Contributors

Amanda A. Achberger Department of Biological Sciences, Louisiana State University, Baton Rouge, LA, USA

Habibu Aliyu Centre for Microbial Ecology and Genomics, University of Pretoria, Pretoria, South Africa

Ianina Altshuler Department of Natural Resource Sciences, McGill University, Ste-Anne-de-Bellevue, QC, Canada

Corien Bakermans Altoona College, The Pennsylvania State University, Altoona, PA, USA

Mário Barroca Centre of Molecular and Environmental Biology (CBMA), Department of Biology, University of Minho, Braga, Portugal

Vladislav Bezrukov State Institute of Gerontology of National Academy of Medical Sciences of Ukraine, Kiev, Ukraine

Erik Borchert School of Microbiology, University College Cork, National University of Ireland, Cork, Ireland

John P. Bowman School of Land and Food, University of Tasmania, Hobart, TAS, Australia

Odd Gunnar Brakstad SINTEF Ocean, Marine Environmental Technology, Trondheim, Norway

Evgeny Brenner Institute of Chemical Biology and Fundamental Medicine of the Siberian Branch of the Russian Academy of Science, Novosibirsk, Russia

Anatoli Brouchkov Tyumen State University, Tyumen, Russia

Lomonosov Moscow State University, Moscow, Russia

Pietro Buzzini Department of Agricultural, Food and Environmental Sciences, University of Perugia, Perugia, Italy

Karen A. Cameron Aberystwyth University Interdisciplinary Centre for Environmental Microbiology (AU iCEM), Aberystwyth University, Aberystwyth, UK

Angela Casillo Department of Chemical Sciences, University of Naples Federico II, Naples, Italy

Brent C. Christner Department of Microbiology and Cell Science, Biodiversity Institute, University of Florida, Gainesville, FL, USA

Tony Collins Department of Biology, Centre of Molecular and Environmental Biology (CBMA), University of Minho, Braga, Portugal

Maria Michela Corsaro Department of Chemical Sciences, University of Naples Federico II, Naples, Italy

Donald A. Cowan Centre for Microbial Ecology and Genomics, University of Pretoria, Pretoria, South Africa

Lewis Cuthbertson Department of Applied Biology, Faculty of Health and Life Sciences, Northumbria University, Newcastle upon Tyne, UK

Jody W. Deming School of Oceanography, University of Washington, Seattle, WA, USA

M. J. Dempsey School of Research, Enterprise and Innovation; Faculty of Science and Engineering, Manchester Metropolitan University and Advanced Bioprocess Development Ltd., Manchester, UK

Alan D.W. Dobson School of Microbiology, University College Cork, National University of Ireland, Cork, Ireland

Environmental Research Institute, University College Cork, National University of Ireland, Cork, Ireland

Arwyn Edwards Institute of Biological, Rural and Environmental Sciences, Aberystwyth University, Aberystwyth, UK

Josef Elster Centre for Polar Ecology, Faculty of Science, University of South Bohemia, České Budějovice, Czech Republic

Institute of Botany, Academy of Sciences of the Czech Republic, Třeboň, Czech Republic

Haluk Ertan School of Biotechnology and Biomolecular Sciences, The University of New South Wales, Sydney, NSW, Australia

Department of Molecular Biology and Genetics, Istanbul University, Istanbul, Turkey

Nobuyoshi Esaki Institute for Chemical Research, Kyoto University, Uji, Kyoto, Japan

Sara Filippucci Department of Agricultural, Food and Environmental Sciences, University of Perugia, Perugia, Italy

Andrew G. Fountain Department of Geology, Portland State University, Portland, OR, USA

Nadesha Fursova State Research Center for Applied Microbiology and Biotechnology, Obolensk, Moscow Region, Russia

Oksana Fursova Lomonosov Moscow State University, Moscow, Russia

Charles Gerday Laboratory of Biochemistry, University of Liège, Institute of Chemistry, Liège, Belgium

Jacqueline Goordial Department of Natural Resource Sciences, McGill University, Macdonald Campus, Sainte-Anne-de-Bellevue, QC, Canada

Bigelow Laboratory for Ocean Sciences, Deep Biosphere Lab, East Boothbay, ME, USA

Charles W. Greer Energy, Mining and Environment, National Research Council Canada, Montreal, QC, Canada

Gea Guerriero Environmental Research and Innovation (ERIN), Luxembourg Institute of Science and Technology (LIST), Esch/Alzette, Luxembourg

Amanda Hopes School of Environmental Sciences, University of East Anglia, Norwich, UK

Sergei Ignatov State Research Center for Applied Microbiology and Biotechnology, Obolensk, Moscow Region, Russia

Stephen A. Jackson School of Microbiology, University College Cork, National University of Ireland, Cork, Ireland

David F. Juck Energy, Mining and Environment, National Research Council Canada, Montreal, QC, Canada

Anne D. Jungblut Life Sciences Department, Natural History Museum, London, UK

Ludmila Kalenova Tyumen State University, Tyumen, Russia

Tyumen Scientific Center, Siberian Branch, Russian Academy of Science, Tyumen, Russia

Hidehisa Kawahara Department of Biotechnology, Faculty of Engineering, Kansai University, Osaka, Japan

Jun Kawamoto Institute for Chemical Research, Kyoto University, Uji, Kyoto, Japan

Michael Kuhn Institute of Atmospheric and Cryospheric Sciences (formerly Institute of Meteorology and Geophysics), University of Innsbruck, Innsbruck, Austria

Tatsuo Kurihara Institute for Chemical Research, Kyoto University, Uji, Kyoto, Japan

Jana Kvíderová Centre for Polar Ecology, Faculty of Science, University of South Bohemia, České Budějovice, Czech Republic

Synnøve Lofthus Department of Biotechnology, NTNU, Trondheim, Norway

Pieter De Maayer School of Molecular and Cell Biology, University of the Witwatersrand, Johannesburg, South Africa

Vladimir Melnikov Tyumen State University, Tyumen, Russia

Tyumen Scientific Center, Siberian Branch, Russian Academy of Science, Tyumen, Russia

Alexander B. Michaud Center for Geomicrobiology, Department of Bioscience, Aarhus University, Aarhus C, Denmark

Thomas Mock School of Environmental Sciences, University of East Anglia, Norwich, UK

Shinjini Mukherjee MEM-Group, Department of Biosciences, University of Helsinki, Helsinki, Finland

Khachik Muradian State Institute of Gerontology of National Academy of Medical Sciences of Ukraine, Kiev, Ukraine

Tahria Najnin Institute for Glycomics, Griffith University, Southport, QLD, Australia

Roman Netzer SINTEF Ocean, Marine Environmental Technology, Trondheim, Norway

Yuichi Nogi Research and Development Center for Marine Biosciences, Japan Agency for Marine-Earth Science and Technology, Yokosuka, Japan

Fergal O'Gara School of Microbiology, University College Cork, National University of Ireland, Cork, Ireland

Biomerit Research Centre, University College Cork, National University of Ireland, Cork, Ireland

School of Biomedical Sciences, Curtin Health Innovation Research Institute, Curtin University, Perth, WA, Australia

Ermegilda Parrilli Department of Chemical Sciences, University of Naples Federico II, Naples, Italy

David A. Pearce Department of Applied Biology, Faculty of Health and Life Sciences, Northumbria University, Newcastle upon Tyne, UK

Gennady Pogorelko Iowa State University, Ames, IA, USA

Vasiliy Potapov State Research Center for Applied Microbiology and Biotechnology, Obolensk, Moscow Region, Russia

John C. Priscu Department of Land Resources and Environmental Sciences, Montana State University, Bozeman, MT, USA

Benjamin Pushparaj ISE-CNR Istituto per lo Studio degli Ecosistemi, Sesto Fiorentino, Italy

Sara M.E. Rassner Department of Geography and Earth Sciences, Aberystwyth University, Aberystwyth, UK

Isabelle Raymond-Bouchard Department of Natural Resource Sciences, McGill University, Ste-Anne-de-Bellevue, QC, Canada

Deni Ribicic Department of Cancer Research and Molecular Medicine, NTNU, Trondheim, Norway

Ciro Sannino Department of Agricultural, Food and Environmental Sciences, University of Perugia, Perugia, Italy

Gustavo Santos Centre of Molecular and Environmental Biology (CBMA), Department of Biology, University of Minho, Braga, Portugal

Ahsan Mushir Shemsi Center for Environment and Water, Research Institute, King Fahd University of Petroleum and Minerals (KFUPM), Dhahran, Saudi Arabia

Satya P. Shukla Aquatic Environment Management Division, Central Institute of Fisheries Education, Mumbai, India

Khawar Sohail Siddiqui Life Sciences Department, King Fahd University of Petroleum and Minerals (KFUPM), Dhahran, Saudi Arabia

Timo P. Sipilä MEM-Group, Department of Biosciences, University of Helsinki, Helsinki, Finland

Sara Sjöling School of Natural Sciences, Technology and Environmental Studies, Sodertorn University, Huddinge, Sweden

Mark L. Skidmore Department of Earth Sciences, Montana State University, Bozeman, MT, USA

Taha School of Biotechnology and Biomolecular Sciences, The University of New South Wales, Sydney, NSW, Australia

Giorgia Tasselli Department of Agricultural, Food and Environmental Sciences, University of Perugia, Perugia, Italy

David N. Thomas School of Ocean Sciences, College of Natural Sciences, Bangor University, Anglesey, UK

Finnish Environment Institute (SYKE), Helsinki, Finland

Martyn Tranter Bristol Glaciology Centre, School of Geographical Sciences, University of Bristol, Bristol, UK

Benedetta Turchetti Department of Agricultural, Food and Environmental Sciences, University of Perugia, Perugia, Italy

Maria Luisa Tutino Department of Chemical Sciences, University of Naples Federico II, Naples, Italy

Trista J. Vick-Majors Département des Sciences Biologiques, Université du Quèbec à Montrèal, Montréal, QC, Canada

Warwick F. Vincent Centre for Northern Studies (CEN), Laval University, Quebec City, QC, Canada

Department of Biology, Laval University, Quebec City, QC, Canada

Lyle G. Whyte Department of Natural Resource Sciences, McGill University, Ste-Anne-de-Bellevue, QC, Canada

Jodi N. Young School of Oceanography, University of Washington, Seattle, WA, USA

Kim Yrjälä MEM-Group, Department of Biosciences, University of Helsinki, Helsinki, Finland

Part I
Boundary Conditions for Microbial Life in the Cold

Chapter 1
The Climate of Snow and Ice as Boundary Condition for Microbial Life

Michael Kuhn and Andrew G. Fountain

Abstract The microclimate and structure of snow and ice are a boundary condition as well as a matrix for a large spectrum of microbial life under alpine and polar conditions. Biological activity critically depends on the supply of energy, water and nutrients, with solar radiation as the prime source of energy, varying with latitude and altitude. The energy balance at the snow or ice surface provides the boundary condition for the fluxes of energy and water to the snow and ice, with important latitudinal differences from the temperate to the polar regions. The extreme situations of sunlit rocks surrounded by snow and the environment of Antarctic cryoconite holes, where ice, water, solar radiation and nutrients interact in particular ways, closes this review on ice and its effect on microbial life.

Contents

M. Kuhn (✉)
Institute of Atmospheric and Cryospheric Sciences, University of Innsbruck, Innrain 52, 6020 Innsbruck, Austria
e-mail: michael.kuhn@uibk.ac.at

A.G. Fountain
Department of Geology, Portland State University, Portland, OR 97201, USA

R. Margesin (ed.), *Psychrophiles: From Biodiversity to Biotechnology*,
DOI 10.1007/978-3-319-57057-0_1

1.1 The Source of Energy: Solar Radiation

At the Earth's surface, the biosphere, the atmosphere, the hydrosphere and the lithosphere interact most extensively. In mountains and polar regions, the cryosphere adds more facets to this multiple interaction. Biogeochemical cycles at the Earth's surface are driven by the vertical exchange of energy and water locally and by the horizontal motion of air, trace gases and water in the global circulation. The energy absorbed from the incident solar radiation is used to heat the ground, snow or water, which in turn heats the overlying air by turbulent convection; to evaporate water, melt or sublimate ice and in part is re-emitted as infrared radiation.

Solar radiation, the prime energy source for climatic and biotic processes, has a strong daily and seasonal variation in mid and high latitudes. This is best illustrated by its reference value, the theoretical amount of energy that would be received if there were no atmosphere. Daily sums of this radiation are displayed in Fig. 1.1 in response to geographical latitude and time of the year. While the tropics have the highest annual sums, the polar regions reach the highest daily totals in their respective summers, with Antarctica receiving more than the Arctic because the Earth is closest to the Sun in the Austral summer.

The presence of gases, aerosols and clouds in the atmosphere reduces solar intensity through absorption, scattering and reflection and varies spatially and with altitude. Global irradiance, the sum of direct and diffuse solar radiation,

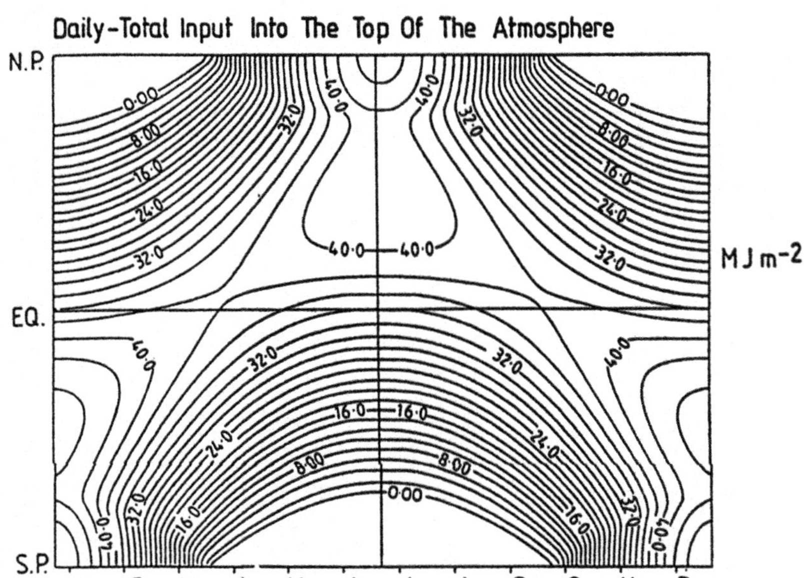

Fig. 1.1 Daily sums of extraterrestrial solar irradiance, the reference amount of energy that would be received without atmosphere. Values are given in MJ m^{-2} day^{-1} computed for a solar constant of 1368 W m^{-2}. Figure courtesy of C. Fröhlich

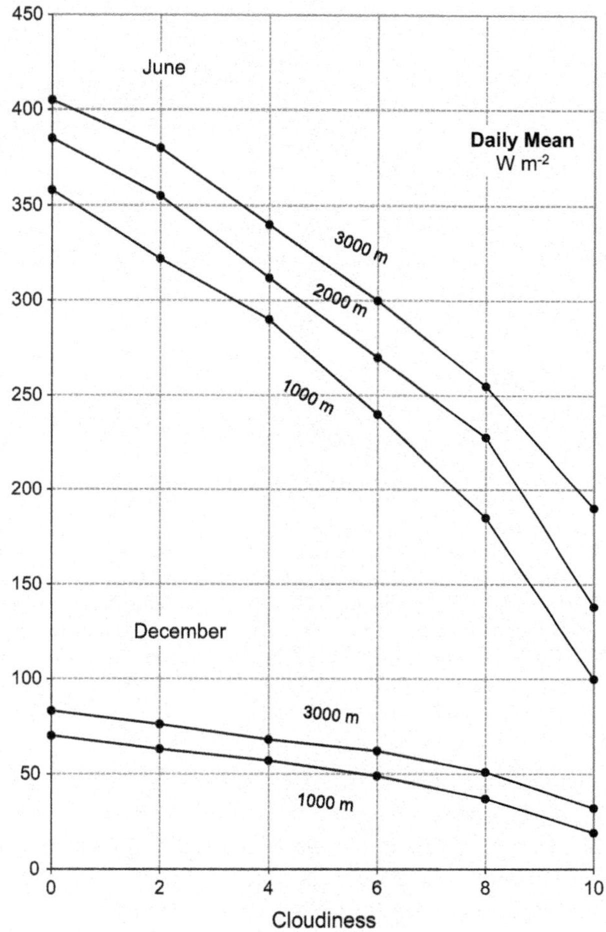

Fig. 1.2 Daily averages of global irradiance at eastern alpine stations, according to altitude and cloudiness, based on data by Dirmhirn (1964)

was compiled from records at Austrian stations in Fig. 1.2, as function of cloudiness and altitude. Results show an increase of global irradiance of the order of 1% per 100 m altitude at mean cloudiness and a decrease by 50% when comparing cloudless and cloud covered sky at an altitude of 3000 m.

The maximum daily average of 400 W m^{-2} at 3000 m (Fig. 1.2) is associated with an instantaneous maximum of ca. 1000 W m^{-2} at noon. The equivalent values in the Dry Valleys of Antarctica, close to sea level, are nearly 300 W m^{-2} daily average and 400 W m^{-2} at noon (Hoffman et al. 2008). This daily average amounts to 83% of the extraterrestrial irradiance in alpine conditions, a fraction similar to that found in the central Antarctic during summer solstice. It is obvious from Figs. 1.1 and 1.2 that this fraction decreases at lower solar elevations.

A large part of this incident solar radiation is reflected back to the atmosphere where clouds, aerosols and air molecules reflect it back to the snow surface, thus increasing the global solar radiation. Higher surface reflectance (albedo) thereby

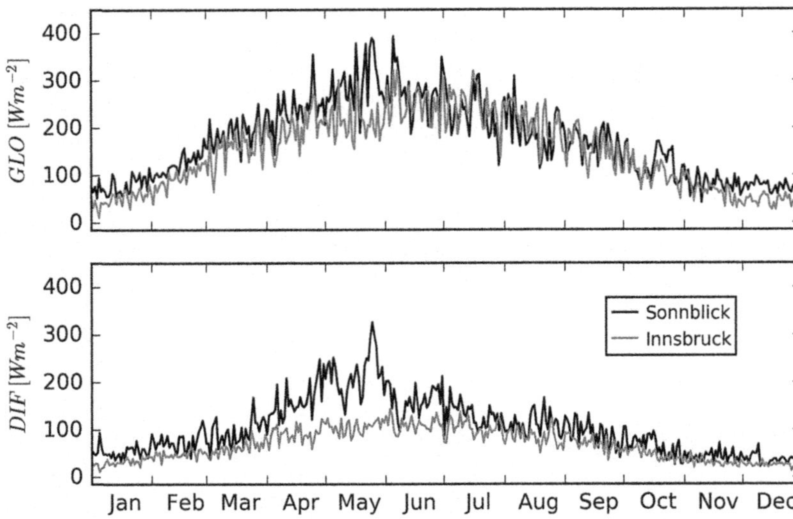

Fig. 1.3 Fluxes of solar radiation reaching the surface at Sonnblick Observatory at 3150 m elevation and the valley station of Innsbruck at 580 m in the Austrian Alps. Average values of 2012–2015. Diffuse radiation DIF responds to the amount of atmospheric aerosols and clouds and to the multiple scattering between surface and atmosphere, particularly obvious in the spring months March–May, when snow covered Sonnblick receives significantly more diffuse radiation than snow free Innsbruck. In a similar way, these effects appear in global radiation GLO, the sum of direct and diffuse radiation. Data provided by the ARAD Project (Olefs et al. 2016)

increases the diffuse part of incoming solar radiation. This is obvious in a comparison of radiation fluxes received at the Sonnblick Observatory in the Austrian Alps at 3110 m elevation and the valley station of Innsbruck at 580 m a.s.l. (Fig. 1.3). Whenever the landscape surrounding this mountain station is snow covered, these multiple reflections significantly increase the diffuse component of the global solar radiation compared with the valley station.

The spatial and temporal variability of both snow cover and clouds thereby affect the seasonal and short term variability to the direct and diffuse solar radiation and their sum, the global radiation that is not apparent in the smoothed values of Fig. 1.2.

The broad band albedo of dry alpine or polar snow exceeds 80%, reaching 90% in the visible and UV parts of the spectrum and dropping to less than 20% in the near-infrared; in the thermal infrared, snow is nearly a perfect absorber/emitter of radiation with an emissivity close to 0.98 (Warren 1982). The angle of incident solar radiation also affects albedo. At low solar angles, typical of polar regions, forward scattering in the snow increases albedo to >90%. The albedo of snow and ice also changes with snow grain size, decreasing with increasing grain size; it also decreases with increasing liquid water content. Clean alpine snow, initially with an albedo exceeding 80%, displays albedo values between 60 and 70% by the end of summer. The presence of dust, biologic materials or other particulates reduces the albedo further.

The albedo of ice depends largely on the presence of cracks and air bubbles: typical clean ice of alpine glaciers reflects about 40% and dust and dirt covered ice may reflect as little as the surrounding rocks, i.e. 15–20%. The so-called blue ice of Antarctica has emerged at the surface after having been subjected to very high pressure in the deepest part of its trajectory: 1100 m of ice exert a hydrostatic pressure of 100 bars (100 atmospheres), a pressure under which air bubbles become dissolved in the crystal lattice of the ice. This bubble free, blue ice has the darkest appearance of any naturally occurring ice.

1.2 Distribution of Energy: The Energy Balance of Snow and Ice

Solar radiation is the prime source of energy for planet Earth. It supplies a global, annual average of 240 W m^{-2}. Geothermal heat supplied by the hot interior of the Earth and by radioactive decay amounts to only 60 mW m^{-2}, negligible compared to solar radiation, but of vital importance at the base of ice sheets.

The principle of the energy balance is most clearly and generally demonstrated for the case of the snow surface, the cryosphere/atmosphere interface. Here, incoming solar (or shortwave) radiation $S\downarrow$ is supplemented by incoming atmospheric infrared (longwave) radiation $L\downarrow$ emitted by greenhouse gases, clouds and aerosols, while part is reflected and emitted back to the atmosphere from the snow surface, $S\uparrow$ and $L\uparrow$. The sum of the four fluxes is called the radiation balance. The amount of other energy interacting with the surface is distributed in four ways:

1. Heat flux, C, to/from the surface by conduction in the snow (internal heating of the snow may occur by penetration of solar radiation and by air convection in the pore space of the snow which may also cool the snow)
2. Turbulent transfer of sensible heat H to/from the atmosphere
3. Turbulent transfer of latent heat of evaporation, sublimation, or condensation LE
4. The latent heat of melting or refreezing LM

Water in contact with the ice of lakes and glacial ice shelves may supply additional sensible and latent heat.

All fluxes are defined positive if they deliver energy to the surface so that, at the surface, their total must be zero.

$$S\downarrow + S\uparrow + L\downarrow + L\uparrow + C + H + \mathrm{LE} + \mathrm{LM} = 0$$

These quantities are usually expressed as energy flux densities in W m^{-2}. As they depend on atmospheric variables that are not locally determined, we first need information on the local climatic boundary conditions.

1.3 Air Temperature: Effects of Altitude and Latitude

The change of air temperature and other climatic conditions with altitude in mid-latitude mountains has often been compared to their change with latitude: a 1000 m higher altitude in the Alps may roughly be equivalent to a 1000 km move northward. In the case of temperature, however, the reasons for the decrease with altitude are basically different from those for the decrease with latitude. If a parcel of dry air is moved upward, it loses pressure, expands and thereby cools at a rate of −1 °C per 100 m altitude; in the case of moist air, condensation may reduce this figure to as much as −0.6 °C per 100 m. In both cases, the cooling is the consequence of vertical motion. The decrease of temperature with increasing latitude, on the other hand, follows from the decreasing annual supply of solar radiation (Fig. 1.1).

From five pairs of mountain and valley stations in the Eastern Alps, situated respectively above 1800 m and below 800 m, typical values of altitudinal temperature gradients are given in Fig. 1.4.

The smaller negative values that prevail in alpine winter are primarily due to temperature inversions above the valley stations. The largest negative values approaching saturated adiabatic conditions occur in spring with intense vertical mixing of the atmosphere. Values in Fig. 1.4 are valid for near surface air temperatures; they differ from gradients in the free atmosphere. Although based on records from the Alps, they are representative for many mountain areas around the world.

Mountain stations generally have smaller diurnal and annual temperature ranges (Table 1.1). This is a ventilation effect due to higher wind speeds and greater mixing at mountain peaks, topographic effect rather than due to altitude.

The decrease of temperature with increasing latitude φ is obvious in Fig. 1.5. The station Decepción is situated on an island at 63° S, at sea level. It displays the

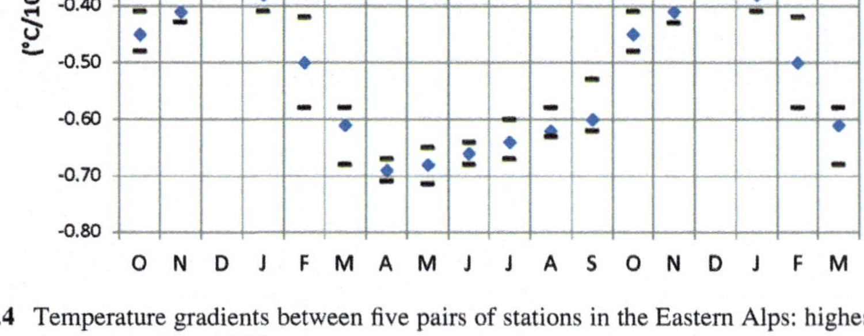

Fig. 1.4 Temperature gradients between five pairs of stations in the Eastern Alps: highest, mean and lowest values in °C per 100 m

Table 1.1 Mean diurnal and annual air temperature ranges at an alpine valley station (Gastein, 1100 m above sea level) and a mountain station (Sonnblick, 3200 m)

Location	Parameter	January	July	Annual range
Gastein	T_{min} (°C)	−7.4	12.5	19.9
	T_{max} (°C)	−1.5	21.1	22.6
	Daily range (°C)	5.9	8.6	
Sonnblick	T_{min} (°C)	−14.2	0.5	14.7
	T_{max} (°C)	−12.7	2.7	15.4
	Daily range (°C)	1.5	2.2	

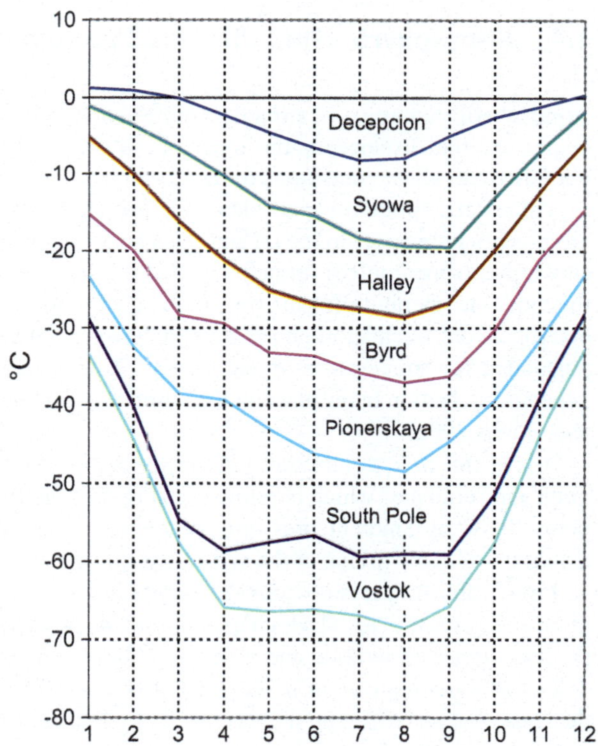

Fig. 1.5 Mean monthly air temperature at Antarctic stations, determined by a mix of latitude and altitude of the locations. Note the coreless winters at South Pole and Vostok. From data by Schwerdtfeger in Orvig (1970)

low annual variation of temperature typical for oceanic situations. The highest latitude of South Pole (90° S, 2800 m) does not yield the coldest temperatures; it is exceeded by Vostok Station, which is 600 m higher at 3400 m. The annual range of temperatures increases with latitude and altitude due to solar geometry decreasing cloudiness at increasingly continental stations.

The daily range of temperature, on the other hand, is determined by the daily range of solar elevation which is $2(90 - \varphi)$, limited by a maximum value at $90 - \varphi + \delta$ where φ is latitude and δ is the solar declination. Consequently no daily range of solar elevation, and therefore temperature, occurs at the poles (90° N, 90° S).

The change from polar day to polar night at latitudes beyond the polar circles (66.5°) causes an asymmetry of solar forcing and an asymmetry in the annual variation of temperature. Figure 1.5 shows how the annual variation changes from a nearly sinusoidal temperature at Decepción to a so-called coreless winter in which temperature reaches low values in April and then slowly decreases to an August minimum at the inland stations South Pole and Vostok.

1.4 Atmospheric Humidity and Precipitation

Atmospheric humidity is strongly controlled by air temperature and to a lesser degree by the distance to the open ocean. Table 1.2 gives figures relevant to humidity and phase transitions in the cryosphere.

Values of saturation vapour density (absolute humidity) and saturation pressure are given with respect to ice (i) and to supercooled water (w). Cloud droplets may stay liquid (supercooled) far below 0 °C. Statistically, −15 to −20 °C seems to be the modal temperature for the transitions from liquid water to ice in clouds due to heterogeneous nucleation on aerosols (ice nuclei). Supercooled droplets have been postulated for temperatures down to −38 °C in a pure atmosphere where homogeneous nucleation takes place in the absence of ice nuclei (see Chap. 6 in Wallace and Hobbs 2006).

It is useful here to introduce the terms "temperate ice" which is at the melting point and "cold ice" which is below freezing. The melting point of ice can be cooled below 0 °C by ambient pressure, by chemical admixtures and by the radius of curvature of snow grains in the sub-millimetre range.

From Table 1.2, it is obvious that saturation vapour pressure at 0 °C is 47 times as large as that at −40 °C which explains the decrease of atmospheric humidity and precipitation with latitude and altitude. While mean annual snow accumulation is 1–2 m of water equivalent in the Alps, it is about 0.5 m at the Antarctic coast and decreases below 0.03 m on the East Antarctic Plateau. A similar, but less impressive decrease goes from subarctic mountains to the centre of the Arctic Ocean (Orvig 1970; Rudolf and Rubel 2005).

The change of precipitation in the Alps and other mountain ranges is controlled by altitude and may increase by a factor of three from the dry, screened interior to the wet, exposed margins at either side of the Alpine range. Screening effects are best developed in mid-latitude mountain ranges of N–S extent: Scandinavia, Pacific Coastal Ranges in North America, Chilean Andes and Southern New Zealand.

Table 1.2 The change with temperature T of density ρ, saturation vapour density ρ_v*, saturation vapour pressure e*, specific heat c and latent heat L

	Density ρ (kg m^{-3})		Saturation vapour density ρ_v* (g m^{-3})		Saturation vapour pressure e* (hPa)		Specific heat c (J kg^{-1} K^{-1})				Latent heat L (MJ kg^{-1})	
T (°C)	ρ_i	ρ_a	ρ^*_{vi}	ρ^*_{vw}	e^*_i	e^*_w	c_i	c_w	c_a	c_v	L_v	L_m
+10		1.24		9.39		12.27		4192			2.477	
0	916.4	1.29	4.85	4.85	6.11	6.11	2105	4218	1005	1846	2.501	0.334
−10	917.4	1.34	2.14	2.36	2.60	2.86	2030	4271			2.525	0.312
−20	918.3	1.39	0.88	1.07	1.03	1.25	1959	4354			2.549	0.289
−30	921.0	1.45	0.39	0.45	0.38	0.51	1884	4520			2.574	0.264
−40	922.0	1.51	0.12	0.18	0.13	0.19	1812	4772			2.602	0.236

Indices i stand for ice, a for air, w for liquid water, v for vapour, m for melting [data from Linke and Baur (1970) and Brutsaert (1982)]

1.5 The Cryosphere: A Matrix for Life

The total extent, or an inventory of the cryosphere, is of little importance to microbiology; rather it is the availability of solar radiation, liquid water and nutrients (Kuhn 2001; Psenner et al. 2003). These conditions in turn depend on altitude, latitude and on the cryospheric feature of concern, as there are: seasonal and perennial snow, glaciers (including ice caps and ice sheets), floating ice (including lake, river and sea ice) and ground ice (including permafrost and transient ground ice).

These four groups differ primarily in their structure and in their connection to other parts of the biosphere like water and soil or in their response to climatic change. For example, the formation and decay of a seasonal snow pack is an immediate product of atmospheric forcing, the base of ice sheets may be the last to respond to its effects. The structure of the cryospheric features determines the transport of energy, water and nutrients, where both energy and nutrient fluxes may be connected to liquid water.

Of the four groups, lake, sea and ground ice are frozen terrestrial water (congelation ice) while snow and glacier ice are of atmospheric origin (meteoric ice). Airborne ice crystals and snowflakes, which form in a large variety of shapes controlled by temperature and humidity, precipitate and accumulate on the Earth's surface. The original, delicate crystals immediately start changing into rounded grains by the transfer of water molecules from tips and convex sites on the crystal surface to the groves and concave sites. The various stages of transformation are very well illustrated in the International Classification of Snow on the Ground (Colbeck et al. 1990; Fierz et al. 2009). The granular shape makes for a more efficient and denser packing of snow, a density of about 300 kg m^{-3} being typical for old, dry alpine and polar snow (Pomeroy and Jones 1996).

Early winter snow packs can be heated from the ground beneath and lose heat from the top layer to the atmosphere, which may result in temperature gradients of several degrees per meter. Associated with the temperature gradient is a gradient of vapour pressure in the pore space which induces upward diffusion of water vapour (from the warmer to cooler layers) and the build-up of facetted crystals several mm in diameter. In severe situations, a facetted layer of large crystals is mechanically weak given the few bonds between crystals creating hazardous avalanche conditions.

In polar snow, in the absence of melting and with weak temperature gradients, the further growth of snow grains proceeds slowly. Under temperate alpine conditions with frequent melt-freeze cycles, the grains become repeatedly surrounded by liquid water films and snow grains grow rapidly. In that stage of metamorphism, the melting point of small grains is lower than that of the larger ones and the latter grow at the expense of the smaller grains. The modal size of the larger grains achieves diameters of 1–2 mm.

The snow matrix in the most general case thus consists of ice, liquid water, water vapour and air. It has peculiar properties being permeable to both air and

Table 1.3 Thermal conductivity λ and thermal diffusivity K of snow and ice at various bulk densities ρ

	Bulk density (ρ; kg m^{-3})				
	100	200	300	500	917
Thermal conductivity (λ; W m^{-1} K^{-1})	0.0003	0.12	0.27	0.74	2.47
Thermal diffusivity (K; 10^{-6} m^2 s^{-1})	0.0014	0.28	0.42	0.70	1.28

water flows at bulk snow densities of up to 830 kg m^{-3} and transmissive of shortwave radiation at any density. Its pore space, the volume not filled by ice, may contain water or air and convection may transport latent and sensible heat. A net radiative transfer of longwave (infrared) radiation from one pore wall to the opposite wall will occur in a temperature gradient. There is, of course, molecular conduction of heat through snow grains in physical contact.

In summary, there are fluxes of air and liquid water through the snow pack, both including gaseous, soluble and insoluble impurities (Kuhn 2001). Electromagnetic radiation penetrates the snow pack, directly in the shortwave range and wall-to-wall in the infrared. The molecular conduction becomes increasingly important in the deeper layers with increasing bulk density. It is usually formulated as

$$C = \lambda\, dT/dz$$

where the depth z is positive into the ground and C is positive when directed towards the surface. The thermal conductivity λ and the thermal diffusivity K effectively include all effects mentioned and thus strongly depend on density and associated permeability as shown in Table 1.3.

The extinction of solar radiation in snow and ice follows an exponential relationship,

$$S\downarrow(z) = S\downarrow(0)\, e^{-\kappa z}$$

where the extinction coefficient κ for shallow layers of dry snow has typical values of about 10–20 m^{-1} for spectrally broadbands (Meirold-Mautner 2004), corresponding to a reduction of the radiation below 10 cm of snow to a fraction of 37–14% of the value incident at the surface. The value of κ strongly depends on wavelength so that broadband extinction coefficients are not constant with depth. Extinction is least for the blue part of the spectrum, giving a blue shade to light that returns upward from deeper layers or crevasses. Actually it is the spectrum of the incident radiation, the spectral extinction coefficient and the spectral sensitivity of the observer's eye that together give snow at depth and ice its blue colour.

1.6 Liquid Water in the Cryosphere

As liquid water is an essential asset of the biosphere, it is of particular interest to investigate the conditions under which it may occur in an environment of 0 °C or colder. Water can cool to temperatures below 0 °C (super-cooling), and in terrestrial environments, such as lakes and rivers, super-cooling is limited to a few tenths of a degree at most because the water readily freezes in contact with geologic or biologic materials. In the atmosphere, however, super-cooling of cloud droplets can reach much colder temperatures as discussed in Sect. 1.4 and by Sattler et al. (2001).

The ice molecules within the surface layer of ice, only several molecules thick, are much more mobile compared to those deeper in the ice lattice due to the lack of surrounding bonds. For this reason, the surface layer is considered 'quasi-liquid'. Closer to 0 °C this surface layer becomes liquid-like allowing for a much higher rate of transfer of molecules from the convex to the concave sites on a crystal than would be possible by diffusion in the vapour phase alone. This process rounds the snow grains (destructive metamorphism) and reduces the total surface free energy.

Surface melting supplies most of the water found in a snow pack and on a glacier surface. At atmospheric pressure of 1013 hPa, nominally sea level, and with an energy supply of 334 kJ kg^{-1}, ice melts when it reaches 0 °C, independent of ambient air temperature. Ice may in fact melt or remain at 0 °C at temperatures of the overlaying air as low as −5 °C and may stay frozen at temperatures of the overlaying air as high as 5 °C, depending on micrometeorological conditions (Kuhn 1987).

If one were to walk down from the highest elevations of a glacier that spans the thermal regime from cold (<10 °C) at the top to temperate (0 °C) at the terminus, one first walks on the dry facies, where melting never occurs. This is followed by a zone where snow melts at the surface but refreezes at depth within the annual layer. Further down-glacier, melt water percolates the entire annual layer either forming a transient layer of water on the impermeable ice or refreezes as superimposed ice. This form of sensible and latent heat transport is the reason why most alpine glaciers below 3400 m are temperate in spite of mean annual ambient air temperatures below freezing. Finally, below that zone melting and net mass loss of glacier ice prevails. Melt water produced at the glacier surface penetrates the ice in cracks, in crevasses and in the funnels of moulins and leaves the glacier in a system of braided channels beyond the glacier front.

Surface melting may produce extended areas of water-soaked snow and firn where algal growth is promoted and further enhances melting due to its low albedo (Meirold-Mautner 2004). When the snow or firn layer on a sloping surface becomes saturated with water, the mechanical stability may fail; the layer suddenly accelerates forming a slush flow. On near-horizontal surfaces or in depressions on the impermeable ice surface, supraglacial lakes may form. They have a lower albedo than the surrounding ice and absorb more solar radiation. Solar heating of the water may warm temperatures above freezing and enhance local melting. That the

Fig. 1.6 Cryoconite holes in the ablation zone of an alpine glacier. The hole on the right has a length of about 20 cm

maximum density of water occurs at 4 °C and the thermodynamic equilibrium at the ice-water contact is 0 °C guarantees a continual convection of water enhancing local ice melting.

A variation of this thermodynamic situation is displayed in the cryoconite holes, a sample of which is shown in Fig. 1.6. The biological importance of these features was first described by Steinböck (1936) of the University of Innsbruck and has recently received revived interest (Margesin et al. 2002). When insoluble organic and inorganic particles concentrate into patches on the ice surface of a glacier (cryoconite), the darker albedo preferentially absorbs solar radiation heating the sediment and it melts into the ice forming a pool of water. A small patch of cryoconite or a small stone will melt into the ice; larger bodies like the one displayed in Fig. 1.6 will have an oblique melt channel of decimetre depth. As the holes deepen, the solar radiation, which is transmitted through the ice to the hole, weakens due to solar absorption in the ice. The hole depth reaches an equilibrium when the rate of deepening equals the ablation rate on the glacier surface (Wharton et al. 1985).

Yet another thermodynamic situation is seen in the lakes of the Antarctic Dry Valleys, which have a permanent ice cover of 3–6 m thickness (McKay et al. 1985). Here, water is kept liquid under an ice cover that separates it from air of mean annual temperature of −20 °C. This is accomplished by energy gain from solar radiation penetrating the ice and warming the water (Wilson and Wellman 1962; Hoare et al. 1965) and by summer influx of melt water from the ice-free surroundings (Lewis et al. 1998). Similarly, cryoconite holes on the glaciers in this region have an ice cap. The surface energy balance maintains the frozen glacier surface while a solid-state greenhouse effect melts the ice around the entombed patch of cryoconite (Fountain et al. 2008; Hoffman et al. 2008) renewing its downward migration each spring.

At the base of glaciers or ice sheets, the load of overlying ice exerts a pressure at the rate of 1 bar (10^5 Pa) per 11 m of ice. Elevated pressure in turn reduces the melting point of ice by 0.76×10^{-6} per Pa so that the pressure melting point at the base of a 4000 m thick ice sheet is −2.8 °C. The depressed freezing point, the geothermal heat flux (global average of 60 mW m^{-2}) and the insulation of the overlying ice provide sufficient conditions to melt 7 mm of ice a year despite average surface air temperatures colder than −20 °C. This basally generated water can move under the ice sheet forming subglacial lakes (Siegert et al. 2001; Fricker et al. 2007). Similar basal melt rates occur under temperate glaciers, but that melt is very small compared to the influx of surface melt water (meters) that reaches the bed.

1.7 Hot Spots in the Ice

The differential absorption of solar radiation by dark rocks may create isolated spots of temperatures far above freezing. While cryoconite, sand or small stones absorb more solar radiation than ice, heat up and melt into the ice, large rocks absorb similar amounts of energy per unit surface area do not melt into the ice. Their larger thickness and smaller contact area with the ice reduces temperature gradients in the rock, and less heat is conducted into the ice. Their energy gain is thus used to raise their temperature, and under alpine summer conditions this may result in significantly elevated rock surface temperatures. Measurements with an infrared thermometer in early summer of one rock, 2 m diameter, protruding 1 m above the snow surface, at 3000 m above sea level, showed temperatures of 42 °C in the early afternoon on its sunlit side.

A profile of surface temperatures from the glacier tongue of Hintereisferner across the moraine was recorded in late May, with peak temperatures again exceeding 40 °C. The values given in Fig. 1.7 show that wet sand is colder than dry sand as is to be expected due to the loss of latent heat; rocks in the vicinity of the

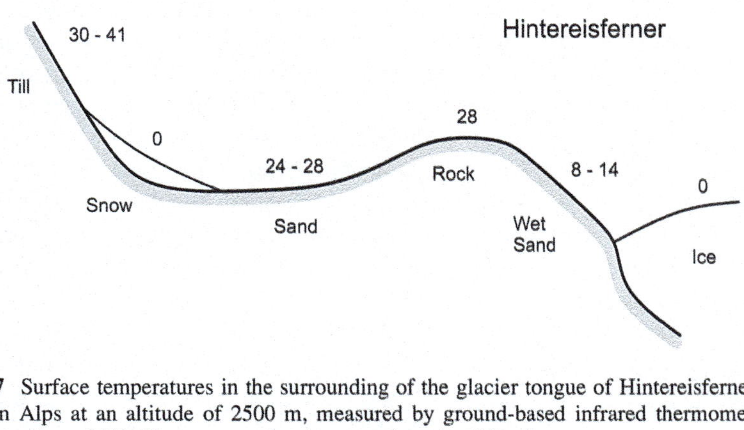

Fig. 1.7 Surface temperatures in the surrounding of the glacier tongue of Hintereisferner in the Austrian Alps at an altitude of 2500 m, measured by ground-based infrared thermometer at a solar elevation of 60°. The cross section is approximately 100 m long

glacier are cooler than those farther up due to the cooling effect of the katabatic glacier wind.

1.8 Ice as a Component of the Biosphere

Life inhabits all components of the cryosphere, and in turn many parts of the cryosphere play important roles within the biosphere (Fountain et al. 2012). For example, sea ice provides both a refugium for small organisms and forging platforms for others. Similarly, snow provides a protective habitat for some animals while it reduces foraging efficiency for others. For microbial life, that part of the cryosphere that includes water is no barrier to biologic activity. Melting alpine snow packs commonly exhibit 'snow algae' (e.g. microalgae, phytoflagelates) (Jones et al. 2001). On the ice surface of glacial ablation zones, melt pools and cryoconite holes may sequester nutrients from the surrounding melting environment as well as from the atmosphere (Tranter et al. 2004). The degree of biological activity on snow and ice surfaces, in terms of both diversity and abundance, is partly controlled by aeolian deposition of biologic materials, usually from local sources (Jones 1999; Hodson 2006; Bagshaw et al. 2013). The diversity of microbial life in cryoconite holes is relatively large, compared to other snow and ice surfaces, and dominated by cyanobacteria and phytoflagellates, algae and fungi, among other biota including diatoms, tardigrades and rotifers (Steinböck 1936; Porazinska et al. 2002; Christner et al. 2003). For algal mats in these pools, their rates of photosynthesis approach that of polar marine waters (Vincent et al. 2000). Surficial streams convey nutrients and biota across the ice surfaces and eventually into the glacier itself via crevasses and moulins. Consequently, the melt of snow and ice inoculates subglacial environments (Tranter et al. 2005).

Within and under glaciers, the biological activity is poorly known. Without sunlight the narrow (10^{-3} to 10^{0} mm) watery veins found between crystals (Cuffey and Patterson 2010), where three crystals meet, are probably dominated by heterotrophic and chemoautotrophic bacteria (Price 2000). In subglacial environments, where glacier ice contacts the lithologic substrate, viable microbial life may be ubiquitous, as long as water is present. This microbial activity exerts a strong influence on the chemistry of subglacial waters and therefore stream flow from glaciers (Tranter et al. 2002; Skidmore et al. 2005). The microbial diversity is probably controlled by the hydraulic conditions that determine the flow speed and residence time of the water and whether the environment is aerated or anaerobic (Tranter 2005). Aerated waters are derived from waters draining from the surface while the anaerobic waters are either not sourced from the surface or in long residence time at the bed. Indeed, under small valley glaciers carbonate dissolution is controlled by sulfide oxidation and microbial CO_2 (Wadham et al. 2010). Under larger glaciers and particularly ice sheets, where water residence and isolation times are much greater, anoxic conditions prevail due to microbial oxidation of sulfide minerals and organic carbon creating ideal conditions for methanogenesis (Wadham et al. 2010).

This enhances silicate mineral dissolution providing nutrients critical to the microbial communities. Under large part of Antarctica, methanogenic archaea may have degraded organic carbon forming a globally significant accumulation of methane in subglacial sedimentary basins (Wadham et al. 2012).

1.9 Conclusions

Contrary to common expectations, the cryosphere harbours abundant microbial life. From the perspective of the biosphere, this represents successful adaptive strategies. From the perspective of the geosphere, it supports environmental conditions that supply light and energy, provides shelter in and under the snow and ice matrix and enables the circulation of nutrients and water.

Acknowledgements We are grateful to F. Pellet and M. Olefs for their assistance in producing the figures.

References

Brutsaert W (1982) Evaporation into the atmosphere. Reidel Publishing Company, Dordrecht, 299 pp

Bagshaw EA, Tranter M, Fountain AG, Welch K, Basagic HJ, Lyons WB (2013) Do cryoconite holes have the potential to be significant sources of C, N, and P to downstream depauperate ecosystems of Taylor Valley, Antarctica? Arct Antarct Alp Res 45:440–454

Colbeck S, Akitaya E, Armstrong R, Gubler H, Lafeuille J, Lied K, McClung D, Morris E (eds) (1990) The international classification for seasonal snow on the ground. International Commission on Snow and Ice, Springfield, VA, 23 pp

Christner BC, Kvitko BH, Reeve JN (2003) Molecular identification of bacteria and eukarya inhabiting an Antarctic cryoconite hole. Extremophiles 7:177–183

Cuffey K, Paterson WSB (2010) The physics of glaciers, 4th edn. Elsevier, Cambridge

Dirmhirn I (1964) Das Strahlungsfeld im Lebensraum. Akademische Verlagsgesellschaft, Frankfurt am Main, 426 pp

Fierz C, Armstrong RL, Durand Y, Etchevers P, Greene E, McClung DM, Nishimura K, Satyawali PK, Sokratov SA (2009) The international classification for seasonal snow on the ground. IHP-VII Technical Documents in Hydrology N°83, IACS Contribution N°1, UNESCO-IHP, Paris, 80 pp

Fountain AG, Campbell JL, Schuur EAG, Stammerjohn SE, Williams MW, Ducklow HW (2012) The disappearing cryosphere: impacts and ecosystem responses to rapid cryosphere loss. BioScience 62(4):405–415

Fountain AG, Nylen TH, Tranter M, Bagshaw E (2008) Temporal variations in physical and chemical features of cryoconite holes on Canada Glacier, McMurdo Dry Valleys, Antarctica. J Geophys Res 113:G01S92

Fricker HA, Scambos T, Bindschadler R, Padman L (2007) An active subglacial water system in West Antarctica mapped from space. Science 315:1544–1548

Hoare RA, Popplewell KB, House DA, Henderson RA, Prebble WM, Wilson AT (1965) Solar heating of Lake Fryxell, a permanently ice-covered Antarctic lake. J Geophys Res 70: 1555–1558

Hodson A (2006) Biogeochemistry of snowmelt in an Antarctic glacial ecosystem. Water Resour Res 42:W11406

Hoffman MJ, Fountain AG, Liston GE (2008) Surface energy balance and melt thresholds over 11 years at Taylor Glacier, Antarctica. J Geophys Res 113:F04014

Jones HG (1999) The ecology of snow-covered systems: a brief overview of nutrient cycling and life in the cold. Hydrol Proc 13:2135–2147

Jones HG, Pomeroy JW, Walker DA, Hoham RW (eds) (2001) Snow ecology: an interdisciplinary examination of snow-covered ecosystems. Cambridge University Press, Cambridge

Kuhn M (1987) Micro-meteorological conditions for snowmelt. J Glaciol 33:24–26

Kuhn M (2001) The nutrient cycle through snow and ice. Aquatic Sci 63:150–167

Lewis K, Fountain A, Dana G (1998) Surface energy balance and meltwater production for a Dry Valley glacier, Taylor Valley, Antarctica. Annals Glaciol 27:603–609

Linke F, Baur F (1970) Meteorologisches Taschenbuch. Akademische Verlagsgesellschaft Geest & Portig, Leipzig, 712 pp

Margesin R, Zacke G, Schinner F (2002) Characterization of heterotrophic microorganisms in alpine glacier cryoconite. Arct Antarct Alp Res 34:88–93

McKay C, Clow G, Wharton R, Squyres S (1985) Thickness of ice on perennially frozen lakes. Nature 313:561–562

Meirold-Mautner I (2004) A physical snow-radiation model: measurements, model development and applications to the ecosystem snow. Doctoral thesis, Institute of Meteorology and Geophysics, University of Innsbruck, Austria, 134 pp

Olefs M, Baumgartner DJ, Obleitner F, Bichler C, Foelsche U, Pietsch H, Rieder HE, Weiss PH, Geyer F, Haider T, Schöner W (2016) The Austrian radiation monitoring network ARAD – best practice and added value. Atmos Meas Tech 9:1513–1531

Orvig S (ed) (1970) Climates of the polar regions, World survey of climatology, vol 14. Elsevier, Amsterdam, 370 pp

Pomeroy JW, Jones HG (1996) Wind-blown snow: sublimation, transport and changes to polar snow. In: Wolff E, Bales RC (eds) Chemical exchange between the atmosphere and polar snow. Springer, Berlin, pp 453–489

Porazinska DL, Fountain AG, Nylen TH, Tranter M (2002) The biodiversity and biogeochemistry of cryoconite holes from McMurdo Dry Valley Glaciers, Antarctica. Arct Antarct Alp Res 54:495–505

Price PB (2000) A habitat for psychrophiles in deep, Antarctic ice. Proc Natl Acad Sci (USA) 97:1247–1251

Psenner R, Wille A, Priscu J, Felip M, Wagenbach D, Sattler B (2003) Extremophiles: ice ecosystems and biodiversity. In: Gerday C (ed) Knowledge for sustainable development. An insight into the Encyclopaedia of Life Support Systems, vol III. UNESCO Publishing – EOlSS Publishers, Oxford, pp 573–598. (updated 2007)

Rudolf B, Rubel F (2005) Global precipitation. In: Hantel M (ed) Observed global climate. Landolt-Börnstein new series geophysics, vol 6. Springer, Berlin

Sattler B, Puxbaum H, Psenner R (2001) Bacterial growth in supercooled cloud droplets. Geophys Res Lett 28:239–242

Siegert M, Ellis-Evans JC, Tranter M, Mayer C, Petit J-R, Salamatin A, Priscu J (2001) Physical, chemical and biological processes in Lake Vostok and other Antarctic subglacial lakes. Nature 414:603–609

Skidmore ML, Anderson SP, Sharp M, Foght J, Lanoil BD (2005) Comparison of microbial community compositions of two subglacial environments reveals a possible role for microorganisms in chemical weathering processes. Appl Environ Microbiol 71:6986–6997

Steinböck O (1936) Über Kryokonitlöcher und ihre biologische Bedeutung. Z Gletscherkd 24: 1–21

Tranter M (2005) Geochemical weathering in glacial and proglacial environments. In: Holland HD, Turekian KK (eds) Treatise on geochemistry, vol 5. Elsevier, London, pp 189–205

Tranter M, Sharp MJ, Lamb H, Brown GH, Hubbard BP, Willis IC (2002) Geochemical weathering at the bed of Haut Glacier d'Arolla, Switzerland: a new model. Hydrol Proc 16:959–993

Tranter M, Fountain A, Fritsen C, Lyons B, Priscu J, Statham P, Welch K (2004) Extreme hydrochemical conditions in natural microcosms entombed within Antarctic ice. Hydrol Proc 18: 379–387

Tranter M, Skidmore ML, Wadham JL (2005) Hydrological controls on microbial communities in subglacial environments. Hydrol Proc 19:996–998

Vincent WF, Gibson JAE, Pienitz R, Villeneuve V (2000) Ice shelf microbial ecosystems in the high Arctic and implications for life on Snowball Earth. Naturwissenschaften 87:137–141

Wadham JL, Tranter M, Skidmore ML, Hodson AJ, Priscu J, Lyons WB, Sharp M, Wynn P, Jackson M (2010) Biogeochemical weathering under ice: size matters. Global Biogeochem Cycles 24:GB3025

Wadham JL, Arndt S, Tulaczyk S, Stibal M, Tranter M, Telling J, Lis JP, Lawson E, Ridgwell A, Dubmick A, Sharp MJ, Anesio AM, Butler CEH (2012) Potential methane reservoirs beneath Antarctica. Nature 488:633–637

Wallace JM, Hobbs PV (2006) Atmospheric science, an introductory survey, 2nd edn. Elsevier, Amsterdam, 483 pp

Warren SG (1982) Optical properties of snow. Rev Geophys 20:67–89

Wharton RA, McKay CP, Simmons GM, Parker BC (1985) Cryoconite holes on glaciers. Bioscience 35:499–503

Wilson AT, Wellman HW (1962) Lake Vanda: an Antarctic lake: Lake Vanda as a solar energy trap. Nature 196:1171–1173

Chapter 2
Determining the Limits of Microbial Life at Subzero Temperatures

Corien Bakermans

Abstract Determining the low-temperature limit of life is a challenge complicated by the reduced availability of liquid water as water freezes and by the low rates of diffusion and reaction brought on by low temperatures. And yet, many microorganisms are able to grow at temperatures of −2 °C to 4 °C and many also survive much lower temperatures of −80 °C to −196 °C. A variety of approaches for determining the low-temperature limit of life are examined in this chapter and relevant data are reported. Theoretical approaches investigate the presence of liquid water at temperatures below 0 °C as well as universal laws of biology which may inform the low-temperature limits of life. Both reductionist and holistic experimental approaches reveal the known limits of cellular components, processes, and whole cells with multiple lines of evidence suggesting that cell reproduction occurs down to temperatures of −20 °C. Finally, observational studies of microorganisms in low-temperature environments of the polar regions expose how the low-temperature limit of life is entangled with other factors (perhaps inextricably) and that time at low temperatures may limit evolution and cold adaptation of terrestrial life.

Contents

C. Bakermans (✉)
Penn State Altoona, The Pennsylvania State University, 3000 Ivyside Park, Altoona, PA 16601, USA
e-mail: cub21@psu.edu

R. Margesin (ed.), *Psychrophiles: From Biodiversity to Biotechnology*,
DOI 10.1007/978-3-319-57057-0_2

2.1 Introduction

The low-temperature limit of life likely depends on many things: the presence of liquid water, the function of different types of molecules, the coordinated interactions of these molecules, an adequate energy and nutrient supply (amount and flux), the absence of damaging conditions, exposure to low temperatures, and the time to evolve adaptations to low-temperature conditions. How do we extract the different effects of these components to discover the true low-temperature limit of life and determine what the critical factor is for that limit (if there is a single limit or a single critical factor)? This question can be tackled via theoretical, observational, and experimental frameworks that are reductionist or holistic. And it is likely that we will find the answer through the application of combinations of these approaches, not a single approach. Considering that many phyla (i.e., Proteobacteria, Firmicutes, Euryarchaeota, Cyanobacteria, Bacteroidetes, and Actinobacteria) contain species whose optimum growth temperature is at or below 20 °C (Corkrey et al. 2016), it should be no surprise that one technique or one answer will not arise as to what defines the low-temperature limit of life. The intent of this chapter is not to provide an exhaustive review but to serve as a guide to specific reviews and to highlight new or previously overlooked studies that inform the low-temperature limit of life.

2.2 Theoretical Approaches

One theoretical limit to life is the presence of liquid water: without liquid water there is no solvent for the biochemistry that makes life possible. At subfreezing temperatures, liquid water exists as films on minerals [theoretically down to −56 °C in glacial ice (Price 2000)] and as brines. On Earth, the $CaCl_2$ rich brines of Don Juan Pond in the McMurdo Dry Valleys of Antarctica (Fig. 2.1a) remain liquid year round even at temperatures of −50 °C (Beaty et al. 2006). However, the water in these brines is not accessible to microorganisms given the very low water activity ($a_w = 0.45$) which is below the demonstrated a_w limit of life (about 0.61) and, indeed, recent studies do not unambiguously detect microbial processes in Don Juan Pond (Samarkin et al. 2010; Peters et al. 2014). The importance of liquid water as a prerequisite for life cannot be understated, and an examination of the many conditions in which liquid water can exist is beyond the scope of this chapter; for a review of the known, and potential for, microbial activity in brines, thin films, and other sources of water, see Stevenson et al. (2015).

As temperature drops below the freezing point of water, one important limit on liquid water inside cells is the temperature at which vitrification of the cytoplasm occurs. Intracellular ice rarely forms (except in specialized habitats like the leaf surface) because temperatures drop slowly in the environment and the cytoplasm is densely packed; instead, dehydration is the main stress on cells as extracellular ice

Fig. 2.1 Environments on Earth where temperature conditions limit life. (**a**) Don Juan Pond, Antarctica (photo courtesy of National Science Foundation); (**b**) Yedoma (ice- and organic-rich syngenetic permafrost with large ice wedges) exposed in the 35-m-high eroding bank of the Itkillik

forms (Fonseca et al. 2006). Even in the laboratory, intracellular ice only forms when cells are cooled very rapidly without cryoprotectants [>100 °C min^{-1} for bacteria (Fonseca et al. 2006) and >20 °C min^{-1} for yeasts (Seki et al. 2009)]. Instead of freezing, vitrification of the cytoplasm occurs when the cytoplasm becomes a noncrystalline amorphous solid at the glass transition temperature (T_g). Because metabolism will be stopped by vitrification, the T_g may define the lower temperature limit for active life. A recent study demonstrated that the T_g of microorganisms varied from −12 °C to −26 °C in the five bacteria, two algae, and two fungi examined (Clarke et al. 2013). The lowest T_g were seen in *Arthrobacter arilaitensis* (−26 °C), *Corynebacterium variabile* (−25.6 °C), and *Chlamydomonas nivalis* (−24.2 °C). Naturally, the T_g will depend on the composition of proteins and small molecules in the cytoplasm which will be affected by physicochemical conditions. The lowest possible T_g of cells in the environment has yet to be determined; however, *Lactobacillus delbrueckii* ssp. *bulgaricus* CFL1 was shown to have a T_g of −19 °C which was significantly reduced in the presence of 0.58 M cryoprotectants such as sucrose, glycerol, and DMSO (Fonseca et al. 2016). DMSO addition resulted in the lowest T_g of −51 °C. While such high concentrations of cryoprotectants are unlikely to be found in the environment, this work demonstrates what is possible.

Furthermore, vitrification of the cytoplasm is likely a key to survival of very low temperatures and the return of metabolism once cells warm back up (Clarke et al. 2013). Extracellular ice formation and the associated increase in extracellular solutes should induce solute dehydration of the cytoplasm [as long as cell membranes remain functional (Mazur 2004)] leading to protective vitrification of the cytoplasm rather than crystallization. This argument suggests that cells may promote extracellular ice formation to induce vitrification and a potential function of ice binding proteins (for reviews of ice binding proteins see Dolev et al. 2016; Vrielink et al. 2016). Survival in a vitrified state is therefore essential to the low-temperature limit of life given the daily, seasonal, and annual temperature fluctuations experienced by organisms in many low-temperature environments.

Theoretical approaches to the low-temperature limit of life commonly look to the basic physical and chemical laws that govern the universe and living systems. This physical chemistry approach can be used to understand the behavior of biomolecules (for a recent review see Gruebele and Thirumalai 2013). For example, Williams and Frausto da Silva outlined how the kinetic and thermodynamic properties of the chemical elements lead to the organization and behavior of molecules and systems (including cells) that exist today (Williams and Frausto da Silva 1996). In a different approach, entropy is seen as the driving force for the organization of self-replicating systems like cells (England 2013). In these examples, physical chemistry concepts are being applied to the study of life generally, but are not yet

---

Fig. 2.1 (continued) River, northern Alaska (image source: Mikhail Kanevskiy, University of Alaska Fairbanks); (**c**) University Valley, Antarctica (photo credit Margarita Marinova, NASA Ames Research Center)

being applied to the question of the limits of life. (Some physical chemistry approaches that are being used to study temperature limits are mentioned below.) And in another theoretical approach, the analysis of the information flow and management within cells, organisms, and ecosystems is predicted to reveal underlying universal laws of biology (Davies and Walker 2016). Perhaps in the future, some of the general theories on life will lead to predictions about the low-temperature (or other) limit of life.

2.3 Reductionist Approach: Limits of Cellular Components

Reductionist work seeks to describe and explain the whole as the sum of its parts, and thus often begins with a description and explanation of the parts. What can reductionist approaches to the function of biological molecules tell us about the limits of life? Is there one particular component of cells that limits life at low temperatures regardless of the functionality of other parts? Some interesting results from reductionist approaches are highlighted below with a focus on proteins and membranes whose functionality is key to metabolism.

2.3.1 Proteins

Without functional enzymes, cells cannot perform the chemical reactions necessary to sustain life. As temperature decreases, reaction rates decrease while the rigidity of proteins increases. To combat these effects, the molecular structure of proteins from psychrophiles is altered by decreasing stabilizing interactions to increase the disorder of the molecule thereby maintaining function (Feller 2013). For example, in a study that compared 47 proteins from psychrophiles with 814 proteins from mesophiles and 269 proteins from thermophiles, all interactions between residues were weakened in proteins from psychrophiles concomitant with a decrease in uncharged polar, charged, and aromatic residues and an increase in hydrophobic residues (Goldstein 2007). As a result of these alterations, which generally result in less compact structures, cold-adapted proteins have a high specific activity (due to a lower energy of activation) and a low thermal stability relative to proteins from mesophiles (Marx et al. 2007). During growth at low temperatures, protein folding is also aided by an increase in the expression of chaperones like trigger factor (Mykytczuk et al. 2011; Cipolla et al. 2012). For a detailed review of the general properties, activity, stability, thermodynamic stability, folding, and engineering of cold-adapted proteins, see Gerday (2013) and Chap. 10. Gerday suggests that the cold adaptation of psychrophilic proteins is not complete given that the k_{cat} at the optimum temperature for activity (T_{opt}) of psychrophilic proteins is lower than the k_{cat} of mesophilic and thermophilic proteins at their T_{opt}. Lastly, given the many ways that proteins could be altered to effect an increase in disorder, predicting the specific cold adaptations in any one protein is exceedingly difficult.

A recent study modeling the temperature dependence of enzyme-catalyzed reactions identified a "psychrophilic trap" which restricts the activity of enzymes with low optimum temperatures (≤20 °C) to a narrow temperature range of about 20°C (Arcus et al. 2016). In this model, psychrophilic enzymes must decrease $\Delta C_p^{\ddagger}$ (the change in heat capacity between the transition and ground state) to have a lower T_{opt} than mesophilic enzymes (note that in this model enzymes with similar T_{opt} should have similar $\Delta C_p^{\ddagger}$ regardless of the reaction catalyzed). Indeed, the increased K_m (which lowers the binding affinity for the substrate) and k_{cat} (which raises the binding affinity for the transition state) of psychrophilic enzymes (Feller 2013) effectively decreases the $\Delta C_p^{\ddagger}$. For psychrophilic enzymes to have $T_{opt} \leq 20$ °C, the required $\Delta C_p^{\ddagger}$ (e.g., −18.0 kJ mol^{-1} K^{-1} for a $T_{opt} = 10$ °C) results in high curvature in this model of the "macromolecular rate theory" and a narrow temperature range of activity. Avoiding this narrow temperature range of activity may help explain why no psychrophilic enzymes are found with $T_{opt} \leq 20$ °C (Cipolla et al. 2012). In addition, the model suggests that larger enzymes are required for "harder" chemical reactions (those with a higher $k_{cat}/k_{noncatalyzed}$) because the higher heat capacity is required to achieve a similar $\Delta C_p^{\ddagger}$. Perhaps psychrophilic enzymes would also benefit from being larger considering that at a given low temperature psychrophilic enzymes have a higher k_{cat} than mesophilic enzymes (Gerday 2013). Indeed, some psychrophilic enzymes have loops which contribute to cold adaptation and are longer than their mesophilic homologs (Sonan et al. 2007; Helland et al. 2009; Yusof et al. 2015; Miao et al. 2016). More studies are needed to support or refute this model of enzyme interaction which has interesting implications for the limits of psychrophilic enzymes.

Remarkably, enzymes can remain functional at extreme low temperatures if the solvent remains liquid. Both catalase (from beef liver) and alkaline phosphatase (from calf intestine) remained active and did not deviate significantly from Arrhenius behavior over the temperature range of +20 °C to −97 °C and −100 °C, respectively, with a seven log decrease in activity (Bragger et al. 2000). Liquid conditions were maintained using the cryosolvent methanol:ethylene glycol:water (70:10:20). While these experiments do not mimic conditions seen in nature, they do demonstrate what is possible for enzymes that can be maintained in the native state with an appropriate solvent.

2.3.2 Membranes

At low temperatures, cell membranes will undergo a phase change from a liquid to a gel. As a gel, membranes are no longer functional becoming too rigid to allow protein movement and hence function. To maintain fluidity at lower temperatures, most cells increase the amount of unsaturated fatty acids while some bacteria can also increase the amount of methyl branched fatty acids or shorter fatty acid chains

(Russell 2007, 2008; Gerday 2011). These modified fatty acids have lower melting temperatures (T_m); for example, palmitoleic acid (C16:1ω7) has a T_m of 0 °C while linoleic acid (C18:2ω9,12) has a T_m of −7 °C (Guendouzi and Mekelleche 2012). These changes in the membrane have been observed in pure culture and in the environment. For example, more unsaturated fatty acids were detected at deeper depths of the active layer of permafrost and after incubation in microcosms at 4 °C versus incubation at 20 °C (Mangelsdorf et al. 2009). Long chain polyunsaturated fatty acids (PUFA) are also common in bacteria and may contribute to membrane fluidity at low temperatures given their very low T_m (Taha et al. 2013). For example, in the membrane of strain 651 isolated from Antarctic sea ice, eicosapentaenoic acid (EPA; C20:5 n−3) was shown to increase from 2% to 12% as the growth temperature decreased from 15 °C to 2 °C, suggesting a contribution to membrane fluidity (Nichols et al. 1997). Furthermore, EPA and docosahexaenoic acid (DHA; C22:6 n−3) are common in marine bacteria with DHA ($T_m = -45$ °C) more common at lower temperatures (Hamamoto et al. 1995; Yoshida et al. 2016); however, the contribution of PUFA to membrane fluidity has yet to be unambiguously demonstrated and confirmed. Individually, UFA and PUFA can have very low T_m, but how do they affect the T_m of a cell membrane? While the T_m of membranes of psychrophiles has not been routinely examined, the T_m of membranes of mesophiles have been examined. For example, while T_m is dependent on growth conditions, the cell membranes of the mesophilic *Lactobacillus delbrueckii* ssp. *bulgaricus* CFL1 that had more unsaturated fatty acids had lower phase transition temperatures ($T_s = -8$ °C and $T_m = -2$ °C) and were less rigid when "frozen" (Gautier et al. 2013). A functional membrane is essential to metabolism at low temperatures, will ensure protective vitrification of the cytoplasm occurs, and may define the low-temperature limit of life.

2.4 Systems Approach: Limits of Whole Cells

Of course, cellular components must interact to become a whole cell that can function under a variety of environmental conditions and reproduce. Moving from components to whole cells introduces a large amount of complexity, particularly if attempting to model networks of interactions (for reviews of recent advances in network and systems biology. see Barabasi and Oltvai 2004; Radde and Hutt 2016). What are whole cells capable of at low temperatures? Ideally, the emergent property of cell division should be measured to understand the low-temperature limits, since life without reproduction can survive, but may not evolve. Measuring cell division in the environment or in communities is difficult, particularly when cell division may proceed extraordinarily slowly. Therefore, only one aspect of cellular metabolism is often measured even for pure cultures with the underlying assumption that the more enzymes that must interact the more relevant the measurement is to the assessment of active metabolism that could lead to reproduction. However, the typically low rates make it difficult to evaluate if metabolism supportive of reproduction is occurring. Next, both pure culture work and microcosm studies are examined for data germane to the low-temperature limit of life.

2.4.1 Pure Cultures

Reproduction (cell division) has been documented at temperatures of −10 °C to −12 °C in several bacteria (Bakermans et al. 2003, 2006; Breezee et al. 2004). But the bacterium with the lowest demonstrated growth temperature to date is *Planococcus halocryophilus* Or1 which was isolated from the active layer (<1 m depth) of permafrost on Ellesmere Island, Canada (Mykytczuk et al. 2012). *P. halocryophilus* Or1 is capable of cell division with a generation time of 50 days at −15 °C with NaCl and glycerol added to maintain the liquidity of the medium (Mykytczuk et al. 2013). Growth at −15 °C is characterized by increased hydrophobicity of the cell wall and "distinct extracellular encrustations" composed of 20% calcium carbonate, 50% peptidoglycan, and 29% choline (Mykytczuk et al. 2016). The authors suggest that choline serves as an osmo- or cryo-protectant and that carbonate may also have some protective role. *P. halocryophilus* Or1 may be able to reproduce at lower temperatures considering that it can mineralize ^{14}C-acetate (1% after 210 days) in permafrost microcosms at −25 °C (Mykytczuk et al. 2013). The cold adaptations found in *P. halocryophilus* Or1 are not unique, but common to many psychrophilic organisms. Numerous fungi can also reproduce at subfreezing temperatures (Hassan et al. 2016). The yeast *Rhodotorula glutinis* has the lowest demonstrated growth temperature with a generation time of 34 days at −18 °C when inoculated onto frozen peas (Collins and Buick 1989).

In addition, a variety of metabolic processes have been demonstrated to continue in pure cultures at low temperatures (for review, see Rummel et al. (2014), recent work is briefly summarized here). The ability to synthesize DNA suggests that cells may be dividing or, if not sustained, that cells may be repairing DNA. DNA synthesis has been documented at −5 °C in frozen cultures of two glacial ice isolates (Bakermans and Skidmore 2011a) and at −15 °C in frozen cultures of the permafrost isolates *Psychrobacter arcticus* 273-4 and *Psychrobacter cryohalolentis* K5 (Amato et al. 2010). Rates at −15 °C were high enough to offset the expected DNA damage from radiation in permafrost (Amato et al. 2010). Bacteria can also repair DNA at low temperatures. Double strand break (DSB) repair of DNA was demonstrated by *Psychrobacter arcticus* 273-4 over 505 days at −15 °C while frozen, but no reproduction was evident (Dieser et al. 2013). The average rate of repair was estimated at eight DSBs per year which is far greater than the estimated rate of damage due to radiation (one DSB every 14,500 years), and the authors concluded that DSBs should not be an issue for long-term survival if energy is available.

Other metabolic processes that cannot clearly differentiate between reproduction or maintenance have also been documented in pure cultures. For example, protein synthesis has been documented at −5 °C in frozen cultures of two glacial ice isolates (Bakermans and Skidmore 2011a), at −15 °C in frozen cultures of the permafrost isolates *P. arcticus* 273-4 and *P. cryohalolentis* K5 (Amato et al. 2010), and at −20 °C in frozen cultures of the sea ice isolate *Colwellia psychroerythraea* 34H (Junge et al. 2006). In *P. cryohalolentis* K5, ATP pools increased as temperatures decreased from 22 °C to −20 °C (Amato and Christner 2009), probably to compensate for decreased reaction rates. Furthermore, respiration ($^{14}CO_2$ production) has been documented at −15 °C (Bakermans and Skidmore 2011b), −25 °C (Mykytczuk et al. 2013), and −33 °C (Bakermans and Skidmore 2011b). And ammonia oxidation has been demonstrated in frozen cultures of *Nitrosomonas cryotolerans* at −32 °C (Miteva et al. 2007). These activities require the coordinated function of more than one enzyme suggesting that complex metabolism is possible at low temperatures and that other metabolism likely occurs to support these reactions.

2.4.2 Microcosms

Recent microcosm studies continue to confirm previous reports of low-temperature metabolism and push the lower temperature limit of metabolism to new lows in more complex and realistic situations. While direct evidence of reproduction is difficult to measure directly in microcosms, bacteria from sea ice increased in numbers during incubation at −12 °C in brine over 200 h (an increase in bacteria was only evident in one of three samples from one sea ice core; Wells and Deming 2006). Many microcosm studies in frozen soils and permafrost have demonstrated respiration down to temperatures of −18 °C (for a review, see Nikrad et al. 2016). DNA synthesis (^{13}C-acetate incorporation into DNA) has been demonstrated in microcosms of Alaskan permafrost at temperatures from 0 °C to −20 °C (Tuorto et al. 2014) over a 6 month incubation. Furthermore, this study demonstrated that most of the organisms present were active as 80% of OTUs in untreated permafrost were evident as labeled DNA in SIP microcosms. Follow-up work has similarly demonstrated the incorporation of ^{13}C from a variety of sources (including acetate, cellobiose, and algal biomass) into DNA occurs in microcosms of Arctic Finnish tundra soils during incubations at 0 °C, −4 °C, and −16 °C and substrate preference changed with temperature (P. Gadkari and M. Haggblom, personal communication). As before, many diverse species were active. These data are highly suggestive of actively dividing cells (at temperatures to −20 °C) and not just DNA repair, particularly given the low rate of expected DNA damage at low temperatures (Hoehler and Jorgensen 2013). Other types of metabolism that have been documented at low temperatures in microcosm studies support the conclusion that active metabolism is possible at low subfreezing temperatures and include: methanogenesis in permafrost at −10 °C (Rivkina et al. 2007), protein synthesis in Lake Vida brine at −13.5 °C (Murray et al. 2012), lipid synthesis in permafrost at

−20 °C (Rivkina et al. 2000), and sulfate reduction in sediments from Lost Hammer Spring hypersaline reducing brine at −20 °C (Lamarche-Gagnon et al. 2015).

2.5 Observational Approach: Limits Determined from Cold Environments

Given that life at low temperatures could proceed extraordinarily slowly, where on Earth do we find cold environments that have persisted for many years for observational studies? Predictably, the polar regions contain some of Earth's oldest and coldest habitats that include permafrost and ice. With the onset of glacial conditions at the end of the Pliocene (about 2.5 million years ago) permafrost formation could begin in the Arctic with a large amount formed during the middle and upper Pleistocene. In North America, the oldest ice dates to ca. 740,000 years ago (Froese et al. 2008); while in Eurasia, 2−3 million year old permafrost exists in northeastern Siberia (Vishnivetskaya et al. 2000, 2006). While the active layer of permafrost is subject to more extreme low temperatures, at depth, permafrost is stable at temperatures from 0 °C to −10 °C in Alaska and −13 °C in Siberia (although it may be hard to constrain exact historical temperatures beyond knowing that these soils have remained below freezing). Microorganisms have been isolated from permafrost environments in many studies, demonstrating that survival of these long-term low-temperature conditions is frequently realized (Gilichinsky and Wagener 1995; Vishnivetskaya et al. 2006; Johnson et al. 2007). However, debate continues over how active microorganisms are while trapped in permafrost.

The high lability of carbon upon thawing of the organic rich permafrost of the Arctic suggests that permafrost carbon has not been noticeably decomposed while frozen (Schuur et al. 2008; Waldrop et al. 2010). For example, when the organic rich Yedoma permafrost from northeast Siberia (Fig. 2.1b) was examined over an age range of 30,100 to >55,000 years old there was no correlation between any indicator of organic matter quality (higher plant fatty acids, C/N ratio, $\delta^{13}C$ values of TOC, or carbon preference index) and depth or age indicating that no significant decomposition had occurred over this time frame (Strauss et al. 2015). Does this suggest that microorganisms are not capable of being active? Or that their metabolism is extraordinarily slow? Certainly permafrost is limited not just by low temperature, but also by diffusion through a frozen matrix. Perhaps organisms could be active at the low temperatures, but are limited by a lack of nutrients (although nutrients are present, they diffuse so slowly that microorganisms cannot be sustained) or inhibited by an accumulation of waste (waste does not diffuse away and poisons cells). Indeed, the long-term storage of carbon in permafrost was more accurately predicted during modeling of carbon respiration in soils when a limit on substrate availability due to low diffusion rates through thin films in frozen soils was included in the model (Schaefer and Jafarov 2016). Perhaps, 25,000 years is not a sufficient amount of time for significant decomposition to be detected.

For comparison, consider that temperature does not limit growth in deep sea sediments (temperatures are above 0 °C), energy limits growth due to low rates of diffusion (Hoehler and Jorgensen 2013). Even so, growth is possible at the low rates of diffusion in subseafloor sediments and may be sustained over millions of years with biomass turnover times of 200–4000 years (Biddle et al. 2006; Lomstein et al. 2012). And deep subsurface rates are 10-fold faster than required to offset damage due to racemization and 1000-fold faster than required to offset damage due to depurination. Not surprisingly then, most cells in the deep subsurface were found to be viable and potentially active, not dormant (Morono et al. 2011).

Since diffusion should be slower in permafrost than in subseafloor sediments, growth should also be slower—so slow that degradation of carbon over the "short" time period that some permafrost has existed would not be evident (below the detection limits of current methodologies). Certainly, analysis of carbon in soils is exceedingly complex and while proxies have been developed for organic matter quality, these proxies are relatively gross measures of carbon degradation. Alternately, the low diffusion rates could be below the necessary energy flux (i.e., the "basal power requirement") that cells require to remain viable (Hoehler and Jorgensen 2013; Lever et al. 2015). Although the basal power requirement of cells in cold environments would be lower due to lower rates of damage (mostly racemization of amino acids) at low temperatures, and would allow for a higher abundance of organisms and longer survival under energy-limited conditions (Hoehler and Jorgensen 2013). Many cells do survive long-term burial in permafrost and likely require intermittent maintenance and repair to remain viable; indeed, long-term burial in permafrost appears to select against endospores which have no maintenance or repair metabolism (Johnson et al. 2007). In addition, various metabolic activities have been demonstrated in microcosms (see above) at low temperatures, also suggesting the ability for long-term survival that requires metabolism in permafrost. Intriguingly, an increase in concentration of acetate with depth in Yedoma permafrost from Siberia that ranged in age from 2000 to 134,000 year old was interpreted as evidence for ongoing fermentation of carbon under anoxic conditions (Ewing et al. 2015). Continued study is needed to resolve what is limiting carbon degradation and the extent of metabolic activity in permafrost.

At the south pole, Antarctica comprises a diversity of low-temperature environments that includes sea ice, water, and sediments; dry, wet, and frozen soils; freshwater, saline, and ice covered lakes; subglacial lakes; cryoconites; and alpine areas (Pearce 2012). Antarctica has a longer history of low temperatures: ice sheets began forming with continental cooling at the start of the Oligocene 34 million years ago, but ice sheets did not become permanent until 15 million years ago (Barrett 2003). Some of the oldest permafrost in Antarctica could be 5–8.1 million years old (Bidle et al. 2007; Gilichinsky et al. 2007). Many Antarctic habitats contain viable microbial communities as demonstrated by the many studies documenting microbial diversity and activity therein (Chong et al. 2015; Chown et al. 2015). Sometimes these microbial communities are surprisingly active. Soil bacteria were found to respond quickly during a seal carcass transplant experiment in the Miers Valley taking only 2 years for bacterial communities under newly covered soils to become similar to those found under 250 year old seals (Tiao et al. 2012). Furthermore, carbon fixation in wetted and arid soils of the McMurdo Dry Valleys had remarkably fast turnover times of 7–140 days during the austral summer (Niederberger et al. 2015). Not surprisingly, metabolic activity is dependent on abiotic factors such as water and temperature (Stomeo et al. 2012).

The McMurdo Dry Valleys of Antarctica are the coldest and driest place on Earth and thus a good place for observing the low-temperature limits of life. University Valley (UV), in the stable upland zone, is one of the coldest places in the Dry Valleys with mean annual air temperatures of −23 °C (Fig. 2.1c). Given that the climate conditions that establish these cold temperatures in UV have probably not changed for the last ca. 150,000 years, UV is a good candidate environment for examining microorganisms at extremes of temperature (and dryness). Recent investigations in UV sought to characterize active communities in both the upper dry layers of permafrost and the deeper ice-cemented permafrost of the valley floor which range in age from 10^4 to 10^5 years old (Goordial et al. 2016a, b). Intriguingly, data did not demonstrate the presence of active microorganisms. Microcosms of valley floor permafrost did not respire ^{14}C-acetate at −5 °C even after 600 days of incubation (Goordial et al. 2016b), although microcosms of cryptoendoliths that inhabit the sandstone walls of UV respired ^{14}C-acetate at temperatures from −5 °C to −20 °C. Further experiments demonstrated that neither a lack of N or P nor soil toxicity was inhibiting respiration in UV permafrost. In addition, cell counts were very low (ca. 10^3 g^{-1} soil) and only six isolates were obtained from over 1000 agar plates in 2 years of trials. A few samples (but not all) of UV permafrost respired ^{14}C-acetate at +5 °C, which was attributed to cells that survived, but were not active, in the permafrost. Metagenomic analyses revealed an enrichment in dormancy and sporulation genes in UV permafrost, corroborating the idea that any cells present are likely endospores or dormant cells. These multiple lines of evidence led the authors to conclude that University Valley is too cold and too dry to sustain microbial life. With a mean annual air temperature of −23 °C, temperatures at the surface only rise above 0 °C for a few hours a day totaling less than 80 h per year. These low temperatures also contribute to a lack of liquid water in UV permafrost where water exists primarily as interfacial water; bulk water (as thin films) is likely only present for about 74 h a year when temperatures in surface soils rise to −1 °C.

2.6 Conclusions

Many studies have demonstrated that reproduction by microorganisms is possible at temperatures of −10 °C to −20 °C and that metabolism continues to even lower temperatures of about −30 °C. Interestingly, these lowest temperatures of activity correspond with the glass transition temperature (T_g) of the cytoplasm of several microorganisms suggesting that T_g may define the low-temperature limit of active metabolizing life. More generally, the lack of liquid water at low temperatures appears to be the true limit for life at low temperatures, and not low temperature per se. The presence of liquid water is of course inextricable tied to temperature. If solvent conditions can be maintained at low temperatures, metabolism may be maintained (enzymes can remain functional to −90 °C as described above). To more fully understand how biomolecules behave at extreme low temperatures, more laboratory studies are needed, especially of more complex enzymes, membranes, and whole cells. More theoretical studies are also needed to better model metabolism at low rates and to discover universal laws of biology that might inform the low-temperature limit of life.

On Earth, low-temperature conditions (and lack of liquid water) appear to limit life in a few places like Don Juan Pond and University Valley in Antarctica. Microorganisms may be active in Arctic permafrost, but metabolism could be so exceptionally slow as to escape detection. Given the slow metabolism expected and the short geological age of permafrost (3 million years at the longest), there is probably insufficient time for microorganisms to evolve to become better adapted to live at subfreezing temperatures. Indeed, terrestrial life may never have the chance to evolve to exploit the low-temperature capabilities of its biomolecules (like the low T_m of PUFA) in view of the limited times at which cold environments persist on Earth. Even the Cryogenian period of near global glaciation only lasted about 85 million years and was followed by warmer conditions in which low-temperature refugia may not have existed to preserve low-temperature adaptations. Instead, the inhabitants of modern low-temperature environments (regularly below 0 °C) likely evolved from mesophiles to survive extreme low-temperature conditions and exploit more clement conditions when they arise (such as the 80 h a year when surface soil temperatures rise above 0 °C in University Valley). If life at low temperatures primarily persists with this discontinuous lifestyle, observational studies will be needed to better understand how much activity is needed in clement conditions for long-term survival and even evolution to extreme low-temperature environments. Lastly, the low-temperature limit of life might be better examined elsewhere in the solar system such as Mars or Europa where low-temperature conditions have existed for much longer.

References

Amato P, Christner BC (2009) Energy metabolism response to low-temperature and frozen conditions in *Psychrobacter cryohalolentis*. Appl Environ Microbiol 75(3):711–718. doi:10.1128/aem.02193-08

Amato P, Doyle SM, Battista JR, Christner BC (2010) Implications of subzero metabolic activity on long-term microbial survival in terrestrial and extraterrestrial permafrost. Astrobiology 10 (8):789–798. doi:10.1089/ast.2010.0477

Arcus VL, Prentice EJ, Hobbs JK, Mulholland AJ, Van der Kamp MW, Pudney CR, Parker EJ, Schipper LA (2016) On the temperature dependence of enzyme-catalyzed rates. Biochemistry (Moscow) 55(12):1681–1688. doi:10.1021/acs.biochem.5b01094

Bakermans C, Skidmore M (2011a) Microbial metabolism in ice and brine at -5°C. Environ Microbiol 13(8):2269–2278. doi:10.1111/j.1462-2920.2011.02485.x
Bakermans C, Skidmore ML (2011b) Microbial respiration in ice at subzero temperatures (-4 to -33°C). Environ Microbiol Rep 3(6):774–782. doi:10.1111/j.1758-2229.2011.00298.x
Bakermans C, Tsapin AI, Souza-Egipsy V, Gilichinsky DA, Nealson KH (2003) Reproduction and metabolism at -10°C of bacteria isolated from Siberian permafrost. Environ Microbiol 5 (4):321–326
Bakermans C, Ayala-del-Río HL, Ponder MA, Vishnivetskaya T, Gilichinsky D, Thomashow MF, Tiedje JM (2006) *Psychrobacter cryohalolentis* sp. nov. and *Psychrobacter arcticus* sp. nov. isolated from Siberian permafrost. Int J Syst Evol Microbiol 56(6):1285–1291
Barabasi AL, Oltvai ZN (2004) Network biology: understanding the cell's functional organization. Nat Rev Genet 5(2):101–U115. doi:10.1038/nrg1272
Barrett P (2003) Palaeoclimatology – Cooling a continent. Nature 421(6920):221–223
Beaty D, Buxbaum K, Meyer M, Barlow N, Boynton W, Clark B, Deming J, Doran PT, Edgett K, Hancock S, Head J, Hecht M, Hipkin V, Kieft T, Mancinelli R, McDonald E, McKay C, Mellon M, Newsom H, Ori G, Paige D, Schuerger AC, Sogin M, Spry JA, Steele A, Tanaka K, Voytek M (2006) Findings of the Mars special regions science analysis group. Astrobiology 6 (5):677–732
Biddle JF, Lipp JS, Lever MA, Lloyd KG, Sorensen KB, Anderson R, Fredricks HF, Elvert M, Kelly TJ, Schrag DP, Sogin ML, Brenchley JE, Teske A, House CH, Hinrichs KU (2006) Heterotrophic Archaea dominate sedimentary subsurface ecosystems off Peru. Proc Natl Acad Sci USA 103(10):3846–3851. doi:10.1073/pnas.0600035103
Bidle KD, Lee S, Marchant DR, Falkowski PG (2007) Fossil genes and microbes in the oldest ice on Earth. Proc Natl Acad Sci USA 104(33):13455–13460
Bragger JM, Dunn RV, Daniel RM (2000) Enzyme activity down to −100°C. Biochim Biophys Acta – Protein Struct Mol Enzymol 1480(1–2):278–282. doi:10.1016/S0167-4838(00)00081-9
Breezee J, Cady N, Staley JT (2004) Subfreezing growth of the sea ice bacterium "*Psychromonas ingrahamii*". Microb Ecol 47(3):300–304
Chong C-W, Pearce DA, Convey P (2015) Emerging spatial patterns in Antarctic prokaryotes. Front Microbiol 6(1058). doi:10.3389/fmicb.2015.01058
Chown SL, Clarke A, Fraser CI, Cary SC, Moon KL, McGeoch MA (2015) The changing form of Antarctic biodiversity. Nature 522(7557):427–434. doi:10.1038/nature14505
Cipolla A, Delbrassine F, Da Lage JL, Feller G (2012) Temperature adaptations in psychrophilic, mesophilic and thermophilic chloride-dependent alpha-amylases. Biochimie 94 (9):1943–1950. doi:10.1016/j.biochi.2012.05.013
Clarke A, Morris GJ, Fonseca F, Murray BJ, Acton E, Price HC (2013) A low temperature limit for life on Earth. PLoS ONE 8(6):e66207. doi:10.1371/journal.pone.0066207
Collins MA, Buick RK (1989) Effect of temperature on the spoilage of stored peas by *Rhodotorula glutinis*. Food Microbiol (London) 6(3):135–142. doi:10.1016/s0740-0020(89)80021-8
Corkrey R, McMeekin TA, Bowman JP, Ratkowsky DA, Olley J, Ross T (2016) The biokinetic spectrum for temperature. PLoS ONE 11(4):e0153343. doi:10.1371/journal.pone.0153343
Davies PCW, Walker SI (2016) The hidden simplicity of biology. Rep Prog Phys 79(10):18. doi:10.1088/0034-4885/79/10/102601
Dieser M, Battista JR, Christner BC (2013) DNA double-strand break repair at -15 degrees C. Appl Environ Microbiol 79(24):7662–7668. doi:10.1128/aem.02845-13
Dolev MB, Braslavsky I, Davies PL (2016) Ice-binding proteins and their function. Annu Rev Biochem 85(1):515–542. doi:10.1146/annurev-biochem-060815-014546
England JL (2013) Statistical physics of self-replication. J Chem Phys 139(12):121923. doi:10.1063/1.4818538
Ewing SA, O'Donnell JA, Aiken GR, Butler K, Butman D, Windham-Myers L, Kanevskiy MZ (2015) Long-term anoxia and release of ancient, labile carbon upon thaw of Pleistocene permafrost. Geophys Res Lett 42(24):10730–10738. doi:10.1002/2015gl066296

Feller G (2013) Psychrophilic enzymes: from folding to function and biotechnology. Scientifica (Cairo). doi:10.1155/2013/512840

Fonseca F, Marin M, Morris GJ (2006) Stabilization of frozen *Lactobacillus delbrueckii* subsp. *bulgaricus* in glycerol suspensions: freezing kinetics and storage temperature effects. Appl Environ Microbiol 72(10):6474–6482. doi:10.1128/aem.00998-06

Fonseca F, Meneghel J, Cenard S, Passot S, Morris GJ (2016) Determination of intracellular vitrification temperatures for unicellular microorganisms under conditions relevant for cryopreservation. PLoS ONE 11(4):e0152939. doi:10.1371/journal.pone.0152939

Froese DG, Westgate JA, Reyes AV, Enkin RJ, Preece SJ (2008) Ancient permafrost and a future, warmer Arctic. Science 321(5896):1648–1648. doi:10.1126/science.1157525

Gautier J, Passot S, Pénicaud C, Guillemin H, Cenard S, Lieben P, Fonseca F (2013) A low membrane lipid phase transition temperature is associated with a high cryotolerance of *Lactobacillus delbrueckii* subspecies *bulgaricus* CFL1. J Dairy Sci 96(9):5591–5602. doi:10.3168/jds.2013-6802

Gerday C (2011) Life at the extremes of temperature. In: Bacterial stress responses, 2nd edn. ASM Press, Washington, DC

Gerday C (2013) Psychrophily and catalysis. Biology (Basel) 2(2):719–741. doi:10.3390/biology2020719

Gilichinsky D, Wagener S (1995) Microbial life in permafrost: a historical review. Permafrost Periglac 6:243–250

Gilichinsky DA, Wilson GS, Friedmann EI, Mckay CP, Sletten RS, Rivkina EM, Vishnivetskaya TA, Erokhina LG, Ivanushkina NE, Kochkina GA, Shcherbakova VA, Soina VS, Spirina EV, Vorobyova EA, Fyodorov-Davydov DG, Hallet B, Ozerskaya SM, Sorokovikov VA, Laurinavichyus KS, Shatilovich AV, Chanton JP, Ostroumov VE, Tiedje JM (2007) Microbial populations in Antarctic permafrost: biodiversity, state, age, and implication for astrobiology. Astrobiology 7(2):275–311

Goldstein RA (2007) Amino-acid interactions in psychrophiles, mesophiles, thermophiles, and hyperthermophiles: insights from the quasi-chemical approximation. Protein Sci 16 (9):1887–1895. doi:10.1110/ps.072947007

Goordial J, Davila A, Greer CW, Cannam R, DiRuggiero J, McKay CP, Whyte LG (2016a) Comparative activity and functional ecology of permafrost soils and lithic niches in a hyper-arid polar desert. Environ Microbiol. doi:10.1111/1462-2920.13353

Goordial J, Davila A, Lacelle D, Pollard W, Marinova MM, Greer CW, DiRuggiero J, McKay CP, Whyte LG (2016b) Nearing the cold-arid limits of microbial life in permafrost of an upper dry valley, Antarctica. ISME J 10(7):1613–1624. doi:10.1038/ismej.2015.239

Gruebele M, Thirumalai D (2013) Perspective: reaches of chemical physics in biology. J Chem Phys 139(12):121701. doi:10.1063/1.4820139

Guendouzi A, Mekelleche SM (2012) Prediction of the melting points of fatty acids from computed molecular descriptors: a quantitative structure-property relationship study. Chem Phys Lipids 165(1):1–6. doi:10.1016/j.chemphyslip.2011.10.001

Hamamoto T, Takata N, Kudo T, Horikoshi K (1995) Characteristic presence of polyunsaturated fatty acids in marine psychrophilic vibrios. FEMS Microbiol Lett 129(1):51–56

Hassan N, Rafiq M, Hayat M, Shah AA, Hasan F (2016) Psychrophilic and psychrotrophic fungi: a comprehensive review. Rev Environ Sci Bio/Technol 15(2):147–172. doi:10.1007/s11157-016-9395-9

Helland R, Larsen RL, Asgeirsson B (2009) The 1.4 angstrom crystal structure of the large and cold-active Vibrio sp. alkaline phosphatase. Biochim Biophys Acta – Proteins Proteom 1794 (2):297–308. doi:10.1016/j.bbapap.2008.09.020

Hoehler TM, Jorgensen BB (2013) Microbial life under extreme energy limitation. Nat Rev Microbiol 11(2):83–94. doi:10.1038/nrmicro2939

Johnson S, Hebsgaard M, Christensen T, Mastepanov M, Nielsen R, Munch K, Brand T, Gilbert M, Zuber M, Bunce M, Ronn R, Gilichinsky D, Froese D, Willerslev E (2007) Ancient bacteria show evidence of DNA repair. Proc Natl Acad Sci 104(36):14401–14405

Junge K, Eicken H, Swanson BD, Deming JW (2006) Bacterial incorporation of leucine into protein down to -20°C with evidence for potential activity in sub-eutectic saline ice formations. Cryobiology 52(3):417–429

Lamarche-Gagnon G, Comery R, Greer CW, Whyte LG (2015) Evidence of in situ microbial activity and sulphidogenesis in perennially sub-0 degrees C and hypersaline sediments of a high Arctic permafrost spring. Extremophiles 19(1):1–15. doi:10.1007/s00792-014-0703-4

Lever MA, Rogers KL, Lloyd KG, Overmann J, Schink B, Thauer RK, Hoehler TM, Jorgensen BB (2015) Life under extreme energy limitation: a synthesis of laboratory- and field-based investigations. FEMS Microbiol Rev 39(5):688–728. doi:10.1093/femsre/fuv020

Lomstein BA, Langerhuus AT, D'Hondt S, Jorgensen BB, Spivack AJ (2012) Endospore abundance, microbial growth and necromass turnover in deep sub-seafloor sediment. Nature 484 (7392):101–104. doi:10.1038/nature10905

Mangelsdorf K, Finsel E, Liebner S, Wagner D (2009) Temperature adaptation of microbial communities in different horizons of Siberian permafrost-affected soils from the Lena Delta. Chemie Erde Geochem 69(2):169–182. doi:10.1016/j.chemer.2009.02.001

Marx JC, Collins T, D'Amico S, Feller G, Gerday C (2007) Cold-adapted enzymes from marine Antarctic microorganisms. Mar Biotechnol (NY) 9(3):293–304. doi:10.1007/s10126-006-6103-8

Mazur P (2004) Principles of cryobiology. In: Fuller BJ, Lane N, Benson EE (eds) Life in the frozen state. CRC Press, Boca Raton, FL, pp 3–65. doi:10.1201/9780203647073.ch1

Miao LL, Hou YJ, Fan HX, Qu J, Qi C, Liu Y, Li DF, Liu ZP (2016) Molecular structural basis for the cold adaptedness of the psychrophilic beta-glucosidase BglU in *Micrococcus antarcticus*. Appl Environ Microbiol 82(7):2021–2030. doi:10.1128/aem.03158-15

Miteva V, Sowers T, Brenchley J (2007) Production of N_2O by ammonia oxidizing bacteria at subfreezing temperatures as a model for assessing the N_2O anolmalies in the Vostok Ice Core. Geomicrobiol J 24:451–459

Morono Y, Terada T, Nishizawa M, Ito M, Hillion F, Takahata N, Sano Y, Inagaki F (2011) Carbon and nitrogen assimilation in deep subseafloor microbial cells. Proc Natl Acad Sci 108 (45):18295–18300. doi:10.1073/pnas.1107763108

Murray AE, Kenig F, Fritsen CH, McKay CP, Cawley KM, Edwards R, Kuhn E, McKnight DM, Ostrom NE, Peng V, Ponce A, Priscu JC, Samarkin V, Townsend AT, Wagh P, Young SA, Yung PT, Doran PT (2012) Microbial life at −13 °C in the brine of an ice-sealed Antarctic lake. Proc Natl Acad Sci 109(50):20626–20631. doi:10.1073/pnas.1208607109

Mykytczuk NC, Trevors JT, Foote SJ, Leduc LG, Ferroni GD, Twine SM (2011) Proteomic insights into cold adaptation of psychrotrophic and mesophilic *Acidithiobacillus ferrooxidans* strains. Antonie Van Leeuwenhoek 100(2):259–277. doi:10.1007/s10482-011-9584-z

Mykytczuk NCS, Wilhelm R, Whyte LG (2012) *Planococcus halocryophilus* sp. nov.; an extreme subzero species from high Arctic permafrost. Int J Syst Evol Microbiol 62(8):1937–1944. doi:10.1099/ijs.0.035782-0

Mykytczuk NCS, Foote SJ, Omelon CR, Southam G, Greer CW, Whyte LG (2013) Bacterial growth at -15C; molecular insights from the permafrost bacterium *Planococcus halocryophilus* Or1. ISME J 7(6):1211–1226. doi:10.1038/ismej.2013.8

Mykytczuk NCS, Lawrence JR, Omelon CR, Southam G, Whyte LG (2016) Microscopic characterization of the bacterial cell envelope of *Planococcus halocryophilus* Or1 during subzero growth at −15 °C. Pol Biol 39(4):701–712. doi:10.1007/s00300-015-1826-5

Nichols DS, Brown JL, Nichols PD, McMeekin TA (1997) Production of eicosapentaenoic and arachidonic acids by an Antarctic bacterium: response to growth temperature. FEMS Microbiol Lett 152(2):349–354. doi:10.1016/s0378-1097(97)00224-3

Niederberger TD, Sohm JA, Gunderson T, Tirindelli J, Capone DG, Carpenter EJ, Cary SC (2015) Carbon-fixation rates and associated microbial communities residing in arid and ephemerally wet Antarctic Dry Valley soils. Front Microbiol 6:9. doi:10.3389/fmicb.2015.01347

Nikrad MP, Kerkhof LJ, Haggblom MM (2016) The subzero microbiome: microbial activity in frozen and thawing soils. FEMS Microbiol Ecol 92(6). doi:10.1093/femsec/fiw081

Pearce DA (2012) Extremophiles in Antarctica: life at low temperatures. In: Stan-Lotter H, Fendrihan S (eds) Adaption of microbial life to environmental extremes: novel research results and application. Springer, Vienna, pp 87–118. doi:10.1007/978-3-211-99691-1_5

Peters B, Casciotti KL, Samarkin VA, Madigan MT, Schutte CA, Joye SB (2014) Stable isotope analyses of NO2−, NO3−, and N2O in the hypersaline ponds and soils of the McMurdo Dry Valleys, Antarctica. Geochim Cosmochim Acta 135:87–101. doi:10.1016/j.gca.2014.03.024

Price PB (2000) A habitat for psychrophiles in deep Antarctic ice. Proc Natl Acad Sci 97 (3):1247–1251

Radde NE, Hutt MT (2016) The physics behind systems biology. EPJ Nonlinear Biomed Phys 4:19. doi:10.1140/epjnbp/s40366-016-0034-8

Rivkina EM, Friedmann EI, McKay CP, Gilichinsky DA (2000) Metabolic activity of permafrost bacteria below the freezing point. Appl Environ Microbiol 66(8):3230–3233

Rivkina E, Shcherbakova V, Laurinavichius K, Petrovskaya L, Krivushin K, Kraev G, Pecheritsina S, Gilichinsky D (2007) Biogeochemistry of methane and methanogenic archaea in permafrost. FEMS Microbiol Ecol 61(1):1–15

Rummel JD, Beaty DW, Jones MA, Bakermans C, Barlow NG, Boston PJ, Chevrier VF, Clark BC, de Vera J-PP, Gough RV, Hallsworth JE, Head JW, Hipkin VJ, Kieft TL, McEwen AS, Mellon MT, Mikucki JA, Nicholson WL, Omelon CR, Peterson R, Roden EE, Sherwood Lollar B, Tanaka KL, Viola D, Wray JJ (2014) A new analysis of Mars "special regions": findings of the second MEPAG special regions science analysis group (SR-SAG2). Astrobiology 14 (11):887–968. doi:10.1089/ast.2014.1227

Russell NJ (2007) Psychrophiles: membrane adaptations. In: Physiology and biochemistry of extremophiles, p 155–164

Russell N (2008) Membrane components and cold sensing. In: Margesin R, Schinner F, Marx J-C, Gerday C (eds) Psychrophiles: from biodiversity to biotechnology. Springer, Berlin, pp 177–190

Samarkin VA, Madigan MT, Bowles MW, Casciotti KL, Priscu JC, McKay CP, Joye SB (2010) Abiotic nitrous oxide emission from the hypersaline Don Juan Pond in Antarctica. Nat Geosci 3 (5):341–344. http://www.nature.com/ngeo/journal/v3/n5/suppinfo/ngeo847_S1.html

Schaefer K, Jafarov E (2016) A parameterization of respiration in frozen soils based on substrate availability. Biogeosciences 13(7):1991–2001. doi:10.5194/bg-13-1991-2016

Schuur EAG, Bockheim J, Canadell JG, Euskirchen E, Field CB, Goryachkin SV, Hagemann S, Kuhry P, Lafleur PM, Lee H, Mazhitova G, Nelson FE, Rinke A, Romanovsky VE, Shiklomanov N, Tarnocai C, Venevsky S, Vogel JG, Zimov SA (2008) Vulnerability of permafrost carbon to climate change: implications for the global carbon cycle. Bioscience 58 (8):701–714. doi:10.1641/b580807

Seki S, Kleinhans FW, Mazur P (2009) Intracellular ice formation in yeast cells vs. cooling rate: predictions from modeling vs. experimental observations by differential scanning calorimetry. Cryobiology 58(2):157–165. doi:10.1016/j.cryobiol.2008.11.011

Sonan GK, Receveur-Brechot V, Duez C, Aghajari N, Czjzek M, Haser R, Gerday C (2007) The linker region plays a key role in the adaptation to cold of the cellulase from an Antarctic bacterium. Biochem J 407:293–302. doi:10.1042/bj20070640

Stevenson A, Burkhardt J, Cockell CS, Cray JA, Dijksterhuis J, Fox-Powell M, Kee TP, Kminek G, McGenity TJ, Timmis KN, Timson DJ, Voytek MA, Westall F, Yakimov MM, Hallsworth JE (2015) Multiplication of microbes below 0.690 water activity: implications for terrestrial and extraterrestrial life. Environ Microbiol 17(2):257–277. doi:10.1111/1462-2920. 12598

Stomeo F, Makhalanyane TP, Valverde A, Pointing SB, Stevens MI, Cary CS, Tuffin MI, Cowan DA (2012) Abiotic factors influence microbial diversity in permanently cold soil horizons of a maritime-associated Antarctic Dry Valley. FEMS Microbiol Ecol 82(2):326–340. doi:10.1111/ j.1574-6941.2012.01360.x

Strauss J, Schirrmeister L, Mangelsdorf K, Eichhorn L, Wetterich S, Herzschuh U (2015) Organic-matter quality of deep permafrost carbon – a study from Arctic Siberia. Biogeosciences 12 (7):2227–2245. doi:10.5194/bg-12-2227-2015

Taha A, Ahmed RZ, Motoigi T, Watanabe K, Kurosawa N, Okuyama H (2013) Lipids in cold-adapted microorganisms. In: Yumoto I (ed) Cold-adapted microorganisms. Caister Academic Press, Wymondham

Tiao G, Lee CK, McDonald IR, Cowan DA, Cary SC (2012) Rapid microbial response to the presence of an ancient relic in the Antarctic Dry Valleys. Nat Commun 3:8. doi:10.1038/ncomms1645

Tuorto SJ, Darias P, McGuinness LR, Panikov N, Zhang T, Haggblom MM, Kerkhof LJ (2014) Bacterial genome replication at subzero temperatures in permafrost. ISME J 8(1):139–149. doi:10.1038/ismej.2013.140

Vishnivetskaya T, Kathariou S, McGrath J, Gilichinsky D, Tiedje JM (2000) Low-temperature recovery strategies for the isolation of bacteria from ancient permafrost sediments. Extremophiles 4:165–173

Vishnivetskaya TA, Petrova MA, Urbance J, Ponder M, Moyer CL, Gilichinsky DA, Tiedje JM (2006) Bacterial community in ancient siberian permafrost as characterized by culture and culture-independent methods. Astrobiology 6(3):400–414

Vrielink ASO, Aloi A, Olijve LLC, Voets IK (2016) Interaction of ice binding proteins with ice, water and ions. Biointerphases 11(1):17. doi:10.1116/1.4939462

Waldrop MP, Wickland KP, White R, Berhe AA, Harden JW, Romanovsky VE (2010) Molecular investigations into a globally important carbon pool: permafrost-protected carbon in Alaskan soils. Glob Change Biol 16(9):2543–2554. doi:10.1111/j.1365-2486.2009.02141.x

Wells LE, Deming JW (2006) Modelled and measured dynamics of viruses in Arctic winter sea-ice brines. Environ Microbiol 8(6):1115–1121

Williams R, Frausto da Silva J (1996) The natural selection of the chemical elements: the environment and life's chemistry. Oxford University Press, New York, NY

Yoshida K, Hashimoto M, Hori R, Adachi T, Okuyama H, Orikasa Y, Nagamine T, Shimizu S, Ueno A, Morita N (2016) Bacterial long-chain polyunsaturated fatty acids: their biosynthetic genes, functions, and practical use. Mar Drugs 14(5):23. doi:10.3390/md14050094

Yusof NY, Abu Bakar FD, Mahadi NM, Raih MF, Murad AMA (2015) Structure prediction of Fe (II) 2-oxoglutarate dioxygenase from a psychrophilic yeast *Glaciozyma antarctica* PI12. In: Ahmad A, AbdKarim NH, Hassan NI et al. (eds) 2015 The UKM FST Postgraduate Colloquium. AIP Conference Proceedings, vol 1678. doi:10.1063/1.4931235

Part II
Microbial Diversity and Activity in Cold Ecosystems

Chapter 3
Aeromicrobiology

Lewis Cuthbertson and David A. Pearce

Abstract The atmosphere presents one of the most challenging environments on Earth with extreme psychrophilic and oligotrophic conditions. Cloud temperatures are typically below 0 °C, and stratospheric temperatures have been measured as low as −100 °C, colder than any other environment in the biosphere. Despite these challenges, the diversity of microbial life in the atmosphere has been found to be high, with over 100 bacterial genera recorded. Viable bacteria have been found in the stratosphere at altitudes of 77 km and bacteria collected from cloud water have been shown to be capable of both growth and reproduction at 0 °C suggesting the existence of psychrophilic bioaerosols. The role that microorganisms perform in processes such as ice nucleation, cloud formation, nitrogen processing, and the degradation of organic carbon-based compounds is becoming progressively clearer. However, the lack of a consensus on the most suitable bioaerosol sampling techniques makes progress challenging. Hence, although aerobiological studies date back as far as the mid nineteenth century, our understanding of microbial life in the atmosphere is still relatively limited. As the importance of understanding microbial biogeography continues to grow, particularly with regard to biogeography, long-range atmospheric dispersal, human health, and agriculture, more research is required to better understand the functional role of psychrophiles in the atmosphere.

Contents

L. Cuthbertson • D.A. Pearce (✉)
Department of Applied Sciences, Faculty of Health and Life Sciences, University of Northumbria at Newcastle, Northumberland Road, Newcastle upon Tyne NE1 8ST, UK
e-mail: david.pearce@northumbria.ac.uk

R. Margesin (ed.), *Psychrophiles: From Biodiversity to Biotechnology*,
DOI 10.1007/978-3-319-57057-0_3

3.1 Introduction

The atmosphere forms part of the biosphere and surrounds and interacts with every habitat and organism on Earth, either directly or indirectly. It is essential to life. However, life in the atmosphere itself is challenging, as organisms must cope with extreme psychrophilic and oligotrophic conditions. Although recent methodological developments in both the collection and analysis of aerial samples have accelerated our understanding of microorganisms inhabiting the atmosphere, many questions still remain. In this chapter, we present a brief overview of aerobiological studies to date, describe the atmosphere as a habitat for microorganisms, review what is known about bioaerosols with regard to dispersal, diversity, and atmospheric processes, and discuss the relevance of these organisms to atmospheric function.

3.2 History of Aerobiology

The term aerobiology was first established in the 1930s by Meier during his expeditions with Charles Lindbergh and was defined as the passive transport of biological particles through the atmosphere and its effect on living systems and the environment (Meier and Lindbergh 1935). The first indirect reference to the atmosphere as a source of microbial life came in the early nineteenth century when Louis Pasteur demonstrated that microorganisms exist all around us invisible to the naked eye (Pasteur 1860). Published research articles specifically focused on microbes in the atmosphere started to emerge relatively soon after this discovery (Airy 1874; Dyar 1894), but it wasn't until the mid-twentieth century when medical research about microorganisms began to take place that research into aerobiology gained momentum with studies proving the aerial transmission of tuberculosis (TB), influenza, and streptococcal infections (Lurie 1930; Brown and Allison 1937; Andrewes and Glover 1941).

Early aerobiological studies employed basic passive techniques such as plate fall assays for the collection of airborne microorganisms. These techniques are still very much in use today, as they result in a microbial culture which can be subjected to physiological investigations, unlike many of the molecular-based methods, which may only generate DNA sequence data. This was followed by the development of active sampling techniques in the early twentieth century, using simple impactor and impinger designs. These early designs inspired the Hirst spore trap and

Anderson air sampler, devised during the 1950s, and derivatives of these, such as the multiphase Anderson sampler (Bourdillon et al. 1941; Wells 1943; Hirst 1952; Andersen 1958). Modern techniques have not diverged greatly from these founding methods in terms of the basic principle, besides the inevitable improvements that come with technological advancement in manufacturing precision and effectiveness of the equipment. Typical of many microbiological studies, early aerobiological techniques relied solely on culture dependent methodologies, restricting observations to the relatively small culturable fraction of the atmospheric biota, which lead to the belief that the atmosphere had considerably lower levels of diversity, a belief that has changed following the development of microscopy and molecular techniques.

3.3 Modern Sampling Methods

Today, there is a much wider range of techniques used in aerobiological studies which rely on either the impaction, impingement, membrane filtration, cyclonic, or plate fall methods (Table 3.1).

3.4 Atmospheric Environment

As an environment for microbial life, the atmosphere can be regarded as harsh as any on Earth with the lowest temperatures, highest ultraviolet radiation, and extreme oligotrophic conditions. The structure of the atmosphere has been described in detail elsewhere, but in summary (adapted from NASA 2013), it can be divided into five main layers. The first is the Troposphere which is the layer at ground level extending up to 14.5 km; it is the densest layer of the atmosphere containing the majority of the Earth's life. It also contains over 99% of the water vapor in the atmosphere and is where almost all familiar weather conditions occur. The second layer is the Stratosphere which begins directly after the Troposphere and extends up to 50 km; the Stratosphere contains the ozone layer. Above that comes the Mesosphere which extends further up to 85 km, and temperatures in this layer reach lows of −100 °C, lower than any temperature ever recorded on the surface of the Earth. This is the highest atmospheric layer in which microscopic life has been found to date (Imshenetsky et al. 1977). Above the Mesosphere comes the Thermosphere and then the Exosphere, respectively, which extend beyond 600 km above the Earth's surface.

The majority of the atmosphere consists of dry air, which provides a challenging environment for microbial life. Clouds are a combination of condensed water droplets and ice crystals; they provide an important refuge for bioaerosols as they contain liquid water and levels of organic acids and alcohols comparable to fresh water lakes (Sattler et al. 2001).

Table 3.1 Summary of the available aerobiological sampling methods

Method	Example of sampler	Flow rates (L min^{-1})	Collection media	Advantages	Disadvantages	Analysis techniques
Impaction	SAS SUPER 100/180/ DUO360 (Bioscience International, Rockville, MD, USA)	<530	Contact plates, petri dishes, dry vessel, membrane filters (e.g. cellulose nitrate, cellulose acetate, and *PTFE*)	– High flow rate/short sampling – Portable – Multiple collection media – Multiple downstream analysis options	– High cost – Desiccation – High flow rate collection bias	– Microscopy – Molecular – Culture
Membrane filtration	Welch WOB-L vacuum pump (Welch, Mt. Prospect, IL, USA) with a Sartorius filtration unit (Göttingen, Germany)	~30	Membrane filters (e.g. cellulose nitrate, cellulose acetate, and *PTFE*)	– Long sample periods – Multiple downstream analysis options – Wide range of filters – Easy sample storage – Duration does not affect viability – Low cost	– Not portable – Low flow rate/long sampling durations – Self assembly	– Microscopy – Molecular – Culture
Impingement	SKC Biosampler (SKC Inc., Eighty Four, PA, USA)	~30	H_2O, *PBS*, mineral oil	– No desiccation – Portable – Viable samples – Multiple downstream analysis options – Multiple collection media	– High cost – Poor in cold environments	– Microscopy – Molecular – Culture
Drop plates	*N/A*	*N/A*	Petri dish	– Very low cost – Wide variety of sampling media – Viability shown	– Low proportion of microbes shown – Limited analysis options	– Culture – Limited microscopy – Limited molecular

Cyclonic (wet and dry)	Bertin Coriolis µ (Bertin Technologies, Montigny-le-Bretonneux, France)	<300	H_2O, *PBS*, dry vessel wall	– Very high flow rate/short sample period – Increased viability (wet) – Portable	– Very high cost – Desiccation (dry)	– Microscopy – Molecular – Culture

N/A not available, *PTFE* polytetrafluoroethylene, *PBS* phosphate-buffered saline

3.5 Microorganisms in the Atmosphere

Microorganisms found in the atmosphere are commonly referred to as bioaerosols; the term bioaerosol refers to all living components of the atmosphere. There are multiple types of bioaerosol found in the atmosphere including bacteria, archaea, algae, fungi, viruses, smaller eukaryotes, and pollen, and these can be viable, dead, dormant, or a combination of the three.

3.5.1 Aerosolization

It has been estimated that a bacterial biomass of 40–1800 Gg is aerosolized annually (Burrows et al. 2009b). Microorganisms enter the atmosphere from exposed terrestrial and aquatic surfaces. In terrestrial locations, bioaerosols are released from plant, soil, and other surfaces when drying reduces bonding forces, and this loosely bonded material is disturbed by strong air movements (Jones and Harrison 2004). Studies have shown vertical bacterial fluxes in terrestrial locations supporting this theory (Lighthart and Shaffer 1994). Recently a specific mechanism by which bioaerosols are formed from soil was discovered, whereby bubbles formed inside raindrops disperse micro-droplets containing bacteria during raindrop impingement, aerosolizing as much as 0.01% of soil surface bacteria (Joung et al. 2017). Aquatic bioaerosol formation is also directly related to air movement, as the majority of aquatic aerosols are released by either evaporation or a bubble bursting process (Blanchard and Syzdek 1982). Theories regarding the aerosolization of microbes are supported in the literature by findings which show direct relationships between the microbiota of the atmosphere and nearby surface level sources. However, contradictory studies also exist which show the potential for long-range atmospheric transport (Griffin 2007).

3.5.2 Dispersal

Once aerosolized, there are two main mechanisms by which bacteria are transported through the atmosphere: free-floating and attached to larger particles. Free-floating particles in the atmosphere are unlikely to come into contact with other microorganisms frequently, but particle-associated bacteria living in close quarters and subjected to stress while suspended in the atmosphere might be subjected to increased horizontal gene flow (Stewart 2013). It is this hypothetical horizontal gene transfer and the abundance of bacteria within the atmosphere that has drawn attention to the environment as a potential source of new antibiotics (Weber and Werth 2015). While airborne, it is estimated that bacteria have a residence time of between 2.2 and 188.1 days (Burrows et al. 2009a). Average generation times of bioaerosols have been measured to be between 3.6 and 19.5 days, a generation time comparable to that of many marine organisms (Sattler et al. 2001; Burrows et al. 2009b). Bioaerosols have been shown to undergo cross continental transport in plumes of desert dust in Asia (Griffin 2007) and across the Pacific Ocean with organisms of Asian origin detected in North America (Smith et al. 2012). Bacteria from the atmosphere can be deposited by two key mechanisms: dry deposition and wet deposition. Dry deposition is the process of bioaerosols adhering to plants, water, and other ground surfaces with which they come into contact (Jones and Harrison 2004), while wet deposition describes the process by which bioaerosols are deposited through precipitation (rain, snow, and hail) (Deguillaume et al. 2008).

3.6 Microbial Activity in the Atmosphere

Despite the long-held belief that the atmosphere acts solely as a reservoir or conduit for microbial life, there are compelling arguments that life in the atmosphere should also be considered a functioning ecosystem. Suspended microbes have been shown to be both metabolically active and capable of reproduction (Dimmick et al. 1975), performing multiple functions such as ice nucleation (Vali 1971), cloud formation (Bauer et al. 2002), the degradation of organic carbon-based compounds (Ariya et al. 2002), nitrogen processing (Hill et al. 2007), sulfur oxidation and reduction (Deguillaume et al. 2008), and photosynthesis (Després et al. 2012). It is likely that the atmosphere works as both a conduit and a functioning ecosystem based on this evidence (Womack et al. 2010); however, further investigation is required before this can be proven definitively.

The majority of metabolic activities in the atmosphere take place within clouds. Bacterial concentrations in cloud water have been described within the range of 10^3–10^5 bacteria mL^{-1} (Margesin and Miteva 2011). The majority of cloud condensation nuclei (CCN) and ice-nucleating (IN) bacteria such as *Pseudomonas syringae*, *Pseudomonas fluorescens*, and *Psychrobacter* sp. are psychrophilic. Bioaerosols like standard aerosols (e.g. mineral dust, sea salt) aid the formation of clouds, by acting as both CCN in suitable warm conditions where relative humidity conditions exceed saturation and IN at temperatures of −2 °C and lower (Möhler et al. 2007); CCN- and IN-associated bacteria also play a role in the initiation of precipitation events (Möhler et al. 2007). The role of CCN and IN bacteria in precipitation has been termed bioprecipitation, a mechanism describing a feedback cycle which enables the wide dispersal of bacteria by wet deposition (Morris et al. 2014). These psychrophilic ice nucleators have been shown to respond to environmental triggers such as changes in humidity and are ubiquitously present in precipitation at abundances of 4–490 ice nuclei per L^{-1} (Margesin and Miteva 2011).

3.7 Aerial Biodiversity

While the atmospheric environment is relatively extreme, thriving diverse bacterial communities have also been found in other challenging environments such as hot springs and deserts (Womack et al. 2010). Microorganisms within the atmosphere are diverse, with airborne microbial communities above both terrestrial and aquatic environments having been shown to contain more than 100 genera of bacteria, a level of generic diversity comparable with that of soil and aquatic environments (Fahlgren et al. 2010; Nonnenmann et al. 2010; Madsen et al. 2015). One recent study by Barberán et al. (2015), collated over 1000 sampling events, and found more than 110,000 different species of airborne bacteria in the USA alone, with more than 55,000 species of fungi. This diversity stems from a multitude of adaptions to atmospheric life such as cryopreservation (Kochkina et al. 2001), spore-forming ability (Moeller et al. 2009), cell pigmentation (Griffin et al. 2011), and DNA repair mechanisms (Johansson et al. 2011).

Bacterial populations can be seen to decrease in number by as much as half with increasing altitude; however, the following viable bacteria and fungi have been found in the Stratosphere at altitudes as high as 77 km: *Mycobacterium luteum, Micrococcus albus, Aspergillus niger, Penicillium notatum, Circinella muscae, Papulaspora anomala,* (Imshenetsky et al. 1977), *Bacillus simplex, Staphylococcus pasteuri,* and *Engyodontium album* (Wainwright et al. 2003).

The majority of these diverse bioaerosol communities are largely comprised of four main bacterial groups which are the Actinobacteria, Bacteroidetes, Firmicutes, and Proteobacteria, an observation consistent in both aquatic and terrestrial environments (Madsen et al. 2015; Seifried et al. 2015). At the genus level, however, there is more variation, dependent upon environmental conditions, such as proximity to agricultural sites and weather (Nonnenmann et al. 2010). Variation is also directly influenced by season; although, seasonal influence varies by location (Bowers et al. 2012; Dong et al. 2016). Concentrations of bacteria in the atmosphere generally range from 10^4 to 10^8 cells per m^3 (Bowers et al. 2011). However, these concentrations are known to vary significantly across all four calendar seasons and can also be affected by weather (wind direction, wind speed, temperature, fog, etc.) (Burrows et al. 2009b; Bowers et al. 2012).

3.8 Psychrophilic Bioaerosols

The aerosolization of microbes from cold environments such as the polar regions propels psychrophilic organisms directly into the atmosphere. Psychrophilic bacteria are better suited than most bacteria to atmospheric life as they are already adapted to survive freezing temperatures of higher altitudes. Bacteria collected from clouds have been shown to be capable of growing and reproducing at 0 °C (Sattler et al. 2001) suggesting the existence of psychrophilic bioaerosols. With cloud temperatures often well below 0 °C, any bacterial species residing there, such as the recently discovered novel bacteria *Deinococcus aethius* and *Bacillus stratosphericus* (Margesin and Miteva 2011), should be considered psychrophilic. Psychrophilic bacteria not only reside in the atmosphere but also play a key role in atmospheric processes, for example, the psychrophilic plant pathogen *Pseudomonas syringae* is involved in ice nucleation in clouds.

3.9 Polar Aerobiology

3.9.1 Arctic

Aerobiological studies in the Arctic date back to the coining of aerobiology as a discipline, with Meier and Lindbergh collecting aerial samples in flight above the region (Meier and Lindbergh 1935); this work was followed up by Polunin et al. in the late 1940s (Polunin et al. 1947), though studies of this nature are sparse. The most recent terrestrial study of bioaerosols in the Arctic was carried out by Harding et al. (2011) on Ward Hunt Island. This study reported the communities in the air to have considerable similarities with communities found in studies of the surrounding Arctic Ocean, drawing the conclusion that local sources contribute a large proportion of communities. The study also found organisms not normally associated with the high Canadian Arctic, from other sources and locations in the Arctic as well as some organisms associated with the Antarctic, potentially supporting the theory of long distance atmospheric dispersal. These findings are consistent with those of previous studies that have stated the dominant groups of bacteria in cold ecosystems to be Alpha-, Beta-, and Gamma-Proteobacteria, Firmicutes, Bacteroidetes, Cyanobacteria, and Actinobacteria (Amato et al. 2007; Møller et al. 2013).

3.9.2 Antarctic

Some of the first ecology-based aerobiological studies took place in the Antarctic in the early 1900s; yet, despite a considerable sampling effort by Marshall, there have only been 12 Antarctic aerobiology studies published since 1996 (Marshall 1996, 1998; Marshall and Chalmers 1997). However, critically the transfer of biological material into Antarctica by atmospheric transport has been demonstrated (Vincent 2000; Herbold et al. 2014). Despite these findings, the small range of studies means that little is still known about the viability, duration of suspension and process of colonization, and establishment of these organisms (Pearce et al. 2016). Bacterial genera that are common in both aerial and Antarctic literature are *Staphylococcus, Bacillus, Corynebacterium, Micrococcus, Streptococcus, Neisseria*, and *Pseudomonas*. Commonly encountered fungal genera include *Penicillium, Aspergillus,*

Cladosporium, Alternaria, Aureobasidium, Botryotrichum, Botrytis, Geotrichum, Staphylotrichum, Paecilomyces, and *Rhizopus* (Pearce et al. 2009).

Bacterial atmospheric residence times in the Antarctic are predicted to be longer than in other environments which implies that long-range transport is more likely in the region (Burrows et al. 2009a). Evidence currently suggests there is an endemic population of bioaerosols in the atmosphere which are in part, but not entirely, related to the surrounding maritime and terrestrial conditions (Bottos et al. 2014). These results are further supplemented by findings suggesting other characteristics of the Antarctic such as sea ice area may have a negligible impact on local biodiversity of atmospheric microbes (Pearce et al. 2010). Along with long-range atmospheric transport, one of the other key inputs of airborne microbes into the Antarctic atmosphere is human activity. The results of aerial studies taken from research stations such as Halley V Research Station, Concordia, and Rothera Point have suggested the potential for input from human-derived sources while marine input into terrestrial samples is low; the most striking comparison across the majority of Antarctic studies, however, is that the biodiversity is markedly different (Hughes et al. 2004; Van Houdt et al. 2009; Pearce et al. 2010; Bottos et al. 2014). Although contradictory studies exist, another feature of aerial input into the region is that it might be directly affected by seasonality, which has potentially been correlated to an increase of keratinous material in summer regions due to increased bird and seal activity (Marshall 1998).

3.10 Biogeography

Microbial dispersal in the atmosphere has been considered ubiquitous in line with the hypothesis that "Everything is everywhere but the environment selects" (Baas Becking 1934). Initially, a substantial amount of evidence supported this theory, where organisms with comparable phylotypes were shown to be present in similar but geographically separated environments (Finlay and Clarke 1999). However, recent developments in modern technologies and an increase in the number of studies of microbial communities across considerable spatial and temporal scales have led to the concept of large-scale microbial biogeography, which would contradict Baas Becking's theory, and refers to patterns in the spatial distribution of microbial life from local to continental scales shaped by processes such as dispersal, speciation, and extinction (Fierer 2008). Studies have since provided evidence for microbial endemism at local scales in Antarctica, for example, in isolated Antarctic habitats (Alger 1999; Pearce et al. 2010; Vyverman et al. 2010).

The atmosphere is key to microbial biogeography, particularly in the cold biosphere, as dispersal provides one of the main exogenous inputs into geographically isolated environments such as the polar regions; however, factors such as dispersal, colonization, and survival rates during atmospheric transport are poorly understood (Fierer 2008; Pearce et al. 2016). Little attention has been given to microbial diversity patterns in the atmosphere as the environment has been disregarded as a conduit rather than a habitat (Womack et al. 2010). Whilst recent studies have begun addressing these issues in the atmosphere, showing, for example that marine bioaerosol communities can be distinct from those found in adjacent terrestrial locations (Lovejoy et al. 2007; Barberán et al. 2015). Whether microbial biogeography in the atmosphere exists at all is still open to question and requires further research (Martiny et al. 2006). Patterns in diversity have been observed in the atmosphere with genera such as *Polaribacter* sp. and *Psychrobacter* sp. being observed in both Arctic and Antarctic studies (Pearce et al. 2010; Harding et al. 2011) and the discovery that bioaerosols over urban environments contain typically higher diversities than those seen in remote locations (Brodie et al. 2007).

While clear patterns are beginning to emerge regarding the atmospheric dispersal of microorganisms and the viability of a number of organisms over extended periods of time in the atmosphere even under the pressures of the environment (Jones and Harrison 2004), studies still consistently fail to consider the viability of these source colonists upon arrival in their new environments (Fierer 2008). These colonists have the potential to interact with the microbiomes of the environments in which they are deposited both positively and negatively, for example bacteria suspended and deposited in low nutrient locations can provide nutrients by recycling, a mechanism which can benefit the ecosystem (Gallisai et al. 2014); conversely, this situation can also disrupt ecosystems causing events such as algal blooms to occur which can be devastating to the native community (Giddings et al. 2014). Migrating bioaerosols pose a considerable pathogenic threat to agriculture due to the heterogeneity of modern day crops (Smith et al. 2011). Human pathogens such as *Mycoplasma pneumonia, Mycobacterium tuberculosis, Corynebacterium diphtheria, Bordetella pertussis*, and influenza virus also utilize the atmosphere as a conduit to spread from host to host. Incidences of both influenza and meningococcal meningitis have been shown to be associated with long-range transport during dust storms (Molesworth et al. 2003; Chen et al. 2010), highlighting the relevance of the atmosphere in the spread of human disease.

3.11 Conclusions

Psychrophilic microorganisms are present in the atmosphere as bioaerosols. Viable bacteria have been found at stratospheric altitudes of 77 km and bacteria collected from cloud water have been shown to be capable of growing and reproducing at 0 °C (Imshenetsky et al. 1977; Sattler et al. 2001). The role bioaerosols play in atmospheric processes, such as cloud condensation nuclei and ice nuclei, is becoming better understood. Biogeographic patterns of psychrophilic bacteria such as *Polaribacter* sp. and *Psychrobacter* sp. are beginning to become apparent (Pearce et al. 2010; Harding et al. 2011). Pathogenic bioaerosols can travel across continents attached to dust particles (Griffin 2007) meaning that psychrophilic plant pathogens such as *Pseudomonas syringae* have the capability to impact agriculture worldwide. These factors highlight the importance of a better understanding of the psychrophilic microbial processes taking place in the atmosphere.

References

Airy H (1874) Pollen-grains in the air. Nature 10(253):355

Alger AS (1999) Diatoms of the McMurdo dry valleys, Antarctica: a taxonomic appraisal including a detailed study of the genus Hantzschia. Thesis, University of Michigan

Amato P, Hennebelle R, Magand O, Sancelme M, Delort AM, Barbante C, Boutron C, Ferrari C (2007) Bacterial characterization of the snow cover at Spitzberg, Svalbard. FEMS Microbiol Ecol 59(2):255–264

Andersen AA (1958) New sampler for the collection, sizing and enumeration of viable airborne particles. J Bacteriol 76(5):471–484

Andrewes CH, Glover RE (1941) Spread of infection from the respiratory tract of the ferret. I. Transmission of influenza A virus. Br J Exp Pathol 22(2):91–97

Ariya PA, Nepotchatykh O, Ignatova O, Amyot M (2002) Microbiological degradation of atmospheric organic compounds. Geophys Res Lett 29(22):34-1–34-4

Baas Becking LGM (1934) Geobiologie of inleiding tot de milieukunde. W.P. Van Stockum & Zoon, Den Haag

Barberán A, Ladau J, Leff JW, Pollard KS, Menninger H, Dunn RR, Fierer N (2015) Continental-scale distributions of dust-associated bacteria and fungi. Proc Natl Acad Sci 112 (18):5756–5761

Bauer H, Kasper-Giebl A, Löflund M, Giebl H, Hitzenberger R, Zibuschka F, Puxbaum H (2002) The contribution of bacteria and fungal spores to the organic carbon content of cloud water, precipitation and aerosols. Atmos Res 64(1–4):109–119

Blanchard DC, Syzdek LD (1982) Water-to-air transfer and enrichment of bacteria in drops from bursting bubbles. Appl Environ Microbiol 43(5):1001–1005

Bottos EM, Woo AC, Zawar-Reza P, Pointing SB, Cary SC (2014) Airborne bacterial populations above desert soils of the McMurdo Dry Valleys, Antarctica. Microb Ecol 67(1):120–128

Bourdillon RB, Lidwell OM, Thomas JC (1941) A slit sampler for collecting and counting air-borne bacteria. J Hyg 41(2):197–224

Bowers RM, McLetchie S, Knight R, Fierer N (2011) Spatial variability in airborne bacterial communities across land-use types and their relationship to the bacterial communities of potential source environments. ISME J 5(4):601–612

Bowers RM, McCubbin IB, Hallar AG, Fierer N (2012) Seasonal variability in airborne bacterial communities at a high-elevation site. Atmos Environ 50:41–49

Brodie EL, DeSantis TZ, Parker JP, Zubietta IX, Piceno YM, Andersen GL (2007) Urban aerosols harbor diverse and dynamic bacterial populations. Proc Natl Acad Sci U S A 104(1):299–304

Brown WA, Allison VD (1937) Infection of the air of scarlet-fever wards with *Streptococcus* pyogenes. J Hyg 37(1):1–13

Burrows SM, Butler T, Jöckel P, Tost H, Kerkweg A, Pöschl U, Lawrence MG (2009a) Bacteria in the global atmosphere – Part 2: Modeling of emissions and transport between different ecosystems. Atmos Chem Phys 9(23):9281–9297

Burrows SM, Elbert W, Lawrence MG, Pöschl U (2009b) Bacteria in the global atmosphere – Part 1: Review and synthesis of literature data for different ecosystems. Atmos Chem Phys 9 (23):9263–9280

Chen PS, Tsai FT, Lin CK, Yang CY, Chan CC, Young CY, Lee CH (2010) Ambient influenza and avian influenza virus during dust storm days and background days. Environ Health Perspect 118(9):1211–1216

Deguillaume L, Leriche M, Amato P, Ariya PA, Delort AM, Pöschl U, Chaumerliac N, Bauer H, Flossmann AI, Morris CE (2008) Microbiology and atmospheric processes: chemical interactions of primary biological aerosols. Biogeosciences 5(4):1073–1084

Després VR, Huffman JA, Burrows SM, Hoose C, Safatov AS, Buryak G, Fröhlich-Nowoisky J, Elbert W, Andreae MO, Pöschl U, Jaenicke R (2012) Primary biological aerosol particles in the atmosphere: a review. Tellus Ser B Chem Phys Meteorol 64:15598

Dimmick RL, Straat PA, Wolochow H, Levin GV, Chatigny MA, Schrot JR (1975) Evidence for metabolic activity of airborne bacteria. J Aerosol Sci 6(6):387–393
Dong L, Qi J, Shao C, Zhong X, Gao D, Cao W, Gao J, Bai R, Long G, Chu C (2016) Concentration and size distribution of total airborne microbes in hazy and foggy weather. Sci Total Environ 541:1011–1018
Dyar HG (1894) XII.—On certain bacteria from the air of New York City. Ann N Y Acad Sci 8:322–380
Fahlgren C, Hagstrom A, Nilsson D, Zweifel UL (2010) Annual variations in the diversity, viability, and origin of airborne bacteria. Appl Environ Microbiol 76(9):3015–3025
Fierer N (2008) Microbial biogeography: patterns in microbial diversity across space and time. In: Zengler K (ed) Accessing uncultivated microorganisms: from the environment to organisms and back. ASM Press, Washington, DC, pp 95–115
Finlay BJ, Clarke KJ (1999) Ubiquitous dispersal of microbial species. Nature 400(6747):828–828
Gallisai R, Peters F, Volpe G, Basart S, Baldasano JM (2014) Saharan dust deposition may affect phytoplankton growth in the Mediterranean Sea at ecological time scales. PLoS ONE 9(10): e110762
Giddings SN, MacCready P, Hickey BM, Banas NS, Davis KA, Siedlecki SA, Trainer VL, Kudela RM, Pelland NA, Connolly TP (2014) Hindcasts of potential harmful algal bloom transport pathways on the Pacific Northwest coast. J Geophys Res Oceans 119(4):2439–2461
Griffin DW (2007) Atmospheric movement of microorganisms in clouds of desert dust and implications for human health. Clin Microbiol Rev 20(3):459–477
Griffin DW, Gonzalez C, Teigell N, Petrosky T, Northup DE, Lyles M (2011) Observations on the use of membrane filtration and liquid impingement to collect airborne microorganisms in various atmospheric environments. Aerobiologia 27(1):25–35
Harding T, Jungblut AD, Lovejoy C, Vincent WF (2011) Microbes in high Arctic snow and implications for the cold biosphere. Appl Environ Microbiol 77(10):3234–3243
Herbold CW, Lee CK, McDonald IR, Cary SC (2014) Evidence of global-scale aeolian dispersal and endemism in isolated geothermal microbial communities of Antarctica. Nat Commun 5:3875
Hill KA, Shepson PB, Galbavy ES, Anastasio C, Kourtev PS, Konopka A, Stirm BH (2007) Processing of atmospheric nitrogen by clouds above a forest environment. J Geophys Res 112: D11301
Hirst JM (1952) An automatic volumetric spore trap. Ann Appl Biol 39(2):257–265
Hughes KA, McCartney HA, Lachlan-cope TA, Pearce DA (2004) A preliminary study of airborne microbial biodiversity over peninsular Antarctica. Cell Mol Biol 50(5):537–542
Imshenetsky AA, Lysenko SV, Kasakov GA, Ramkova NV (1977) Resistance of stratospheric and mesospheric micro-organisms to extreme factors. Life Sci Space Res 15:37–39
Johansson E, Adhikari A, Reponen T, Yermakov M, Grinshpun SA (2011) Association between increased DNA mutational frequency and thermal inactivation of aerosolized *Bacillus* spores exposed to dry heat. Aerosol Sci Technol 45(3):376–381
Jones AM, Harrison RM (2004) The effects of meteorological factors on atmospheric bioaerosol concentrations – a review. Sci Total Environ 326(1–3):151–180
Joung YS, Ge Z, Buie CR (2017) Bioaerosol generation by raindrops on soil. Nat Commun 8:14668
Kochkina GA, Ivanushkina NE, Karasev SG, Gavrish EY, Gurina LV, Evtushenko LI, Spirina EV, Vorob'eva EA, Gilichinskii DA, Ozerskaya SM (2001) Survival of micromycetes and actinobacteria under conditions of long-term natural cryopreservation. Microbiology 70 (3):356–364
Lighthart B, Shaffer BT (1994) Bacterial flux from chaparral into the atmosphere in mid-summer at a high desert location. Atmos Environ 28(7):1267–1274
Lovejoy C, Vincent WF, Bonilla S, Roy S, Martineau M-J, Terrado R, Potvin M, Massana R, Pedrós-Alió C (2007) Distribution, phylogeny, and growth of cold-adapted picoprasinophytes in Arctic seas. J Phycol 43(1):78–89

Lurie MB (1930) Experimental epidemiology of tuberculosis. J Exp Med 51(5):743
Madsen AM, Zervas A, Tendal K, Nielsen JL (2015) Microbial diversity in bioaerosol samples causing ODTS compared to reference bioaerosol samples as measured using Illumina sequencing and MALDI-TOF. Environ Res 140:255–267
Margesin R, Miteva V (2011) Diversity and ecology of psychrophilic microorganisms. Res Microbiol 162(3):346–361
Marshall WA (1996) Aerial dispersal of lichen soredia in the maritime Antarctic. New Phytol 134 (3):523–530
Marshall WA (1998) Aerial transport of keratinaceous substrate and distribution of the fungus *Geomyces pannorum* in Antarctic soils. Microb Ecol 36(2):212–219
Marshall WA, Chalmers MO (1997) Airborne dispersal of antarctic terrestrial algae and cyanobacteria. Ecography 20(6):585–594
Martiny JBH, Bohannan BJM, Brown JH, Colwell RK, Fuhrman JA, Green JL, Horner-Devine MC, Kane M, Krumins JA, Kuske CR, Morin PJ, Naeem S, Ovreas L, Reysenbach A-L, Smith VH, Staley JT (2006) Microbial biogeography: putting microorganisms on the map. Nat Rev Microbiol 4(2):102–112
Meier FC, Lindbergh CA (1935) Collecting micro-organisms from the Arctic atmosphere. Sci Monthly 40:5–20
Moeller R, Setlow P, Reitz G, Nicholson WL (2009) Roles of small, acid-soluble spore proteins and core water content in survival of *Bacillus subtilis* spores exposed to environmental solar UV radiation. Appl Environ Microbiol 75(16):5202–5208
Möhler O, DeMott PJ, Vali G, Levin Z (2007) Microbiology and atmospheric processes: the role of biological particles in cloud physics. Biogeosciences 4(6):1059–1071
Molesworth AM, Cuevas LE, Connor SJ, Morse AP, Thomson MC (2003) Environmental risk and meningitis epidemics in Africa. Emerg Infect Dis 9(10):1287–1293
Møller AK, Søborg DA, Al-Soud WA, Sørensen SJ, Kroer N (2013) Bacterial community structure in High-Arctic snow and freshwater as revealed by pyrosequencing of 16S rRNA genes and cultivation. Polar Res 32:17390
Morris CE, Conen F, Alex Huffman J, Phillips V, Pöschl U, Sands DC (2014) Bioprecipitation: a feedback cycle linking Earth history, ecosystem dynamics and land use through biological ice nucleators in the atmosphere. Glob Change Biol 20(2):341–351
NASA (2013) Earth's atmospheric layers. https://www.nasa.gov/mission_pages/sunearth/science/atmosphere-layers2.html. Accessed 8 Feb 2017
Nonnenmann MW, Bextine B, Dowd SE, Gilmore K, Levin JL (2010) Culture-independent characterization of bacteria and fungi in a poultry bioaerosol using pyrosequencing: a new approach. J Occup Environ Hyg 7(12):693–699
Pasteur L (1860) Nouvelles expériences relatives aux générations dites spontanées. C R Acad Sci 51:348–359
Pearce DA, Bridge PD, Hughes KA, Sattler B, Psenner R, Russell NJ (2009) Microorganisms in the atmosphere over Antarctica. FEMS Microbiol Ecol 69(2):143–157
Pearce DA, Hughes KA, Lachlan-Cope T, Harangozo SA, Jones AE (2010) Biodiversity of air-borne microorganisms at Halley Station, Antarctica. Extremophiles 14(2):145–159
Pearce DA, Alekhina IA, Terauds A, Wilmotte A, Quesada A, Edwards A, Dommergue A, Sattler B, Adams BJ, Magalhaes C, Chu WL, Lau MC, Cary C, Smith DJ, Wall DH, Eguren G, Matcher G, Bradley JA, de Vera JP, Elster J, Hughes KA, Cuthbertson L, Benning LG, Gunde-Cimerman N, Convey P, Hong SG, Pointing SB, Pellizari VH, Vincent WF (2016) Aerobiology over Antarctica – a new initiative for atmospheric ecology. Front Microbiol 7:16
Polunin N, Pady SM, Kelly CD (1947) Arctic aerobiology. Nature 160(4077):876–877
Sattler B, Puxbaum H, Psenner R (2001) Bacterial growth in supercooled cloud droplets. Geophys Res Lett 28(2):239–242
Seifried JS, Wichels A, Gerdts G (2015) Spatial distribution of marine airborne bacterial communities. Microbiology 4(3):475–490

Smith DJ, Griffin DW, Jaffe DA (2011) The high life: transport of microbes in the atmosphere. EOS Trans Am Geophys Union 92(30):249–250
Smith DJ, Jaffe DA, Birmele MN, Griffin DW, Schuerger AC, Hee J, Roberts MS (2012) Free tropospheric transport of microorganisms from Asia to North America. Microb Ecol 64 (4):973–985
Stewart FJ (2013) Where the genes flow. Nat Geosci 6(9):688–690
Vali G (1971) Quantitative evaluation of experimental results and the heterogeneous freezing nucleation of supercooled liquids. J Atmos Sci 28(3):402–409
Van Houdt R, De Boever P, Coninx I, Le Calvez C, Dicasillati R, Mahillon J, Mergeay M, Leys N (2009) Evaluation of the airborne bacterial population in the periodically confined Antarctic base Concordia. Microb Ecol 57(4):640–648
Vincent WF (2000) Evolutionary origins of Antarctic microbiota: invasion, selection and endemism. Antarct Sci 12(3):374–385
Vyverman W, Verleyen E, Wilmotte A, Hodgson DA, Willems A, Peeters K, Van de Vijver B, De Wever A, Leliaert F, Sabbe K (2010) Evidence for widespread endemism among Antarctic microorganisms. Polar Sci 4(2):103–113
Wainwright M, Wickramasinghe NC, Narlikar JV, Rajaratnam P (2003) Microorganisms cultured from stratospheric air samples obtained at 41 km. FEMS Microbiol Lett 218(1):161–165
Weber CF, Werth JT (2015) Is the lower atmosphere a readily accessible reservoir of culturable, antimicrobial compound-producing Actinomycetales? Front Microbiol 6:802
Wells WF (1943) Bacteriologic procedures in sanitary air analysis: with special reference to air disinfection. J Bacteriol 46(6):549–557
Womack AM, Bohannan BJM, Green JL (2010) Biodiversity and biogeography of the atmosphere. Phil Trans Roy Soc London B: Biol Sci 365(1558):3645–3653

Chapter 4
Microbial Life in Supraglacial Environments

Arwyn Edwards and Karen A. Cameron

Abstract Supraglacial environments occupy 11% of Earth's surface area and represent a critical interface between climate and ice. This century has brought a renewed appreciation that glacier surfaces represent a collective of diverse microbial niches which occur wherever sufficient liquid water is available to support microbial activity: even at the microscopic scales of ice crystal boundaries within the crystalline matrices of snow or glacial ice. Within this chapter, we review the range of microbial habitats associated with snowpacks, the glacial ice photic zone, and phototrophic microbial biofilms formed by supraglacial algae or by the darkening of microbe–mineral aggregates known as cryoconite. In summary, glacier surfaces are home to surprisingly biodiverse and active microbial communities despite their low temperatures and austere conditions. Consequently, microbial communities and their processes are interposed between climate and ice and merit urgent consideration in the light of the effects of climate warming on Earth's supraglacial environments.

Contents

A. Edwards (✉) • K.A. Cameron
Institute of Biological, Rural and Environmental Sciences, Aberystwyth University, Cledwyn Building, Aberystwyth, UK

Aberystwyth University Interdisciplinary Centre for Environmental Microbiology (AU iCEM), Aberystwyth University, Aberystwyth, UK
e-mail: aye@aber.ac.uk

R. Margesin (ed.), *Psychrophiles: From Biodiversity to Biotechnology*,
DOI 10.1007/978-3-319-57057-0_4

4.1 Introduction

In the twenty-first century, approximately 198,000 glaciers and 2 ice sheets cover roughly 11% of Earth's surface area, sequestering 70% of Earth's freshwater (Shiklomanov 1993; Pfeffer et al. 2014). Within the recent geologic past, glacial ice covered ca. 30% of the Earth's surface during the last glacial maximum (Boyd et al. 2010), and the extent of ice is thought to have approached total coverage during so-called Snowball Earth events during the Cryogenian period 600–700 million years ago (Hoffman et al. 1998). However, within the near future, anthropogenic climate change will contribute to the decimation of Earth's glacial ice (Meier et al. 2007; Joughin et al. 2014; Pachauri et al. 2014). In the event of severe climate warming, the loss of up to 85% of mountain glaciers is predicted within this century (Pachauri et al. 2014). Moreover, climatic warming spurs the destabilization of Greenland's ice sheet and risks commitment to the loss of Antarctica's ice sheet (e.g., Joughin et al. 2014). Consequently, ice melt is set to be a major contributor to rising sea levels with impacts upon the habitability of coastal areas (FitzGerald et al. 2008) and the food and water security of densely populated regions nourished by glacial meltwaters (Edwards et al. 2014a; Hodson 2014). The importance of glaciers and ice sheets within the Earth system at both geologic timescales as well as the future of our contemporary society is therefore readily apparent and merits considerable investment in the study of both the fundamental properties of glacial systems and predictions of the rates and extents of their responses to climatic warming.

In contrast, a neglected facet of glacial systems is that they are also microbial habitats. In volumetric terms, glaciers and ice sheets represent Earth's largest freshwater ecosystem (Edwards et al. 2014a). Only in this century has the paradigm of glaciers and ice sheets as truly inhospitable environments been displaced as a result of experimental and conceptual advances spawning the field of glacier ecology, summarized in the seminal review of glacial ecosystems by Andrew Hodson and his colleagues in 2008 (Hodson et al. 2008). We now recognize that glacial systems represent both a deep frozen archive of microbial biodiversity (Bidle et al. 2007) and loci of globally significant biogeochemical processes mediated by microbes (Anesio et al. 2009; Boyd et al. 2010; Wadham et al. 2012; Hawkings et al. 2014), despite the unfavorable effects of their prevailing low temperatures (Rodrigues and Tiedje 2008). Indeed, microbial communities at the glacier–atmosphere interface constitute a supraglacial ecosystem and assume considerable importance as modulators of both biogeochemical cycling and glacial system response to melting (Anesio et al. 2009; Lutz et al. 2016b).

Within this chapter, we focus upon the microbial communities within the supraglacial ecosystem and their interactions within this habitat, addressing the key concepts, processes, and approaches relevant to the study of microbial life in supraglacial environments.

4.2 The Physical Properties of Glacial Systems

An understanding of the fundamental properties of glacial systems is an essential prerequisite to their consideration as microbial ecosystems. Readers from a non-glaciological background embarking upon the study of glacial ecosystems are advised to solicit the guidance offered by popular scholarly texts on the theory (e.g., Benn and Evans 2014) and practice (e.g., Hubbard and Glasser 2005) of glaciology to enrich their multidisciplinary endeavors. Within this section, we provide a synopsis of the properties of glacial systems pertinent to a discussion of supraglacial ecosystems.

As large persistent masses of ice, glaciers and ice sheets represent major features of the terrestrial cryosphere. These ice masses form where the prolonged accumulation of ice mass exceeds its loss by ablation, resulting in the metamorphosis of snow crystals to glacial ice. Glacial ice itself is a non-Newtonian fluid which is deformed under the pressure created by its own mass, resulting in the gravitational flow of glacial ice from higher to lower elevations once a critical thickness of glacial ice cover is achieved. Moreover, thick accumulations of glacial ice can sufficiently elevate pressures at the glacier bed to incur pressure-induced melting of glacial ice, resulting in an admixture of ice above its pressure melting point and influencing the *thermal regime* of a glacial system.

Typically, the distribution of ice mass across an elevational gradient incurs variation in the rates of ice accumulation and ablation as a consequence of altitudinal effects on local temperature from adiabatic cooling. Colder, higher elevations where the accumulation of ice mass directly from snowfall (or indirectly from wind-redistributed snow or avalanches) annually exceeds its ablation represent zones of net accumulation. Meanwhile, warmer lower elevations experience net ablation since the loss of ice mass (via sublimation, the generation of meltwater, or the direct loss of ice mass by the formation of icebergs) annually exceeds its accumulation. This property is known as the *mass balance gradient* of a glacial system and pivots upon an elevational contour known as the *equilibrium line altitude* where accumulation equals ablation on an annual basis. Glacial systems well-nourished by precipitation and residing in colder climates will tend towards the net accumulation of ice mass and consequently will advance as a result of their *positive mass balance* while glacial systems experiencing net ablation of ice mass will exhibit *negative mass balance* and exhibit recession. As such, glacial systems can be thought of as climatic thermometers and thermostats, and indeed the properties of ice within a glacial system can provide a proxy for the reconstruction of past climate (e.g., Petit et al. 1999).

Glacial systems take a range of forms. Glaciers are entities comprised of flowing glacial ice constrained by the topography of the landscape in which they reside while the accumulation of glacial ice present in ice sheets and ice caps override the underlying topography of their catchments. The division between ice caps and ice sheets is a function of their scale; ice sheets are defined as possessing surface areas greater than 50,000 km^2 (Benn and Evans 2014; Gokul et al. 2016) and presently include the ice sheet on Greenland and the Antarctic Ice Sheet which is divided into two sections, East and West, by the Transantarctic mountains. At the opposite end of the size spectrum, some glacial systems exhibit unconventional flow modes or, indeed, no detectable flow modes at all. These include ice patches, which are small, localized accumulations of glacial ice typically nourished by the wind-driven accumulation of snow (Andrews and MacKay 2012) and debris covered or buried ice masses (Franzetti et al. 2013). The scale of the ice mass and its relationship with the parent catchment has implications for cognate ecosystems in terms of the sources, residence time, and dynamics of microbiota.

Finally, glacial systems entrain both water and debris, affording resident microbes with solvent, nutritional solutes, habitats, and dispersal modes. In terms of debris accumulation transport, the burden of debris associated with glacial systems can be considerable, ranging from fine grained aeolian debris (McGee et al. 2010) to in extremis debris covered glaciers which are extensively covered in rocks, affording shelter and nutrition to lithotrophic microbial communities (Franzetti et al. 2013). Glacial melt can be stored and transported within glacial systems as water percolating in intercrystalline spaces (e.g., in water-saturated decomposing snowpacks, firn aquifers, the near-surface ice itself, or in saturated basal sediments); it can be captured by compartmentalized storage systems (e.g., in supraglacial lakes, englacial chambers, or subglacial lakes) or it can move rapidly though channelized drainage through surficial, englacial, or subglacial drainage networks. The interplay between glaciers and melt leading to the evolution of glacier hydraulic systems is multifaceted since multiple interactions that occur between mass balance and thermal regime (Irvine-Fynn et al. 2011b) complicate the evolution, storage, and flux of meltwater. Nevertheless, considering life's cardinal requirement is for liquid water, it is safe to conclude the relationship between melt and microbes is necessarily intimate within glacial systems.

The glacier surface itself represents the interface between the glacier and the atmosphere. Accordingly, there is a continual exchange of matter and energy, with the surface energy budget of a glacier being represented by Eq. (1) (Paterson 1994):

$$Q_{\mathrm{M}} + Q_{\mathrm{SW}} + Q_{\mathrm{LW}} + Q_{\mathrm{S}} + Q_{\mathrm{L}} + Q_{\mathrm{P}} + Q_{\mathrm{C}} = 0 \left(\mathrm{w\,m^2}\right) \tag{1}$$

where Q_{M} is the energy accessible for melting, Q_{SW} and Q_{LW} represent shortwave and longwave radiation, respectively, Q_{S} and Q_{L} are sensible and latent heat, while Q_{P} is energy from precipitation, and Q_{C} is energy conducted into the glacial system. Conditions at the supraglacial environment become most conducive for microbial life when the influx of energy, predominantly as shortwave radiation, is sufficient to incur melting. As a consequence, life can thrive at the glacier surface thanks to the genesis of liquid water and liberation of nutrients within snow and ice melt concomitant with the onslaught of photosynthetically available radiation (Anesio and Laybourn-Parry 2012) and elevated ambient temperatures.

4.3 Snow Habitats

Every glacier starts with a single snowflake. While the recognition of microbial processes in snow dates to Aristotle (Hell et al. 2013), and Van Leeuwenhoek found microbes within snowmelt among the first samples he viewed with a microscope (van Leewenhoeck 1677), the study of microbial life in supraglacial snow is a recent endeavor (Kol 1942). In spite of the low temperature and limited supply of nutrients typical of snow, microbial abundances in the range of 2×10^2 to 7.2×10^5 cells ml^{-1} of melted snow are reported from supraglacial snow (Carpenter et al. 2000; Amato et al. 2007; Irvine-Fynn et al. 2012). Considering that the 11% of Earth surface overridden by glacial ice is seasonally or perennially covered by snow, it is clear that supraglacial snowpacks represent globally extensive repositories of microbial assemblages, with considerable scope for variation in abundance, diversity, and activity. The abundance and activity of microbes within snow is influenced by their origins and the onset of melt (Xiang et al. 2009), so we consider distinct stages in the interactions between snow and its microbiota.

4.3.1 Depositional Modes

Snowflakes themselves, as the micro-scaled crystalline precipitates of water ice, represent microbial habitats (Sattler et al. 2001; Temkiv et al. 2011). While a detailed consideration of snow microbiology while aloft lies within the domain of atmospheric microbiology (see Chap. 3) and is thus beyond the scope of this chapter, the windborne redistribution of poorly consolidated supraglacial snow raises the potential for metabolic activity by microbes in ice precipitation (Temkiv et al. 2013). Moreover, microbial processes contribute to the nucleation of ice crystals, including fresh snowfall on glacier surfaces (Christner et al. 2008). Therefore, snow and ice precipitates both nourish the mass balance of glacial systems and inoculate glacial systems with microbiota (Harding et al. 2011; Cameron et al. 2014), representing *wet* modes of microbial deposition (Xiang et al. 2009). In contrast, *dry* modes of microbial deposition entail the passive sedimentation of airborne microbiota and in particular the co-deposition of microbial biomass with aeolian particulates (Xiang et al. 2009). Necessarily, the wet or dry mode of deposition will influence the origins, rates, viability, and composition of microbial inoculants to the snowpack (Hell et al. 2013; Cameron et al. 2014). Both locally derived material advected to the glacier surface (Hell et al. 2013) and (co-) deposition of microbiota following long-range transport from source environments are possible (Harding et al. 2011; Cameron et al. 2014; Nagatsuka et al. 2014; Wunderlin et al. 2016). As such, the supraglacial ecosystem can be connected with distant regions of the global cryosphere via atmospheric transport (Pearce et al. 2009) prompting the inoculation of globally ubiquitous microbial taxa (Darcy et al. 2011) to the supraglacial environment.

4.3.2 Microbial Life in the Dynamic Snowpack

Snowpacks represent complex habitats at the microbial scale. Although intrinsically defined by the presence of crystals of water ice, changes in the availability of water and impurities (including both nutrients and pollutants) are driven by the consolidation and melting of the snowpack, affecting the fundamental niches occupied by snow microbiota as the snowpack metamorphoses over time.

While dry, cold (i.e., subzero, non-melting) snowpacks are thought to represent a low end-member in the spectrum of activity and biomass associated with snow, they are not necessarily microbiologically inert. Initial evidence for in situ microbial activities in dry snow was offered (Carpenter et al. 2000) following low, but nonzero, rates of macromolecular synthesis via radioisotope inclusion in incubations of snow collected at the South Pole and incubated at temperatures representing in situ conditions. The validity of the inferences drawn from the experimental work was questioned within the literature (Warren and Hudson 2003) on the grounds of limited water availability and the eventual entombment of snow at depth, at temperatures of −50 °C. However, more recent work lends tacit support for the argument that microbes are active within dry polar snow. Firstly, sophisticated analyses of CRISPR repeat loci in shotgun metagenomic datasets of surface snow from the Antarctic plateau (Lopatina et al. 2016) infer a selection process is at play within the dry snow microbiota. While the stage (i.e., pre- or post-depositional) at which the selective pressures are manifest is less clear, the recovery of different community profiles from the sequencing of reverse-transcribed cDNA from 16S rRNA versus the sequencing of 16S rRNA genes from the same dry Antarctic snow samples (Lopatina et al. 2013) mitigates for the presence of translational machinery consistent with microbial activity (Blazewicz et al. 2013) in dry snow leading to interannual variability in communities within the snow (Lopatina et al. 2013). Moreover, Amoroso et al. (2009) concluded that microbial activities within dry, dark snowpacks must be responsible for nitrification occurring under conditions which do not permit abiotic photolysis of reactive nitrogen species. Considering the vast scale of dry, cold snowpacks, future work should examine these potential microbial habitats closely, since the cumulative effect of processes occurring at marginal rates across expansive habitats is nontrivial.

Seasonal melt transforms the snowpack habitat. Evidently, melt promotes the availability of liquid water. Furthermore, melt mobilizes of ionic impurities derived from aerosol deposits (e.g., from sea spray and atmospheric pollution) from the snowpack as they are rejected from decaying ice crystals (Davies et al. 1987; Kuhn 2001). Conservative, non-nutrient ions are eluted rapidly from the decaying snowpack, changing the boundary conditions for microbial life. Meanwhile, nutrient ions are sequestered within the snowpack, evidencing their biogeochemical transformations and identifying the melting snowpack as the locus of an active microbial ecosystem as well as the modulator of snowmelt composition, affecting nutrient availability in downstream habitats (Hodson et al. 2005).

The snowpack microbial community interacts closely with these complex changes in conditions (Larose et al. 2013; Maccario et al. 2014). In the first instance, the bacterial community of the snowpack has been shown to shift in its structure rapidly as melt progresses. Hell et al. (2013) examined the snowpack of a High Arctic glacier; in line with earlier studies, (Amato et al. 2007; Larose et al. 2010) a diverse bacterial community was identified in the snowpack. Notably, surface layers of snow harbor discrete communities of bacteria relative to melting snow and water-saturated snow (slush), indicating the role of post-depositional changes in community. While the class Betaproteobacteria retained its dominance of the community over the course of a week, the genus *Polaromonas* was able to adapt to the changing conditions. *Polaromonas* itself is ubiquitous within the global cryosphere (Darcy et al. 2011; Franzetti et al. 2013) and is characterized as a metabolically flexible genus, with the ability to degrade complex xenobiotics (Mattes et al. 2008). More recently, Franzetti et al. (2016) raised the prospect that supraglacial *Polaromonas* may act as anoxygenic phototrophs from the oxidation of carbon monoxide arising from photolytic degradation of supraglacial organic matter. While this study focused upon cryoconite environments, which are, at depth, attenuated in their exposure to high energy UV and possess an abundance of bioavailable organic carbon sources (Anesio et al. 2009), the potential for *Polaromonas*-mediated mixotrophy in the photochemically reactive (Amoroso et al. 2009) snowpack is unexplored.

Considering that multiple studies indicate the bacterial community is responsive to melt at week- to seasonal timescales (Hell et al. 2013; Maccario et al. 2014), the impacts of contemporary climate change on the timing, rate, duration, spatial extent, and hydrological flowpaths of seasonal snowmelt and the cognate bacterial community must be considered (Fig. 4.1). Spatially expansive yet transient melting

Fig. 4.1 Liquid water saturating a shallow firn core in the accumulation zone of the Greenland Ice Sheet (Photo credit: Sara Penrhyn-Jones)

episodes have been observed already, most notably during August 2012 ca. 97% of the surface area of the Greenland Ice Sheet (GrIS) exhibited melting, albeit for a duration of less than a week (Nghiem et al. 2012). While Betaproteobacteria may respond and influence the cycling of nitrogen pollutants deposited in the snowpack at such timescales on valley glaciers (Hell et al. 2013), whether the microbiota of the GrIS snowpack (Cameron et al. 2014) responded to this event is open to question. Stibal et al. (2015a) identified elevated microbial cell concentrations present in the refrozen 2012 melt layer relative to un-melted layers of the snowpack at a site high upon the GrIS, but the limited numbers of samples available precluded further detailed investigation. Moreover, whether the percolative redistribution of snowmelt into the perennial, near-surface firn aquifer of the GrIS accumulation zone (Forster et al. 2013) permits microbial community activities in a potential habitat four times the size of Wales is entirely unknown. While the empirical evidence for bacterial activities of biogeochemical significance in melting snowpacks is clear, it is evident that there are significant lacunae in our understanding of these processes in the context of rapid changes in the cryosphere.

Beyond the bacterial community, the snowpack is home to other types of microbiota. The presence of Archaea is variable, notably absent in some studies (Hell et al. 2013; Lazzaro et al. 2015) but detected independently by others (Cameron et al. 2014; Lutz et al. 2015). Over 30 Archaeal taxa were identified (Choudhari et al. 2013) in avalanche debris cone snow, paralleling an average number of 30 Archaeal taxa per sample across GrIS snow samples (Cameron et al. 2014). Archaea from the class Nitrosphaerales, associated with ammonia oxidation, predominate in both GrIS and Icelandic snow samples harboring Archaea (Cameron et al. 2014; Lutz et al. 2015), implying a potential role for Archaea in supraglacial nitrification, although further evidence for archaeal contributions to nitrogen cycling in glacial habitats is limited (Boyd et al. 2011).

In contrast to the enigmatic Archaea, the microeukaryotic component of the snowpack is readily apparent in the form of charismatic snow algae. Typically described as *Chlamydomonas nivalis*, snow algae are comprised of several algal lineages within the Chlorophyceae, in particular the Chlamydomonadaceae genera *Chloromonas, Chlamydomonas,* and *Raphidonema* (Spijkerman et al. 2012; Lutz et al. 2016a). Characteristic of these algae are the presence of unicellular motile cells within the isothermal snowpack, which form either green snow or red snow dependent upon reaching a quiescent phase in which the cells accumulate carotenoid pigments such as astaxanthin (Remias et al. 2005) as a means of photoprotection from high levels of UV penetrating the atrophied snowpack later in the melting season. Algal communities in red snow are apparent upon glaciers worldwide (Kol 1942; Yoshimura et al. 1997; Lutz et al. 2016b), and the community composition of red snow is remarkably uniform at the interspecific level across the European and Greenlandic Arctic (Lutz et al. 2016b). Interestingly, within defined, regional-scale catchments, patches of red snow appear exclusively dominated by differing single haplotypes (Brown et al. 2016) consistent with dispersal limitation incurred ecological priority effects. As such, reconciling these contrasts in the biogeography of glacial colonization by algae remains a challenge. Similarly, blooms of snow algae are associated with a diverse community of microbial heterotrophs, including bacteria, algae, and fungi (Weiss 1983; Lutz et al. 2015, 2016a). While some bacteria are intimately associated with snow algae (Weiss 1983), whether co-occurring microbes in general simply share a habitat with the algae or whether algal colonization facilitates their presence via the transfer of autochthonous carbon (Brown et al. 2015) is open to question; bacterial communities in algal blooms appear structured by locally varying lithological factors (Lutz et al. 2016a). If the translocation of carbon from algal photosynthesis itself is not pivotal to the assembly of snow algae associated communities, it may be that the evolution of meltwater from algal-mediated albedo depression (Lutz et al. 2016b) promotes the development of the cognate assemblage of microbes within snow algal blooms.

4.4 Supraglacial Ice and Meltwater Habitats

Within this section, we focus upon supraglacial ice and meltwater as active microbial habitats; discrete biofilms found in association with bare ice zones will be discussed in Sect. 4.5.

4.4.1 Life in Glacial Ice

As the snowline of a glacier recedes to higher elevations in summer, the extent of bare glacial ice increases in its prominence within the supraglacial environment. Glacial ice is typically considered as an archive of microbial biomass (Willerslev et al. 1999; Biddle et al. 2007; Castello and Rogers 2005) which may serve to seed the re-emergent glacial ice surface and downstream fluvial habitats with its microbiota. While this source of microbiota may be particularly important for environments characterized by low dust fluxes and limited surface melting, for example, blue ice ecosystems within the Antarctic (Hodson et al. 2013), microbial biomass in supraglacial ice is also sourced from both wet and dry deposition and the translocation of cells from the supraglacial snowpack itself (Irvine-Fynn et al. 2012; Hell et al. 2013; Björkman et al. 2014). Furthermore, arguments that englacial ice itself represents active microbial habitats within intercrystalline vein junctions, ice–mineral interfaces and even intracrystalline spaces have been advanced (Mader et al. 2006; Rohde and Price 2007). Therefore, while glacial ice may archive microbial biomass, the immured assemblage of microbiota is likely subject to post-depositional changes (Xiang et al. 2009; Liu et al. 2016) occurring over prolonged residence times (10^2–10^4 years).

4.4.2 Life in the Glacial Photic Zone

Bare ice itself represents an active microbial habitat. The glacial ice extent exposed by seasonal melting (excepting Antarctica) is estimated at a maximum of 7.5×10^5 km^2 (Anesio et al. 2009; Irvine-Fynn and Edwards 2013). The microbial communities of bare ice vary spatially, likely the result of localized inoculation sources and the physicochemical conditions presented to them (Edwards et al. 2013c; Cameron et al. 2014; Stibal et al. 2015a). Similarly, the total abundance of microbiota on surface ice is spatially variable and has been reported to range from 10^3 cells ml^{-1} in the zone of accumulation to 10^6 cells ml^{-1} in the ablation zone (Irvine-Fynn et al. 2012; Stibal et al. 2015a).

The physical processes of ice melt make it a viable microbial habitat. Incident shortwave radiation, which affects glacier surface energy balance (Eq. 1), penetrates ablating glacial ice to incur subsurface melt. The depth of subsurface melt is influenced by the extinction coefficient of the ice matrix, which will vary but can reach 2–20 m in optically clear ice (Hodson et al. 2013). Dissipation of incident radiation penetrating the surface ice incurs subsurface melt, decaying ice crystal structure and expanding interstitial spaces. This genesis of near-surface melt and its subsequent percolation at very low velocities further perpetuates the evolution of a highly porous near-surface ice layer, termed the weathering crust which stores meltwater in a perched aquifer atop nonporous englacial ice (Müller and Keeler 1969). As such, glacial ice surfaces in receipt of solar shortwave radiation represent a meteorologically controlled, seasonally evolving three-dimensional porous ice matrix which harbors a trifecta of decaying ice crystals, the percolative interstitial transfer of melt, and a decay gradient of incident radiation, all occurring to a depth which is delimited by the transmission of incident radiation and the corresponding inversely increasing density of ice (Cook et al. 2015b). Irvine-Fynn and Edwards (2013) recognized that these physical parameters amount to a glacial ice photic zone in which the liberation of melt commingled with nutrient, cell, and particulate mobilization and the availability of photosynthetically available radiation distributed in its intensity across a spectrum of photosynthetic optima promotes biological activity. Empirical support for a glacial ice photic zone is derived from the flow cytometric quantification of cell and particulate budgets of the weathering crust of a Svalbard glacier (Irvine-Fynn et al. 2012). When the rates of microbial biomass aeolian input, storage within ice, and fluvial release are considered, considerable microbial biomass is accumulated within the glacial ice photic zone under typical melting conditions (Irvine-Fynn et al. 2012) and thus the perched aquifer of the weathering crust modulates the fluvial delivery of cells to proglacial habitats.

The potential for physical redistribution and substantial concentration of cells in the interstitial spaces of the ice matrix (Mader et al. 2006; Irvine-Fynn et al. 2012) is coupled with the in situ activities of the microbial community. While the overall abundance of cells discharged from the weathering crust was consistent, specific subpopulations (Irvine-Fynn et al. 2012) were seasonally eluted, indicating a temporal shift in the microbial community's structure. Moreover, measurements of bacterial and abundance carbon production (Rassner et al. 2016) indicate the potential community doubling time in the weathering crust is considerably briefer than the hydrological residence time of cells within percolating melt. Finally, Irvine-Fynn et al. (2012) identified nucleic acid bearing cell populations with a median size of 0.5 μm as well as smaller nucleic acid-rich particles consistent with virus-like particles. Rassner et al. (2016) amended meltwater with organic carbon and nutrients to observe the interactions of supraglacial meltwater bacteria and their viruses, finding a community dominated by the Betaproteobacteria genus *Janthinobacterium* which was resilient to viral predation. Therefore, it appears likely that the glacial ice photic zone presents a locus of microbial interactions.

The implications of a glacial ice photic zone for glacial systems and their biogeochemistry are threefold. Firstly, it appears to be an extensive yet poorly characterized habitat. Irvine-Fynn and Edwards (2013) extrapolated measurements of the cellular concentrations within glacial ice photic zones with the potential scale of the glacial ice photic zone on ablating glacial ice. Excluding Antarctic ice, the calculations suggest 1×10^{21} to 1×10^{26} cells reside within the global glacial ice photic zone. This is comparable to the abundance of Archaea and Bacteria within the global oceanic photic zone (4×10^{25} cells; Whitman et al. 1998) which is both greater in surface area and volume (reaching 200 m deep) than the glacial ice photic zone. The pivotal role of marine microbes within the global carbon cycle and biosphere function is well known (Fuhrman and Steele 2008); what, therefore, constrains the impact of microbes from the enigmatic glacial ice photic zone? At present we do not know.

Secondly, the net accumulation of microbes, noncellular, and inorganic particulates within the glacial ice photic zone is enhanced at high melt rates. Indeed, Irvine-Fynn et al. (2012) observed an inverse nonlinear relationship between melt discharge and the accumulation of microbial biomass. As such, this prompts a positive feedback in which the contact rates between the glacial ice photic zone microbiota and inorganic particulates are promoted. Moreover, noting the tendency for *Janthinobacterium* as a dominant taxon within the glacial ice photic zone (Rassner et al. 2016) to exude biofilms and extracellular material (Pantanella et al. 2007), it is likely that cell–mineral aggregates are initiated within the glacial ice photic zone (e.g., Simon et al. 2009) further accelerating near-surface melt. Irvine-Fynn et al. (2012) identified this retention of biomass and its confection with inorganic particulates as a mechanism for the "biological darkening" of ice surfaces, accelerating ice melt.

Finally, the evolution of the glacial ice photic zone and its ability to modulate the flux of melt, biomass, and particulate matter influences the properties of proximal and downstream ecosystems. Cook et al. (2015b) used cryoconite holes (discussed below; Sect. 4.5.2) as naturally occurring piezometers dwelling within the weathering crust to explore the hydrology of the glacial ice photic zone of a second Svalbard glacier. Diurnal fluctuations in the storage and transmission of meltwater were identified, corresponding to changes in the porosity of the weathering crust. While the porosity of the crust is insufficient to permit the mobility of microbe–mineral aggregates, the crust permits the interconnection of supraglacial microbial habitats, buffering against local variations in soluble nutrient availability. Moreover, microbial productivity contributes to the organic carbon budget of glacial systems (Hood et al. 2015). The porous ice of the glacial ice photic zone permits the connection of supraglacial loci of productivity with downstream habitats by the transmission of runoff to supraglacial channels, releasing bioavailable carbon (Lawson et al. 2014; Feng et al. 2016), nutrients, and microbiota to depauperate proglacial habitats including proglacial streams, glacial forefields, and coastal oceans (Hood et al. 2009; Singer et al. 2012; Wilhelm et al. 2013, 2014; Hood et al. 2015).

4.5 From Bioalbedo to Biocryomorphology: Microbial Biofilms on the Ice Surface

Each year, as winter turns to summer, snow cover on lower altitude glacial surfaces melts away as a result of elevated air temperatures and solar irradiation. This event exposes bare ice surfaces and brings about notable changes in the physical, chemical, and biological properties of supraglacial ecosystems. This area is extensive (Anesio et al. 2009; Irvine-Fynn and Edwards 2013) and growing in scale. On the GrIS, the maximum area of snow-free ice that is exposed each year has increased steadily since passive microwave satellite observations began in 1979 (Fettweis et al. 2007), one of many indicators that this region is responding to changes in climate. Bare ice has a higher spectral absorption (Warren et al. 2006) and is denser than snow, and impurities within the ice help to contribute towards a topographically uneven surface (Irvine-Fynn et al. 2014). The melting of snow gives rise to a pulse of nutrient release, including SO_4^{2-}, NH_4^+, NO_3^-, Ca^{2+}, Cl^-, and Na^+ (reviewed in Kuhn 2001). Bioavailable nitrogen, in the form of ammonia, may be sequestered and mineralized within the sediment material of cryoconite holes (Wynn et al. 2007). However, the remaining nutrients released from melting snowpacks are likely evacuated from the surface ice environment by supraglacial meltwater rivers. Therefore, the chemistry of ice is predominantly dependant on the era and location in which it was laid down as snow, a feature that can be used when analyzing ice cores to recreate historical records of atmospheric chemistry.

4.5.1 Algal Biofilms

The most visually notable biological feature of bare ice surfaces is that they are plastered with Zygnematophyceae green algae. In addition to the lutein, violaxanthin, chlorophyll a and b, and β-carotene primary pigments that these cells contain, they also carry pigment-filled vacuoles, which are responsible for coloring the cells into shades of dark brown and purple (Uetake et al. 2010; Remias et al. 2012b; Yallop et al. 2012). Populations of these cells are so dense that they darken glacial surfaces: a "bloom" phenomenon which can be observed both on the ground with the naked eye (Fig. 4.2) and from space in satellite imagery. Surface ice algal abundance has been reported in the range of 10^5 cells ml^{-1} (Yallop et al. 2012); however, as these cells coat the exposed solid ice crystals of this crusty surface, expressing abundance as a liquid concentration is an experimentally and conceptually challenging task. Three ice algae species have been found to dominate polar and alpine glaciers: *Cylindrocystis brébissonii*, *Mesotaenium berggrenii*, and *Ancylonema nordenskiöldii* (Remias et al. 2009, 2012a, b; Remias 2012; Yallop et al. 2012). Analysis of vacuole pigments from Alpine *M. berggrenii* reveals the

Fig. 4.2 The ablation zone of the Greenland Ice Sheet. Surface ice is darkened by Zygnematophyceae green algae, in contrast to englacial ice, as highlighted by the crevasse opening (Photo credit: Sara Penrhyn-Jones)

presence of a phenolic compound called purpurogallin carboxylic acid-6-O-β-D-glucopyranoside. This compound likely functions as an ultraviolet and visible radiation photoprotectant, due to its broad spectral absorbance capacity. In addition, it has been suggested that these vacuoles may act as chemical deterrents against grazers or as energy sinks when temperature or nutrient availability limits cell function (Remias et al. 2012b). While little is known about the specific ecological activities of these communities, their darkening of glacial surfaces, and therefore their ability to reduce solar reflectance and consequently enhance surface melt through bioalbedo reduction (Yallop et al. 2012; Lutz et al. 2014), has become a key consideration for glacial surface mass balance studies in recent years (Tedesco et al. 2016).

4.5.2 Cryoconite Ecosystems

While algal populations may bloom on bare ice, the ice surface is also home to more stable microbial habitats. Key among these are cryoconite ecosystems, formed as microbial communities biofilm aeolian dusts and organic matter (e.g., wind-blown material from proximal habitats or surficial algal necromass) (see the recent review by Cook et al. (2016a) and references therein). The ensuing microbial–mineral aggregate is termed cryoconite, which darkens the ice surface through localized reduction of the albedo through the accumulation of dark humic substances (Takeuchi et al. 2001a, b; Takeuchi 2002). This contributes to the evolution and storage of surface meltwater and the formation of quasi-circular holes within the ice surface (Wharton et al. 1985). The thermodynamic evolution of so-called cryoconite holes occurs in three dimensions. Firstly, depth evolution proceeds until an equilibrium depth is maintained, at which the melting rate of ice in contact with the dark cryoconite debris is equivalent to the melting rate of the adjacent bare ice (Gribbon 1979). Secondly, lateral re-equilibration of cryoconite debris and hence cryoconite hole morphology can occur as sediment loads shift as the result of coalescence or evacuation of cryoconite holes as the ablation season proceeds. In this case, thick layers of cryoconite are redistributed to laterally widen the cryoconite hole (Cook et al. 2016b). In milder climates typical of Arctic and mountain glacial surfaces, cryoconite holes are typically open (or at least transiently lidded by a thin layer of ice on a diurnal cycle) to the atmosphere and reside within the porous ice of the glacial ice photic zone (Cook et al. 2015b) which likely permits hydraulic connectivity of cryoconite ecosystems within the same supraglacial catchments (Edwards et al. 2011). However, in the colder climate of Antarctica, attainment of equilibrium depth may mean the cryoconite hole is sufficiently deep to permit re-freezing of its surface, isolating the cryoconite hole from the exchange of biomass, gases and nutrients for extended periods, up to a decade (Tranter et al. 2004). In the most extreme conditions of Antarctica, where strong winds scour bare ice clean to form areas of blue ice, it is thought that the entombed cryoconite holes can only be colonized by microbes liberated from the melt-out of ancient ice and entrapped dusts (Hodson et al. 2013). As such cryoconite ecosystems occupy glacial ice surfaces across the world and endure conditions over an extended gradient of climatic and geochemical conditions. It is thought cryoconite ecosystems may have contributed to the demise (Abbot and Pierrehumbert 2010) of the pervasive Neoproterozoic glaciation ("Snowball Earth"; Hoffman et al. 1998) and the survival of microbial eukaryotes despite (near-)global ice cover (Hoffman 2016).

Indeed, cryoconite holes (Fig. 4.3) are considered "ice-cold hot spots" of microbial activity and diversity on contemporary ice surfaces (Edwards et al. 2013b, 2014b). It appears the principal agent in the formation of cryoconite is the action of filamentous microbial phototrophs, typically cyanobacteria such as *Phormidesmis pristleyi* (Edwards et al. 2011; Chrismas et al. 2015, 2016; Gokul et al. 2016) which exude photosynthetic carbon as extracellular polymeric substances which aggregate biomass and particulate matter (Hodson et al. 2010; Langford et al. 2010, 2014). Consequently, cyanobacteria are thought of as "ecosystem engineers" of the cryoconite ecosystem (Edwards et al. 2014b). As a result of such activity, cryoconite is a locus of carbon and macronutrient biogeochemical cycling on glacial ice the world over (Smith et al. 2016) exhibiting surprisingly high rates of net carbon fixation during the melt season (Anesio et al. 2009), as well as the cycling of nitrogen (Cameron et al. 2012a) through fixation (Telling et al. 2011, 2012b) and both nitrification and denitrification (Segawa et al. 2014). Where

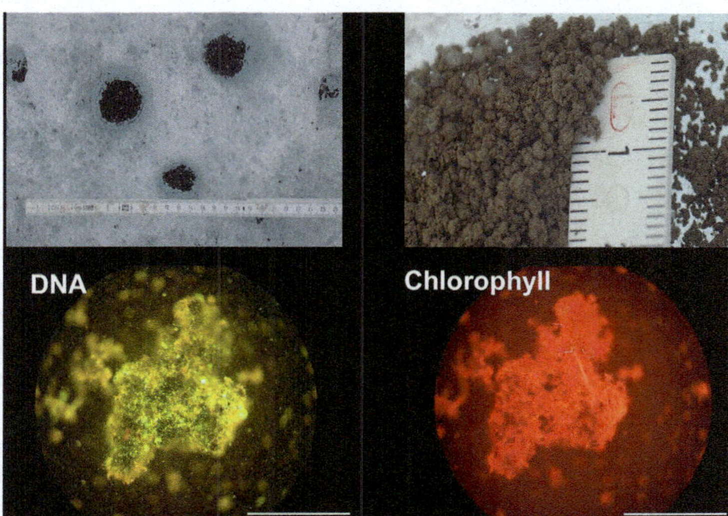

Fig. 4.3 Cryoconite and cryoconite holes: *Top panel*: A cryoconite hole in plan view (*left*), with close up on granular cryoconite typical of Arctic cryoconite sediments (*right*). *Lower panel*: Epifluorescence microscopy of individual cryoconite granules reveals the abundance of microbial life stained for nucleic acid with SYBR Gold (*left*) and chlorophyll *red* autofluorescence (*right*). Scale bars = 200 μm (Top panel photo credit: Nozomu Takeuchi; Lower panel photo credit: Arwyn Edwards)

cryoconite ecosystems are hydraulically connected to proximal habitats or open to the atmosphere the impacts of these supraglacial "bioreactors" can be disseminated further, fertilizing downstream environments (Feng et al. 2016).

The biogeochemical activities of cryoconite are the concerted activity of a diverse, microbially dominated biotic community which ranges from viruses to meiofauna. Along with the cyanobacterial ecosystem engineers, other microbial phototrophs include green algae and diatoms (Stibal et al. 2006; Yallop and Anesio 2010; Stanish et al. 2013; Vonnahme et al. 2015). Conglomeration of cryoconite and moss can result in the development of moss-dominated surface aggregates (Uetake et al. 2014). Notable cryoconite heterotrophs include bacteria, protozoal grazers, fungi and meiofauna such as tardigrades and rotifers (Desmet and Vanrompus 1994; Säwström et al. 2002; Edwards et al. 2013a); Archaea appear to be a minor and variable component of the cryoconite community, but more prevalent within Antarctic and some alpine cryoconite (Cameron et al. 2012b; Hamilton et al. 2013). Perhaps most is known of the bacterial community of cryoconite, which is typically dominated by Proteobacteria, with an apparent divide between Alphaproteobacteria in the Arctic and Betaproteobacteria within alpine cryoconite (Edwards et al. 2014b). Other notable groups include the Actinobacteria, in the form of taxa associated with soil humus (Gokul et al. 2016), Bacteroidetes (Edwards et al. 2013b) and Fibrobacteres (Ransom-Jones et al. 2014) associated with the cycling of complex carbohydrates. The composition of the cryoconite bacterial community is regionally variable at a range of scales from the local (Gokul et al. 2016), to ice sheet (Stibal et al. 2015b) and between global regions (Cameron et al. 2012b; Edwards et al. 2014b). While lithological and hydrologic factors are relevant (Edwards et al. 2011; Lutz et al. 2016a) the ability of the cryoconite community to sequester organic matter (Edwards et al. 2011, 2014b) is closely coupled to the structure and function of the bacterial community.

Accordingly, cryoconite ecosystems represent microbial habitats where the range and rate of microbial processes and the taxa present interact closely within the habitable confines of the cryoconite holes. The receipt and transfer of solar energy as both photosynthetically available radiation and heat for melting is pivotal. The equilibrium depth of seasonally open cryoconite holes coincides with the photosynthetic maxima of phototrophs resident within the cryoconite; as such the flux of photosynthetically available radiation is maintained at optimal levels (Cook et al. 2010, 2012, 2016b). As noted above, the thickening of cryoconite sediments as the consequence of the overloading of cryoconite holes incurs the lateral redistribution of sediments towards single-granule layers and expansion of the cryoconite hole's profile. Since thicker sediment layers of cryoconite tend towards net heterotrophy through the self-shading of cryoconite granules (Cook et al. 2010; Telling et al. 2012a) this restores net autotrophy (Cook et al. 2016a). Disturbing this process artificially incurs a profound metabolomic stress response and impact upon carbon cycling (Cook et al. 2016b). As such the floors of cryoconite holes are sensitively adjusted in three dimensions to provide optimal conditions for the accumulation of organic matter within cryoconite, which in turn enhances the melting of ice and the shaping of the cryoconite hole. Cook et al.

(2015a) defined this multi-lateral network of interactions between cryoconite biota and the ice surface as "biocryomorphology." Considering the ubiquity and mobility of cryoconite holes (Irvine-Fynn et al. 2011a), the potential for microbial processes to shape the ice surface topography by modulating micro- to meso-scale roughness and further influence surface melt rates is raised (Cook et al. 2015a). Understanding how the concerted actions of microbial players within the varied range of biofilms on ice surfaces—from algal blooms to cryoconite holes—influence glacier processes from bioalbedo to biocryomorphology therefore presents a vital research priority.

4.6 Conclusions

Glacier surfaces are a tripwire for climate change since their response to climatic warming is to form meltwater which contributes to sea level rise. They also represent a spatially expansive collective of microbial habitats where life finds niches in the voids between ice crystals or forms biofilms which promote the melting of the ice. As always, the abundance and activity of life is closely intertwined with liquid water, and glacier surfaces are no exception in this regard. In turn glacial microbial processes accelerate melting and are exported in meltwater. Given the contemporary trends for climate warming and consequent wastage of glacial ice, the interactions between life and the surfaces of Earth's glacier ice will continue to merit close study by scientists at the interface of microbiology and glaciology.

References

Abbot DS, Pierrehumbert RT (2010) Mudball: surface dust and snowball Earth deglaciation. J Geophys Res 115(D3):D03104. doi:10.1029/2009jd012007

Amato P, Parazols M, Sancelme M, Laj P, Mailhot G, Delort A-M (2007) Microorganisms isolated from the water phase of tropospheric clouds at the Puy de Dôme: major groups and growth abilities at low temperatures. FEMS Microbiol Ecol 59(2):242–254. doi:10.1111/j.1574-6941.2006.00199.x

Amoroso A, Domine F, Esposito G, Morin S, Savarino J, Nardino M, Montagnoli M, Bonneville JM, Clement JC, Ianniello A, Beine HJ (2009) Microorganisms in dry polar snow are involved in the exchanges of reactive nitrogen species with the atmosphere. Environ Sci Technol 44 (2):714–719. doi:10.1021/es9027309

Andrews TD, MacKay G (2012) The archaeology and paleoecology of alpine ice patches: a global perspective. Arctic 65(5):iii–ivi

Anesio AM, Laybourn-Parry J (2012) Glaciers and ice sheets as a biome. Trends Ecol Evol 27 (4):219–225

Anesio AM, Hodson AJ, Fritz A, Psenner R, Sattler B (2009) High microbial activity on glaciers: importance to the global carbon cycle. Glob Change Biol 15(4):955–960. doi:10.1111/j.1365-2486.2008.01758.x

Benn D, Evans DJ (2014) Glaciers and glaciation. Routledge, Abingdon

Bidle KD, Lee S, Marchant DR, Falkowski PG (2007) Fossil genes and microbes in the oldest ice on Earth. Proc Natl Acad Sci 104(33):13455–13460. doi:10.1073/pnas.0702196104

Björkman MP, Zarsky JD, Kühnel R, Hodson A, Sattler B, Psenner R (2014) Microbial cell retention in a melting high Arctic snowpack, Svalbard. Arct Antarct Alp Res 46(2):471–482

Blazewicz SJ, Barnard RL, Daly RA, Firestone MK (2013) Evaluating rRNA as an indicator of microbial activity in environmental communities: limitations and uses. ISME J 7 (11):2061–2068. doi:10.1038/ismej.2013.102

Boyd ES, Skidmore M, Mitchell AC, Bakermans C, Peters JW (2010) Methanogenesis in subglacial sediments. Environ Microbiol Rep 2(5):685–692. doi:10.1111/j.1758-2229.2010.00162.x

Boyd ES, Lange RK, Mitchell AC, Havig JR, Hamilton TL, Lafrenière MJ, Shock EL, Peters JW, Skidmore M (2011) Diversity, abundance, and potential activity of nitrifying and nitrate-reducing microbial assemblages in a subglacial ecosystem. Appl Environ Microbiol 77 (14):4778–4787. doi:10.1128/aem.00376-11

Brown SP, Olson BJ, Jumpponen A (2015) Fungi and algae co-occur in snow: an issue of shared habitat or algal facilitation of heterotrophs? Arct Antarct Alp Res 47(4):729–749

Brown SP, Ungerer MC, Jumpponen A, Graham LE (2016) A community of clones: snow algae are diverse communities of spatially structured clones. Int J Plant Sci 177(5):432–439

Cameron K, Hodson AJ, Osborn AM (2012a) Carbon and nitrogen biogeochemical cycling potentials of supraglacial cryoconite communities. Polar Biol 35(9):1375–1393. doi:10.1007/s00300-012-1178-3

Cameron KA, Hodson AJ, Osborn AM (2012b) Structure and diversity of bacterial, eukaryotic and archaeal communities in glacial cryoconite holes from the Arctic and the Antarctic. FEMS Microbiol Ecol 82:254–267. doi:10.1111/j.1574-6941.2011.01277.x

Cameron KA, Hagedorn B, Dieser M, Christner BC, Choquette K, Sletten R, Crump B, Kellogg C, Junge K (2014) Diversity and potential sources of microbiota associated with snow on western portions of the Greenland Ice Sheet. Environ Microbiol 17:594–609

Carpenter EJ, Lin S, Capone DG (2000) Bacterial activity in South Pole snow. Appl Environ Microbiol 66(10):4514–4517

Castello JD, Rogers SO (2005) Life in ancient ice. Princeton University Press, Princeton, NJ

Choudhari S, Smith S, Owens S, Gilbert JA, Shain DH, Dial RJ, Grigoriev A (2013) Metagenome sequencing of prokaryotic microbiota collected from Byron Glacier, Alaska. Genome Announc 1(2):e0009913. doi:10.1128/genomeA.00099-13

Chrismas NAM, Anesio A, Sanchez-Baracaldo P (2015) Multiple adaptations to polar and alpine environments within cyanobacteria: a phylogenomic and Bayesian approach. Front Microbiol 6:1070

Chrismas NA, Barker G, Anesio AM, Sánchez-Baracaldo P (2016) Genomic mechanisms for cold tolerance and production of exopolysaccharides in the Arctic cyanobacterium *Phormidesmis priestleyi* BC1401. BMC Genom 17(1):533

Christner BC, Morris CE, Foreman CM, Cai R, Sands DC (2008) Ubiquity of biological ice nucleators in snowfall. Science 319(5867):1214. doi:10.1126/science.1149757

Cook J, Hodson A, Telling J, Anesio A, Irvine-Fynn T, Bellas C (2010) The mass-area relationship within cryoconite holes and its implications for primary production. Ann Glaciol 51 (56):106–110. doi:10.3189/172756411795932038

Cook JM, Hodson AJ, Anesio AM, Hanna E, Yallop M, Stibal M, Telling J, Huybrechts P (2012) An improved estimate of microbially mediated carbon fluxes from the Greenland ice sheet. J Glaciol 58(212):1098–1108

Cook J, Edwards A, Hubbard A (2015a) Biocryomorphology: integrating microbial processes with ice surface hydrology, topography and roughness. Front Earth Sci 3:78. doi:10.3389/feart.2015.00078

Cook JM, Hodson AJ, Irvine-Fynn TDL (2015b) Supraglacial weathering crust dynamics inferred from cryoconite hole hydrology. Hydrol Process 30:433–443. doi:10.1002/hyp.10602

Cook J, Edwards A, Takeuchi N, Irvine-Fynn T (2016a) Cryoconite: the dark biological secret of the cryosphere. Prog Phys Geogr 40(1):66–111

Cook J, Edwards A, Bulling M, Mur L, Cook S, Gokul J, Cameron K, Sweet M, Irvine-Fynn T (2016b) Metabolome-mediated biocryomorphic evolution promotes carbon fixation in Greenlandic cryoconite holes. Environ Microbiol 18(12):4674–4686. doi:10.1111/1462-2920.13349

Darcy JL, Lynch RC, King AJ, Robeson MS, Schmidt SK (2011) Global distribution of *Polaromonas* phylotypes – evidence for a highly successful dispersal capacity. PLoS ONE 6 (8):e23742

Davies TD, Brimblecombe P, Tranter M, Tsiouris S, Vincent CE, Abrahams P, Blackwood IL (1987) The removal of soluble ions from melting snowpacks. In: Jones HG, Orville-Thomas WJ (eds) Seasonal snowcovers: physics, chemistry, hydrology. Springer Netherlands, Dordrecht, pp 337–392. doi:10.1007/978-94-009-3947-9_20

Desmet WH, Vanrompus EA (1994) Rotifera and Tardigrada from some cryoconite holes on a Spitsbergen (Svalbard) Glacier. Belg J Zool 124(1):27–37

Edwards A, Anesio AM, Rassner SM, Sattler B, Hubbard B, Perkins WT, Young M, Griffith GW (2011) Possible interactions between bacterial diversity, microbial activity and supraglacial hydrology of cryoconite holes in Svalbard. ISME J 5:150–160. doi:10.1038/ismej.2010.100

Edwards A, Douglas B, Anesio AM, Rassner SM, Irvine-Fynn TDL, Sattler B, Griffith GW (2013a) A distinctive fungal community inhabiting cryoconite holes on glaciers in Svalbard. Fung Ecol 6(2):168–176. doi:10.1016/j.funeco.2012.11.001

Edwards A, Pachebat JA, Swain M, Hegarty M, Hodson A, Irvine-Fynn TDL, Rassner SME, Sattler B (2013b) A metagenomic snapshot of taxonomic and functional diversity in an alpine glacier cryoconite ecosystem. Environ Res Lett 8(3):035003

Edwards A, Rassner SM, Anesio AM, Worgan H, Irvine-Fynn T, Williams HW, Sattler B, Griffith GW (2013c) Contrasts between the cryoconite and ice-marginal bacterial communities of Svalbard glaciers. Polar Res 32:19468

Edwards A, Irvine-Fynn T, Mitchell AC, Rassner SME (2014a) A germ theory for glacial systems? Wiley Interdiscip Rev Water 1:331–340. doi:10.1002/wat2.1029

Edwards A, Mur LAJ, Girdwood S, Anesio A, Stibal M, Rassner SM, Hell K, Pachebat JA, Post B, Bussell J, Cameron SJ, Griffith GW, Hodson AJ, Sattler B (2014b) Coupled cryoconite ecosystem structure-function relationships are revealed by comparing bacterial communities in Alpine and Arctic glaciers. FEMS Microbiol Ecol 89:222–237

Feng L, Xu J, Kang S, Li X, Li Y, Jiang B, Shi Q (2016) Chemical composition of microbe-derived dissolved organic matter in cryoconite in Tibetan Plateau glaciers: insights from Fourier transform ion cyclotron resonance mass spectrometry analysis. Environ Sci Technol 50 (24):13215–13223

Fettweis X, van Ypersele JP, Gallée H, Lefebre F, Lefebvre W (2007) The 1979–2005 Greenland ice sheet melt extent from passive microwave data using an improved version of the melt retrieval XPGR algorithm. Geophys Res Lett 34(5):L05502

FitzGerald DM, Fenster MS, Argow BA, Buynevich IV (2008) Coastal impacts due to sea-level rise. Annu Rev Earth Planet Sci 36:601–647

Forster RR, Box JE, van den Broeke MR, Miege C, Burgess EW, van Angelen JH, Lenaerts JTM, Koenig LS, Paden J, Lewis C, Gogineni SP, Leuschen C, McConnell JR (2013) Extensive liquid meltwater storage in firn within the Greenland ice sheet. Nat Geosci 7:95–98

Franzetti A, Tatangelo V, Gandolfi I, Bertolini V, Bestetti G, Diolaiuti G, D'Agata C, Mihalcea C, Smiraglia C, Ambrosini R (2013) Bacterial community structure on two alpine debris-covered glaciers and biogeography of *Polaromonas* phylotypes. ISME J 7:1483–1492

Franzetti A, Tagliaferri I, Gandolfi I, Bestetti G, Minora U, Mayer C, Azzoni RS, Diolaiuti G, Smiraglia C, Ambrosini R (2016) Light-dependent microbial metabolisms drive carbon fluxes on glacier surfaces. ISME J 10:2984–2988

Fuhrman J, Steele J (2008) Community structure of marine bacterioplankton: patterns, networks, and relationships to function. Aquat Microb Ecol 53(1):69–81. doi:10.3354/ame01222

Gokul JK, Hodson AJ, Saetnan ER, Irvine-Fynn TD, Westall PJ, Detheridge AP, Takeuchi N, Bussell J, Mur LA, Edwards A (2016) Taxon interactions control the distributions of cryoconite bacteria colonizing a High Arctic ice cap. Mol Ecol 25:3752–3767

Gribbon PWF (1979) Cryoconite holes on Sermikavask, West Greenland. J Glaciol 22 (86):177–181

Hamilton TL, Peters JW, Skidmore ML, Boyd ES (2013) Molecular evidence for an active endogenous microbiome beneath glacial ice. ISME J 7:1402–1412

Harding T, Jungblut AD, Lovejoy C, Vincent WF (2011) Microbes in High Arctic snow and implications for the cold biosphere. Appl Environ Microbiol 77(10):3234–3243. doi:10.1128/aem.02611-10

Hawkings JR, Wadham JL, Tranter M, Raiswell R, Benning LG, Statham PJ, Tedstone A, Nienow P, Lee K, Telling J (2014) Ice sheets as a significant source of highly reactive nanoparticulate iron to the oceans. Nat Commun 5:3929. doi:10.1038/ncomms4929

Hell K, Edwards A, Zarsky J, Podmirseg SM, Girdwood S, Pachebat JA, Insam H, Sattler B (2013) The dynamic bacterial communities of a melting high Arctic glacier snowpack. ISME J 7 (9):1814–1826

Hodson AJ (2014) Understanding the dynamics of black carbon and associated contaminants in glacial systems. Wiley Interdiscip Rev Water 1(2):141–149. doi:10.1002/wat2.1016

Hodson AJ, Mumford PN, Kohler J, Wynn PM (2005) The high Arctic glacial ecosystem: new insights from nutrient budgets. Biogeochemistry 72(2):233–256. doi:10.1007/s10533-004-0362-0

Hodson A, Anesio AM, Tranter M, Fountain A, Osborn M, Priscu J, Laybourn-Parry J, Sattler B (2008) Glacial ecosystems. Ecol Monogr 78(1):41–67

Hodson A, Cameron K, Bøggild C, Irvine-Fynn T, Langford H, Pearce D, Banwart S (2010) The structure, biogeochemistry and formation of cryoconite aggregates upon an Arctic valley glacier; Longyearbreen, Svalbard. J Glaciol 56(196):349–362

Hodson A, Paterson H, Westwood K, Cameron K, Laybourn-Parry J (2013) A blue-ice ecosystem on the margins of the East Antarctic ice sheet. J Glaciol 59(214):255–268

Hoffman PF (2016) Cryoconite pans on Snowball Earth: supraglacial oases for Cryogenian eukaryotes? Geobiology 14(6):531–542. doi:10.1111/gbi.12191

Hoffman PF, Kaufman AJ, Halverson GP, Schrag DP (1998) A neoproterozoic snowball Earth. Science 281(5381):1342–1346. doi:10.1126/science.281.5381.1342

Hood E, Fellman J, Spencer RG, Hernes PJ, Edwards R, D'Amore D, Scott D (2009) Glaciers as a source of ancient and labile organic matter to the marine environment. Nature 462 (7276):1044–1047

Hood E, Battin TJ, Fellman J, O'Neel S, Spencer RGM (2015) Storage and release of organic carbon from glaciers and ice sheets. Nat Geosci 8(2):91–96. doi:10.1038/ngeo2331

Hubbard B, Glasser NF (2005) Field techniques in glaciology and glacial geomorphology. Wiley, Chichester

Irvine-Fynn TDL, Edwards A (2013) A frozen asset: the potential of flow cytometry in constraining the glacial biome. Cytometry A 85(1):3–7. doi:10.1002/cyto.a.22411

Irvine-Fynn TDL, Bridge JW, Hodson AJ (2011a) In situ quantification of supraglacial cryoconite morpho-dynamics using time-lapse imaging: an example from Svalbard. J Glaciol 57:651–657

Irvine-Fynn TDL, Hodson AJ, Moorman BJ, Vatne G, Hubbard AL (2011b) Polythermal glacier hydrology: a review. Rev Geophys 49(4):RG4002. doi:10.1029/2010rg000350

Irvine-Fynn TDL, Edwards A, Newton S, Langford H, Rassner SM, Telling J, Anesio AM, Hodson AJ (2012) Microbial cell budgets of an Arctic glacier surface quantified using flow cytometry. Environ Microbiol 14(11):2998–3012. doi:10.1111/j.1462-2920.2012.02876.x

Irvine-Fynn TD, Sanz-Ablanedo E, Rutter N, Smith MW, Chandler JH (2014) Measuring glacier surface roughness using plot-scale, close-range digital photogrammetry. J Glaciol 60 (223):957–969

Joughin I, Smith BE, Medley B (2014) Marine ice sheet collapse potentially under way for the Thwaites Glacier Basin, West Antarctica. Science 344(6185):735–738. doi:10.1126/science.1249055

Kol E (1942) The snow and ice algae of Alaska. Smithsonian Miscellaneous Collections 101:1–36

Kuhn M (2001) The nutrient cycle through snow and ice, a review. Aquat Sci 63(2):150–167

Langford H, Hodson A, Banwart S, Bøggild C (2010) The microstructure and biogeochemistry of Arctic cryoconite granules. Ann Glaciol 51(56):87–94

Langford HJ, Irvine-Fynn TDL, Edwards A, Banwart SA, Hodson AJ (2014) A spatial investigation of the environmental controls over cryoconite aggregation on Longyearbreen glacier, Svalbard. Biogeosciences 11(19):5365–5380. doi:10.5194/bg-11-5365-2014

Larose C, Berger S, Ferrari C, Navarro E, Dommergue A, Schneider D, Vogel T (2010) Microbial sequences retrieved from environmental samples from seasonal Arctic snow and meltwater from Svalbard, Norway. Extremophiles 14(2):205–212. doi:10.1007/s00792-009-0299-2

Larose C, Dommergue A, Vogel TM (2013) The dynamic arctic snow pack: an unexplored environment for microbial diversity and activity. Biology 2(1):317–330

Lawson EC, Wadham JL, Tranter M, Stibal M, Lis GP, Butler CE, Laybourn-Parry J, Nienow P, Chandler D, Dewsbury P (2014) Greenland Ice Sheet exports labile organic carbon to the Arctic oceans. Biogeosciences 11(14):4015–4028

Lazzaro A, Wismer A, Schneebeli M, Erny I, Zeyer J (2015) Microbial abundance and community structure in a melting alpine snowpack. Extremophiles 19(3):631–642

Liu Y, Priscu JC, Yao T, Vick-Majors TJ, Xu B, Jiao N, Santibáñez P, Huang S, Wang N, Greenwood M (2016) Bacterial responses to environmental change on the Tibetan Plateau over the past half century. Environ Microbiol 18(6):1930–1941

Lopatina A, Krylenkov V, Severinov K (2013) Activity and bacterial diversity of snow around Russian Antarctic stations. Res Microbiol 164(9):949–958

Lopatina A, Medvedeva S, Shmakov S, Logacheva MD, Krylenkov V, Severinov K (2016) Metagenomic analysis of bacterial communities of Antarctic surface snow. Front Microbiol 7:398

Lutz S, Anesio AM, Villar SEJ, Benning LG (2014) Variations of algal communities cause darkening of a Greenland glacier. FEMS Microbiol Ecol 89(2):402–414

Lutz S, Anesio AM, Edwards A, Benning LG (2015) Microbial diversity on Icelandic glaciers and ice caps. Front Microbiol 6:307

Lutz S, Anesio AM, Edwards A, Benning LG (2016a) Linking microbial diversity and functionality of arctic glacial surface habitats. Environ Microbiol. doi:10.1111/1462-2920.13494

Lutz S, Anesio AM, Raiswell R, Edwards A, Newton RJ, Gill F, Benning LG (2016b) The biogeography of red snow microbiomes and their role in melting arctic glaciers. Nat Commun 7:11968

Maccario L, Vogel TM, Larose C (2014) Potential drivers of microbial community structure and function in Arctic spring snow. Front Microbiol 5:413

Mader HM, Pettitt ME, Wadham JL, Wolff EW, Parkes RJ (2006) Subsurface ice as a microbial habitat. Geology 34(3):169–172. doi:10.1130/G22096.1

Mattes TE, Alexander AK, Richardson PM, Munk AC, Han CS, Stothard P, Coleman NV (2008) The Genome of *Polaromonas* sp. strain JS666: insights into the evolution of a hydrocarbon- and xenobiotic-degrading bacterium, and features of relevance to biotechnology. Appl Environ Microbiol 74(20):6405–6416. doi:10.1128/aem.00197-08

McGee D, Broecker WS, Winckler G (2010) Gustiness: the driver of glacial dustiness? Quat Sci Rev 29(17–18):2340–2350

Meier MF, Dyurgerov MB, Rick UK, O'Neel S, Pfeffer WT, Anderson RS, Anderson SP, Glazovsky AF (2007) Glaciers dominate eustatic sea-level rise in the 21st century. Science 317(5841):1064–1067

Müller F, Keeler CM (1969) Errors in short-term ablation measurements on melting ice surfaces. J Glaciol 8(52):91–105

Nagatsuka N, Takeuchi N, Nakano T, Shin K, Kokado E (2014) Geographical variations in Sr and Nd isotopic ratios of cryoconite on Asian glaciers. Environ Res Lett 9(4):045007

Nghiem SV, Hall DK, Mote TL, Tedesco M, Albert MR, Keegan K, Shuman CA, DiGirolamo NE, Neumann G (2012) The extreme melt across the Greenland ice sheet in 2012. Geophys Res Lett 39:L20502. doi:10.1029/2012gl053611

Pachauri RK, Allen MR, Barros V, Broome J, Cramer W, Christ R, Church J, Clarke L, Dahe Q, Dasgupta P (2014) Climate change 2014: synthesis Report. Contribution of working groups I, II and III to the fifth assessment report of the intergovernmental panel on climate change. IPCC, Geneva, Switzerland

Pantanella F, Berlutti F, Passariello C, Sarli S, Morea C, Schippa S (2007) Violacein and biofilm production in *Janthinobacterium lividum*. J Appl Microbiol 102(4):992–999. doi:10.1111/j.1365-2672.2006.03155.x

Paterson W (1994) The Physics of Glaciers. Butterworth-Heinemann, Burlington, MA

Pearce DA, Bridge PD, Hughes KA, Sattler B, Psenner R, Russell NJ (2009) Microorganisms in the atmosphere over Antarctica. FEMS Microbiol Ecol 69(2):143–157. doi:10.1111/j.1574-6941.2009.00706.x

Petit JR, Jouzel J, Raynaud D, Barkov NI, Barnola JM, Basile I, Bender M, Chappellaz J, Davis M, Delaygue G, Delmotte M, Kotlyakov VM, Legrand M, Lipenkov VY, Lorius C, Pepin L, Ritz C, Saltzman E, Stievenard M (1999) Climate and atmospheric history of the past 420,000 years from the Vostok ice core, Antarctica. Nature 399(6735):429–436

Pfeffer WT, Arendt AA, Bliss A, Bolch T, Cogley JG, Gardner AS, Hagen J-O, Hock R, Kaser G, Kienholz C, Miles ES, Moholdt G, Mölg N, Paul F, Radic V, Rastner P, Raup BH, Rich J, Sharp MJ, Consortium R (2014) The Randolph glacier inventory (2014): a globally complete inventory of glaciers. J Glaciol 60:221. doi:10.3189/2014JoG13J176

Ransom-Jones E, Jones DL, Edwards A, McDonald JE (2014) Distribution and diversity of members of the bacterial phylum *Fibrobacteres* in environments where cellulose degradation occurs. Syst Appl Microbiol 37(7):502–509

Rassner SME, Anesio A, Girdwood SE, Hell K, Gokul J, Whitworth DE, Edwards A (2016) Can the bacterial community of a high Arctic glacier surface escape viral control? Front Microbiol 7:956

Remias D (2012) Cell structure and physiology of alpine snow and ice algae. In: Lütz C (ed) Plants in Alpine regions. Springer, New York, NY, pp 175–185

Remias D, Lütz-Meindl U, Lütz C (2005) Photosynthesis, pigments and ultrastructure of the alpine snow alga *Chlamydomonas nivalis*. Eur J Phycol 40(3):259–268

Remias D, Holzinger A, Lütz C (2009) Ultrastructure and physiological characterization of the ice alga *Mesotaenium berggrenii* (Zygnemaphyceae, Chlorophyta) from glaciers in the European alps. Phycologia 48:302–312

Remias D, Holzinger A, Aigner S, Lütz C (2012a) Ecophysiology and ultrastructure of *Ancylonema nordenskiöldii* (Zygnematales, Streptophyta), causing brown ice on glaciers in Svalbard (high arctic). Polar Biol 35(6):899–908

Remias D, Schwaiger S, Aigner S, Leya T, Stuppner H, Lütz C (2012b) Characterization of an UV- and VIS-absorbing, purpurogallin-derived secondary pigment new to algae and highly abundant in *Mesotaenium berggrenii* (Zygnematophyceae, Chlorophyta), an extremophyte living on glaciers. FEMS Microbiol Ecol 79(3):638–648

Rodrigues DF, Tiedje JM (2008) Coping with our cold planet. Appl Environ Microbiol 74 (6):1677–1686. doi:10.1128/aem.02000-07

Rohde RA, Price PB (2007) Diffusion-controlled metabolism for long-term survival of single isolated microorganisms trapped within ice crystals. Proc Natl Acad Sci 104 (42):16592–16597. doi:10.1073/pnas.0708183104

Sattler B, Puxbaum H, Psenner R (2001) Bacterial growth in supercooled cloud droplets. Geophys Res Lett 28(2):239–242. doi:10.1029/2000gl011684

Säwström C, Mumford P, Marshall W, Hodson A, Laybourn-Parry J (2002) The microbial communities and primary productivity of cryoconite holes in an Arctic glacier (Svalbard 79°N). Polar Biol 25(8):591–596. doi:10.1007/s00300-002-0388-5

Segawa T, Ishii S, Ohte N, Akiyoshi A, Yamada A, Maruyama F, Li Z, Hongoh Y, Takeuchi N (2014) The nitrogen cycle in cryoconites: naturally occurring nitrification-denitrification granules on a glacier. Environ Microbiol 16(10):3250–3262. doi:10.1111/1462-2920.12543

Shiklomanov I (1993) World freshwater resources. In: Gleick PH (ed) Water in crisis: a guide to the world's fresh water resources. Oxford University Press, New York, NY, pp 13–14

Simon C, Wiezer A, Strittmatter AW, Daniel R (2009) Phylogenetic diversity and metabolic potential revealed in a glacier ice metagenome. Appl Environ Microbiol 75(23):7519–7526. doi:10.1128/aem.00946-09

Singer GA, Fasching C, Wilhelm L, Niggemann J, Steier P, Dittmar T, Battin TJ (2012) Biogeochemically diverse organic matter in Alpine glaciers and its downstream fate. Nat Geosci 5(10):710–714

Smith HJ, Schmit A, Foster R, Littman S, Kuypers MM, Foreman CM (2016) Biofilms on glacial surfaces: hotspots for biological activity. NPJ Biofilms Microbiomes 2:16008

Spijkerman E, Wacker A, Weithoff G, Leya T (2012) Elemental and fatty acid composition of snow algae in Arctic habitats. Front Microbiol 3:380

Stanish LF, Bagshaw EA, McKnight DM, Fountain AG, Tranter M (2013) Environmental factors influencing diatom communities in Antarctic cryoconite holes. Environ Res Lett 8(4):045006

Stibal M, Sabacka M, Kastovska K (2006) Microbial communities on glacier surfaces in Svalbard: Impact of physical and chemical properties on abundance and structure of cyanobacteria and algae. Microb Ecol 52(4):644–654. doi:10.1007/s00248-006-9083-3

Stibal M, Gözdereliler E, Cameron KA, Box JE, Stevens IT, Gokul JK, Schostag M, Zarsky JD, Edwards A, Irvine-Fynn TD (2015a) Microbial abundance in surface ice on the Greenland Ice Sheet. Front Microbiol 6:225

Stibal M, Schostag M, Cameron KA, Hansen LH, Chandler DM, Wadham JL, Jacobsen CS (2015b) Different bulk and active bacterial communities in cryoconite from the margin and interior of the Greenland ice sheet. Environ Microbiol Rep 7(2):293–300

Takeuchi N (2002) Optical characteristics of cryoconite (surface dust) on glaciers: the relationship between light absorency and the property of organic matter contained in the cryoconite. Ann Glaciol 34:409–414

Takeuchi N, Kohshima S, Goto-Azuma K, Koerner R (2001a) Biological characteristics of dark colored material (cryoconite) on Canadian Arctic glaciers (Devon and Penny ice caps). Memoirs of the National Institute of Polar Research 54:495–505

Takeuchi N, Kohshima S, Seko K (2001b) Structure, formation, and darkening process of albedo-reducing material (cryoconite) on a Himalayan glacier: a granular algal mat growing on the glacier. Arct Antarct Alp Res 33(2):115–122

Tedesco M, Doherty S, Fettweis X, Alexander P, Jeyaratnam J, Noble E, Stroeve J (2016) The darkening of the Greenland ice sheet: trends, drivers and projections (1981–2100). Cryosphere 10:477–496

Telling J, Anesio AM, Tranter M, Irvine-Fynn T, Hodson A, Butler C, Wadham J (2011) Nitrogen fixation on Arctic glaciers, Svalbard. J Geophys Res-Biogeosci 116:G03039. doi:10.1029/2010jg001632

Telling J, Anesio AM, Tranter M, Stibal M, Hawkings J, Irvine-Fynn T, Hodson A, Butler C, Yallop M, Wadham J (2012a) Controls on the autochthonous production and respiration of organic matter in cryoconite holes on high Arctic glaciers. J Geophys Res 117(G1):G01017. doi:10.1029/2011jg001828

Telling J, Stibal M, Anesio AM, Tranter M, Nias I, Cook J, Lis G, Wadham JL, Sole A, Nienow P, Hodson A (2012b) Microbial nitrogen cycling on the Greenland ice sheet. Biogeosci Discuss 9:2431–2442. doi:10.5194/bgd-8-10423-2011

Temkiv TŠ, Finster K, Hansen BM, Nielsen NW, Karlson UG (2011) The microbial diversity of a storm cloud as assessed by hailstones. FEMS Microbiol Ecol 81(3):684–695. doi:10.1111/j.1574-6941.2012.01402.x

Temkiv TŠ, Finster K, Hansen BM, Pašić L, Karlson UG (2013) Viable methanotrophic bacteria enriched from air and rain can oxidize methane at cloud-like conditions. Aerobiologia 29 (3):373–384. doi:10.1007/s10453-013-9287-1

Tranter M, Fountain AG, Fritsen CH, Berry Lyons W, Priscu JC, Statham PJ, Welch KA (2004) Extreme hydrochemical conditions in natural microcosms entombed within Antarctic ice. Hydrol Process 18(2):379–387. doi:10.1002/hyp.5217

Uetake J, Naganuma T, Hebsgaard MB, Kanda H, Kohshima S (2010) Communities of algae and cyanobacteria on glaciers in west Greenland. Polar Sci 4(1):71–80

Uetake J, Tanaka S, Hara K, Tanabe Y, Samyn D, Motoyama H, Imura S, Kohshima S (2014) Novel biogenic aggregation of moss gemmae on a disappearing African glacier. PLoS ONE 9 (11):e112510

van Leewenhoeck A (1677) Observations, Communicated to the Publisher by Mr. Antony van Leewenhoeck, in a Dutch Letter of the 9th of Octob. 1676. Here English'd: concerning little animals by him observed in rain-well-sea and snow water; as also in water wherein pepper had lain infused. Phil Trans 12(133–142):821–831. doi:10.1098/rstl.1677.0003

Vonnahme T, Devetter M, Žárský J, Šabacká M, Elster J (2015) Controls on microalgal community structures in cryoconite holes upon high Arctic glaciers, Svalbard. Biogeosci Discuss 12:11751–11795

Wadham JL, Arndt S, Tulaczyk S, Stibal M, Tranter M, Telling J, Lis GP, Lawson E, Ridgwell A, Dubnick A, Sharp MJ, Anesio AM, Butler CEH (2012) Potential methane reservoirs beneath Antarctica. Nature 488(7413):633–637

Warren SG, Hudson SR (2003) Bacterial activity in South Pole snow is questionable. Appl Environ Microbiol 69(10):6340–6341

Warren SG, Brandt RE, Grenfell TC (2006) Visible and near-ultraviolet absorption spectrum of ice from transmission of solar radiation into snow. Appl Optics 45(21):5320–5334

Weiss RL (1983) Fine Structure of the snow algae (*Chlamydamonas nivalis*) and associated bacteria. J Phycol 19(2):200–204

Wharton RA, Mckay CP, Simmons GM, Parker BC (1985) Cryoconite holes on glaciers. Bioscience 35(8):499–503

Whitman WB, Coleman DC, Wiebe WJ (1998) Prokaryotes: the unseen majority. Proc Natl Acad Sci 95(12):6578–6583

Wilhelm L, Singer GA, Fasching C, Battin TJ, Besemer K (2013) Microbial biodiversity in glacier-fed streams. ISME J 7:1651–1660

Wilhelm L, Besemer K, Fasching C, Urich T, Singer GA, Quince C, Battin TJ (2014) Rare but active taxa contribute to community dynamics of benthic biofilms in glacier-fed streams. Environ Microbiol 16(8):2514–2524. doi:10.1111/1462-2920.12392

Willerslev E, Hansen AJ, Christensen B, Steffensen JP, Arctander P (1999) Diversity of Holocene life forms in fossil glacier ice. Proc Natl Acad Sci 96:8017–8021

Wunderlin T, Ferrari B, Power M (2016) Global and local-scale variation in bacterial community structure of snow from the Swiss and Australian Alps. FEMS Microbiol Ecol 92(9):fivq32

Wynn PM, Hodson AJ, Heaton TH, Chenery S (2007) Nitrate production beneath a High Arctic glacier, Svalbard. Chem Geol 244(1):88–102

Xiang S-R, Shang T-C, Chen Y, Yao T-D (2009) Deposition and postdeposition mechanisms as possible drivers of microbial population variability in glacier ice. FEMS Microbiol Ecol 70 (2):165–176. doi:10.1111/j.1574-6941.2009.00759.x

Yallop M, Anesio A (2010) Benthic diatom flora in supraglacial habitats: a generic-level comparison. Ann Glaciol 51(56):15–22

Yallop ML, Anesio AM, Perkins RG, Cook J, Telling J, Fagan D, MacFarlane J, Stibal M, Barker G, Bellas C, Hodson A, Tranter M, Wadham J, Roberts NW (2012) Photophysiology and albedo-changing potential of the ice algal community on the surface of the Greenland ice sheet. ISME J 6:2302–2313

Yoshimura Y, Kohshima S, Ohtani S (1997) A community of snow algae on Himalayan glacier: change of algal biomass and community structure with altitude. Arct Alp Res 29(1):126–137. doi:10.2307/1551843

Chapter 5
Microbiology of Subglacial Environments

Amanda M. Achberger, Alexander B. Michaud, Trista J. Vick-Majors, Brent C. Christner, Mark L. Skidmore, John C. Priscu, and Martyn Tranter

Abstract The abundance of water at the base of glaciers and polar ice sheets forms lacustrine features and habitats in the saturated sediments of subglacial hydrological systems. Nutrients and energy sources may be made available through mineralization of stored organic matter or through glacial processes (e.g., bedrock comminution) that provide redox couples for microbial life. The logistical challenges of accessing subglacial environments has limited direct observations to a small number of locations, but microorganisms and associated microbial activities have been found in all subglacial environments examined to date (i.e., basal ice and sediment cores, subglacial lakes, and subglacial outflows at glacial margins). Molecular and biogeochemical data imply that the microbial clades common in subglacial environments are utilizing reduced iron, sulfur, and nitrogen compounds

A.M. Achberger
Department of Biological Sciences, Louisiana State University, 202 Life Sciences Building, Baton Rouge, LA 70803, USA

A.B. Michaud
Department of Bioscience, Center for Geomicrobiology, Aarhus University, 8000, Aarhus C Aarhus, Denmark

T.J. Vick-Majors
Départment des Sciences Biologiques, Université du Quèbec à Montrèal, C.P. 8888 succursale Centre-ville, Montréal, QC, Canada, H3C 3P8

B.C. Christner (✉)
Department of Microbiology and Cell Science, Biodiversity Institute, University of Florida, 1355 Museum Drive, Gainesville, FL 32611, USA
e-mail: xner@ufl.edu

M.L. Skidmore
Department of Earth Sciences, Montana State University, 200 Traphagen Hall, Bozeman, MT 59717, USA

J.C. Priscu
Department of Land Resources and Environmental Sciences, Montana State University, 334 Leon Johnson Hall, Bozeman, MT 59717, USA

M. Tranter
Bristol Glaciology Centre, School of Geographical Sciences, University of Bristol, Bristol BS8 1SS, UK

R. Margesin (ed.), *Psychrophiles: From Biodiversity to Biotechnology*,
DOI 10.1007/978-3-319-57057-0_5

as energy sources to fuel primary production at the glacial bed. Here, we review the latest information on the diversity of subglacial environments and discuss how interactions between physical and biogeochemical processes affect microbial ecosystems and processes at the glacier bed.

Contents

5.1 Introduction

Glaciers exist where the annual temperature remains cold enough to allow snowfall to accumulate for an extended period of time and where conditions allow subsequent metamorphosis to ice. Glacial ice forms expansive continental ice sheets in the polar regions (e.g., in Antarctica and Greenland), and at lower latitudes, ice fields (valley or alpine glaciers) and ice caps (if a volcano or mountain range is completely glaciated) exist globally at high altitude. Temperate glaciers comprise $<4\%$ of the glacial ice on the planet, but are important freshwater reservoirs and are often the sources for major rivers vital for irrigation, industry, and providing millions of people with drinking water (e.g., Severskiy 2004). The Greenland and Antarctic ice sheets currently cover ~10% of the terrestrial surface ($>1.5 \times 10^7$ km^2) and contain ~75% of the freshwater on Earth (Paterson 1994). The Antarctic ice sheet alone contains ~90% of the planet's ice and, if melted, would result in a sea level rise of ~65 m (The National Snow and Ice Data Center; http://nsidc.org/).

Evidence for liquid water in the basal zones of polar ice sheets (e.g., Ueda and Garfield 1970), the discovery of more than 400 subglacial lakes in Antarctica (Wright and Siegert 2012), and expanding perspectives on the tenacity of life under cold conditions (e.g., Priscu and Christner 2004; Skidmore 2011; Doyle et al. 2013) have motivated research to gain new insights about microorganisms and active biogeochemical reactions in the subglacial environment. The presence of viable microorganisms has been documented in deep glacier ice (Abyzov et al. 1998; Christner et al. 2000, 2003, 2006; Miteva et al. 2004), basal ice (Skidmore et al. 2000; Sheridan et al. 2003; Foght et al. 2004; Miteva et al. 2004; Doyle et al. 2013), subglacial waters (Sharp et al. 1999; Mikucki et al. 2004; Boyd et al. 2011; Hamilton et al. 2013; Dieser et al. 2014), subglacial sediments (Skidmore et al. 2005; Lanoil et al. 2009), and subglacial lakes and accreted ice (Karl et al. 1999; Priscu et al. 1999; Christner et al. 2001, 2006, 2014; Gaidos et al. 2004; Achberger et al. 2016). Here, we review current information on the diversity of subglacial environments and discuss how interactions between physical and biogeochemical processes affect microbial ecosystems at the glacier bed.

5.2 Liquid Water Beneath Ice Masses

5.2.1 Water and Life

Water is required as a solvent in biochemical reactions, for mass transfer (i.e., the physical transport of molecules), and to establish electrochemical gradients (e.g., proton motive force). Nutrient-containing (i.e., C, N) surface melt waters enter the basal zone of warm and polythermal alpine glaciers, and studies of subglacial outflow and sediments have provided information on the biogeochemistry and microbial diversity of these environments (e.g., Sharp et al. 1999; Foght et al. 2004; Tranter et al. 2005). The occurrence of subglacial water under polar ice sheets was first discovered in Antarctica (Ueda and Garfield 1970) and, more recently, in Greenland (Anderson et al. 2004; Palmer et al. 2013). In the presence of liquid water and chemical (inorganic or organic) energy sources, subglacial microbial communities cycle nutrients under some of the most extreme conditions in the biosphere.

5.2.2 Liquid Water in Arctic and Alpine Subglacial Environments

Ice masses may be classified into three categories based on the temperature regime of the ice: cold, polythermal, and temperate (Paterson 1994). Cold-based ice masses consist of ice with temperatures below freezing throughout, and liquid water is only present in the veins between ice crystals (Nye 1992), with no significant water layer at the glacier bed. Alpine glaciers in the McMurdo Dry Valleys of Antarctica and smaller glaciers in the Canadian High Arctic are examples of cold-based ice masses. Polythermal ice masses are largely found at latitudes above the Arctic Circle, whereas temperate ice masses (i.e., ice is at the melting point from the surface to the base) are located at low and mid-latitudes typically as valley glaciers. Sections of polythermal ice masses are frozen to the bed especially beneath the thinner margins and termini. However, the bed is temperate beneath the thicker, inner zones of the ice mass. Melting can occur at the bed of the glacier due to geothermal and frictional heating, and a layer of liquid water between the basal ice and the bed substrate results (Alley et al. 1997). Temperate-based ice masses have ice at the bed which is at the pressure-melting point throughout, and thus the entire glacier bed is wet based. The surface of the ice remains snow covered during the winter, and, during the summer ablation season, surface meltwater is delivered to the glacier bed via crevasses and moulins from surface snow and ice melt in both polythermal (Skidmore and Sharp 1999) and temperate ice masses (Nienow et al. 1998).

A number of different subglacial drainage system configurations are possible where the glacier bed is temperate, for example, linked cavity systems, canals, sheet flow, and channelized drainage [for reviews of temperate glacier hydrology and subglacial processes, see Hubbard and Nienow (1997), Fountain and Walder (1998), Clarke (2005), and Chu (2014)]. Where basal water is present, microbes have been documented in the subglacial environment of both temperate glaciers (Sharp et al. 1999; Foght et al. 2004; Skidmore et al. 2005; Boyd et al. 2011, 2014; Hamilton et al. 2013) and polythermal glaciers (Skidmore et al. 2000; Wadham et al. 2004; Bhatia et al. 2006). Tranter et al. (2005) review how subglacial hydrological flowpaths control the connectivity of chemical weathering environments, access to atmospheric oxygen, and the redox potential (Eh) of the environments that microbes colonize (see also Sect. 5.4.1).

5.2.3 Liquid Water Under the Greenland Ice Sheet

Several ice core drilling projects have been carried out in Greenland since the mid 1960s: Camp Century, the Greenland Ice Sheet Project (GISP and GISP2), the Greenland Ice Core Project (GRIP), and the North Greenland Ice Core Project (NGRIP). In the deepest portion of these ice cores, referred to as "silty ice," the ice contains numerous organic and inorganic inclusions, which are thought to originate from the subglacial environment rather than from aeolian deposition in snowfall (Gow and Meese 1996). Anderson et al. (2004) encountered pink-colored basal water when drilling at NGRIP in 2003, which entered the borehole and raised the fluid level 45 m, signifying that liquid water exists in certain locations at the bed of the Greenland Ice Sheet. Analysis of the refrozen material from NGRIP revealed dissolved CH_4 and H_2 concentrations approximately 60 and 700-fold higher, respectively, than air-equilibrated freshwater (Christner et al. 2012). If a large reservoir of organic matter exists beneath the Greenland Ice Sheet (Wadham et al. 2012) and/or rock crushing provides sufficient H_2 (Telling et al. 2015), subglacial carbon and methane biogeochemical cycling could be widespread at the bed.

Using airborne radio echo sounding (RES), Palmer et al. (2013) identified two small ($\leq$10 km^2) lakes beneath the western margin of the Greenland Ice Sheet, providing the first evidence for subglacial lakes in this region. More recently, additional locations have been identified (Palmer et al. 2015; Willis et al. 2015) that exhibit characteristics of active lakes (i.e., periodically undergo filling and draining cycles). The proximity of these subglacial lakes to supraglacial features (e.g., lakes, crevasses, and moulins) implies that these hydrologic systems are sustained through the conveyance of surface melt water to the base of the ice sheet (Palmer et al. 2013, 2015). Despite the limited number of reports for subglacial lakes in Greenland, high resolution digital elevation models have identified several regions where hydrologic sinks exist (Lindbäck et al. 2015), indicating that small (<1 km^2) lacustrine features may be more prevalent than previously thought. Due to their small size and shallow depths, many sensing techniques (e.g., ice penetrating radar and laser altimetry) must be used in tandem to help identify these features beneath the Greenland Ice Sheet (Lindbäck et al. 2015). At present, the volume and retention time of water at the base of the Greenland Ice Sheet is poorly constrained but fluctuates seasonally due to surface melt inputs (e.g., Rennermalm et al. 2013; Chu 2014).

5.2.4 Discovery of Subglacial Water Beneath Antarctica's Ice Sheets

The Antarctic polar plateau has a mean annual surface temperature of −37 °C and was first explored in an expedition led by Robert Falcon Scott in 1902. During the 1950s, geothermal heat flow models were developed that predicted an increase in temperature with depth in the ice, implying that, if the ice sheet is thick enough, then the basal ice in contact with the bedrock is at the pressure-melting point. Theoretical predictions, together with seismic measurements of the ice sheet thickness, supported the idea that liquid water existed at the base of the ice sheet in central portions of the Antarctic continent (Zotikov 2006).

The first deep drilling operation in Antarctica was initiated in 1966 at Byrd Station. An ice core was successfully recovered at this site, and bedrock was reached at a depth of 2164 m below the ice surface in January 1968. The drilling team did not expect to encounter liquid water at the ice–bedrock interface; however, water entered the borehole and raised the drilling fluid level 55 m. Due to the resulting "hydrostatic unbalance" and movement of the aqueous basal fluid into colder portions of the borehole, further coring and collection of subglacial materials were suspended to prevent loss of the drill by freezing (Ueda and Garfield 1970; Zotikov 2006). Shortly after this discovery, Oswald and Robin (1973) used RES to survey portions of East Antarctica and provided evidence for the existence of water under the ice sheet by identifying 17 subglacial lakes. Large subglacial lakes create flat surface depressions in the overlying ice sheet, and some of these features were used by pilots with the Soviet Antarctic Expedition in the late 1950s as navigation markers (Zotikov 2006). Robin et al. (1977) later conducted many flights in the vicinity of Vostok Station and discovered that a large lake (Subglacial Lake Vostok) exists beneath the Russian base, which was further supported by satellite radar altimetry (Ridley et al. 1993) and seismic data (Kapitsa et al. 1996).

Over the last 20 years, RES and satellite laser altimetry data have expanded the inventory of known Antarctic subglacial lakes to greater than 400 (Wright and Siegert 2012; Siegert et al. 2016). Subglacial lakes are frequently located near ice divides, with a majority (~53%) within <300 km of a divide (Wright and Siegert 2011). There is evidence that more than a quarter of Antarctic subglacial lakes are active due to rapid fluctuations in the surface elevation of the overlying ice, indicative of water movement at the base of the ice sheet. The transport of water between lakes occurs along hydropotential gradients (Fricker and Scambos 2009), and based on studies of the Aurora Subglacial Basin, East Antarctica, subglacial water may flow unimpeded from interior lakes located over 1000 km from the ice margin (Wright et al. 2012). Thus, many subglacial lakes are not isolated systems, and it is likely that hydrologic transfer disperses microorganisms, carbon, and nutrients between lakes and ultimately to the Southern Ocean. The volume of known Antarctic subglacial lakes is estimated at 13,000 km^3 (Wright and Siegert 2011), which represents ~20% of all freshwater on non-glaciated continents and enough to cover the whole Antarctic continent with a uniform water layer thickness of ~1 m (Priscu et al. 2008).

5.3 Subglacial Aquatic Environments

5.3.1 Sampling Subglacial Environments

Over the last decade, significant strides have been made towards understanding the microbial ecosystems that exist beneath ice masses. The majority of amassed data on the diversity and physiology of microbial life in subglacial environments has been derived from the examination of subglacial outflows collected at the margins of polar and alpine glaciers. However, much less is known about the communities that exist in the large internal areas of ice masses, primarily due to the substantial logistical effort needed to directly access and sample such locations (e.g., lakes; Doran et al. 2008; Siegert et al. 2012). These efforts are further challenged with the need to introduce minimal chemical and microbial contamination when exploring these pristine environments (Priscu et al. 2003, 2013; Inman 2005; National Research Council 2007; Siegert et al. 2007). Despite these obstacles, several projects have successfully drilled into and sampled subglacial lakes beneath the Antarctic ice sheet and Icelandic ice caps.

To minimize disturbance and the introduction of surface contamination during subglacial lake access, several groups have adopted the method of hot water drilling (HWD; e.g., Thorsteinsson et al. 2008; Rack et al. 2014; Makinson et al. 2016), which involves heating water sourced from melted snow to 80–90 °C and using the thermal energy from the water to melt a hole through the ice mass. Water within the borehole is circulated back to the surface, reheated, and subsequently pumped through a deployable hose to further the drilling process at the base of the borehole (e.g., Fig. 5.1a). Using water as a drilling fluid in HWD has many advantages, including the ability to efficiently purge cells via inline filtration and sterilize the cleaned effluent with ultraviolet irradiation (e.g., Doran et al. 2008; Fig. 5.1b). HWD technology was implemented successfully to gain access to several Icelandic caldera lakes (Gaidos et al. 2004, 2009; Marteinsson et al. 2013) and Subglacial Lake Whillans, West Antarctica (Priscu et al. 2013; Christner et al. 2014; Rack et al. 2014). Although unsuccessful (Makinson et al. 2016; Pearce et al. 2016), HWD was also the approach adopted by the Lake Ellsworth Consortium to enter Subglacial Lake Ellsworth, West Antarctica, when attempted in December 2012 (Siegert et al. 2007, 2012).

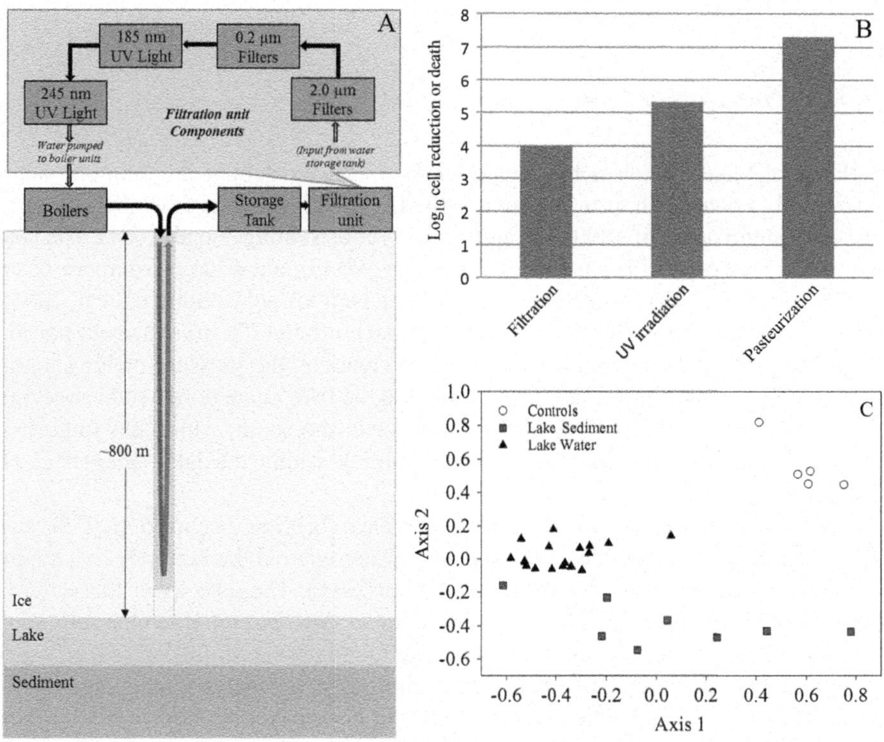

Fig. 5.1 (**a**) Schematic representation of the clean access hot water drill system used at Subglacial Lake Whillans. Filtration unit components consist of 2.0 and 0.2 μm filters that remove cells and particulates. Following filtration, water is irradiated with ultraviolet light (185 and 245 nm) before pasteurization in the boiler unit. (**b**) The efficiency of each stage of the HWD was evaluated for the removal of cells and reduction in cell viability, the latter of which was determined by counts of colony forming units of *Escherichia coli* and adenylate triphosphate concentrations of aquatic communities passed through the system. (**c**) Non-metric multidimensional scaling analysis of the microbial community (based on 16S rRNA gene datasets) of the Subglacial Lake Whillans (SLW) water column and sediments compared to control water samples collected from the drilling system and borehole

Due to the enormous energy requirements, HWD has only been used to create boreholes ≤2.5 km in depth (Benson et al. 2014; Makinson et al. 2016). Thus, it has been difficult to implement this technology to gain entry into lakes under thick ice such as Subglacial Lake Vostok, East Antarctica, which lies beneath ~3.7 km of ice. To access Subglacial Lake Vostok, the Russian Antarctic Expedition took advantage of a pre-existing mechanically drilled borehole (borehole 5G) at Vostok Station that was originally created to obtain ice cores for climatic records. The 5G borehole was drilled to a depth of 3623 m in 1998 by a coordinated Russian, French, and American effort. Owing to concerns of contaminating the lake environment with the kerosene-based fluid in the borehole, drilling was terminated in a zone of accretion ice ~120 m above the water–ice interface. In 2006, a solely Russian drilling effort resumed coring at this site, and in 2012, Subglacial Lake Vostok was initially penetrated (e.g., Bulat 2016).

5.3.2 Subglacial Lake Vostok

Subglacial Lake Vostok is by far the largest of all the subglacial Antarctic lakes identified to date, with a surface area >14,000 km^2, volume of 5400 ± 1600 km^3, and maximum depth of ~800 m (Kapitsa et al. 1996; Studinger et al. 2004). The lake consists of a northern basin (water depth of ~500 m) and a larger southern basin (~800 m water depth), which are separated by a bedrock sill (Studinger et al. 2004). The variation in ice sheet thickness between the north (~4200 m) and south portion (~3900 m) of the lake produces a 0.3 °C difference in the pressure-melting point. This gradient results in glacial ice melting into the lake in the north, and refreezing (i.e., accretion) to the bottom of the ice sheet in the south, which has important repercussions for circulation and vertical mixing within the lake (Siegert et al. 2001).

Entry into Lake Vostok via electrochemical drilling occurred in February 2012 at a borehole depth of ~3769 m, when water entered the borehole to a height of approximately 363 m above the lake (Bulat 2016). The lake water that entered the borehole was allowed to freeze, and a 32 m ice core was retrieved the following season. Although the recovered samples were found to be an emulsion of lake water and microbially contaminated kerosene drilling fluid, an analysis conducted on the refrozen water estimated the surface lake water contained only tens of cells per ml (Bulat 2016). In January 2015, Lake Vostok was accessed for a second time and more refrozen lake water was similarly recovered (Bulat 2016). Analysis of this material has yet to be published.

Perhaps the most reliable data on conditions in Lake Vostok are derived from molecular, microbiological, and geochemical analysis of the basal portion (3539–3623 m) of the 5G borehole (Karl et al. 1999; Priscu et al. 1999; Christner et al. 2001, 2006; Bulat et al. 2004). Based on ice sheet flow (Bell et al. 2002) and the concentration of particle inclusions within the accretion ice (Souchez et al. 2002; Royston-Bishop et al. 2005; Christner et al. 2006), ice cores recovered between 3539 and 3609 m (referred to as Type I accretion ice) are inferred to have formed from lake water that accreted in a shallow embayment in the southwestern portion of the lake, whereas accretion ice between 3610 and 3623 m (Type II accretion ice) has a much lower debris content and probably formed over the deep central portion of the lake's southern basin (de Angelis et al. 2004). Thus, the accretion ice profile captured in the Vostok ice core represents a transect of surface waters ranging from the shallow depths in the east to the deep waters in the vicinity of Vostok Station.

Heterotrophic activity has been reported within melted samples of the accretion ice (Karl et al. 1999; Christner et al. 2006) while metagenomic and small subunit (16S) rRNA gene sequencing of extracted DNA, RNA, and isolated cultures imply the lake is dominated by bacteria related to the Proteobacteria (alpha, beta, gamma, and delta subdivisions), Firmicutes, Actinobacteria, and Bacteroidetes (Priscu et al. 1999; Christner et al. 2001, 2006; Bulat et al. 2004) with few archaeal or eukaryotic members (Rogers et al. 2013). Using a quantitative decontamination protocol (Christner et al. 2005; Fig. 5.2a, b), Christner et al. (2006) examined 20 depths in the Types I and II accretion ice (between 3540 and 3623 m) and predicted that the average concentration of organic carbon, prokaryotic cells, and total dissolved solids in surface waters of the shallow embayment and open lake are 86 and 160 μM, 150 and 460 cells ml^{-1}, and 1.5 and 34 mM, respectively. The input of organic carbon from the ice sheet has been estimated to be insufficient to support reproductive growth of the entire lake community, and a sustained ecosystem would likely require a supplemental chemical energy source (Christner et al. 2006). Supplemental energy needed to support a sustained chemolithotrophic-based ecosystem (Fig. 5.3) may originate from sulfide and iron mineral substrates in subglacial debris (e.g., Bottrell and Tranter 2002; Tranter et al. 2002) entering the lake (Christner et al. 2006) and perhaps by geothermal input from deep faults within the bottom of the lake (Bulat et al. 2004).

5.3.3 Subglacial Lake Whillans

Subglacial Lake Whillans (SLW) is a small, active lake (maximum area of ~60 km^2) located along the Siple Coast of West Antarctica, beneath ~800 m of ice of the Whillans Ice Stream (WIS; Fricker and Scambos 2009; Christianson et al. 2012). Recent observations in this region have shown that water flows between several interconnected lake systems during episodic flooding events (Fricker and Scambos 2009; Carter et al. 2013; Siegfried et al. 2016). As one of the terminal lakes in this hydrologic network, SLW receives input from the upper WIS and the neighboring Kamb Ice Stream (Smith et al. 2009; Wright and Siegert 2012). Since 2003, SLW has filled and drained on three occasions, with the outflow eventually flowing to the Ross Sea beneath the Ross Ice Shelf (Fricker and Scambos 2009; Siegfried et al. 2016). The short residence time of water in SLW (Fricker et al. 2007) meant that any drilling-related impact to the subglacial aquatic system should be minor and

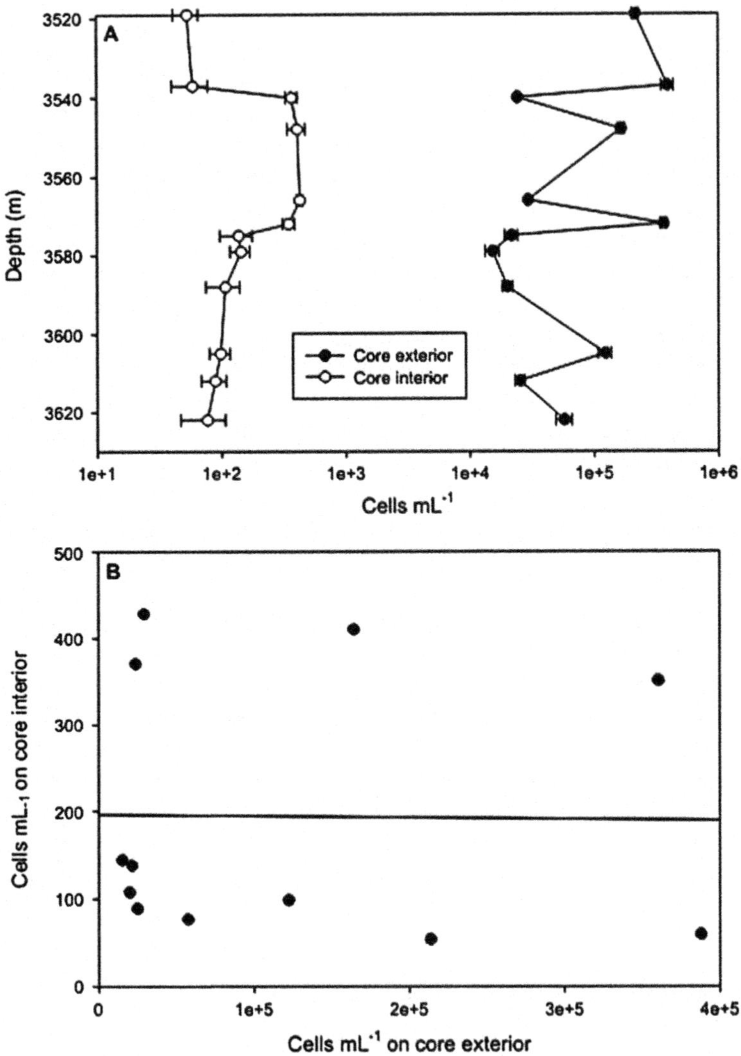

Fig. 5.2 (**a**) Concentration of cells on the exterior and interior of ice cores from the bottom ~100 m (i.e., accretion ice) of the Vostok 5G ice core. (**b**) Cell densities on the inside versus the outside of the ice core are statistically different ($r = 0.016$) and the data do not co-vary with depth (paired t-test, $p < 0.050$). The line is a regression plot of the data points. These data are not consistent with cells penetrating the ice core as a result of drilling (i.e., through microfractures in the ice) or via the drilling fluid, supporting the notion that bona fide cell concentrations were detected within the ice core interior. For detail on the decontamination protocol and the cell enumeration method, see Christner et al. (2005, 2006)

transitory (Fricker et al. 2011). This combined with close proximity to the United States research base (~1000 km from McMurdo Station) made SLW an ideal candidate for exploration.

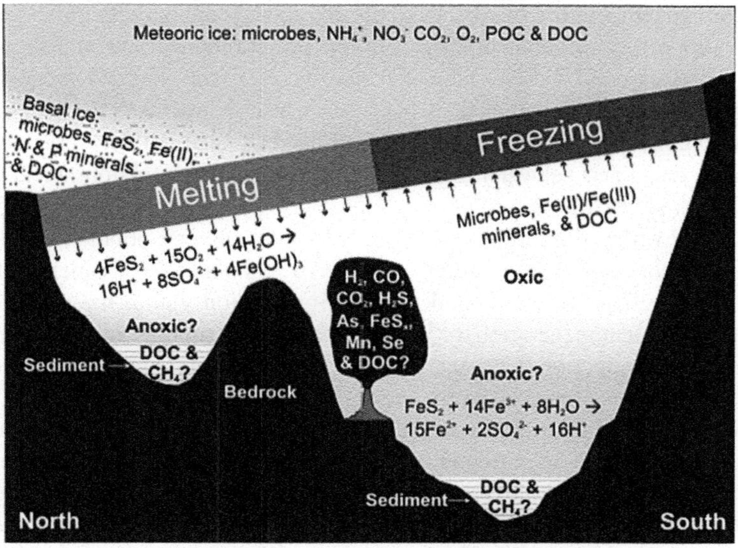

Fig. 5.3 Hypothetical scenario for chemically driven biogeochemical reactions that could be used for bioenergetics in Lake Vostok. Inputs to the system (northern portion of the lake; see text) are through the melting of basal ice, which contains crushed sulfide and iron minerals and organic material from the bedrock, and glacial ice, which provides a constant supply of oxidants (O_2 and NO_3^-), nutrients, and organic material. Microbes, minerals, and organic carbon are removed from the lake via the accretion ice (southern portion of the lake). Shown are oxic and anoxic chemolithotrophic reactions (i.e., metal sulfide oxidation) that have been documented in Alpine and Arctic subglacial environments (Bottrell and Tranter 2002; Tranter et al. 2002; Wadham et al. 2004). Fault bound vents may be present in the shallow embayment of the lake (Bulat et al. 2004), which could introduce significant amounts of thermal energy, geochemical energy, and organic carbon to the lake. If biotic and/or abiotic oxygen sinks exist in the lake, then the deep waters and sediments would be expected to be anaerobic. POC, particulate organic carbon; DOC, dissolved organic carbon

In January 2013, the Whillans Ice Stream Subglacial Access Research Drilling (WISSARD) project used microbiologically clean HWD technology (Priscu et al. 2013) to create a ~800 m deep access borehole into the SLW water column. Upon breakthrough, lake water entered the borehole, raising the water level in the borehole by 28 m (Tulaczyk et al. 2014), indicating that significant mixing of the borehole water with the lake body did not occur. Over the course of 3 days, measurements and sample collection were conducted in the water column and sediments (0–40 cm); see Tulaczyk et al. (2014) for details of the scientific operations.

During sampling, SLW was at low stand following a drainage event, and the water column depth at the borehole location was ~2.2 m (Tulaczyk et al. 2014; Siegfried et al. 2016). Analysis of water samples as well as *in situ* temperature, conductivity, and depth (CTD) profiles revealed that the lake water was aerobic with a temperature of −0.49 °C, a pH of 8.1, and a conductivity of 720 µS cm^{-1} (Christner et al. 2014). These conditions were distinct from those of the borehole (i.e., average temperature of −0.17 °C, pH of 5.4, and conductivity of 5.3 μS cm^{-1}) further supporting minimal mixing between the borehole water and lake during entry. Analysis of $\delta^{18}O\text{-}H_2O$ indicated that the lake was sourced predominantly from glacial melt (Christner et al. 2014). The surface sediments of the lake bed were largely composed of homogenous glacial till (Hodson et al. 2016), while the deeper sediments contained seawater signatures of past marine intrusions in the region (Michaud et al. 2016).

The SLW water column harbored morphologically diverse populations of microbial cells at concentrations of approximately 10^5 cells ml^{-1} (Christner et al. 2014; Purcell et al. 2014). Rates of dark primary production within the water column revealed sufficient generation of organic carbon to support measured rates of heterotrophic production (Christner et al. 2014). Relatively, high dissolved organic carbon (DOC) concentrations in the SLW water column (221 ± 55 μmol L^{-1}) may also support the heterotrophic activity observed, and the lake was found to be biologically deficient in nitrogen with respect to phosphorous (Christner et al. 2014; Vick-Majors et al. 2016).

Molecular analysis of 16S rRNA genes and transcripts revealed the microbial communities associated with SLW water and sediments were distinct from water sampled in the borehole and HWD system, thus enabling a confident evaluation of the members of the SLW ecosystem (Fig. 5.1c; Achberger et al. 2016). SLW was dominated by bacterial species, primarily Proteobacteria (gamma, beta, and delta subdivisions) and Actinobacteria, with a minor archaeal component (Thaumarchaeota and Euryarchaeota) and little evidence for eukaryotic species (Achberger 2016). The water column community was primarily composed of taxa most closely related to species of *Polaromonas*, *Sideroxydans*, *Albidiferax*, *Candidatus* Nitrotoga, and *Nitrosoarchaeum* (Fig. 5.4; Achberger et al. 2016). Within the sediment profile, species of *Methylobacter* and *Thiobacillus* were prevalent at the sediment–water interface (0–2 cm), while the deeper sediments were inhabited by poorly classified members of the Actinobacteria, Nitrospirae, and Candidate division JS1 phyla (Fig. 5.4; Achberger et al. 2016). Based on the physiology of the nearest cultured relatives, the taxa abundant in SLW are predicted to rely on the lithotrophic oxidation of reduced iron, sulfur, and nitrogen compounds as pathways for energy generation. Additionally, the abundance of *Methylobacter* species also suggests the potential for methane oxidation to occur beneath the Antarctic Ice Sheet (Michaud 2016).

5.3.4 Subglacial Caldera Lakes

Smaller subglacial lakes have also been found in volcanic calderas beneath an Icelandic ice cap. Hydrothermal vent activity results in melting of the overlying ice and filling of the subglacial caldera with water to form a subglacial lake. The overlying ice on Vatnajokull is ~250–300 m thick, and water depths in the lakes vary from 20 to 140 m (Gaidos et al. 2004). These lakes typically drain

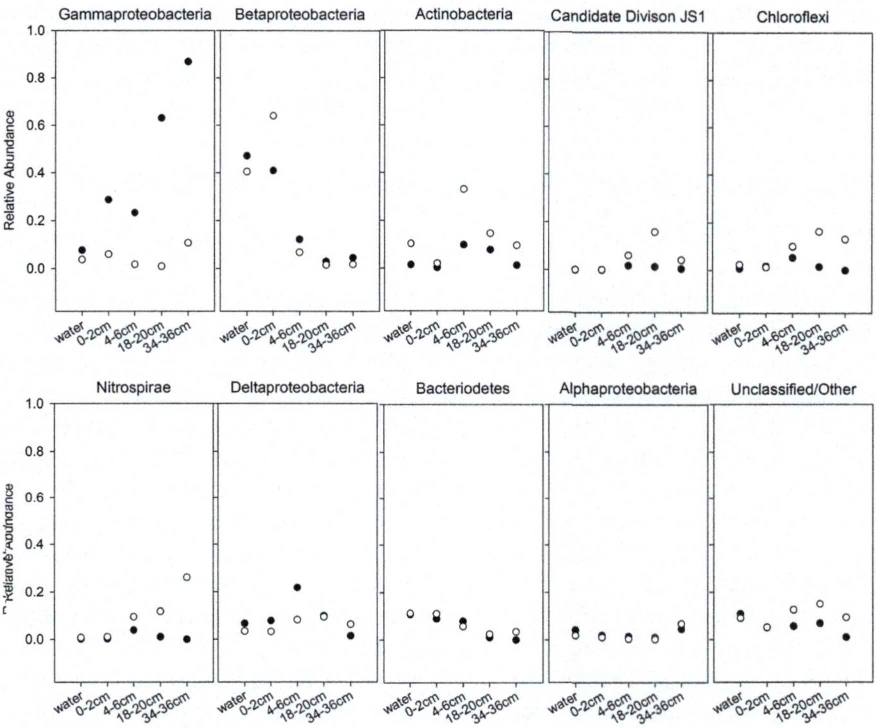

Fig. 5.4 The relative abundance of the dominant phyla detected in the water column and sediments of Subglacial Lake Whillans based on 16S rRNA gene (*white*) and transcript (*black*) analysis

catastrophically beneath the ice resulting in an outburst flood, termed a jökulhlaup. The draining and filling cycle for these subglacial lakes over the past four decades has ranged from one to less than ten years (T. Thorsteinsson, personal communication).

Three Icelandic subglacial lakes, one in the Grímsvötn Caldera and the two Skaftárkatlar Cauldrons, have been accessed via HWD. Molecular analysis of the Grímsvötn water column samples indicates that the lake contains a viable microbial assemblage that is distinct from that in the borehole water (before lake penetration) and overlying ice and snow (Gaidos et al. 2004). Sequencing of amplified 16S rRNA gene fragments from samples of the lake water revealed phylotypes with high identity to psychrophilic and thermophilic organisms (Gaidos et al. 2004). The uptake of ^{14}C-labeled bicarbonate in dark, low-temperature incubations of lake water samples also indicated the presence of chemoautotrophs (Gaidos et al. 2004). At the time of sampling, the water column in the Skaftárkatlar caldera lakes was sulfidic (Gaidos et al. 2009; Marteinsson et al. 2013). Both lakes were bacterially dominated with no detectable archaea (Gaidos et al. 2009; Marteinsson et al. 2013). Many of the abundant organisms within the lakes, as identified through molecular analysis of 16S rRNA genes, were related to autotrophic species that use hydrogen or sulfur species as electron donors (e.g., *Acetobacterium*, *Sulfurospirillum*, *Sulfuricurvum*, and *Desulfosporosinus*; Gaidos et al. 2009; Marteinsson et al. 2013).

5.3.5 Glacial Margin Environments: Insights to Subglacial Processes

Due to the logistical challenges associated with drilling into and directly accessing subglacial aquatic environments (Sect. 5.3.1), numerous investigations have instead focused on examining subglacial water outflows and freshly exposed proglacial sediments collected at the margin of glaciers as a proxy for conditions at the bed. Studies of these environments have provided a wealth of information concerning the biogeochemical characteristics beneath Arctic and Alpine glaciers. For example, Robertson Glacier, in the Canadian Rockies, has been extensively studied and found to harbor diverse archaeal, bacterial, and eukaryotic communities (Hamilton et al. 2013). Comparative surveys of 16S and 18S rRNA genes and transcripts recovered from glacial sediment, cryoconite, and snow samples revealed the presence of taxa shared between the subglacial and supraglacial environments, indicative of exogenous surface input to the glacial bed (Hamilton et al. 2013). However, among the organisms found to be highly prevalent in the glacial sediments were methanogenic archaea and species of *Sideroxydans* and *Sulfurihydrogenibium*, which can derive energy from iron and sulfur oxidation (Hamilton et al. 2013). Transcripts for genes involved in ammonia oxidation were identified that were related to species within the order Nitrosomonadales (Boyd et al. 2011), providing evidence for their involvement in subglacial primary production and nitrogen cycling. The presence of active methanogenic and chemolithoautotrophic organisms in the Robertson Glacier subglacial environment is further supported by geochemical data (e.g., Boyd et al. 2010, 2014; Telling et al. 2015), ^{14}C-labeled bicarbonate uptake experiments (e.g., Boyd et al. 2014), and microcosom/culture-based studies (e.g., Boyd et al. 2010; Harrold et al. 2016).

At Russell Glacier, an outlet glacier along the western margin of the Greenland Ice Sheet (GrIS), subglacial water collected in an outflow over the melt season exhibited a range in electrical conductivity (EC) and dissolved oxygen (DO) that was attributed to variable mixing of supraglacial–subglacial melt water and biogeochemical processes (Dieser et al. 2014). Relatively high dissolved methane concentrations (2.7–83 μM), isotopically light ($\delta^{13}C$) methane, detectable methane oxidation rates (0.32 μM per day), detection of particulate methane monooxygenase (*pmoA*) transcripts, and the presence of methanotrophic and methanogenic taxa in the outflows supported a proposition that active methane cycling, and more generally an active carbon cycle, occurs beneath this portion of the ice sheet (Dieser et al. 2014).

Blood Falls, a subglacial outflow feature found at Taylor Glacier (TG) in the McMurdo Dry Valleys of Antarctica, is so named because anoxic, iron-rich water emanating from the glacier terminus turns an orange/red color when oxidized on exposure to the atmosphere (Mikucki et al. 2004). Airborne transient electromagnetic sensing has revealed the source of the subglacial outflow to be hypersaline brine saturated sediments (~1.5 km^3) thought to be derived from the concentration of ancient marine waters that inundated the region during the Miocene (Mikucki et al. 2015). Although TG is considered to be a cold-based glacier, the salinity of the subglacial brine beneath TG is sufficient to remain unfrozen at temperatures around −6 °C (atmospheric pressure), providing a stable aquatic habitat for microbial life (Mikucki et al. 2015).

Samples of the outflow from Blood Falls contain ~6×10^4 microbial cells ml^{-1}, a portion of which are metabolically active and capable of heterotrophic and autotrophic metabolisms (Mikucki et al. 2004; Mikucki and Priscu 2007). The microbial assemblages characterized at Blood Falls are largely composed of Proteobacteria (beta, delta, and gamma subdivisions) and Bacteroidetes, with most of the taxa phylogenetically related to marine species, consistent with the geological history of the region (Mikucki and Priscu 2007). Furthermore, several of the dominant phylotypes are related to taxa capable of deriving energy from iron and sulfur compounds (e.g., *Thiomicrospira* and *Desulfocapsa*), metabolisms that would be well suited for an environment where sulfur and iron cycling are thought to be tightly coupled (Mikucki and Priscu 2007; Mikucki et al. 2009).

5.4 Biogeochemical Processes and Microorganisms in the Subglacial Environment

5.4.1 Chemolithoautotrophy in Subglacial Environments

Glacier flow results in two factors that promote habitability and production of biomass via chemolithoautotrophic metabolisms within subglacial environments. First, the comminution of the underlying bedrock produces mechanically weathered rock with a high surface area to volume ratio. These crushed mineral grains provide fresh, reactive surfaces for microorganisms to access redox active elements for microbial metabolism (Tranter et al. 2005; Telling et al. 2015). Second, frictional heating provides liquid water to subglacial environments. The presence of finely crushed rock with large mineral surface areas and water promotes rock–water interactions which efficiently solubilize cations and anions present in the bedrock. Of particular interest for microorganisms is the release of redox active elements (i.e., sulfur, iron, manganese) and H_2 from crushed bedrock (Tranter et al. 2005; Telling et al. 2015). Products from rock–water interactions may be used as electron donors or acceptors based on the redox state of the subglacial environment to drive microbial metabolism and formation of biomass through chemolithoautotrophic metabolic pathways. Microbial metabolism in subglacial environments is dependent on ice–water–bed interactions providing redox active elements as well as the availability of oxygen (Tranter et al. 2005; Skidmore 2011; Mitchell et al. 2013).

The availability of oxygen at the glacier bed is controlled by water transported from the supraglacial to subglacial environment through crevasses and moulins or through the melting of basal ice which contains atmospheric gases. In ice masses where water is transported to the bed (i.e., temperate and polythermal glaciers and GrIS), the water typically enters the bed with dissolved oxygen near saturation. Once oxygen reaches the bed of the ice mass, the size of the glacier bed and overall structure of the hydrologic drainage system determines the water retention time and, thus, controls the temporal and spatial distribution of microbial oxygen consumption (Fountain and Walder 1998; Tranter et al. 2005). A channelized subglacial drainage system, resembling a stream network, provides an efficient path for water to move towards the glacier terminus. These channels do not house water throughout the year and atmospheric gases can fill the channels when water is not present (Fountain and Walder 1998; Tranter et al. 2005). A distributed drainage system can also be present beneath an ice mass and led to longer water retention times as the water flows slower through many smaller, circuitous branches, which are typically water filled throughout the year. These two different routes for water through a subglacial environment permit differing oxygen dynamics that favor aerobic microorganisms (channelized drainage system) and anaerobic microorganisms (distributed drainage system).

The geochemical signatures of metabolism vary given the oxygen supplied by contrasting subglacial drainage systems. Microorganisms directly influence geochemistry through their metabolism (e.g., produce sulfate from sulfide oxidation) and indirectly through generating metabolic end products that enhance mineral weathering (e.g., H^+ from sulfide oxidation). Subglacial hydrology and drainage will influence the delivery of oxygen and types of metabolisms possible at the bed, which may be inferred by examining geochemical end products that are exported from the terminus of the glacier. For example, sulfide oxidation and organic carbon remineralization at the glacier bed consume oxygen, eventually creating anoxia. Tranter et al. (2002) reported excesses of sulfate relative to that theoretically possible by oxidizing FeS_2 in O_2-saturated water, indicating the microbial oxidation of metal sulfides in glacial flour also occurs under anoxic conditions, perhaps using Fe(III) as an oxidant (Bottrell and Tranter 2002; Tranter et al. 2002). Recently, a cold-adapted, autotrophic bacteria was isolated from subglacial sediments that couples the oxidation of thiosulfate to the reduction of oxygen, nitrate, or nitrite, linking the sulfur and nitrogen cycle beneath glaciers (Harrold et al. 2016). Most sulfide- and iron-oxidizing bacteria are members of the β- and γ-Proteobacteria, and taxa related to autotrophic species of *Thiobacillus* and *Sideroxydans* are found in many subglacial environments, where they may provide fixed carbon to the ecosystem (Foght et al. 2004; Skidmore et al. 2005; Mitchell et al. 2013; Boyd et al. 2014; Achberger et al. 2016; Harrold et al. 2016).

Christner et al. (2006) suggest that sulfide and iron oxidation could serve as the basis for a chemolithotrophic food web in Lake Vostok (Fig. 5.2). Electron acceptors such as oxygen and nitrate are continually introduced into the lake through the melting of basal ice, and sulfate is produced through the chemical weathering of sulfide minerals in the bedrock. There has been speculation regarding geothermal energy input from high-enthalpy mantle processes or seismotectonic activity (Bulat et al. 2004), which could introduce significant amounts of geochemical energy and CO_2 and could support an ecosystem similar to those found in deep-sea hydrothermal vents. However, the data support the role of glaciological processes in supplying subglacial lake ecosystems with nutrient and redox couples for microbial metabolism, and therefore the search for viable subglacial communities need not be exclusive to environments with geothermal input.

Although most of the evidence for subglacial chemolithoautotrophy is derived from studies of smaller valley glaciers (e.g., Skidmore et al. 2005; Boyd et al. 2014), geochemical and microbiological data imply that similar physiologies (i.e., sulfide and iron oxidation) are active in aquatic environments beneath the West Antarctic Ice Sheet (WAIS) (Lanoil et al. 2009; Skidmore et al. 2010; Purcell et al. 2014; Achberger et al. 2016). The production of excess sulfate in the upper 15 cm of sediments beneath SLW, concurrent with evidence for O_2 depletion, and 16S rRNA transcripts closely related to *Thiobacillus* and *Sideroxydans* spp., in the upper 6 cm provides additional support for active sulfide-oxidizing microorganisms (Achberger et al. 2016; Michaud et al. 2016). Further, the metabolic activity of *Thiobacillus* and *Sideroxydans* spp. could indirectly impact mineral weathering through the production of protons during sulfide oxidation, as shown by a peak in Si concentration coinciding with a peak in their rRNA abundance (Achberger et al. 2016; Michaud et al. 2016).

It has been hypothesized that the oxic conditions in SLW, which favor the observed sulfide- and iron-oxidizing microorganisms, may be are maintained by high geothermal heat flux, which induces melting at the ice sheet bed releasing atmospheric oxygen trapped during ice formation at the surface (Fisher et al. 2015; Vick-Majors et al. 2016). The oxygen concentrations then depend on the biological and chemical oxygen demand of the system. For example, the estimated biological demand of SLW exceeds oxygen inputs. Based on dissolved O_2 concentrations in the waters at the time of sampling, the lake would go anoxic in ~4 years without additional inputs of oxygen (Vick-Majors et al. 2016). The concentration profile of redox sensitive species in SLW sediments (such as V) indicate oxygen is consumed within the upper 15 cm of sediment and horizons below this depth are anoxic (Michaud et al. 2016). These anoxic sediments beneath SLW should be conducive to anaerobic microorganisms, notably methanogenic archaea that have been identified in the deepest horizons of SLW sediment available (Achberger et al. 2016) and in other subglacial environments (Boyd et al. 2010; Wadham et al. 2012).

5.4.2 Heterotrophic Activity in Subglacial Environments

Microbial activities in subglacial environments are most likely limited by some combination of: the permanent darkness (i.e., no photosynthetic activity), low temperature, availability and quality of organic substrates, and low concentrations of N relative to C and P (Boyd et al. 2011; Christner et al. 2014; Vick-Majors et al. 2016). In spite of this, microbial activity has been detected in many subglacial environments. For example, samples collected from the outflow of Blood Falls (Sect. 5.3.5) that contained relatively high dissolved organic carbon (DOC; 711 μM) concentrations were positive for both thymidine and leucine incorporation (proxies for heterotrophic microbial activity; 1.63–12.5 nM thymidine per hour and 0.51–1.0 nM leucine per hour; Mikucki et al. 2004). Data from Lake Vostok accretion ice suggest that the lake is oligotrophic (≤250 μM DOC, ≤2.50 μM total N), but that 75–99% of microorganisms present are potentially viable and capable of metabolizing glucose and acetate (Karl et al. 1999; Priscu et al. 1999; Christner et al. 2006). There is also evidence that microbial activity occurs in the unfrozen liquid vein network in ice at −4 °C to −33 °C, a temperature range that is similar to those at frozen glacier beds (e.g., −15 °C at Taylor Glacier), albeit at rates indicative of maintenance metabolism (Bakermans and Skidmore 2011; Doyle et al. 2013). Heterotrophic activity in the subglacial environment must ultimately be supported by a combination of organic matter generated by chemolithotrophic primary production (e.g., Boyd et al. 2014), and that stored in the sediments (e.g., Bhatia et al. 2010), or transported to the bed (i.e., in regions where the supraglacial and subglacial hydrological system are directly connected; Willis et al. 2015).

There is taxonomic and potential functional coherence among subglacial heterotrophic taxa. For example, taxa related to the *Polaromonas* comprised one of the most abundant groups in the SLW water column (Achberger et al. 2016). *Polaromonas* can degrade a variety of carbon compounds and are commonly found in cryospheric and high latitude environments (Lanoil et al. 2009; Boyd et al. 2011; Hell et al. 2013; Hodson et al. 2015). Taxa related to *Albidiferax ferrireducens,* which grows heterotrophically on acetate using iron as an electron acceptor, have also been found to be numerically abundant in subglacial environments including SLW and beneath northern hemisphere glaciers (Skidmore et al. 2005; Achberger et al. 2016).

The availability of organic carbon substrates to support heterotrophic growth has been documented in a number of subglacial environments, with carbon pools characterized as labile microbial carbon (e.g., Wright Glacier, Antarctica; Stibal et al. 2012), more recalcitrant, relict plant material (e.g. Russell Glacier, Greenland; Stibal et al. 2012), and more recalcitrant, relict marine material mixed with recently microbially derived material (e.g., Subglacial Lake Whillans, Antarctica; Christner et al. 2014; Vick-Majors 2016). Concentrations of organic carbon reported for subglacial environments are low compared to more productive surface environments: Lake Vostok (17–250 μM, estimated from the accretion ice: Priscu et al. 1999; Christner et al. 2006), subglacial meltwaters and basal ice in the high Arctic (8–100 μM: Skidmore et al. 2000; Barker et al. 2006), glacial meltwaters of the Canadian Rocky Mountains (13–64 μM: Lafreniere and Sharp 2004), and the water column of SLW (220 μM; Christner et al. 2014). In addition, the lack of photosynthetically produced carbon may mean that significant fractions of the DOC are of poor quality (Ogawa et al. 2001).

Heterotrophic metabolic rates in subsurface environments are typically low, such as those found in deep sea sediments, (10^{-5} to 10^{-3} fmol C per cell per day) compared to surface environments (0.1–10 fmol C per cell per day; (Jørgensen 2011); similarly low rates ($2–8 \times 10^{-3}$ fmol C per cell per day) were found in the water column at SLW (Vick-Majors 2016). Nutrient amendments (C, N, and P) failed to consistently stimulate rates of heterotrophic activity in SLW, and productivity increased by 5% per °C of temperature increase to a maximum of 0.6 nmol C L^{-1} per day at 10 °C, with a community Q_{10} of 2.13, indicating that the community was not temperature-limited (Vick-Majors et al. 2016). The availability of freshly produced carbon substrate did not immediately explain the low heterotrophic rates. Most of the heterotrophic carbon demand in SLW was apportioned to respiration rather than biomass production (92%; Vick-Majors et al. 2016), and when this was considered, chemolithoautotrophic production exceeded the heterotrophic carbon demand by a factor of 1.5. Thermodynamic calculations in terms of energy density (LaRowe and Amend 2014) imply that the energy available to support heterotrophic activity was 20-fold lower than that to support chemolithoautotrophic metabolism (Vick-Majors et al. 2016), suggesting that energy limitation may constrain heterotrophic metabolism in SLW. Whether this is characteristic of all subglacial environments remains to be seen, as few studies have reported rates of both heterotrophic and chemolithoautotrophic activity. One such study at Blood Falls reported a ratio of chemolithoautotrophic to heterotrophic production as ~0.5 (Mikucki et al. 2004, Mikucki and Priscu 2007), suggesting that a range of ecological conditions (net heterotrophic vs. net autotrophic) may be encountered in disparate subglacial environments.

5.4.3 Do Subglacial Environments Harbor Endemic Microbial Species?

Microbiological investigations of glacial environments (i.e., cryoconite holes, subsurface glacial ice, accreted ice from Lake Vostok, subglacial sediment, and subglacial outflow and streams) in polar and non-polar locations indicate that strong phylogenetic relationships exist between bacteria from geographically distant environments (e.g., Priscu and Christner 2004). In plants and animals, allopatric speciation can occur when a geographically isolated population diverges from the parent population, resulting in the emergence of a genetically distinct species. Biogeographical relationships exist between cyanobacteria in hot springs (e.g., Papke and Ward 2004). However, the question remains as to whether geographically separated glacial environments possess endemic or cosmopolitan species.

Figure 5.5 shows the phylogenetic relationships (based on 16S rRNA gene analysis) between Alphaproteobacteria of the genera *Methylobacterim* and *Sphingomonas,* the Gammaproteobacterium *Methylobacter*, and the Betaproteobactia genera of *Polaromonas*, *Sideroxydans*, *Thiobacillus*, and Candidatus *Nitrotoga* isolated from glacial and subglacial environments in

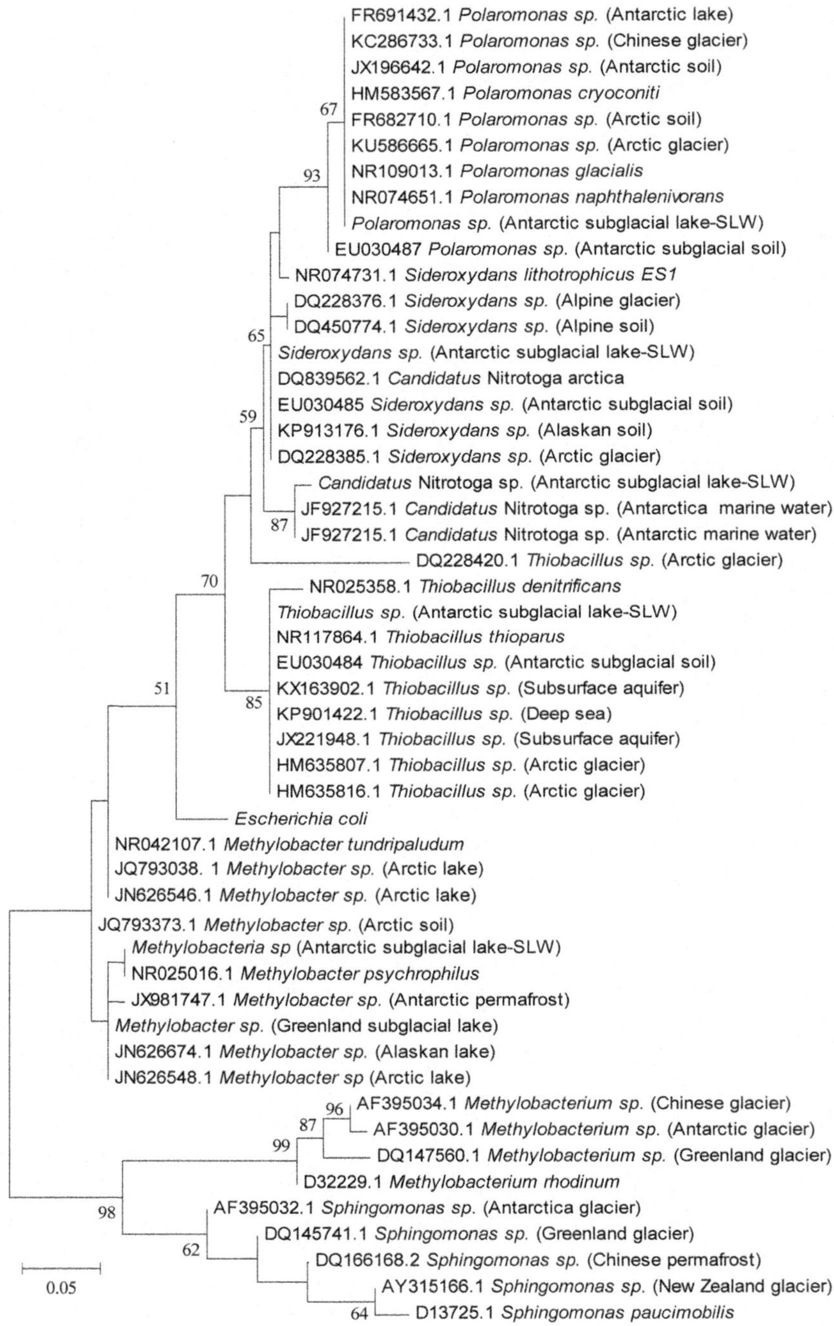

Fig. 5.5 Phylogenetic analysis of small subunit (16S) rRNA gene sequences from species frequently isolated from and identified in glacial and subglacial environments. The phylogenetic tree was generated using the maximum likelihood method. The *scale bar* represents 0.05 fixed substitutions per nucleotide position

Antarctica (glacial ice, Subglacial Lake Whillans, Kamb Ice Stream sediments), Greenland (glacial ice and subglacial outflow), China (glacial ice), and New Zealand (subglacial sediment). Also included in Fig. 5.5 are related strains and cloned sequences recovered globally from the deep sea vents, Antarctic marine waters, subsurface aquifers, frozen Arctic lakes, and permafrost, polar, and alpine soils. The distribution of related bacteria in worldwide glacial and subglacial environments implies that some members of these genera evolved under cold conditions and likely possess similar strategies to survive freezing and metabolize at low temperatures. While the phylogeny of a single gene (i.e., 16S rRNA; Fig. 5.5) is not sufficient to resolve fine scale evolutionary relationships, analysis of multiple loci and recent advances in genomic sequencing technology make these types of experiments now feasible and cost-effective. Due to the isolated nature of subglacial environments, these systems may represent promising evolutionary models for investigating bacterial endemism and to test theory-based species concepts (e.g., Cohan 2002).

5.5 Conclusions

Despite the fact that >80% of the biosphere (by volume) is permanently below 5 °C and most of the biomass is microbial (Priscu and Christner 2004), very little is known about the biology of microorganisms inhabiting permanently cold environments. Biologists have studied life on the margins and surfaces of glaciers for nearly a century, but until recently, the subglacial environment was thought to be inhospitable for life. The discovery of active microbial assemblages beneath glaciers and realization that large quantities of liquid water exist beneath polar ice sheets has resulted in a new paradigm in the study of life in the cryosphere.

Knowledge of microbial life in subglacial ecosystems is limited due to sparse data and the technological, financial, and environmental challenges associated with sampling of such cold and remote subsurface environments. Considerable progress has been made over the last 10 years in the exploration and study of subglacial environments, permitting a glimpse of the microbial life that exists under conditions of cold temperature, low nutrient input, and no sunlight. Priscu et al. (2008) estimate that the number of cells and organic carbon content in Earth's glaciers and subglacial environments (4×10^{29} cells and 10 Pg C) exceeds that reported for the Earth's surface freshwater lakes and rivers (1.3×10^{26} cells and 0.5 Pg C) and is close to that for the open ocean. These tentative estimates imply that glaciated environments contain a considerable pool of cells and organic carbon, and the deep cold biosphere may represent a significant and previously unknown global source of CO_2 and CH_4 (Sharp et al. 1999). As such, biogeochemical cycling models assuming zero rates of microbial mineralization in glacially overridden soils may underestimate the flux of CO_2 and CH_4 released to the atmosphere during glacial to interglacial transitions.

The study of ecosystems in the cold deep biosphere also has implications for the natural history and evolution of life on Earth, as well as on icy planets and moons in the solar system. Geological evidence indicates a number of periods during Earth history where there have been lengthy episodes of glaciation (Eyles 2008). One notable example of pervasive low latitude glaciation occurred during the late Proterozoic, referred to as a "Snowball Earth" (Kirschvink 1992). Hoffman et al. (1998) argue that during the "Snowball Earth" period, the planet was completely covered in ice for at least 10 million years, and liquid water only existed in the ocean under a thick ice cover. If this scenario is accurate, such a long period of global freeze would have had drastic consequences on ecosystems established prior to this event, and subglacial environments may have provided an important refuge for life during such an extended ice age. Polar ice caps composed of water ice exist on Mars; there is evidence for glaciers at lower latitudes during times of higher obliquity (Head et al. 2005), and the Jovian moon Europa is thought to maintain a 50-100 km-deep liquid ocean under a 3–4 km-thick ice shell (Turtle and Pierazzo 2001). Thus, the study of cold, dark, subglacial environments on Earth will provide insight as to the likelihood of microbial life surviving and persisting in icy extra-terrestrial environments. Furthermore, the challenge of identifying appropriate extraterrestrial sites for exploration and developing technology to sample icy subsurface environments will directly benefit from the experience gained by studying earthly analogs.

Subglacial environments remain one of the last unexplored frontiers on our planet. While the study of microbial communities that function near the freezing point of water is inherently interesting, these ecosystems are also clearly relevant to determining the boundaries for life in the biosphere, biogeochemical cycling, the natural history of life on Earth, and astrobiology. We can therefore expect subglacial exploration to be at the forefront of cryospheric research in the future and the years to follow should prove to be an interesting time of discovery.

References

Abyzov SS, Mitskevich IN, Poglazova MN (1998) Microflora of the deep glacier horizons of central Antarctica. Microbiology (Moscow) 67:66–73

Achberger AM (2016) Structure and functional potential of microbial communities in Subglacial Lake Whillans and at the Ross Ice Shelf grounding zone, West Antarctica. PhD Thesis, Louisiana State University

Achberger AM, Christner BC, Michaud AB, Priscu JC, Skidmore ML, Vick-Majors TJ (2016) Microbial community structure of Subglacial Lake Whillans, West Antarctica. Front Microbiol 7:1457

Alley RB, Cuffey KM, Evenson EB, Strasser JC, Lawson DE, Larson GJ (1997) How glaciers entrain and transport basal sediment: physical constraints. Q Sci Rev 16:1017–1038

Anderson KK and 49 others (2004) High-resolution record of Northern Hemisphere climate extending into the last interglacial period. Nature 431:47–51

Bakermans C, Skidmore M (2011) Microbial respiration in ice at subzero temperatures (-4°C to -33°C). Environ Microbiol Rep 3:774–782

Barker JD, Sharp MJ, Fitzsimons RJT (2006) Abundance and dynamics of dissolved organic carbon in glacier systems. Arct Antarct Alp Res 38:193–172

Bell RE, Studinger M, Tikku AA, Clarke GKC, Gutner MM, Meertens C (2002) Origin and fate of Lake Vostok water frozen to the base of the East Antarctic ice sheet. Nature 416:307–310

Benson T, Cherwinka J, Duvernois M, Elcheikh A, Feyzi F, Greenler L et al (2014) IceCube enhanced hot water drill functional description. Ann Glaciol 55:105–114

Bhatia MP, Charette MA, Bhatia MP, Das SB, Das SB, Longnecker K, Kujawinski EB (2010) Molecular characterization of dissolved organic matter associated with the Greenland ice sheet. Geochim Cosmochim Acta 74:3768–3784

Bhatia M, Sharp MJ, Foght JM (2006) Distinct bacterial communities exist beneath a High Arctic polythermal glacier. Appl Environ Microbiol 72:5838–5845

Bottrell S, Tranter M (2002) Sulphide oxidation under partially anoxic conditions at the bed of Haut Glacier d'Arolla, Switzerland. Hydrol Process 16:2363–3468

Boyd ES, Hamilton TL, Havig JR, Skidmore ML, Shock EL (2014) Chemolithotrophic primary production in a subglacial ecosystem. Appl Environ Microbiol 80:6146–6153

Boyd ES, Lange RK, Mitchell AC et al (2011) Diversity, abundance, and potential activity of nitrifying and nitrate-reducing microbial assemblages in a subglacial ecosystem. Appl Environ Microbiol 77:4778–4787

Boyd ES, Skidmore M, Mitchell AC, Bakermans C, Peters JW (2010) Methanogenesis in subglacial sediments. Environ Microbiol Rep 2:685–692

Bulat SA (2016) Microbiology of the subglacial Lake Vostok: first results of borehole-frozen lake water analysis and prospects for searching for lake inhabitants. Philos Trans R Soc A 374:20140292

Bulat SA, Alekhina IA, Blot M, Petit JR, Vasileva LP, de Angelis M, Wagenbach D, Lipenkov VY, Vasilyeva LP, Wloch D, Raynaud D, Lukin VV (2004) DNA signature of thermophilic bacteria from the aged accretion ice of Lake Vostok, Antarctica: implications for searching for life in extreme icy environments. Int J Astrobiol 3:1–12

Carter SP, Fricker HA, Siegfried MR (2013) Evidence of rapid subglacial water piracy under Whillans Ice Stream, West Antarctica. J Glaciol 59:1147–1162

Christianson K, Jacobel RW, Horgan HJ, Anandakrishnan S, Alley RB (2012) Subglacial Lake Whillans—ice-penetrating radar and GPS observations of a shallow active reservoir beneath a West Antarctic ice stream. Earth Planet Sci Lett 331:237–245

Christner BC, Mikucki JA, Foreman CM, Denson J, Priscu JC (2005) Glacial ice cores: a model system for developing extraterrestrial decontamination protocols. Icarus 174:572–584

Christner B, Montross G, Priscu J (2012) Dissolved gases in frozen basal water from the NGRIP borehole: implications for biogeochemical processes beneath the Greenland Ice Sheet. Polar Biol 35:1735–1741

Christner BC, Mosley-Thompson E, Thompson LG, Reeve JN (2001) Isolation of bacteria and 16S rDNAs from Lake Vostok accretion ice. Environ Microbiol 3:570–577

Christner BC, Mosley-Thompson E, Thompson LG, Reeve JN (2003) Bacterial recovery from ancient ice. Environ Microbiol 5:433–436

Christner BC, Mosley-Thompson E, Thompson LG, Zagorodnov V, Sandman K, Reeve JN (2000) Recovery and identification of viable bacteria immured in glacial ice. Icarus 144:479–485

Christner BC, Priscu JC, Achberger AM et al (2014) A microbial ecosystem beneath the West Antarctic ice sheet. Nature 512:310–313

Christner BC, Royston-Bishop G, Foreman CM, Arnold BR, Tranter M, Welch KA, Lyons WB, Tsapin AI, Studinger M, Priscu JC (2006) Limnological conditions in subglacial Lake Vostok, Antarctica. Limnol Oceanogr 51:2485–2501

Chu VW (2014) Greenland ice sheet hydrology. A review. Prog Phys Geogr 38:19–54

Clarke GKC (2005) Subglacial processes. Annu Rev Earth Planet Sci 33:247–276

Cohan FM (2002) What are bacterial species? Annu Rev Microbiol 56:457–487

de Angelis M, Petit J-R, Savarino J, Souches R, Thiemens MH (2004) Contributions of an ancient evaporitic-type reservoir to subglacial Lake Vostok chemistry. Earth Planet Sci Lett 222:751–765

Dieser M, Broemsen EL, Cameron KA, King GM, Achberger A, Choquette K et al (2014) Molecular and biogeochemical evidence for methane cycling beneath the western margin of the Greenland Ice Sheet. ISME J 8:2305–2316
Doran PT, Fritsen CH, Murray AE, Kenig F, McKay CP, Kyne JD (2008) Entry approach into pristine ice-sealed lakes—Lake Vida, East Antarctica, a model ecosystem. Limnol Oceanogr Methods 6:542–547
Doyle S, Montross S, Skidmore M, Christner B (2013) Characterizing microbial diversity and the potential for metabolic function at $-15°$ C in the basal ice of Taylor Glacier, Antarctica. Biology 2:1034–1053
Eyles N (2008) Glacio-epochs and the supercontinent cycle after ~3.0 Ga: Tectonic boundary conditions for glaciation. Palaeogeogr Palaeoclimatol Palaeoecol 258:89–129
Fisher AT, Mankoff KD, Tulaczyk SM, Tyler SW, Foley N (2015) High geothermal heat flux measured below the West Antarctic Ice Sheet. Sci Adv 1:e1500093
Foght J, Aislabie J, Turner S, Brown CE, Ryburn J, Saul DJ, Lawson W (2004) Culturable bacteria in subglacial sediments and ice from two southern hemisphere glaciers. Microb Ecol 47:329–340
Fountain AG, Walder JS (1998) Water flow through temperate glaciers. Rev Geophys 36:299–328
Fricker HA, Powell R, Priscu J, Tulaczyk S, Anandakrishnan S, Christner B et al (2011) Siple coast subglacial aquatic environments: The Whillans Ice Stream Subglacial Access Research Drilling Project. Geophys Monogr Ser 194:199–219
Fricker HA, Scambos T (2009) Connected subglacial lake activity on lower Mercer and Whillans ice streams, West Antarctica, 2003–2008. J Glaciol 55:303–315
Fricker HA, Scambos T, Bindschadler R, Padman L (2007) An active subglacial water system in West Antarctica mapped from space. Science 315:1544–1548
Gaidos E, Lanoil B, Thorsteinsson T, Graham A, Skidmore ML, Han S-K, Rust T, Popp B (2004) A viable microbial community in a subglacial volcanic crater lake, Iceland. Astrobiology 4:327–344
Gaidos E, Marteinsson V, Thorsteinsson T, Johannesson T, Rúnarsson ÁR, Stefansson A et al (2009) An oligarchic microbial assemblage in the anoxic bottom waters of a volcanic subglacial lake. ISME J 3:486–497
Gow AJ, Meese DA (1996) Nature of basal debris in the GISP2 and Byrd ice cores and its relevance to bed processes. Ann Glaciol 22:134–140
Hamilton TL, Peters JW, Skidmore ML, Boyd ES (2013) Molecular evidence for an active endogenous microbiome beneath glacial ice. ISME J 7:1402–1412
Harrold ZR, Skidmore ML, Hamilton TL, Desch L, Amada K, van Gelder W et al (2016) Aerobic and anaerobic thiosulfate oxidation by a cold-adapted, subglacial chemoautotroph. Appl Environ Microbiol 82:1486–1495
Head JW, Neukum G, Jaumann R, Hiesinger H, Hauber E, Carr M, Masson P, Foing B, Hoffmann H, Kreslavsky M, Werner S, Milkovich S, van Gasselt S, The HRSC Co-Investigator Team (2005) Tropical to mid-latitude snow and ice accumulation, flow and glaciation on Mars. Nature 434:346–351
Hell K, Insam H, Edwards A, Žárský J, Podmirseg SM, Girdwood S, Pachebat JA, Sattler B (2013) The dynamic bacterial communities of a melting High Arctic glacier snowpack. ISME J 7:1814–1826
Hodson A, Brock B, Pearce D, Laybourn-Parry J, Tranter M (2015) Cryospheric ecosystems: a synthesis of snowpack and glacial research. Environ Res Lett 10:110201
Hodson TO, Powell RD, Brachfeld SA, Tulaczyk S, Scherer RP (2016) Physical processes in Subglacial Lake Whillans, West Antarctica: inferences from sediment cores. Earth Planet Sci Lett 444:56–63
Hoffman PF, Kaufman AJ, Halverson GP, Schrag DP (1998) A neoproterozoic snowball Earth. Science 281:1342–1346
Hubbard B, Nienow PW (1997) Alpine glacier hydrology. Quat Sci Rev 16:939–955
Inman M (2005) The plan to unlock Lake Vostok. Science 310:611–612

Jørgensen BB (2011) Deep subseafloor microbial cells on physiological standby. Proc Natl Acad Sci 108:18193–18194

Kapitsa AP, Ridley JK, Robin GD, Siegert MJ, Zotikov IA (1996) A large deep freshwater lake beneath the ice of central East Antarctica. Nature 381:684–686

Karl DM, Bird DF, Björkman K, Houlihan T, Shackelford R, Tupas L (1999) Microorganisms in the accreted ice of Lake Vostok, Antarctica. Science 286:2144–2147

Kirschvink JL (1992) Late Proterozoic low-latitude global glaciation: the snowball earth. In: Schopf JW, Klein C, Des Marais D (eds) The Proterozoic biosphere: a multidisciplinary study. Cambridge University Press, Cambridge, UK, pp 51–52

Lafreniere MJ, Sharp MJ (2004) The concentration and fluorescence of dissolved organic carbon (DOC) in glacial and nonglacial catchments: interpreting hydrological flow routing and DOC sources. Arct Antarct Alp Res 36:156–165

Lanoil B, Skidmore M, Priscu JC, Han S, Foo W. Vogel SW, Tulaczyk S, Engelhardt H (2009) Bacteria beneath the West Antarctic Ice Sheet. Environ Microbiol 11:609–615

LaRowe D, Amend J (2014) Energetic constraints on life in marine deep sediments. In: Kallmeyer J, Wagner D (eds) Microbial life of the deep biosphere. DE Gruyter, Berlin, pp 279–302

Lindbäck K, Pettersson R, Hubbard AL, Doyle SH, As D, Mikkelsen AB, Fitzpatrick AA (2015) Subglacial water drainage, storage, and piracy beneath the Greenland ice sheet. Geophys Res Lett 42:7606–7614

Makinson K, Pearce D, Hodgson DA, Bentley MJ, Smith AM, Tranter M et al (2016) Clean subglacial access: prospects for future deep hot-water drilling. Philos Trans R Soc A 374:20140304

Marteinsson VT, Runarsson A, Stefánsson A, Thorsteinsson T, Johannesson T, Magnusson SH et al (2013) Microbial communities in the subglacial waters of the Vatnajökull ice cap, Iceland. ISME J 7:427–437

Michaud AB (2016) Microbially mediated biogeochemical cycles in polar ice covered lakes. PhD Thesis, Montana State University

Michaud AB, Skidmore ML, Mitchell AC, Vick-Majors TJ, Barbante C, Turetta C et al (2016) Solute sources and geochemical processes in Subglacial Lake Whillans, West Antarctica. Geology 44:347–350

Mikucki J, Auken E, Tulaczyk S, Virginia R, Schamper C, Sørensen K et al (2015) Deep groundwater and potential subsurface habitats beneath an Antarctic dry valley. Nat Commun 6:6831

Mikucki JA, Foreman CM, Sattler B, Lyons WB, Priscu JP (2004) Geomicrobiology of Blood Falls: an iron-rich saline discharge at the terminus of the Taylor Glacier, Antarctica. Aquat Geochem 10:199–220

Mikucki JA, Pearson A, Johnston DT, Turchyn AV, Farquhar J, Schrag DP et al (2009) A contemporary microbially maintained subglacial ferrous "ocean". Science 324:397–400

Mikucki JA, Priscu JC (2007) Bacterial diversity associated with blood falls, a subglacial outflow from the Taylor Glacier, Antarctica. Appl Environ Microbiol 73:4029–4039

Mitchell AC, Lafrenière MJ, Skidmore ML, Boyd ES (2013) Influence of bedrock mineral composition on microbial diversity in a subglacial environment. Geology 41:855–858

Miteva VI, Sheridan PP, Brenchley JE (2004) Phylogenetic and physiological diversity of microorganisms isolated from a deep Greenland glacier ice core. Appl Environ Microbiol 70:202–213

National Research Council (2007) Exploration of Antarctic subglacial aquatic environments: environmental and scientific stewardship. National Academies Press, Washington DC

Nienow PW, Sharp M, Willis IC (1998) Seasonal changes in the morphology of the subglacial drainage system, Haut Glacier d'Arolla, Switzerland. Earth Surf Process Landf 23:105–133

Nye JF (1992) Water veins and lenses in polycrystalline ice. In: Maeno N, Hondoh T (eds) Physics and chemistry of ice. Hokkaido University Press, Sapporo, Japan, pp 200–205

Ogawa H, Amagai Y, Koike I, Kaiser K, Benner R (2001) Production of refractory dissolved organic matter by bacteria. Science 292:917–920

Oswald GKL, Robin GQ (1973) Lakes beneath the Antarctic ice sheet. Nature 245:251–254
Palmer SJ, Dowdeswell JA, Christoffersen P, Young DA, Blankenship DD, Greenbaum JS et al (2013) Greenland subglacial lakes detected by radar. Geophys Res Lett 40:6154–6159
Palmer S, McMillan M, Morlighem M (2015) Subglacial lake drainage detected beneath the Greenland ice sheet. Nat Commun 6:8408
Papke RT, Ward DM (2004) The importance of physical isolation to microbial diversification. FEMS Microbiol Ecol 48:293–303
Paterson WSB (1994) The physics of glaciers, 3rd edn. Elsevier Science, Tarrytown, New York
Pearce D, Magiopoulos I, Mowlem M, Tranter M, Holt G, Woodward J et al (2016) Microbiology: lessons from a first attempt at Lake Ellsworth. Philos Trans R Soc A 374:20140291
Priscu JC, Achberger AM, Cahoon JE, Christner BC, Edwards RL, Jones WL et al (2013) A microbiologically clean strategy for access to the Whillans Ice Stream subglacial environment. Antarct Sci 25:637–647
Priscu JC, Adams EE, Lyons WB, Voytek MA, Mogk DW, Brown RL, McKay CP, Takacs CD, Welch KA, Wolf CF, Kirschtein JD, Avci R (1999) Geomicrobiology of subglacial ice above Lake Vostok, Antarctica. Science 286:2141–2144
Priscu JC, Bell RE, Bulat SA, Ellis-Evans JC, Kennicutt MC, Lukin VV, Petit JR, Powell RD, Siegert MJ, Tabacco I (2003) An international plan for Antarctic subglacial lake exploration. Polar Geogr 27:69–83
Priscu JC, Christner BC (2004) Earth's icy biosphere In: Bull AT (ed) Microbial biodiversity and bioprospecting. American Society for Microbiology Press, Washington, DC, pp 130–145
Priscu JC, Tulaczyk S, Studinger M, Kennicutt M, Christner BC, Foreman CM (2008) Antarctic subglacial water: origin, evolution, and ecology. In: Laybourn-Parry J, Vincent W (eds) Polar lakes and rivers: limnology of Arctic and Antarctic aquatic ecosytems. Oxford University Press, Oxford, pp 119–135
Purcell AM, Mikucki JA, Achberger AM, Alekhina IA, Barbante C, Christner BC et al (2014) Microbial sulfur transformations in sediments from Subglacial Lake Whillans. Front Microbiol 5:594
Rack FR, Duling D, Blythe D, Burnett J, Gibson D, Roberts G et al (2014) Developing a hot-water drill system for the WISSARD project: 1. Basic drill system components and design. Ann Glaciol 55:285–297
Rennermalm AK, Smith LC, Chu VW, Box JE, Forster RR, Van den Broeke MR, Van As D, Moustafa SE (2013) Evidence of meltwater retention within the Greenland ice sheet. Cryosphere 7:1433–1445
Ridley JK, Cudlip W, Laxon SW (1993) Identification of subglacial lakes using ERS-1 radar altimeter. J Glaciol 39:625–634
Robin GQ, Drewry DJ, Meldrum DT (1977) International studies of ice sheet and bedrock. Philos Trans R Soc Lond 279:185–196
Rogers SO, Shtarkman YM, Koçer ZA, Edgar R, Veerapaneni R, D'Elia T (2013) Ecology of subglacial Lake Vostok (Antarctica), based on metagenomic/metatranscriptomic analyses of accretion ice. Biology 2:629–650
Royston-Bishop G, Priscu JC, Tranter M, Christner BC, Siegert MJ, Lee V (2005) Incorporation of particulates into accreted ice above subglacial Lake Vostok, Antarctica. Ann Glaciol 40:145–150
Severskiy IV (2004) Water-related problems of Central Asia: some results of the (GIWA) International Water Assessment Program. Ambio 33:52–62
Sharp M, Parkes J, Cragg B, Fairchild IJ, Lamb H, Tranter M (1999) Widespread bacterial populations at glacier beds and their relationship to rock weathering and carbon cycling. Geology 27:107–110
Sheridan PP, Miteva VI, Brenchley JE (2003) Phylogenetic analysis of anaerobic psychrophilic enrichment cultures obtained from a Greenland glacier ice core. Appl Environ Microbiol 69:2153–2160

Siegert MJ, 31 others (2007) Exploration of Ellsworth Subglacial Lake: a concept paper on the development, organisation and execution of an experiment to explore, measure and sample the environment of a West Antarctic subglacial lake. Rev Environ Sci Biotechnol 6:161–179

Siegert MJ, Clarke RJ, Mowlem M, Ross N, Hill CS, Tait A et al (2012) Clean access, measurement, and sampling of Ellsworth Subglacial Lake: A method for exploring deep Antarctic subglacial lake environments. Rev Geophys 50:RG1003

Siegert MJ, Ellis-Evans JC, Tranter M, Mayer C, Petit JR, Salamatin A, Priscu JC (2001) Physical, chemical and biological processes in Lake Vostok and other Antarctic subglacial lakes. Nature 414:603–609

Siegert MJ, Ross N, Le Brocq AM (2016) Recent advances in understanding Antarctic subglacial lakes and hydrology. Philos Trans R Soc A 374:20140306

Siegfried MR, Fricker HA, Carter SP, Tulaczyk S (2016) Episodic ice velocity fluctuations triggered by a subglacial flood in West Antarctica. Geophys Res Lett 43:2640–2648

Skidmore M (2011) Microbial communities in Antarctic subglacial aquatic environments (SAE). In: Siegert MJ, Kennicutt MC (eds) Antarctic subglacial aquatic environments. American Geophysical Union Press, Washington, DC, pp 61–68

Skidmore ML, Anderson SP, Sharp MJ, Foght JM, Lanoil BD (2005) Comparison of microbial community compositions of two subglacial environments reveals a possible role for microbes in chemical weathering processes. Appl Environ Microbiol 71:6986–6997

Skidmore ML, Foght JM, Sharp MJ (2000) Microbial life beneath a High Arctic glacier. Appl Environ Microbiol 66:3214–3220

Skidmore M, Sharp MJ (1999) Drainage system behaviour of a high Arctic polythermal glacier. Ann Glaciol 28:209–215

Skidmore M, Tranter M, Tulaczyk S, Lanoil B (2010) Hydrochemistry of ice stream beds—evaporitic or microbial effects? Hydrol Process 24:517–523

Smith BE, Fricker HA, Joughin IR, Tulaczyk S (2009) An inventory of active subglacial lakes in Antarctica detected by ICESat (2003–2008). J Glaciol 55:573–595

Souchez R, Jean-Baptiste P, Petit JR, Lipenkov VY, Jouzel J (2002) What is the deepest part of the Vostok ice core telling us? Earth Sci Rev 60:131–146

Stibal M, Hasan F, Wadham JL, Sharp MJ, Anesio AM (2012) Prokaryotic diversity in sediments beneath two polar glaciers with contrasting organic carbon substrates. Extremophiles 16:255–265

Studinger M, Bell RE, Tikku AA (2004) Estimating the depth and shape of subglacial Lake Vostok's water cavity from aerogravity data. Geophys Res Lett 31:L12401

Telling J, Boyd E, Bone N, Jones E, Tranter M, MacFarlane J et al (2015) Rock comminution as a source of hydrogen for subglacial ecosystems. Nat Geosci 8:851–855

Thorsteinsson T, Elefsen SO, Gaidos E, Lanoil B, Jóhannesson T, Kjartansson V, Marteinsson V, Stefánsson A, Thorsteinsson T (2008) A hot water drill with built-in sterilization: design, testing and performance. Jökull 57:71–82

Tranter M, Sharp MJ, Lamb HR, Brown GH, Hubbard BP, Willis IC (2002) Geochemical weathering at the bed of Haut Glacier d'Arolla, Switzerland—a new model. Hydrol Process 16:959–993

Tranter M, Skidmore M, Wadham J (2005) Hydrological controls on microbial communities in subglacial environments. Hydrol Process 19:995–998

Tulaczyk S, Mikucki JA, Siegfried MR, Priscu JC, Barcheck CG, Beem LH et al (2014) WISSARD at Subglacial Lake Whillans, West Antarctica: scientific operations and initial observations. Ann Glaciol 55:51–58

Turtle EP, Pierazzo E (2001) Thickness of a Europan ice shell from impact crater simulations. Science 294:1326–1328

Ueda HT, Garfield DE (1970) Deep core drilling at Byrd Station Antarctica. In: Gow AJ (ed) International symposium on Antarctic glaciological exploration (ISAGE), Hanover, New Hampshire, USA, 3–7 September 1968, Cambridge, UK, pp 53–62

Vick-Majors T (2016) Biogeochemical processes in Antarctic aquatic environments: linkages and limitations. Dissertation, Montana State University

Vick-Majors TJ, Mitchell AC, Achberger AM et al (2016) Physiological ecology of microorganisms in Subglacial Lake Whillans. Front Microbiol 7:1457–1416

Wadham J, Arndt S, Tulaczyk S, Stibal M, Tranter M, Telling J, Lis G, Lawson E, Ridgwell A, Dubnick A, Sharp M, Anesio A, Butler (2012) Potential methane reservoirs beneath Antarctica. Nature 488:633-637

Wadham JL, Bottrell SH, Tranter M, Raiswell R (2004) Stable isotope evidence for microbial sulphate reduction at the bed of a polythermal high Arctic glacier. Earth Planet Sci Lett 219:341–355

Willis MJ, Bevis MG, Bell RE, Herried BG, Willis MJ (2015) Recharge of a subglacial lake by surface meltwater in northeast Greenland. Nature 518:223–227

Wright A, Siegert MJ (2011) The identification and physiographical setting of Antarctic subglacial lakes: an update based on recent discoveries. Geophys Monogr Ser 192:9–26

Wright A, Siegert M (2012) A fourth inventory of Antarctic subglacial lakes. Antarct Sci 24:659–664

Wright AP, Young DA, Roberts JL, Schroeder DM, Bamber JL, Dowdeswell JA, Young NW, Le Brocq AM, Warner RC, Payne AJ, Blankenship DD (2012) Evidence of a hydrological connection between the ice divide and ice sheet margin in the Aurora Subglacial Basin, East Antarctica. J Geophys Res Earth Surf 117:F01033

Zotikov IA (2006) The Antarctic subglacial Lake Vostok: glaciology, biology, and planetology. Springer, Chichester, UK

Chapter 6
Viruses in Glacial Environments

Sara M.E. Rassner

Abstract Viruses play a central role in glacial microbial communities. Prokaryotes in glacial environments support surprisingly large viral communities, which, in turn, have a considerable impact on the prokaryotic communities. Through the lysis of host cells and by lowering the growth efficiency of prokaryotic communities, viruses substantially alter the carbon cycling in glacial environments. Despite many similarities with viruses in other habitats, the unique characteristics of glacial environments have accentuated certain features in glacial viruses and their interactions with their hosts, e.g. low viral decay rates in supraglacial viruses as a mechanism for overcoming low host contact rates in systems with low prokaryotic abundances, virus-specific temperature adaptation that differ from that of the host, and virus-mediated transfer of CRISPR arrays that confer immunity against superinfection. Current literature suggests that viral communities in glacial environments are as genetically diverse as those in other environments and, with recent technological advances in environmental genomics and bioinformatics, we are posed to tackle the next great challenge in viral ecology of identifying and quantifying the dynamics of individual virus–host pairs in environmental samples.

Contents

S.M.E. Rassner (✉)
Department of Geography and Earth Sciences, Aberystwyth University, Llandinam Building, Aberystwyth SY23 3DB, UK
e-mail: skr@aber.ac.uk

R. Margesin (ed.), *Psychrophiles: From Biodiversity to Biotechnology*,
DOI 10.1007/978-3-319-57057-0_6

6.1 Introduction

Viruses are integral to microbial ecosystems and glacial environments are no exception. In fact, viruses are both as numerous and diverse, if not more so, in microbially dominated polar aquatic habitats as in the oceans (Aguirre de Cárcer et al. 2015; Cavicchioli 2015).

For every group of organisms, there are viruses that infect them (i.e. algal, archaeal, protozoan and other eukaryotic viruses, such as mycoviruses, as well as cyano- and bacteriophages). As can be expected from this diverse range of hosts, there is a great diversity among viruses, in terms of their size (18–1500 nm; Legendre et al. 2014; Tu et al. 2015), morphology (e.g. icosahedral, filamentous, caudate; Ackermann 2007), genome size (3.2 kb–2.5 Mb; Kay and Zoulim 2007; Philippe et al. 2013), genetic material (e.g. single- and double-stranded RNA and DNA; Weinbauer 2004) and mode of action (e.g. lytic, lysogenic, chronic infection; Weinbauer 2004).

Although they come in a plethora of shapes and sizes, it is usually not possible to identify viruses without co-culturing with their host or by sequencing of their genome. It is therefore common to group all viruses found in environmental samples and simply refer to them as 'viruses' or 'bacteriophages' (on the assumption that a majority of the viruses found in a bacteria-dominated habitat have been produced in bacteria). In addition, because of their small size, electron microscopy is required to obtain sufficiently detailed images of their structure (Weinbauer 2004). As a result, many studies rely on epifluorescence microscopy of nucleic acid stained samples at ca. 1000× magnification (Noble and Fuhrman 1998). Using this method, viruses show up as bright pin pricks of green or yellow light against a black background (Patel et al. 2007) and since it is not possible to see the structure of the viruses, they are referred to as *virus-like-particles* (VLPs).

6.1.1 History of the Study of Glacial Viruses

The study of environmental viruses in glacial environments broadly mirrors and follows on from that of the more well-studied marine viruses and, similarly to that of marine viruses, the field has developed through three fairly distinct phases:

Initially, the focus was on determining whether viruses were present in a variety of habitats and to enumerate viruses and compare their abundance to that of the potential host (e.g. bacterial or algal) community in the same habitat. Efforts were also made to estimate frequency of infection, burst size and proportion of lysogens. A majority of these studies were included in the review by Säwström et al. (2008), in which the authors showed that despite their typically low abundances of bacteria (Table 6.1), polar glacial and lentic habitats supported higher-than-expected numbers of VLPs, as evidenced by high virus-to-bacterium ratios (VBRs). However, the large variability in the data collated and presented by Säwström and co-workers highlights one of the difficulties facing polar microbiologists, namely the patchiness of microbial communities in glacial environments. During this time period, the focus of environmental viral ecology was on the role of viruses in bacterial mortality and their effect on bacterial production and carbon cycling (Säwström et al. 2008). Due to technical limitations, bacterial and viral communities were (mainly) treated as whole units, masking any differences in response between subpopulations.

During the second phase, the focus shifted from treating microbial communities as large uniform units, and more emphasis was put on the genetic variety within each community (e.g. Cottrell and Kirchman 2012). It had been previously acknowledged that the overall response of a specific microbial community was likely to hide contrasting and sometimes opposite changes in populations; however,

Table 6.1 Reported ranges of virus-like-particle (VLP) and bacterial abundances, and virus-to-bacterium ratios (VBR), in glacial environments, after type of habitat

	VLP ($\times 10^6$)	Bacteria ($\times 10^6$)	VBR	*n*	Reference
Arctic					
Supraglacial	0.05–1.19	0.007–0.07	7.4–31.2	5	Säwström et al. (2008) and references therein
	≤1.012	≤0.017	≤58.1	1	Rassner et al. (2016)
	562–2450	*268–1960*	*1.2–4.0*	20	Bellas et al. (2013)
Englacial	0.01–0.56		0.7–74.0		Anesio et al. (2007)
Subglacial	≤0.9	≤0.12	≤7.3	1	Säwström et al. (2008) and references therein
Lakes	0.92–28.9	0.38–2.06	1.6–25.2	17	Säwström et al. (2008) and references therein
	0.62–2.61	0.50–1.59	1.21–2.47	8	Gorniak et al. (2016)
Alpine					
Lake	0.438–12.5	0.068–0.610	ca. 4–105	16	Drewes et al. (2016)
Antarctic					
Lakes	0.02–120	0.08–5.0	0.1–126.7	30	Säwström et al. (2008) and references therein

Italicised values are expressed as *per gram dry mass*; all other values are expressed as mL^{-1}. Sampling effort (*n*) reflects both number of sites and number of sampling occasions in the case of repeat sampling. Values for lakes are given for comparative purposes and the data presented here are representative rather than exhaustive

it was not until the development and decreasing costs of high-throughput sequencing that this issue could be addressed in sufficient detail. From 2008 to ca. 2013, a number of papers on the metaviromes of various glacial environments were published, thereby beginning to describe the viral diversity of these habitats (Bellas and Anesio 2013; Bellas et al. 2015).

In the current third phase, a move has been made to try to understand the intricacies of the biotic interactions between viruses and their hosts (Anesio and Bellas 2011; Yau et al. 2011), such as CRISPR-Cas systems (Bellas et al. 2015; Sanguino et al. 2015), host defence-mechanisms and temperature dependence of virus–host interactions (Wells and Deming 2006a). A fascinating array of virus–host interactions have been identified and their mechanisms of action have started to be unravelled.

6.2 Viruses in Glacial Habitats

The term *glacial environments* encompasses a large variety of habitats, which present vastly different challenges to the microbes and viruses inhabiting them (Boetius et al. 2015). Conditions at the surface of glaciers, ice caps and ice sheets are very different from those within or underneath the same bodies of ice. Communities at the surface are exposed to fluctuations in solar radiation and temperature, but experience excellent conditions for photosynthesis (at the poles, during summer only). Communities at the bed of glacial ice have to contend with high pressure, high ion concentrations and complete darkness and are thus reliant on chemoautotrophy (Skidmore et al. 2005) and a limited influx of organic matter.

In the same year as the review by Säwström et al. (2008) was published, Hodson et al. (2008) published a seminal paper consolidating the concept of a 'glacial ecosystem' and providing a framework for the field of glacial microbiology. An additional benefit has been that by researchers distinguishing between the different types of glacial habitats (e.g. supra-, en- and subglacial, snowpack, meltponds, proglacial lakes) when reporting results, it has become easier to use data from the literature to make predictions about specific microbial communities.

6.2.1 Limited Focus on Viruses

In the majority of studies of glacial microbial communities, the focus has been predominantly on the prokaryotes, and bacteria and cyanobacteria, in particular (see Chap. 4). The number of studies of the viral component of these communities is small and typically limited to measuring specific parameters in relation to the prokaryotic community in question. As a result, we do not have a very cohesive picture of the role of viruses in these communities.

Most studies of glacial viruses are from studies of supraglacial habitats, in particular cryoconite and cryoconite holes (e.g. Castello et al. 1999; Anesio et al. 2007, 2009; Säwström et al. 2007b; Bellas and Anesio 2013; Ji et al. 2015). Arctic aquatic supraglacial habitats typically have low bacterial (0.007–0.07 $\times$ 10^6 cells mL^{-1}) and viral abundances (0.05–1.19 $\times$ 10^6 VLPs mL^{-1}) (Table 6.1). However, when viewed in relation to the size of the bacterial community, it becomes apparent that these habitats can sustain remarkably high virus (VBR 7.4–58.1) (Säwström et al. 2008; Rassner et al. 2016). In contrast, the values reported for supraglacial sediments (cryoconite) from Svalbard and the Greenland Ice Sheet suggest that viruses are less prominent in these particle-associated habitats: 2.68–19.6 $\times$ 10^8 cells g^{-1} dry mass, 5.62–24.5 $\times$ 10^8 VLPs g^{-1} dry mass, mean VBR 1.2-4.0 (Bellas et al. 2013).

The studies of englacial and subglacial environments are much more limited. In a study of viruses in cryoconite holes on Svalbard, Anesio et al. (2007) also took some shallow ice cores (0–90 cm depth), which showed that at least the viral abundance of the upper layer of ice (0.01–0.56 $\times$ 10^6 VLPs mL^{-1}) was comparable to that of supraglacial meltwater from the same glacier, but the range of VBRs were a lot larger (0.7–74). Säwström et al. (2007b) reportedly found comparatively high numbers of bacteria ($\leq$0.12 $\times$ 10^6 cells mL^{-1}) and viruses ($\leq$0.9 $\times$ 10^6 VLPs mL^{-1}; VBR $\leq$7.3) in subglacial water emerging at the front of a Svalbard glacier. Whereas Górniak et al. (2016) could not detect any viruses in the subglacial outflow from another Svalbard glacier, despite finding seven times as many bacteria (0.85 $\times$ 10^6 cells mL^{-1}).

6.2.2 *Reliance on Comparisons with Studies of Polar Lakes*

The number of papers concerning (mainly or partly) viruses in glacial ecosystems is rather limited and a literature search will only result in a little over 40 papers. In contrast, the studies of viruses in Arctic and Antarctic lakes are more numerous [e.g. Cavicchioli (2015) and references therein]. As a result, the research community has to rely on these data despite there being a not insignificant difference in physical parameters between large water bodies like lakes and most glacial habitats (e.g. cryoconite, ice, meltwater streams and ponds). One notable difference is the potential for using studies of permanently ice-covered Antarctic lakes to make inferences about subglacial lakes (Filippova et al. 2016).

In their review, Säwström et al. (2008) included data compiled from 17 Arctic lakes and 30 Antarctic lakes. The Arctic lakes generally supported larger bacterial and viral communities: 0.38–2.06 $\times$ 10^6 cells mL^{-1}; 0.92–28.9 $\times$ 10^6 VLPs mL^{-1}; VBR 1.6-25.2. The data from the Antarctic lakes highlighted the large variations in bacterial (0.08–5.0 $\times$ 10^6 cells mL^{-1}) and viral abundances (0.02–120 $\times$ 10^6 VLPs mL^{-1}; VBR 0.10–126.7). Because of the large variation in morphology, nutrient status and water chemistry, it is perhaps not surprising that

Antarctic lakes display a much larger range of bacterial and viral abundance than both Arctic lakes and glacial habitats.

6.2.3 Temperature Ranges

Similarly to all other biological entities, viruses seem to have preferred temperature ranges, each with their specific optimal temperature range within which infection results in larger number of progeny. In a study of marine psychrophilic virus–host systems, Wells and Deming (2006a) defined *cold-active bacteriophages* as viruses that can successfully infect host bacterial cells and produce progeny at temperatures lower than +4 °C. They also argued that there is a distinction between the host being psychrophilic and the virus being cold-active and showed that the production of a bacteriophage varied with temperature semi-independently to the host. The cold-active bacteriophage in Wells and Deming's study was incapable of producing progeny at temperatures above 4–8 °C.

In contrast, two recent culture-based studies of glacier-derived phage–host pairs obtained from Mingyong glacier in China showed that viral production was at its highest at ca +15 °C (host: *Janthinobacterium* sp.; Li et al. 2016) and suggested that a lower burst size and extended latent period was indicative of an adaptation to low temperatures (host: *Bacillus cereus*; Ji et al. 2015). Neither *B. cereus* nor *Janthinobacterium* are true psychrophiles (both being soil bacteria also found in temperate climates), but rather psychrotolerant species, and the bacteriophages in these studies were clearly capable of producing progeny at temperatures well above +4 °C (e.g. +20 °C). However, unlike the two host species, the bacteriophages were not able to cope with higher temperatures. Similar results were reported by Sillankorva et al. (2004) who linked the success of phage infection and production to temperature-dependent changes in types of cell wall receptors present on the host bacterium.

Thus, in those glacial environments where the temperature fluctuates around or above 0 °C, there is a potential for several hosts and viruses with overlapping but very different temperature ranges to co-exist, thereby possibly increasing the complexity of the interactions taking place.

6.2.4 Host Specificity

Viruses have traditionally been considered very host specific, i.e. each virus is limited to a small number of host species (Wommack and Colwell 2000). However, transplantation experiments have shown that some viruses are capable of infecting prokaryotes from completely different environments (Sano et al. 2004). Anesio et al. (2007) found that viruses from a supraglacial habitat of a Svalbard glacier could infect bacteria taken from a lake in the glacier's forefield. However, the lack of genomic data makes it impossible to determine whether the same bacterial species were present in both habitats.

6.3 Diversity of Glacial Viruses

There are a limited number of studies of the genomic diversity of viruses in glacial environments and a combination of factors has contributed to this. The typically low abundances of viruses in glacial habitats means that viruses must be concentrated from large sample volumes (e.g. hundreds of litres of glacial meltwater) in order to obtain enough material for downstream applications, be that electron microscopy or genome sequencing. This often presents not only a methodological challenge, such as concentrating viruses from large volumes of water or ice (e.g. clogging of filters, ultracentrifugation of large volumes; Wells and Deming 2006a; Rassner et al. 2016) and dislodging particle-associated viruses (e.g. from cryoconite or clay particles; Allen et al. 2010; Bellas et al. 2015), but also a logistical one, as many sampling sites are in remote location without adequate laboratory facilities and requiring prolonged transport of samples. The development of methods for analysis of assemblages of viruses from environmental samples has also lagged behind that of methods for prokaryotic communities, but is now catching up (Rose et al. 2016). Ongoing adaptation of analytic and sequencing techniques to accommodate the size and diversity of viral genomes has opened up for high-resolution analysis of the diversity of environmental viral communities. However, the research community is still tackling obstacles, such as the absence of a universally present conserved region similar to the 16S rRNA gene in bacteria and archaea, a lack of communal genomic sequence repositories and frequent misidentification of viral and prophage DNA as prokaryotic DNA (Bruder et al. 2016; Rose et al. 2016).

In the studies that exist, a variety of approaches has been used. Bellas et al. (2013) used capsid protein *g23* to investigate the diversity of T4-type (a subgroup of the *Myoviridae*) bacteriophages in samples from cryoconite, supraglacial meltwater, proglacial lake water and near-coastal seawater and found the different habitats to harbour distinct and diverse T4 communities. The phages isolated from Mingyong glacier by Li et al. (2016) and Ji et al. (2015) were identified as *Siphoviridae* and *Podoviridae*, respectively, based on morphology using TEM, in combination with SDS-PAGE of structural proteins. The genome of the *Podoviridae* phage has recently been sequenced by Qin et al. (2016). With the development of high-throughput sequencing, it is now becoming possible to study the metaviromes of glacial habitats. In a study of dsDNA viromes of cryoconite from the Greenland Ice Sheet and two Svalbard glaciers, Bellas et al. (2015) found large amounts of viral dsDNA, with a majority of matches belonging to the *Caudovirales* (caudate bacteriophages), mainly *Siphoviridae* but also *Myoviridae* and *Podoviridae*, and they succeeded in generating a total of 54 consensus genomes.

However, once again, more data are available for the viral diversity of polar lakes (López-Bueno et al. 2009, 2015; Yau et al. 2011; Wilkins et al. 2013; Cavicchioli and Erdmann 2015; Aguirre de Cárcer et al. 2016; and references therein). From these, it seems that despite the similarities in the viruses present in Antarctic and Arctic freshwaters at higher taxonomic levels, there are distinct differences at the lower taxonomic levels, with each polar area having its set of unique taxa, although some taxa are bipolar in their distribution (Aguirre de Cárcer et al. 2015, 2016). Because of the variations in the type of nucleic acid used by viruses, there is an obvious problem with reporting bias, especially in favour of dsDNA viruses. When other types of viruses have been included, a higher abundance of ssDNA viruses than dsDNA viruses is generally reported (López-Bueno et al. 2009; Aguirre de Cárcer et al. 2015, 2016) and diverse communities of RNA viruses have also been found (López-Bueno et al. 2015). The viral communities of polar area have also proven to be as taxon rich and diverse as those of lower latitudes (Aguirre de Cárcer et al. 2015, 2016) and may be one of the drivers of the high diversity in polar microbial communities (Anesio and Bellas 2011).

6.4 Activity of Glacial Viruses

Due to viruses' reliance on host cells for reproduction, their effects on the environment are to an extent governed by the nature of their interactions with their host (Weinbauer 2004).

6.4.1 Viral Life Cycles

There are four main viral life cycles:

1. *Lytic infection*, where the virus genome enters the host cell and causes the cell to produce new virus particles, which are then released to the outside through lysis of the cell;
2. *Chronic infection*, in which the cell is not lysed, but the new viruses are released through budding;
3. *Lysogenic infection*, where the viral genome fuses with the host genome to become a provirus or prophage, which is duplicated at each cell division (i.e. each daughter cell contains a copy of the provirus) until an induction event causes the lysogenic virus to enter the lytic pathways; and
4. *Pseudolysogeny*, where the viral genome remains free in the cell cytoplasm as a circular or linear plasmid and is passed to one of the daughter cells upon cell division (Weinbauer 2004).

In cases where there are more than one copy of the virus plasmid within the cell, such as during a chronic infection, these can be divided between daughter cells either unevenly (by chance) or evenly with the help of a partition system (Sengupta et al. 2010).

There is little consensus on the extent and drivers of lysogeny among viruses in glacial environments. Säwström et al. (2008) found that data for Antarctic lakes showed very large seasonal variations in the occurrence of lysogens (up to 71%), although no obvious seasonal trend was evident, but concluded that, in contrast to Antarctic habitats, there was no or very little evidence of lysogens in Arctic freshwater systems. However, in an earlier paper, Anesio et al. (2007) reported finding 25% lysogens in samples of supraglacial water from a glacier on Svalbard. Later papers on studies carried out in the Arctic have continued to report contradicting results. Bellas et al. (2015) found evidence of lysogenic phages in several of their viral genome scaffolds, as well as evidence of viral plasmids indicative of chronic infection, whereas Bellas et al. (2013) were unable to induce any lysogens using the common induction agent mitomycin C. In a study of viral genes incorporated into bacterial metagenomes from polar marine environments, Cottrell and Kirchman (2012) found that the prevalence of prophages was higher in Arctic bacterial metagenomes than in temperate waters, whereas it was much lower in Antarctic metagenomes.

6.4.2 Viral Decay and Loss of Infectivity

Viruses lack the ability to move through their environment under their own steam and are therefore reliant on passive encounters with potential hosts. On contact with a cell, the virus will recognise and bind to any binding sites (e.g. cell wall proteins) of the correct type if they are present, after which the virus can initiate infection (Weinbauer 2004). However, because the virus cannot actively swim away from an encountered particle, it can be trapped if attached to non-host cells or other particular organic matter (POM). Viruses are affected by electrical forces and can be reversibly, or irreversibly, bound to high-molecular mass dissolved organic matter (DOM) and non-organic particles, such as clay minerals (Suttle and Chen 1992; Cattaneo et al. 2010; Drewes et al. 2016). This binding to non-hosts causes virus particles to be in effect removed from the water column (or other environment), either permanently or temporarily. Viruses can also be removed from the environment through ingestion by heterotrophic nanoflagellates and other predators. Irrespective of the mechanism, the removal of active, free viruses from the environment is referred to as *viral decay*.

Depending on the method used to determine changes in viral abundance (e.g. microscopy counts of stained virus particles, plaque forming units (PFUs), virus titres), the term viral decay is also, perhaps more accurately, used to describe a loss of infectivity (Suttle and Chen 1992; Noble and Fuhrman 1997). Viruses can lose their ability to infect cells through a variety of mechanisms, for instance by damage to the proteins involved in attachment to and injection into the host (Wells and Deming 2006b). Despite their sturdy protective capsid, viral genomes are also susceptible to damage from, for instance, UV radiation, high temperatures and extracellular enzymes (Suttle and Chen 1992; Noble and Fuhrman 1997; Mojica and Brussaard 2014). Such damage is likely to lower the effectiveness of the virus upon infection or completely incapacitate the virus. Thus, not only are viruses removed from the pool of active viruses, but a proportion of free viruses in this pool are damaged and thereby unable to successfully infect potential hosts upon contact. This inactivation, as opposed to destruction, of viruses is probably contributing to the many of the reports of solar radiation and temperature having no, or little, impact on viral decay (e.g. Guixa-Boixereu et al. (2002), Rassner et al. (2016); but see Madan et al. (2005)).

Reports of viral decay rates for glacial environments are somewhat scarce. Rassner et al. (2016) calculated viral decay rates of 0.0007 h^{-1} for supraglacial meltwater and 0.0082 h^{-1} for a lake in the forefield of a Svalbard glacier. Although not reports of actual viral decay rates, Anesio et al. (2007) calculated that the virus–host encounter rate was two orders of magnitude lower in the supraglacial meltwater of a Svalbard glacier than in a nearby forefield lake (2.46 and 466.5 encounters L^{-1} h^{-1}, respectively), while Collins and Deming (2011) estimated that bacterial cells in sea ice in the Admunsen Gulf encountered between approximately 1.04 and 33.33 viruses per hour. In comparison, Guixa-Boixereu et al. (2002) reported viral decay rates ranging from 0.006 h^{-1} to 0.308 h^{-1} for Antarctic marine waters, and Heldal and Bratbak (1991) reported viral decay values of 0.26–0.55 and 0.33–0.64 for a Norwegian fjord and lake, respectively.

6.4.3 Dynamic Virus–Host Interactions

The presence of agents or factors that cause viral decay (e.g. clay particles, high levels of UV-radiation) can severely alter the virus–host dynamics (Mojica and Brussaard 2014; but see Drewes et al. 2016). However, the success rate of infection of environmental viruses is varied and dependent on several factors (Weinbauer 2004). Because of their reliance on passively encountering hosts, the contact rate between virus and host is greatly affected by the abundance of hosts. The more host cells present in a system, the higher the contact rate and thus the higher the production and release of new viruses into the environment. However, in a natural environment, the host population is only one of many populations, as there are multiple co-existing virus–host systems, and the presence of high numbers of non-hosts means a reduced contact rate, and may also lead to an increased decay rate for a virus population, as the viruses attach (reversibly or irreversibly) to these non-hosts. Thus, in any environment where there are multiple co-existing virus–host systems, any one virus population will benefit from a high abundance of its host population, whereas any one prokaryote population will experience an increased viral predation pressure as it increases in relative abundance.

Simultaneously, the production of new lytic viruses is intimately linked to the growth rate of its prokaryotic host, and host populations with higher growth rates are likely to produce ever more virus particles (Thingstad 2000; Weinbauer 2004). Therefore, it seems reasonable to assume that as any one prokaryotic population gets a particularly competitive advantage and becomes dominant, its associated viruses will increase in abundance, and as the virus becomes more abundant and exerts a higher pressure on its hosts, it will ultimately cause the population of its host to decrease until the viral pressure again is in line with that of other populations. This mechanism was formulated in the '*kill-the-winner' theory* by Thingstad and Lignell (1997). Since then, evidence that either support or refute the theory has been put forward and several modifications and alternatives to the theory have been suggested (e.g. 'Piggyback-the-winner', Knowles et al. (2016); 'Red Queen', Clarke et al. (1994); Zhao et al. (2013); 'King of the Mountain', Giovannoni et al. (2013); 'Cheshire Cat', Frada et al. (2008)).

These, more recent, papers highlight the complexity of virus–host interactions in the natural environment (Rohwer and Thurber 2009) and the need for further work untangling the impacts of different factors on the interplay between viruses and their hosts before we can accurately predict the responses of these dynamics to external pressures. For instance, Rassner et al. (2016) reported that the bacterial community of supraglacial meltwater from a Svalbard glacier was dominated by *Janthinobacterium* sp. and that despite an increase in total viral abundance, the dominance of *Janthinobacterium* sp. only strengthened over a 2-week microcosm experiment. Initially, this seems to be completely contrary to the kill-the-winner theory; however, they propose a mechanism that explains this apparent contradiction. Electron microscopy revealed that a proportion of the bacteria present in the community produced outer membrane vesicles (Biller et al. 2014), something that *Janthinobacterium* is known to do (Hornung et al. 2013). Because these vesicles consist of cell membrane and associated proteins, a proportion of the vesicles released by a prokaryote will contain the binding sites recognised by the viruses of that prokaryote. Since a virus that binds to these vesicles and inject their genome into the vesicle will not reproduce, the overall viral pressure on the *Janthinobacterium* population will be less than if it did not produce vesicles. In effect, the vesicles act as decoys and the actual viral pressure on the bacterial population is reduced.

6.5 Viral Impacts on Host Communities

Regardless of the uncertainties in the host ranges of viruses in natural environments, each species of virus can only infect a certain range of organisms and thus each virus population is likely to only be able to infect a subset of the microbial community present in a given habitat. In environments with diverse viral and prokaryotic communities, one can therefore imagine that the various viral populations may be competing for suitable host organisms and that mechanisms that give an advantage over other viruses may evolve. Some polar aquatic viruses are, for instance, able to modify proteins on the surface of infected host cells, thereby reducing the risk of subsequent infections by other viruses (so-called superinfection exclusion) (Cavicchioli 2015; Tschitschko et al. 2015).

6.5.1 Virus-Mediated Horizontal Gene Transfer

The packaging of viral genomes into progeny virus capsids is far from perfect (Hanks et al. 1988; Jiang and Paul 1998). As a result, far from all of the virus particles that are released from a cell are identical copies of their 'parent' virus. Packing errors include the inclusion of partial or incomplete viral genomes, empty capsids (a complete lack of nucleic acids) and inclusion of plasmid or host DNA. Inclusion of host DNA is particularly common in the case of lysogenic viruses, as host DNA on either side of the prophage is prone to be included during the transcription of the prophage.

Transduction, or horizontal gene transfer mediated by a virus, occurs when a virus particle carrying prokaryotic DNA injects this DNA into another prokaryotic the cell and the DNA is incorporated into the genome (or plasmid) of the new host. If the transferred prokaryotic DNA contains an intact gene, its expression may confer new traits on the receiving prokaryote, such as antibiotic resistance or new metabolic pathways (Ochman et al. 2000). Estimates of the frequency of transduction events in natural environments are still insufficient, but reported ranges are in the order of 10^{-8} to 10^{-5} transduction events PFU^{-1}, which, for example, would equate 10^{13} and 10^{14} transduction events per year for the Mediterranean Sea and Tampa Bay, respectively (Weinbauer and Rassoulzadegan 2004). Thus, transduction is a very important consequence of viral infection, since as genes are transferred between individual prokaryotes, traits spread through the prokaryotic community (Ochman et al. 2000).

6.5.2 CRISPRs Afford Immunity

Despite any benefits that it may confer, viral infection is a lethal threat to single-cell organisms and prokaryotes have developed mechanisms that interrupt viral infections (Labrie et al. 2010; Goldfarb et al. 2015), including CRISPRs (clustered regularly interspaced short palindromic repeats) which are common throughout the archaea and also found in many bacteria (Barrangou et al. 2007). In essence, CRISPRs consist of snippets of different viral or plasmid DNA, called *spacers*, that are sandwiched between identical segments of host DNA, called *direct repeats*. The CRISPR array provides a record of previous infections that allows the cell to identify new infection attempts by the same virus species, and in combination with an associated nuclease (e.g. CRISPR-associated protein *Cas9*), the cell can render the injected viral DNA harmless by cleaving it.

The CRISPR arrays are not only useful for the prokaryotes. By searching for direct repeats in prokaryotic metagenomes, CRISPRs can be located. The spacers in these CRISPRs can be used to identify viruses that have previously infected that particular prokaryote. In combination with taxonomic data (e.g. 16S rDNA sequence data), it is possible to identify new potential virus–prokaryote systems from environmental samples without culturing (Sanguino et al. 2015).

The extent of CRISPR arrays in glacial environments has not yet been determined. However, Sanguino et al. (2015) found a larger number of CRISPRs in samples from Arctic glacial ice than in samples from Arctic soil. Bellas et al. (2015) reported finding the genome of a lysogenic phage containing three CRISPR arrays, with a total of 59 spacers, and several *Cas* genes in a sample from the Greenland Ice Sheet. If incorporated into a prokaryotic genome, this prophage would thus have the potential to impart immunity against superinfection by a large number of other phages to its host. Similarly, Tschitschko et al. (2015) found several CRISPRs and *Cas* genes in haloarchaea in Deep Lake, Antarctica, and Chénard et al. (2016) reportedly found a CRISPR array in a phage of the cyanobacterium *Nostoc*, which is commonly found in polar environments (Jungblut et al. 2009; Stibal et al. 2015; Edwards et al. 2016; Gokul et al. 2016). The fact that one of the more recently acquired spacers in the CRISPR array found by Bellas et al. (2015) was a complete match to another phage also present in the sample highlights how prophage-mediated horizontal gene transfer can provide a virus with a competitive edge in the competition between virus populations.

6.6 Viral Impacts on the Ecosystem

When discussing viruses in glacial environments with people from outside the field, the greatest concern raised is often whether the warming of the polar regions and the retreat of glaciers will lead to the release of viable pathogenic viruses (Edwards 2015). This question is the focus for a range of scientific publications (e.g. Pearce and Wilson 2003; Legendre et al. 2014; Dudley et al. 2015). Many of these have been more speculative in nature (Rogers et al. 2004; Smith et al. 2004), but there are examples of studies where viable viruses have been extracted from glacial environments (Castello et al. 1999; Zhang et al. 2006). For instance, Filippova et al. (2016) found and revived bacteriophages from a 25,000–40,000 year old ice wedge in Northeast Siberia.

6.6.1 The Viral Shunt

However, this focus on pathogenic viruses detracts attention from the much more important impact of viruses on the environment, in form of the *viral shunt* (Wilhelm and Suttle 1999). As the prokaryotic host cell is lysed to release the virus progeny, organic matter and energy that could have made its way into higher trophic levels, through the *microbial loop* (Pomeroy 1974; Azam et al. 1983), are released to the environment as DOM and POM.

Viral lysis causes a very varied but often substantial part of prokaryotic mortality in natural environments (e.g. 0–97.3%; Wommack and Colwell 2000; Wilhelm and Matteson 2008), and there are few estimates of virus-induced mortality in glacial environments (Pearce and Wilson 2003; Säwström et al. 2008). Bellas et al. (2013) reported very large estimates (342–2606%) of virus-induced mortality for bacteria in cryoconite sediment from two glaciers on Svalbard and one glacier on Greenland. Whereas, Rassner et al. (2016) reported a more moderate estimate of virus-induced mortality equating 35.5% of bacterial secondary production, or 6.1 ng C L^{-1} h^{-1}, for supraglacial meltwater on a Svalbard glacier. In comparison, studies of Antarctic lakes have found a big range and large variations in virus-induced mortality (38–251%; Säwström et al. 2007a, 2008), whereas reports from marine environments tend to be lower (Pearce and Wilson 2003). For example, Evans et al. (2009) estimated that viral lysis was responsible for the release of 17–56% of bacterial production, or 0.9–3.6 μg C L^{-1} day^{-1}, in an area of the Southern Ocean adjacent to the Antarctic continent, while Middelboe et al. (2012) reported that viral lysis contributed to the loss of 4–36% of bacterial production (12–88 mg C m^{-2} day^{-1}) in a fjord in Greenland.

6.6.2 Impact of Viral Lysis on the Fate of Carbon

Dissolved organic carbon (DOC) taken up by bacteria can either be used to synthesise new biomass (*bacterial production*, BP) or to produce energy through remineralisation of the DOC to CO_2 (*bacterial respiration*, BR). A proportion of DOC will always be used for cell maintenance, but the relative allocation of DOC to production and respiration, respectively, says a lot about the prosperity of a community. Therefore, measured bacterial production is often expressed relative to bacterial respiration as *bacterial growth efficiency* (BGE; BGE = BP/(BP+BR)). When a majority of the DOC is used for bacterial production (high BGE), the bacterial community as a whole will increase in abundance or at least compensate for losses due to, for example, grazing and viral lysis, and the carbon will be locked away in the biomass. In contrast, when most of the DOC is used for respiration (low BGE), the bacteria are mainly maintaining existing cells, and a majority of the carbon is released to the environment as CO_2.

Viral lysis leads to the release of cellular contents rich in nitrogen, phosphorus and labile DOC into the environment, thereby providing the surviving members of the community with high-quality and readily accessible substrates, such as DNA fragments, proteins and carbohydrates (Middelboe et al. 1996). It is therefore not unreasonable to assume that uninfected bacteria in a nutrient-poor habitat would benefit from the presence and activity of lytic viruses, as long as the benefits from the increased availability of nutrients outweighed the risk of infection. However, several studies of marine bacteria show that even if bacterial production may increase as a result of viral lysis, the BGE is more likely to be negatively affected by viral activity (Middelboe et al. 1996; Bonilla-Findji et al. 2008; Motegi et al. 2009). Middelboe et al. (1996) suggested that the decrease in BGE may be caused by the higher energetic cost of extracellular degradation of complex organic molecules, while Motegi et al. (2009) highlighted the importance of the relative impact of viral activity by showing that the larger the proportion of bacterial production that is lysed by viruses, the lower the BGE.

In a more recent study, Middelboe et al. (2012) found that the very low BGE (9–10%) of the bacterial community in a Greenland fjord did increase in response to addition of DOC, although this community was C limited at the time. Interestingly, these are similar to values of BGE reported for polar freshwater environments (2–5%; Säwström et al. 2007a), and the authors suggested that the low impact of the viral community (4–36% of bacterial production lost through lysis, again similar to some of the lower values reported for glacial environments) allowed the bacterial community to profit from the increase in available DOC. Given the large variation in both bacterial production rates and organic carbon concentrations in glacial environments (Anesio et al. 2010), the response in BGE to any changes in DOC concentration and quality is very likely to be highly habitat and location specific. Considering the additional uncertainty introduced by large variations in virus-induced mortality, it is very difficult to make any general statements on the impact of viral lysis on the BGE of glacial environments. However, since the lysis of prokaryotic cells by viruses can not only divert the flow of energy and carbon through the food web but also alter the fate of carbon in the system and may ultimately have an impact on the CO_2 fluxes to and from the system, it is paramount that future studies of microbial communities in glacial environments should seek to include a viral component.

6.7 Conclusions

Viruses play a central role in glacial microbial communities, yet glacial viruses have hitherto not been given the same attention as those in other high latitude/high altitude aquatic systems. Although viral abundances are lower in glacial environments than in many other environments, glacial prokaryotic communities can support surprisingly large viral communities, as evident by the large VBRs found in many glacial habitats. A common problem in studies of viral ecology is the difficulty in identifying and quantifying the dynamics of individual virus–host pairs in environmental samples. However, with current technological advances in environmental genomics and bioinformatics, new possibilities are opening up and the existing literature suggests that glacial viral communities are as diverse as those in other habitats.

Despite many similarities with viruses in other habitats, the unique characteristics of glacial environments have accentuated certain features in glacial viruses and their interactions with their hosts, such as the longevity of supraglacial viruses, the possibility of temperature adaptation separate from that of the host and the potential importance of transduction (virus-mediated horizontal gene transfer). Low viral decay rate is a mechanism for overcoming low host contact rates in systems with low prokaryotic abundances, while a difference in infectivity between psychrotolerant and psychrophilic viruses at increased temperatures is likely to influence the often-discussed impact of glacial viruses on downstream habitats—a common concern in the discussion of increased glacial melting. The initial evidence of transduction altering the virus–host dynamics of glacial microbial communities by the transfer of CRISPR arrays that confer immunity against superinfection is also a very exciting development for glacial viral ecology.

Although fascinating in their own right, the most crucial aspect of glacial viruses is arguably their impact on the flow of organic carbon. By lowering bacterial growth efficiency and through the lysis of bacterial cells, viruses substantially alter the carbon cycling in glacial environments.

Acknowledgements This work was supported by the Freshwater Biological Association through the Hugh Cary Gilson Award 2013 and by the Welsh Government and HEFCW through the *Sêr Cymru* National Research Network for Low Carbon, Energy and the Environment.

References

Ackermann H-W (2007) 5500 phages examined in the electron microscope. Arch Virol 152 (2):227–243. doi:10.1007/s00705-006-0849-1

Aguirre de Cárcer D, López-Bueno A, Alonso-Lobo JM, Quesada A, Alcami A (2016) Metagenomic analysis of lacustrine viral diversity along a latitudinal transect of the Antarctic Peninsula. FEMS Microbiol Ecol 92(6). doi:10.1093/femsec/fiw074

Aguirre de Cárcer D, López-Bueno A, Pearce DA, Alcamí A (2015) Biodiversity and distribution of polar freshwater DNA viruses. Sci Adv 1(5):e1400127. doi:10.1126/sciadv.1400127

Allen B, Willner D, Oechel WC, Lipson D (2010) Top-down control of microbial activity and biomass in an Arctic soil ecosystem. Environ Microbiol 12(3):642–648. doi:10.1111/j.1462-2920.2009.02104.x

Anesio AM, Bellas CM (2011) Are low temperature habitats hot spots of microbial evolution driven by viruses? Trends Microbiol 19(2):52–57. doi:10.1016/j.tim.2010.11.002

Anesio AM, Hodson AJ, Fritz A, Psenner R, Sattler B (2009) High microbial activity on glaciers: importance to the global carbon cycle. Glob Chang Biol 15(4):955–960. doi:10.1111/j.1365-2486.2008.01758.x

Anesio AM, Mindl B, Laybourn-Parry J, Hodson AJ, Sattler B (2007) Viral dynamics in cryoconite holes on a high Arctic glacier (Svalbard). J Geophys Res Biogeosci 112(G4):10. doi:10.1029/2006jg000350

Anesio AM, Sattler B, Foreman C, Telling J, Hodson A, Tranter M, Psenner R (2010) Carbon fluxes through bacterial communities on glacier surfaces. Ann Glaciol 51(56):32–40

Azam F, Fenchel T, Field JG, Gray J, Meyer-Reil L, Thingstad F (1983) The ecological role of water-column microbes in the sea. Mar Ecol Prog Ser 10:257–263

Barrangou R, Fremaux C, Deveau H, Richards M, Boyaval P, Moineau S, Romero DA, Horvath P (2007) CRISPR provides acquired resistance against viruses in prokaryotes. Science 315 (5819):1709–1712. doi:10.1126/science.1138140

Bellas CM, Anesio AM (2013) High diversity and potential origins of T4-type bacteriophages on the surface of Arctic glaciers. Extremophiles 17(5):861–870. doi:10.1007/s00792-013-0569-x
Bellas CM, Anesio AM, Barker G (2015) Analysis of virus genomes from glacial environments reveals novel virus groups with unusual host interactions. Front Microbiol 6:14. doi:10.3389/fmicb.2015.00656
Bellas CM, Anesio AM, Telling J, Stibal M, Tranter M, Davis S (2013) Viral impacts on bacterial communities in Arctic cryoconite. Environ Res Lett 8(4):9. doi:10.1088/1748-9326/8/4/045021
Biller SJ, Schubotz F, Roggensack SE, Thompson AW, Summons RE, Chisholm SW (2014) Bacterial vesicles in marine ecosystems. Science 343(6167):183–186. doi:10.1126/science.1243457
Boetius A, Anesio AM, Deming JW, Mikucki JA, Rapp JZ (2015) Microbial ecology of the cryosphere: sea ice and glacial habitats. Nat Rev Microbiol 13(11):677–690. doi:10.1038/nrmicro3522
Bonilla-Findji O, Malits A, Lefèvre D, Rochelle-Newall E, Lemée R, Weinbauer MG, Gattuso J-P (2008) Viral effects on bacterial respiration, production and growth efficiency: consistent trends in the Southern Ocean and the Mediterranean Sea. Deep-Sea Res II Top Stud Oceanogr 55(5):790–800
Bruder K, Malki K, Cooper A, Sible E, Shapiro JW, Watkins SC, Putonti C (2016) Freshwater metaviromics and bacteriophages: a current assessment of the state of the art in relation to bioinformatic challenges. Evol Bioinforma 12:25–33. doi:10.4137/ebo.s38549
Castello JD, Rogers SO, Starmer WT, Catranis CM, Ma LJ, Bachand GD, Zhao YH, Smith JE (1999) Detection of tomato mosaic tobamovirus RNA in ancient glacial ice. Polar Biol 22 (3):207–212. doi:10.1007/s003000050411
Cattaneo R, Rouviere C, Rassoulzadegan F, Weinbauer MG (2010) Association of marine viral and bacterial communities with reference black carbon particles under experimental conditions: an analysis with scanning electron, epifluorescence and confocal laser scanning microscopy. FEMS Microbiol Ecol 74(2):382–396. doi:10.1111/j.1574-6941.2010.00953.x
Cavicchioli R (2015) Microbial ecology of Antarctic aquatic systems. Nat Rev Microbiol 13 (11):691–706. doi:10.1038/nrmicro3549
Cavicchioli R, Erdmann S (2015) The discovery of Antarctic RNA viruses: a new game changer. Mol Ecol 24(19):4809–4811. doi:10.1111/mec.13387
Chénard C, Wirth JF, Suttle CA (2016) Viruses infecting a freshwater filamentous cyanobacterium (*Nostoc* sp.) encode a functional CRISPR array and a proteobacterial DNA polymerase B. mBio 7(3):e00667-16. doi:10.1128/mBio.00667-16
Clarke DK, Duarte EA, Elena SF, Moya A, Domingo E, Holland J (1994) The red queen reigns in the kingdom of RNA viruses. Proc Natl Acad Sci 91(11):4821–4824
Collins RE, Deming JW (2011) Abundant dissolved genetic material in Arctic sea ice. Part II: Viral dynamics during autumn freeze-up. Polar Biol 34(12):1831–1841. doi:10.1007/s00300-011-1008-z
Cottrell MT, Kirchman DL (2012) Virus genes in Arctic marine bacteria identified by metagenomic analysis. Aquat Microb Ecol 66(2):107–116. doi:10.3354/ame01569
Drewes F, Peter H, Sommaruga R (2016) Are viruses important in the plankton of highly turbid glacier-fed lakes? Sci Report 6:24608. doi:10.1038/srep24608
Dudley JP, Hoberg EP, Jenkins EJ, Parkinson AJ (2015) Climate change in the North American Arctic: a one health perspective. EcoHealth 12(4):713–725. doi:10.1007/s10393-015-1036-1
Edwards A (2015) Coming in from the cold: potential microbial threats from the terrestrial cryosphere. Front Earth Sci 3:12
Edwards A, Debbonaire AR, Sattler B, Mur LA, Hodson AJ (2016) Extreme metagenomics using nanopore DNA sequencing: a field report from Svalbard, 78 N. bioRxiv. doi:10.1101/073965
Evans C, Pearce I, Brussaard CPD (2009) Viral-mediated lysis of microbes and carbon release in the sub-Antarctic and polar frontal zones of the Australian Southern Ocean. Environ Microbiol 11(11):2924–2934. doi:10.1111/j.1462-2920.2009.02050.x

Filippova SN, Surgucheva NA, Sorokin VV, Akimov VN, Karnysheva EA, Brushkov AV, Andersen D, Gal'chenko VF (2016) Bacteriophages in Arctic and Antarctic low-temperature systems. Microbiology 85(3):359–366. doi:10.1134/s0026261716030048

Frada M, Probert I, Allen MJ, Wilson WH, de Vargas C (2008) The "Cheshire Cat" escape strategy of the coccolithophore *Emiliania huxleyi* in response to viral infection. Proc Natl Acad Sci 105 (41):15944–15949. doi:10.1073/pnas.0807707105

Giovannoni S, Temperton B, Zhao Y (2013) Giovannoni et al. reply. Nature 499(7459):E4–E5. doi:10.1038/nature12388

Gokul JK, Hodson AJ, Saetnan ER, Irvine-Fynn TDL, Westall PJ, Detheridge AP, Takeuchi N, Bussell J, Mur LAJ, Edwards A (2016) Taxon interactions control the distributions of cryoconite bacteria colonizing a High Arctic ice cap. Mol Ecol 25(15):3752–3767. doi:10.1111/mec.13715

Goldfarb T, Sberro H, Weinstock E, Cohen O, Doron S, Charpak-Amikam Y, Afik S, Ofir G, Sorek R (2015) BREX is a novel phage resistance system widespread in microbial genomes. EMBO J 34(2):169–183. doi:10.15252/embj.201489455

Gorniak D, Marszalek H, Jankowska K, Dunalska J (2016) Bacterial community succession in an Arctic lake-stream system (Brattegg Valley, SW Spitsbergen). Boreal Environ Res 21 (1-2):115–133

Guixa-Boixereu N, Vaqué D, Gasol JM, Sánchez-Cámara J, Pedrós-Alió C (2002) Viral distribution and activity in Antarctic waters. Deep Sea Res II Top Stud Oceanogr 49(4–5):827–845. doi:10.1016/S0967-0645(01)00126-6

Hanks MC, Newman B, Oliver IR, Masters M (1988) Packaging of transducing DNA by bacteriophage P1. Mol Gen Genet MGG 214(3):523–532. doi:10.1007/bf00330490

Heldal M, Bratbak G (1991) Production and decay of viruses in aquatic environments. Mar Ecol Prog Ser 72(3):205–212

Hodson A, Anesio AM, Tranter M, Fountain A, Osborn M, Priscu J, Laybourn-Parry J, Sattler B (2008) Glacial ecosystems. Ecol Monogr 78(1):41–67. doi:10.1890/07-0187.1

Hornung C, Poehlein A, Haack FS, Schmidt M, Dierking K, Pohlen A, Schulenburg H, Blokesch M, Plener L, Jung K, Bonge A, Krohn-Molt I, Utpatel C, Timmermann G, Spieck E, Pommerening-Röser A, Bode E, Bode HB, Daniel R, Schmeisser C, Streit WR (2013) The *Janthinobacterium* sp. HH01 genome encodes a homologue of the *V. cholerae* CqsA and *L. pneumophila* LqsA autoinducer synthases. PLoS One 8(2):e55045. doi:10.1371/journal.pone.0055045

Ji XL, Zhang CJ, Fang Y, Zhang Q, Lin LB, Tang B, Wei YL (2015) Isolation and characterization of glacier VMY22, a novel lytic cold-active bacteriophage of *Bacillus cereus*. Virol Sin 30 (1):52–58. doi:10.1007/s12250-014-3529-4

Jiang SC, Paul JH (1998) Gene transfer by transduction in the marine environment. Appl Environ Microbiol 64(8):2780–2787

Jungblut AD, Lovejoy C, Vincent WF (2009) Global distribution of cyanobacterial ecotypes in the cold biosphere. ISME J 4(2):191–202. http://www.nature.com/ismej/journal/v4/n2/suppinfo/ismej2009113s1.html

Kay A, Zoulim F (2007) Hepatitis B virus genetic variability and evolution. Virus Res 127 (2):164–176. doi:10.1016/j.virusres.2007.02.021

Knowles B, Silveira CB, Bailey BA, Barott K, Cantu VA, Cobián-Güemes AG, Coutinho FH, Dinsdale EA, Felts B, Furby KA, George EE, Green KT, Gregoracci GB, Haas AF, Haggerty JM, Hester ER, Hisakawa N, Kelly LW, Lim YW, Little M, Luque A, McDole-Somera T, McNair K, de Oliveira LS, Quistad SD, Robinett NL, Sala E, Salamon P, Sanchez SE, Sandin S, Silva GGZ, Smith J, Sullivan C, Thompson C, Vermeij MJA, Youle M, Young C, Zgliczynski B, Brainard R, Edwards RA, Nulton J, Thompson F, Rohwer F (2016) Lytic to temperate switching of viral communities. Nature 531(7595):466–470. doi:10.1038/nature17193

Labrie SJ, Samson JE, Moineau S (2010) Bacteriophage resistance mechanisms. Nat Rev Microbiol 8(5):317–327

Legendre M, Bartoli J, Shmakova L, Jeudy S, Labadie K, Adrait A, Lescot M, Poirot O, Bertaux L, Bruley C, Couté Y, Rivkina E, Abergel C, Claverie J-M (2014) Thirty-thousand-year-old distant relative of giant icosahedral DNA viruses with a pandoravirus morphology. Proc Natl Acad Sci 111(11):4274–4279. doi:10.1073/pnas.1320670111

Li MY, Wang JL, Zhang Q, Lin LB, Kuang AX, Materon L, Ji XL, Wei Y (2016) Isolation and characterization of the lytic cold-active bacteriophage MYSP06 from the Mingyong Glacier in China. Curr Microbiol 72(2):120–127. doi:10.1007/s00284-015-0926-3

López-Bueno A, Rastrojo A, Peiro R, Arenas M, Alcami A (2015) Ecological connectivity shapes quasispecies structure of RNA viruses in an Antarctic lake. Mol Ecol 24(19):4812–4825. doi:10.1111/mec.13321

López-Bueno A, Tamames J, Velázquez D, Moya A, Quesada A, Alcamí A (2009) High diversity of the viral community from an Antarctic lake. Science 326(5954):858–861. doi:10.1126/science.1179287

Madan NJ, Marshall WA, Laybourn-Parry J (2005) Virus and microbial loop dynamics over an annual cycle in three contrasting Antarctic lakes. Freshw Biol 50(8):1291–1300. doi:10.1111/j.1365-2427.2005.01399.x

Middelboe M, Glud RN, Sejr MK (2012) Bacterial carbon cycling in a subarctic fjord: a seasonal study on microbial activity, growth efficiency, and virus-induced mortality in Kobbefjord, Greenland. Limnol Oceanogr 57(6):1732–1742

Middelboe M, Jorgensen N, Kroer N (1996) Effects of viruses on nutrient turnover and growth efficiency of noninfected marine bacterioplankton. Appl Environ Microbiol 62(6):1991–1997

Mojica KDA, Brussaard CPD (2014) Factors affecting virus dynamics and microbial host–virus interactions in marine environments. FEMS Microbiol Ecol 89(3):495–515. doi:10.1111/1574-6941.12343

Motegi C, Nagata T, Miki T, Weinbauer MG, Legendre L, Rassoulzadegan F (2009) Viral control of bacterial growth efficiency in marine pelagic environments. Limnol Oceanogr 54 (6):1901–1910

Noble RT, Fuhrman JA (1997) Virus decay and its causes in coastal waters. Appl Environ Microbiol 63(1):77–83

Noble RT, Fuhrman JA (1998) Use of SYBR Green I for rapid epifluorescence counts of marine viruses and bacteria. Aquat Microb Ecol 14(2):113–118

Ochman H, Lawrence JG, Groisman EA (2000) Lateral gene transfer and the nature of bacterial innovation. Nature 405(6784):299–304

Patel A, Noble RT, Steele JA, Schwalbach MS, Hewson I, Fuhrman JA (2007) Virus and prokaryote enumeration from planktonic aquatic environments by epifluorescence microscopy with SYBR Green I. Nat Protoc 2(2):269–276

Pearce DA, Wilson WH (2003) Viruses in Antarctic ecosystems. Antarct Sci 15(3):319–331. doi:10.1017/s0954102003001330

Philippe N, Legendre M, Doutre G, Couté Y, Poirot O, Lescot M, Arslan D, Seltzer V, Bertaux L, Bruley C, Garin J, Claverie J-M, Abergel C (2013) Pandoraviruses: Amoeba viruses with genomes up to 2.5 Mb reaching that of parasitic eukaryotes. Science 341(6143):281–286. doi:10.1126/science.1239181

Pomeroy LR (1974) The Ocean's food web, a changing paradigm. Bioscience 24(9):499–504. doi:10.2307/1296885

Qin KH, Cheng BX, Zhang ST, Wang N, Fang Y, Zhang Q, Kuang AX, Lin LB, Ji XL, Wei YL (2016) Complete genome sequence of the cold-active bacteriophage VMY22 from *Bacillus cereus*. Virus Genes 52(3):432–435. doi:10.1007/s11262-016-1300-7

Rassner SME, Anesio AM, Girdwood SE, Hell K, Gokul JK, Whitworth DE, Edwards A (2016) Can the bacterial community of a high Arctic glacier surface escape viral control? Front Microbiol 7:956. doi:10.3389/fmicb.2016.00956

Rogers SO, Starmer WT, Castello JD (2004) Recycling of pathogenic microbes through survival in ice. Med Hypotheses 63(5):773–777. doi:10.1016/j.mehy.2004.04.004

Rohwer F, Thurber RV (2009) Viruses manipulate the marine environment. Nature 459 (7244):207–212

Rose R, Constantinides B, Tapinos A, Robertson DL, Prosperi M (2016) Challenges in the analysis of viral metagenomes. Virus Evol 2(2):vew022. doi:10.1093/ve/vew022

Sanguino L, Franqueville L, Vogel TM, Larose C (2015) Linking environmental prokaryotic viruses and their host through CRISPRs. FEMS Microbiol Ecol 91(5). doi:10.1093/femsec/fiv046

Sano E, Carlson S, Wegley L, Rohwer F (2004) Movement of viruses between biomes. Appl Environ Microbiol 70(10):5842–5846. doi:10.1128/AEM.70.10.5842-5846.2004

Säwström C, Anesio MA, Graneli W, Laybourn-Parry J (2007a) Seasonal viral loop dynamics in two large ultraoligotrophic Antarctic freshwater lakes. Microb Ecol 53(1):1–11. doi:10.1007/s00248-006-9146-5

Säwström C, Laybourn-Parry J, Graneli W, Anesio AM (2007b) Heterotrophic bacterial and viral dynamics in Arctic freshwaters: results from a field study and nutrient-temperature manipulation experiments. Polar Biol 30(11):1407–1415. doi:10.1007/s00300-007-0301-3

Säwström C, Lisle J, Anesio AM, Priscu JC, Laybourn-Parry J (2008) Bacteriophage in polar inland waters. Extremophiles 12(2):167–175. doi:10.1007/s00792-007-0134-6

Sengupta M, Nielsen HJ, Youngren B, Austin S (2010) P1 plasmid segregation: accurate redistribution by dynamic plasmid pairing and separation. J Bacteriol 192(5):1175–1183. doi:10.1128/JB.01245-09

Sillankorva S, Oliveira R, Vieira MJ, Sutherland I, Azeredo J (2004) *Pseudomonas fluorescens* infection by bacteriophage ΦS1: the influence of temperature, host growth phase and media. FEMS Microbiol Lett 241(1):13–20. doi:10.1016/j.femsle.2004.06.058

Skidmore M, Anderson SP, Sharp M, Foght J, Lanoil BD (2005) Comparison of microbial community compositions of two subglacial environments reveals a possible role for microbes in chemical weathering processes. Appl Environ Microbiol 71(11):6986–6997. doi:10.1128/AEM.71.11.6986-6997.2005

Smith AW, Skilling DE, Castello JD, Rogers SO (2004) Ice as a reservoir for pathogenic human viruses: specifically, caliciviruses, influenza viruses, and enteroviruses. Med Hypotheses 63 (4):560–566. doi:10.1016/j.mehy.2004.05.011

Stibal M, Schostag M, Cameron KA, Hansen LH, Chandler DM, Wadham JL, Jacobsen CS (2015) Different bulk and active bacterial communities in cryoconite from the margin and interior of the Greenland ice sheet. Environ Microbiol Rep 7(2):293–300. doi:10.1111/1758-2229.12246

Suttle CA, Chen F (1992) Mechanisms and rates of decay of marine viruses in seawater. Appl Environ Microbiol 58(11):3721–3729

Thingstad TF (2000) Elements of a theory for the mechanisms controlling abundance, diversity, and biogeochemical role of lytic bacterial viruses in aquatic systems. Limnol Oceanogr 45 (6):1320–1328. doi:10.4319/lo.2000.45.6.1320

Thingstad TF, Lignell R (1997) Theoretical models for the control of bacterial growth rate, abundance, diversity and carbon demand. Aquat Microb Ecol 13(1):19–27

Tschitschko B, Williams TJ, Allen MA, Paez-Espino D, Kyrpides N, Zhong L, Raftery MJ, Cavicchioli R (2015) Antarctic archaea-virus interactions: metaproteome-led analysis of invasion, evasion and adaptation. ISME J 9(9):2094–2107. doi:10.1038/ismej.2015.110

Tu M, Liu F, Chen S, Wang M, Cheng A (2015) Role of capsid proteins in parvoviruses infection. Virol J 12:114. doi:10.1186/s12985-015-0344-y

Weinbauer MG (2004) Ecology of prokaryotic viruses. FEMS Microbiol Rev 28(2):127–181. doi:10.1016/j.femsre.2003.08.001

Weinbauer MG, Rassoulzadegan F (2004) Are viruses driving microbial diversification and diversity? Environ Microbiol 6(1):1–11. doi:10.1046/j.1462-2920.2003.00539.x

Wells LE, Deming JW (2006a) Characterization of a cold-active bacteriophage on two psychrophilic marine hosts. Aquat Microb Ecol 45(1):15–29

Wells LE, Deming JW (2006b) Effects of temperature, salinity and clay particles on inactivation and decay of cold-active marine bacteriophage 9A. Aquat Microb Ecol 45(1):31–39. doi:10.3354/ame045031

Wilhelm SW, Matteson AR (2008) Freshwater and marine virioplankton: a brief overview of commonalities and differences. Freshw Biol 53(6):1076–1089. doi:10.1111/j.1365-2427.2008.01980.x

Wilhelm SW, Suttle CA (1999) Viruses and nutrient cycles in the sea: viruses play critical roles in the structure and function of aquatic food webs. Bioscience 49(10):781–788. doi:10.2307/1313569

Wilkins D, Yau S, Williams TJ, Allen MA, Brown MV, DeMaere MZ, Lauro FM, Cavicchioli R (2013) Key microbial drivers in Antarctic aquatic environments. FEMS Microbiol Rev 37(3):303–335. doi:10.1111/1574-6976.12007

Wommack KE, Colwell RR (2000) Virioplankton: viruses in aquatic ecosystems. Microbiol Mol Biol Rev 64(1):69–114

Yau S, Lauro FM, DeMaere MZ, Brown MV, Thomas T, Raftery MJ, Andrews-Pfannkoch C, Lewis M, Hoffman JM, Gibson JA, Cavicchioli R (2011) Virophage control of antarctic algal host–virus dynamics. Proc Natl Acad Sci 108(15):6163–6168. doi:10.1073/pnas.1018221108

Zhang G, Shoham D, Gilichinsky D, Davydov S, Castello JD, Rogers SO (2006) Evidence of influenza A virus RNA in Siberian lake ice. J Virol 80(24):12229–12235. doi:10.1128/jvi.00986-06

Zhao Y, Temperton B, Thrash JC, Schwalbach MS, Vergin KL, Landry ZC, Ellisman M, Deerinck T, Sullivan MB, Giovannoni SJ (2013) Abundant SAR11 viruses in the ocean. Nature 494(7437):357–360. http://www.nature.com/nature/journal/v494/n7437/abs/nature11921.html#supplementary-information

Chapter 7
Microbial Life in the Deep Sea: Psychropiezophiles

Yuichi Nogi

Abstract Psychropiezophiles are microorganisms specialized for living in the deep-sea environment. Piezophiles display maximum growth at high pressure. Some can also grow at atmospheric pressure; those that cannot are referred to as obligatory piezophiles. A temperature change affects deep-sea psychropiezophiles more than a pressure change. Therefore, sample collection and cultivation temperature should be kept low. The preferred method for the long-term preservation and storage of psychropiezophiles is freezing in the vapor phase of liquid nitrogen. Initially, cultures of deep-sea psychropiezophilic bacteria were only species affiliated with one of five genera within the *Gammaproteobacteria* subgroup: *Shewanella*, *Photobacterium*, *Colwellia*, *Moritella*, and *Psychromonas*. However, more recently, species classified as *Alphaproteobacteria* and *Firmicutes* have also been found. The genome of several of these bacteria has been analyzed, which revealed characteristic features of these microorganisms. Psychropiezophiles contain unsaturated fatty acids in their cell membrane layers, but the presence of polyunsaturated fatty acids, like eicosapentaenoic acid and docosahexaenoic acid, is not obligatory for growth under high pressure. In the future, along with the development of culture methods and isolation techniques, a variety of other psychropiezophilic species will be discovered and the relationship between pressure and growth of psychropiezophiles will be clarified.

Contents

Y. Nogi (✉)
Research and Development Center for Marine Biosciences, Japan Agency for Marine-Earth Science and Technology, 2-15 Natsushima-cho, Yokosuka 237-0061, Japan
e-mail: nogiy@jamstec.go.jp

R. Margesin (ed.), *Psychrophiles: From Biodiversity to Biotechnology*,
DOI 10.1007/978-3-319-57057-0_7

7.1 Introduction

Despite the fact that up to 60% of the Earth's surface is covered by seas with depths exceeding 1000 m, the study of microorganisms in the deep sea is very incomplete. The deep sea is regarded as an extreme environment with high hydrostatic pressures (because pressure in the ocean increases by about 0.1 megapascal [MPa, 1 Pa $=1$ kg/(m·s^2) $= 9.87 \times 10^{-6}$ atm] for each 10 meters in depth up to 110 MPa), predominantly low temperatures (2–4 °C) (Fig. 7.1) but with occasional regions of extremely high temperatures (up to 370 °C) at hydrothermal vents, darkness, and low nutrient availability, although with sufficient dissolved oxygen (Fig. 7.1).

Until the exploratory voyage of the *Challenger* in the late nineteenth century, the deep sea was considered devoid of life. In the early twentieth century, only inconsequential organisms were considered to inhabit the deep sea. In 1949, research started on the effects of hydrostatic pressure on microbial activities (ZoBell and Johnson 1949). The word "barophilic" was first used by ZoBell and Johnson (1949), and it is defined today as optimal growth at pressure greater than 0.1 MPa or by a requirement for increased pressure for growth. The term "piezophile" was proposed as a replacement for "barophile" as the Greek translations of the prefixes "baro" and "piezo" mean "weight" and "pressure," respectively (Yayanos 1995). Thus, the word piezophile is more suitable than barophile to describe bacteria that grow better at high pressure than at atmospheric pressure. Therefore, researchers have opted to use the term "piezophilic bacteria" meaning high-pressure-loving bacteria. The growth patterns of piezotolerant and piezophilic bacteria are shown in Fig. 7.2. Piezophiles display maximum growth at high pressure. Some can also grow at atmospheric pressure; those that cannot are referred to as obligatory piezophiles. Piezotolerant bacteria grow best at atmospheric pressure but can sustain growth at high pressure.

7.2 Deep-Sea Psychropiezophiles

Bacteria living in the deep sea have several unusual characteristics that allow them to grow in their extreme environment. In 1979, the first pure culture isolate of a piezophilic bacterium was reported (Yayanos et al. 1979). The spirillum-like bacterial strain CNPT-3 had a rapid doubling rate at 50 MPa but did not grow at atmospheric pressure. However, no public culture collections are maintained and no

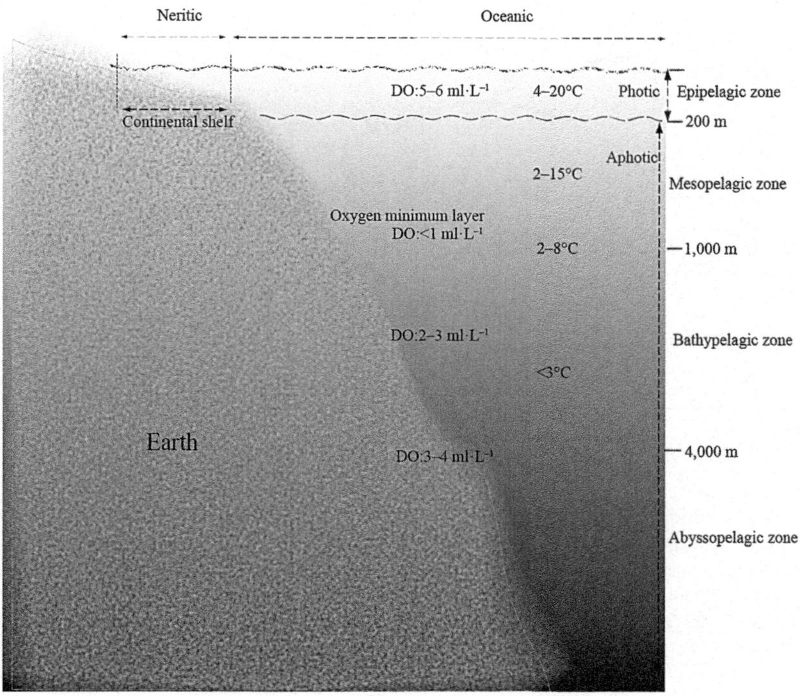

Fig. 7.1 Layers of the ocean. The oceans are divided into two broad realms, the pelagic and the benthic. These layers, known as "zones," extend from the surface to the most extreme depths where light can no longer penetrate. Biologists divide the pelagic zone into the epipelagic (less than 200 m, where photosynthesis can occur), the mesopelagic (200–1000 m, the "twilight" zone with faint sunlight but no photosynthesis), the bathypelagic (1000–4000 m), the abyssopelagic (4000–6000 m), and the deepest zone, the hadopelagic (trenches more than 6000 m deep). Thermoclines vary in thickness from a few hundred meters to nearly 1000 m. The temperature of the epipelagic zone in the tropics is usually higher than 20 °C. From the base of the epipelagic, the temperature drops to 2–8 °C at 1000 m. It continues to decrease to the bottom, but at a much slower rate. Below 1000 m, the water is isothermal between 2 °C and 4 °C. The cold water stems from the sinking of heavy surface water in the polar regions. *DO*, dissolved oxygen

name has been added to strain CNPT-3. The first psychropiezophile to be named was *Shewanella benthica* (Table 7.1). *S. benthica* strain W I45 was isolated from the holothurian at a depth of 4575 m in the South Atlantic Ocean (Deming et al. 1984). Thereafter, we isolated and characterized numerous piezophilic and piezotolerant bacteria from cold deep-sea sediments at depths ranging from 2500 m to 11,000 m using sterilized sediment samplers (Fig. 7.3) on the submersibles SHINKAI 6500 and KAIKO systems operated by the Japan Agency for Marine-Earth Science and Technology (JAMSTEC) (Kato et al. 1995; Nogi and Kato 1999; Nogi et al. 2007). Groups at the Scripps Research Institute collected deep-sea organisms such as amphipods using traps and isolated psychropiezophiles among them (Lauro et al. 2007; Cao et al. 2014). Most isolated strains are not only

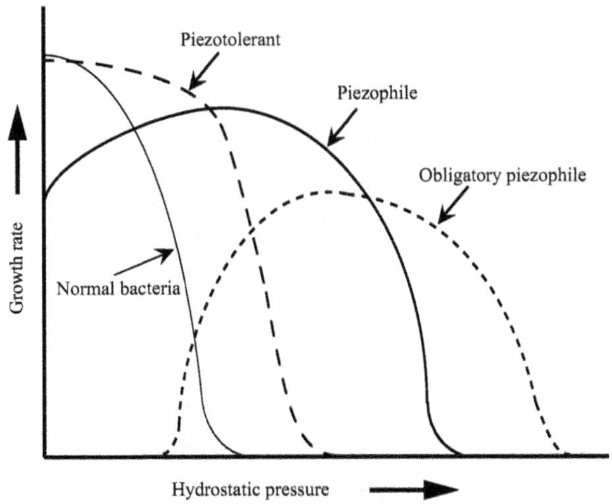

Fig. 7.2 Definition of piezophilic growth properties. This figure is a conceptual chart, and the *x*-axis value does not indicate specific pressure. Piezophiles display maximum growth at high pressure. They may or may not grow at atmospheric pressure, and in the latter case are obligatory piezophiles. Piezotolerant bacteria grow optimally at atmospheric pressure, but can sustain growth at high pressure (about 30–50 MPa), whereas normal bacteria (piezosensitive) stop growing at about 30–50 MPa

piezophilic but also psychrophilic (psychropiezophilic) and cannot be cultured at temperatures higher than 20 °C.

7.3 Isolation and Preservation of Deep-Sea Psychropiezophiles

The isolation of deep-sea psychropiezophiles requires the maintenance of samples at low temperature (0–4 °C). Furthermore, the culture medium used for isolation must also be cooled in advance. In the polar zone where the temperature of surface seawater is low, deep-sea samples can be collected relatively easily at the optimum temperature. However, in higher surface seawater temperature areas, special samplers are required. The cell walls of bacteria allow easy entry and exit of water, which do not destroy the cell if an instantaneous pressure change does not occur. A temperature change affects deep-sea psychropiezophiles more than a pressure change. JAMSTEC researchers use the sampler shown in Fig. 7.3 to collect deep-sea microbes (Ikemoto and Kyo 1993; Kato et al. 1995), which ensures only a minimal change in the sample temperature. It is possible to culture many psychropiezophiles other than obligatory piezophiles on agar plates as well as in a liquid medium under atmospheric pressure.

Table 7.1 Growth properties and genetic information of cultivated psychropiezophilic bacteria

Strain	Culture collection no.	Opt. temp. (°C)	Opt. pressure (MPa)	Isolation source (depth [m])	16S rRNA Genbank accession no.	Complete genome	Reference
Gammaproteobacteria							
Colwelliaceae							
Colwellia hadaliensis BNL-1^T		10	90	Puerto Rico Trench (7410)			Deming et al. (1988)
Colwellia piezophila Y223G^T	JCM 11831^T	10	60	Japan Trench, sediment (6278)	NR 024805	NZ KB905161	Nogi et al. (2004)
Colwellia sp. strain MT41		8	103	Mariana Trench, decaying amphipod (10,476)	DQ027051	NZ CP013145	Yayanos et al. (1981)
Moritellaceae							
Moritella abyssi 2693^T	JCM 11436^T	10	30	Atlantic Ocean, sediment (2815)	AJ252022		Xu et al. (2003b)
Moritella japonica DSK1^T	JCM 10249^T	15	50	Japan Trench, sediment (6356)	D21224		Nogi et al. (1998a)
Moritella profunda 2674^T	JCM 11435^T	6	30	Atlantic Ocean, sediment (2815)	AJ252023		Xu et al. (2003b)
Moritella yayanosii DB21MT-5^T	JCM 10263^T	10	80	Mariana Trench, sediment (10,898)	AB008797		Nogi and Kato (1999)
Moritella sp. strain PE36	ATCC BAA-1251	10	41	North Pacific Ocean, amphipod trap water (3584)	DQ027053		Yayanos (1986)
Vibrionaceae							
Photobacterium profundum DSJ4^T	JCM 10084^T	10	10	Ryukyu Trench, sediment (5110)	D21226		Nogi et al. (1998c)
Photobacterium profundum SS9	ATCC BAA-2589	15	28	Sulu Trough, amphipod homogenate (2551)	AB003191	NC 006370.1	DeLong et al. (1997)
Oceanospirillaceae							
"*Profundimonas piezophila*" YC-1^T	ATCC BAA-2591	8	50	Puerto Rico Trench, water column (6000)	HQ230045		Cao et al. (2014)

(continued)

Table 7.1 (continued)

Strain	Culture collection no.	Opt. temp. (°C)	Opt. pressure (MPa)	Isolation source (depth [m])	16S rRNA Genbank accession no.	Complete genome	Reference
Psychromonadaceae							
Psychromonas hadalis K41G^T	JCM 11830^T	6	60	Japan Trench, sediment (7542)	AB094413	NZ_ATUO01000000	Nogi et al. (2007)
Psychromonas kaikoae JT7304^T	JCM 11054^T	10	50	Japan Trench, cold-seep sediment (7434)	AB052160		Nogi et al. (2002)
Psychromonas profunda 2825^T	JCM 11437^T	10	25	Atlantic Ocean sediment (2770)	AJ416756		Xu et al. (2003a)
Psychromonas sp. strain CNPT3		12	52	Central North Pacific, decaying amphipod (5800)	DQ027056	NC 020802.1	Yayanos et al. (1979)
Shewanellaceae							
Shewanella benthica W145^T	DSM 8812^T	10–15	50	Intestine, holothurian, South Atlantic Ocean (4575)	X82131		Deming et al. (1984)
Shewanella benthica KT99	ATCC BAA-2590	~2	~98	Kermadec Trench, amphipod homogenate (9856)	DQ027058	NZ_ABIC01000000	Lauro et al. (2007)
Shewanella piezotolerans WP3^T	JCM 13877^T	15–20	20	West Pacific, deep-sea sediment (1914)	AJ551090	NC 011566.1	Xiao et al. (2007)
Shewanella psychrophila WP2^T	JCM 13876^T	10–15	20	West Pacific, deep-sea sediment (1914)	AJ551089		Xiao et al. (2007)
Shewanella violacea DSS12^T	JCM 10179^T	10	30	Ryukyu Trench, sediment (5110)	D21225	NC 014012.1	Nogi et al. (1998b)
Alphaproteobacteria							
Rhodobacterales bacterium PRT1		10	80	Puerto Rico Trench, seawater (8350)	JF303756		Eloe et al. (2011)
Firmicutes							
Carnobacterium sp. AT7		20	20	Aleutian Trench, water column (2500)	DQ027061	NZ_ABHH00000000.1	Lauro et al. (2007)

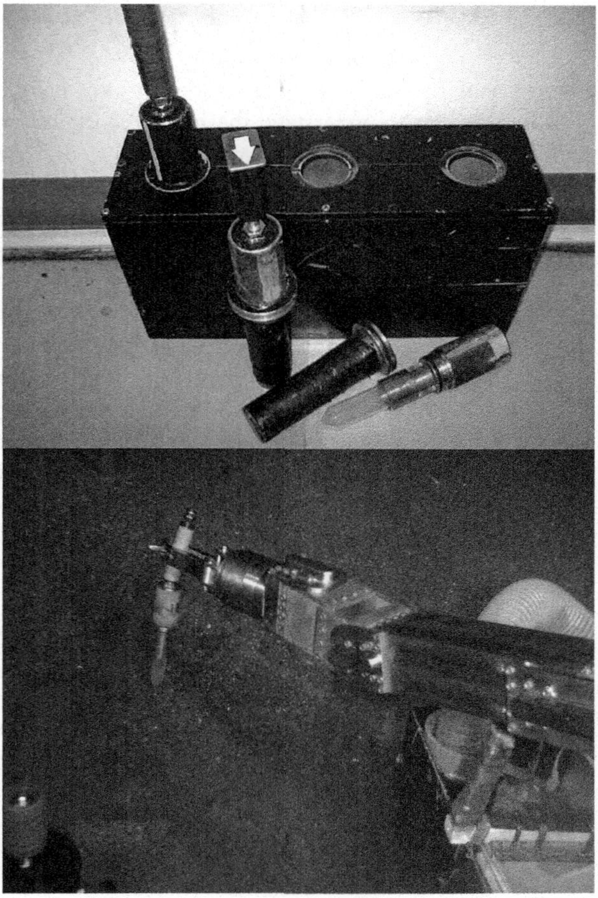

Fig. 7.3 Sterilized sediment sampler (*upper photograph*) and sediment sampling (*lower photograph*)

The preferred method for the long-term preservation and storage of psychropiezophiles is freezing in the vapor phase of liquid nitrogen (–130 °C and lower). Long survival of more than 20 years and good recovery rates can be achieved with this method. Cultures to be stored in liquid nitrogen are usually grown to the late-log growth phase and mixed with a cryopreservative (10% glycerol or 5% DMSO). Sample cultures can also be preserved at –70 to –80 °C. However, since the survival and recovery rates decrease with this method, it is necessary to use a preservative for dense suspensions of cells. Since almost all psychropiezophilic strains die after freeze-drying, this method cannot be used.

7.4 Taxonomy of the Psychropiezophiles

Numerous deep-sea piezophilic bacterial strains have been isolated and characterized in an effort to understand the interaction between the deep-sea environment and its microbial inhabitants (Yayanos et al. 1979; Kato et al. 1998; Margesin and Nogi 2004; Lauro et al. 2007; Eloe et al. 2011). Approximately 30 years after psychropiezophiles were first isolated from the deep sea, they were assigned to the gamma-subgroup of the *Proteobacteria* according to phylogenetic classifications based on 5S and 16S ribosomal RNA (rRNA) sequence information (DeLong et al. 1997; Kato 1999; Margesin and Nogi 2004; Nogi et al. 2007). Prior to the reports by the JAMSTEC group, only two deep-sea piezophilic bacterial species had been described; they were named *Shewanella benthica* (Deming et al. 1984; MacDonell and Colwell 1985) and *Colwellia hadaliensis* (Deming et al. 1988). We identified several novel piezophilic species within these genera based on the results of chromosomal DNA–DNA hybridization studies and several other taxonomic properties. Both previously described and novel species of bacteria were identified among the piezophilic bacterial isolates. Nogi et al. (2002) reported that cultivated psychropiezophilic deep-sea bacteria are represented by the genera *Colwellia*, *Moritella*, *Psychromonas*, and *Shewanella* within the *Alteromonadaceae* family of the *Gammaproteobacteria*, and that the genus *Photobacterium* is assigned to the *Vibrionaceae* family within the *Gammaproteobacteria* (Table 7.1). Subsequently, Lauro et al. (2007) reported that a psychropiezophile isolated from seawater of the Aleutian Trench was classified as the genus *Carnobacterium* of *Firmicutes* based on 16S rRNA analysis. In addition, Eloe et al. (2011) isolated a psychropiezophile from seawater of the Puerto Rico Trench which was classified from 16S rRNA analysis as a clade of the *Rhodobacterales* of the *Alphaproteobacteria* subgroup (Fig. 7.4). However, the complete identification of these two psychropiezophiles was not performed, and they were not deposited in a culture collection. More recently, Cao et al. (2014) isolated psychropiezophiles from seawater of the Puerto Rico Trench which were classified from 16S rRNA analysis as the genus *Profundimonas* in the *Oceanospirillales* clade of the *Gammaproteobacteria* subgroup (Table 7.1). In the future, it is expected that a wide variety of other psychropiezophilic genera will be isolated and identified.

7.4.1 The Genus Colwellia

Species of the genus *Colwellia* are defined as facultatively anaerobic and psychrophilic bacteria, and the type species of this genus is *Colwellia psychroerythrus* (Deming et al. 1988). This genus belongs to the *Gammaproteobacteria*. At the time of writing, the genus comprised 15 species with validly published names. *Colwellia hadaliensis*, *Colwellia piezophila*, and *Colwellia* sp. strain MT41 are the only known members of the genus *Colwellia* showing psychropiezophilic growth

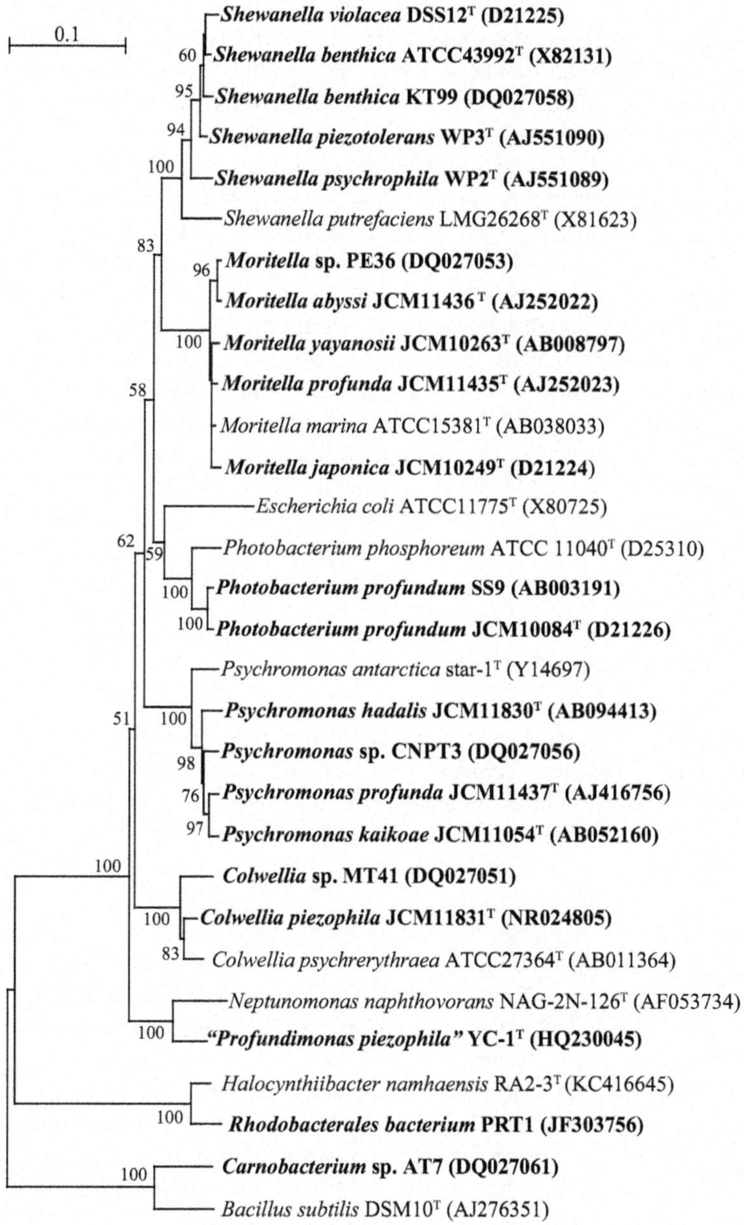

Fig. 7.4 Phylogenetic tree showing the relationships between isolated deep-sea piezophilic bacteria (in *bold*) determined by comparing 16S rDNA sequences using the neighbor-joining method (references for species description are indicated in the *text*). The scale represents the average number of nucleotide substitutions per site. Bootstrap values (%) are calculated from 1000 trees and shown for frequencies above the threshold of 50%

properties (Deming et al. 1988; Nogi et al. 2004). The taxonomic data for *C. hadaliensis* have been published, although the 16S rRNA gene sequence information has not, and it has not been deposited in public culture collections. Furthermore, the taxonomic data for *Colwellia* sp. strain MT41 have not been published and only the 16S rRNA gene sequence information has been registered. The other species, *C. piezophila*, has been isolated as an obligatory psychropiezophilic strain from the sediment of the deep-sea fissure of the Japan Trench (Nogi et al. 2004). Bowman et al. (1998) reported that *Colwellia* species produce the long-chain polyunsaturated fatty acid (PUFA) docosahexaenoic acid (DHA). However, *C. piezophila* does not produce eicosapentaenoic acid (EPA) or DHA in the membrane layer, although high levels of unsaturated fatty acids ($C_{16:1}$) are produced. This suggests that the production of long-chain PUFAs should not be a requirement for classification as a piezophilic bacterium, although the production of unsaturated fatty acids could be a common property of psychropiezophiles. In addition, genome analysis has been performed for *C. psychroerythrus* 34H^T, *C. piezophila* Y223G^T, and *Colwellia* sp. strain MT41 (Methé et al. 2005; Stelling et al. 2014). As a result, further progress in the analysis of psychropiezophiles is also expected.

7.4.2 The Genus Moritella

The type strain of the genus *Moritella* is *Moritella marina* (Urakawa et al. 1998), previously known as *Vibrio marinus* (Colwell and Morita 1964), and is one of the most common psychrophilic organisms isolated from marine environments. At the time of writing, the genus *Moritella* consisted of seven species. Many species of the genus *Moritella* are psychropiezophilic, but *M. marina* is not a piezophilic bacterium.

Strain DSK1, a moderately psychropiezophilic bacterium isolated from the Japan Trench, was identified as *Moritella japonica* (Nogi et al. 1998a). It was the first piezophilic species identified in the genus *Moritella*. Production of the long-chain PUFA DHA is a characteristic property of the genus *Moritella*. The obligatory psychropiezophilic bacterial strain DB21MT-5 isolated from the world's deepest sea bottom, the Mariana Trench Challenger Deep, at a depth of 10,898 m was also identified as a *Moritella* species and designated *Moritella yayanosii* (Nogi and Kato 1999). The optimal pressure for the growth of *M. yayanosii* strain DB21MT-5 is 80 MPa; this strain is unable to grow at pressures of less than 50 MPa but grows well at pressures as high as 100 MPa (Kato et al. 1998). The fatty acid composition of psychropiezophilic strains changes as a function of pressure, and in general greater amounts of PUFAs are synthesized at higher growth pressures. Approximately 70% of the membrane lipids in *M. yayanosii* are unsaturated fatty acids, which is a finding consistent with its adaptation to very high pressures (Nogi and Kato 1999; Fang et al. 2000). Two other species of the genus *Moritella*, *Moritella abyssi* and *Moritella profunda*, were isolated from a depth of

2815 m off the West African coast (Xu et al. 2003b), and strain PE36 was isolated from a depth of 3584 m in the North Pacific Ocean (Yayanos 1986); they are moderately piezophilic and their growth properties are similar to those of *M. japonica*.

7.4.3 The Genus Photobacterium

The genus *Photobacterium* was one of the earliest known bacterial taxa (Beijerinck 1889), and the type species of this genus is *Photobacterium phosphoreum*. Phylogenetic analyses based on 16S rRNA gene sequences showed that this genus falls within the *Gammaproteobacteria*, is particularly closely related to the genus *Vibrio* (Nogi et al. 1998c), and is typical of marine bacterial genera.

Photobacterium profundum, a novel species, was identified through studies of the moderately psychropiezophilic strains DSJ4 and SS9 (Nogi et al. 1998c), and *Photobacterium frigidiphilum* was reported to be slightly piezophilic; its optimal pressure for growth is 10 MPa (Seo et al. 2005). About 25 *Photobacterium* species have been isolated, but *P. profundum* and *P. frigidiphilum* are the only species within this genus known to display piezophily and to produce the long-chain PUFA EPA. No other known species of *Photobacterium* produces EPA (Nogi et al. 1998c). *P. profundum* strain SS9 has been extensively studied and subjected to genome sequencing and expression analysis. The genome consists of a 4.1-Mbp circular chromosome, a 2.2-Mbp minor circular chromosome, and an 80-kbp circular plasmid (Vezzi et al. 2005). A study of strain SS9 showed that several stress response genes are upregulated in response to atmospheric pressure, including *htpG*, *dnaK*, *dnaJ*, and *groEL* (Vezzi et al. 2005). In addition, studies were conducted in relation to the pressure regulation of the outer membrane proteins OmpH and OmpL (Bartlett and Welch 1995).

7.4.4 The Genus Psychromonas

The genus *Psychromonas* is composed of psychrophilic bacteria; it also belongs to the *Gammaproteobacteria* and is closely related to the genera *Shewanella* and *Moritella* on the basis of 16S rRNA gene sequence data. The type species of the genus *Psychromonas*, *Psychromonas antarctica*, was isolated as an aerotolerant anaerobic bacterium from a high-salinity pond in Antarctica (Mountfort et al. 1998). This strain does not display piezophilic properties. At the time of writing, the genus comprised 14 species with validly published names.

Psychromonas kaikoae and *Psychromonas hadalis* are novel obligatory psychropiezophilic bacteria (Nogi et al. 2002, 2007). *P. kaikoae* was isolated from sediment collected from the deepest cold-seep environment (sometimes called a cold vent, an area of the ocean floor where hydrogen sulfide, methane, and other hydrocarbon-rich fluid seepage occurs) in the Japan Trench at a depth of 7434 m, where chemoautotrophic animal communities were also found. The optimal temperature and pressure for the growth of *P. kaikoae* are 10 °C and 50 MPa, respectively, and both EPA and DHA are produced in the membrane layer. *P. hadalis* was isolated from sediment collected from the bottom of the Japan Trench at a depth of 7542 m. The optimal temperature and pressure for the growth of *P. hadalis* are 6 °C and 60 MPa, respectively, and DHA is produced in the membrane layer. *Psychromonas profunda* is a moderately piezophilic bacterium isolated from deep Atlantic sediments at a depth of 2770 m (Xu et al. 2003a). In contrast, *P. profunda* is similar to the piezosensitive strain *P. antarctica* and does not produce either EPA or DHA in its membrane layer. The piezosensitive strains *Psychromonas marina* and *Psychromonas ossibalaenae* (Miyazaki et al. 2008) also produce small amounts of DHA. In the genus *Psychromonas*, only *P. kaikoae* produces both EPA and DHA. *Psychromonas* strain CNPT-3 proved to be closely related to *Psychromonas* species based on 16S rRNA sequence information, and therefore it was assumed that strain CNPT-3 should be included in the genus *Psychromonas*.

7.4.5 The Genus Shewanella

The genus *Shewanella* comprises Gram-negative, aerobic, and facultatively anaerobic *Gammaproteobacteria* (MacDonell and Colwell 1985) and is typical of deep-sea bacterial genera (DeLong et al. 1997). The genus includes psychrophilic and mesophilic species that are widely distributed in marine environments. The type species of this genus is *Shewanella putrefaciens*, a bacterium formerly known as *Pseudomonas putrefaciens* (MacDonell and Colwell 1985; Owen et al. 1978). About 60 *Shewanella* species have been isolated and described.

Prior to the present report, *S. benthica, Shewanella piezotolerans, Shewanella psychrophila*, and *Shewanella violacea* were the only known members of the genus *Shewanella* showing psychropiezophilic growth properties (Nogi et al. 1998b; Xiao et al. 2007). The psychrophilic and piezophilic *Shewanella* strains, including *S. benthica, S. piezotolerans, S. psychrophila*, and *S. violacea*, produce EPA, and thus the production of such long-chain PUFAs is a property shared by many deep-sea bacteria to maintain cell-membrane fluidity under conditions of extreme cold and high hydrostatic pressure (Fang et al. 2003). *S. violacea* strain DSS12 has been studied extensively, particularly with respect to its molecular mechanisms of adaptation to high pressure (Kato et al. 2000; Nakasone et al. 1998, 2002). As there are only a few differences in the growth characteristics of strain DSS12 under different pressure conditions, this strain is a very convenient deep-sea bacterium for the study of the mechanisms of adaptation to high-pressure environments. In terms of respiratory proteins (Tamegai et al. 2005), the RNA polymerase subunit (Kawano et al. 2009), dihydrofolate reductase (Ohmae et al. 2015), and isopropylmalate dehydrogenase (De Poorter et al. 2004), piezophilic proteins are unique for adaptation to high-pressure environments, and some are notably more stable and active under higher-pressure conditions. Therefore, genome analysis of strain DSS12 was performed as a model deep-sea psychropiezophilic bacterium. This strain contains 4.96 Mbp, a single chromosome, and no known plasmids. It has 4346 protein genes and 169 RNA genes (Aono et al. 2010).

7.4.6 *The Genus* Profundimonas

The genus *Profundimonas* is a Gram-negative, facultatively anaerobic heterotroph within the family *Oceanospirillaceae*, closely related to the uncultured symbiont of the deep-sea whale bone-eating worms of the genus *Osedax* (Cao et al. 2014). This taxonomic name has been effectively published but not validly published under the rules of the International Code of Nomenclature of Bacteria.

Profundimonas piezophila, the type species of the genus *Profundimonas*, was isolated from deep seawater collected from the Puerto Rico Trench at a depth of 6000 m as novel obligatory psychropiezophilic bacterium. The optimal temperature and pressure for the growth of *P. piezophila* are 8 °C and 50 MPa, respectively, and it does not produce either EPA or DHA in its membrane layer.

7.4.7 *The Genus* Carnobacterium

The genus *Carnobacterium* is a Gram-positive, facultatively anaerobic, heterofermentative, psychrotolerant, rod-shaped lactic acid bacteria that produces L-lactic acid from glucose, within the family *Leuconostocaceae* of *Firmicutes*. The type species of this genus is *Carnobacterium divergens*. It is found in vacuum-packed meat and is capable of growing in products stored at low temperatures, including refrigerated food (Collins et al. 1987). About 12 *Carnobacterium* species have been isolated and described. *Carnobacterium* sp. strains AT7 and AT12 were isolated from seawater collected from a depth of 2500 m in the Aleutian Trench and are novel psychropiezophilic bacteria. These strains are closely related to the recently isolated *Carnobacterium pleistocenium* (Lauro et al. 2007), but this is the first report of a piezophilic isolate of this species as well as the first Gram-positive piezophile ever identified. The pressure range for growth of strains AT7 and AT12 is 0.1–60 MPa, with the optimum at 15 MPa (Yayanos and DeLong 1987). Detailed data on this strain have not yet been reported. However, since genome analysis has been carried out, a detailed report is expected in the future.

7.4.8 *The Order* Rhodobacterales

Strain PRT1 was isolated from hadal seawater collected from the Puerto Rico Trench at a depth of 8350 m as a novel obligatory psychropiezophilic bacterium within the *Roseobacter* clade of the order *Rhodobacterales* within the *Alphaproteobacteria*. The optimal temperature and pressure for the growth of this strain are 10 °C and 80 MPa, respectively. This is the first report of a piezophilic isolate of *Alphaproteobacteria*. Strain PRT1 is the slowest growing (minimal doubling time, 36 h) and lowest cell density-producing (maximal densities of 5.0 $\times 10^6$ cells ml^{-1}) (Eloe et al. 2011). Therefore, taxonomic studies appear to be difficult. However, it is expected that it will be proposed as a novel genus after further analysis.

7.5 Fatty Acid Composition of Psychropiezophiles

The psychropiezophilic *Shewanella* and *Photobacterium* strains produce EPA (Nogi et al. 1998b, c), *Moritella* strains produce DHA (Nogi et al. 1998a; Nogi and Kato 1999), and *P. kaikoae* produces both EPA and DHA (Nogi et al. 2002), but *C. piezophila* does not produce such PUFAs (Nogi et al. 2004). The fatty acid composition of these psychropiezophilic strains is dependent on the taxonomic affiliation (genus); high levels of unsaturated fatty acids (about 40–70%), including EPA or DHA, are commonly found in their membrane layer. Generally, species included in these genera other than piezophilic bacteria tend to have a high ratio of unsaturated fatty acids (Table 7.2). However, the ratios of unsaturated fatty acids of obligatory psychropiezophilic bacteria are particularly high (60% or more). This indicates that it is important for psychropiezophilic bacteria to contain high ratios of unsaturated fatty acids.

The fatty acid composition of psychropiezophilic strains also changes as a function of pressure, and, in general, greater amounts of PUFAs are synthesized under high-pressure conditions for growth (DeLong and Yayanos 1985, 1986). All pychropiezophilic bacteria were believed to produce one of the long-chain PUFAs, either EPA or DHA, but this does not appear to be obligatory. For example, Allen et al. (1999) reported that monounsaturated fatty acids, but not PUFAs, are required for the growth of the psychropiezophilic bacterium *P. profundum* SS9 based on the analysis of pressure-sensitive mutants. In their mutant experiments, the $C_{18:1}$ fatty acid proved to be necessary for growth under low-temperature and/or high-pressure conditions. In the case of *C. piezophila* Y223G^{T} and *P. profunda* 2825^{T}, either EPA or DHA and the $C_{18:1}$ fatty acid are absent but the strain produces a large amount of the fatty acid $C_{16:1}$ in the cell membrane (Table 7.2). All psychropiezophilic bacteria analyzed so far have the $C_{16:1}$ fatty acid, and thus this fatty acid appears to be one of the important components required for high-pressure growth.

Table 7.2 Whole-cell fatty acid compositions of psychropiezophiles and related strains

Strain	Char.	Whole-cell fatty acid composition (%) of:																						UFA (%)
		$C_{11:0}$	$C_{12:0}$	$C_{13:0}$	$C_{14:0}$	$C_{15:0}$	$C_{16:0}$	$C_{17:0}$	$C_{18:0}$	iso-$C_{13:0}$	iso-$C_{14:0}$	iso-$C_{15:0}$	iso-$C_{16:0}$	$C_{14:1}$	$C_{15:1}$	$C_{16:1}$	$C_{17:1}$	$C_{18:1}$	$C_{20:5}$ ω3	$C_{22:6}$ ω3	$C_{10:0}$ 3OH	$C_{12:0}$ 3OH	$C_{14:0}$ 3OH	iso-$C_{13:0}$ 3OH	
C. psychroerythraea ATCC27364[T]	P				7	6	30	1	1					6		40	1	1	1	7					56
C. piezophila Y223G[T]	OPP		1		3	3	31							9	2	50						1			61
M. marina	P		1		16	1	21						1	1		45		2		12					60
M. abyssi 2693[T]	PP				22		11							10		46				6		6			62
M. japonica DSK1[T]	PP				18	1	21							2		50		2		6					60
M. profunda 2674[T]	PP				16		13							10		45				11		5			65
M. yayanosii DB21MT-5[T]	OPP				15	1	13							6		53		1		9					69
Pho. phosphoreum ATCC11040[T]	M		6		11		26		1					2		40		3				9	3		44
Pho. profundum DSJ4[T]	PP		2		3	1	9		1	2	4	2	15	3		32		9	13			5			57
Pho. profundum SS9	PP		4		10	1	22		2	1			6	5		30		7	7			6	1		48
"*Pro. piezophila*" YC-1[T]	OPP						7									36		46			7				81
Psy. antarctica DSM 10704[T]	P		1				24							8		58		3					6		69
Psy. hadalis K41G[T]	OPP		1		1		31							17		37				8			3		62
Psy. kaikoae JT7304[T]	OPP		1		6	1	15							10		54		2	2	2		2	4		76

(continued)

Table 7.2 (continued)

Strain	Char.	Whole-cell fatty acid composition (%) of:																							UFA (%)
		$C_{11:0}$	$C_{12:0}$	$C_{13:0}$	$C_{14:0}$	$C_{15:0}$	$C_{16:0}$	$C_{17:0}$	$C_{18:0}$	iso-$C_{13:0}$	iso-$C_{14:0}$	iso-$C_{15:0}$	iso-$C_{16:0}$	$C_{14:1}$	$C_{15:1}$	$C_{16:1}$	$C_{17:1}$	$C_{18:1}$	$C_{20:5}$ ω3	$C_{22:6}$ ω3	$C_{10:0}$ 3OH	$C_{12:0}$ 3OH	$C_{14:0}$ 3OH	iso-$C_{13:0}$ 3OH	
Psy. profunda 2825[T]	PP						31							15		44							10		59
S. putrefaciens IAM12079[T]	M		5	1	3	6	10	1		11		13		1	1	31	13	3							49
S. benthica ATCC43992[T]	PP	2	5		17		15			11		5				37		1	8						46
S. benthica DB21-MT2	OPP	1	2		3	1	15			1		3				35		13	25						73
S. piezotolerans WP3[T]	PP		3	1	3	3	10	2	2	8		10				24	6	12	13			1		1	54
S. psychrophila WP2[T]	PP		6		8		13		2	7		4				38	2	9	7					2	56
S. violacea DSS12[T]	PP	2	4		6	7	15	1		8		14		1		19	1	5	15						41

Char., strain characteristics: P, psychrophile; PP, psychropiezophile; OPP, obligatory psychropiezophile; M, mesophile, UFA, the proportion of unsaturated fatty acids

7.6 Conclusions

Initially, cultures of deep-sea psychropiezophilic bacteria were only species affiliated with one of five genera within the *Gammaproteobacteria* subgroup: *Shewanella*, *Photobacterium*, *Colwellia*, *Moritella*, and *Psychromonas*. However, more recently, species classified as *Alphaproteobacteria* and *Firmicutes* have also been found. In the future, a variety of other species will no doubt be discovered with the development of culture methods and isolation techniques. These psychropiezophiles are characterized by containing unsaturated fatty acids in their cell membrane layers but PUFAs, like EPA and DHA, are not obligatory for growth under high pressure. Subsequent fatty acid analysis of psychropiezophilic species of *Alphaproteobacteria* and *Firmicutes* will clarify the relationship between psychropiezophilic bacteria and unsaturated fatty acids. Progress in genome research will enable researchers to elucidate numerous details, such as the pressure response of psychropiezophilic bacteria.

References

Allen EE, Facciotti D, Bartlett DH (1999) Monounsaturated but not polyunsaturated fatty acids are required for growth of the deep-sea bacterium *Photobacterium profundum* SS9 at high pressure and low temperature. Appl Environ Microbiol 65:1710–1720

Aono E, Baba T, Ara T, Nishi T, Nakamichi T, Inamoto E, Toyonaga H, Hasegawa M, Takai Y, Okumura Y, Baba M, Tomita M, Kato C, Oshima T, Nakasone K, Mori H (2010) Complete genome sequence and comparative analysis of *Shewanella violacea*, a psychrophilic and piezophilic bacterium from deep sea floor sediments. Mol Biosyst 6:1216–1226

Bartlett DH, Welch TJ (1995) OmpH gene expression is regulated by multiple environmental cues in addition to high pressure in the deep-sea bacterium *Photobacterium* species strain SS9. J Bacteriol 177:1008–1016

Beijerinck MW (1889) Le *Photobacterium luminosum,* bactérie luminosum de la Mer Nord. Arch Néerl Sci 23:401–427 (in French)

Bowman JP, Gosink JJ, McCammon SA, Lewis TE, Nichols DS, Nichols PD, Skerratt JH, Staley JT, McMeekin TA (1998) *Colwellia demingiae* sp. nov., *Colwellia hornerae* sp. nov., *Colwellia rossensis* sp. nov. and *Colwellia psychrotropica* sp. nov.: psychrophilic Antarctic species with the ability to synthesize docosahexaenoic acid (22:6w3). Int J Syst Bacteriol 48:1171–1180

Cao Y, Chastain RA, Eloe EA, Nogi Y, Kato C, Bartlett DH (2014) Novel psychropiezophilic Oceanospirillales species *Profundimonas piezophila gen. nov., sp. nov.*, isolated from the deep-sea environment of the Puerto Rico Trench. Appl Environ Microbiol 80:54–60

Collins MD, Farrow JAE, Phillips BA, Ferusu S, Jones D (1987) Classification of *Lactobacillus divergens*, *Lactobacillus piscicola*, and some catalase-negative, asporogenous, rod-shaped bacteria from poultry in a new genus, *Carnobacterium.* Int J Syst Bacteriol 37:310–316

Colwell RR, Morita RY (1964) Reisolation and emendation of description of *Vibrio marinus* (Russell) Ford. J Bacteriol 88:831–837

DeLong EF, Yayanos AA (1985) Adaptation of the membrane lipids of a deep-sea bacterium to changes in hydrostatic pressure. Science 228:1101–1103

DeLong EF, Yayanos AA (1986) Biochemical function and ecological significance of novel bacterial lipids in deep-sea prokaryotes. Appl Environ Microbiol 51:730–737

DeLong EF, Franks DG, Yayanos AA (1997) Evolutionary relationship of cultivated psychrophilic and barophilic deep-sea bacteria. Appl Environ Microbiol 63:2105–2108

Deming JW, Hada H, Colwell RR, Luehrsen KR, Fox GE (1984) The nucleotide sequence of 5S rRNA from two strains of deep-sea barophilic bacteria. J Gen Microbiol 130:1911–1920

Deming JW, Somers LK, Straube WL, Swartz DG, Macdonell MT (1988) Isolation of an obligately barophilic bacterium and description of a new genus, *Colwellia* gen. nov. System Appl Microbiol 10:152–160

De Poorter LMI, Suzaki Y, Sato T, Tamegai H, Kato C (2004) Effects of pressure on the structure and activity of isopropylmalate dehydrogenases from deep-sea *Shewanella* species. Mar Biotechnol 6:s190–s194

Eloe EA, Malfatti F, Gutierrez J, Hardy K, Schmidt WE, Pogliano K, Pogliano J, Azam F, Bartlett DH (2011) Isolation and characterization of a psychropiezophilic Alphaproteobacterium. Appl Environ Microbiol 77:8145–8153

Fang JS, Barcelona MJ, Nogi Y, Kato C (2000) Biochemical implications and geochemical significance of novel phospholipids of the extremely barophilic bacteria from the Marianas Trench at 11,000 m. Deep-Sea Res Part I 47:1173–1182

Fang JS, Chan O, Kato C, Sato T, Peeples T, Niggemeyer K (2003) Phospholipid FA of piezophilic bacteria from the deep sea. Lipids 38:885–887

Ikemoto E, Kyo M (1993) Development of microbiological compact mud sampler. Jpn Mar Sci Technol Res 30:1–16

Kato C (1999) Barophiles (piezophiles). In: Horikoshi K, Tsujii K (eds) Extremophiles in deep-sea environments. Springer, Tokyo, pp 91–111

Kato C, Sato T, Horikoshi K (1995) Isolation and properties of barophilic and barotolerant bacteria from deep-sea mud samples. Biodiv Conserv 4:1–9

Kato C, Li L, Nakamura Y, Nogi Y, Tamaoka J, Horikoshi K (1998) Extremely barophilic bacteria isolated from the Mariana Trench, Challenger Deep, at a depth of 11,000 meters. Appl Environ Microbiol 64:1510–1513

Kato C, Nakasone K, Qureshi MH, Horikoshi K (2000) How do deep-sea microorganisms respond to changes in environmental pressure? In: Storey KB, Storey JM (eds) Cell and molecular response to stress, vol 1. Environmental stressors and gene responses. Elsevier Science BV, Amsterdam, pp 277–291

Kawano H, Takahashi H, Abe F, Kato C, Horikoshi K (2009) Identification and characterization of two alternative sigma factors of RNA polymerase in the deep-sea piezophilic bacterium *Shewanella violacea*, strain DSS12. Biosci Biotechnol Biochem 73:200–202

Lauro FM, Chastain RA, Blankenship LE, Yayanos AA, Bartlett DH (2007) The unique 16S rRNA genes of piezophiles reflect both phylogeny and adaptation. Appl Environ Microbiol 73:838–845

MacDonell MT, Colwell RR (1985) Phylogeny of the Vibrionaceae, and recommendation for two new genera, *Listonella* and *Shewanella*. Syst Appl Microbiol 6:171–182

Margesin R, Nogi Y (2004) Psychropiezophilic microorganisms. Cell Mol Biol 50:429–436

Methé BA, Nelson KE, Deming JW, Momen B, Melamud E, Zhang X, Moult J, Madupu R, Nelson WC, Dodson RJ, Brinkac LM, Daugherty SC, Durkin AS, DeBoy RT, Kolonay JF, Sullivan SA, Zhou L, Davidsen TM, Wu M, Huston AL, Lewis M, Weaver B, Weidman JF, Khouri H, Utterback TR, Feldblyum TV, Fraser CM (2005) The psychrophilic lifestyle as revealed by the genome sequence of *Colwellia psychrerythraea* 34H through genomic and proteomic analyses. Proc Natl Acad Sci USA 102:10913–10918

Miyazaki M, Nogi Y, Fujiwara Y, Horikoshi K (2008) P*sychromonas japonica* sp. nov., *Psychromonas aquimarina* sp. nov., *Psychromonas macrocephali* sp. nov. and *Psychromonas ossibalaenae* sp. nov., psychrotrophic bacteria isolated from sediment adjacent to sperm whale carcasses off Kagoshima, Japan. Int J Syst Evol Microbiol 58:1709–1714

Mountfort DO, Rainey FA, Burghardt J, Kasper F, Stackebrant E (1998) *Psychromonas antarcticus* gen. nov., sp. nov., a new aerotolerant anaerobic, halophilic psychrophile isolated from pond sediment of the McMurdo ice shelf, Antarctica. Arch Microbiol 169:231–238

Nakasone K, Ikegami A, Kato C, Usami R, Horikoshi K (1998) Mechanisms of gene expression controlled by pressure in deep-sea microorganisms. Extremophiles 2:149–154

Nakasone K, Ikegami A, Kawano H, Usami R, Kato C, Horikoshi K (2002) Transcriptional regulation under pressure conditions by the RNA polymerase s54 factor with a two component regulatory system in *Shewanella violacea*. Extremophiles 6:89–95

Nogi Y, Kato C (1999) Taxonomic studies of extremely barophilic bacteria isolated from the Mariana Trench, and *Moritella yayanosii* sp. nov., a new barophilic bacterial species. Extremophiles 3:71–77

Nogi Y, Kato C, Horikoshi K (1998a) *Moritella japonica* sp. nov., a novel barophilic bacterium isolated from a Japan Trench sediment. J Gen Appl Microbiol 44:289–295

Nogi Y, Kato C, Horikoshi K (1998b) Taxonomic studies of deep-sea barophilic *Shewanella* species, and *Shewanella violacea* sp. nov., a new barophilic bacterial species. Arch Microbiol 170:331–338

Nogi Y, Masui N, Kato C (1998c) *Photobacterium profundum* sp. nov., a new, moderately barophilic bacterial species isolated from a deep-sea sediment. Extremophiles 2:1–7

Nogi Y, Kato C, Horikoshi K (2002) *Psychromonas kaikoae* sp. nov., a novel piezophilic bacterium from the deepest cold-seep sediments in the Japan Trench. Int J Syst Evol Microbiol 52:1527–1532

Nogi Y, Hosoya S, Kato C, Horikoshi K (2004) *Colwellia piezophila* sp. nov., isolation of novel piezophilic bacteria from the deep-sea fissure sediments of the Japan Trench. Int J Syst Evol Microbiol 54:1627–1631

Nogi Y, Hosoya S, Kato C, Horikoshi K (2007) *Psychromonas hadalis* sp. nov., a novel piezophilic bacterium isolated from the bottom of the Japan Trench. Int J Syst Evol Microbiol 57:1360–1364

Ohmae E, Gekko K, Kato C (2015) Environmental adaptation of dihydrofolate reductase from deep-sea bacteria. In: Akasaka K, Matsuki H (eds) High pressure bioscience—basic concepts, applications and frontiers. Springer, Berlin, pp 423–442

Owen R, Legros RM, Lapage SP (1978) Base composition, size and sequence similarities of genome deoxyribonucleic acids from clinical isolates of *Pseudomonas putrefaciens*. J Gen Microbiol 104:127–138

Seo HJ, Bae SS, Lee J-H, Kim S-J (2005) *Photobacterium frigidiphilum* sp. nov., a psychrophilic, lipolytic bacterium isolated from deep-sea sediments of Edison Seamount. Int J Syst Evol Microbiol 55:1661–1666

Stelling SC, Techtmann SM, Utturkar SM, Alshibli NK, Brown SD, Hazen TC (2014) Draft genome sequence of *Thalassotalea* sp. strain ND16A isolated from eastern Mediterranean Sea water collected from a depth of 1,055 meters. Genome Announcement 2:e01231-14

Tamegai H, Kawano H, Ishii A, Chikuma S, Nakasone K, Kato C (2005) Pressure-regulated biosynthesis of cytochrome bd in piezo- and psychrophilic deepsea bacterium *Shewanella violacea* DSS12. Extremophiles 9:247–253

Urakawa H, Kita-Tsukamoto K, Steven SE, Ohwada K, Colwell RR (1998) A proposal to transfer *Vibrio marinus* (Russell 1891) to a new genus *Moritella* gen. nov. as *Moritella marina* comb. nov. FEMS Microbiol Lett 165:373–378

Vezzi A, Campanaro S, D'Angelo M, Simonato F, Vitulo N, Laauro FM, Cestaro A, Malacrida G, Simionati B, Cannata N, Romualdi C, Bartlett DH, Valle G (2005) Life at depth: *Photobacterium profundum* genome sequence and expression analysis. Science 307:1459–1461

Xiao X, Wang P, Zeng X, Bartlett DH, Wang F (2007) *Shewanella psychrophila* sp. nov. and *Shewanella piezotolerans* sp. nov., isolated from west Pacific deep-sea sediment. Int J Syst Evol Microbiol 57:60–65

Xu Y, Nogi Y, Kato C, Liang Z, Rüger H-J, Kegel DD, Glansdorff N (2003a) *Psychromonas profunda* sp. nov., a psychropiezophilic bacterium from deep Atlantic sediments. Int J Syst Evol Microbiol 53:527–532

Xu Y, Nogi Y, Kato C, Liang Z, Rüger H-J, Kegel DD, Glansdorff N (2003b) *Moritella profunda* sp. nov. and *Moritella abyssi* sp. nov., two psychropiezophilic organisms isolated from deep Atlantic sediments. Int J Syst Evol Microbiol 53:533–538

Yayanos AA (1986) Evolutional and ecological implications of the propertiesof deep-sea barophilic bacteria. Proc Natl Acad Sci U S A 83:9542–9546

Yayanos AA (1995) Microbiology to 10,500 meters in the deep sea. Annu Rev Microbiol 49:777–805

Yayanos AA, DeLong EF (1987) Deep-sea bacterial fitness to environmental temperatures and pressure. In: Jannasch HW, Marquis RE, Zimmerman AM (eds) Current perspectives in high pressure biology. Academic, Toronto, pp 17–32

Yayanos AA, Dietz AS, Van Boxtel R (1979) Isolation of a deep-sea barophilic bacterium and some of its growth characteristics. Science 205:808–810

Yayanos AA, Dietz AS, Van Boxtel R (1981) Obligately barophilic bacterium from the Mariana Trench. Proc Natl Acad Sci U S A 78:5212–5215

ZoBell CE, Johnson FH (1949) The influence of hydrostatic pressure on the growth and viability of terrestrial and marine bacteria. J Bacteriol 57:179–189

Chapter 8
Microbial Life in Permafrost

Ianina Altshuler, Jacqueline Goordial, and Lyle G. Whyte

Abstract Permafrost is a hostile environment that harbors a diverse and active microbial community. Next generation sequencing studies have demonstrated a wide diversity of microorganisms present in Arctic, Antarctic and high altitude permafrost soils. In situ activity of these microorganisms has been demonstrated through multiple lines of evidence. Radiolabeled studies and stable isotope probing have established that active respiration and DNA replication occur in permafrost soils under frozen conditions. Furthermore, microorganisms capable of subzero growth have been isolated from permafrost samples. These isolates have adapted to the permafrost environment through a multitude of molecular changes, such as increased expression of cold shock and metabolite transport proteins, reduced fatty acid saturation in the membrane, and presence of temperature specific isozymes. Recent studies have focused on permafrost thaw due to anthropogenic climate change. The subsequent thaw of frozen organic carbon stores in permafrost is thought to increase microbial activity and emissions of greenhouse gases to the atmosphere. As the permafrost thaws, the microbial community changes in terms of diversity and functional potential in response to warmer temperatures, and increased carbon and water availability.

Contents

I. Altshuler • L.G. Whyte (✉)
Department of Natural Resource Sciences, McGill University, Macdonald Campus, 21111 Lakeshore Road, Sainte-Anne-de-Bellevue, QC, Canada, H9X 3V9
e-mail: lyle.whyte@mcgill.ca

J. Goordial
Department of Natural Resource Sciences, McGill University, Macdonald Campus, 21111 Lakeshore Road, Sainte-Anne-de-Bellevue, QC, Canada, H9X 3V9

Bigelow Laboratory for Ocean Sciences, Deep Biosphere Lab, 60 Bigelow Drive, East Boothbay, ME 04555, USA

R. Margesin (ed.), *Psychrophiles: From Biodiversity to Biotechnology*,
DOI 10.1007/978-3-319-57057-0_8

8.1 Introduction

Permafrost represents a large and extremely challenging environment for microbial life, covering 27% of the terrestrial surface on Earth, in which microorganism must cope with multiple environmental stressors (Goordial et al. 2013). Freezing temperatures, low kinetic energy, and low water and carbon availability limit microbial growth (Nikrad et al. 2016). Despite this, permafrost harbors a diverse microbial community that is viable and metabolically active. Furthermore, there has been a surge of interest into the diversity and activity of permafrost microorganisms as the permafrost thaws due to anthropogenic climate change.

Permafrost microorganisms are generally characterized as cryophiles, organisms that can sustain growth and reproduction at low temperatures ranging from −17 °C to +10 °C (D'Amico et al. 2006; De Maayer et al. 2014). In addition to permanently subfreezing temperatures, any microbial life within permafrost must be able to survive the often oligotrophic conditions, background radiation on geological timescales, and limited liquid water activity; any liquid water present in permafrost is thought to exist in saline brine veins or in special saline niche environments such as cryopegs (Gilichinsky et al. 2003). As such, any organisms which are able to survive these conditions are often polyextremophilic. While many microorganisms are either non-viable or dormant under these harsh conditions, there is clear evidence that globally, diverse and abundant microbial communities within permafrost can be viable and active in situ. In addition to in situ measurements indicative of microbial growth, psychrophilic organisms have been successfully isolated from permafrost environments, with subzero growth observed in the laboratory. This chapter discusses the diversity of microorganisms found in permafrost globally, as well as evidence for viability and activity within the permafrost environment. Finally, we discuss the issue of climate change and how microbial communities in permafrost are expected to respond, with an emphasis on methane dynamics based on research to date.

8.2 Permafrost Environments

Permafrost, defined as ground material that is at ≤0 °C for two or more consecutive years (van Everdingen 1998), is found primarily in the Arctic, sub-Arctic, and Antarctic regions, as well as in alpine regions (for example in the Qinghai-Tibet Plateau, South America, and Sweden) (Bockheim and Munroe 2014). Perennially frozen permafrost is most often overlaid by a seasonally thawed active layer, the depth of which is dependent on air temperatures, moisture content, vegetation, and snow cover (Tarnocai 1980). Permafrost can be hundreds of meters thick (e.g., over 500 m in Siberia), while the active layers can range between a few centimeters to several meters in depth (Tarnocai 1980). A transition zone exists between permafrost and the active layer, which acts as a temperature buffer and fluctuates between being seasonally frozen and perennially frozen over decadal time scales (Shur et al. 2005). Permafrost is a heterogeneous environment due to frost heave, front sorting, and cryoturbation, processes which disrupt and mix different soil horizons, sometimes transferring organic carbon from surface to deeper layers and often creating patterned ground (van Everdingen 1998; Hugelius and Kuhry 2009).

Permafrost soils can also contain different geomorphological features such as ice-wedges, taliks, cryopegs, massive ground ice, frost boils, thermokarst lakes, organic accumulations, and broken soil horizons, each providing a unique habitat for microbial growth (Fig. 8.1). Ice-wedge polygons are formed when the frozen ground contracts during the cold winter months and cracks the surface, dividing it into polygonal blocks (Kerfoot 1972). During the spring, these cracks are filled with snow melt water, which freezes and forms ice-wedges. As the temperature rises, the ground between the cracks expands and elevates, forming the polygon center (Kerfoot 1972). The ice-wedges are overlaid with active layer soil, creating the polygon trough (Shur et al. 2005). Over subsequent years of freeze and thaw cycles,

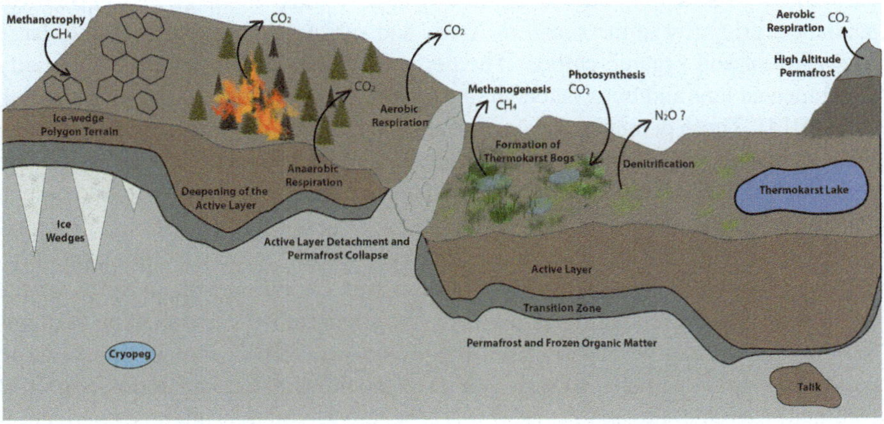

Fig. 8.1 Schematic diagram of permafrost, including microbial and abiotic processes involved in permafrost thaw and GHG emissions

the ice-wedges grow, resulting in high centered polygon surrounded by lower troughs. Taliks are unfrozen masses of ground soil found within the permafrost; often they are located under lakes due to the water's ability to transfer heat (Shur and Jorgenson 2007). Cryopegs are supercooled groundwater brine lenses that remain liquid at below 0 °C due to their high salt content (Gilichinsky et al. 2003). Along with a very steep temperature gradient, soluble nutrients, such as nitrogen, phosphorus, calcium, magnesium, and potassium, can also potentially form a gradient in the soil. As the nutrients are solubilized in meltwater, they move along the thermal gradient, enriching the top frozen permafrost layer (Kokelj and Burn 2005; Tarnocai 2009). This results in the upper layer of permafrost becoming a sink of soluble materials and nutrients during permafrost formation and can then conversely become a source of these nutrients during permafrost degradation (Kokelj and Burn 2005).

Arctic permafrost overall is a heterogeneous environment with both mineral and organic soils. The polar deserts of High Arctic permafrost tend to harbor mineral permafrost, without an organic layer over the mineral horizons (Steven et al. 2006; Shur and Jorgenson 2007; Tarnocai et al. 2009). However, the Arctic does contain large stores of organic carbon frozen in permafrost. Pleistocene Yedoma permafrost deposits in the North East Siberia are rich in organic carbon and were formed by windblown dust and sediment deposits that were subsequently frozen during the glacial age (Zimov et al. 2006; Tarnocai 2009; Vonk et al. 2013). These deposits contain large ice-wedges and well-preserved organic material, which constitutes roughly one-third of organic carbon stored in permafrost globally (Vonk et al. 2013). Due to the cold climate during the formation of the deposits, the organic carbon is well preserved and readily available for biological degradation (Schuur et al. 2008; Vonk et al. 2013). A prevalent feature of the Yedoma permafrost deposits is the high water content present in the form of ice veins and large ice-wedges that account for ~50% of the soil content (Vonk et al. 2013). Due to the high ice content, the Yedoma permafrost deposits are highly susceptible to permafrost degradation due to climate change (Vonk et al. 2013). Permafrost peatlands, primarily in the southern Arctic and sub-Arctic regions, also hold large reservoirs of soil organic carbon. The permafrost deposits in these areas initially developed as fens and bogs that contained woody plants, mosses, and sages (Routh et al. 2014). These perennially frozen peatlands are vulnerable to climate warming in the Arctic and sub-Arctic regions; the melting of ice in the permafrost will lead to degradation of the peatlands and biodegradation of the currently sequestered carbon (Tarnocai 2006).

Antarctic permafrost is less studied than its Arctic counterpart; however, Antarctic permafrost represents a much more extreme environment in terms of the combined freezing temperatures, aridity, and oligotrophy. The Antarctic contains 37% of the world's permafrost (Bockheim and Hall 2002), though compared to Arctic and alpine permafrost, relatively little is known about Antarctic permafrost communities (Goordial and Whyte 2014). The majority of the Antarctic continent is snow and ice covered, with only 0.35% of exposed ground on which permafrost occurs (Campbell and Claridge 2009). The McMurdo Dry Valleys' polar desert represents the largest ice-free area in the Antarctic and has been the focus of most permafrost studies to date (Gilichinsky et al. 2007; Goordial et al. 2016). Trough-like depressions may be underlain by sand wedges instead of ice, though ice veins and ice lenses may also be found within sand wedge structures (Bockheim et al. 2009). The McMurdo Dry Valleys receive very low annual precipitation and is the only known place on Earth where dry permafrost is found (defined as permafrost which contains less than $<3\%$ water by mass) which forms from sublimation of ice-cemented permafrost over time (Bockheim et al. 2007). Ice-cemented permafrost is found primarily on the coastal areas and younger surfaces where the geography facilitates drainage and therefore results in wetter soils. Dry permafrost is found overlaying ice-cemented permafrost and is found at higher elevations and in older inland arid areas (Campbell and Claridge 2006). In some higher elevation valleys, there is an absence of any active layer that rises above 0 °C seasonally (Marinova et al. 2013). The majority of the water in the permafrost is frozen and the possibility for brine veins or thin films of water depends on temperature, solute concentration, and distance from the coast. Coastal and lower elevation valleys receive higher salt influx and have higher soil chlorine concentrations, as well as have a larger number of thaw days throughout the year (Goordial et al. 2016). As a result, higher elevation valleys, such as University Valley, do not have sufficient solute concentrations or temperatures to form thin films of water for more than a few hours a year (Goordial et al. 2016).

In addition to the poles, permafrost is also present at high altitudes. The largest amount of high altitude (alpine) permafrost is located in China (Ran et al. 2012) with the Tibetan plateau being the largest alpine permafrost region (Chen et al. 2016). Temperate mountain permafrost soils have well-drained coarse sediments, steep slopes that have higher spatial and geothermal variability, more variable snow distribution, lower influence of vegetation, and warmer mean annual temperatures compared to Arctic and Antarctic permafrost (Haeberli and Gruber 2009; Frey et al. 2016). Furthermore, incoming solar radiation is moderated by the slope and shading (Etzelmüller 2013). In more coastal areas, alpine permafrost tends to be located above the tree line; however in continental areas, forests may promote permafrost development (Etzelmüller 2013). Recently, high altitude permafrost has garnered more attention since, in the last decade, monitoring of alpine permafrost has shown warming across the globe, but particularly in colder regions, with unknown consequences on geotechnical stability (Haeberli and Gruber 2009; Etzelmüller 2013).

8.3 Microbial Diversity and Abundance in Permafrost

Despite being a hostile environment, permafrost does harbor a microbial community with 10^5–10^9 cells g^{-1} in Arctic permafrost, 10^{10}–10^8 in alpine permafrost, and a lower amount of 10^3–10^6 cells g^{-1} in Antarctic permafrost (Vishnivetskaya et al. 2006; Gilichinsky et al. 2007; Hansen et al. 2007; Steven et al. 2007; Blanco et al. 2012; Hu et al. 2015; Goordial et al. 2016). Across studies in alpine and polar regions, the dominant groups present in permafrost soils tend to be Actinobacteria, Bacteroidetes, Proteobacteria (Alpha- and Beta-primarily but also Delta- and Gamma-Proteobacteria), Firmicutes, Chloroflexi, and Acidobacteria (Steven et al. 2008; Yergeau et al. 2010; Wilhelm et al. 2011; Deng et al. 2015; Stackhouse et al. 2015) (Table 8.1). The fermentative members of Chloroflexi and Bacterioidetes increase with soil depth indicating anaerobic carbon degradation in permafrost (Deng et al. 2015). Archaea and fungi are present, though 200–1000 times lower in abundance (Yergeau et al. 2010; Stackhouse et al. 2015; Frey et al. 2016). Abundant archaeal groups in permafrost are often related to halophilic archaea, part of Euryarchaeota, likely due to the brine veins thought to host active microbial life within permafrost soils (Steven et al. 2007). However, Crenarchaeota appear to dominate acidic wetland permafrost (Wilhelm et al. 2011). Under frozen permafrost conditions, based on RNA data, the most transcriptionally active organisms appear to be the same as the dominant phyla: Proteobacteria, Firmicutes, Acidobacteria, and Actinobacteria, as well as Euryarchaeota and ascomycetous fungi (Coolen and Orsi 2015) (Table 8.1). Furthermore, permafrost contains active sulfate and Iron (III) reducers, nitrifying bacteria, methanotrophs, and methanogens (Yergeau et al. 2010; Stackhouse et al. 2015). This is reflected in the presence of genes, transcripts, and proteins involved in sulfate reduction, Iron(III) reduction, nitrogen cycle, methanogenesis, and methane oxidation found in the permafrost (Yergeau et al. 2010; Mackelprang et al. 2011; Hultman et al. 2015). Functional groups of bacteria in high altitude permafrost also include ammonia-oxidizing bacteria and archaea, methane-oxidizing bacteria, nitrifying bacteria, nitrogen-fixing rhizobial symbiont bacteria, sulfur- and sulfate-reducing bacteria, and thiosulfate-oxidizing bacteria (Wu et al. 2012; Zhang et al. 2009; Yun et al. 2014; Hu et al. 2016). A unique study of the northern slope of Mount Everest showed an interesting comparison of ammonia oxidizers above and below 5800 m a.s.l.: the soil above this threshold is permafrost and is characterized by increasingly colder temperatures, stronger radiation, lower oxygen concentration, and lower nutrients with increase in altitude (Zhang et al. 2009). The increase in altitude also decreased the abundance of both archaeal and bacterial ammonia oxidizers; however, at lower elevation, the soils were dominated by archaeal ammonia oxidizers, but were replaced by bacterial ammonia oxidizers in the higher altitude permafrost soils (Zhang et al. 2009). Overall, caution should be used when taking the results of higher altitude permafrost microbial abundance and function studies at face value, since authors in these papers often confusingly refer to active layer soils as permafrost.

Overall, permafrost contains genes involved in response to stressors, pathogenicity, toxicity, degradation of carbon compounds, methanogenesis, methane oxidation, denitrification, nitrogen fixation, ammonia assimilation, and sulfate reduction (Yergeau et al. 2010; Hultman et al. 2015) and have shown presence of proteins involved in chemotaxis and motility (Hultman et al. 2015). This result is also supported by the isolation of motile permafrost microbes from brine veins containing flagella (Shcherbakova et al. 2005). Compared to the overlaying active layer, the permafrost contains lower abundance of transporter proteins and transcripts, but higher presence of cold-shock proteins and other stress response genes

Table 8.1. Table outlining the microbial members of permafrost community present, active, and cultivable in Arctic, Antarctic and in high altitude permafrost

Permafrost microbial member	(sub)Arctic				Antarctic				High altitude			
	Present[a]	In situ active[b]	Cultivable isolates	References	Present[a]	In situ active[b]	Cultivable isolates	References	Present[a]	In situ active[b]	Cultivable isolates	References
Proteobacteria	+	+	+	1–15	+		+	16–19	+		+	20–24
Alpha	+	+	+	1, 4, 9, 13, 14, 15	+		+	16, 17	+		+	20–23
Beta	+	+	+	1, 4, 9, 12, 14, 15	+		+	16, 17	+		+	20–23
Gamma	+		+	3, 4, 9–15	+		+	16, 17, 25	+		+	20–23
Delta	+			3,4,6,9,13–15					+			20, 21
Firmicutes	+	+	+	1,3,5-9, 11,13,14,26,27	+		+	16-18	+			20–23
Acidobacteria	+	+		1, 2, 5, 7, 8, 14, 15	+			19	+			21, 24
Actinobacteria	+	+	+	1-3, 5, 7, 8, 9, 12-15, 28	+		+	16, 17, 19	+			20–24
Chloroflexi	+	+		6, 7, 13–15	+			19	+			21, 24
Gemmatimonadetes	+			2, 4, 7, 14, 15	+			19	+			24
Verrucomicrobia	+	+		2, 7, 14, 15	+			19	+			21, 24
Bacteroidetes	+		+	1, 3-7, 9, 13–15	+			16, 17, 19	+		+	22–24, 29
Planctomycetes	+	+		4, 7	+			19	+			21, 24
OP11 (Microgenomates)	+			7					+			24
OP5 (Caldiserica)									+			24
OD1 (Parcubacteria)									+			24
TM7 (Saccharibacteria)	+			9					+			24
GN02 (Gracilibacteria)									+			24
AD3 candidate division	+			15								
Spirochaetes	+	+		7								
Chlorobi	+	+		7								

(continued)

Table 8.1. (continued)

Permafrost microbial member	(sub)Arctic				Antarctic				High altitude			
	Present[a]	In situ active[b]	Cultivable isolates	References	Present[a]	In situ active[b]	Cultivable isolates	References	Present[a]	In situ active[b]	Cultivable isolates	References
Cyanobacteria									+			21
Nitrospirae									+			21, 24
Euryarchaeota	+	+	+	1, 3, 4, 7, 8, 30–32, 34	+			17	+			20, 33
Crenarchaeota	+			2, 4					+			33
Thaumarchaeota	+			34					+			20
Bathyarchaeota	+			34								
Woesearchaeota	+			34								

[a]DNA based evidence
[b]RNA based evidence
1, Yergeau et al. (2010); 2, Wilhelm et al. (2011); 3, Steven et al. (2008); 4, Steven et al. (2007); 5, Stackhouse et al. (2015); 6, Deng et al. (2015); 7, Hultman et al. (2015); 8, Coolen and Orsi (2015); 9, Hansen et al. (2007); 10, Bakermans et al. (2003); 11, Ponder et al. (2005); 12, Panikov and Sizova (2007); 13, Gittel et al. (2014a); 14, Gittel et al. (2014b); 15, Taş et al. (2014); 16, Gilichinsky et al. (2007); 17, Goordial et al. (2016); 18, Tamppari et al. (2012); 19, Bakermans et al. (2014); 20, Hu et al. (2016); 21, (Yun et al. (2014); 22, Bai et al. (2006); 23, Zhang et al. (2007); 24, Frey et al. (2016); 25, Kim et al. (2012); 26, Shcherbakova et al. (2005); 27, Mykytczuk et al. (2013); 28, Finster et al. (2009); 29, Zhao et al. (2011); 30, (Rivkina et al. (2007); 31, Shcherbakova et al. (2011); 32, Mackelprang et al. (2011); 33, Wei et al. (2014); 34, Shcherbakova et al. (2016)

(Yergeau et al. 2010; Mackelprang et al. 2011; Hultman et al. 2015). Microbiota in mineral permafrost horizons have lower carbon degradation capacity compared to organisms from organic active layers; this can be due to lower taxonomic diversity in oligotrophic conditions found in the permafrost (Ernakovich and Wallenstein, 2015; Hultman et al. 2015). However, the permafrost environment does contain genes for degradation of sugar alcohols; aminosugars; mono-, di-, and oligosaccharides; starch; lignocellulose; chitin; trehalose; and cellulose (Yergeau et al. 2010; Hultman et al. 2015). Furthermore, Actinobacteria is often a dominant bacterial phyla present in permafrost soils (Table 8.1). This phyla is known to contain members that are facultative anaerobic degraders of complex soil organic matter and are adapted to low carbon availability. In addition, permafrost soils have demonstrated potential enzymatic activities of hydrolytic (cellobiohydrolase, endochitinase, *N*-acetylglucosaminidase, and leucine aminopeptidase) and oxidative (phenoloxidase and peroxidase) enzymes (Gittel et al. 2014a, b).

Compared to permafrost soils, ice-wedges and ground ice are relatively low in diversity and abundance of microorganisms (Steven et al. 2008), they reflect the community of the surrounding permafrost soils and include members of Proteobacteria, Actinobacteria, Acidobacteria, Bacterioidetes, and Firmicutes (Steven et al. 2008; Wilhelm et al. 2012).These are habitable cryoenvironments which show evidence of in situ heterotrophic activity based on occluded gas measurements (Katayama et al. 2007; Lacelle et al. 2011; Wilhelm et al. 2012) and contain cultivable microbiota, despite low viable cell counts (Gilichinsky et al. 1995; Katayama et al. 2007; Lacelle et al. 2011). Soil particles suspended in the ice-wedges and brine veins are thought to serve as refugia for microorganisms and protect the cells from ice crystals.

Specifically in the Antarctic, the microbial community consists of Proteobacteria (Gammaproteobacteria) (Gilichinsky et al. 2007) (as well as members of *Nitrospina*, green non-sulfur bacteria and relatives, *Fibrobacter*, *Acidobacterium*, and the Flexibacter–Cytophaga–Bacteroides group) and low levels of anaerobes including denitrifying bacteria, methanogens, and sulfate reducers. Culturable isolates belonged to Alpha-, Beta-, and Gamma-Proteobacteria; Actinobacteria; Firmicutes; and Methylobacterium; as well as mycelial fungi and yeast groups (Gilichinsky et al. 2007). Interestingly, unicellular green algae including *Mychonastes* sp. (Chlorellaceae), *Chlorococcum* sp. (Chlorococcaceae), and *Chlorella* sp. (Chlorellaceae) have also been isolated from Antarctic permafrost (Gilichinsky et al. 2007). The McMurdo Dry Valleys' permafrost ranges from cold and dry in the coastal Taylor Valley (TV; subxerous; 1–75 mm precipitation; summer temperatures sometimes above 0 °C in the Valley) to extremely cold, extremely dry, and highly oligotrophic in high inland University Valley (UV; ultraxerous; average summer temperature −5 °C in the Valley; 0.01−0.05% total carbon, undetectable to 0.09% total nitrogen) (Tamppari et al. 2012; Goordial et al. 2016). The diversity of the microbiota is also reflected between the two valleys. The wetter TV permafrost contains a higher abundance of microbes compared to the drier and colder UV (Tamppari et al. 2012). For example, TV permafrost showed presence of both anaerobic and aerobic microbial groups, whereas in the UV permafrost the anaerobic microbial groups were undetected based on PLFA data; however, later sequencing data did show presence of anaerobes in UV permafrost as well; this is likely due to TV permafrost being "wetter" and therefore potentially having anoxic conditions near the permafrost table (Tamppari et al. 2012; Goordial et al. 2016). Furthermore, the dominant microbial groups in TV are Proteobacteria, sulfate-reducing bacteria, anaerobic metal reducers, Acidobacteria, Gemmatimonadetes, and Firmicutes, whereas in UV the dominant groups are Gamma- and Beta-Proteobacteria, though Firmicutes, Actinobacteria (including Actinomycetes), and Bacteroidetes were also present (Tamppari et al. 2012; Bakermans et al. 2014; Goordial et al. 2016).

The permafrost microbial community in high elevation permafrost appears to be dominated by bacterial phyla of Proteobacteria and Actinobacteria and in one case Patescibacteria superphyla (Hu et al. 2016; Frey et al. 2016), although Actinobacteria, Proteobacteria, and Chloroflexi dominate high altitude wetland permafrost (Yun et al. 2014). Arachaeal diversity is dominated by Thaumarchaeota (Hu et al. 2016). Culturable isolates in high altitude permafrost belong to Arthrobacter; Pseudomonas; Alpha-, Beta-, and Gamma-Proteobacteria; Firmicutes; and CFB (Bai et al. 2006; Zhang et al. 2007; Hu et al. 2015). Interestingly, the majority of the phylotypes of bacterial and archaeal origin were less than 97% similar to previously isolated strains (Hu et al. 2016). Frey et al. (2016) conducted an elegant study on alpine permafrost and found that the permafrost compared to the nearby non-permafrost soils was highly enriched in uncultured bacteria. The uncultured members were part of candidate phyla OD1 (proposed Parcubacteria), TM7 (Saccharibacteria), GN02 (Gracilibacteria), OP11 (Microgenomates), OP5 (Caldiserica), SR1, MVP-21, WS5, and Kazan-3B-28. The candidate phyla OD1, TM7, GN02, and OP11 are part of a proposed superphylum Patescibacteria (Frey et al. 2016). Members of Patescibacteria have very limited biosynthesis abilities of key macronutrients and are hypothesized to lead a semi parasitic or symbiotic lifestyle by attaching to the surface of other cells. These bacteria can be characterized by their low C–G content, small genome size, and are found in anoxic environments (Frey et al. 2016). Such features may embed a selective advantage to this superphylum in the harsh, oxygen-limited permafrost environment.

8.4 Are They Livin' or Just Chillin'?

It may be intuitive to imagine that the microbes in the harsh frozen permafrost environments are in a state of dormancy and stasis. However, microbes in permafrost are biologically active. The active layer of the permafrost does boast higher microbial activity and microbial diversity; however, despite the harsh permafrost conditions, active microbes are still found within the permafrost (Yergeau et al. 2010; Coolen and Orsi 2015; Mackelprang et al. 2016). There are several complimentary lines of evidence for this: subzero respiration under aerobic and anaerobic conditions measured, isotope incorporation and mineralization under frozen conditions in permafrost soils, isolation of permafrost microbes capable of subzero growth, and evidence from RNA based studies

8.4.1 Soil Respiration and Laboratory Incorporation Studies

Respiration and subsequent release of gases has been measured in frozen soils in situ and in laboratory studies. Some of the early evidence that microbes are living and are active in permafrost soils came from measurements of CO_2 and CH_4 release to the atmosphere from wintertime frozen tundra and peat bog soils (Fahnestock et al. 1999; Panikov and Dedysh 2000; Elberling and Brandt 2003). However, these in situ soil respiration studies should be interpreted with caution since the gas emissions could be due to releases of trapped gas. Laboratory experiments using permafrost soils have also demonstrated respiration at freezing temperatures as low as −2 °C (Michaelson and Ping 2003), −4 °C (Larsen et al. 2002), −16 °C (Panikov and Dedysh 2000), −18 °C (Elberling and Brandt 2003), −39 °C (controversial) (Panikov et al. 2006), and in Antarctic soils down to −5 °C (Bakermans et al. 2014). Overall, respiration in frozen soils is dependent on water availability, temperature, and carbon content of the soils (Panikov and Dedysh 2000; Michaelson and Ping 2003; Öquist et al. 2009).

Respiration and heterotrophic activity under subzero conditions on environmental permafrost samples has also been shown in studies using highly sensitive radiolabeling microbial activity assays. For example, respiration in permafrost soils, active layer soil, and ice-wedges has been demonstrated using C^{14} labeled glucose and acetate as substrates (Rivkina et al. 2000; Steven et al. 2008; Wilhelm et al. 2012). The briny habitat existing within cryopegs is also thought to host active microorganisms as activity down to −15 °C was demonstrated via uptake of C^{14} labeled glucose (Gilichinsky et al. 2003). Massive ground ice on the other hand was not found to host an actively mineralizing community (Steven et al. 2008). Similarly, the highly unique permafrost in the high elevation Dry Valleys do not appear to respire under frozen conditions in laboratory or in situ, as demonstrated in University Valley soils (Goordial et al. 2016). In this instance, the combination of cold temperature (mean annual temp −23 °C with no days above freezing), aridity (<10 mm snow melt equivalent per year), and low salinity to facilitate brine veins within the permafrost (Goordial et al. 2016) all contribute to an inactive community. In contrast, the permafrost in the more coastal, and therefore wetter and warmer, Taylor Valley of McMurdo Dry Valleys does show active respiration at subzero conditions (Bakermans et al. 2014). These incorporation studies highlight environmental samples and conditions which favor active microbial life—namely the presence of liquid water, facilitated through the presence of solutes or salt. While these studies show microbial activity under ambient permafrost conditions, they do not show active DNA replication and growth, and do not reveal which microbial community members are active in the subzero permafrost environment.

8.4.2 Stable Isotope Probing in Permafrost Soils

Stable isotope probing (SIP) is used to label the DNA of actively dividing microbes. In an elegantly designed study, Tuorto et al. (2014) were able to demonstrate microbial genome replication in permafrost soils in temperatures ranging from 0 °C to −20 °C using C^{13}-acetate. The active community members that were able to perform DNA replication in frozen permafrost conditions were part of Acidobacteria, Actinobacteria, Chloroflexi, Gemmatimonadetes and Proteobacteria phyla and were distantly subzero related to isolated psychrophilic strains that are able to grow at subzero temperatures (Tuorto et al. 2014). Firmicutes were not detected by SIP in this study, suggesting that spore-forming members of this phyla are dormant and non-metabolically active in an ambient permafrost environment, and their near ubiquitous detection in molecular studies may be due to the increased longevity of spores. Some members of the permafrost community were actively growing across all temperatures, whereas some were limited to specific niche temperature ranges; for example, uncultured members of Actinobacteria and Proteobacteria were only able to synthesize DNA between −9 °C and −20 °C. This could be due to changes in solute concentrations and available water, permitting only the growth of microorganisms that are adapted to very specific niche conditions (Tuorto et al. 2014).

8.4.3 RNA/DNA Ratios

Ratios of RNA/DNA can be used to infer metabolic activity of the community or a particular group of microbes. Microbial community members with higher RNA/DNA ratios are thought to be more metabolically active in the environment, as the active microbes would be synthesizing higher amounts of RNA per cell compared to inactive ones, while the DNA content would remain static regardless of activity (Eriksson et al. 2001; Blazewicz et al. 2013). Hultman et al. (2015) looked at the ratios of RNA transcripts in metatranscriptomes (MT) to the DNA in metagenomes (MG) of permafrost soils to determine that the most active groups of microbes in permafrost were Proteobacteria, Acidobacteria, and Firmicutes, suggesting that these members of the community were more acclimated to life at subzero temperatures (Hultman et al. 2015). Furthermore, the study showed that permafrost had a higher ration of methanogenesis and methane oxidation transcripts (RNA) to genes (DNA) than the active layer, suggesting that the permafrost had relatively higher ratio of active methanogens and methanotrophs compared to the active layer. However, the active layer appeared to have a higher RNA/DNA ratio of *nif* genes, suggesting that it harbors a more active community of nitrogen fixers (Hultman et al. 2015). Schostag et al. (2015) also performed a DNA and RNA based analysis though only on the active layer of permafrost; their study spanned the winter and the summer season allowing the comparison of the same soils in frozen and thawed conditions. The copy number of rRNA genes and transcripts did not fluctuate between the two seasons, suggesting that a similar relative abundance of microbes continue to be active during the winter with the soil temperatures below −10 °C (Schostag et al. 2015).

8.4.4 Subzero Growth of Permafrost Isolates

Further evidence that microbes in permafrost are active is the ability of isolated permafrost microorganisms to grow at below freezing temperatures. There have been many permafrost isolates, mainly belonging to the Firmicutes, Actinobacteria, Proteobacteria, and Bacterioidetes phyla (Goordial et al. 2016; Jansson and Taş 2014). However, few are able to sustain subzero growth and those that do have evolved key adaptive strategies. For example, *Psychrobacter cryopegella* is able to grow down to −10 °C and up to 28 °C, with maximum growth at 22 °C. It was isolated from saline cryopegs buried within 40,000 years old Siberian permafrost (Bakermans et al. 2003). The authors demonstrated that growth yield of the isolate peaked at 4 °C; at this temperature, the microbial cells needed the least amount of RNA and proteins to divide. At freezing temperatures, the isolate likely needed to produce more cold acclimation and cold shock proteins as well as initiate other cold adaptive changes; above this temperature, the isolate needed to produce more RNA and proteins due to higher turnover and degradation rate caused by higher temperatures (Bakermans and Nealson 2004). Other Siberian permafrost isolates include a Gram-positive *Exiguobacterium sibiricum* and a Gram-negative *Psychrobacter* sp. 273-4. These isolates were able to grow at –2.5 °C and are thought to cope with the subzero temperatures by lowering fatty acid saturation and chain length in their membranes, changing the composition of exopolysaccharides, and increasing their ice-nucleation activity which is thought to reduce damaging intracellular ice accumulation (Ponder et al. 2005). An obligate anaerobic, spore forming, bacterial isolate that was capable of subzero growth was isolated from a water brine from the Kolyma Lowland region; *Clostridium algoriphilum* grows down to –5 °C, with optimal growth at +5 °C (Shcherbakova et al. 2005). From Alaskan permafrost, Panikov and Sizova (2007) isolated bacterial and fungal members capable of subzero growth. The bacterial isolates *Pseudomonas* sp. 3-2005 and *Arthrobacter* sp. 9-2 grew at temperatures down to −17 °C; *Polaromonas* sp. strain *hydrogenovorans* grew at −1 °C. The fungal members *Leucosporidium* spp. MS-1, -2 and *Geomyces* spp. FMCC-1, -2, -3, -4 were able to grow down to –35 °C; *Mrakia* sp. MS-2 grew down to −12 °C. However, in these isolates, growth between −18 °C and –35 °C was transient and ceased after three weeks, though normal growth dynamics were sustained in subzero growth above −18 °C (Panikov and Sizova 2007). An Antarctic *Psychrobacter* sp. PAMC 21119 isolate capable of subzero growth was isolated from permafrost soil on Barton Peninsula (Kim et al. 2012). The microbe is able to grow down to at least –5 °C, partially due to an increase in production of proteins involved in metabolite transport, proper protein folding, and membrane fluidity (Koh et al. 2016). Two other Antarctic isolates (*Rhodotorula* and *Rhodococcus* spp.) are also able to grow at −10 °C and −5 °C, respectively, and were isolated from extremely dry University Valley Antarctic permafrost (Goordial et al. 2016).

Another bacterial permafrost isolate capable of subzero growth of note is *Planococcus halocryophilus* Or1; it was isolated from high Arctic permafrost and is able to grow down to −15 °C and sustain low level of metabolic activity down to −25 °C (Mykytczuk et al. 2013). Cells grown under the colder temperatures counter-intuitively contained higher levels of saturated fatty acids over branched ones and developed a crust of dense nodular material (Mykytczuk et al. 2013). The genome of the isolate showed adaptations to cold and osmotic stress including cold shock proteins, chaperones, genes involved in regulation, repair mechanisms, osmolyte uptake, and membrane alteration. Genome redundancy in *P. halocryophilus* would also suggest the presence of isozymes that may be adapted to specific temperatures (Mykytczuk et al. 2013). However, despite our ability to culture subzero permafrost isolates, it appears that the uncultivable native consortia of permafrost microorganisms are more adapted to subzero temperatures compared to individual permafrost members currently isolated, based on comparisons of growth yield and utilization of C^{14} labeled ethanol between permafrost isolates and permafrost soils (Panikov and Sizova 2007).

8.4.5 DNA Repair

Another piece of evidence that microorganisms are likely alive, active, and growing in the permafrost environment is their need and ability to repair DNA damage from background ionizing radiation. Natural background radiation can damage a cell's DNA over long geological timescales, such as those encountered in permafrost (Johnson et al. 2007). Completely dormant and inactive bacteria frozen in permafrost would continue to accumulate DNA damage and eventually lose viability; thus any viable microbes recovered from an ancient permafrost environment would need to have been at least minimally metabolically active in order to repair DNA damage during their tenure in the subzero permafrost (Price and Sowers 2004). Unchecked DNA damage will cross link the DNA and/or reduce it into 100 bp fragments within 100,000 to 1 million years in frozen conditions (Poinar et al. 1996; Hansen et al. 2006). To date, there have been several permafrost bacteria isolated from ancient permafrost; for example, *Exiguobacterium sibiricum* was isolated and cultured at subzero temperatures from a depth of 43.6 m from a 2–3 million years old Siberian permafrost (Ponder et al. 2005). Furthermore, experiments on *Psychrobacter cryohalolentis* K5 and *P. arcticus* 273-4 showed that the microbes are able to grow under −15 °C conditions while simultaneously exposed to ionizing radiation (Amato et al. 2010). These microbes were metabolically active, as demonstrated by [^{3}H] thymidine incorporation, and showed that microbes were able to sustain enough metabolic activity to repair DNA damage in the permafrost environment

(Amato et al. 2010). Indeed, RNA sequencing of permafrost under frozen and thawed conditions has also demonstrated a relative overexpression of genes involved in DNA repair mechanisms in frozen permafrost soils (Coolen and Orsi 2015). In addition, Johnson et al. (2007) demonstrated bacterial survival for at least half a million years in sealed permafrost. The authors amplified a long 4 Kbp DNA fragment from ancient permafrost samples (400,000–500,000 years). The fragment was 20-fold longer than ancient DNA from dead plant/animal samples (max amplicon lengths of 100–500 bp) of a similar age, suggesting that an active DNA repair mechanism must have been present to yield such a large DNA fragment from the ancient permafrost (Johnson et al. 2007). Sequence diversity greatly decreased with permafrost age, suggesting select few microorganisms are able to sustain long-term survival in permafrost (Johnson et al. 2007). In younger permafrost (5000–30,000 years), endospore forming bacterial members were shown to accumulate DNA damage; in older permafrost samples (400,000–600,000 years), there was no presence of bacteria with capacity for dormancy; instead, members related to non-spore forming *Arthrobacter* (Actinobacteria) were dominant. In addition, authors were able to show active respiration in these older permafrost samples at ambient permafrost temperatures. Together, these results suggest that microbes in ancient permafrost can sustain viability by maintaining low levels of metabolic activity and DNA repair; bacterial members that are capable of this strategy may outperform bacterial members capable of dormancy in very old permafrost environments (Johnson et al. 2007).

8.5 Live Microbes or Ancient DNA

The advent of next generation sequencing technologies has greatly propelled the study of microbial diversity, giving us a wealth of information and greater insight into the microbial processes that take place in permafrost environments. However, one uncertainty of studying microbial life and ecology of permafrost through novel molecular means is whether the nucleic acids that are isolated from the permafrost represent the current active microbial community in the ground, DNA within cryopreserved cells, spores, or just extracellular DNA and RNA molecules (i.e., eDNA) adsorbed to soil particles (Willerslev et al. 2004a, b; Pietramellara et al. 2009). Do nucleic acids extracted from permafrost represent a frozen snapshot of past life or does it represent current life adapted to a harsh environment? Dry conditions, low temperatures, and salinity promote the persistence of nucleic acids in the environment, with the rate of DNA degradation decreasing by a level of magnitude for every 10 °C drop in temperature (Smith et al. 2001; Willerslev et al. 2004a; Hebsgaard and Willerslev 2009). However, rates of nucleic acid degradation under different environmental factors are not well understood (Hebsgaard and Willerslev 2009), and permafrost conditions are thought to be favorable for long-term microbial and nucleic acid persistence (Johnson et al. 2007). Though DNA fragments (100–500 bp) may not persist in colder environments for more than 10^5 years (Lindahl 1993; Briggs and Summons 2014). DNA in the environment, resting cells, and endospores degrade over time because of chemical hydrolysis and oxidation, eventually becoming non-viable and non-amplifiable due to lack of active DNA repair. However, even minimally metabolically active cells may retain a functioning DNA repair mechanism and persist in the permafrost environment over longer time periods (Price and Sowers 2004; Johnson et al. 2007).

Finally, contamination can never be ruled out, when working with such low quantities and quality of nucleic acids, as in the case of permafrost, the sensitivity of PCR to contaminants becomes problematic. In addition, when we are studying the current microbial community in the permafrost soils, extracellular eDNA from dead cells can obscure the microbial diversity recovered (Carini et al. 2016). On the other hand, when trying to use ancient nucleic acids as fossil molecules for paleodiversity studies and to reconstruct past ecosystems, there is risk of currently active microbes interfering with the results (Bellemain et al. 2013; Briggs and Summons 2014). Bellemain et al. (2013) run into this problem in their fungal paleodiversity study of two Siberian Pleistocene aged permafrost samples. The authors used permafrost samples to infer past ecology and environment of the area and through metabarcoding detected presence of plant-associated fungal taxa and fungal insect pathogens; based on this, they concluded that these fungi were an active component of the Pleistocene environments. However, psychrophilic and psychrotolerant fungal taxa were also detected in their molecular analysis, suggesting presence of metabolically active taxa in the permafrost that were potentially interfering with an accurate paleo-reconstruction (Bellemain et al. 2013). So how is it possible to differentiate between old nucleic acids preserved in the environment and a potentially metabolically active microbial community?

8.5.1 *Strategies for Differentiating Between Old Biomarkers and an Active Microbial Community*

As DNA degrades over time, it is increasingly hard to amplify long stretches of preserved DNA. Designing primers that flank large portions of 16S or other genes of interest is one way to ensure that only recent DNA is amplified (Johnson et al. 2007). RNA molecules on the other hand are more susceptible to degradations and do not persist in the environment as free molecules like DNA, due to their single stranded nature allowing direct cleavage of the phosphodiester bonds. Because of this, it is thought that molecular reconstruction of current microbial communities through RNA extractions and sequencing is more reliable. However, RNA extractions are notoriously difficult in permafrost soils; this is either due to low biomass in the samples or lack of actual RNA molecules altogether. The latter would imply lack of active cells. It is prudent to note that some studies have been able to extract or show evidence of long-term RNA preservation in permafrost, ice, snow, and other environments aged 50–140,000 years; for now, these studies are limited to viral and plant RNA, though this does not exclude the possibility of microbial and fungal RNA preservation (Guy 2014). Contamination may be an even bigger hurdle in permafrost RNA studies compared to DNA ones, due to ubiquitous prevalence of RNases. Another option is the propidium monoazide (PMA) treatment, which is able to differentiate between live cells' DNA and eDNA/DNA within dead cells. PMA binds to DNA and inhibits the PCR reaction; however, it is not able to penetrate intact membranes of live cells (Bae and Wuertz 2009). Therefore, permafrost studies that aim to describe the current viable microbial community can use PMA treatment on their samples (Yergeau et al. 2010). However, these studies are limited to amplicon sequencing or any design that requires a PCR amplification step in the protocol. The advantage of this is that differences between the "live" and total DNA can be used to elucidate the portion of DNA that is environmental and preserved in the permafrost. Though ancient DNA would provide us with the most comprehensive phylogenetic information of past microbial communities, other biomolecules can be more persistent in the environment and may serve better for paleodiversity studies (Briggs and Summons 2014).

8.6 Warming Climate and Permafrost

Permafrost contains large amounts of frozen ancient carbon stores, in the range of 25–50% of the total soil organic carbon (Tarnocai et al. 2009). These pools are currently mostly inaccessible to microbial metabolism (Mackelprang et al. 2016). However, as the climate warms due to anthropogenic climate change, these pools of frozen carbon thawing are becoming available for heterotrophic microbial decomposition (Schuur et al. 2015). As the carbon pools are degraded, the microbes release greenhouse gases (GHG; CO_2, CH_4, and N_2O) into the atmosphere which can potentially increase the rate of climate change via a positive GHG feedback loop (Marushchak et al. 2011; Graham et al. 2012). Auxiliary effects of climate change such as wildfires further increase permafrost degradation and active layer deepening (Taş et al. 2014) (Fig. 8.1). Active layer detachment and permafrost collapse due to thawing further expose formerly buried permafrost and also increase microbial activity and degradation of previously unavailable soil organic matter (Pautler et al. 2010).

In laboratory warming experiments, permafrost thaw induced an increased CO_2 production in both Arctic and high altitude permafrost soils (Stackhouse et al. 2015; Mu et al. 2016). While carbon emissions from permafrost at subzero temperatures are present, rates of CH_4 and CO_2 emissions generally significantly increase with permafrost thaw under aerobic and anaerobic conditions (Song et al. 2014). Furthermore, N_2O emissions have been reported in laboratory permafrost core melting experiments (Elberling et al. 2010). Interestingly, the initial permafrost thaw released minimal N_2O; however, a cycle of drying and rewetting with the meltwater induced high rates of N_2O production in wetland permafrost (Elberling et al. 2010).

Understanding the extent of ancient carbon degradation and resulting GHG emissions from permafrost is challenging due to different factors that affect the thaw of permafrost and the heterogeneity of permafrost environments, such as vegetation cover, latitude, soil composition, hydrology and geology of the area, precipitation, types of organic carbon stored, C:N ratios, and permafrost depth (Chen et al. 2016). Indeed, deeper permafrost soils upon thaw may release less CO_2 due to these older soils containing higher proportions of recalcitrant carbon and lower enzyme (amylase and cellulose) concentrations and activity (Song et al. 2014). Thus, characterizing the current microbial community and activity in permafrost and understanding how that will shift with permafrost thaw and deepening of the active layer is one of the key current questions of environmental microbiology and the study of permafrost life. Indeed, studies that not only consider the current microbial biodiversity of permafrost, but that directly explore the changes associated with permafrost thaw by looking at environments that are currently undergoing this transition or by directly thawing permafrost samples in the laboratory, are starting to provide us with real insight into what the microbial community, its metabolic potential, and its effects on further GHG production may be, once permafrost thaws.

Coolen and Orsi (2015) looked at the transcriptional response of a microbial community to permafrost thaw in a moist acidic Arctic tundra location. Thawing permafrost conditions stimulated the growth of Firmicutes, Bacteroidetes, Euryarchaeota, Chloroflexi, Crenarchaeota, and ascomycetous fungi (Coolen and Orsi 2015). However, other studies reported that permafrost melt increased abundance of Actinobacteria upon thawing short term (Mackelprang et al. 2011) and long term (Deng et al. 2015). In mineral permafrost horizons, thaw increased the abundance of Actinomycetales that degrade complex recalcitrant carbon sources, as well as *Chitinophaga* (a chitinolytic genus) and Sphingomonadales (degraders of aromatics compounds) (Deng et al. 2015). Gene expression upon thaw tended towards overexpression of genes involved in amino acid transport and metabolism; energy production; and DNA repair, replication, and recombination (Coolen and Orsi 2015). However, genes encoding for biofilm formation, virulence, and horizontal gene transfer were higher expressed under frozen conditions compared to thawed soil (Coolen and Orsi 2015). Biofilm formation in the frozen soils could be restricted to the liquid brine vein microhabitat that surrounds the frozen permafrost soil particles (Gilichinsky et al. 2003; Coolen and Orsi 2015). As the permafrost thaws, genes that code for translation, ribosomal structure, and biogenesis are upregulated, as well as genes involved in extracellular protein degradation, anaerobic metabolism, and the uptake, transport, and degradation of carbohydrates and are thus, likely contributing to permafrost soil organic carbon degradation. However, genes coding for hydrolases responsible for the cleavage of complex carbon polymers into C1 and C2 substrates were expressed in both frozen and thawed permafrost (Coolen and Orsi 2015).

Wildfires in upland Boreal Alaska stimulate near surface permafrost thaw; this thaw can shift the microbial community in the soils (Taş et al. 2014). Taş et al. (2014) conducted a novel study looking at the effects of fire and subsequent permafrost thaw on the microbial community and its metabolic potential for further

GHG fluxes. The fire burned the majority of the top organic layer, thawed the permafrost to at least 1m in depth and lowered the C, N, DOC, and moisture content of the soils, but increased the pH and percentage of aromaticity. While fire and the subsequent thawing permafrost had a negative impact on the abundance of Verrumicrobia and Chloroflexi, it had a positive impact on the abundance of candidate phyla AD5, which was one of the most abundance phyla in the deeper thawed and frozen permafrost soils. This is likely due to lowered C, N, and moisture content and a higher pH, as the abundance of AD5 was also correlated to these soil parameters (Taş et al. 2014). It is likely that this candidate phyla, with no known culturable isolates, thrives in nutrient and moisture poor environments, as it was shown to increase in abundance with both depth and fire disturbance. Furthermore, the permafrost thawed soils had a different functional potential compared to intact permafrost. The thawed permafrost contained more genes for the hydrogenotrophic methanogenesis compared to aceticlastic methanogenesis in intact permafrost (Taş et al. 2014). However, regardless of the metabolic pathways, anaerobic incubations of permafrost soils demonstrated an overall reduction in CH_4 production in burned soils; this is potentially due to a reduced moisture content and thus reduced microbial activity and anaerobic niches (Taş et al. 2014). Nitrogen cycling genes were overall more abundant in the thawed permafrost compared to intact permafrost, including genes for nitrate assimilation and denitrification. However, genes coding for nitrite and nitrous oxide reductases were lower in the thawed permafrost soils; this suggests a potential for incomplete denitrification and release of N_2O upon thaw (Elberling et al. 2010; Taş et al. 2014).

Effects of Antarctic permafrost thaw on the microbial community and GHG emissions have not been studied. However, Buelow et al. (2016) have looked at the potential effects of permafrost thaw on McMurdo Dry Valley soils by simulating addition of water and organic matter to the arid Antarctic topsoils. Soils were dominated by Actinobacteria, Firmicutes, and Proteobacteria phyla (Buelow et al. 2016). Water and organic matter supplements increased bacterial abundance but decreased diversity suggesting that the region may experience loss of endemic dry-adapted oligotrophic taxa and replacement by generalist taxa with increasing temperatures due to climate warming (Buelow et al. 2016).

8.6.1 Methane Dynamics

Methane is ~20 times more potent that CO_2 as a GHG. Currently, it is uncertain the portion of stored organic carbon that will be released as CH_4 versus CO_2 (Mackelprang et al. 2016). The flux of methane from the soils is governed by the equilibrium of methanotrophs and methanogens. Methanogens are anaerobic archaea (part of Euryarchaeota) that are responsible for the biogenic production of CH_4. Methanotrophs are characterized by their ability to oxidize CH_4 and assimilate it as organic carbon (Hanson and Hanson 1996), belonging to the phyla Verrucomicrobia and Proteobacteria and are further classified as either

Type I methanotrophs belonging to the Gamaproteobacteria or Type II methanotrophs belonging to the Alphaproteobacteria (Conrad 2007). Furthermore, anaerobic oxidation of CH_4 is also possible via reverse methanogenesis by a group of ANME archaea related to methanogens (Knittel and Boetius 2009). Currently, wetland areas of the Arctic that are acting as methane sources and upland polar desert soils are acting as methane sinks (Christiansen et al. 2015; Lau et al. 2015). It is thought the emissions of CH_4 in peat and wetlands are currently offset by CH_4 uptake in upland dryer soils (Emmerton et al. 2014; Christiansen et al. 2015). However, this may not hold up as the climate warms and permafrost degrades. As the permafrost thaws, it not only introduces organic carbon into the deepening active layer, but also causes land surface collapse, changes in soil hydrology, and the formation of thermokarst bogs and wetlands with anoxic conditions (Johansson et al. 2006; Graham et al. 2012). These anoxic conditions promote the growth of methanogens and are favorable conditions for CH_4 production. So the question arises: what is the potential of permafrost to act as a CH_4 sink or source once it thaws?

Permafrost does contain an active community of methanotrophs and methanogens (Yergeau et al. 2010; Mackelprang et al. 2011; Allan et al. 2014; Deng et al. 2015), though there are conflicting results in terms of diversity of these organisms and no consensus as to how their abundance will be affected with permafrost thaw. Overall, the permafrost methanogen community is dominated by Methanococcales and Methanomicrobiales in mineral permafrost and *Methanosarcina*, *Methanoregula*, and *Methanobacterium* in methane containing permafrost (Allan et al. 2014; Shcherbakova et al. 2016). In contrast, active layer soil microbial community is dominated by Methanobacteriales, Methanosarcinales, Methanocellaceae, and Methanomicrobiaceae (Ganzert et al. 2007; Barbier et al. 2012). Methanogens' diversity and abundance increases with soil depth in permafrost, likely due to the anaerobic conditions in the subsurface (Stackhouse et al. 2015; Shcherbakova et al. 2016). Furthermore, methanogens increase in abundance and diversity with permafrost thaw and creation of wetlands; this may unfortunately increase the ratio of CH_4 to CO_2 flux from microbial decomposition of stored organic matter (Allan et al. 2014; McCalley et al. 2014). By looking at habitats at different stages of permafrost thaw, Hultman et al. (2015) observed higher abundance of methanogens in the thawed sites, possibly suggesting that even if the current permafrost soil do not harbor a large number of methanogens, they can be colonized by these organisms as the permafrost thaws. In addition to abundance, the thaw also stimulates methanogen activity, likely due to increase in temperature (Allan et al. 2014); this is reflected in a higher abundance of *mcrA* gene transcripts, a gene that catalyzes the final methanogenesis step (Coolen and Orsi 2015). Furthermore, upon thaw there is an increase in transcripts of the *fhs* gene which is part of the acetogenic fermentation pathway; this corresponded with an increase in the genes involved in the acetoclassic methanogenesis by *Methanosarcina barkeri* (able to utilize C1 and C2 compounds including acetate) (Coolen and Orsi 2015). This supported a previous finding by McCalley et al. (2014), who demonstrated that increased CH_4 emissions due to peatland permafrost thaw are associated with a switch from a hydrogenotrophic to acetoclastic methanogenesis. This was also reflected in the shift of dominant novel methanogen taxa from

Methanoflorens stordalenmirensis in partially thawed sites to members of *Methanosaeta* genus in fully thawed sites (McCalley et al. 2014; Mondav et al. 2014). However, this is in contrast to two studies that found higher abundance of hydrogenotrophic Methanobacteria members in the upper permafrost horizons compared to acetoclastic Methanomicrobia members in the lower permafrost horizons (Barbier et al. 2012; Deng et al. 2015). This could have been also due to a shift from organic more aerobic upper soils to more anoxic mineral lower permafrost soils (Mondav et al. 2014; Deng et al. 2015). In addition to thawing permafrost, there is evidence that methanogenesis is able to occur in frozen permafrost as well (Rivkina et al. 2007; Yergeau et al. 2010). When methane is not able to diffuse through the frozen soil, permafrost ends up accumulating methane that may be released rapidly into the atmosphere as the permafrost thaws (Rivkina et al. 2007; Mackelprang et al. 2011).

Whether permafrost thaw will have a significant impact on release of CH_4 into the atmosphere is also dependent on the response of methanotrophs. Permafrost and active layer soils harbor Type I, Type II, and Verrucomicrobia methanotrophs; however, these organisms are not ubiquitous across all permafrost soils (Yergeau et al. 2010; Christiansen et al. 2015, Stackhouse et al. 2015). The relative abundance of methanotrophs decreases with depth, possibly due to an increasingly anoxic environment (Stackhouse et al. 2015). The dominant methanotroph organisms in permafrost soils are members of Methylococcaceae, Methylocystaceae, and *Methylocapsa*, *Methylocella*, and *Methylacidiphilum* (Verrucomicrobia) (Yergeau et al. 2010; Deng et al. 2015; Stackhouse et al. 2015). Currently, organic rich permafrost and permafrost affected soils act as a methane source and mineral permafrost soils act as a methane sink, possibly due to the activity of high affinity methanotrophs (Christiansen et al. 2015; Lau et al. 2015). Though many studies predict increases in CH_4 releases with permafrost thaw (Mackelprang et al. 2016), Emmerton et al. (2014) suggests that the future changes in temperature and soil moisture content may increase CH_4 consumption in mineral well-drained permafrost affected soils, but at the same time will have minimal effects on increasing CH_4 production (Emmerton et al. 2014).

8.7 Conclusions

Despite permafrost being an inhospitable environment, it harbors an active microbial community in the Arctic, Antarctic, and at high altitudes. Next generation sequencing and -omics techniques have allowed us to fully grasp the diversity of microbial life in permafrost. This unique environment has also provided us with many microbial isolates that help us understand the physiological limits of life and the necessary adaptations needed to survive in such an extreme environment. In the more recent years, with the warming climate, it has become imperative to determine the effects of permafrost degradation on the microbial community. Understanding how the current microbial community is responding to this warming and subsequent GHG release is the focus of many current permafrost studies.

References

Allan J, Ronholm J, Mykytczuk N et al (2014) Methanogen community composition and rates of methane consumption in Canadian High Arctic permafrost soils. Environ Microbiol Rep 6:136–144

Amato P, Doyle SM, Battista JR et al (2010) Implications of subzero metabolic activity on long-term microbial survival in terrestrial and extraterrestrial permafrost. Astrobiology 10:789–798

Bae S, Wuertz S (2009) Discrimination of viable and dead fecal Bacteroidales bacteria by quantitative PCR with propidium monoazide. Appl Environ Microbiol 75:2940–2944

Bai Y, Yang D, Wang J et al (2006) Phylogenetic diversity of culturable bacteria from alpine permafrost in the Tianshan Mountains, northwestern China. Res Microbiol 157:741–751

Bakermans C, Nealson KH (2004) Relationship of critical temperature to macromolecular synthesis and growth yield in *Psychrobacter cryopegella*. J Bacteriol 186:2340–2345

Bakermans C, Tsapin AI, Souza-Egipsy V et al (2003) Reproduction and metabolism at −10°C of bacteria isolated from Siberian permafrost. Environ Microbiol 5:321–326

Bakermans C, Skidmore ML, Douglas S et al (2014) Molecular characterization of bacteria from permafrost of the Taylor Valley, Antarctica. FEMS Microbiol Ecol 89:331–346

Barbier BA, Dziduch I, Liebner S et al (2012) Methane-cycling communities in a permafrost-affected soil on Herschel Island, Western Canadian Arctic: active layer profiling of mcrA and pmoA genes. FEMS Microbiol Ecol 82:287–302

Bellemain E, Davey ML, Kauserud H et al (2013) Fungal palaeodiversity revealed using high-throughput metabarcoding of ancient DNA from arctic permafrost. Environ Microbiol 15:1176–1189

Blanco Y, Prieto-Ballesteros O, Gómez MJ et al (2012) Prokaryotic communities and operating metabolisms in the surface and the permafrost of Deception Island (Antarctica). Environ Microbiol 14:2495–2510

Blazewicz SJ, Barnard RL, Daly RA et al (2013) Evaluating rRNA as an indicator of microbial activity in environmental communities: limitations and uses. ISME J 7:2061–2068

Bockheim JG, Hall KJ (2002) Permafrost, active-layer dynamics and periglacial environments of continental Antarctica. S Afr J Sci 98:82–90

Bockheim JG, Munroe JS (2014) Organic carbon pools and genesis of alpine soils with permafrost: a review. Arct Antarct Alp Res 46:987–1006

Bockheim JG, Campbell IB, Mcleod M (2007) Permafrost distribution and active-layer depths in the McMurdo Dry valleys, Antarctica. Permafr Periglac Process 18:217–227

Bockheim JG, Kurz MD, Soule SA et al (2009) Genesis of active sand-filled polygons in lower and central Beacon Valley, Antarctica. Permafr Periglac Process 20:295–308

Briggs DE, Summons RE (2014) Ancient biomolecules: their origins, fossilization, and role in revealing the history of life. BioEssays 36:482–490

Buelow HN, Winter AS, Van Horn DJ et al (2016) Microbial community responses to increased water and organic matter in the arid Soils of the McMurdo Dry Valleys, Antarctica. Front Microbiol 7:1040

Campbell IB, Claridge GG (2006) Permafrost properties, patterns and processes in the Transantarctic Mountains region. Permafr Periglac Process 17:215–232

Campbell IB, Claridge GG (2009) Antarctic permafrost soils. In: Margesin R (ed) Permafrost soils, Springer, Heidelberg, pp 17-31

Carini P, Marsden PJ, Leff JW, et al (2016) Relic DNA is abundant in soil and obscures estimates of soil microbial diversity. bioRxiv:043372

Chen L, Liang J, Qin S et al (2016) Determinants of carbon release from the active layer and permafrost deposits on the Tibetan Plateau. Nat Commun 7:13046

Christiansen JR, Romero AJB, Jørgensen NO et al (2015) Methane fluxes and the functional groups of methanotrophs and methanogens in a young Arctic landscape on Disko Island, West Greenland. Biogeochemistry 122:15–33

Conrad R (2007) Microbial ecology of methanogens and methanotrophs. Adv Agron 96:1–63

Coolen MJ, Orsi WD (2015) The transcriptional response of microbial communities in thawing Alaskan permafrost soils. Front Microbiol 6:197
D'Amico S, Collins T, Marx JC et al (2006) Psychrophilic microorganisms: challenges for life. EMBO Rep 7:385–389
De Maayer P, Anderson D, Cary C et al (2014) Some like it cold: understanding the survival strategies of psychrophiles. EMBO Rep 15:508–517
Deng J, Gu Y, Zhang J et al (2015) Shifts of tundra bacterial and archaeal communities along a permafrost thaw gradient in Alaska. Mol Ecol 24:222–234
Elberling B, Brandt KK (2003) Uncoupling of microbial CO2 production and release in frozen soil and its implications for field studies of arctic C cycling. Soil Biol Biochem 35:263–272
Elberling B, Christiansen HH, Hansen BU (2010) High nitrous oxide production from thawing permafrost. Nat Geosci 3:332–335
Emmerton CA, St Louis V, Lehnherr I et al (2014) The net exchange of methane with high Arctic landscapes during the summer growing season. Biogeosciences 11:3095–3106
Eriksson M, Ka J-O, Mohn WW (2001) Effects of low temperature and freeze-thaw cycles on hydrocarbon biodegradation in Arctic tundra soil. Appl Environ Microbiol 67:5107–5112
Ernakovich JG, Wallenstein MD (2015) Permafrost microbial community traits and functional diversity indicate low activity at in situ thaw temperatures. Soil Biol Biochem 87:78–89
Etzelmüller B (2013) Recent advances in mountain permafrost research. Permafr Periglac Process 24:99–107
Fahnestock JT, Jones MH, Welker JM (1999) Wintertime CO2 efflux from arctic soils: implications for annual carbon budgets. Glob Biogeochem Cycles 13:775–779
Finster KW, Herbert RA, Kjeldsen KU et al (2009) *Demequina lutea* sp. nov. isolated from a high Arctic permafrost soil. Int J Syst Evol Microbiol 59:649–653
Frey B, Rime T, Phillips M et al (2016) Microbial diversity in European alpine permafrost and active layers. FEMS Microbiol Ecol 92:fiw018
Ganzert L, Jurgens G, Münster U et al (2007) Methanogenic communities in permafrost-affected soils of the Laptev Sea coast, Siberian Arctic, characterized by 16S rRNA gene fingerprints. FEMS Microbiol Ecol 59:476–488
Gilichinsky D, Wagener S, Vishnevetskaya T (1995) Permafrost microbiology. Permafr Periglac Process 6:281–291
Gilichinsky D, Rivkina E, Shcherbakova V et al (2003) Supercooled water brines within permafrost-an unknown ecological niche for microorganisms: a model for astrobiology. Astrobiology 3:331–341
Gilichinsky D, Wilson G, Friedmann E et al (2007) Microbial populations in Antarctic permafrost: biodiversity, state, age, and implication for astrobiology. Astrobiology 7:275–311
Gittel A, Bárta J, Kohoutová I et al (2014a) Distinct microbial communities associated with buried soils in the Siberian tundra. ISME J 8:841–853
Gittel A, Bárta J, Kohoutová I et al (2014b) Site-and horizon-specific patterns of microbial community structure and enzyme activities in permafrost-affected soils of Greenland. Front Microbiol 5:541
Goordial J, Lamarche-Gagnon G, Lay C-Y, Whyte LG (2013) Left out in the cold: life in cryoenvironments. In: Seckbach J, Oren A, StanLotter H (eds) Polyextremophiles, vol 27, Cellular origin, life in extreme habitats and astrobiology. Springer, Amsterdam, pp 335–363
Goordial J, Whyte L (2014) Microbial life in Antarctic permafrost environments. In: Cowan D (ed) Antarctic terrestrial microbiology. Springer, Heidelberg, pp 217–232
Goordial J, Davila A, Lacelle D et al (2016) Nearing the cold-arid limits of microbial life in permafrost of an upper dry valley, Antarctica. ISME J 10:1613–1624
Graham DE, Wallenstein MD, Vishnivetskaya TA et al (2012) Microbes in thawing permafrost: the unknown variable in the climate change equation. ISME J 6:709–712
Guy PL (2014) Prospects for analyzing ancient RNA in preserved materials. Wiley Interdiscip Rev RNA 5:87–94

Haeberli W, Gruber S (2009) Global warming and mountain permafrost. In: Margesin R (ed) Permafrost soils. Springer, Heidelberg, pp 205–218
Hansen AA, Herbert RA, Mikkelsen K et al (2007) Viability, diversity and composition of the bacterial community in a high Arctic permafrost soil from Spitsbergen, Northern Norway. Environ Microbiol 9:2870–2884
Hansen AJ, Mitchell DL, Wiuf C et al (2006) Crosslinks rather than strand breaks determine access to ancient DNA sequences from frozen sediments. Genetics 173:1175–1179
Hanson RS, Hanson TE (1996) Methanotrophic bacteria. Microbiol Rev 60:439–471
Hebsgaard MB, Willerslev E (2009) Very old DNA. In: Margesin R (ed) Permafrost soils. Springer, Heidelberg, pp 47–57
Hu W, Zhang Q, Tian T et al (2015) The microbial diversity, distribution, and ecology of permafrost in China: a review. Extremophiles 19:693–705
Hu W, Zhang Q, Tian T et al (2016) Characterization of the prokaryotic diversity through a stratigraphic permafrost core profile from the Qinghai-Tibet Plateau. Extremophiles 20:337–349
Hugelius G, Kuhry P (2009) Landscape partitioning and environmental gradient analyses of soil organic carbon in a permafrost environment. Glob Biogeochem Cycles 23:GB3006
Hultman J, Waldrop MP, Mackelprang R et al (2015) Multi-omics of permafrost, active layer and thermokarst bog soil microbiomes. Nature 521:208–212
Jansson JK, Taş N (2014) The microbial ecology of permafrost. Nat Rev Microbiol 12:414–425
Johansson T, Malmer N, Crill PM et al (2006) Decadal vegetation changes in a northern peatland, greenhouse gas fluxes and net radiative forcing. Glob Chang Biol 12:2352–2369
Johnson SS, Hebsgaard MB, Christensen TR et al (2007) Ancient bacteria show evidence of DNA repair. PNAS 104:14401–14405
Katayama T, Tanaka M, Moriizumi J et al (2007) Phylogenetic analysis of bacteria preserved in a permafrost ice wedge for 25,000 years. Appl Environ Microbiol 73:2360–2363
Kerfoot DE (1972) Thermal contraction cracks in an arctic tundra environment. Arctic 25:142–150
Kim SJ, Shin SC, Hong SG et al (2012) Genome sequence of a novel member of the genus *Psychrobacter* isolated from Antarctic soil. J Bacteriol 194:2403–2403
Knittel K, Boetius A (2009) Anaerobic oxidation of methane: progress with an unknown process. Annu Rev Microbiol 63:311–334
Koh HY, Park H, Lee JH et al (2016) Proteomic and transcriptomic investigations on cold-responsive properties of the psychrophilic *Antarctic bacterium* sp. PAMC 21119 at subzero temperatures. Environ Microbiol. doi:10.1111/1462-2920.13578
Kokelj S, Burn C (2005) Geochemistry of the active layer and near *Psychrobacter*-surface permafrost, Mackenzie delta region, Northwest Territories, Canada. Can J Earth Sci 42:37–48
Lacelle D, Radtke K, Clark ID et al (2011) Geomicrobiology and occluded O_2–CO_2–Ar gas analyses provide evidence of microbial respiration in ancient terrestrial ground ice. Earth Planet Sci Lett 306:46–54
Larsen KS, Jonasson S, Michelsen A (2002) Repeated freeze–thaw cycles and their effects on biological processes in two arctic ecosystem types. Appl Soil Ecol 21:187–195
Lau M, Stackhouse B, Layton A et al (2015) An active atmospheric methane sink in high Arctic mineral cryosols. ISME J 9:1880–1891
Lindahl T (1993) Instability and decay of the primary structure of DNA. Nature 362:709–715
Mackelprang R, Waldrop MP, Deangelis KM et al (2011) Metagenomic analysis of a permafrost microbial community reveals a rapid response to thaw. Nature 480:368–371
Mackelprang R, Saleska SR, Jacobsen CS et al (2016) Permafrost meta-omics and climate change. Annu Rev Earth Planet Sci 44:439–462
Marinova MM, Mckay CP, Pollard WH et al (2013) Distribution of depth to ice-cemented soils in the high-elevation Quartermain Mountains, McMurdo Dry Valleys, Antarctica. Antarct Sci 25:575–582
Marushchak M, Pitkämäki A, Koponen H et al (2011) Hot spots for nitrous oxide emissions found in different types of permafrost peatlands. Glob Chang Biol 17:2601–2614

Mccalley CK, Woodcroft BJ, Hodgkins SB et al (2014) Methane dynamics regulated by microbial community response to permafrost thaw. Nature 514:478–481

Michaelson G, Ping C (2003) Soil organic carbon and CO_2 respiration at subzero temperature in soils of Arctic Alaska. J Geophys Res 108:8164

Mondav R, Woodcroft BJ, Kim E-H et al (2014) Discovery of a novel methanogen prevalent in thawing permafrost. Nat Commun 5:3212

Mu C, Zhang T, Zhang X et al (2016) Sensitivity of soil organic matter decomposition to temperature at different depths in permafrost regions on the northern Qinghai-Tibet Plateau. Eur J Soil Sci 67:773–781

Mykytczuk NC, Foote SJ, Omelon CR et al (2013) Bacterial growth at−15°C; molecular insights from the permafrost bacterium *Planococcus halocryophilus* Or1. ISME J 7:1211–1226

Nikrad MP, Kerkhof LJ, Häggblom MM (2016) The subzero microbiome: Microbial activity in frozen and thawing soils. FEMS Microbiol Ecol 92:fiw081

Öquist MG, Sparrman T, Klemedtsson L et al (2009) Water availability controls microbial temperature responses in frozen soil CO_2 production. Glob Chang Biol 15:2715–2722

Panikov NS, Dedysh S (2000) Cold season CH_4 and CO_2 emission from boreal peat bogs (West Siberia): Winter fluxes and thaw activation dynamics. Glob Biogeochem Cycles 14:1071–1080

Panikov NS, Sizova MV (2007) Growth kinetics of microorganisms isolated from Alaskan soil and permafrost in solid media frozen down to -35°C. FEMS Microbiol Ecol 59:500–512

Panikov N, Flanagan P, Oechel W et al (2006) Microbial activity in soils frozen to below−39°C. Soil Biol Biochem 38:785–794

Pautler BG, Simpson AJ, Mcnally DJ et al (2010) Arctic permafrost active layer detachments stimulate microbial activity and degradation of soil organic matter. Environ Sci Technol 44:4076–4082

Pietramellara G, Ascher J, Borgogni F et al (2009) Extracellular DNA in soil and sediment: fate and ecological relevance. Biol Fertil Soils 45:219–235

Poinar HN, Hoss M, Bada JL et al (1996) Amino acid racemization and the preservation of ancient DNA. Science 272:864

Ponder MA, Gilmour SJ, Bergholz PW et al (2005) Characterization of potential stress responses in ancient Siberian permafrost psychroactive bacteria. FEMS Microbiol Ecol 53:103–115

Price PB, Sowers T (2004) Temperature dependence of metabolic rates for microbial growth, maintenance, and survival. PNAS 101:4631–4636

Ran Y, Li X, Cheng G et al (2012) Distribution of permafrost in China: an overview of existing permafrost maps. Permafr Periglac Process 23:322–333

Rivkina E, Friedmann E, Mckay C et al (2000) Metabolic activity of permafrost bacteria below the freezing point. Appl Environ Microbiol 66:3230–3233

Rivkina E, Shcherbakova V, Laurinavichius K et al (2007) Biogeochemistry of methane and methanogenic archaea in permafrost. FEMS Microbiol Ecol 61:1–15

Routh J, Hugelius G, Kuhry P et al (2014) Multi-proxy study of soil organic matter dynamics in permafrost peat deposits reveal vulnerability to climate change in the European Russian Arctic. Chem Geol 368:104–117

Schostag M, Stibal M, Jacobsen CS et al (2015) Distinct summer and winter bacterial communities in the active layer of Svalbard permafrost revealed by DNA-and RNA-based analyses. Front Microbiol 6:399

Schuur EA, Bockheim J, Canadell JG et al (2008) Vulnerability of permafrost carbon to climate change: implications for the global carbon cycle. Bioscience 58:701–714

Schuur E, Mcguire A, Schädel C et al (2015) Climate change and the permafrost carbon feedback. Nature 520:171–179

Shcherbakova V, Rivkina E, Pecheritsyna S et al (2011) *Methanobacterium arcticum* sp. nov. a methanogenic archaeon from Holocene Arctic permafrost. Int J Syst Evol Microbiol 61:144–147

Shcherbakova V, Chuvilskaya N, Rivkina E et al (2005) Novel psychrophilic anaerobic spore-forming bacterium from the overcooled water brine in permafrost: description of *Clostridium algoriphilum* sp. nov. Extremophiles 9:239–246

Shcherbakova V, Yoshimura Y, Ryzhmanova Y et al (2016) Archaeal communities of Arctic methane-containing permafrost. FEMS Microbiol Ecol 92:fiw135
Shur Y, Jorgenson M (2007) Patterns of permafrost formation and degradation in relation to climate and ecosystems. Permafr Periglac Process 18:7–19
Shur Y, Hinkel KM, Nelson FE (2005) The transient layer: implications for geocryology and climate-change science. Permafr Periglac Process 16:5–17
Smith CI, Chamberlain AT, Riley MS et al (2001) Neanderthal DNA: not just old but old and cold? Nature 410:771–772
Song C, Wang X, Miao Y et al (2014) Effects of permafrost thaw on carbon emissions under aerobic and anaerobic environments in the Great Hing'an Mountains, China. Sci Total Environ 487:604–610
Stackhouse BT, Vishnivetskaya TA, Layton A et al (2015) Effects of simulated spring thaw of permafrost from mineral cryosol on CO_2 emissions and atmospheric CH_4 uptake. J Geophys Res Biogeosci 120:1764–1784
Steven B, Leveille R, Pollard WH et al (2006) Microbial ecology and biodiversity in permafrost. Extremophiles 10:259–267
Steven B, Briggs G, Mckay CP et al (2007) Characterization of the microbial diversity in a permafrost sample from the Canadian high Arctic using culture-dependent and culture-independent methods. FEMS Microbiol Ecol 59:513–523
Steven B, Pollard WH, Greer CW et al (2008) Microbial diversity and activity through a permafrost/ground ice core profile from the Canadian high Arctic. Environ Microbiol 10:3388–3403
Tamppari L, Anderson R, Archer P et al (2012) Effects of extreme cold and aridity on soils and habitability: McMurdo Dry Valleys as an analogue for the Mars Phoenix landing site. Antarct Sci 24:211–228
Tarnocai C (1980) Summer temperatures of cryosolic soils in the north-central Keewatin, NWT. Can J Soil Sci 60:311–327
Tarnocai C (2006) The effect of climate change on carbon in Canadian peatlands. Glob Planet Chang 53:222–232
Tarnocai C (2009) Arctic permafrost soils. In: Margesin R (ed) Permafrost soils. Springer, pp 3–16
Tarnocai C, Canadell J, Schuur E et al (2009) Soil organic carbon pools in the northern circumpolar permafrost region. Glob Biogeochem Cycles 23:GB2023
Taş N, Prestat E, Mcfarland JW et al (2014) Impact of fire on active layer and permafrost microbial communities and metagenomes in an upland Alaskan boreal forest. ISME J 8:1904–1919
Tuorto SJ, Darias P, Mcguinness LR et al (2014) Bacterial genome replication at subzero temperatures in permafrost. ISME J 8:139–149
Van Everdingen R (ed) (1998) Multi-language glossary of permafrost and related ground-ice terms. (revised May 2005) National Snow and Ice Center, Boulder, Colarado
Vishnivetskaya TA, Petrova MA, Urbance J et al (2006) Bacterial community in ancient Siberian permafrost as characterized by culture and culture-independent methods. Astrobiology 6:400–414
Vonk J, Mann P, Dowdy K et al (2013) Dissolved organic carbon loss from Yedoma permafrost amplified by ice wedge thaw. Environ Res Lett 8:035023
Wei S, Cui H, He H et al (2014) Diversity and distribution of Archaea community along a stratigraphic permafrost profile from Qinghai-Tibetan Plateau, China. Archaea 2014:240817
Wilhelm RC, Niederberger TD, Greer C et al (2011) Microbial diversity of active layer and permafrost in an acidic wetland from the Canadian High Arctic. Can J Microbiol 57:303–315
Wilhelm RC, Radtke KJ, Mykytczuk NC et al (2012) Life at the wedge: the activity and diversity of Arctic ice wedge microbial communities. Astrobiology 12:347–360
Willerslev E, Hansen AJ, Poinar HN (2004a) Isolation of nucleic acids and cultures from fossil ice and permafrost. Trends Ecol Evol 19:141–147
Willerslev E, Hansen AJ, Rønn R et al (2004b) Long-term persistence of bacterial DNA. Curr Biol 14:R9–R10

Wu X, Zhang W, Liu G et al (2012) Bacterial diversity in the foreland of the Tianshan No. 1 glacier, China. Environ Res Lett 7:014038

Yergeau E, Hogues H, Whyte LG et al (2010) The functional potential of high Arctic permafrost revealed by metagenomic sequencing, qPCR and microarray analyses. ISME J 4:1206–1214

Yun J, Ju Y, Deng Y et al (2014) Bacterial community structure in two permafrost wetlands on the Tibetan Plateau and Sanjiang Plain, China. Microb Ecol 68:360–369

Zhang G, Niu F, Ma X et al (2007) Phylogenetic diversity of bacteria isolates from the Qinghai-Tibet Plateau permafrost region. Can J Microbiol 53:1000–1010

Zhang L-M, Wang M, Prosser JI et al (2009) Altitude ammonia-oxidizing bacteria and archaea in soils of Mount Everest. FEMS Microbiol Ecol 70:208–217

Zhao Q, Bai Y, Zhang G et al (2011) *Chryseobacterium xinjiangense* sp. nov. isolated from alpine permafrost. Int J Syst Evol Microbiol 61:1397–1401

Zimov SA, Schuur EA, Chapin Iii FS (2006) Permafrost and the global carbon budget. Science 312:1612–1613

Chapter 9
Cyanobacteria in Polar and Alpine Ecosystems

Anne D. Jungblut and Warwick F. Vincent

Abstract Cyanobacteria are commonly found in freshwaters, soils and glacial environments in polar and alpine regions. Studies to date indicate these cold-dwelling phototrophs are psychrotolerant rather than psychrophilic, with temperature optima for growth that lie well above the temperature ranges of their ambient environment. Cyanobacterial mats occur at the bottom of lakes, ponds and streams and within meltwater habitats on glaciers and ice shelves. They can accumulate large biomass stocks and may account for the dominant fraction of total ecosystem productivity in such environments. Certain taxa in these benthic communities are known to produce cyanotoxins, including microcystins. Planktonic cyanobacteria are also found in many high latitude lakes, specifically picocyanobacteria, but they are conspicuously absent or poorly represented in polar seas, probably as a result of their minimal growth rates in extreme cold. Cyanobacteria also occur in a variety of nonaquatic habitats in the cold regions, including on and within rocks, and as a major constituent of soil crusts in polar and alpine deserts. The nitrogen-fixing capabilities of some cyanobacteria make them especially important for the natural enrichment of soils that have been newly exposed after glacial retreat. The evolution and biogeography of cyanobacterial ecotypes in the cold biosphere is a current focus of genomic analysis and pole-to-pole comparisons, and these studies are providing insights into how microbial ecosystems survived prolonged periods of cold and freeze-up on early Earth.

Contents

A.D. Jungblut (✉)
Life Sciences Department, Natural History Museum, London SW7 5BD, UK
e-mail: a.jungblut@nhm.ac.uk

W.F. Vincent
Centre d'Études Nordiques (CEN), Quebec City, QC G1V 0A6, Canada

Department of Biology, Laval University, Quebec City, QC G1V 0A6, Canada

R. Margesin (ed.), *Psychrophiles: From Biodiversity to Biotechnology*,
DOI 10.1007/978-3-319-57057-0_9

9.1 Introduction

Perennially cold environments in which temperatures remain below 5 °C are common throughout the biosphere (Margesin and Häggblom 2007). In these habitats, cold temperatures are often accompanied by freeze-thaw cycles, extreme fluctuations in irradiance including ultraviolet radiation and large variations in nutrient supply and salinity. As a result of these constraints, polar and alpine environments contain a reduced biodiversity of plants and animals, but they are rich in microscopic life, including cyanobacteria.

Cyanobacteria are phototrophic, oxygenic Gram-negative bacteria. They are of special importance for freshwater and terrestrial ecosystems in polar and alpine regions, where they are major primary producers, key taxa in food webs and drivers of carbon and nitrogen cycling. Their presence in the polar regions was first observed during early expeditions to the Arctic and Antarctica. In 1870, the Swedish-Finnish explorer Adolf Erik Nordenskiöld described the extensive growth of cyanobacteria in the cryosphere (the ensemble of cold environments containing snow and ice) across the Greenland Ice Cap (Leslie 1879). Somewhat later in Antarctica, James Murray, the biologist on the Shackleton's 1907–1909 expedition to Ross Island, also gathered detailed information on cyanobacterial diversity and their importance (Murray 1910). Current research shows a diverse range of cyanobacteria in polar and alpine habitats and that these organisms have broad tolerances to the abiotic stresses that prevail in these environments. This has made them of great interest for the reconstruction of microbial life, diversification and geobiological weathering processes on early Earth, and polar cyanobacterial mats and their ice-based habitats have been identified as potential analogues for biotopes present during the major glaciation events of the Precambrian (Vincent and

Howard-Williams 2000; Vincent et al. 2004a, b; Sumner et al. 2015; Hoffman 2016).

In this chapter, we first introduce the taxonomic status and general characteristics of cyanobacteria. We examine their ecophysiological traits that allow them to survive and often thrive in such cold environments. Cyanobacterial diversity in freshwater, soil and cryosphere ecosystems in the Arctic, Antarctic and alpine regions is presented, and we conclude this review by consideration of the biogeographical distribution and evolution of polar cyanobacteria.

9.2 Taxonomy and Diversity

The classification of Cyanobacteria has been controversial for centuries, and several taxonomic schemes have been proposed. These organisms were initially described as algae in the eighteenth century, and the first classification system was based on the International Code of Botanical Nomenclature as described by Oren (2004). In the botanical taxonomy, two major works are especially noteworthy. Firstly, Geitler (1932) produced a flora that compiled all European taxa, which already encompassed 150 genera and 1500 species based on the morphology. Secondly, the recent revisions by Anagnostidis and Komárek (e.g. Komárek and Anagnostidis 2005) aimed to define more consistent genera, still based on the morphology. After the prokaryotic nature of cyanobacteria became more obvious on the basis of ultrastructural and molecular studies, it was proposed that their nomenclature should be governed by the International Code for Nomenclature of Bacteria (Stanier et al. 1978). Currently, the bacterial phylum Cyanobacteria encompasses five subsections (corresponding to the five orders in the botanical classification) in Bergey's Manual of Systematic Bacteriology (Castenholz 2001): *Chroococcales* (unicellular), *Pleurocapsales* (large cells subdividing into smaller baeocysts), *Oscillatoriales* (simple filamentous), *Nostocales* (filamentous, non-branching heterocyst-forming) and *Stigonematales* (filamentous, branching, heterocyst-forming).

In recent years, several other classifications have been suggested. Hoffmann et al. (2005a, b) recommended the recognition of four subclasses considering both phylogenetic inferences and morphological features: *Nostocophycideae*, *Oscillatoriophycideae*, *Synechococcophycideae* and *Gloeobacteriophycidae* with nine orders *Nostocales*, *Chroococcidiopsidales*, *Spirulinales*, *Pleurocapsales*, *Chroococcales*, *Oscillatoriales*, *Synechococcales* and *Gloeobacteriales* (Hoffman et al. 2005b; Komárek et al. 2014). It is the first system that recognises that coccoid and filamentous morphologies are polyphyletic. This is also the classification that has been adopted by the Greengenes 16S rRNA gene reference database (DeSantis et al. 2006) that is commonly used for high-throughput sequencing community structure analysis.

To date, only a few names of cyanobacterial taxa have been validly published according to bacterial rules, reflecting not only technical difficulties but also the confusion due to the existence of two nomenclature systems (Oren 2004). Current taxonomic studies on cyanobacteria are now adopting a polyphasic approach, which combines genotypic studies with morphological and phenotypic analyses.

Cyanobacteria often have quite simple morphologies and some of these characters exhibit plasticity with environmental parameters, so that their taxonomic usefulness can be limited. Moreover, a number of botanical taxa have been delimited based on minute morphological differences (e.g. sheath characteristics, slight deviations in cell dimensions or form), and many authors have shown that the genetic diversity does often not coincide with that based on morphology (e.g. Rajaniemi et al. 2005; Taton et al. 2006b). To address these problems, studies on environmental samples (natural mixed assemblages of microorganisms) are typically using molecular taxonomic markers, most often the 16S rRNA gene. The obtained 16S rRNA sequences are compared and often grouped into OTUs (operational taxonomic units) or phylotypes on the basis of their similarity and taxonomic assignment performed based on either phylogenetic interference or percentage similarity match to public databases (Taton et al. 2006a, b; Komárek et al. 2012; Martineau et al. 2013; Jancusova et al. 2016). Another taxonomic marker is the internal transcribed spacer regions (ITS). It is highly variable and therefore not suitable for phylogenetic interference of distinctly related taxa or even between genera. However, within species and 16S rRNA genotypes, ITS-based analyses have shown that the resolution of the 16S rRNA gene is often not sufficient to distinguish between ecotypes as demonstrated for the non-polar marine cyanobacterium *Prochlorococcus* (Rocap et al. 2002).

9.3 General Characteristics and Ecophysiology

Cyanobacteria are oxygenic photosynthetic bacteria and possess photosystems I and II, which are located on thylakoid membranes (except in the genus *Gloeobacter*). The cells usually have a characteristic blue-green coloration due to the proteins phycocyanin (blue) and allophycocyanin (blue). They additionally have chlorophyll *a*, and some species also contain the protein phycoerythrin that colours the cells red. The protein pigments are organised into specialised light-capturing structures called phycobilisomes, which transfer their absorbed energy to the reaction centres of photosystem II. In a few taxa, other chlorophylls have been observed, including chlorophylls *b*, *d*, and *f* (Miyashita et al. 1996; Chen and Scheer 2013). Some cyanobacteria are also able to fix atmospheric nitrogen, and several species produce toxic secondary compounds. Cyanobacteria have various storage bodies for carbon, nitrogen, phosphate and the enzyme ribulose 1,5-biphosphate carboxylase/oxygenase (RubisCO) (Castenholz 2001).

9.3.1 Cold Temperatures

In general, high latitude and high altitude cyanobacteria tend to be cold tolerant (psychrotolerant), with suboptimal growth at low temperatures, rather than psychrophiles that grow optimally at low temperature (Tang et al. 1997; Tang and Vincent 1999; Nadeau et al. 2001). They have a variety of mechanisms that allows them to tolerate and continue to grow, albeit often at slow rates, in the cold and to tolerate freeze-thaw conditions (Vincent 2007). To maintain membrane fluidity at low temperatures, polyunsaturated fatty acids with decreased chain lengths are incorporated into the membrane. In addition, the production of compatible solutes (e.g. trehalose) and update of choline and betaine help to reduce the freezing point of the intracellular fluid. On the cellular level, genomic and metagenomic analysis of the cyanobacterium *Phormidium priestleyi* and cyanobacterial-based microbial mats in the Arctic and Antarctic identified several genes potentially implicated in the adaptation to cold environments linked to DNA replication, a translation initiation factor for protein biosynthesis and chaperones for protein folding (Varin et al. 2012; Chrismas et al. 2016).

Cyanobacteria must withstand prolonged seasonal desiccation and freezing. *Phormidium*-dominated microbial mats in the Arctic have been shown to be perennial and appear to survive the winter in the vegetative state as they were metabolically active shortly before and after freezing (Tashyreva and Elster 2016). Similarly, Antarctica microbial mats have been shown to resume photosynthesis within minutes to hours after re-thawing (Vincent 2007). However, the tolerance to desiccation varies between genera. While taxa belonging to *Nostoc* are anhydrobiotic, *Microcoleus* was found not to be truly anhydrobiotic as it survived extensive dehydration (to 0.23 g water g^{-1} dry mass) but not complete desiccation (to 0.03 g water g^{-1} dry mass). Laboratory experiments also suggested that nitrogen starvation prior to desiccation increased the tolerance to reduced water availability in Arctic *Microcoleus* stains (Tashyreva and Elster 2016).

9.3.2 Salt Stress

Typical hypersaline environments are saline ponds and lakes in terrestrial polar ecosystems (Vincent 1988) in permanently cold environments, and salinity can be an important driver of community assemblages (Sabbe et al. 2004; Jungblut et al. 2006). Sudden increases in salt concentration are counterbalanced by a rapid accumulation of salts to maintain the osmotic equilibrium. Long-term survival strategies involve uptake of inorganic ions to balance the extracellular ion concentrations, as well as the production of organic osmolytes (Oren 2000) and similar mechanisms as to protect from desiccation. Studies on microbial mats in Ward Hunt Lake, Canadian High Arctic, have shown that the photosynthetic communities are highly resistant to the major increases in salinity that may be associated with solute concentration during freeze-up (Lionard et al. 2012).

9.3.3 High and Low Irradiance

UV radiation and high-energy photosynthetically active radiation (PAR) can induce photoinhibition, phycobiliprotein degradation, chlorophyll bleaching and DNA damage or the production of reactive oxygen species, and the net damage may be exacerbated at low temperatures (Vincent 2007). Cyanobacteria have evolved a variety of DNA repair mechanisms, such as excision repair and photoreactivation, to cope with UV-induced DNA damage (Castenholz and Garcia-Pichel 2000). However, these processes are reduced at lower temperatures. Furthermore, the cyanobacteria produce photoprotective screening (gloeocapsin, scytonemin, mycosporine) and quenching pigments (carotenoids), and many Antarctic cyanobacteria are able to avoid UV radiation exposure by migrating to deeper layers within the microbial mats (Castenholz and Garcia-Pichel 2000). High concentrations of scytonemin can lead to a black coloration in many cyanobacterial mats and soil crusts (Vincent 2007). Conversely, phototrophs in polar and alpine regions must also contend with low irradiances caused by prolonged snow and ice cover. The cyanobacteria utilise highly efficient light-capturing complexes, with photosynthetic quantum yields close to the theoretical maximum (Hawes and Schwarz 2001; Vincent 2007).

9.3.4 Cyanobacterial Toxin Production

Many cyanobacteria can produce toxic substances, known collectively as cyanotoxins. The mechanisms of toxicity include hepatotoxicity, neurotoxicity, dermatotoxicity, tumour promotion and genotoxicity and are therefore of great importance for public health and freshwater ecology. Several studies have also found cyanotoxins in cyanobacteria-based microbial mats from freshwater ecosystems in polar regions. Microcystins are hepatotoxins and inhibit protein phosphatases, and acute exposure can lead to liver failure and death (Chorus and Bartram 1999). Microcystins have been detected in meltwater ponds on the McMurdo Ice Shelf, Miers Valley (McMurdo Dry Valleys), and Adelaide Island, Antarctica, as well as in thermokarst ponds on Bylot Island, Canadian High Arctic. Concentrations ranged from 1 ng g^{-1} to 16 μg g^{-1} dry mass (Hitzfield et al. 2000; Jungblut et al. 2006; Wood et al. 2008; Kleinteich et al. 2012, 2013). Culture studies with Arctic mats showed that warming caused changes in species composition and an increase in toxins that could have been the result of increased cellular toxin production or a shift in community structure towards more toxic species (Kleinteich et al. 2012).

Common planktonic microcystin producers, such as *Microcystis*, are absent from polar freshwater ecosystems, and benthic *Nostoc* populations have been proposed as the most likely potential microcystin producers based on morphological and DNA analysis (Jungblut et al. 2006; Wood et al. 2008). In addition, Kleinteich et al. (2013) also detected cylindrospermopsin in freshwater ponds on Adelaide Island, Antarctica, but the genus *Cylindrospermosis*, the main producer of this toxin in temperate and tropical environments, was not found. The ecological or physiological functions of toxins in polar ecosystems are not fully understood at present. For microcystins, potential roles include protection against grazers, quorum sensing, gene regulation or iron scavenging, or they may be ancestral relicts of past pathways and functions (Kaebernick and Neilan 2001; Moffitt and Neilan 2004; Rantala et al. 2004).

9.4 Polar Inland Waters

Cyanobacteria often form cohesive, highly pigmented biofilms and mats, from mm- to dm-scale thicknesses that coat the benthic environments in ponds and streams in the Arctic and Antarctic (e.g. Tanabe et al. 2010; Andersen et al. 2011; Lionard et al. 2012; de los Ríos et al. 2015). Due to the high biomass accumulations of cyanobacteria-based microbial mats with rich heterotrophic and microbial eukaryote communities in terrestrial aquatic ecosystems, they are therefore seen as hotspots of biodiversity in comparison to soil environments in the cold, arid polar desert in permanently ice-free regions.

9.4.1 Benthic Communities in Shallow Ponds

In Antarctica, a great variety of shallow meltwater ponds are present that span a wide range of environmental conditions. During the austral summer they are ice-free and brightly lit habitats, whereas they can fully freeze during winter but can also experience high variability in irradiance and temperatures including daily freeze-thaw cycling, depending on weather conditions. These aquatic ecosystems are characterised by high benthic biomass accumulations of cyanobacteria-based microbial mats. Some of these microbial mats can have vertical zonation of organisms and functions in response to light gradients. Many of these cyanobacteria-based microbial mats have an orange, carotenoid-rich upper layer that likely reduces UV and oxidative stress, underlain by green deep chlorophyll maximum layer, which in turn may be further underlain by an anoxic bottom layer that contains anoxygenic phototrophs such as purple bacteria. At the molecular level, these ponds can show a large cyanobacterial diversity (Taton et al. 2003; Kleinteich et al. 2014; Archer et al. 2015); for example, Kleinteich et al. (2014)

identified 274 cyanobacterial OTUs from 30 meltwater ponds from Adelaide Island, Antarctic Peninsula, using 16S rRNA gene high-throughput sequencing.

9.4.2 Benthic Communities in Lakes

Lakes similarly span a wide range of environmental conditions in the Arctic and Antarctic. Many of them are covered with ice for most of the year or even have a perennial ice cover. Studies on perennially ice-covered Lake Hoare in the McMurdo Dry Valleys have shown that PAR irradiance exerts an overall control on microbial photosynthetic production, composition and mat structure (Vopel and Hawes 2006). Other characteristics such as nutrients and salinity also influence the cyanobacterial richness and community structure (Sabbe et al. 2004; Jungblut et al. 2012; Zhang et al. 2015). The most spectacular cyanobacteria-based mats have been described to date from the McMurdo Dry Valley lakes, such as Lakes Vanda, Hoare, Fryxell and Joyce, and Lake Untersee in Dronning Maud Land, East Antarctica, with striking differences in shape and form. Microbial mats can range from prostrate (flat) to decimetre thick, three-dimensional branched pinnacle and cone-shaped growth morphologies that form through accumulation of annual layers during the summer months (Anderson et al. 2011; Jungblut et al. 2016; Sumner et al. 2016).

It is still not well understood what triggers the three-dimensional growth, but Sumner et al. (2016) proposed for Lake Vanda that first small tufts initiate from random irregularities in prostrate mat and that these then grow into pinnacles over the course of several years. As pinnacles increase in size and age, their interiors become colonised by a more diverse community of cyanobacteria with high photosynthetic potential. Biomass accumulation within this subsurface community causes pinnacles to swell, expanding laminae thickness and creating distinctive cylindrical bases and cuspate tops. This change in shape suggests that pinnacle morphology emerges from a specific distribution of biomass accumulation that depends on multiple microbial communities fixing carbon in different parts of pinnacles. However, local habitat conditions such as physical and chemical variables must also play some role as microbial mat structures can vary greatly among lakes.

Differences along vertical lake profiles can also be observed. In Lake Fryxell (Taylor Valley), cyanobacterial mat morphologies and cyanobacterial assemblages and microbial mat communities appear to be a result of local habitat conditions likely defined by irradiance and oxygen and sulphide concentrations. Dissolved oxygen falls from 20 mg L^{-1} to undetectable over one vertical metre from 8.9 to 9.9 m depth, and three macroscopic mat morphologies were associated with different parts of the oxygen gradient. Cuspate pinnacles occurred in the upper hyperoxic zone, which displayed complex topography, and were dominated by phycoerythrin-rich cyanobacteria attributable to the genus *Leptolyngbya* and a diverse but sparse assemblage of pennate diatoms. A less topographically complex "ridge-pit" mat was located immediately above the oxygen limit and contained *Leptolyngbya* and an increasing abundance of diatoms. Finally, flat prostrate mats occurred in the upper anoxic zone, dominated by a green cyanobacterium phylogenetically identified as *Phormidium pseudopriestleyi* and a single diatom, *Diadesmis contenta* (Jungblut et al. 2016).

In contrast, in Lake Joyce, Taylor Valley (McMurdo Dry Valleys), the water is supersaturated with calcium carbonate (Green et al. 1988), and this leads to the formation of three-dimensional microbial mats with well-developed calcite skeletons associated with the active microbial mats (Mackey et al. 2015). Interestingly, in Lake Joyce, depth-specific distributions of these calcitic microbialites, their organic carbon, photosynthetic pigments and photosynthetic potential cannot be explained by current growth conditions; it was therefore proposed that they are a legacy of past lake conditions before a 7-m lake level rise between 1973 and 2009, thus representing sentinels of environmental change (Hawes et al. 2011).

At the other pole, cyanobacteria play an important role in Arctic lakes, ponds and streams and have been well studied in the Canadian High Arctic (Bonilla et al. 2005). The benthic microbial mats in lakes often have a cohesive layering, which is established through an extra polysaccharide matrix and often have a characteristic pigment stratification (Bonilla et al. 2005; Jungblut et al. 2010; Lionard et al. 2012). The most common groups are *Oscillatoriales* and *Nostocales*, with some *Chroococcales*; however, elaborate microbialite structures have not yet been found in Arctic lake ecosystems.

9.4.3 Benthic Communities in Flowing Water Ecosystems

Streams, runnels and seepages are other aquatic features in permanently cold ecosystems in the Arctic and Antarctic. They are more ephemeral than lakes and meltwater ponds and only have liquid water during the warmer summer months from glacial and snow meltwater. Benthic microbial mats grow in these environments (Vincent 1988; Elster et al. 1997). These communities remain dormant during the winter, then rapidly resume biological activity when water wets the dry channels (Vincent et al. 1993; Tashyreva and Elster 2016). Filamentous cyanobacteria, such as *Phormidium*, *Leptolyngbya* and *Pseudanabaena*, are important for formation of the microbial mat matrix structure as found in ponds and lakes in the McMurdo Dry Valleys. Stream mats are usually less well developed and can be made of two macroscopically different layers with a more pigmented upper layer and more photosynthetically active lower layers. Other components of the communities are heterotrophic bacteria and diatoms and significant co-occurrence patterns have been identified between the different taxonomic groups, which suggests ecological interaction between these groups in stream mat communities in the McMurdo Dry Valleys (Stanish et al. 2013). The stream water also contains cyanobacteria in the plankton, which are similar taxa in the benthic stream mats, and they likely originate from dispersed benthic mat material rather than represent a distinct planktonic community (Jungblut unpublished).

Microbial mats also grow in water-filled depressions in the soil, such as the ice-free central plateau of Byers Peninsula, Livingston Island (South Shetland Islands), which form meltwater catchment areas for lakes and can cover extensive surface areas of up to several hundred square metres. These microbial mats are typically bi-layered, with an upper layer of non-active biomass and the sheath pigment scytonemin and a basal layer that contains most of the active photosynthetic microbiota. Cyanobacteria in these mats were *Leptolyngbya*, *Phormidium*, *Anabaena*, *Tychonema*, *Synechococcus*, *Oscillatoria*, *Gloeobacter*, *Calothrix*, *Microcoleus*, *Pseudanabaena and Tolypothrix*, and a comparison of samples from spring, summer and autumn found differences suggesting microbial succession within these communities (Velázques et al. 2017).

Interestingly, when looking at the trophic interactions of these mats, a Bayesian mixing model suggested that carbon flow from cyanobacteria to upper trophic levels was limited to tardigrades and rotifers, whereas fungal and bacterial activity were likely the main connectors between consumers and producers via a heterotrophic loop. This suggests that a homeostatic state displayed in freshwater microbial mats from maritime Antarctica provides stability to the microbial mats under the fluctuating environmental conditions commonly found in permanently cold shallow terrestrial aquatic ecosystems in Antarctica (Velázques et al. 2017).

Cyanobacterial mat communities are less well described from the Arctic. High biomass accumulations of *Nostoc* can be found in runnels, but further work is required to better understand the trophic interactions and relationship with the environmental conditions of these microbial mats. A detailed analysis by a range of microscopy techniques of a black pigmented stream mat on Ellesmere Island in High Arctic Canada showed that it contained *Gloeocapsa*, *Nostoc*, cf. *Tolypothrix*, *Calothrix* and cf. *Tychonema*, with the filaments binding together aggregates of calcium carbonate and other mineral particles; a pink coloured mat in the same stream also contained abundant populations of *Phormidium*, *Pseudanabaena* and *Leptolyngbya* (de los Ríos et al. 2015). A study on shallow seeps on Svalbard found extensive mats dominated by *Phormidium* that maintained large, viable overwintering populations (Tashyreva and Elster 2016).

9.4.4 Planktonic Communities in Lakes

Bloom-forming cyanobacteria are mostly absent from polar and alpine waters, reflecting the typically oligotrophic (low nutrient) status of these aquatic environments as well as their low temperatures and the lack of water column stability in summer that would favour gas-vacuolate taxa (Vincent and Quesada 2012). However, picoplanktonic forms of cyanobacteria can be present in high abundance in lakes. These usually have a lower contribution to total biomass and primary production than phototrophic microbial eukaryotes such as chrysophytes, dinoflagellates and green algal flagellates and may be a food supply for mixotrophic algae. Picocyanobacteria can also be grazed by crustacean zooplankton in high latitude waters, although they are a relatively poor quality food source for growth (Przytulska et al. 2015).

The abundance of planktonic picocyanobacteria is dependent on nutrient availability and light (Vincent 2000a) and more often nutrient limited than benthic cyanobacterial mat communities, based on in situ growth experiments in the High Arctic (Bonilla et al. 2005). In a High Arctic lake, improved irradiance conditions combined with entrainment of nutrient-rich deeper waters into the surface by mixing after loss of its ice cover resulted in a threefold increase in concentrations of picocyanobacteria (Veillette et al. 2011).

In relation to the composition, a study on lakes in the Vestfold Hills (Ace Lake, Pendant Lake and Clear Lake) found 16S rRNA sequences of *Synechococcus*-like picocyanobacteria but distinct from other *Synechococcus* genotypes, such as *Synechococcus* PS840 from the Russian marine coast (Waleron et al. 2007). In Lakes Vanda and Hoare, cyanobacteria-specific 16S rRNA gene clone analysis revealed sequences with highest BLAST similarity up to 99% to *Limnothrix redekei* and *Chamaesiphon subglobosus*, which are also genotypes found in the microbial mats (Jungblut unpublished). Interestingly, although picocyanobacteria are often the most common photosynthetic cell type in many lakes in the polar regions, they are absent or only poorly represented in polar seas (Vincent 2000a; Vincent and Quesada 2012).

9.5 Ice-Based Habitats in the Polar Regions

Similar to lakes, ponds and streams, cyanobacteria can also form large biomass accumulations in Arctic and Antarctic ice-based ecosystems (Vincent 1988; Quesada and Vincent 2012) and dominate many microbial consortia formed in ice-based habitats, such as cryoconite holes on glaciers and meltwater ponds on ice shelves both in the Arctic and Antarctic.

9.5.1 Glacial Ecosystems

Cryoconite (literally "cold rock dust") gives rise to vertical, cylindrically formed holes in the ice surface that contain a thin layer of sediment overlain by water. The formation of these habitats is initiated through the absorption of solar radiation by the sediment and the subsequent ablation of the surrounding ice (Wharton et al. 1985). Two different types of cryoconite holes can be found depending on summer conditions, such as those that are open to the air annually and those that remain closed by a perennial ice lid. Typically cryoconite holes on glaciers in the Arctic and Greenland Ice Sheet open up every summer, whereas in the Antarctic, there are also closed cryoconite holes (Webster-Brown et al. 2015).

Studies of cryoconite have described diverse microbial communities of cyanobacteria, heterotrophic bacteria and microbial eukaryotes (Cameron et al. 2012; Mueller et al. 2003; Edwards et al. 2014; Stibal et al. 2015; Webster-Brown et al. 2015) from regions such as Svalbard, Greenland Ice Shelf, Canadian Arctic, southern Victoria Land and the McMurdo Dry Valleys in Antarctica. In closed cryoconite hole communities on the Koettlitz, Diamond and Wright glaciers in southern Victoria Land, Webster-Brown et al. (2015) found 73 cyanobacterial OTUs belonging to the genera *Chamaeosiphon*, *Leptolyngbya*, *Pseudanabaena*, *Phormidesmis*, *Phormidium*, *Hormoscilla*, *Nostoc* and *Oscillatoria* using a combination of high-throughput sequencing and 16S rRNA gene clone library-based survey. Some of these cyanobacterial taxa were also identified by Christner et al. (2003) in a 16S rRNA clone library survey of open cryoconite holes from Canada Glacier, Taylor Valley in the McMurdo Dry Valleys, Antarctica, and by Edwards et al. (2011) in Svalbard. Environmental samples and isolates from four glaciers in Svalbard included the morphotypes *Chlorogloea*, *Microcoleus* and *Pseudophormidium* (Stibal et al. 2006). Many cyanobacterial genera appear to be present in supraglacial ecosystems in the Arctic and Antarctic but with large differences among sites. Variations in community structure among several glaciers in southern Victoria Land, Antarctica, were attributed to both initial wind-borne inocula from local and regional sources as well as to differences in habitat properties, such as pH of the meltwaters in the cryoconite holes (Webster-Brown et al. 2015).

Food webs in these glacial systems tend to be simple, with algae and cyanobacteria as the primary producers and heterotrophic bacteria and fungi as microbial recycling agents in cryoconite holes (Porazinska et al. 2004; Foreman et al. 2007). Cyanobacteria are not only key taxa for photosynthesis in these environments but also aggregate and bind together wind-blown debris leading to the formation of granular cryoconites. Filamentous, non-heterocystous genera, such as *Phormidium, Phormidesmis* and *Leptolyngbya,* appear to be especially important in this process. It has therefore been suggested that cyanobacteria are habitat engineers of granular microbial-mineral cryoconite aggregates in glacial ecosystems (Hodsion et al. 2010; Langford et al. 2010; Cook et al. 2015; Gokul et al. 2016).

9.5.2 Ice Shelf Ecosystems

Another important class of ice-based habitats is represented by meltwater ponds that form on ice shelves in the Arctic and Antarctic. These contain liquid water during the summer months but completely freeze over the winter. The biota of these habitats must therefore contend with extreme temperature changes, freezing and desiccation stress and high salinities. In Antarctica, the ponds on the McMurdo Ice

Shelf have low nutrient concentrations, especially nitrogen, due to the marine origin of the sediments (Hawes et al. 1993); their characteristics are described by Wait et al. (2006).

Thick benthic cyanobacterial mats coat the base of ice shelf melt ponds and are diverse communities of *Chroococcales*, *Nostocales* and *Oscillatoriales* as well as other bacterial phyla and microeukaryotes, as revealed by microscopy, culture isolation, environmental sequencing and lipid biomarker analysis (Howard-Williams et al. 1989; Nadeau et al. 2001; Jungblut et al. 2005, 2010). Salt content, measured as conductivity, can vary greatly from 100 to 70,000 μS cm^{-1}, from freshwater to brines that are more saline than sea water, and salt concentrations appear to be a key driver of cyanobacterial community assemblages in these benthic ecosystems (Jungblut et al. 2005). These meltwater ponds are also usually limited in nitrogen and cyanobacteria, especially heterocyst-forming genera such as *Nostoc*, are thought to play a more important role in the nitrogen budget than heterotrophic bacteria. Filamentous, oscillatorian cyanobacteria dominate the microbial mats in these ice shelf meltwater ponds (Fernandez-Valiente et al. 2007; Jungblut and Neilan 2010), but nitrogen-fixing *Nostoc* colonies are also present.

Similar to the south polar region, ice shelves as well as glaciers provide a variety of habitats for cyanobacteria in addition to other biota in the Arctic (Säwström et al. 2002; Varin et al. 2010; Jungblut et al. 2017). However, the total area of ice shelves is lower than that of Antarctica, where 40% of coastline is fringed by ice shelves. Furthermore, the recent breakup of the Ward Hunt Ice Shelf (Mueller et al. 2003) and the loss of the Markham Ice Shelf (Mueller et al. 2008) signal the massive reduction of these habitats through global warming.

Microbial mats in the Arctic can be prolific at some locations (Vincent et al. 2004a, b) but are generally less developed than in Antarctica; this may be due to the increased grazing pressure and also a lesser habitat stability as meltwater ponds are more ephemeral on Arctic ice shelves (Vincent 2000a). As in Antarctica, the north polar cyanobacteria communities are also dominated by oscillatorian morphotypes and genotypes such as *Leptolyngbya*, *Phormidesmis*, *Phormidium* and *Oscillatoria*. Jungblut et al. (2010) found overlaps in the abundance of genera in Arctic and Antarctic ice shelf mats and also identified Arctic sequences that had similarities up to 99% to *Leptolyngbya antarctica* that was thought to be endemic to Antarctica.

More unicellular genera were found on Arctic ice shelves than in Antarctica, including *Chamaesiphon*, *Gloeocapsa*, *Merismopedia*, cf. *Chlorogloea* sp., cf. *Aphanocapsa* sp., *Chroococcus* sp. and *Synechococcus*. This may be due to the wind-blown material from the nearby terrestrial lakes and streams (Harding et al. 2011). *Nostoc* and *Tolypothrix* can also be found; however, the genus *Nodularia* was not detected using 16S rRNA gen clone library sequencing. In total, 10 ribotypes and 19 morphotypes were identified, but this is still an underestimation, and high-throughput analyses are required for the more complete understanding of the cyanobacterial richness in ice shelf ecosystems.

9.6 Polar Soil and Rock Habitats

Cyanobacteria are often the primary colonisers of permafrost soils in areas where meltwater flushes occur through snowmelt or glacial meltwater (Pointing et al. 2015). Colonisation by cyanobacteria can increase soil stability and contribute to nutrient concentrations. Niederberger et al. (2015) showed that cyanobacteria and diverse groups of heterotrophic bacteria contributed to nitrogen fixation in wetted soils in the McMurdo Dry Valleys by comparing DNA, transcript *nifH* genes and acetylene reduction assays. However, in the Dry Valleys of Antarctica, the soils are old and weathered and have low carbon and nutrient concentrations (Vincent 1988; Cary et al. 2010), and cyanobacteria do not seem to be ubiquitously present in soil samples in the McMurdo Dry Valleys as cyanobacteria were only detected in some of the studied soil samples using 16S rRNA gene high-throughput sequencing (Van Horn et al. 2013) and ARISA (Magalhaes et al. 2012). Wood et al. (2008) also concluded that moisture content of Antarctic Dry Valley soils was a poor indicator of cyanobacteria distribution and that other variables such as soil elemental composition may play a role in influencing edaphic cyanobacterial habitat suitability.

Terrestrial dark crusts are found throughout the Arctic and Antarctica and are commonly dominated by cyanobacteria. A study performing 16S rRNA gene high-throughput sequencing on dark soil crusts growing in water tracks in the Ward Hunt Lake catchment, Canadian high Arctic (Stevens et al. 2013), found communities dominated by *Oscillatoriales* and OTUs grouping within *Nostocales*, *Chroococcales* and *Acaryochloris*, as well as a diverse heterotrophic community that included *Acidobacteria*, *Proteobacteria*, *Bacteroides* and *Verrucumicrobia*. Although there was no clear relationship between overall community structure and water availability, cyanobacteria were present in higher abundance in the water tracks with more regular water.

Cyanobacteria are also often identified in biofilms below and within the rocks where the microclimate gives protection against environmental stresses, such as high UV radiation, temperature extremes, desiccation and physical removal by wind in hyperarid polar deserts. They can be found in depth below the rock surface, depending on the optical characteristics of the rocks and the level of available PAR. Depending on the spatial location of the communities, they are hypolithic (beneath rocks), endolithic (in pore spaces of rocks), chasmoendolithic (in cracks and fissures of rocks) or cryptoendolithic (in the pore space between mineral grains forming sedimentary rocks) (Pointing et al. 2009; Makhalanyane et al. 2013; Stomeo et al. 2013).

Hypolithic communities under translucent rocks can be found in Antarctica, for example, in the Miers Valley of the McMurdo Dry Valleys. These microbial assemblages differ from soil communities, and cyanobacteria appear particularly during the first stage of developmental succession. The increased nutrient enrichment of the hypolithic microenvironment resulting from cyanobacterial colonisation could set the stage for subsequent heterotrophic fungal colonisation and the eventual development of bryophyte-based communities (Cowan et al. 2010). A phylogenetic inference analysis found that hypolithic 16S rRNA cyanobacterial sequences formed clades with Oscillatoriales, such as *Phormidium priestleyi* and *Leptolyngbya frigida*, and Nostocales such as *Nostoc* (Pointing et al. 2009; Khan et al. 2011) in hypoliths in the McKelvey and Miers valleys. In the Arctic, hypolithic cyanobacteria are also commonly observed under opaque rocks subjected to periglacial movements (Cockell and Stokes 2004); however there is still a lack of 16S rRNA gene community structure analysis and phylogenetic analysis for a biogeographic and ecological comparison.

Chasmoliths and endoliths have also been studied in the McKelvey Valley in Antarctica and appear to have bacterial, eukaryotic and functional gene communities that are distinct from hypolithic and soil bacterial communities (Pointing et al. 2009; Chan et al. 2013). 16S rRNA gene cyanobacterial communities were dominated by *Chroococcidiopsis*, which had already been identified by morphology in sandstones of the Dry Valleys by Friedmann and Ocampo (1976). This genus is remarkably resistant to desiccation and has close relatives in hot deserts (Fewer et al. 2002). Other 16S rRNA sequences identified in this habitat have included *Acaryochloris* and *Tolypothrix* (Miyashita et al. 1996; de los Ríos et al. 2007; Pointing et al. 2009), and microscopic analysis has identified *Gloeocapsa*-like cyanobacteria in other types of continental Antarctic rocks such as sandstones. In the Arctic, cryptoendolithic communities are also common in sandstone outcrops of Eureka, Ellesmere Island, and consist of similar cyanobacterial morphotypes as in Antarctic rocks (Omelon et al. 2006); however, there is still a lack of molecular analysis.

9.7 Marine Ecosystems

The abundance of picocyanobacteria decreases markedly from temperate latitudes to polar regions (Marchant et al. 1987; Fouilland et al. 1999; Vincent and Quesada 2012). This decrease is likely due to temperature-limited growth rates (Vincent 2000a; Marchant 2005) in combination with continuous losses due to grazing, advection and mixing (Vincent 2000a; Vincent and Quesada 2012). In a 16S rRNA gene study of bacterial communities in the Beaufort Sea and Amundson Gulf, cyanobacteria were not part of the top 50 most abundant bacterial taxa (Comeau et al. 2011) and represented only 0.5% of the relative abundance in the epishelf lake in Milne Fjord, Canadian High Arctic, whereas phototrophic microbial eukaryotes were more abundant (Thaler et al. 2017). Interestingly a study of their molecular diversity in the Beaufort Sea identified picocyanobacteria using a cyanobacteria-specific amplification protocol, but they were affiliated with freshwater and brackish *Synechococcus* lineages and not with oceanic ones (Waleron et al. 2007). This was further confirmed by Blais et al. (2012), who also identified cyanobacterial nitrogenase reductase genes (*nifH*) from the Beaufort Sea, but again had highest match to cyanobacteria in freshwater or terrestrial habitats, such as the heterocyst-forming and nitrogen-fixing genus *Nostoc*. Their origin, therefore, seemed allochthonous for the Arctic Ocean as it is much influenced by large riverine inputs. Similar findings have also resulted from Antarctic marine studies (e.g. Koh et al. 2012) where cyanobacteria were rare in the Antarctic ice communities and most closely related to freshwater *Cyanobacteria* from neighbouring terrestrial sources. Interestingly, Wilmotte et al. (2002) detected 16S rRNA sequences of picocyanobacteria that were closely related to temperate oceanic *Synechococcus*, such as WH8103 and WH7803 (Wilmotte et al. 2002) from the Subantarctic Front (51°S), suggesting that some marine genotypes may occur in polar waters.

9.8 Alpine Habitats

9.8.1 Ice-Based Supraglacial Ecosystems

Glaciers are a key feature of alpine landscapes, and supraglacial ecosystems, such as cryoconite hole communities, can be found from European and New Zealand Alps to the Himalayas. Therefore, it is not surprising that similarities have been found across supraglacial microbial diversity and communities and *Cyanobacteria* and *Proteobacteria* are important taxa in cryoconite assemblages. Although there are similarities in the cryosphere ecosystems across Svalbard, Greenland and Austria, a comparative study also found distinct geographic signals in the communities suggesting that local conditions play a role in shaping the assemblages (Edwards et al. 2014).

9.8.2 Streams and Lakes

In alpine streams, water chemistry, geochemical conditions, hydraulic conditions and permanence of flow are the key factors defining taxonomic diversity. Cyanobacteria have been found as part of microbial mats, epiphytic on mosses and endosymbiotic in lichens in stream habitats of many alpine regions, but there are no specific studies on their molecular diversity or adaptations (McClintic et al. 2003; Rott et al. 2006). Nutrient concentrations show large variations during the year with peaks in late winter and autumn. PAR and UV radiation also range from low levels in the presence of ice and snow cover to high levels during summer months, creating a need for protective mechanisms to survive. Rott et al. (2006) have described different colonisation patterns for several cyanobacterial morphotypes in alpine streams.

Cyanobacteria in alpine freshwater lakes can be found as benthic and planktonic communities. Plankton communities are mostly comprised of *Synechococcus* morphotypes, and their abundance is correlated to nutrient availability, particularly nitrogen and phosphorus. Benthic communities were studied by Mez et al. (1998) and Sommaruga and Garcia-Pichel (1999).

9.8.3 *Rocks and Soils*

Cyanobacteria are also a component of alpine soil crusts and of soils from recently deglaciated areas and rock-associated communities, as described for endolithic communities of dolomite rocks in the Swiss Alps (Sigler et al. 2003). Diverse cyanobacteria, such as *Microcoleus*, *Nodosilinea*, *Nostoc* and Chroococcales, were found, based on morphological and 16S rRNA gene phylogenetic analysis in soil crust altitudinal gradients (4600–5900 m) in the Himalayas (Janatková et al. 2013). The presence and role of cyanobacteria in microbial and plant succession in the Andes has also been examined (Schmidt et al. 2008). Cyanobacterial sequences in Andean soils (Nemergut et al. 2007) included those related to the *Chamaesiphon* PCC7430 (96% sequence similarity), diverse *Nostoc* strains (up to 98.5% sequence similarity with the Antarctic ANT.L52B.1), the Antarctic *Leptolyngbya frigida* ANT.LH52.2 (98.5%) and ANT.LH52B.3 (99.6%), or differed from database sequences and thus represented a novel diversity.

Cyanobacteria have also been described from rock habitats ranging from European alpine dolomite layers to rocks in cold desert environment, such as stony desert pavements in the Qaidam Basin in China where communities contained in particular *Chroococcidiopsis*, based on 16S rRNA gene phylogenies, and cyanobacterial communities in comparison to hot desert environments were spatially aggregated at multiple scales in patterns distinct from the underlying rock pattern. Site-level differences in cyanobacterial spatial pattern (e.g. mean inter-patch distance) were linked with rainfall, whereas patchiness within sites was correlated with local geology (greater colonisation frequency of large rocks) and biology (dispersal during rainfall) (Warren-Rhodes et al. 2007).

9.9 Evolution and Biogeography

The origin and biogeography of cyanobacteria found in permanently cold environments, especially in the Arctic and Antarctic, remain topics of much debate. Physiological studies of polar cyanobacterial strains found that they were mostly psychrotolerant and had temperature optima for growth rates in the range 15–20 °C, well above their habitat temperatures, and it was therefore suggested that they likely had their evolutionary origins within temperate latitudes (Tang et al. 1997; Nadeau et al. 2001) and subsequently colonised perennial cold habitats. Phylogenetic interference based on a limited number of 16S rRNA genes of these first polar cyanobacteria isolates showed that they grouped with clades across the cyanobacterial Tree of Life, implying that psychrotolerant phenotypes evolved several times (Nadeau et al. 2001). A far more comprehensive study based on a 16S rRNA gene tree with a large-scale multi gene analysis (135 proteins and 2 ribosomal RNAs) also identified clades that are common to Arctic, Antarctic and alpine sites in all major lineages (Chrismas et al. 2015). Bayesian ancestral state reconstruction was also performed with 16S rRNA gene sequences from these three cold region environments. This analysis of 270 sequences of cyanobacteria identified 20 clades with each having common ancestors that were likely capable of surviving in the cold (Chrismas et al. 2015).

Linked to the question of where and what evolutionary mechanisms led to the radiation of cyanobacteria with an ability to dealing with extreme cold is the ongoing discussion about cyanobacterial biogeography. A long-standing theory of microbial distribution is that "everything is everywhere, but the environment selects" and that local habitats select for specific microbiota that are globally distributed (Baas-Becking 1934). Castenholz (1992) noted the slow rates of speciation in the cyanobacteria together with their large dispersal abilities, and this in combination with the relatively young age of most polar ice-free environments suggests that endemism is likely to be rare among polar cyanobacteria. On the other hand, several features of Antarctica suggest that microbial endemism may be possible there (Vincent 2000b): (1) Antarctica has been more isolated than other parts of the world for several million years; (2) dispersal processes which favour local species are more efficient than long-range dispersal processes; and (3) there has probably been strong environmental selection for adaptive strategies. Morphological identifications seemed to support this hypothesis, but such characterisation is limited due to morphological plasticity. There are also several phylogenetic studies based on 16S rRNA gene sequences from isolated cyanobacteria and from environmental samples that have identified taxa which appeared to be restricted to specific geographic sites and which may suggest the presence of endemicity (Taton et al. 2006a, b; Strunecky et al. 2011; Komárek et al. 2012). However, comparisons with other geographic locations are still unsatisfactory due to the limited number of reference strains in culture and the limited sampling over the vast polar and alpine regions.

The first comprehensive 16S rRNA gene surveys of cyanobacterial mat assemblages from the High Arctic indicated the presence of cyanobacterial ecotypes that were 99% similar, based on 16S rRNA gene analyses, to taxa previously thought to be endemic to Antarctica. More than 68% of all identified ribotypes at each site matched only cyanobacterial sequences from perennially cold terrestrial ecosystems and were less than 97.5% similar to sequences from warmer environments. This implied the global distribution of low-temperature cyanobacterial ecotypes throughout the cold terrestrial biosphere (Jungblut et al. 2010).

Although several studies have shown the potentially cosmopolitan distribution of cold ecotypes, other work has indicated that there are dispersal barriers on local to global scales leading to distinct populations within 16S rRNA gene genotypes and ecotypes. At the local scale, Novis and Smissen (2006) showed with amplified fragment length polymorphism (AFLP) analysis that *Nostoc commune* formed distinct genetic groups according to the habitat in the McMurdo Dry Valleys,

Antarctica. At the global scale, barriers of dispersal have also been identified between 16S rRNA gene ribotypes of desert-dwelling *Chroococcidiopsis*, as molecular clock analysis of 16S RRNA genotypes from cold arid and hot deserts suggested no evidence of recent interregional gene flow. The temporal phylogeny suggests that the time of most recent common ancestry to all contemporary variants was 2.5 Ga (range: 3.1–1.9 Ga), indicating populations have not shared common ancestry since before the formation of modern continents (Bahl et al. 2011). This suggests that the biogeography of polar cyanobacteria is complex and may vary across taxonomic species and ecotypes. Therefore, further studies are needed with better species sampling and using multilocus gene and genomic analysis for adequate resolution.

9.10 Conclusions

Cyanobacteria evolved under the harsh, highly variable temperature regimes of the Precambrian and their modern representatives retain a remarkable ability to adapt to and survive within extreme conditions. They are among the microbial dominants in soil ecosystems, the hydrosphere and the cryosphere of Arctic, Antarctic and alpine regions. They are keystone taxa as they play a major ecological role in primary colonisation of substrates, primary production, nitrogen input and even as habitat engineers in these permanently cold ecosystems. The application of molecular tools, in particular high-throughput sequencing and metagenomics, in combination with laboratory experimentation has begun to provide new insights into the full diversity of cyanobacteria, their community dynamics and their biogeographical distribution in cold environments on a global scale. However, more studies and especially comparative genomic evaluations are needed to unravel the enigmatic origins and evolution of cyanobacteria in the cold biosphere. This is an urgent task in view of the climatic changes that have already begun to alter the structure and functioning of polar and alpine microbial ecosystems.

Acknowledgements We thank our funding support including from the Natural History Museum, the Centre d'études nordiques (CEN), the Natural Sciences and Research Council of Canada (NSERC) and the Network of Centres of Excellence ArcticNet and our logistics support in the polar regions from Polar Continental Shelf Program (Canada), British Antarctic Survey, Antarctica New Zealand and the US Antarctic Program.

References

Andersen DT, Sumner DY, Hawes I, Webster-Brown J, Mackay CP (2011) Discovery of large conical stromatolites in Lake Untersee, Antarctica. Geobiology 9:280–293

Archer SDJ, McDonald IR, Herbold CW, Lee CK, Cary CS (2015) Benthic microbial communities of coastal terrestrial and ice shelf Antarctic meltwater ponds. Front Microbiol 6:485

Baas-Becking LGM (1934) Geobiologie of inleiding tot de milieukunde. van Stockum WP, Zoon NV, The Hague
Bahl J, Lau MCY, Smith GJD, Vijaykrishna D, Cary C, Lacap DC, Lee CK, Papke RT, Warren-Rhodes KA, Wong FKY, McKay CP, Pointing SB (2011) Ancient origins determine global biogeography of hot and cold desert cyanobacteria. Nat Commun 2:163
Blais M, Tremblay J-É, Jungblut AD, Gagnon J, Martin J, Thaler M, Lovejoy C (2012) Nitrogen fixation and identification of potential diazotrophs in the Canadian Arctic. Glob Biogeochem Cycles 26:GB3022. doi:10.1029/2011GB004096
Bonilla S, Villeneuve V, Vincent WF (2005) Benthic and planktonic algal communities in a high arctic lake: pigment structure and contrasting responses to nutrient enrichment. J Phycol 41:1120–1130
Cameron KA, Hodson AJ, Osborn AM (2012) Structure and diversity of bacteria, eukaryotic and archaeal communities in glacial cryoconite holes from the Arctic and the Antarctic. FEMS Microbiol Ecol 91:254–267
Cary SC, McDonald IR, Barrett JE, Coawn DA (2010) On the rocks: the microbiology of Antarctic Dry Valley soils. Nat Rev Microbiol 8:129–138
Castenholz RW (1992) Species usage, concept, and evolution in the Cyanobacteria (blue-green algae). J Phycol 28:737–745
Castenholz RW (2001) Phylum BX. Cyanobacteria. Oxygenic photosynthetic bacteria. In: Boone DR, Castenholz RW, Garrity GM (eds) Bergey's manual of systematic bacteriology (the archaea and the deeply branching and phototrophic bacteria), vol 1. Springer, New York, NY, pp 473–599
Castenholz RW, Garcia-Pichel F (2000) Cyanobacterial responses to UV-radiation. In: Whitton BA, Potts M (eds) The ecology of cyanobacteria. Kluwer, Dordrecht, pp 591–611
Chen M, Scheer H (2013) Extending the limits of natural photosynthesis and implications for technical light harvesting. J Porphyrins Phthalocyanines 17:1–15
Chan Y, Van Nostrand JD, Zhou J, Pointing SB, Farrell RL (2013) Proc Natl Acad Sci USA 110:8990–8995
Chorus I, Bartram J (1999) Toxic cyanobacteria in water. A guide to their public health consequences, monitoring and management. World Health Organization, London
Chrismas N, Anesio AM, Sanches-Baracaldo P (2015) Multiple adaptations to polar and alpine environments within cyanobacteria: a phylogenomic and Bayesian approach. Front Microbiol 6:1070
Chrismas N, Barker G, Anesio AM, Sanches-Baracaldo P (2016) Genomic mechanisms for cold tolerance and production of exopolysaccharides in the Arctic cyanobacterium *Phormidesmis priestleyi* BC1401. BMC Genom 17:533
Christner BC, Kvitoko BH, Reeve JN (2003) Molecular identification of Bacteria and Eukarya inhabiting an Antarctic cryoconite hole. Extremophiles 7:177–183
Cockell CS, Stokes MD (2004) Widespread colonization by polar hypoliths. Nature 431:414
Comeau AM, Li WKW, Tremblay J-E, Carmack EC, Lovejoy C (2011) Changes in Arctic Ocean microbial community structure following the 2007 record sea ice minimum. PLoS ONE 6: e27492
Cook JM, Edwards A, Takeuchi N, Irvine-Fynn TDL (2015) Cryoconite: the dirty biological secret of the cryosphere. Prog Phys Geogr 40:66–111
Cowan DA, Khan N, Pointin SB, Cary SG (2010) Diversity of hypolithic refuge communities in the McMurdo Dry Valleys. Antarct Sci 22:714–720
de los Ríos A, Grube M, Sancho LG, Ascaso C (2007) Ultrastructural and genetic characteristics of endolithic cyanobacterial biofilms colonizing Antarctic granite rocks. FEMS Microbiol Ecol 59:386–395
de los Ríos A, Ascaso C, Wierzchos J, Vincent WF, Quesada A (2015) Microstructure and cyanobacterial composition of microbial mats from the High Arctic. Biodivers Conserv 24:841–886

DeSantis TZ, Hugenholtz P, Larsen N et al (2006) Greengenes, a chimera-checked 16S rRNA gene database and workbench compatible with ARB. Appl Environ Microbiol 72:5069–5072

Edwards A, Anesio AM, Rassner SM et al (2011) Possible interactions between bacterial diversity, microbial activity and supraglacial hydrology of cryoconite holes in Svalbard. ISME J 5:150–160

Edwards A, Mur LAJ, Girdwood SE et al (2014) Coupled cryoconite ecosystem structure-function relationships are revealed by comparing bacterial communities in alpine and Arctic glaciers. FEMS Microbiol Ecol 89:222–237

Elster J, Syobody J, Komárek J, Marvan P (1997) Algal and cyanoprocaryote communities in a glacial stream, Sverdrup Pass, 79N, Central Ellesmere Island, Canada. Arch Hydrobiol Suppl 119:57–93

Fernandez-Valiente E, Camacho A, Rochera C, Rico E, Vincent WF, Quesada A (2007) Community structure and physiological characterization of microbial mats in Byers Peninsula, Livingston Island (South Shetland Islands, Antarctica). FEMS Microbiol Ecol 59:377–385

Fewer D, Friedl T, Büdel B (2002) *Chroococcidiopsis* and heterocyst-differentiating cyanobacteria are each other's closest living relatives. Mol Phylogenet Evol 23:82–90

Foreman CM, Sattler B, Mikucki J et al (2007) Metabolic activity and diversity of cryoconites in the Taylor Valley, Antarctica. J Geophys Res 112:1–11

Fouilland E, Descolas-Gros C, Courties C, Pons V (1999) Autotrophic carbon assimilation and biomass from size-fractionated phytoplankton in the surface waters across the subtropical frontal zone (Indian Ocean). Polar Biol 21:90–96

Friedmann EI, Ocampo R (1976) Endolithic blue-green algae in the Dry Valleys: primary producers in the Antarctic ecosystem. Science 193:1247–1249

Geitler L (1932) Cyanophyceae. In: Kolkwitz R (ed) Rabenhorst's Kryptogamenflora von Deutschland, Osterreich und der Schweiz. Akademische Verlagsgesellschaft, Leibzig

Gokul JK, Hodson AJ, Saetnan ER, Irvine-Fynn TDL, Westall PJ, Detheride AP, Takeychi N, Bussell J, Mur LAJ, Edwards E (2016) Taxon interactions control the distributions of cryoconite bacteria colonizing a High Arctic ice cap. Mol Ecol 25:3752–3767

Green WJ, Angle MP, Chave KE (1988) The geochemistry of Antarctic streams and their role in the evolution of four lakes of the McMurdo Dry Valleys. Geochim Cosmochim Acta 52:1265–1274

Harding T, Jungblut AD, Lovejoy C, Vincent WF (2011) Microbes in High Arctic snow and implications for the cold biosphere. Appl Environ Microbiol 77:3234–3243

Hawes I, Schwarz AM (2001) Absorption and utilization of irradiance by cyanobacterial mats in two ice-covered Antarctic lakes with contrasting light climates. J Phycol 37:5–15

Hawes I, Howard-Williams C, Pridmore RD (1993) Environmental control of microbial biomass in the ponds of the McMurdo Ice Shelf, Antarctica. Arch Hydrobiol 127:271–287

Hawes I, Sumner DY, Andersen DT, Mackey TJ (2011) Legacies of recent environmental change in the benthic communities of Lake Joyce, a perennially ice-covered Antarctic lake. Geobiology 9:394–410

Hitzfeld B, Lampert CS, Spaeth N, Mountfort D, Kaspar H, Dietrich DR (2000) Toxin production in cyanobacterial mats from ponds on the McMurdo Ice Shelf, Antarctica. Toxicon 38:1731–1748

Hodson A, Cameron K, Bøggild C et al (2010) The structure, biogeochemistry and formation of cryoconite aggregates upon an Arctic valley glacier; Longyearbreen, Svalbard. J Glaciol 56:349–362

Hoffman PF (2016) Cryoconite pans on Snowball Earth: supraglacial oases for Cryogenian eukaryotes? Geobiology 14:531–542

Hoffmann L, Komárek J, Kaštovský J (2005a) System of cyanoprokaryotes (cyanobacteria) – state in 2004. Algol Stud 117:95–115

Hoffmann L, Komárek J, Kaštovský J (2005b) Proposal of cyanobacterial system – 2004. In: Büdel B, Krienitz L, Gärtner G, Schagerl M (eds) Süsswasserflora von Mitteleuropa 19/2. Elsevier/Spektrum, Heidelberg, pp 657–660

Howard-Williams C, Pridmore RD, Downes MT, Vincent WF (1989) Microbial biomass, photosynthesis and chlorophyll a related pigments in the ponds of the McMurdo Ice Shelf, Antarctica. Antarct Sci 1:125–131

Janatková K, Řeháková K, Doležal J, Šimek M, Chlumská Z, Dvorský M, Kopecký M (2013) Community structure of soil phototrophs along environmental gradients in arid Himalaya. Environ Microbiol 15:2505–2516

Jancusova M, Kovacik L, Batista Periera A, Dusinsky R, Wilmotte A (2016) Polyphasic characterization of 10 selected ecologically relevant filamentous cyanobacterial strains from the South Shetland Islands, Maritime Antarctica. FEMS Microbiol Ecol 92:fiw100

Jungblut AD, Neilan BA (2010) NifH-gene diversity and expression in a microbial mat community on the McMurdo Ice Shelf, Antarctica. Antarct Sci 22:177–122

Jungblut AD, Hawes I, Mountfort D, Hitzfeld B, Dietrich DR, Burns BP, Neilan BA (2005) Diversity within cyanobacterial mat communities in variable salinity meltwater ponds of McMurdo Ice Shelf, Antarctica. Environ Microbiol 7:519–529

Jungblut AD, Hoeger SJ, Mountford D, Hitzfeld BC, Dietrich DR, Neilan BA (2006) Characterization of microcystin production in an Antarctic cyanobacterial mat community. Toxicon 47:271–278

Jungblut AD, Lovejoy C, Vincent WF (2010) Global distribution of cyanobacterial ecotypes in the cold biosphere. ISME J 4:191–202

Jungblut AD, Wood SA, Hawes I, Webster-Brown J, Harris C (2012) The Pyramid Trough Wetland: environmental and biological diversity in a newly created Antarctic protected area. FEMS Microbiol Ecol 82:356–366

Jungblut AD, Hawes I, Mackey TJ, Krusor M, Doran PT, Sumner DY, Eisen JA, Hillman C, Goroncy AK (2016) Microbial mat communities along an oxygen gradient in a perennially ice-covered Antarctic lake. Appl Environ Microbiol 82:620–630

Jungblut AD, Mueller D, Vincent WF (2017) Arctic ice shelf ecosystems. In: Copland L, Mueller DR (eds) Arctic Ice Shelves and Ice Islands. Springer SBM, Dordrecht

Kaebernick M, Neilan BA (2001) Ecological and molecular investigations of cyanotoxin production. FEMS Microbiol Ecol 35:1–9

Khan N, Tuffin M, Stafford W, Cary C, Lacap DC, Pointing SB et al (2011) Hypolithic microbial communities of quartz rocks from Miers Valley, McMurdo Dry Valleys, Antarctica. Polar Biol 34:1657–1668

Kleinteich J, Wood SA, Kupper FC, Camacho A, Quesada A, Frickey T, Dietrich DR (2012) Temperature-related changes in polar cyanobacterial mat diversity and toxin production. Nat Clim Chang 2:356–360

Kleinteich J, Wood SA, Puddick J, Schleheck D, Kupper FC, Dietrich D (2013) Potent toxins in Arctic environments: presence of saxitoxins and an unusual microcystin variant in Arctic freshwater ecosystems. Chem Biol Interact 206:423–431

Kleinteich J, Hildebrand F, Wood SA, Cirs S, Agha R, Quesada A, Pearce D, Convey P, Küpper F, Dietrich DR (2014) Diversity of toxin and non-toxin containing cyanobacterial mats of meltwater ponds on the Antarctic Peninsula: a pyrosequencing approach. Antarct Sci 26:521–532

Koh EY, Cowie ROM, Simpson AM, O'Toole R, Ryan KG (2012) The origin of cyanobacteria in Antarctic sea ice: marine or freshwater? Environ Microbiol Rep 4:479–483

Komárek J, Anagnostidis K (2005) Cyanoprokaryota 2. Teil Oscillatoriales. Spektrum Akademischer Verlag, Heidelberg

Komárek J, Nedbalová L, Hauer T (2012) Phylogenetic position and taxonomy of three heterocytous cyanobacteria dominating the littoral of deglaciated lakes, James Ross Island, Antarctica. Polar Biol 35:759–774

Komárek J, Kaštovský J, Mareš J, Johansen JR (2014) Taxonomic classification of cyanoprokaryotes (cyanobacterial genera) 2014, using a polyphasic approach. Preslia 86:295–335

Langford H, Hodson A, Banwart S, Bøggild C (2010) The microstructure and biogeochemistry of Arctic cryoconite granules. Ann Glaciol 51:87–94
Leslie A (1879) The Arctic voyages of Adolf Erik Nordenskiöld. MacMillan, London
Lionard M, Péquin B, Lovejoy C, Vincent WF (2012) Benthic cyanobacterial mats in the High Arctic: multi-layer structure and fluorescence responses to osmotic stress. Front Aquat Microbiol 3:140
Mackey TJ, Sumner DY, Hawes I, Jungblut A-D, Andersen DT (2015) Branched stromatolites in Lake Joyce, Antarctica. Geobiology 13:373–390
Magalhaes C, Stevens MI, Cary SC, Ball BA, Storey BC, Wall DH, Tuerk R, Ruprecht U (2012) At limits of life: multidisciplinary insights reveal environmental constraints on biotic diversity in continental Antarctica. PLoS ONE 7:e44578
Makhalanyane TP, Valverde A, Lacap DC, Pointing SB, Tuffin M, Cowan DA (2013) Evidence of species recruitment and development of hot desert hypolithic communities. Environ Microbiol Rep 5:219–224
Marchant HJ (2005) Cyanophytes. In: Scott FJ, Marchant HJ (eds) Antarctic marine protists. Australian Biological Resources Study, Canberra, pp 324–325
Marchant HJ, Davidson AT, Wright SW (1987) The distribution and abundance of chroococcoid cyanobacteria in the Southern Ocean. Proc NIPR Symp Polar Biol 1:1–9
Margesin R, Häggblom M (2007) Thematic issue: microorganisms in cold environments. FEMS Microbiol Ecol 59:215–216
Martineau E, Wood SA, Miller MR, Jungblut AD, Hawes I, Webster-Brown J, Packer MA (2013) Characterisation of Antarctic cyanobacteria and comparison with New Zealand strains. Hydrobiology 711:139–154
McClintic AS, Casamatta DA, Vis ML (2003) A survey of algae from montane cloud forest and alpine streams in Bolivia: macroalgae and associated microalgae. Nova Hedwigia 76:363–379
Mez K, Hanselmann K, Preisig HR (1998) Environmental conditions in high mountain lakes containing toxic benthic cyanobacteria. Hydrobiologia 368:1–15
Miyashita H, Ikemoto H, Kurano N, Adachi K, Chihara M, Miyashi S (1996) Chlorophyll *d* as a major pigment. Nature 383:402
Moffitt CM, Neilan AB (2004) Characterization of the nodularin synthetase gene cluster and proposed evolution of cyanobacterial hepatotoxins. Appl Environ Microbiol 70:6353–6362
Mueller DR, Vincent WF, Jeffries MO (2003) Break-up of the largest Arctic ice shelf and associated loss of an epishelf lake. Geophys Res Lett 30:2031
Mueller DR, Copland L, Hamilton A, Stern DR (2008) Examining Arctic ice shelves prior to 2008 breakup. EOS Trans Am Geophys Union 89:502–503
Murray J (1910) British Antarctic expedition 1907–1909. Reports on the scientific investigations. Biology 1(Part V):83–187
Nadeau TL, Milbrandt EC, Castenholz RW (2001) Evolutionary relationships of cultivated Antarctic Oscillatoriaceans (cyanobacteria). J Phycol 37:650–654
Nemergut DR, Anderson SP, Cleveland CC, Martin AP, Miller AE, Seimon A, Schmidt SK (2007) Microbial community succession in an unvegetated, recently deglaciated soil. Microb Ecol 53:110–122
Niederberger TD, Sohm JA, Gunderson TE, Parker AE, Tirindelli J, Capone DG, Carpenter EK, Cary SC (2015) Microbial community composition of transiently wetted Antarctic Dry Valley soils. Front Microbiol 6:9
Novis PM, Smissen RD (2006) Two genetic and ecological groups of *Nostoc commune* in Victoria Land, Antarctia, revealed by AFLP analysis. Antarct Sci 18:573–581
Omelon CR, Pollard WH, Ferris FG (2006) Environmental controls on microbial colonization of high Arctic cryptoendolithic habitats. Polar Biol 30:19–29
Oren A (2000) Salt and brines. In: Whitton BA, Potts M (eds) The ecology of cyanobacteria. Kluwer Academic Press, Dordrecht, pp 281–306
Oren A (2004) A proposal for further integration of the cyanobacteria under the Bacteriological Code. Int J Syst Evol Microbiol 54:1895–1902

Pointing SB, Chan Y, Lacap DC, Lau MC, Jurgens JA, Farrell RL (2009) Highly specialized microbial diversity in hyper-arid polar desert. Proc Natl Acad Sci U S A 106:19964–19969

Pointing SB, Büdel B, Convey P, Gillman L, Körner C, Leuzinger S, Vincent WF (2015) Biogeography of photoautotrophs in the high polar biome. Front Plant Sci 6:692

Porazinska DL, Fountain AG, Nylen TH et al (2004) The biodiversity and biogeochemistry of cryoconite holes from McMurdo Dry Valley Glaciers, Antarctica. Arct Antarct Alp Res 36:84–91

Przytulska A, Bartosiewicz M, Rautio M, Dufresne F, Vincent WF (2015) Climate effects on Arctic *Daphnia* via food quality and thresholds. PLoS ONE 10:e0126231

Quesada A, Vincent WF (2012) Cyanobacteria in the cryosphere: snow, ice and extreme cold. In: Whitton BA (ed) Ecology of cyanobacteria II: their diversity in space and time. Springer, New York, NY, pp 387–399

Rajaniemi P, Hrouzek P, Kastovska K, Willame R, Rantala A, Hoffmann L, Komarek J, Sivonen K (2005) Phylogenetic and morphological evaluation of the genera *Anabaena, Aphanizomenon, Trichormus* and *Nostoc* (Nostocales, Cyanobacteria). Int J Syst Evol Microbiol 55:11–26

Rantala A, Fewer D, Hisburges M, Rouhiainen L, Vaitomaa J, Börner T (2004) Phylogenetic evidence for the early evolution of the microcystin synthesis. Proc Natl Acad Sci U S A 101:568–573

Rocap G, Distel DL, Waterbury JB, Chisholm SW (2002) Resolution of *Prochlorococcus* and *Synechococcus* ecotypes by using 16S-23S ribosomal DNA internal transcribed spacer sequences. Appl Environ Microbiol 68(3):1180–1191

Rott E, Cantonati M, Füreder L, Pfister P (2006) Benthic algae in high altitude streams of the Alps-a neglected component of the aquatic biota. Hydrobiology 562:195–216

Sabbe K, Hodgson DA, Verlyen E, Taton A, Wilmotte A, Vanhoutte KD, Vyverman W (2004) Salinity, depth and the structure and composition of microbial mats in continental Antarctic lakes. Freshw Biol 49:296–319

Säwström C, Mumford P, Marshall W, Hodson A, Laybourn-Parry J (2002) The microbial communities and primary productivity of crycoconite holes in an Arctic glacier (Svalbard 79°N). Polar Biol 25:591–596

Schmidt SK, Reed SC, Nemergut DR, Grandy AS, Cleveland CC, Weintraub MN, Hill AW, Costello EK, Meyer AF, Neff JC, Martin AM (2008) The earliest stages of ecosystem succession in high-elevation (5000 metres above sea level), recently deglaciated soils. Proc R Soc Lond B Biol Sci 275:2793–2802

Sigler WV, Bachofen R, Zeier J (2003) Molecular characterization of endolithic cyanobacteria inhabiting exposed dolomite in central Switzerland. Environ Microbiol 5:618–627

Sommaruga R, Garcia-Pichel F (1999) UV-absorbing mycosporine-like compounds in planktonic and benthic organisms from a high-mountain lake. Arch Hydrobiol 144:255–269

Stanier RY, Sistrom WR, Hansen TA, Whitton BA, Castenholz RW, Pfennig N, Gorlenko VN, Kondratieva EN, Eimhjellen KE, Whittenbury R, Gherma RL, Trüper HG (1978) Proposal to place nomenclature of cyanobacteria (blue-green-algae) under rules of International Code of Nomenclature of Bacteria. Int J Syst Bacteriol 28:335–336

Stanish LF, O'Neil SP, Gonzales A, Legg TM, Knelman J, McKnight DM, Spaulding S, Nemergut DR (2013) Bacteria and diatom co-occurrence patterns in microbial mats from polar desert streams. Environ Microbiol 15:1115–1131

Stevens B, Lionard M, Kuske CR, Vincent WF (2013) High bacterial diversity of biological soil crusts in water tracks over permafrost in the High Arctic polar desert. PLoS ONE 8:e71489

Stibal M, Sabacká M, Kastovská K (2006) Microbial communities on glacier surfaces in Svalbard: impact of physical and chemical properties on abundance and structure of cyanobacteria and algae. Microb Ecol 52:644–654

Stibal M, Schostag M, Cameron KA et al (2015) Different bulk and active bacterial communities in cryoconite from the margin and interior of the Greenland ice sheet. Environ Microbiol Rep 7:293–300

Stomeo F, Valverde A, Pointing SB, McKay CP, Warren-Rhodes KA, Tuffin MI et al (2013) Hypolithic and soil microbial community assembly along an aridity gradient in the Namib Desert. Extremophiles 17:329–337

Strunecky O, Elster J, Komarek J (2011) Taxonomic revision of the freshwater cyanobacterium *Phormidium murrayi* = *Wilmottia murrayi*. Fottea 11:57–71

Sumner DY, Hawes I, Mackey TJ, Jungblut AD, Doran PT (2015) Antarctic microbial mats: a modern analogue for Archean lacustrine oxygen oases. Geology 43:887–890

Sumner D, Jungblut AD, Hawes I, Andersen DT, Mackey TJ, Wall K (2016) Growth of elaborate microbial pinnacles in Lake Vanda, Antarctica. Geobiology 14:556–574

Tanabe Y, Ohtan S, Kasamatsu N, Fukuchi M, Kudoh S (2010) Photophysiological responses of phytobenthic communities to the strong light and UV in Antarctic shallow lakes. Polar Biol 33:85–100

Tang EPY, Vincent WF (1999) Strategies of thermal adaptation by high latitude cyanobacteria. New Phytol 142:315–323

Tang EPY, Tremblay R, Vincent WF (1997) Cyanobacterial dominance of polar freshwater ecosystems: are high latitude mat-formers adapted to the low temperature environment? J Phycol 33:171–181

Tashyreva D, Elster J (2016) Annual cycles of two cyanobacterial mat communities in hydro-terrestrial habitats of the High Arctic. Microb Ecol 71:887–900

Taton A, Grubisic S, Brambilla E, De Wit R, Wilmotte A (2003) Cyanobacterial diversity in natural and artificial microbial mats of Lake Fryxell (McMurdo Dry Valleys, Antarctica): a morphological and molecular approach. Appl Environ Microbiol 69:5157–5169

Taton A, Grubisic S, Balhasart P, Hodgson DA, Laybourn-Parry J, Wilmotte A (2006a) Biogeographical distribution and ecological ranges of benthic cyanobacteria in East Antarctic lakes. FEMS Microbiol Ecol 57:272–289

Taton A, Grubisic S, Ertz D, Hodgson DA, Piccardi R, Biondi N, Tredici MR, Mainini M, Losi D, Marinelli F, Wilmotte A (2006b) Polyphasic study of Antarctic cyanobacterial strains. J Phycol 42:1257–1270

Thaler M, Vincent WF, Lionard M, Hamilton AK, Lovejoy C (2017) Microbial community structure and interannual change in the last epishelf lake ecosystem in the North Polar Region. Front Mar Sci 3:275

Van Horn DJ, Van Horn ML, Barret JE, Gooseff MN, Altrichter E, Geyer KM, Zeglin LH, Takacs-Vesbach CD (2013) Factors controlling soil microbial biomass and bacterial diversity and community composition in a cold desert ecosystem: Role of geographic scale. PLoS ONE 8: e66103

Varin T, Lovejoy C, Jungblut AD, Vincent WF, Corbeil J (2010) Metagenomic profiling of Arctic microbial mat communities as nutrient scavenging and recycling systems. Limnol Oceanogr 55:1901–1911

Varin T, Lovejoy C, Jungblut AD, Vincent WF, Corbeil J (2012) Metagenomic analysis of stress genes in microbial mat communities from Antarctica and the High Arctic. Appl Environ Microbiol 78:549–559

Veillette J, Martineau M-J, Antoniades D, Sarrazin D, Vincent WF (2011) Effects of loss of perennial lake ice on mixing and phytoplankton dynamics: Insights from High Arctic Canada. Ann Glaciol 51:56–70

Velázques D, Jungblut AD, Rochera C, Camacho A, Quesada A, Rico E (2017) Seasonal dynamics and trophic interactions of a microbial mat in maritime Antarctic. Polar Biol. doi:10.1007/s00300-016-2039-2

Vincent WF (1988) Microbial ecosystems of Antarctica. Cambridge University Press, Cambridge, 304 pp

Vincent WF (2000a) Cyanobacterial dominance in the polar regions. In: Whitton BA, Potts M (eds) The ecology of Cyanobacteria. Kluwers Academic Press, Dordrecht, pp 321–340

Vincent WF (2000b) Evolutionary origins of Antarctic microbiota: invasion, selection and endemism. Antarct Sci 12:374–385

Vincent WF (2007) Cold tolerance in cyanobacteria and life in the cryosphere. In: Seckbach J (ed) Algae and cyanobacteria in extreme environments. Springer, Heidelberg, pp 287–301
Vincent WF, Howard-Williams C (2000) Life on snowball Earth. Science 287:2421v
Vincent WF, Quesada A (2012) Cyanobacteria in high latitude lakes, rivers and seas. In: Whitton BA (ed) Ecology of cyanobacteria II: their diversity in space and time. Springer, New York, NY, pp 371–385
Vincent WF, Castenholz RW, Downes MT, Howard-Williams C (1993) Antarctic cyanobacteria: light, nutrients, and photosynthesis on the microbial mat environments. J Phycol 29:745–755
Vincent WF, Mueller D, Van Hove P, Howard-Williams C (2004a) Glacial periods on early Earth and implications for the evolution of life. In: Seckbach J (ed) Origins: genesis, evolution and diversity of life. Kluwer Academic Publishers, Dordrecht, pp 481–501
Vincent WF, Mueller DR, Bonilla S (2004b) Ecosystems on ice: the microbial ecology of Markham Ice Shelf in the High Arctic. Cryobiology 48:103–112
Vopel K, Hawes I (2006) Photosynthetic performance of benthic microbial mats in Lake Hoare, Antarctica. Limnol Oceanogr 51:1801–1812
Wait BR, Webster-Brown JG, Brown KR, Healy M, Hawes I (2006) Chemistry and stratification of Antarctic meltwater ponds I: coastal ponds near Bratina Island, McMurdo Ice Shelf. Antarct Sci 18:515–524
Waleron M, Waleron K, Vincent WF, Wilmotte A (2007) Allochthonous inputs of riverine picocyanobacteria to coastal waters in the Arctic Ocean. FEMS Microbiol Ecol 59:356–365
Warren-Rhodes KA, Rhodes KL, Boyle LN, Pointing SB, Chen Y et al (2007) Cyanobacterial ecology across environmental gradients and spatial scales in China's hot and cold deserts. FEMS Microbiol Ecol 61:470–482
Webster-Brown JG, Hawes I, Jungblut AD, Wood SA, Christenson HK (2015) The effects of entombment on water chemistry and bacterial assemblages in closed cryoconite holes on Antarctic glaciers. FEMS Microbiol Ecol 91:fiv144
Wharton RA, McKay CP, Simmons GM, Parker BC (1985) Cryoconite holes on glaciers. Bioscience 35:499–503
Wilmotte A, Demonceau C, Goffart A, Hecq J-H, Demoulin V, Crossley AC (2002) Molecular and pigment studies of the picophytoplankton in a region of Southern Ocean (42-54°S, 141-144°E) in March 1998. Deep-Sea Res II 49:3351–3363
Wood SA, Mountfort D, Selwood AI, Holland PT, Puddick J, Cary SC (2008) Widespread distribution and identification of eight novel microcystins in Antarctic cyanobacterial mats. Appl Environ Microbiol 74:7243–7251
Zhang L, Jungblut AD, Hawes I, Andersen DT, Sumner DY, Mackey TJ (2015) Cyanobacterial diversity in benthic mats of the McMurdo Dry Valley lakes, Antarctica. Polar Biol 38:1097–1110

Part III
Molecular and Physiological Adaptations to Cold Habitats

Chapter 10
Enzyme Catalysis in Psychrophiles

Tony Collins and Charles Gerday

Abstract Cold-active enzymes are produced by organisms, known as psychrophiles, adapted to permanently cold habitats. Low temperatures have an exponential deleterious effct on reaction rates, and thus psychrophilic enzymes have to be adapted to secure appropriate reaction rates in their environment. These enzymes have a high specific activity at low temperatures, in any case higher than that of their mesophilic and thermophilic counterparts, and display a shift of the apparent optimum temperature for activity towards low temperatures as well as a reduced thermal stability and increased flexibility. The increased flexibility may be global, involving the overall edifice, or local, involving only those zones crucial for activity, be they near or distant from the active site. The reduced thermodynamic stability of cold-adapted enzymes is illustrated by a significantly lower stabilisation energy as compared to that of their mesophilic and thermophilic counterparts, yet maximum stability occurs at similar temperatures in all cases. The comparison of their three-dimensional structures with higher temperature-adapted homologues, in conjunction with various mutagenesis studies, has shown that their high activity results from rather discrete molecular changes that tend to decrease the stability of the molecular edifice. Each cold-adapted enzyme however adopts a specific strategy. There is apparently a continuum in the adaptation, with some enzymes showing extremely acute cold adaptation, as illustrated by a severe shift of the activity towards low temperatures, whereas others appear to cover a broader range of temperatures. This probably depends on the specific evolutionary history of the organisms which produce them.

Contents

T. Collins
Centre of Molecular and Environmental Biology (CBMA), Department of Biology, University of Minho, Campus de Gualtar, 4710-057 Braga, Portugal

C. Gerday (✉)
Laboratory of Biochemistry, Institute of Chemistry, University of Liège, B6, Sart-Tilman, 4000 Liège, Belgium
e-mail: ch.gerday@ulg.ac.be

R. Margesin (ed.), *Psychrophiles: From Biodiversity to Biotechnology*,
DOI 10.1007/978-3-319-57057-0_10

10.1 Introduction

Enzymes are proteins that increase the rate of chemical reactions without modifying the equilibrium constant of the reaction. Their efficiency is notably exponentially dependent on temperature and follows the Arrhenius law (Arrhenius 1889):

$$k_{cat} = A\mathrm{e}^{-\mathrm{Ea}/RT} \tag{10.1}$$

in which k_{cat} is the reaction rate; A the pre-exponential term, also known as the frequency factor, which is related to the frequency of collision of the reactants and to the probability of the reactants being in the appropriate orientation to react; Ea the activation energy, R the universal gas constant (8.314 J mol^{-1} K^{-1}); and T the temperature in Kelvin.

Similarly, the temperature dependence of a reaction can be described by the following equation derived from the transition state theory (Eyring 1935):

$$k_{cat=}\kappa(\mathrm{k_B}T/h)\mathrm{e}^{\left(-\Delta G^{\#}/RT\right)} \tag{10.2}$$

where k_B is the Boltzmann constant (1.38×10^{-23} J K^{-1}), h the Planck constant (6.63×10^{-34} J s) and $\Delta G^{\#}$ the free energy of activation, that is, the difference in Gibbs energy between the activated and ground states. K, the transmission coefficient, takes into account possible re-crossing (activated state returning to the ground state) and tunnelling (complexes with lower free energy than the activation energy giving rise to product) effects as well as deviation from the equilibrium distribution (Garcia-Viloca et al. 2004). Importantly, κ is negatively influenced by the viscosity of the medium (Garcia-Viloca et al. 2004; Siddiqui et al. 2004) and is particularly relevant here as the viscosity of the reaction medium increases significantly at low temperatures (Mastro and Keith 1984; Demchenko et al. 1989; Homchaudhuri et al. 2006). Hence, enzyme reaction rates are presented with a double challenge at low temperatures, being negatively affected by both the decreased temperature and the increased viscosity resultant of this decreased temperature.

Cold-active enzymes, which have successfully overcome the low-temperature challenge, are produced by a group of organisms known as the psychrophiles, a term introduced by Schmidt-Nielsen in 1902. One knows, from the preceding chapters in this book, that psychrophiles are quite diversified and have successfully colonised various environments that experiment temperatures close to or below 0 °C. Interestingly, due to the fact that their apparent optimum growth temperature is generally above 20 °C, these organisms were initially considered as cold-tolerant, or psychrotolerant, rather than cold-loving, or psychrophiles (Ingraham and Stokes 1959). Indeed, this confusion still persists today as some investigators erroneously consider the so-called optimum temperature to correspond to the best possible temperature for the organism. It was the merit of Hess (1934) to point out that numerous bacteria, even with maximum growth rates around 20 °C, were true psychrophiles simply because in broth cultures the largest cell populations were obtained at 5 °C rather than at 20 °C. This has been confirmed by more recent studies, such as that of Margesin (2009). Furthermore, it has also been shown that the production of extracellular enzymes is highest at temperatures close to that of the environment of the microorganisms and far below the so-called optimum temperature (Feller et al. 1994). Thus, it is the habitat temperature of the organism and not the temperature enabling highest growth rates that is most relevant, and psychrophiles are found to thrive in environments characterised by temperatures close to or below the freezing point of water.

Despite the negative effects of low temperatures on reaction rates, psychrophiles reproduce and grow well in cold environments. Indeed, the adaptation of psychrophiles to cold habitats, having real growth optima close to those of their environments, indicates that their cell components and in particular their enzymes, which catalyse nearly all chemical reactions occurring in their cells, are perfectly adapted to work at low temperatures. The first significant report related to cold-active enzymes dates back to more than a quarter of a century with the publication of the properties of a heat-labile alkaline phosphatase isolated from an Antarctic bacterium (Kobori et al. 1984). This paper can be considered as a seed paper since the two main properties of cold-active catalysts were already correctly described, i.e. a high specific activity at low temperature and a rather high sensitivity to temperature. The biotechnological potential of this enzyme was also underlined since this cold-active enzyme could be rapidly inactivated by mild heat treatment, contrary to calf intestinal ATPase, and thereby offered a very convenient method for the 5′ end radio-labelling of DNA fragments. Numerous reviews or book chapters devoted to the study of cold-active enzymes have since been published; one can mention, among the latest, Russell (2000), Collins et al. (2002, 2007, 2008), Feller and Gerday (2003), Georlette et al. (2004), D'Amico et al. (2006b), Siddiqui and Cavicchioli (2006), Marx et al. (2007), Feller (2008, 2010), Papaleo et al. (2011), Gerday (2013, 2014), Fields et al. (2015) and Santiago et al. (2016).

10.2 Activity at Low Temperatures

The relation between enzyme activity and temperature can be roughly expressed by the Q_{10} which represents the ratio between the rates of a chemical reaction measured at an interval temperature of 10 °C. The value of Q_{10} depends on the activation energy (Ea) of the enzyme but in a typical mesophilic enzyme is usually close to 3. Therefore, in this case, a temperature shift from 40 °C to 0 °C will decrease the rate of the chemical reactions occurring in a non-adapted organism by a factor of 81 and severely impair the survival of this organism at low temperature. The temperature dependence of enzymatic reactions occurring in psychrophilic and mesophilic organisms is illustrated in Fig. 10.1. Three main differences can be observed:

1. When compared to its mesophilic counterpart, the apparent optimum of the psychrophilic α-amylase is shifted towards low temperatures by more than 30 °C.
2. The cold-adapted enzyme displays a high catalytic efficiency at low and moderate temperatures, contrary to the mesophilic counterpart.
3. The cold-active enzyme is rapidly inactivated at moderately high temperatures.

As far as the temperature shift is concerned, it is worth mentioning that its amplitude strongly depends on the enzymes considered. Also, in many cold-adapted enzymes, the activity recorded at the environmental temperature is somewhat lower than that recorded for homologous mesophilic and thermophilic enzymes at their own environmental temperatures. This possibly reflects an incomplete adaptation to low temperatures in some cold-adapted enzymes.

The modification of enzymatic properties as a function of environmental temperature in ectothermic organisms has been discussed for many years (Hochachka and Somero 1973, 2002; Somero 1977). These authors defined three possible strategies for the adaptation of enzymes to low temperatures: first, the organism could increase the concentration of enzymes in the cell or secreted; second, the activity could be modulated by increasing the substrate binding ability; and third, the organism could modify in a permanent way the catalytic efficiency of the enzyme. The first strategy is, at first sight, improbable due to its energy cost in a cold environment; the second strategy will not be efficient in the case of enzymes working at saturated substrate concentration, which probably concerns most of the

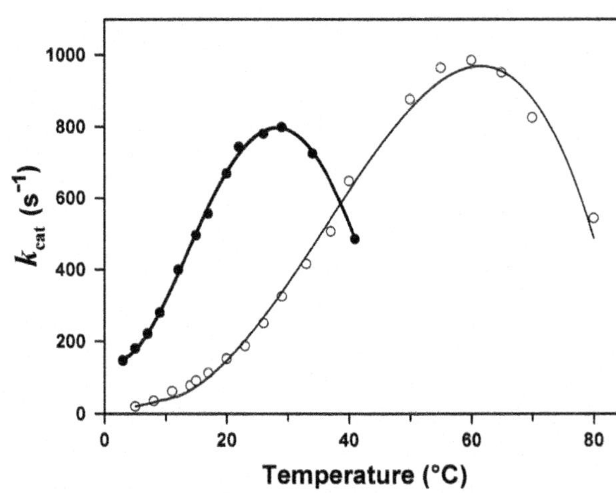

Fig. 10.1 Temperature dependence of activity. The temperature dependence of activity for a cold-adapted α-amylase isolated from the Antarctic bacterium *Pseudoalteromonas haloplanktis* (*filled circles*) and a closely related mesophilic enzyme isolated from pig pancreas (*open circles*) are shown (from Feller et al. 1992)

extracellular enzymes. In the case of intracellular enzymes, it has been shown that the K_m values for phosphoenolpyruvate of fish pyruvate kinases adapted to different temperatures were lowest in the range of temperatures usually experimented by these organisms, and, at the respective environmental temperatures, the substrate binding affinities were found to be similar (Somero 1977). In fact, only the third strategy, i.e. the production of enzymes specifically adapted to low temperatures, seemed to be the most appropriate.

The initial discoveries as well as an easier accessibility to Arctic and Antarctic sites stimulated an interest in cold enzymes, and in the 1990s several groups were formed, thanks notably to the support of the European Union, to focus on psychrophilic organisms with the aim of shedding light on the fundamental aspects of their adaptation and their biotechnological potential. In addition to other factors, these studies enabled significant progress in the understanding of the molecular adaptations to cold of cold-active enzymes. Numerous cold-adapted enzymes were notably isolated; one can mention recent investigations carried out on a cold-active cellulase from the Antarctic bacterium *Pseudoalteromonas haloplanktis* (Garsoux et al. 2004; Sonan et al. 2007), the hormone-sensitive lipase from the Antarctic strain *Psychrobacter* sp. TA144 (De Santi et al. 2010), the periplasmic nitrate reductase from the Antarctic bacterium *Shewanella gelidimarina* (Simpson and Codd 2011), the serine hydroxymethyltransferase from a cold-adapted *Psychromonas ingrahamii* (Angelaccio et al. 2012), the cold-adapted thermolysin-like protease from *Halobacillus* sp. SCSIO isolated from marine sediment (Yang et al. 2013), the cold-adapted pullulanase from *Shewanella arctica* 40-3 (Qoura et al. 2014), the psychrophilic acetate kinase from *Shewanella* sp. AS-11 (Tang et al. 2014), the cold-adapted α-amylase from a metagenomics library (Vester et al. 2015), the N-acetylneuraminic acid synthase from a psychrophilic *Moritella viscosa* (Berg et al. 2015), the ectoine synthase from the marine bacterium *Sphingopyxis alaskensis* (Kobus et al. 2015) and the psychrophilic β-glucosidase from *Micrococcus antarcticus* (Miao et al. 2016). All these newly isolated enzymes show the typical characteristics of cold adaptation, i.e. a higher specific activity than their mesophilic counterparts at low and moderate temperatures (see also Table 10.1) and a higher sensitivity to moderately high temperatures. Some cold-adapted enzymes show hyperpsychrophilic character (a counterpart to hyperthermophilic enzymes), such as the extracellular enzymes of microorganisms isolated from sea ice in the Arctic with apparent temperature optima around 15 °C (Huston et al. 2000), the cold-active dihydrofolate reductase from the deep sea microorganism *Moritella profunda* (Xu et al. 2003a) with a maximal growth rate at 2 °C and the α-amylase isolated from submarine ikaite columns in SW Greenland that displays an apparent optimum at 10–15 °C (Vester et al. 2015). In a limited number of cases, the specific activity of the cold-active enzyme does not exceed that of the mesophilic enzyme studied, as demonstrated by an arginine kinase from *Desulfotalea psychrophila* (Suzuki et al. 2012), an Antarctic chitinase (Lonhienne et al. 2001a) and an isocitrate reductase (Fedoy et al. 2007). Interestingly, in this latter case, the thermal stability also seems higher than that of the mesophilic homologue. Also, in the first case, the data tends to indicate that the arginine kinase

Table 10.1 K_m and k_{cat} values of psychrophilic (P) and mesophilic (M) enzymes

Enzyme/organism	*T* (°C)	K_m	k_{cat} (s^{-1})	Reference
Alpha-amylase				D'Amico et al. (2001)
P: *Pseudoalteromonas haloplanktis*	25	234 μM	294 (10 °C)	
M: Pig pancreatic	25	65 μM	97	
Cellulase				Garsoux et al. (2004)
P: *P. haloplanktis*	4	600 μM	0.18	
M: *Erwinia chrysanthemi*	4	200 μM	0.01	
DNA ligase				Georlette et al. (2000)
P: *P. haloplanktis*	18	0.30 μM	0.034	
M: *Escherichia coli*	18	0.18 μM	0.004	
Endonuclease I				Altermark et al. (2007)
P: *Vibrio salmonicida*	0.5	246 mM	9.41	
M: *Vibrio cholerae*	0.5	118 mM	1.03	
Isocitrate dehydrogenase				Watanabe et al. (2005)
P: *Colwellia maris*	15	62 mM	70.8	
M: *E. coli*	15	3.3 mM	22.0	
Lactate dehydrogenase				Coquelle et al. (2007)
P: *Champsocephalus gunnari*	0	0.16 mM	230	
M: *Squalus acanthias*	0	~0.3 mM	72	
Ornithine transcarbamylase				Xu et al. (2003b)
P: *Moritella abyssi*	30	45 mM	690	
M: *Saccharomyces cerevisiae*	30	0.9 mM	235	
Subtilisin				Narinx et al. (1997)
P: *Bacillus* sp. (Antarctic)	5	26 μM	32	
M: Subtilisin Carlsberg	5	6 μM	18	

from the psychrophilic bacterium *D. psychrophila* is not well adapted to low temperatures. In fact it would be interesting to compare the catalytic properties of other enzymes from *D. psychrophila* to those of homologous enzymes from the mesophilic bacterium *Desulfitobacterium hafniense* to determine whether this apparently incomplete adaptation to cold can be extended to other enzymes and possibly to the whole organism. Finally, in the case of the chitinase, the type of substrate used in the study is probably important since chitins are structurally very different from each other.

Another parameter which can play an important role in chemical reactions controlled by enzymes is the K_m. On a first approximation, this represents the affinity of the enzyme for the substrate, provided that the rate constants that could interfere with the constants directly involved in the true dissociation constant of the enzyme-substrate complex can be neglected. K_m values have been reported for numerous cold-adapted enzymes and compared to those of mesophilic homologues (Table 10.1). With the exception of the lactate dehydrogenases, this table reveals a main feature of cold-adapted enzymes, i.e. their K_m values are higher than that of their mesophilic homologues, indicating a lower affinity for the substrate. Indeed, this loss in substrate binding affinity contributes to fix the limit of the thermal tolerance of living organisms. The increase of K_m in cold-adapted enzymes suggests that the active sites of these have lost part of their initial rigidity and that the enzyme-substrate complex ES is less stable than in mesophilic and a fortiori thermophilic counterparts. Therefore, in the case of poly-substrate enzymes, one can also expect a modification of the substrate specificity following this adaptation drift. This has been clearly illustrated in the case of a few cold-adapted enzymes: elastases (Smalas et al. 2000), alcohol dehydrogenases (Tsigos et al. 1998) and a psychrophilic α-amylase (D'Amico et al. 2001, 2006a). The latter has been compared to its close homologue from pig pancreas, and structural information shows these to have strictly identical substrate binding sites. Nonetheless, the cold-active enzyme has a higher catalytic efficiency towards large substrates such as starch, amylopectin, amylose, dextrin and glycogen, whereas the mesophilic homologue is more efficient on small-sized substrates such as maltopentaose, maltohexaose and maltoheptaose and on a mix (G4–G10) of maltooligosaccharides. Here it is believed that, being less rigid, the active site of the cold-adapted enzyme can more easily accommodate the large macromolecular oligosaccharides.

10.3 Thermal Inactivation and Structural Stability

The relative activities as a function of temperature of psychrophilic, mesophilic and moderately thermophilic α-amylases are shown in the upper diagram of Fig. 10.2, and the unfolding as a function of temperature (D'Amico et al. 2003) is illustrated in the lower diagram. As expected, the optimum temperature (T_{opt}) and the subsequent loss of activity, or inactivation, of the cold-active enzyme occur at temperatures much lower than that of its mesophilic and thermophilic homologues, but a comparison of the two parts of the figure reveals an intriguing property of the cold-active enzyme. The reduction in activity (upper panel) occurs at a lower temperature than that of unfolding (lower panel), whereas in the case of the mesophilic and thermophilic enzymes, the apparent optimum is correlated to the unfolding transition. Such an apparent anomaly has also been observed in other cold-active enzymes, such as DNA ligases (Georlette et al. 2003) and xylanases (Collins et al. 2003), and different hypotheses have been put forward to explain these observations. These include an increased sensitivity of the active site to heat (Fields and Somero 1998) and an increased instability of the ES complex in these enzymes as compared to mesophilic and thermophilic enzymes. An interesting and recent study (Arcus et al. 2016) has possibly given a definite answer to this problem; it implicates the heat capacity change $\Delta C_p^{\#}$ that occurs during the activation process: ES → ES$^{\#}$. Enzymes are characterised by high heat capacities (C) which correspond to the change in internal energy following a change in temperature, and in proteins this is dominated by the absorption of energy by the vibration modes of the

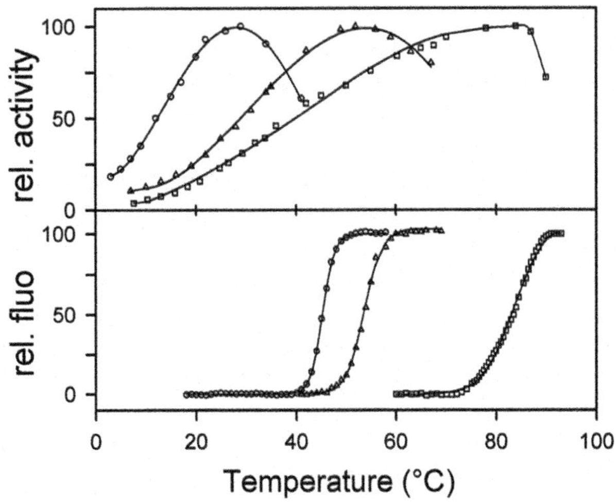

Fig. 10.2 Comparison of the temperature dependence of enzyme activity and thermal unfolding. *Upper diagram*: relative activities of a psychrophilic *Pseudoalteromonas haloplanktis* α-amylase (*open circles*), a mesophilic α-amylase from pig pancreas (*open triangles*) and a thermophilic α-amylase from *Bacillus amyloliquefaciens* (*open squares*) as a function of temperature. *Lower diagram*: unfolding as a function of temperature of psychrophilic, mesophilic and thermophilic α-amylases. The unfolding was monitored by fluorescence measurements at an excitation wavelength of 280 nm and emission wavelength of 350 nm (from D'Amico et al. 2003)

molecule. Interestingly, it was shown (Arcus et al. 2016) that enzyme-catalysed reactions proceed with a generally negative value of $\Delta C_p^\#$, that is, the heat capacity of the ES complex is generally larger than that of the transition state $ES^\#$. This is possibly consequent of a tighter binding of the substrate to the transition state that reduces the frequencies of the vibration modes as compared to the ground state ES. This observation of a non-zero value for $\Delta C_p^\#$ has major consequences for our understanding of the temperature dependence of enzyme-catalysed reactions as described in Eq. 10.3.

$$\Delta G^\# = \{\Delta H^\#_{T0} + \Delta C^\#_p (T - T_0)\} - T\{\Delta S^\#_{T0} + \Delta C^\#_p (ln\, T - ln\, T_0)\} \quad (10.3)$$

It can be seen that with non-zero $\Delta C_p^\#$, the activation enthalpy ($\Delta H^\#$) and activation entropy ($\Delta S^\#$) are actually temperature (T) dependent for enzyme-catalysed reactions and that the Gibbs free energy of activation ($\Delta G^\#$) is curved, with the curvature being a function of the value of $\Delta C_p^\#$ (Daniel and Danson 2010). As a consequence, for $\Delta C_p^\# < 0$ the rate of the catalysed reaction increases with temperature up to T_{opt} and then decreases, and this even in the absence of any unfolding process (Arcus et al. 2016). This is indeed observed in Fig. 10.2 for the psychrophilic α-amylase where a negative $\Delta C_p^\#$ leads to a pronounced curvature and a loss of activity before the onset of unfolding. The magnitude of $\Delta C_p^{\#}$ is much less negative for the mesophilic and thermophilic counterparts, probably as a consequence of fewer vibration low-frequency modes in the ES complex due to the stronger interaction of substrate with these enzymes (lower K_m) in the ground state. This leads to lower curvature in the activity-temperature curve and results in the activity loss corresponding to denaturation as seen in Fig. 10.2.

The above-discussed observations should also be reflected in differences in the activation parameters for the catalytic reaction. From Table 10.2, one can see that the activation energy, $\Delta G^{\#}$, for cold-active enzymes, is systematically lower than that of their mesophilic counterparts. The rather moderate differences result from significant decreases of the activation enthalpies, $\Delta H^{\#}$, partially compensated for by more negative values of the activation entropy term, $T\Delta S^{\#}$. This indicates that a much more important reordering of the ground states is required to reach the activated states and fits perfectly well with the more negative heat capacity changes seen during activation of cold-active enzymes as discussed above. The compensation effect observed between activation enthalpy and entropy could originate from the protein surface mobility (Isaksen et al. 2016). Indeed, at least in the case of psychrophilic and mesophilic trypsins, rigidification of the surface of the psychrophilic enzyme by applying positional restraints from the water-enzyme interface and inward yielded an increase of the activation enthalpy partially compensated by an increase of the activation entropy eventually turning the psychrophilic enzyme into a mesophilic one. In Table 10.2, the activation entropy terms, $T\Delta S^{\#}$, of three mesophilic enzymes, i.e. chitobiase, endonuclease and lysozyme, are positive. This could mean that, contrary to the other enzymes listed, the activated states are characterised by greater disorder than the ground state or, more probably, that the rearrangement of water molecules during the activation process gives rise to a positive value of the activation entropy. A schematic representation of the activation process for a psychrophilic and mesophilic enzyme is shown in Fig. 10.3. The energy change resulting from the interaction of the enzyme with the substrate is associated with a negative energy change that places the enzyme-substrate complex ES in an unfavourable deep energy pit. As the K_m values of cold-active enzymes are higher than their mesophilic counterparts, one can consider that the ground state of the ES complex of the cold-active enzyme has a higher free energy level that is favourable in limiting the energy change necessary to reach the activated state $ES^{\#}$. If less or weaker bonds are implicated in the formation of the ES complex, less energetically costly structural modifications will be necessary to reach the activated state.

Table 10.2 Activation parameters for activity of psychrophilic enzymes (P) as compared to those of their mesophilic counterparts (M)

Enzyme	Type	Temp (°C)	$\Delta G^{\#}$ (kJ mol^{-1})	$\Delta H^{\#}$ (kJ mol^{-1})	$T\Delta S^{\#}$ (kJ mol^{-1})	Reference
Amylase	P	10	57.7	34.7	−23.0	D'Amico et al. (2003)
	M		58.5	46.4	−12.1	
Cellulase	P	4	71.6	46.2	−25.4	Garsoux et al. (2004)
	M		78.2	65.8	−12.4	
Chitobiase	P	15	59.5	44.7	−14.8	Lonhienne et al. (2001b)
	M		63.5	71.5	+8.0	
Endonuclease	P	5	62.8	33.4	−29.4	Altermark et al. (2007)
	M		67.9	74.0	+ 6.1	
Lysozyme	P	25	45.1	31.9	−13.2	Sotelo-Mundo et al. (2007)
	M		46.2	49.4	+3.2	
Subtilisin	P	15	62.0	36.0	−26.5	Davail et al. (1994)
	M		66.0	46.0	−20.2	
Xylanase (bacteria)	P	10	54.0	21.0	−33.0	Collins et al. (2003)
	M		60.0	58.0	−2.0	
Xylanase (yeast)	P	5	52.3	45.3	−7.0	Petrescu et al. (2000)
	M		54.6	49.9	−4.7	

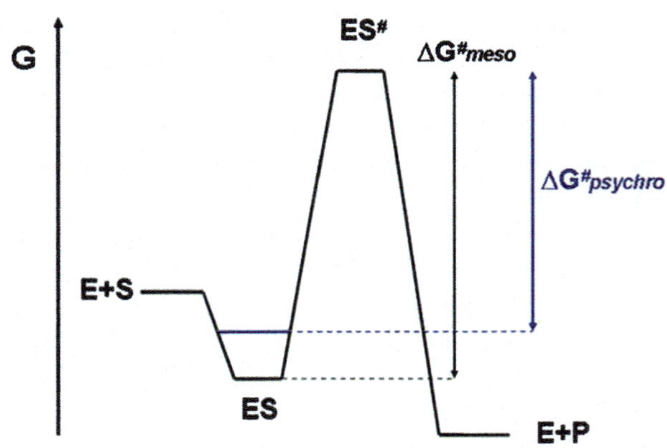

Fig. 10.3 Energetics of the activation of an enzyme-catalysed reaction by psychrophilic and mesophilic enzymes. The activation in psychrophilic enzymes is rendered easier by a decrease of the affinity of the enzyme for the substrate (higher level of ES). It is assumed here that the activated forms, ES$^{\#}$, have identical energy levels (adapted from Gerday 2014)

10.4 Flexibility

The notion of an increased flexibility in relation to the high specific activity and generally low thermal stability of cold-active enzymes has been introduced a long time ago (Somero 1977). This correlation has been demonstrated in numerous cases, but a more refined view of the linkage between the catalytic properties and overall thermal stability of enzymes is now emerging. In fact, the catalytic efficiency of an enzyme is believed to be mainly related to the mobility-flexibility of crucial parts of the molecular edifice. An increase of this flexibility usually leads to an overall or domain-specific decrease of the thermal stability, but in a limited number of cases, this could probably also occur without significant modifications of the thermal stability. Intuitively, one can indeed suggest that, for example, substitutions of amino acids in existing loops could modify the conformational flexibility of regions crucial for catalysis without altering the thermal stability.

The term "flexibility" is rather complex since it can refer to structural modifications as a function of time which can be assimilated to a "respiration" of the molecular edifice or to the amplitude of the conformational changes. Both are probably important for the accommodation of substrates at low temperature and the release of products. The flexibility of proteins has been evaluated by various techniques such as molecular dynamics simulations (MDS), neutron scattering, EPR spectroscopy, crystallographic B-factors, time-resolved fluorescence and quenching of tryptophan fluorescence using acrylamide. MDS have indicated that the higher catalytic efficiencies of cold-active enzymes are correlated to a higher flexibility of regions important for catalysis and a reduced thermal stability. Enzymes studied include subtilisin-like proteases (Tindbaek et al. 2004; Tiberti and Papaleo 2011), chitinases (Ramli et al. 2012), frataxins (Roman et al. 2013), haemoglobins (Stadler et al. 2012), thermolysin-like proteases (Radestock and Gohlke 2011) and beta-tubulins (Chiappori et al. 2012). However, in the case of α-amylases, MDS showed that the overall flexibility of the mesophilic enzyme was higher than that of the psychrophilic counterpart as a result of an insertion of additional loops. It was also shown that the flexibility of orthologous xylanases was approximately the same at their respective apparent optimum temperatures (Spiwok et al. 2007).

EPR spectroscopy has also shown that the catalytic function of a cold-active alkaline phosphatase was closely related to the relative mobility of the helix carrying catalytic site residue cysteine 67 and nucleophilic serine 65 (Heidarsson et al. 2009).

The problem of relative flexibility was also approached by neutron scattering (Tehei et al. 2004). In this study, the mean macromolecular dynamics of whole crude extracts of bacteria adapted to different environmental temperatures were compared. It was shown that, at a given temperature, the resilience (equivalent to rigidity) increased from psychrophiles to thermophiles and that at their respective environmental temperatures, the measured resiliences were similar, in perfect agreement with the "corresponding state" theory (Somero 1995).

The temperature factor, or B-factor, has also been used to evaluate the relative flexibility of proteins. It represents the spread of X-ray crystallography electron densities and can be calculated from known three-dimensional structures. Average B-factors are not systematically correlated with the activity-flexibility relationship (Sun-Yong et al. 1999), but relative B-factors, representing the ratio between the B-factors of local parts of the proteins and the average B-factor, apparently correlate well (Russell et al. 1998; Sun-Yong et al. 1999). A recent systematic comparison of the B-factors of 20 pairs of psychrophilic and mesophilic enzymes demonstrated that psychrophilic enzymes are more flexible in S-turn and β-strand

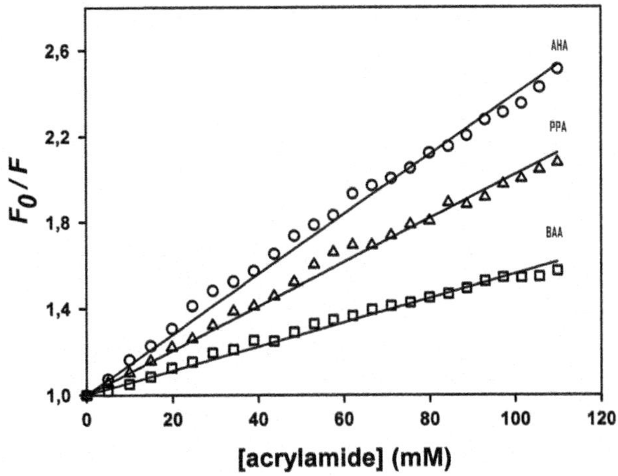

Fig. 10.4 Protein flexibility as monitored by acrylamide quenching of fluorescence. Quenching of protein fluorescence as a function of acrylamide concentration at 37 °C for the cold-active α-amylase AHA from the Antarctic bacterium *Pseudoalteromonas haloplanktis* (*open circles*), the mesophilic α-amylase from pig pancreas PPA (*open triangles*) and the thermophilic α-amylase from *Bacillus amyloliquefaciens* BAA (*open squares*). F_0/F represents the ratio of the fluorescence in the absence (F_0) and presence (F) of acrylamide at various concentrations. The relative steepness of the slopes is related to the ability of the quenching agent to penetrate the respective protein structures (adapted from D'Amico et al. 2003)

secondary structures; they have also larger average cavity sizes, and these cavities are lined with an increased frequency of acidic groups (Paredes et al. 2011).

Fluorescence techniques have demonstrated a clear relationship between the activity and local flexibility of carbonic anhydrases. It was shown that the cold-active enzyme, with a higher specific activity, displays a higher flexibility in the region that controls the folding of the protein (Chiuri et al. 2009). More clear-cut data in relation to the relative flexibility of enzymes have been obtained via comparisons of the quenching of protein fluorescence by acrylamide. It was shown that psychrophilic enzymes are much more permeable to acrylamide than their mesophilic counterparts in the case of Ca^{2+}-Zn^{2+} proteases (Chessa et al. 2000), α-amylases (see Fig. 10.4) (D'Amico et al. 2003; Cipolla et al. 2012), DNA ligases (Georlette et al. 2003), xylanases (Collins et al. 2003), cellulases (Sonan et al. 2007), aminopeptidases (Huston et al. 2008), thermolysins (Xie et al. 2009), subtilisin-like proteases (Sigtryggsdóttir et al. 2014) and cold-adapted esterases (Truongvan et al. 2016).

The relationship between protein instability, high flexibility and activity at low temperature has also been demonstrated in a recent intriguing experiment carried out on a hyperthermophilic archeon, *Aeropyrum pernix* (Schwartz and Pan 2016). It was shown that, when exposed to low-temperature growth conditions, mistranslation leading to substitution of leucine residues with methionine occurred during protein synthesis. A citrate synthase produced at low temperatures with these mistranslations was found to display enhanced low-temperature activity and a significantly increased flexibility of the molecular structure. This is due to the fact that Met side chains, although of a similar volume to that of Leu side chains, are much more flexible and have more degrees of freedom. In line with this observation, in general, psychrophilic proteins have a higher content of Met residues when compared to their mesophilic and thermophilic homologues (Gudmundsdóttir et al. 1996; Marshall 1997; Thomas and Cavicchioli 1998).

10.5 Thermodynamic Stability

The thermodynamic stability of many proteins has been investigated assuming that unfolding is a reversible two-state process defined by $\Delta G_{\text{N-U}} = -RT\ln K_{\text{N-U}}$, where $K_{\text{N-U}} = [\text{U}]/[\text{N}]$. For a protein in the native state, $\Delta G_{\text{N-U}}$ is a positive value and can be defined as the energy necessary to unfold the protein. The thermodynamic stability of enzymes can be evaluated by non-calorimetric techniques making use of chemical denaturants (Pace and Laurents 1989; Talla-Singh and Stites 2008) or more easily in a microcalorimeter in which the heat absorbed during the unfolding of the protein is measured and defined as ΔH_{cal}. This represents the energy necessary to disrupt the weak bonds that stabilise the three-dimensional structure and appears as a peak with the area limited by this peak being equivalent to ΔH_{cal}. ΔH_{cal} has been shown to increase significantly on going from psychrophiles to thermophiles and is often more symmetrical for cold-active enzymes, indicative of a more pronounced cooperativity of unfolding. In contrast, this is often unsymmetrical and flattened and distributed over a rather broad range of temperatures for mesophilic and thermophilic enzymes (Feller 2010).

As a function of temperature, the stabilisation energy or free energy of unfolding is described by an equivalent of the Gibbs-Helmholtz equation:

$$\Delta G_{(T)} = \Delta H_{\text{cal}}(1 - T/T_{\text{m}}) + \Delta C_{\text{p}}(T - T_{\text{m}}) - T\Delta C_{\text{p}} \ln (T/T_{\text{m}}) \qquad (10.4)$$

In this equation, T_{m} is the so-called melting temperature, located at the top of the peak, and defined as the temperature of half denaturation where $[\text{N}] = [\text{U}]$, ΔH_{cal} has been defined above and is equivalent to the van't Hoff enthalpy in a two-state model, T is a given temperature and ΔC_{p} is the heat capacity change occurring during the transition from the native to the unfolded state. The calculation of the stabilisation energy, over a temperature range where the native state is in excess of the unfolded state, gives access to stability curves as shown in Fig. 10.5. These parabola-shaped curves have been established for the cold-active α-amylase AHA, mesophilic α-amylase PPA and thermophilic α-amylase BAA (D'Amico et al. 2003). One can see that over the whole range of temperatures, the cold-active enzyme presents a much lower thermodynamic stability than its mesophilic and thermophilic counterparts, even in the low-temperature range. This has been attributed to the lower number and/or strength of weak bonds that stabilise the structure. A second striking feature is that the maximum stability for the three orthologous

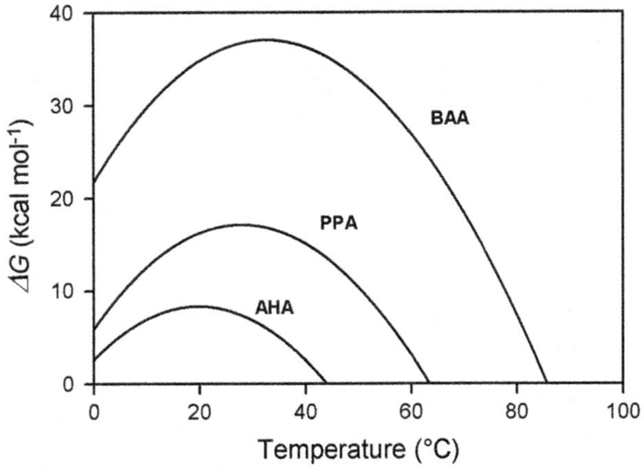

Fig. 10.5 Stabilisation energy *curves* of enzymes adapted to different temperatures. Stabilisation energy as a function of temperature for the psychrophilic α-amylase (AHA) from the Antarctic bacterium *Pseudoalteromonas haloplanktis*, the mesophilic enzyme from pig pancreas (PPA) and the thermophilic counterpart (BAA) from *Bacillus amyloliquefaciens* (adapted from D'Amico et al. 2003)

enzymes is found in the rather narrow range of 20–30 °C, even though the melting temperatures are very different: around 42 °C for AHA, 62 °C for PPA and 85 °C in BAA. This is due to the fact that the hydrophobic effect is maximum around room temperature (Kumar et al. 2002).

Another interesting point displayed by Fig. 10.5 is that the usual environmental temperature of mesophilic and thermophilic enzymes lies on the right-hand side of the curve and therefore does not correspond to the maximum stability of these enzymes. This relatively low stability at the environmental temperature is in fact required to secure an appropriate flexibility of the molecular structures and allow good interaction with the substrate. This plasticity is induced by the increase with temperature of the unfavourable stabilisation entropy, $\Delta S_{\text{N-U}}$. At the maximum stability, $\Delta S_{\text{N-U}} = 0$, and this term becomes negative at low temperatures due to the propensity of hydrophobic groups to favour hydration rather than interaction with other hydrophobic groups and thereby decreases the entropy of the unfolded state (Dias et al. 2010). Therefore, on the right side of the curve, the mesophilic and thermophilic enzymes are stabilised by the enthalpic term. On the contrary, in cold-active enzymes, the environmental temperature (near 0 °C) lies on the left-hand side of the curve which is characterised by a negative value of the entropic and enthalpic terms. The former is therefore stabilising under these conditions, whereas the negative value of the stabilisation enthalpy appears essential to confer to the enzyme a good flexibility. This negative enthalpic value originates from the weakening of hydrophobic and ionic interactions by their preferential hydration at low temperatures, and ultimately, the hydration of these groups leads to cold denaturation, a process which occurs with a release of heat. Interestingly, one can see from Fig. 10.5 that the cold-active enzyme, with a T_{m} predicted at around

$-10\,^{\circ}C$, is more sensitive to low temperatures than the mesophilic and thermophilic homologues. This is probably due to the fact that the cold-adapted structure is stabilised by less weak interactions.

10.6 Determinants of Cold Adaptation

To understand cold adaptation, it is necessary to define the structural changes that are directly or indirectly involved in the adaptation to cold. This is complicated by the high number of neutral mutations resulting from genetic drift that lead to large differences in the amino acid sequence of homologous enzymes produced by related species. The elucidation of the three-dimensional structure of several cold-active enzymes by X-ray crystallography, presently more than 30, and the production of reliable structural models have allowed to detect, in a first approach, the amino acid substitutions that could lead to a decreased thermal stability of cold-active enzymes. Again, only a few of these substitutions are intuitively involved in the adaptation to cold since, in low-temperature environments, the absence of a high selective pressure on thermal stability can enhance the importance of the genetic drift. The first structure of a psychrophilic enzyme, an α-amylase, was solved in the 1990s (Aghajari et al. 1998) shortly followed by that of a citrate synthase (Gerike et al. 1997; Russell et al. 1998). The structural parameters involved in the thermal stability of proteins have been extensively discussed (Vieille and Zeikus 2001), and their implications in cold adaptation have been presented in many reviews, notably in those mentioned in Sect. 10.1.

There is a consensus to agree that the enhanced flexibility of crucial parts of these enzymes is mainly secured through a weakening of intramolecular weak bonds that stabilise the structure of mesophilic counterparts and/or to entropic factors that tend to increase the entropy of the unfolded state. In general, the structural changes leading to cold adaptation are rather discrete since the three-dimensional structures of psychrophilic and mesophilic homologous enzymes can be nearly superimposed. Hydrophobic interactions play important roles in driving the folding of proteins in aqueous media and in stabilising the nascent polypeptide chain. In cold-active enzymes, the number and size of hydrophobic interactions as well as the size of the hydrophobic groups involved can be reduced so as to decrease the compactness of the molecule by creating cavities that can be occupied by water molecules (Paredes et al. 2011). In other cases, there is an increase in the proportion of hydrophobic groups at the surface of the protein; this induces a destabilisation of the molecular edifice through the reduction of the overall entropy of the system and the formation of clathrate-like structures around these groups. The stabilising effect of salt bridges can also be altered through a modification of the orientation of ionised groups and their reduction in number (Papaleo et al. 2007). The presence of ionised groups carrying similar charge, in general negative, can also modify the stability through charge repulsion and by increasing interaction with the solvent (Narinx et al. 1997; Feller et al. 1999; Adekoya et al. 2006). In a cold-active uracil-DNA

N-glycosylase, however, a more positively charged surface near the active site seems to be important for the adaptation to cold (Moe et al. 2004). In citrate synthases, psychrophilic, mesophilic and thermophilic enzymes have an increasingly stronger electrostatic stabilisation of the transition state (Bjelic et al. 2008). Some cold-active enzymes are also poorer in arginine residue content and display a lower Arg/Arg+Lys ratio (Adekoya et al. 2006). This is possibly due to the fact that arginine can develop more electrostatic interactions with surrounding amino acids than lysine. α-Helices can be assimilated to macroscopic electrostatic dipoles that carry a net positive charge at their N-terminus and a net negative charge at their C-terminus (Serrano and Fersht 1989) and are sometimes destabilised in psychrophilic enzymes, such as in cold-active trypsins (Leiros et al. 2000). Also, the hydrogen bonding network is often weakened in psychrophilic enzymes and has been shown to be responsible for the increased flexibility of these enzymes (Xie et al. 2009). The structure of loops connecting secondary structures also seems important in conferring an appropriate flexibility; their length can be increased, their level of interactions with the internal moiety lowered and their flexibility enhanced (Feller et al. 1997; Gudmundsdóttir 2002; Matsuura et al. 2002). Metal binding can also be involved through the reduction of binding affinity or deletion of binding sites (Narinx et al. 1997; Almog et al. 2009). Finally, the deletion of disulphide bridges can also improve the flexibility and activity of cold-active enzymes and has been demonstrated in a cold-active α-amylase and alkaline phosphatase (D'Amico et al. 2003; Asgeirsson et al. 2007; Papaleo et al. 2007). In a recent study, a cold-adapted xylanase was compared to a thermophilic homologue; it was found to contain more flexible loops, an increased number of Gly residues, decreased number of Pro residues and a much higher number of Arg residues at the surface (Zheng et al. 2016). This latter factor can improve the interaction with the solvent and also produce charge-charge repulsions that could cause the destabilisation of the cold-adapted structure.

Following the comparison of the three-dimensional structures of psychrophilic and mesophilic proteins, mutagenesis experiments were carried out in an attempt to transform psychrophilic enzymes into mesophilic counterparts or to try to force mesophilic enzymes to adopt psychrophilic characteristics. Cold-active subtilisin (S41) from an Antarctic *Bacillus* sp. (Davail et al. 1994) was submitted to random mutagenesis and mutant libraries screened for enzymes displaying greater thermal stability without sacrificing low-temperature activity (Miyazaki et al. 2000; Wintrode et al. 2001). These experiments tended to demonstrate that it was quite possible to increase the thermal stability of an enzyme while keeping its high catalytic activity at low temperature; in other words, the two properties were not necessarily inversely related. However, these experiments were carried out using synthetic and small-sized substrates, and later it was shown that the mutants showing a better thermal stability and higher activity than the wild-type enzyme were poorly active on natural large-sized substrates such as proteins. Nevertheless, this does not mean that in vitro it is not possible to increase the thermostability of a cold-active enzyme while preserving or even increasing its catalytic efficiency at low temperature. Indeed, another cold-active subtilisin (S39), highly similar to S41, was also submitted to site-directed mutagenesis with the aim of increasing stability (Narinx et al. 1997). Several mutations were introduced to restore: a salt bridge found in subtilisin BPN′, a hydrophobic interaction found in subtilisin Carlsberg, a disulphide bridge present in aqualysine, a mutation N136S supposed to increase the specific activity towards large substrates and a mutation T85D to exchange a poor Ca^{2+} binding site for a strong one. Most of these mutations had a positive effect on the thermostability of the enzyme, especially the mutation T85D that increased the half-life at 50 °C by more than 50 min. At 5 °C, the specific activity of this mutant with the synthetic substrate s-FAAF-*p*Na was four times higher than that of the wild-type enzyme, but the K_m value was not affected. Interestingly however, the specific activity was only slightly increased when the natural substrate azocasein was used. In another cold-active subtilisin-like proteinase, Ser-Ala and Xaa-Pro mutations were introduced to produce single, double and triple mutants. Most of these mutations, especially Xaa-Pro, led to an increased thermostability with a concomitant decrease of the specific activity (Arnorsdóttir et al. 2007). A thermophilic subtilase was also engineered in order to produce cold-active variants (Zhong et al. 2009). All successful single mutations were located within or near the active site, and a variant carrying four amino acid substitutions showed a sixfold increase of specific activity towards casein in the temperature range of 15–25 °C, a decrease in thermal stability and a shift of its apparent optimum towards low temperature by approximately 15 °C. Interestingly, this multiple mutant was less active towards the synthetic substrate s-AAPF-*p*NA due to a large increase in K_m and decrease in k_{cat}. This again underlines the importance, before drawing robust conclusions, of selecting appropriate substrates when studying multi-substrate enzymes by mutagenesis.

The molecular basis of the adaptation to cold of the psychrophilic α-amylase, already mentioned above, was also investigated by site-directed mutagenesis. Fourteen amino acid substitutions, including double mutations, were introduced with the aim of mimicking specific structural characteristics found in mesophilic and thermophilic homologous enzymes (D'Amico et al. 2001, 2002). These included an introduction of hydrogen bonds, salt bridges and hydrophobic interactions, helix dipole stabilisation and reinsertion of a disulphide bridge that connects domain A and B in mesophilic counterparts. The highest contribution to stability, in terms of both T_m and ΔH_{cal}, was obtained by the introduction of an electrostatic interaction. A double aromatic interaction was found to increase the stabilisation energy by a factor of 2 at 10 °C, whereas the reinforcement of hydrophobic clusters within the hydrophobic core of the enzyme induced the production of multiple calorimetric domains similar to those existing in the mesophilic counterpart from pig pancreas. Most of the introduced mutations that aimed to increase the stability were efficient but led to a decrease in k_{cat} and K_m. Also, any attempts to further increase the catalytic efficiency at low temperature through a decrease in stability gave rise to unfolded forms, suggesting that the cold-active α-amylase had reached a limit in stability, precluding any improvement of the specific activity via a decrease in thermal stability. The incorporation of a disulphide bridge decreased the specific activity of the enzyme by a factor of 2 at 5 °C as well as the K_m and resulted in a microcalorimetric profile similar to that of the mesophilic counterpart with however the appearance of a domain transition at a lower T_m. This indicates that the introduction of this disulphide bridge leads to a stabilisation of some parts of the cold-adapted protein but also induces an unfavourable structural constraint in other parts. Multiple mutants of this cold-active enzyme were also recently produced (Cipolla et al. 2011) and enabled for variants with a reduced specific activity, drastically improved thermal stability and decreased flexibility. It was concluded that these mutants can be considered as structural intermediates between their parent psychrophilic and mesophilic enzymes. The effects of these mutations on the molecular dynamics of the proteins were also recently investigated (Papaleo et al. 2011) and showed a reduced flexibility in various regions, not only near the active site and substrate binding groove but also at long distance from these, even in domain C which didn't contain any mutations.

Several other cold-active enzymes have also been the target of site-directed mutagenesis studies aimed at shedding light on the residues involved in their adaptation. One can notably mention a lipase (Santarossa et al. 2005), an alkaline phosphatase (Koutsiolis et al. 2008) and an isocitrate lyase (Sato et al. 2008). These experiments indicated that in most cases, the high specific activity of cold-active enzymes is related to an improved flexibility of crucial parts of the molecular structure that induces a high thermal sensitivity and a generally better accessibility for the substrate at low temperatures. On the other hand, attempts to try to simultaneously increase the thermostability and activity at low temperature have had little success and directed evolution experiments that have apparently produced more stable and more active variants used essentially multi-substrate enzymes with small artificial substrates.

10.7 Folding of Cold-Active Enzymes

As discussed above, psychrophilic enzymes show a marginal stability at low temperatures. This can complicate the folding process, in particular as this marginal stability is induced by the preferential hydration of hydrophobic and ionic groups that can impair the formation of bonds and also, in certain cases, by the difficulty, at low temperatures, to transform proline *trans* configuration into *cis* configuration. The folding process can also be affected by the high viscosity of the aqueous medium in addition to the macromolecular crowding that characterises the intracellular space (Gershenson and Gierasch 2011).

The protein folding process is most commonly represented by a folding funnel populated by various and transient energy-distinct intermediates converging to a more or less defined native form characterised by the lowest energy level (Leopold et al. 1992). As proposed by D'Amico et al. (2003) psychrophilic, mesophilic and thermophilic proteins display distinct folding funnels as illustrated in Fig 10.6. Cold-active enzymes are less stable than their mesophilic homologues, and thus the energy level of the native state is higher than that of the mesophilic protein. The

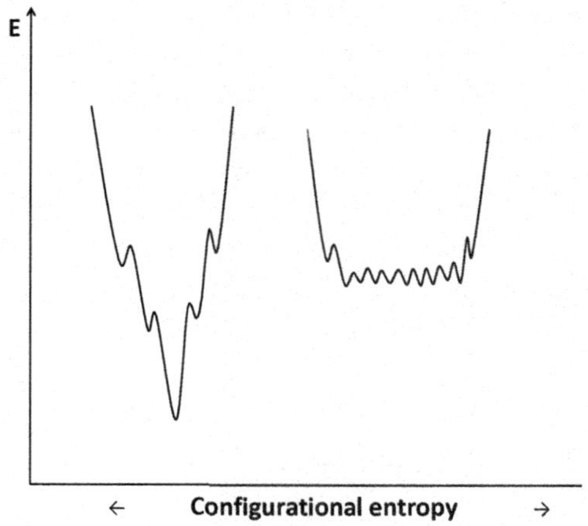

Fig. 10.6 Folding energy landscapes. Schematic representation of the funnel-shaped folding energy landscapes of mesophilic (*left diagram*) and psychrophilic (*right diagram*) enzymes (modified from D'Amico et al. 2003)

proposed dynamical properties of the cold enzyme are reflected by the number of local minima of the native state that show a large population of conformations differing by a low energy level on the free energy landscape. This allows a rapid conversion of one conformation into another (flexibility) and is contrary to the mesophilic counterpart which is characterised by a limited number of local free energy minima with quite distinct energy levels that preclude the interconversion between each other (rigidity). This model was shown to be consistent with metalloproteases (Xie et al. 2009), elastases and uracil DNA glycosylases studied by molecular dynamics (Mereghetti et al. 2010). Finally, the lower free energy level of the unfolded form of psychrophilic enzymes shown in Fig. 10.6 reflects the larger entropy of the unfolded form of these, due to their generally higher proportion of glycine residues and lower content of proline residues.

Chaperones are helper proteins that assist in the efficient folding of proteins into their functional three-dimensional structure and are essential elements in protein folding and protein quality control. Their expression and activity as a function of growth temperature in psychrophilic microorganisms has been the subject of a number of studies. One can mention studies with *Shewanella* sp. (Suzuki et al. 2004), *Methanococcoides burtonii* (Goodchild et al. 2004), *Exiguobacterium sibiricum* (Qiu et al. 2006; Rodrigues et al. 2008), *Psychrobacter arcticus* (Zheng et al. 2007; Kuhn 2012), *Shewanella livingstonensis* (Kawamoto et al. 2007), *Sphingopyxis alaskensis* (Ting et al. 2010), *Pseudoalteromonas haloplanktis* (Piette et al. 2010, 2011), *Shewanella* S1B1 FKBP22 (Budiman et al. 2011), *Acidithiobacillus ferrooxidans* (Mykytczuk et al. 2011), *Saccharomyces cerevisiae* (Naicker et al. 2012), *Psychrobacter* sp. PAMC21119 (Kim et al. 2015) and *Glaciozyma antarctica* (Yusof et al. 2016). Interestingly, no general conclusions can be made from these studies with, depending on the species, chaperone overexpression or downregulation being observed at low temperature. This is possibly related to the fact that in psychrophiles, the hydrophobic clusters that constitute the core of proteins are often weaker than in their mesophilic homologues. Also, at low temperatures, hydrophobic interactions are weakened, and the risk of aggregation and misfolding of the nascent polypeptide chains is lower. Indeed, this idea is corroborated by the fact that cold-adapted proteins often show a spontaneous reversible thermal unfolding (Feller 2010). In contrast to the lack of a general trend with chaperone production at low temperatures, an enhancement at low temperatures of the activity of peptidyl-prolyl *cis/trans* isomerases (PPIases) has been commonly observed. PPIase is critical in the conversion of *cis*-proline into the *trans* form during protein folding, and it has been shown that this increased activity is achieved either through PPIase alone or through a chaperone, known as trigger factor, that carries a PPIase domain.

10.8 Conclusions

Psychrophilic organisms living in permanently cold habitats have developed numerous adaptations that allow them to successfully thrive in low-temperature environments. Their enzymes are a key feature of this adaptation. Cold-enzymes are much more active than their mesophilic counterparts at low and moderate temperatures and display an increased flexibility of the active site and/or of other regions of the structure indirectly involved in activity. The adaptation takes the form of rather discrete structural modifications that are often located at long distance from the active site. These modifications systematically lead to a higher thermal instability of these enzymes. This is often preceded by a faster heat inactivation that results from the more negative heat capacity change occurring during the activation as compared to the mesophilic homologues. Although the general strategy adopted by nature consists of a weakening of the intramolecular forces that stabilise the structure of their mesophilic and thermophilic counterparts, the strategy appears to be specific to each enzyme and can depend on a number of factors. These include the position of the enzyme within a metabolic pathway, its localisation in the intra- or extracellular space, its capacity for structural modification, the environment of the organism and its evolutionary history. The adaptation to cold of the microorganisms living in permanently cold habitats is also highly diversified. Some display growth within a low and narrow temperature range only and could be considered as hyperpsychrophiles. Others grow over a relatively large temperature range which may even partially overlap that of mesophilic organisms and could be simply named psychrophiles. Obviously, there is a continuum in the adaptation to cold as is the case for mesophilic and thermophilic organisms at moderate and high temperatures, respectively.

Acknowledgements Tony Collins is supported by the FCT, the European Social Fund, the Programa Operacional Potencial Humano and the Investigador FCT Programme (IF/01635/2014). The FCT is thanked for their funding through EXPL/BBB-BIO/1772/2013-FCOMP-01-0124-FEDER-041595, the strategic programme UID/BIA/04050/2013 (POCI-01-0145-FEDER-007569), and the ERDF through COMPETE2020—Programa Operacional Competitividade e Internacionalização (POCI). All the technical staff at the CBMA is thanked for the skillful technical assistance.

References

Adekoya OA, Helland R, Willassen NP, Sylte I (2006) Comparative sequence and structure analysis reveal features of cold adaptation of an enzyme in the thermolysin family. Proteins 62:435–439

Aghajari N, Feller G, Gerday C, Haser R (1998) Structures of the psychrophilic *Alteromonas haloplanktis* α-amylase give insights into cold adaptation at a molecular level. Structure 6:1503–1516

Almog O, Gonzalez A, Godin N, de Leeuw M, Mekel MJ, Klein D, Braun S, Shoham G, Walter RL (2009) The crystal structure of the psychrophilic subtilisin S41 and the mesophilic subtilisin Sph reveal the same calcium loaded state. Proteins 74:489–496

Altermark B, Niiranen L, Willassen NP, Smalas AO, Moe E (2007) Comparative studies of endonuclease I from cold-adapted *Vibrio salmonicida* and mesophilic *Vibrio cholerae*. FEBS J 274:252–263

Angelaccio S, Florio R, Consalvi V, Festa G, Pascarella P (2012) Serine hydroxymethyltransferase from the cold-adapted microorganism *Psychromonas ingrahamii*: a low temperature active enzyme with broad specificity. Int J Mol Sci 13:1314–1326

Arcus VL, Prentice EJ, Hobbs JK, Mulholland AJ, Van der Kamp MW, Pudney CR, Parker EJ, Schipper LA (2016) On the temperature dependence of enzyme-catalysed rates. Biochemistry 55:1681–1688

Arnorsdóttir J, Helgadóttir S, Thorbjarnardóttir SH, Eggertsson G, Kristjansson MM (2007) Effect of selected Ser/Ala and Xaa/Pro mutations on the stability and catalytic properties of a cold adapted subtilisin-like serine proteinase. Biochim Biophys Acta 1774:749–755

Arrhenius S (1889) Uber die Reaktionsgeschwindigkeit bei der Inversion von Rohrzuckerdurch Säuren. Z Physic Chem 4:226–248

Asgeirsson B, Adalbjörsson BV, Gylfason GA (2007) Engineered disulfide bonds increase active-site local stability and reduce catalytic activity of a cold-adapted alkaline phosphatase. Biochim Biophys Acta 1774:679–687

Berg TO, Gurung MK, Altermark B, Smalas AO, Raeder H (2015) Characterization of the N-acetylneuraminic acid synthase (NeuB) from the psychrophilic fish pathogen *Moritella viscosa*. Carbohydr Res 402:133–145

Bjelic S, Bransdal BO, Aqvist J (2008) Cold adaptation of enzyme reaction rates. Biochemistry 47:10049–10057

Budiman C, Koga Y, Takano K, Kanaya S (2011) FK 506-binding protein 22 from a psychrophilic bacterium, a cold shock-inducible peptidyl prolyl isomerase with the ability to assist in protein folding. Int J Mol Sci 12:5261–5284

Chess JP, Petrescu I, Bentahir M, Van Beeumen J, Gerday C (2000) Purification, physico-chemical characterization and sequence of the heat-labile alkaline metalloprotease from a psychrophilic *Pseudomonas* species. Biochim Biophys Acta 1479:265–274

Chiappori F, Pucciarelli S, Merelli I, Ballarini P, Miceli C, Milanesi L (2012) Structural thermal adaptation of β-tubulins from the Antarctic psychrophilic protozoan *Euplotes focardii*. Proteins 80:1154–1166

Chiuri R, Majorano G, Rizello A, del Mercato LL, Cingolani R, Rinaldi R, Maffia M, Pompa PP (2009) Exploring local flexibility/rigidity in psychrophilic and mesophilic carbonic anhydrases. Biophys J 96:1586–1596

Cipolla A, Delbrassine F, Da Lage JL, Feller G (2011) Temperature adaptations in psychrophilic, mesophilic and thermophilic chloride-dependent alpha-amylases. Biochimie 94:1943–1950

Cipolla A, D'Amico S, Barumandzadeh R, Matagne A, Feller G (2012) Stepwise adaptation to low temperature as revealed by multiple mutants of psychrophilic α-amylase from Antarctic bacterium. J Biol Chem 286:38348–38355

Collins T, Claverie P, D'Amico S, Georlette D, Gratia E, Hoyoux A, Meuwis MA, Poncin J, Sonan G, Feller G, Gerday C (2002) Life in the cold: psychrophilic enzymes. In: Pandalai SG

(ed) Recent research developments in proteins, vol 1. Transworld Research Network, Trivandrum, pp 13–26
Collins T, Meuwis M-A, Gerday C, Feller G (2003) Activity, stability and flexibility in glycosidases adapted to extreme thermal environments. J Mol Biol 338:419–428
Collins T, D'Amico S, Marx J-C, Feller G, Gerday C (2007) Cold-adapted enzymes. In: Gerday C, Glansdorff N (eds) Physiology and biochemistry of extremophiles. ASM Press, Washington, DC, pp 165–179
Collins T, Roulling F, Piette F, Marx J-C, Feller G, Gerday C, D'Amico S (2008) Fundamentals of cold-adapted enzymes. In: Margesin R, Schinner F, Marx J-C, Gerday C (eds) Psychrophiles: from biodiversity to biotechnology. Springer, Berlin, pp 211–227
Coquelle N, Fioravanti E, Weik M, Vellieux F, Madern D (2007) Activity, stability and structural studies of lactate dehydrogenases adapted to extreme thermal environments. J Mol Biol 374:547–562
D'Amico S, Gerday C, Feller G (2001) Structural determinants of cold adaptation and stability in a large protein. J Biol Chem 276:25791–25796
D'Amico S, Gerday C, Feller G (2002) Dual effects of an extra disulphide bond on the activity and stability of a cold-adapted α-amylase. J Biol Chem 277:46110–46115
D'Amico S, Marx J-C, Gerday C, Feller G (2003) Activity-stability relationship in extremophilic enzymes. J Biol Chem 278:7891–7896
D'Amico S, Sohier J-S, Feller G (2006a) Kinetics and energetics of ligand binding determined by microcalorimetry: insight into active site mobility in a psychrophilic alpha-amylase. J Mol Biol 358:1296–1304
D'Amico S, Collins T, Marx JC, Feller G, Gerday C (2006b) Psychrophilic microorganisms: challenges for life. EMBO Rep 7:385–389
Daniel RM, Danson MJ (2010) A new understanding of how temperature affects the catalytic activity of enzymes. Trends Biochem Sci 35:584–591
Davail S, Feller G, Narinx E, Gerday C (1994) Cold adaptation of proteins. Purification, characterization, and sequence of the heat labile subtilisin from the Antarctic psychrophile *Bacillus* TA 41. J Biol Chem 269:17448–17453
De Santi C, Tutino ML, Mandrich L, Giuliani M, Parilli E, Del Vecchio P, De Pascale D (2010) The hormone sensitive lipase from *Psychrobacter* sp. TA144: new insight in the structure/functional characterization. Biochimie 92:949–957
Demchenko AP, Rusyn OI, Saburova EA (1989) Kinetics of the lactate dehydrogenase reaction in high-viscosity media. Biochim Biophys Acta 998:196–203
Dias CL, Ala-Nissila T, Wong E, Kabut J, Vattulainen I, Grant M, Karttunen M (2010) The hydrophobic effect and its role in cold denaturation. Cryobiology 60:91–99
Eyring H (1935) The activated complex in chemical reactions. J Chem Phys 3:107–115
Fedoy A-E, Yang N, Martinez A, Leiros H-K, Stee H (2007) Structural and functional properties of isocitrate dehydrogenase from the psychrophilic bacterium *Desulfotalea psychrophila* reveal a cold-active enzyme with an unusual high thermal stability. J Mol Biol 372:130–149
Feller G (2008) Enzyme function at low temperatures in psychrophiles. In: Siddiqui KS, Thomas T (eds) Protein adaptation in extremophiles. Nova Science, New York, NY, pp 35–69
Feller G (2010) Protein stability and enzyme activity at extreme biological temperatures. J Phys Condens 22:32101–32118
Feller G, Gerday C (2003) Psychrophilic enzymes: hot topics in cold adaptation. Nat Rev Microbiol 1:200–208
Feller G, Lonhienne T, Deroanne C, Libioulle C, Van Beeumen J, Gerday C (1992) Purification, characterization, and nucleotide sequence of the thermolabile α-amylase from the Antarctic psychrotroph *Alteromonas haloplanktis* A 23. J Biol Chem 267:5217–5221
Feller G, Narinx E, Arpigny JL, Zekhnini Z, Swings J, Gerday C (1994) Temperature dependence of growth, enzyme secretion and activity of psychrophilic Antarctic bacteria. Appl Microbiol Biotechnol 41:477–479

Feller G, Zekhnini Z, Lamotte-Brasseur J, Gerday C (1997) Enzymes from cold-adapted microorganisms. The class C beta-lactamase from the Antarctic psychrophile *Psychrobacter immobilis* A5. Eur J Biochem 244:186–191

Feller G, D'Amico S, Gerday C (1999) Thermodynamic stability of a cold-active α-amylase from the Antarctic bacterium *Alteromonas haloplanktis*. Biochemistry 38:4613–4619

Fields PA, Somero GN (1998) Hot spots in cold adaptation: localized increases in conformational flexibility in lactate dehydrogenases A4 orthologs of Antarctic notothenioid fishes. Proc Natl Acad Sci U S A 95:11476–11481

Fields PA, Dong Y, Meng X, Somero GN (2015) Adaptations of protein structure and function to temperature: there is more than one way to 'skin a cat'. J Exp Biol 2018:1801–1811

Garcia-Viloca M, Gao J, Karplus M, Truhlar DG (2004) How enzymes work: analysis by modern rate theory and computer simulations. Science 303:186–195

Garsoux G, Lamotte-Brasseur J, Gerday C, Feller G (2004) Kinetic and structural optimisation to catalysis at low temperatures in a psychrophilic cellulase from the Antarctic bacterium *Pseudoalteromonas haloplanktis*. Biochem J 384:247–253

Georlette D, Jonsson ZO, Van Petegem F, Chessa J-P, Van Beeumen J, Hubscher U, Gerday C (2000) A DNA ligase from the psychrophile *Pseudoalteromonas haloplanktis* gives insights into the adaptation of proteins at low temperatures. Eur J Biochem 267:3502–3512

Georlette D, Damien B, Blaise V, Depiereux E, Uversky VN, Gerday C, Feller G (2003) Structural and functional adaptations to extreme temperatures in psychrophilic, mesophilic, and thermophilic DNA ligases. J Biol Chem 278:37015–37023

Georlette D, Blaise V, Collins T, D'Amico S, Gratia E, Hoyoux A, Marx J-C, Sonan G, Feller G, Gerday C (2004) Some like it cold: biocatalysis at low temperatures. FEMS Microbiol Rev 28:25–42

Gerday C (2013) Catalysis and protein folding in psychrophiles. In: Yumoto I (ed) Cold-adapted microorganisms. Caister Academic press, Norfolk, pp 137–157

Gerday C (2014) Fundamentals of cold-active enzymes. In: Buzzini P, Margesin R (eds) Cold-adapted yeasts. Springer, Berlin, pp 325–350

Gerike U, Danson MJ, Russell NJ, Hough DW (1997) Sequencing and expression of the gene encoding a cold-active citrate synthase from an Antarctic bacterium strain DS-3R. Eur J Biochem 248:49–57

Gershenson A, Gierasch LM (2011) Protein folding in the cell: challenges and progress. Curr Opin Struct Biol 21:32–41

Goodchild A, Saunders NF, Erlan H, Raftery M, Guilhaus M, Curmi PM, Cavicchioli R (2004) A proteomic determination of cold adaptation in the Antarctic archeon, *Methanococcoides burtonii*. Mol Microbiol 53:309–321

Gudmundsdóttir A (2002) Cold-adapted and mesophilic brachyurins. Biol Chem 383:1125–1131

Gudmundsdóttir E, Spilliaert R, Yang Q, Craik CS, Bjarnason JB, Gudmundsdóttir A (1996) Isolation and characterization of two cDNAs from Atlantic cod encoding two distinct psychrophilic elastases. Comp Biochem Physiol B 113:795–801

Heidarsson PO, Sigurdsson ST, Asgeirsson B (2009) Structural features and dynamics of a cold-adapted alkaline phosphatase studied by EPR spectroscopy. FEBS J 276:2725–2735

Hess E (1934) Effects of low temperatures on the growth of marine bacteria. Contribs Can Biol Fisheries Ser C 8:491–505

Hochachka PW, Somero GN (1973) Strategies of biochemical adaptation. WB Saunders, Philadelphia, PA

Hochachka PW, Somero GN (2002) Temperature. In: Hochachka PW, Somero GN (eds) Biochemical adaptation. Oxford University Press, New York, NY, 190 pp

Homchaudhuri L, Sarma N, Swaminathan R (2006) Effect of crowding by dextrans and ficols on the rate of alkaline phosphatase-catalysed hydrolysis: a size-dependent investigation. Biopolymers 83:477–486

Huston AL, Krieger-Brockett BB, Deming JW (2000) Remarkably low temperature optima for extracellular enzyme activity from Arctic bacteria sea ice. Environ Microbiol 4:383–388

Huston AL, Haeggström JZ, Feller G (2008) Cold adaptation of enzymes: structural, kinetic and microcalorimetric characterizations of an aminopeptidase from the Arctic psychrophile *Colwellia psychrerythraea* and of human leukotriene A (4) hydrolase. Biochim Biophys Acta 1784:1865–1872

Ingraham JL, Stokes JL (1959) Psychrophilic bacteria. Bacteriol Rev 23:97–108

Isaksen GV, Aqvist J, Brandsdal BO (2016) Enzyme surface rigidity tunes the temperature dependence of catalytic rates. Proc Natl Acad Sci U S A 113:7822–7827

Kawamoto J, Kurihara T, Kitagawa M, Kato I, Esaki N (2007) Proteomic studies of an Antarctic cold-adapted bacterium, *Shewanella livingstonensis* AC 10, for global identification of cold-inducible proteins. Extremophiles 10:819-826

Kim HW, Wi AR, Jeon BW, Lee JH, Shin SC, Park H, Jeon SJ (2015) Cold adaptation of a psychrophilic chaperonin from *Psychrobacter* sp. and its application for heterologous protein expression. Biotechnol Lett 37:1887–1893

Kobori H, Sullivan CW, Shizuya H (1984) Heat-labile alkaline phosphatase from Antarctic bacteria: rapid 5′ end-labeling of nucleic acid. Proc Natl Acad Sci U S A 81:6691–6695

Kobus S, Widderich N, Hoeppner A, Bremer E, Smits SH (2015) Overproduction, crystallization and X-ray diffraction data analysis of ectoine synthase from the cold-adapted marine bacterium *Sphingopyxis alaskensis*. Acta Cristallogr F Struct Biol Commun 71:1027–1032

Koutsiolis D, Wang E, Tzanodaskalaki M, Nikiforaki D, Deli A, Feller G, Heikinheimo P, Bouriotis V (2008) Directed evolution on the cold adapted properties of TAB5 alkaline phosphatase. Protein Eng 21:319–327

Kuhn E (2012) Toward understanding life under subzero conditions: the significance of exploring psychrophilic "cold-shock" proteins. Astrobiology 12:1078–1086

Kumar S, Tsai CJ, Nussinov R (2002) Maximal stabilities of reversible two-state proteins. Biochemistry 41:5359–5374

Leiros HK, Willassen NP, Smalas AO (2000) Structural comparison of psychrophilic and mesophilic trypsins. Elucidating the molecular basis of cold-adaptation. Eur J Biochem 267:1039–1049

Leopold PE, Montal M, Onuchic JN (1992) Protein folding funnels: a kinetic approach to the sequence-structure relationship. Proc Natl Acad Sci U S A 89:8721–8725

Lonhienne T, Baise E, Feller G, Bouriotis V, Gerday C (2001a) Enzyme activity determination on macromolecular substrates by isothermal calorimetry: application to mesophilic and psychrophilic chitinases. Biochim Biophys Acta 1545:349–356

Lonhienne T, Zoidakis J, Vorgias E, Feller G, Gerday C, Bouriotis V (2001b) Modular structure, local flexibility and cold-activity of a novel chitobiase from a psychrophilic Antarctic bacterium. J Mol Biol 310:291–297

Margesin R (2009) Effect of temperature on growth parameters of psychrophilic bacteria and yeasts. Extremophiles 13:257–262

Marshall CI (1997) Cold-adapted enzymes. Trends Biotechnol 15:359–364

Marx J-C, Collins T, D'Amico S, Feller G, Gerday C (2007) Cold-adapted enzymes from marine Antarctic microorganisms. Mar Biotechnol 9:293–304

Mastro AM, Keith AD (1984) Diffusion in the aqueous compartment. J Cell Biol 99:180–187

Matsuura A, Yao M, Aizawa T, Koganesawa N, Masaki K, Miyazawa M, Demura M, Tanaka I, Kawano K, Nitta K (2002) Structural analysis of an insect lysozyme exhibiting catalytic efficiency at low temperature. Biochemistry 41:12086–12092

Mereghetti P, Riccardi L, Brandsdal BO, Fantucci P, De Gioia L, Papaleo E (2010) Near native-state conformational landscape of psychrophilic and mesophilic enzymes: probing the folding funnel model. J Phys Chem B 114:7609–7619

Miao LL, Hou YJ, Fan HX, Qu J, Qi C, Liu Y, Li DF, Liu ZP (2016) Molecular structural basis for the cold adaptedness of the psychrophilic β-Glucosidase BgIU in *Micrococcus antarcticus*. Appl Environ Microbiol 82:2021–2030

Miyazaki K, Wintrode PL, Grayling RA, Rubigh DN, Arnold FH (2000) Directed evolution study of temperature adaptation in a psychrophilic enzyme. J Mol Biol 297:1015–1026

Moe E, Leiros I, Rijse EK, Olufsen M, Lanes O, Smalas A, Willassen NP (2004) Optimisation of the surface electrostatics as a strategy for cold adaptation of uracil-DNA N-glycosylase. J Mol Biol 343:1221–1230

Mykytczuk NC, Trevors JT, Foote SJ, Leduc LG, Ferroni GD, Twine SM (2011) Proteomic insight into cold adaptation of psychrotrophic and mesophilic *Acidithiobacillus ferrooxidans* strains. Antonie Van Leeuwenhoek 100:259–277

Naicker MC, Seul JI, Im H (2012) Identification of chaperones in freeze tolerance in *Saccharomyces cerevisiae*. J Microbiol 50:882–887

Narinx E, Baise E, Gerday C (1997) Subtilisin from Antarctic bacteria: characterization and site-directed mutagenesis of residues possibly involved in the adaptation to cold. Protein Eng 10:1271–1279

Pace CN, Laurents DV (1989) A new method for determining the heat capacity change for protein folding. Biochemistry 28:2520–2525

Papaleo E, Olufsen M, De Gioia L, Bransdal BO (2007) Optimization of electrostatics as a strategy for cold-adaptation: a case study of cold- and warm-active elastases. J Mol Graph Model 26:93–103

Papaleo E, Pasi M, Tiberti M, De Gioia L (2011) Molecular dynamics of mesophilic-like mutants of a cold-adapted enzyme: insight into distal effects induced by the mutations. PLoS ONE 6: e24214

Paredes DI, Watters K, Pitman DJ, Bystrff C, Dordick JS (2011) Comparative void-volume analysis of psychrophilic and mesophilic enzymes: structural bioinformatics of psychrophilic enzymes reveals sources of core flexibility. BMC Struct Biol 11:42

Petrescu I, Lamotte-Brasseur J, Chessa J-P, Ntarima P, Claeyssens M, Devreese B, Marino G, Gerday C (2000) Xylanase from the psychrophilic yeast *Cryptococcus adeliae*. Extremophiles 4:137–144

Piette F, D'Amico S, Struvay C, Mazzuchelli G, Remaut J, Tutino ML, Danchin A, Leprince P, Feller G (2010) Proteomics of life at low temperature: trigger factor is the primary chaperone in the Antarctic bacterium *Pseudoalteromonas haloplanktis* TAC 125. Mol Microbiol 76:120–132

Piette F, Struvay C, Feller G (2011) The protein folding in psychrophiles: facts and current issues. Environ Microbiol 13:1924–1933

Qiu Y, Kathariou S, Lubman DM (2006) Proteomic analysis of cold adaptation in a Siberian permafrost bacterium *Exiguobacterium sibiricum* 255-15 by two-dimensional liquid separation coupled with mass spectrometry. Proteomics 6:135–148

Qoura F, Elleuche S, Brueck T, Antranikian G (2014) Purification and characterization of a cold-adapted pullulanase from a psychrophilic bacterial isolate. Extremophiles 18:1095–1102

Radestock S, Gohlke H (2011) Protein rigidity and thermophilic adaptation. Proteins 79:1089–1108

Ramli AN, Mahadi NM, Shamsir MS, Rabu A, Joyce-Tan KH, Murad AM, Illias RM (2012) Structural prediction of a novel chitinase from the psychrophilic *Glaciozyma antarctica* PI12 and an analysis of its structural properties and function. J Comput Aided Mol Des 26:947–961

Rodrigues DF, Ivanova N, He Z, Huebner M, Zhou J, Tiedje JM (2008) Architecture of thermal adaptation in *Exiguobacterium sibiricum* strain isolated from 3 million years old permafrost: a genome and transcriptome approach. BMC Genom 9:547

Roman EA, Faraj SE, Cousido-Siah A, Mitscler A, Podjarny A, Santos J (2013) Frataxin from *Psychromonas ingrahamii* as a model to study stability modulation within the CYaY protein family. Biochim Biophys Acta 1834:1168–1180

Russell NJ (2000) Toward a molecular understanding of cold activity of enzymes from psychrophiles. Extremophiles 4:83–90

Russell RJ, Gerike U, Danson MJ, Hough DW, Taylor GL (1998) Structural adaptations of the cold-active citrate synthase from an Antarctic bacterium. Structure 6:351–361

Santarossa G, Gatti-Lafranconi P, Alquati C, De Gioia L, Alberghina L, Fantucci P, Lotti M (2005) Mutations in the « lid » region affect chain length specificity and thermostability of a *Pseudomonas fragi* lipase. FEBS Lett 579:2383–2386

Santiago M, Ramírez-Sarmiento CA, Zamora RA, Parra LP (2016) Discovery, molecular mechanisms, and industrial applications of cold-active enzymes. Front Microbiol 7:1408

Sato Y, Watanabe S, Yamaoka N, Takada Y (2008) Gene cloning of cold-adapted isocitrate lyase from a psychrophilic bacterium, *Colwellia psychrerythraea*, and analysis of amino acid residues involved in cold adaptation of this enzyme. Extremophiles 12:107–117

Schmidt-Nielsen S (1902) Ueber einige psychrophile Mikrooganismen und ihr Vorkommen. Centr Bakteriol Parasitenk Abt II 9:145–147

Schwartz MH, Pan T (2016) Temperature dependent mistranslation in a hyperthermophile adapts proteins to lower temperature. Nucleic Acids Res 44:294–303

Serrano L, Fersht AR (1989) Capping and alpha-helix stability. Nature 342:296–299

Siddiqui KS, Cavicchioli R (2006) Cold-adapted enzymes. Annu Rev Biochem 75:403–433

Siddiqui KS, Bokhari SA, Afzal AJ, Singh S (2004) A novel thermodynamic relationship based on Kramers Theory for studying enzyme kinetics under high viscosity. IUBMB Life 56:403–407

Sigtryggsdóttir AR, Papaleo E, Thorbjarnardóttir SH, Kristjánsson MM (2014) Flexibility of cold- and-heat adapted subtilisin-like serine proteinase evaluated with fluorescence quenching and molecular dynamics. Biochim Biophys Acta 1844:705–712

Simpson PJL, Codd R (2011) Cold adaptation of the mononuclear molybdoenzyme periplasmic nitrate reductase from the Antarctic bacterium *Shewanella gelidimarina*. Biochem Biophys Res Commun 414:783–788

Smalas AO, Leiros HK, Os V, Willassen NP (2000) Cold-adapted enzymes. Biotechnol Annu Rev 6:1–57

Somero GN (1977) Temperature as a selective factor in protein evolution: the adaptational strategy of compromise. J Exp Zool 194:175–188

Somero GN (1995) Proteins and temperature. Annu Rev Physiol 57:43–68

Sonan GK, Receveur-Brechot V, Duez C, Aghajari N, Czjzek M, Haser R, Gerday C (2007) The linker region plays a key role in the adaptation to cold of the cellulase from an Antarctic bacterium. Biochem J 407:293–302

Sotelo-Mundo RR, Lopez-Zavala AA, Garcia-Orozco KD, Arvizu AA, Velazquez-Contreras EF, Valenzuela-Soto EM, Rojo-Dominguez A, Kanost MR (2007) The lysozyme from insect (*Manduca sexta*) is a cold-adapted enzyme. Protein Pept Lett 14:774–778

Spiwok V, Lipovova P, Skalova T, Duskova J, Dohnalek J, Hasek J, Russell N, Kralova B (2007) Cold-active enzymes studied by comparative molecular dynamics simulation. J Mol Model 13:485–497

Stadler AM, Garvey CJ, Bocahut A, Sacquin-Mora S, Digel I, Schneider GJ, Natali F, Artmann GM, Zaccai G (2012) Thermal fluctuations of haemoglobin from different species: adaptation to temperature via conformational dynamics. J R Soc Interface 7:2845–2855

Sun-Yong K, Kwang-Yeon H, Sung-Hou K, Ha-Chin S, Ye-Sun H, Yunge C (1999) Structural basis for cold adaptation. Sequence, biochemical properties, and crystal structure of malate dehydrogenase from a psychrophile *Aquaspirillium arcticum*. J Biol Chem 274:11761–11767

Suzuki Y, Haruki M, Takano K, Morikawa M, Kanaya S (2004) Possible involvement of an FKBP family member protein from a psychrotrophic bacterium, *Shewanella* sp. SLB1 in cold adaptation. J Biochem 271:1372–1381

Suzuki T, Yamamoto K, Tada H, Uda K (2012) Cold-adapted features of arginine kinase from the deep-sea *Calyptogena kaikoi*. Mar Biotechnol 14:294–303

Talla-Singh D, Stites WE (2008) Refinement of noncalorimetric determination of the change in heat capacity, ΔC_p, of protein unfolding and validation across a wide temperature range. Proteins 71:1607–1616

Tang MA, Motoshima H, Watanabe K (2014) Cold adaptation: structural and functional characterizations of psychrophilic and mesophilic acetate kinase. Protein J 33:313–322

Tehei M, Franzetti B, Madern D, Ginzburg M, Ginzburg BZ, Giudici-Orticoni MT, Bruschi M, Zaccai G (2004) Adaptation to extreme environments: macromolecular dynamics in bacteria compared in vivo by neutron scattering. EMBO Rep 5:66–70

Thomas T, Cavicchioli R (1998) Archaeal cold-adapted proteins: structural and evolutionary analysis of elongation factor 2 proteins from psychrophilic, mesophilic and thermophilic methanogens. FEBS Lett 439:281–286

Tiberti M, Papaleo E (2011) Dynamic properties of extremophilic subtilisin-like serine-proteases. J Struct Biol 174:69–83

Tindbaek N, Svendsen A, Oestezrgaard PR, Draborg H (2004) Engineering a substrate-specific cold-adapted subtilisin. Protein Eng Des Sel 17:149–156

Ting L, Williams TJ, Cowley MJ, Lauro FM, Guilhaus M, Raftery MJ, Cavicchioli R (2010) Cold adaptation in the marine bacterium, *Sphingopyxis alaskensis* assessed using quantitative proteomics. Environ Microbiol 12:2658–2676

Truongvan N, Jang SH, Lee CW (2016) Flexibility and stability trade-off in active site of cold-adapted *Pseudomonas mandelii* esterase EstK. Biochemistry 55:3542–3549

Tsigos I, Velonia K, Smonou I, Bouriotis V (1998) Purification and characterization of an alcohol dehydrogenase from the Antarctic psychrophile *Moraxella* sp. TAE123. Eur J Biochem 254:356–362

Vester JK, Glaring MA, Stougaard P (2015) An exceptionally cold-adapted alpha-amylase from a metagenomics library of a cold and alkaline environment. Appl Microbiol Biotechnol 99:717–727

Vieille C, Zeikus G (2001) Hyperthermophilic enzymes: sources, uses and molecular mechanisms for thermostability. Microbiol Mol Biol Rev 65:1–43

Watanabe S, Yasutake Y, Tanaka I, Takada Y (2005) Elucidation of stability determinants of cold-adapted monomeric isocitrate dehydrogenase from a psychrophilic bacterium, *Colwellia maris*, by construction of chimeric enzymes. Microbiology 151:1083–1094

Wintrode PL, Miyazaki K, Arnold FH (2001) Patterns of adaptation in a laboratory evolved thermophilic enzyme. Biochim Biophys Acta 1549:1–8

Xie BB, Bian F, Chen XL, He HL, Guo J, Gao X, Zeng YX, Chen B, Zhou BC, Zhang YZ (2009) Cold adaptation of zinc metalloprotease in the thermolysin family from deep sea and arctic sea ice bacteria revealed by catalytic and structural properties and molecular dynamics: new insights into relationship between conformational flexibility and hydrogen bonding. J Biol Chem 284:9257–9269

Xu Y, Feller G, Gerday C, Glansdorff N (2003a) Moritella cold-active dihydrofolate reductase: are there natural limits to optimization of catalytic efficiency at low temperature. J Bacteriol 185:5519–5526

Xu Y, Feller G, Gerday C, Glansdorff N (2003b) Metabolic enzymes from psychrophilic bacteria: challenge of adaptation to low temperatures in ornithine carbamoyltransferase from *Moritella abyssi*. J Bacteriol 185:2161–2168

Yang J, Li J, Mai Z, Tian X, Zhang S (2013) Purification, characterization, and gene cloning of a cold-adapted thermolysin-like protease from *Halobacillus* sp. SCSIO 20089. J Biosci Bioeng 115:628–632

Yusof NA, Hashim NH, Beddoe T, Mahadi NM, Ililias RM, Bakar FD, Murad AM (2016) Thermotolerance and molecular chaperone function of an SGT1-like protein from the psychrophilic yeast, *Glaciozyma Antarctica*. Cell Stress Chaperones 4:707–715

Zheng S, Ponder MA, Shih JY, Tiedje JM, Thomashow MF, Lubman DM (2007) A proteomic analysis of *Psychrobacter arcticus* 273-4 adaptation to low temperature and salinity using 2-D liquid mapping approach. Electrophoresis 28:467–488

Zheng Y, Li Y, Liu W, Chen CC, Ko TP, He M, Xu Z, Liu M, Luo H, Guo RT, Yao B, Ma Y (2016) Structural insight into potential cold adaptation mechanism through a psychrophilic glycoside hydrolase family 10 endo-β-1,4-xylanase. J Struct Biol 193:206–211

Zhong CQ, Song S, Fang N, Liang X, Zhu H, Tang XF, Tang B (2009) Improvement of low-temperature caseinolytic activity of a thermophilic subtilase by directed evolution and site directed mutagenesis. Biotechnol Bioeng 104:862–870

Chapter 11
Cryoprotectants and Ice-Binding Proteins

Hidehisa Kawahara

Abstract Some bacteria have developed a variety of strategies to survive and colonize extremely cold environments such as the Antarctic and the Arctic. In these frozen environments, bacteria are exposed to conditions that necessitate the partial removal of water from the intracellular space in order to maintain the structure and function of the cell. To avoid the dehydration condition under frozen environments, bacteria can accumulate small compounds, i.e. glucose, trehalose, etc., for keeping the unfrozen conditions. Also, some bacteria can produce some ice crystal-controlling materials into the intracellular or extracellular space. The ice crystal-controlling materials are ice-nucleating proteins, anti-nucleating materials, and antifreeze proteins. Among these three compounds, ice-nucleating proteins can facilitate the ice-nucleating activity at temperatures more than −3 °C. Also, anti-nucleating materials can inhibit ice-nucleating activity, thereby facilitating the supercooling temperature. Antifreeze proteins can inhibit ice crystal growth by binding with the surface of ice crystal. The production of these materials can diminish the fear of physical action by ice crystals in the intra- or extracellular space. In this chapter, the origin, structure, and functions of the ice crystal-controlling materials are mentioned.

Contents

H. Kawahara (✉)
Department of Biotechnology, Faculty of Engineering, Kansai University, Osaka, Japan, 564-8680
e-mail: kawahara@kansai-u.ac.jp

R. Margesin (ed.), *Psychrophiles: From Biodiversity to Biotechnology*,
DOI 10.1007/978-3-319-57057-0_11

11.1 Introduction

Some bacteria have developed a variety of strategies to survive and colonize extremely cold environments such as the Antarctic and the Arctic (Russell 1997; Chattopadhyay 2006). In these frozen environments, bacteria are exposed to conditions that necessitate the partial removal of water from the intracellular space in order to maintain the structure and function of the cell. As water is essential for the function of various macromolecular structures, i.e. proteins, polysaccharides, lipids, and nucleic acids, any significant deviation in the accessibility of water due to dehydration, desiccation, or the alteration of its physical state, from the aqueous phase to an ice crystal, will pose a severe threat to the normal function and survival of organisms (Beall 1983).

It has been suggested that substances such as trehalose, glycerol, and sorbitol are major cryoprotectants (CRPs) for prokaryotic cells exposed to freezing damage. They allow the maintenance of enzyme activities in vivo (Storey and Storey 1986) and also prevent the cold denaturation of proteins (Phadtare 2004).

As shown in Fig. 11.1, the degree of injury in various intact cells is affected by the freezing condition, i.e. by the freezing rate, and by the recrystallization of ice in the intra- and extracellular space that can occur when the freezing rate is lower than $10\,^{\circ}\mathrm{C\,min}^{-1}$ (Fujikawa 1987). The lipoteichoic acid (LTA), a biopolymer in the cell wall of Gram-positive bacteria, can be added to *Bacillus subtilis* cultures to increase freeze tolerance (Rice et al. 2015).

Usually the ice crystals are initially formed in the extracellular space, while the intracellular freezing of yeast and bacteria requires a freezing rate of $1\text{–}10\,^{\circ}\mathrm{C\,min}^{-1}$ (Mazur 1970). When a freezing rate of over $100\,^{\circ}\mathrm{C\,min}^{-1}$ is applied, ice appears as

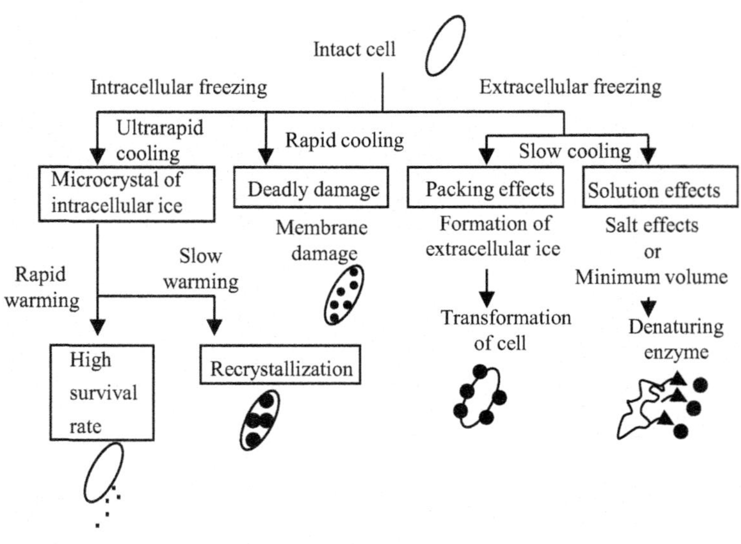

Fig. 11.1 Mechanism of freezing injury against an intact cell

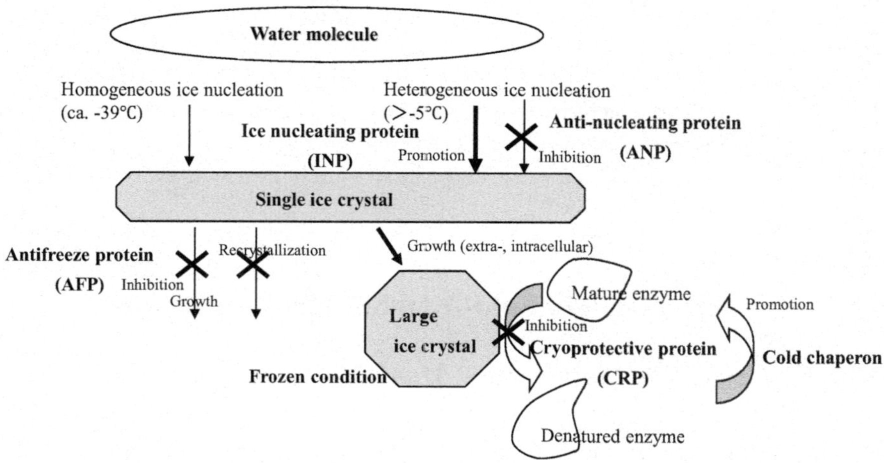

Fig. 11.2 The scheme of action modes of various ice crystal-controlling proteins

microcrystals and is less damaging since freezing rates in the order of 10–100 °C min^{-1} can damage the cytoplasmic membrane.

For many organisms, and especially for some bacteria, specific proteins or/and saccharides can be produced to counteract or minimize the deleterious effects of ice crystal formation. As shown in Fig. 11.2, ice nuclei can be formed by the so-called homogeneous or heterogeneous nucleation in a first step. In a second step, all ice crystals can grow from their ice nuclei due to the binding of free water molecules. The formation of ice nuclei through heterogeneous ice nucleation is promoted by foreign particles, which are called ice nucleation activators. Various types of ice nucleation activators of biogenic origin (Hew and Yang 1992) are known to exist in plant bacteria, insects, intertidal invertebrates, plants, and lichens. The highest level of ice nucleation activation is provided by ice nucleation proteins (INPs) produced by some ice-nucleating bacteria.

Paradoxically, the inhibitors of heterogeneous ice nucleation which can favour supercooling have been poorly studied. These inhibitors can contribute to minimize the threats of intra- or extracellular ice formation. These anti-nucleating materials (ANMs) include proteins and polysaccharides and are also classified as ice crystal-controlling materials.

Other ice crystal-controlling proteins which can play a crucial role in the second step of ice formation are antifreeze proteins (AFPs). The function of the AFP is to inhibit ice formation by suppressing the growth of ice nuclei. In the past two decades, these proteins were found in bacteria originating from various environments. INPs have been discovered in bacteria such as *Pseudomonas*, *Pantoea*, and *Xanthomonas*, in fungi such as *Fusarium*, in insects, and in lichens such as *Umbilicaria esculenta* and *Anaptychia obscurata*. ANMs have been found in bacteria, such as *Acinetobacter* and *Bacillus*, and in coniferous trees. AFPs have been first discovered in fish, plants, insects, and lichens such as *Thamnolia*; in filamentous fungi such as *Penicillium*, *Typhula*, *Coprinus*, and *Flammulina*; in yeasts such as *Leucosporidium*, *Glaciozyma*, and *Rhodotorula*; and in bacteria such as *Pseudomonas*, *Moraxella*, *Flavobacterium*, and *Marinomonas*. Furthermore, a polysaccharide, xylomannan, which is a component of the cell wall in *Flammulina velutipes*, has antifreeze activity like other AFPs. CRPs have been found in bacteria such as *Pseudomonas*, *Pantoea*, and *Lactobacillus*, in plants such as spinach and soybean, in algae such as *Chlorella*, and also in vertebrates such as frogs.

11.2 Cryoprotection in Cold-Adapted Microorganisms

11.2.1 Cryoprotectants of Low Molecular Mass

Cryoprotectants are chemical substances, which are known to accumulate in the body fluids of some wintering frogs and insects (Storey 1983). Examples of these substances include sugars (glucose, fructose, sucrose, trehalose, etc.), sugar alcohol (sorbitol, glycerol), and amino acids (alanine, proline) (Morita et al. 2003). As an example, glycine betaine, which is a well-known osmolyte, can enhance the growth of *Listeria monocytogenes* at low temperatures (Ko et al. 1994). The cryoprotective roles of glycine betaine in bacteria are believed to prevent cold-induced aggregation of proteins and to maintain optimum membrane fluidity at low temperatures (Chattopadhyay 2002). Other examples include ice-nucleating bacteria, such as the genera *Pantoea* (Lindow 1983), *Pseudomonas* (Maki et al. 1974; Obata et al. 1987), and *Xanthomonas* (Kim 1987). They can induce the nucleation of ice in supercooled water at −2 to −3 °C and induce frost damage in many crops (Lindow 1983). The ability of ice-nucleating bacteria to survive on the surface of a leaf after freezing and thawing has been shown to be due to ice-nucleating activity after a temperature decrease (Hirano et al. 1982). This phenomenon is responsible for the high degree of cryotolerance of ice-nucleating bacteria as shown in Fig. 11.3. The temperature change during cold acclimation resembles the temperature change that

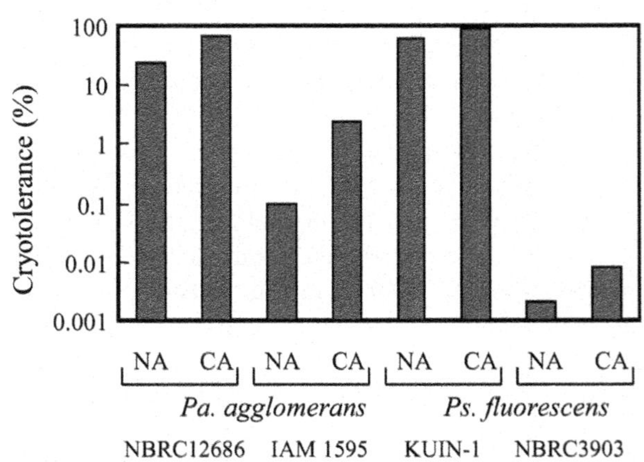

Fig. 11.3 Effects of cold acclimation (CA) on the cryotolerance of ice-nucleating bacteria, *Pantoea agglomerans* NBRC12686 and *Pseudomonas fluorescens* KUIN-1, and non-ice-nucleating bacteria, *Pa. agglomerans* IAM 1595 and *Ps. fluorescens* NBRC3903. *CA* cold-acclimated strains, *NA* non-acclimated strains (Obata et al. 1998; Koda et al. 2000)

Table 11.1 Accumulations of various cryoprotectants in some ice-nucleating bacteria

Strains	Cryoprotectants	Functions	Reference
Pseudomonas fluorescens KUIN-1	COR26	Cryoprotection of enzymes	Obata et al. (1998)
Pantoea agglomerans NBRC12686	CRP	Cryoprotection of enzymes	Koda et al. (2001)
	Uridine phosphorylase	Cryoprotection of enzymes	Obata et al. (2004)
	Ribose-1-phosphate	Cryoprotection of enzymes Depression of freezing points	Obata et al. (2004)
Pantoea ananatis KUIN-3	HSC25	Refolding of denatured enzymes	Kawahara et al. (2000b)
	Glucose	Depression of freezing points	Koda et al. (2002)

causes the frost damage in plants on spring mornings. *Pantoea agglomerans* NBRC12686 and *Pantoea ananatis* KUIN-1 had a high survival rate (65% and 79%, respectively) after cold acclimation at 10 °C (Koda et al. 2002). *Pseudomonas fluorescens* KUIN-1 had the highest rate of survival (94%) after cold acclimation at 4 °C. These high survival rates in ice-nucleating bacteria are partly due to the accumulation of the cold acclimation proteins that possess cryoprotective or chaperone activity and also due to the accumulation of glucose (1.9 mg g^{-1} cell wet mass) in *Pantoea ananatis* KUIN-3 (Table 11.1). Glucose-6-phosphatase, which is related to the glucose accumulation, had a 1.44-fold higher activity (117 mU mg^{-1}) after cold acclimation for 24 h at 10 °C following growth for 12 h at 30 °C (81 mU mg^{-1}). Furthermore, *Pantoea agglomerans* NBRC12686 accumulates ribose-1-phophate and glucose in cells following a shift in temperature (12 °C) from the optimum growth temperature (30 °C) (Table 11.1). Uridine phosphorylase, which is related to the ribose-1-phosphate and glucose accumulation, had a 1.48-fold higher activity after cold acclimation for 48 h at 12 °C following growth for 12 h at 30 °C. Both enzymes play an important role as some of the various regulators for glucose accumulation during cold acclimation to achieve high survival rates in ice-nucleating bacteria.

11.2.2 Cryoprotective Proteins and Cold Chaperones from Ice-Nucleating Bacteria

Many enzymes in low concentrations are easily inactivated during storage, especially frozen storage (Chilson et al. 1965; Anderson et al. 1978). The factors involved in this inactivation, and the relationship between freeze denaturation and conformational changes, have been investigated in detail, particularly for lactate

Table 11.2 Comparison of various cryoprotective proteins

Protein	Molecular mass	CP_{50} (μg ml^{-1})	CP_{50} (nM)	Reference
BSA	66,000	2.8×10^1	4.0×10^2	Lin and Thomasow (1992)
Sucrose	340	2.7×10^5	8.0×10^8	Lin and Thomasow (1992)
COR 15	15,000	8.3×10^{-2}	5.6×10^0	Lin and Thomasow (1992)
COR 85	350,000	1.5×10^1	4.3×10^1	Thomashow (1990)
COR 26	159,000	1.6×10^4	1.0×10^5	Obata et al. (1998)
HIC 6	14,700	1.1×10^{-1}	7.4×10^0	Honjoh et al. (2000)
CRP 29	29,000	8.0×10^{-4}	2.7×10^{-2}	Koda et al. (2001)
CRP 62	62,000	2.0×10^{-1}	3.2×10^0	Neven et al. (1993)
AS 26k	26,000	2.0×10^1	7.7×10^2	Momma et al. (1997)

CP_{50} is the concentration required to protect for 50% of LDH activity

dehydrogenase (LDH) (Seguro et al. 1989). The cryoprotective effects of albumin on common metabolic enzymes including LDH and mutarotase, etc. have been reported (Tamiya et al. 1985). Indeed, albumin (0.1%) prevents the freeze inactivation of LDH in response to cold acclimation at 12 °C in surrounding the enzyme molecules, thereby protecting them at a molar ratio of 1 LDH for 10 BSA (bovine serum albumin) from conformational changes and aggregation when freezing. The polypeptide COR15 (15 kDa) from *Arabidopsis thaliana* is, besides albumin, also known as cryoprotectant (Lin and Thomasow 1992). The gene *cor15* which encodes the cold-regulated polypeptide COR15 is expressed during cold acclimation processes; COR15 has an in vitro cryoprotective activity about 10^2 to 10^3 times more effective than BSA (Table 11.2). The psychrotolerant ice-nucleating bacterium *Pseudomonas fluorescens* KUIN-1 accumulates COR26, which is a 26 kDa polypeptide induced in response to a decrease in temperature (Obata et al. 1998). Also *Pantoea agglomerans* NBRC12686 produces a 29 kDa CRP (Koda et al. 2001). This CRP protects freeze-labile enzymes such as LDH, alcohol dehydrogenase (ADH), and isocitrate dehydrogenase against freezing and thawing denaturation. The activity of this CRP as cryoprotectant was about 3.5×10^4 times better than that of COR26 (Table 11.2) (Tamiya et al. 1985). In contrast to BSA, the mole ratio of protected LDH and CRP was 6.49×10^{-4}, so it is impossible for 1 mole of CRP to surround 1540 moles of LDH. Although this mechanism of cryoprotection remains unexplained, it seems that the association between the CRP molecules and water molecules may play an important role. Obata et al. (2004) reported that uridine phosphorylase (29.7 kDa) from *Pantoea agglomerans* NBRC12686 has also a weak cryoprotective activity (30%) at a concentration of 5.0×10^{-3} mg ml^{-1} (Table 11.2).

Another ice-nucleating bacterium, *Pantoea ananatis* KUIN-3, produces Hsc25 as one of the cold acclimation proteins (Kawahara et al. 2000b). This Hsc25 has a refolding activity similar to the chaperon GroELS; it is composed of eight subunits of 25 kDa each, with a total molecular mass of 200 kDa. It has a refolding activity towards denatured enzymes induced by both heat treatment at 100 °C and guanidinium chloride and also a cryopreservation effect at −20 °C for 24 h. The affinity of Hsc25 for freeze-denatured enzymes is higher than that of GroEL for heat-denatured enzymes. The high rate of survival after freezing/thawing of ice-nucleating bacteria is therefore also achieved by some components of the cytoplasmic space, which are different from the ice-nucleating activity present at the surfaces of their cell membranes. LTA, a biopolymer in the cell wall of Gram-positive bacteria, enables at 1% (w/v) a 50% survival rate, similar to the results obtained with 1% (w/v) glycerol (Rice et al. 2015).

11.3 Ice Crystal-Controlling Proteins

11.3.1 The Structure and Function of Ice Nucleation Proteins

When pure liquid water is cooled at atmospheric pressure, it does not freeze spontaneously at 0 °C. Due to density fluctuations in liquid water, the water molecules form clusters with the same molecular arrangement as ice crystals but remain in a liquid state due to the fluctuation of energy. This is called supercooling. A drop of pure water that is perfectly cleared from all foreign particles can display a supercooling temperature or freezing temperature of about −40 °C. This process has been called 'homogeneous ice nucleation'. In practice, however, impurities or foreign particles that are usually present in water attach water molecules on their surface. As water molecules may be oriented in a way such as to resemble an ice nucleus, these become compatible with the critical dimension of ice nucleation. This process is called 'heterogeneous ice nucleation'; it always occurs at a temperature between −2 °C and −15 °C. Some bacteria produce INPs which can induce ice nucleation at temperatures higher than −3 °C; they are called ice-nucleating bacteria (Fig. 11.2). At least six species of ice-nucleating bacteria have been found; some strains of *Fusarium* and related fungal genera are also active in ice nucleation (Pouleur et al. 1992).

The ice-nucleating components from some lichens have been partially purified (Kawahara et al. 2005; Obata et al. 2006). These ice-nucleating substances have different properties when compared to those of bacterial and fungal INPs. Although almost all microorganisms producing INPs were found at the surface of plants, *Pseudomonas* sp. KUIN-5 was isolated from the marine alga *Monostroma latissum*. This strain produces a bacterial cellulose in the culture broth which can act as an ice-nucleating component (Kawahara et al. 1996a).

Genes conferring ice-nucleating activity have been sequenced from six bacterial strains and all encode INPs (120–150 kDa) with similar primary structures (Green and Warren 1985; Warren et al. 1986; Abe et al. 1989; Warren and Corotto 1989; Zhao and Orser 1990; Michigami et al. 1994). As shown in Table 11.3, all INPs are composed of three domains, the N-, R-, and C-domains. The N-domain (15% of the

Table 11.3 Comparison of the amino acid sequences of various ice nucleation-active proteins

Bacteria	Gene	Total amino acids	Predicted mol. mass	Amino acid residue in each domain			Reference
				N-domain	R-domain	C-domain	
Ps. fluorescens KUIN-1	inaF	1352	132,472	183	1120	49	Kawahara (unpubl.)
Ps. fluorescens	inaW	1210	111,881	164	992	54	Warren et al. (1986)
Ps. syringae	inaZ	1200	118,587	175	976	49	Green and Warren (1985)
Pa. ananatis	inaA	1322	131,094	161	1120	41	Abe et al. (1989)
Pa. ananatis	inaU	1034	103,387	161	832	41	Michigami et al. (1994)
Pa. agglomerans	iceE	1258	125,076	161	1056	41	Warren and Corrotto (1989)
X. campestris	inaX	1567	152,548	219	1280	68	Zhao and Orser (1990)

total sequence) is at least responsible for the binding of lipids, polysaccharides, and INPs (Kozloff et al. 1991); it is relatively hydrophobic and contains a membrane anchor with a mannan-phosphatidylinositol. The R-domains play an important role as a template for ice formation; their length (832–1280 amino acids) is correlated with the amplitude of ice nucleation activity (Table 11.3). They are made (ca. 80%) of contiguous repeats of a consensus octapeptide (-Ala-Gly-Tyr-Gly-Ser-Thr-Leu-Thr-), mainly consisting of hydrophilic amino acids. The C-terminal domain (less than 40–50 residues) is rich in basic amino acid residues and is very hydrophilic. It is necessary for the ice-nucleating activity since, for example, in the INP from *Pantoea ananatis*, the deletion of the last 12 C-terminal residues does not modify the activity, while another mutant with Met29 as new C-terminal amino acid displays a nearly complete loss of activity (Michigami et al. 1995a). It was shown that one of the residues of this domain, Tyr27, is important for the activity, although not exclusively required since the activity was lost to a great extent when this residue was replaced by Gly or Ala, but to a much lesser extent when it was replaced by Leu. These results point to the importance of the secondary and/or tertiary structure of the C-terminal region for the ice nucleation activity.

The characterization of INPs has involved sequencing, truncation, activity measurements, peptide studies, and theoretical modelling (Warren and Wolber 1991; Hew and Yang 1992; Guriansherman and Lindow 1993). Several research groups have previously presented theoretical structures of the highly repetitive region (R-domain) (Mizuno 1989; Warren and Wolber 1991), and one has described models consisting of 8, 16, and 24 residue peptides based on NMR data

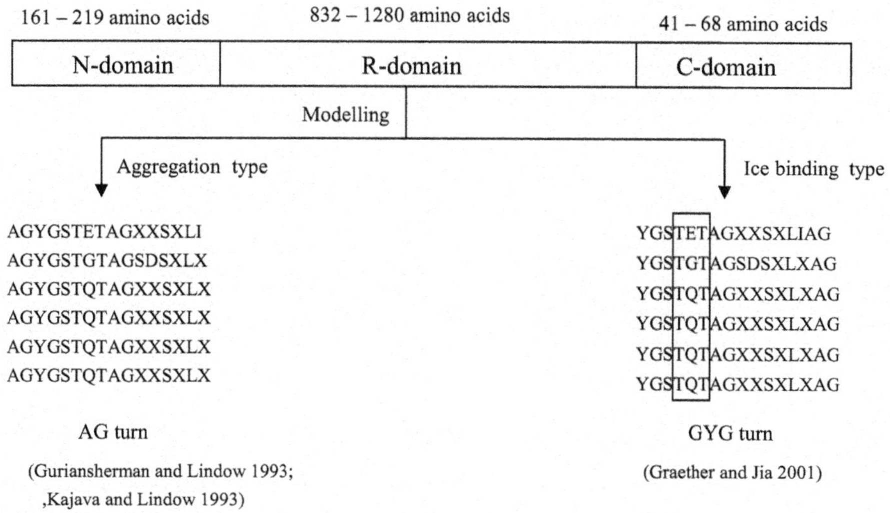

Fig. 11.4 Different models of the central repeat region of ice-nucleating proteins

(Tsuda et al. 1997). The recombinant 96-amino-acid-residue polypeptide corresponding to the sequence Tyr176 to Gly273 of INP from *Pseudomonas syringae* had the ability to shape an ice crystal, whose morphology was highly similar to that of the hexagonal bipyramid generally identified for AFPs (Kobashigawa et al. 2005). Also, Wilson et al. (2006) have suggested that the ice affinity of INPs directs the role of bacteria either towards inhibition of ice recrystallization, ice nucleation, or ice shaping. The ice-controlling protein from *P. syringae* exhibits only ice nucleation activity and has no AFP activity.

As shown in Fig. 11.4, the conserved glycine residues, which are often involved in chain bending, are located at every turn of the proposed R-domain structure, while the highly conserved Ser and Tyr residues are only present in the middle of the beta-strands allowing them to act as an icelike template. They are involved in the aggregation of individual INPs, a process that renders the ice-nucleating properties more efficient (Kajava and Lindow 1993). Based on the ice-binding ability (Kobashigawa et al. 2005), it was suggested that INPs may have a similar beta-helical fold and may interact with water through the repetitive TXT motif (Graether and Jia 2001).

Proteins and other components located on the outer membranes of bacteria are responsible for ice nucleation. *Pseudomonas fluorescens* (Obata et al. 1993), *Pantoea ananatis* (*Erwinia uredovora*) (Kawahara et al. 1993), *Erwinia herbicola* (Phelps et al. 1986), and *Erwinia carotovora* (Fukuoka et al. 1992) have been reported to release the extracellular ice-nucleating materials into the culture fluid. However, purified INPs, contrary to raw material (−3 °C), exhibit an ice nucleation activity at a temperature below −6 °C (Fall and Wolber 1995). This difference is probably due to a requirement for phospholipids (Govindarajan and Lindow 1988).

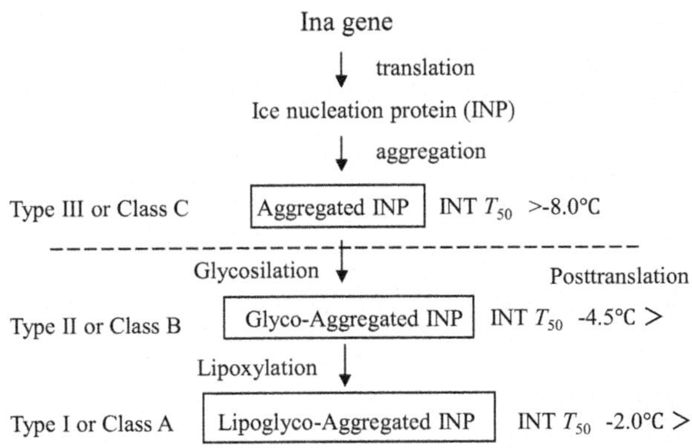

Fig. 11.5 Tentative pathway of the biosynthesis of ice nucleation materials in some ice nucleation bacteria

Bacterial ice activators are in fact complexes of lipids, saccharides, and proteins; they are lipoglycoproteins (Kozloff et al. 1991) and classified into three chemically distinct classes according to their chemical composition: A, lipoglycoprotein; B, glycoprotein; and C, only protein (Turner et al. 1991) (Fig. 11.5). Also, the activation effect of the class C structures is due to the aggregation of INPs giving rise to molecular masses over 1000 kDa (Govindarajan and Lindow 1988). Franks (1985) reported that the factors involved in the formation of ice nuclei have to obey to three conditions: similarity to the ice crystal lattice, paucity of the surface charge, and high hydrophobicity. Class A structures apparently display the full set of conditions required for maximal activity.

The extracellular ice-nucleating material produced by *Erwinia uredovora* KUIN-3 was purified and characterized as a spherical structure (0.2–0.4 μm); it is composed of lipid, protein, saccharide, and polyamine (Kawahara et al. 1993). The polyamine plays a role in the surface charge, in the control of hydrophobicity, and in the stability of the protein conformation of class A and B structures. It is a critical component of these structures (Kawahara et al. 1994). The purified INP from the extracellular ice-nucleating material of *E. uredovora* is a glycoprotein (Michigami et al. 1995b). Not all INPs secreted in the culture broth have the same signal sequences of 15–70 amino acid residues at the N-terminus. Michigami et al. (1995b) have reported that, after the assembly of INP near the inner membrane, the assembly then enters a vesicle just formed on the surface of the outer membrane and leaves the surface. It has been shown that ATP is essential for the secretion of extracellular ice-nucleating material into the culture broth (Kawahara et al. 1999). Considering these results, it appears that the secretion of INPs or extracellular ice-nucleating material might be a unique mechanism.

Almost all ice-nucleating bacteria belong to Gram-negative epiphytic genera and are known as pathogenic bacteria, either mesophilic or psychrotolerant. The isolation and characterization of ice-nucleating bacteria from Antarctica are still in an early state of investigation; that is the reason why we have started a research programme to isolate ice-nucleating bacteria from the bacterial collection stock at

Ross Island in the McMurdo Dry Valley region of Antarctica (Obata et al. 1999). Eleven ice-nucleating bacteria were discovered among 135 strains (ca. 8%). Strain IN-74, with the highest level of ice-nucleating activity, was identified as *Pseudomonas antarctica*. Strain IN-74 also displayed a high level of ice-nucleating activity following induction at 0 °C, and this temperature was significantly higher than that required by other ice-nucleating bacteria. The molecular mass of its INP was over 120 kDa. As *ina U* mRNA transcription was induced at 18 °C (Michigami et al. 1995b), this transcription in strain IN-74 might be activated at 0 °C. This could mean the discovery of a novel ice-nucleating bacterium, which has been found in various environments. What remains to be resolved concerning the ice-nucleating bacteria are the elucidations of various interesting biochemical processes.

11.3.2 The Structure and Function of Anti-nucleating Proteins

Ice-nucleating inhibitors have the ability to lower the supercooling point of water (Fig. 11.6), and this point, expressed in °C, is defined as the anti-nucleation activity. These components have been up to now poorly studied. An enzyme-modified gelatin (EMG-12) has been reported as an ice-nucleating inhibitor of silver iodide (AgI) (Arai and Watanabe 1985). Also hinokitiol, a compound found in the leaves of coniferous trees, has an anti-nucleating activity against *P. fluorescens* (Kawahara et al. 2000a). Furthermore, poly-glycerol, with a molecular mass of 750 Da, has been found to apparently bind and inhibit the ice-nucleating activity of *P. syringae* (Wowk and Fahy 2002).

We have discovered a bacterial strain with anti-nucleating activity (Kawahara et al. 1996b); this strain, KINI-1, was identified as *Acinetobacter calcoaceticus*. The anti-nucleating protein produced by this strain has a molecular mass of 55 kDa. It exhibits a broad specificity with the capability to lower the nucleating activity of a

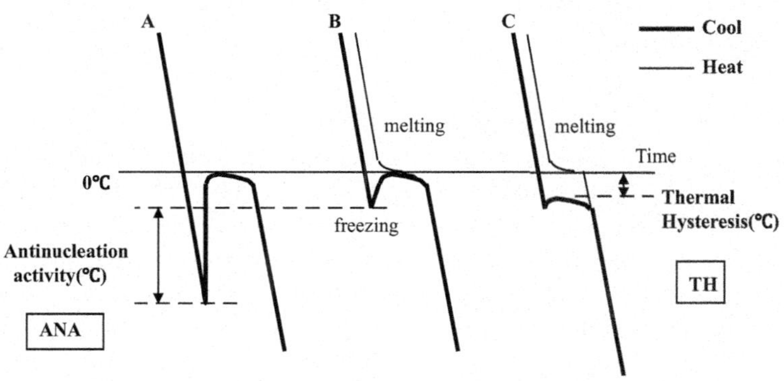

Fig. 11.6 Freezing curves of solutions with anti-nucleation material (A), distilled water (B), and antifreeze protein (C)

wide range of ice nucleators including some bacterial components and AgI. This anti-nucleating protein is, for example, able to shift the supercooling point of water to −5.0 °C in a cell suspension of *E. uredovora*; it inhibits the formation of ice nuclei in a way similar to EMG-8 and EMG-12 (Arai and Watanabe 1985).

Furthermore, a strain of *Bacillus thuringiensis* produces an anti-ice-nucleating polysaccharide having a molecular mass of 130 kDa (Yamashita et al. 2002b). The analysis of this compound by LCMS demonstrated that this polysaccharide consists of a polyacetyl-D-glucosamine structure similar to that of chitin. This polysaccharide has an activity ranging from 0.1 to 4.2 °C against various ice-nucleating materials including AgI; some organic compounds such as metaldehyde, fluoren-9-one, and phenazine; and some ice-nucleating bacteria. It does not have, however, any activity against homogeneous ice nucleation, that is, pure water.

Organ cryopreservation is hindered by ice-inflicted damages. Nonfreezing preservation of livers at subzero temperatures might offer advantages over the current method of preservation. In order to improve the preservation procedure for liver grafts, Matsukawa et al. (2000) have attempted to apply subzero nonfreezing (SZNF) storage methods making use of bacterial anti-nucleating proteins and ascorbic acid 2-glucoside (AA2-G) as an antioxidant. When the liver graft was kept for 24 h at SZNF storage (−3.0 °C) in the above-mentioned solution made of anti-nucleating protein (20 μg ml^{-1}) and AA2-G (100 μg ml^{-1}), both the number of apoptotic cells and the serum level of alanine aminotransferase after 4 h of reperfusion could be greatly diminished. Furthermore, ATP concentrations in grafted liver tissues were significantly greater than those obtained after normal storage at 4 °C. From these results, SZNF liver storage using anti-nucleating proteins has the potential to surpass conventional hypothermic liver storage procedures for long periods. Further studies are carried out to isolate novel bacteria that produce anti-nucleating proteins and characterize their anti-nucleating proteins or polysaccharides.

11.3.3 The Structure and Function of Bacterial Antifreeze Proteins

All AFPs and antifreeze glycoproteins (AFGPs) lower the freezing point of water without altering the melting point (Fig. 11.6). This activity is defined as the difference between both temperatures and is called thermal hysteresis (TH). It is caused by a non-colligative phenomenon, the activity of which can be as high as 500-fold that of colligatively acting substances such as sucrose, glycerol, and trehalose. As shown in Fig. 11.7, the forms of ice crystals in the presence of AFPs at the initiation time of growth during cooling at 1 °C min^{-1} are different from that of pure water. Primarily, needle-like growth along the c-axis occurs with fish AFP and bacterial AFP from *Pseudomonas putida* GR12-2 (Sun et al. 1995), rather than the usual predominantly highly dendritic growth along the a-axis within

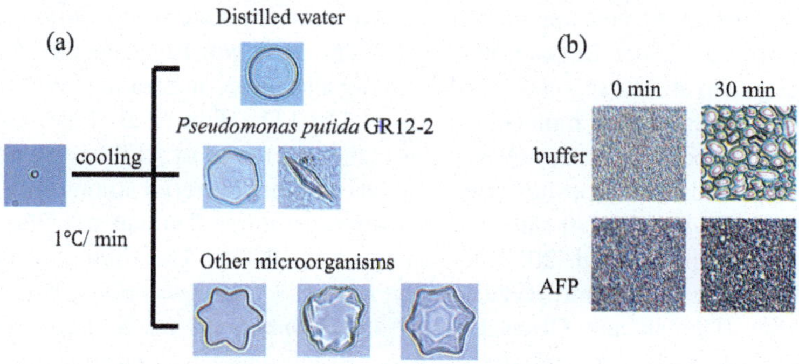

Fig. 11.7 Ice crystal regulation by antifreeze proteins from some microorganisms. (**a**) Ice crystal morphology; (**b**) ice crystal recrystallization inhibition

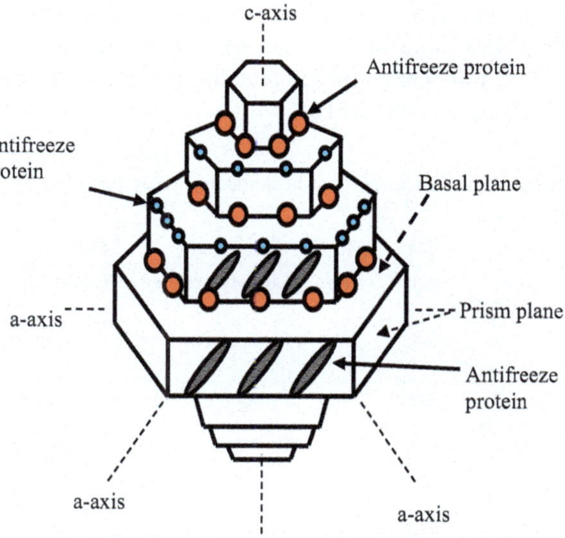

Fig. 11.8 Binding sites of some antifreeze proteins to ice crystal surfaces

the basal plane. Also AFPs from other microorganisms like fungi can form different ice crystal (Fig. 11.7). This may also be affected by other materials including saccharides or inorganic compounds. Recrystallization of frozen solutions can be inhibited at very low concentrations (0.1 μg ml^{-1}) (Fig. 11.7). So far, AFP and AFGP have been identified in polar fish, insects, plants (Hew and Yang 1992), and fungi (Duman and Olsen 1993). Ice-shaping activity, thermal hysteresis (TH) activity, and recrystallization inhibition (RI) are common activities of most AFPs isolated from fishes. A model of the binding sites of AFPs from fishes and insect larvae to ice crystal surfaces is shown in Fig. 11.8. As AFPs isolated from insect larvae have a superactive TH activity (ca. 5 °C), these proteins were named thermal hysteresis proteins (THPs) (Barrett 2001). AFPs possessing only RI activity have been discovered in the Antarctic nematode *Panagrolaimus davidi* and in winter wheat grass and are named ice-active or ice-structuring proteins (ISPs) (Wharton et al. 2005; Regand and Goff 2006). It is worth noting that *P. putida* GR12-2 grown at 5 °C secretes an AFP having also an ice-nucleating activity; it is a lipoglycoprotein with a molecular mass of 164 kDa (Sun et al. 1995; Xu et al. 1998). To the best of our knowledge, this is the first report of AFP having activities directed towards two opposite actions. So far, all other bacterial strains only exhibit antifreeze activity; they all can grow at low temperatures (Duman and Olsen 1993; Mills 1999; Gilbert et al. 2004; Kawahara et al. 2004). The following bacterial representatives showed antifreeze activity: *Rhodococcus erythropolis*, *Micrococcus cryophilus* (Duman and Olsen 1993), *Stenotrophomonas maltophilia*, *Sphingomonas*, *Halomonas*, *Pseudoalteromonas*, *Psychrobacter*, *Enterobacter agglomerans* (Gilbert et al. 2004), *Pseudomonas fluorescens* (Gilbert et al. 2004; Kawahara et al. 2004), and *Marinomonas protea* (Gilbert et al. 2004; Mills 1999). Besides *Marinomonas primoryensis* (Gilbert et al. 2005) and *Flavobacterium xanthum* (Kawahara et al. 2007), almost all other bacteria secrete AFP into the culture broth (Table 11.4). Some AFPs from the genera denoted in bold letters in Table 11.5 were characterized regarding molecular mass, antifreeze activity, and the localization of the AFPs.

Table 11.4 Comparison of various purified bacterial antifreeze proteins

Bacteria	Mol. mass	TH	RI	Compound	Localization	Reference
Pseudomonas putida GR12–2	164 kDa	0.1 °C	++	Lipoglycoprotein	Extracellular	Xu et al. (1998)
Moraxella sp.	52 kDa	0.1 °C	+	Lipoprotein	Extracellular	Yamashita et al. (2002b)
Marinomonas primoryensis	60 kDa	3.5 °C	+++	Protein	Intracellular	Gilbert et al. (2005)

Table 11.5 Taxonomic affiliation of bacteria producing phenotypically verified antifreeze proteins (based on Cid et al. 2016)

Class	Branch 1	Branch 2	Branch 3
Alphaproteobacteria	*Rhizobium* *Ensifer*		*Bradyrhizobium*
Betaproteobacteria			*Herbaspirillum*
Gammaproteobacteria	***Marinomonas*** ***Pseudomonas*** *Xanthomonas*		***Colwellia*** *Pseudomonas* *Serratia*
Deltaproteobacteria			*Stigmatella*
Bacilli		*Paenibacillus* *Bacillus*	
Flavobacteriia			***Flavobacterium***
Fimbriimonadia			*Fimbriimonas*

Bold letters denote the presence of phenotypically verified strains within the bacterial genus

Although the TH activities of most bacterial AFPs are lower than those of THPs from some insect larvae, a lysate supernatant of *M. primoryensis* containing 11 mg protein ml^{-1} had a TH activity of 0.8 °C. The activity of this lysate is lost following extensive dialysis using a 3500 MW cut-off membrane. The addition of 10 mM $CaCl_2$ to the dialysate restores full activity. When concentrated, the crude lysate has a TH activity >2 °C, which is higher than the maximal activity of most fish AFPs. This AFP is a Ca^{2+}-dependent, superactive AFP like THP. Furthermore, ice crystals formed in the presence of this lysate do not have distinct facets. They are typically rounded in shape, and their morphology does not change during the course of the TH measurement. The ice crystal 'burst' occurring at the end point of TH is dendritic with hexagonal symmetry, suggesting growth from primary or secondary prism faces or edges. A DNA fragment from *P. putida* GR12-2 with an open reading frame encoding 473 amino acids was cloned by PCR and inverse PCR, using primers elaborated from the partial amino acid sequence of the isolated AFP (Muryoi et al. 2004). The predicted gene product, AfpA, has a molecular mass of 47.3 kDa, a pI of 3.51, and no previously known function. Although AfpA is a secreted protein, it lacks an N-terminal signal peptide and was shown by sequence analysis to have two possible secretion systems: a hemolysin-like calcium-binding secretion domain and a type V auto-transporter domain typical of Gram-negative bacteria. Also the expression of *afpA* in *Escherichia coli* yields an intracellular 72 kDa protein modified with both sugars and lipids and exhibits lower levels of antifreeze and ice nucleation activities. The deduced amino acid sequence from *afpA* shows 7 potential sites for N-glycosylation, 2 sites for O-glycosylation, and 20 sites for myristoylation. This AFP was secreted into the culture broth; the physiological role of the AFP for the bacterial cell might be the stabilization of the membrane lipids at a low temperature. Following transposon Tn5 mutagenesis of strain GR12-2, five mutants with different freeze resistance were selected (Kawahara et al. 2001). These mutants secreted low amounts of AFP (0.4–2.2 μg ml^{-1}) into the culture broth, compared to the wild type (4.8 μg ml^{-1}). Furthermore, the decreased freeze resistance of three mutants could be partially restored by adding purified AFP to the mutant cell suspensions. We concluded that accumulation of AFPs is one component of the mechanism for freeze resistance in bacteria.

We were also able to select 6 bacteria (ca. 5%) out of 130 strains, isolated from Ross Island in the McMurdo Dry Valley region in Antarctica, capable of producing AFPs (Yamashita et al. 2002a). Strain 82 has the highest antifreeze activity without ice-nucleating activity and was identified as *Moraxella* sp. This AFP is a lipoprotein with a molecular mass of 52 kDa. Its N-terminal amino acid sequence is very similar to that of the outer membrane proteins from *Moraxella (Branhamella) catarrhalis*. Both the isolation of ice-nucleating bacteria and AFP-producing bacteria may provide new insights into the elucidation of the freeze-resistance or cryotolerance mechanisms in bacteria.

Recently, some yeasts isolated from the Antarctic and Arctic exhibited antifreeze activity in the culture broth (Lee et al. 2010; Hashim et al. 2013; Singh et al. 2014). Among these yeasts, *Leucosporidium* sp. (Lee et al. 2010) and *Glaciozyma antarctica* (Hashim et al. 2013) produce extracellular AFPs that are glycoproteins

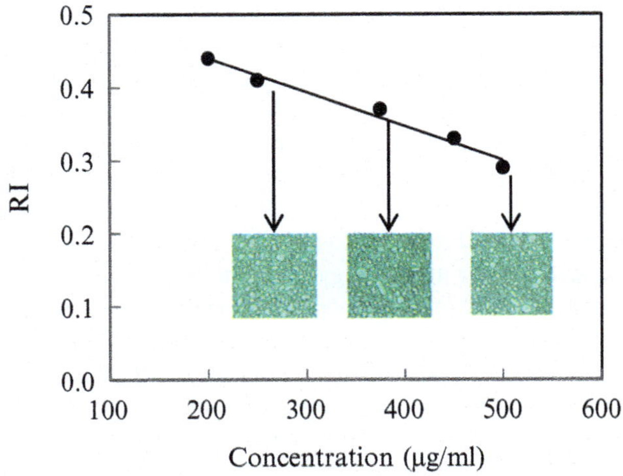

Fig. 11.9 Effect of RI activity on purified xylomannan

or proteins with a molecular mass of 25 and 18 kDa, respectively, and with TH activities of 0.30 (300 μm) and 0.10 (100 μm), respectively. No homology was found between the amino acid sequences of the two AFPs.

Flammulina velutipes (Enokitake) has the highest level of freeze tolerance. Until recently, it has been thought that all biological large molecular mass antifreezes were proteins. Walters et al. (2009) reported the isolation of a novel non-proteinaceous material, a glycolipid having xylomannan as the structural moiety, from the freeze-tolerant beetle *Upis ceramboides*. It was predicted that the xylomannan moiety of this glycolipid might play an important role in the TH activity in this molecule. The cell wall of *F. valutipes* contains xylomannan as one major component (Smiderle et al. 2006). This xylomannan was isolated by the modified Smiderle method using KOH solution (Kawahara et al. 2016). After purificating by gel filtration chromatography, the xylomannan purified from the mycelium exhibited RI activity (RI = 0.30, 0.5 mg ml^{-1}) like various AFPs from various organisms (Fig. 11.9). Also, the RI activity of this polysaccharide was 0.07 ± 0.01 °C at a concentration of 5 mg ml^{-1}. This polysaccharide was composed of mannose and xylose in a 2:1 molar ratio, the molecular mass was 320 kDa. It was predicted that xylomannan in the cell wall has an important role of freeze tolerance in *F. velutipes* (Kawahara et al. 2016).

11.4 Conclusions

Thus far, potential strategies for microbial surviving under extremely cold environments have been investigated. Various ice crystal-controlling proteins and other materials from some bacteria have been isolated and characterized. These proteins have a high potential in biotechnology. For example, ice nucleation activators can be used to regulate both the rate of freezing and the texture of frozen foods (Li and Lee 1998), and AFPs can offer a new way of improving the quality of foods. However, the structure and function of these proteins from various organisms, especially from bacteria, are poorly characterized. The use of ice nucleation activators and AFPs in foods most likely will depend on the cost of their production and also on the safety of the bacteria involved. Currently, our work is oriented towards the search of ice sublimation-controlling proteins from bacteria as a new type of ice-controlling proteins.

Acknowledgement This work was carried out in collaboration with Professor Hitoshi Obata in the Laboratory of Microbial Technology, Department of Biotechnology, Kansai University. I would like to thank Jiro Nishikawa, Department of Applied Biological Science, Science University of Tokyo, for providing me with some microorganisms including bacteria, yeasts, and filamentous fungi from Antarctica. I am also grateful to Drs. BR Glick and M Griffith from the Department of Biology, Waterloo University, for providing me with *P. putida* GR12-2. Also, I like to thank all graduate and undergraduate students, who were studying the mechanism of cryotolerance in bacteria and ice crystal-controlling proteins from some microorganisms.

References

Abe K, Watabe S, Emori Y, Watanabe M, Arai S (1989) An ice nucleation active gene of *Erwinia ananas*. FEBS Lett 258:297–300

Anderson JO, Nath J, Harner EJ (1978) Effect of freeze preservation on some pollen enzyme. II. Freezing and freeze-drying stresses. Cryobiology 15:468–477

Arai S, Watanabe M (1985) The protein that prevents the formation of ice nuclei. Kagaku To Seibutsu 23:363–373. (in Japanese)

Barrett J (2001) Thermal hysteresis proteins. Int J Biochem Cell Biol 33:105–117

Beall PT (1983) States of water in biological systems. Cryobiology 20:324–334

Chattopadhyay MK (2002) The cryoprotective effects of glycine betaine. Trends Microbiol 10:311

Chattopadhyay MK (2006) Mechanism of bacterial adaptation to low temperature. J Biosci 31: 157–165

Chilson OP, Costello LA, Kaplan NO (1965) Effects of freezing on enzymes. Fed Proc 24:55–65

Cid FP, Rillng JI, Graether SP, Bravo LA, Mora M, Jorquera MA (2016) Properties and biotechnological applications of ice-binding proteins in bacteria. FEMS Microbiol Lett 363:1–12

Duman JG, Olsen TM (1993) Thermal hysteresis protein activity in bacteria, fungi, and phylogenetically diverse plants. Cryobiology 30:322–328

Fall R, Wolber PK (1995) Biochemistry of bacterial ice nuclei. In: Lee RE, Jr Warren GJ, Gusta LV (eds) Biological ice nucleation and its applications. APS Press, St. Paul, MN, pp 63–81

Franks F (1985) Biophysics and biochemistry at low temperatures. Japan UNI Agency, Tokyo

Fujikawa S (1987) Mechanisms of freezing injury in cellular level. Trans JAR 14:11–25

Fukuoka S, Kamishima H, Tamiya E, Karube I (1992) Spontaneous release of outer membrane vesicles by *Erwinia carotovora*. Microbios 72:167–173

Gilbert JA, Hill PJ, Dodd CER, Laybourn-Party J (2004) Demonstration of antifreeze protein activity in Antarctic lake bacteria. Microbiology 150:171–180

Gilbert JA, Davies PL, Laybourn-Party J (2005) A hyperactive, Ca^{2+}-dependent antifreeze protein in an Antarctic bacterium. FEMS Microbiol Lett 245:67–72

Govindarajan A, Lindow SE (1988) Phospholipid requirement for expression of ice nuclei in *Pseudomonas syringae* and in vitro. J Biol Chem 263:9333–9338

Graether SP, Jia Z (2001) Modeling *Pseudomonas syringae* ice-nucleation protein as a beta-helical protein. Biophys J 80:1169–1173
Green R, Warren G (1985) Physical and functional repetition in a bacterial ice nucleation gene. Nature 317:645–648
Guriansherman D, Lindow SE (1993) Bacterial ice nucleation: significance and molecular basis. FASEB J 7:1338–1343
Hashim NH, Bharudin I, Nguong DL, Higa S, Bakar FD, Nathan S, Rabu A, Kawahara H, Illias RM, Najimudin N, Mahadi NM, Murad AM (2013) Characterization of Afp1, an antifreeze protein from the psychrophilic yeast *Glaciozyma antarctica*. Extremophiles 17:63–73
Hew CL, Yang DSC (1992) Protein interaction with ice. Eur J Biochem 203:33–42
Hirano SS, Nordheim EV, Arny DC, Upper CD (1982) Longnormal distribution of epiphytic bacterial population of leaf surfaces. Appl Environ Microbiol 44:697–700
Honjoh K, Matsumoto H, Shimizu H, Ooyama K, Tanaka K, Oda Y, Takata R, Joh T, Suga K, Miyamoto T, Iio M, Hatano S (2000) Cryoprotective activities of group 3 late embryogenesis abundant proteins from *Chlorella vulgaris* C-27. Biosci Biotechnol Biochem 64:1656–1663
Kajava AV, Lindow SE (1993) A model of the three-dimensional structure of ice nucleation proteins. J Mol Biol 232:709–717
Kawahara H, Mano Y, Obata H (1993) Purification and characterization of extracellular ice-nucleating matter from *Erwinia uredovora* KUIN-3. Biosci Biotechnol Biochem 57:1429–1432
Kawahara H, Mano Y, Hamada R, Obata H (1994) Role of spermidine in ice-nucleating activity of the EIM from *Erwinia uredovora* KUIN-3. Biosci Biotechnol Biochem 59:2201–2206
Kawahara H, Ikugawa H, Obata H (1996a) Isolation and characterization of a marine ice-nucleating bacterium, *Pseudomonas* sp. KUIN-5, which produces cellulose and secretes it in the culture broth. Biosci Biotechnol Biochem 60:1474–1478
Kawahara H, Nagae I, Obata H (1996b) Purification and characterization of a new anti-nucleating protein isolated from *Acinetobacter calcoaceticus* KINI-1. Biocontrol Sci 1:11–17
Kawahara H, Matsushita M, Yamada K, Obata H (1999) The control of the production and secretion of extracellular ice-nucleating materials of *Erwinia uredovora* KUIN-3. Biocontrol Sci 4:9–16
Kawahara H, Masuda K, Obata H (2000a) Identification of a compound in *Chamaecyparis taiwanensis* inhibiting the ice-nucleating activity of *Pseudomonas fluorescens* KUIN-1. Biosci Biotechnol Biochem 64:2651–2656
Kawahara H, Koda N, Oshio M, Obata H (2000b) A cold acclimation protein with refolding activity on frozen denatured enzymes. Biosci Biotechnol Biochem 64:2668–2674
Kawahara H, Li J, Griffith M, Glick BR (2001) Relationship between antifreeze protein and freezing resistance in *Pseudomonas putida* GR12-2. Curr Microbiol 43:365–370
Kawahara H, Nakano Y, Omiya K, Muryoi N, Nishikawa J, Obata H (2004) Production of two types of ice crystal-controlling proteins in Antarctic bacterium. J Biosci Bioeng 98:220–223
Kawahara H, Takemura T, Obata H (2005) Ice crystal-controlling protein from *Thamnolia vermicularis*. Lichenology 4:97–104
Kawahara H, Iwanaka Y, Higa S, Muryoi N, Sato M, Honda M, Omura H, Obata H (2007) A novel, intracellular antifreeze protein in an antarctic bacterium, *Flavobacterium xanthum*. CryoLett 28:39–49
Kawahara H, Matsuda Y, Sakaguchi T, Arai N, Koide Y (2016) Antifreeze activity of xylomannan from the mycelium and fruit body of *Flammulina velutipes*. Biocontrol Sci 21:153–159
Kim HK (1987) *Xanthomonas campestris* pv. *translucens* strains active in ice nucleation. Plant Dis 71:994–997
Ko R, Smith LT, Smith GM (1994) Glycine betaine confers enhanced osmotolerance and cryotolerance in *Listeria monocytogenes*. J Bacteriol 176:426–431
Kobashigawa Y, Nishimiya Y, Miura K, Ohgiya S, Miura A, Tsuda S (2005) A part of ice nucleation protein exhibits the ice-binding ability. FEBS Lett 579:1493–1497

Koda N, Aoki M, Kawahara H, Obata H (2000) Characterization and properties of intracellular proteins after cold acclimation of ice-nucleating bacterium *Pantoea agglomerans* (*Erwinia herbicola*) IFO12686. Cryobiology 41:195–203

Koda N, Asaeda T, Yamade K, Kawahara H, Obata H (2001) A novel cryoprotective protein (CRP) with high activity from the ice-nucleating bacterium *Pantoea agglomerance* IFO12686. Biosci Biotechnol Biochem 65:888–894

Koda N, Inada Y, Nakayama S, Kawahara H, Obata H (2002) Response of the ice-nucleating bacterium *Pantoea ananas* KUIN-3 during cold acclimation. Biosci Biotechnol Biochem 66:866–868

Kozloff LM, Turner MA, Arellano F, Lute M (1991) Phosphatidylinositol, a phospholipid of ice-nucleating bacteria. J Bacteriol 173:2053–2060

Lee JK, Park KS, Park S, Park H, Song YH, Kang S-H, Kim HJ (2010) An extracellular ice-binding glycoprotein from an Arctic psychrophilic yeast. Cryobiology 60:222–228

Li J, Lee TC (1998) Bacterial extracellular ice nucleator effects on freezing of foods. J Food Sci 63:375–381

Lin C, Thomasow MF (1992) DNA sequence analysis of a complementary DNA for cold-regulated *Arabidopsis* gene *cor15* and characterization of the COR15. Plant Physiol 99:519–525

Lindow SE (1983) The role of bacterial ice nucleation in frost injury to plants. Annu Rev Phytopathol 21:363–384

Maki LR, Galyan EL, Chang-Cheng NM, Caldwell DR (1974) Ice nucleation induced by *Pseudomonas syringae*. Appl Microbiol 28:456–459

Matsukawa H, Yagi Y, Matsuda H, Kawahara H, Yamamoto I, Matsuoka J, Tanaka N (2000) Ascorbic acid 2-glucoside prevents sinusoidal endothelial cell apoptosis in supercooled preserved grafts in rat liver transplantation. Transplant Proc 32:313–317

Mazur P (1970) Cryobiology: the freezing of biological systems. Science 168:939–949

Michigami Y, Watabe S, Abe K, Obata H, Arai S (1994) Cloning and sequencing of an ice nucleation active gene of *Erwinia uredovora*. Biosci Biotechnol Biochem 58:762–764

Michigami Y, Abe K, Obata H, Arai S (1995a) Significance of the C-terminal domain of *Erwinia uredovora* ice nucleation-active protein (Ina U). J Biochem (Tokyo) 118:1279–1284

Michigami Y, Abe K, Iwabuchi K, Obata H, Arai S (1995b) Formation of ice-nucleation-active vesicle in *Erwinia uredovora* at low temperature and transport of InaU molecules into shed vesicles. Biosci Biotechnol Biochem 59:1996–1998

Mills SV (1999) Novel biochemical compounds from Antarctic microorganisms. PhD thesis, Nottingham University, UK

Mizuno H (1989) Prediction of the conformation of ice-nucleation protein by conformational energy calculation. Proteins 5:47–65

Momma M, Haraguchi K, Saito M, Chikuni K, Harada K (1997) Purification and characterization of the acid soluble 26-kDa polypeptide from soybean seeds. Biosci Biotechnol Biochem 61: 1286–1289

Morita Y, Nakamori S, Takagi H (2003) L-Proline accumulation and freeze tolerance in *Saccharomyces cerevisiae* are caused by a mutation in the PRO1 gene encoding gamma-glutamyl kinase. Appl Environ Microbiol 69:212–219

Muryoi N, Sato M, Kaneko S, Kawahara H, Obata H, Yaish MWF, Griffith M, Glick BR (2004) Cloning and expression of afpA, a gene encoding an antifreeze protein from the Arctic plant growth-promoting rhizobacterium *Pseudomonas putida* GR12-2. J Bacteriol 186:5661–5667

Neven LG, Haskell DW, Hong A, Li Q-B, Guy CL (1993) Characterization of a spinach gene responsive to low temperature and water stress. Plant Mol Biol 21:291–305

Obata H, Saeki Y, Tanashita J, Tokuyama T, Hori H, Higashi Y (1987) Identification of an ice-nucleating bacterium KUIN-1 as *Pseudomonas fluorescens* and its nucleation properties. Agric Biol Chem 51:1761–1766

Obata H, Tanaka T, Kawahara H, Tokuyama T (1993) Properties of cell-free ice nuclei from ice nucleation-active *Pseudomonas fluorescens* KUIN-1. J Ferment Bioeng 76:19–24

Obata H, Ishigaki H, Kawahara H, Yamade K (1998) Purification and characterization of a novel cold-regulated protein from an ice-nucleating bacterium, *Pseudomonas fluorescens* KUIN-1. Biosci Biotechnol Biochem 62:2091–2097

Obata H, Muryoi N, Kawahara H, Yamade K, Nishikawa J (1999) Identification of a novel ice-nucleating bacterium of Antarctic origin and its ice nucleation properties. Cryobiology 38:131–139

Obata H, Muryoi N, Kawahara H, Nishiyama A (2004) Purification and characterization of uridine phosphorylase from the ice-nucleating bacterium, *Pantoea agglomerance* NBRC12686. CryoLett 25:195–204

Obata H, Shiga N, Takemura T, Kawahara H, Yamamoto Y (2006) Properties of cell-free ice nuclei from a novel ice nucleation-active *Heterodermia obscurata* (Nyl.) *Trevis*. Lichenology 5:37–44

Phadtare S (2004) Recent developments in bacterial cold-shock response. Curr Issues Mol Biol 6: 125–136

Phelps P, Geddings TH, Prochoda M, Fall R (1986) Release of cell-free ice nuclei by *Erwinia herbicola*. J Bacteriol 167:496–502

Pouleur S, Richard C, Martin JG, Autoun H (1992) Ice nucleation activity in *Fusarium acuminatum* and *Fusarium avenaceum*. Appl Environ Microbiol 58:2960–2964

Regand A, Goff HD (2006) Ice recrystallization inhibition in ice cream as affected by ice structuring proteins from winter wheat grass. J Dairy Sci 89:49–57

Rice CV, Middaugh A, Wickham JR, Friedline A, Rhomas KJ III, Scull E, Johnson K, Zachariah M, Garimella R (2015) Bacterial lipoteichoic acid enhanced cryosurvival. Extremophiles 19:297–305

Russell NJ (1997) Psychrophilic bacteria –molecular adaptations of membrane lipids. Comp Biochem Physiol Physiol 118:489–493

Seguro K, Tamiya T, Tsuchiya T, Matsumoto J (1989) Effect of chemical modifications on freeze denaturation of lactate dehydrogenase. Cryobiology 26:154–161

Singh P, Singh SM, Tsuji M, Prasad GS, Hoshino T (2014) *Rhodotorula svalbardensis* sp. nov., a novel yeast species isolated from cryoconite holes of Ny-Ålesund, Arctic. Cryobiology 68:122–128

Smiderle FR, Carbonero ER, Mellinger CG, Sassaki GL, Gorin PA, Iacomini M (2006) Structural characterization of a polysaccharide and a beta-glucan isolated from the edible mushroom *Flammulina velutipes*. Phytochemistry 67:2189–2196

Storey KB (1983) Metabolism and bound water in overwintering insects. Cryobiology 20:365–379

Storey KB, Storey JM (1986) Freeze tolerance and intolerance as strategies of winter survival in terrestrially-hibernating amphibians. Comp Biochem Physiol 83:613–617

Sun X, Griffith M, Pasternak JJ, Glick BR (1995) Low temperature growth, freezing survival, and production of antifreeze protein by the plant growth promoting rhizobacterium *Pseudomonas putida* GR12-2. Can J Microbiol 44:64–73

Tamiya T, Okahashi N, Sakuma R, Aoyama T, Akahane T, Matsumoto J (1985) Freeze denaturation of enzymes and its prevention with additives. Cryobiology 22:446–456

Thomashow MF (1990) Molecular genetics of cold acclimation in higher plant. Adv Genet 28: 99–131

Tsuda S, Ito A, Matsushima N (1997) A hairpin-loop conformation in tandem repeat sequence of the ice nucleation protein revealed by NMR spectroscopy. FEBS Lett 409:227–231

Turner MA, Arellano F, Kozloff LM (1991) Components of ice nucleation structures of bacteria. J Bacteriol 173:6515–6527

Walters KR, Serianni AS, Sformo T, Barnes BM, Duman JG (2009) A nonprotein thermal hysteresis-producing xylomannan antifreeze in the freezetolerant Alaskan beetle *Upis ceramboides*. Proc Natl Acad Sci 106:20210–20215

Warren G, Corotto L (1989) The consensus sequence of ice nucleation proteins from *Erwinia herbicola*, *Pseudomonas fluorescens*, and *Pseudomonas syringae*. Gene 85:241–244

Warren G, Wolber P (1991) Molecular aspects of microbial ice nucleation. Mol Microbiol 5: 239–243

Warren G, Corotto L, Wolber P (1986) Conserved repeats in diverged ice nucleation structural genes from two species of *Pseudomonas*. Nucleic Acids Res 14:8047–8060

Wharton DA, Barrett J, Goodall G, Marshall CJ, Ramlov H (2005) Ice-active proteins from the Antarctic nematode *Panagrolaimus davidi*. Cryobiology 51:198–207

Wilson SL, Kelley DL, Walker VK (2006) Ice-active characteristics of soil bacteria selected by ice-affinity. Environ Microbiol 8:1816–1824

Wowk B, Fahy GM (2002) Inhibition of bacterial ice nucleation by polyglycerol polymers. Cryobiology 44:14–23

Xu H, Griffith M, Patten CL, Glick BR (1998) Isolation and characterization of an antifreeze protein with ice nucleation activity from the plant growth promoting rhizobacterium *Pseudomonas putida* GR12-2. Can J Microbiol 44:64–73

Yamashita Y, Nakamura N, Omita K, Nishikwa J, Kawahara H, Obata H (2002a) Identification of an antifreeze lipoprotein from *Moraxella* sp. of Antarctic origin. Biosci Biotechnol Biochem 66:239–247

Yamashita Y, Kawahara H, Obata H (2002b) Identification of a novel anti-ice-nucleating polysaccharide from *Bacillus thuringiensis* YY529. Biosci Biotechnol Biochem 66:948–954

Zhao J, Orser CS (1990) Conserved repetition in the ice nucleation gene inaX from *Xanthomonas campestris* pv. *translucens*. Mol Gen Genet 223:163–166

Chapter 12
The Role of Exopolysaccharides in Microbial Adaptation to Cold Habitats

Jody W. Deming and Jodi N. Young

Abstract The cellular exterior that a single-celled microorganism presents to its surroundings marks its first line of defense against environmental pressures, from energy deprivation, shifts in ionic strength, and thermal stress to viral and higher-order attack. The extracellular production of complex sugar compounds (extracellular polysaccharides or exopolysaccharides), whether to provide an individual, multipurpose cell coating or be released for consortial arrangements such as biofilm formation, is a hallmark of microbial life in soil, water, and host (plant and animal)-associated environments. The basic features of exopolysaccharides and their functions pertain to all manner of environments and forms of microbial adaptation, regardless of ambient temperature. At very low temperatures, however, where a phase change comes into play, special considerations arise. In this chapter, which represents an update and reframing of previous work [Krembs and Deming (Psychrophiles: from biodiversity to biotechnology. Springer, 2008)], we pay particular attention to the role of microbially produced exopolysaccharides at subfreezing temperatures. Given recent advances, we have kept the focus on sea ice, that frozen yet brine-filled and exopolysaccharide-rich environment where the physical, chemical, and viral challenges present in the space inhabited by microbes greatly influence their ability to survive and evolve. In the last decade, the study of exopolysaccharides in sea ice has advanced on several fronts, from ocean-scale analyses of their microalgal production and iron-storage capacity to biochemical analyses of their novel ice-binding functions in bacteria. The field has thus moved closer to realizing the pervasive importance of exopolysaccharides in very cold environments, including their biogeochemical, evolutionary, and biotechnological potential.

Contents

J.W. Deming (✉) • J.N. Young
School of Oceanography, University of Washington, Box 357950, Seattle, WA 98195, USA
e-mail: jdeming@uw.edu

R. Margesin (ed.), *Psychrophiles: From Biodiversity to Biotechnology*,
DOI 10.1007/978-3-319-57057-0_12

12.1 Introduction

The cellular exterior that a single-celled microorganism presents to its surroundings marks its first line of defense against environmental pressures, from energy deprivation, shifts in ionic strength, and thermal stress to viral and higher-order attack. The extracellular production of complex sugar compounds (extracellular polysaccharides or exopolysaccharides), whether to provide an individual, multipurpose cell coating or be released for consortial arrangements such as biofilm formation, is a hallmark of microbial life in soil, water, and host (plant and animal)-associated environments. The basic features of exopolysaccharides and their functions pertain to all manner of environments and forms of microbial adaptation, regardless of ambient temperature. At very low temperatures, however, where a phase change comes into play, special considerations arise.

In this chapter, which represents an update and reframing of previous work (Krembs and Deming 2008), we pay particular attention to the role of microbially produced exopolysaccharides in habitats at subfreezing temperatures. Whereas the earlier chapter generalized to exopolymers and used the term exopolymeric substances (and the abbreviation EPS) freely, we have attempted to highlight exopolysaccharides and their unique features, for they comprise the dominant portion of exopolysaccharides in the environment and are targeted by most methods used to quantify naturally occurring exopolysaccharides. In favoring new developments, we relinquished a detailed discussion of the microscale physics of polymer behavior at low temperature (see Krembs and Deming 2008). Given advances over the last decade, however, we have kept the focus on sea ice (touching upon other cold habitats), that frozen yet brine-filled and exopolysaccharide-rich environment where the physical, chemical, and viral challenges present in the space inhabited by microbes greatly influence their ability to survive and evolve. In the last decade, the study of exopolysaccharides in sea ice has advanced on several fronts, from ocean-scale analyses of their microalgal production and iron-storage capacity to biochemical analyses of their novel ice-binding functions in bacteria. The field has thus moved closer to realizing the pervasive importance of exopolysaccharides in very cold habitats and particularly their roles in microbial adaptation and biogeochemical cycles.

12.2 Exopolysaccharides in the Environment

12.2.1 Terminology and Methodology

Although dominant in microbial exudates, complex sugar polymers are not the only organic compounds produced and released into the environment by various microorganisms. Other extracellular macromolecules include proteins (from peptides to large molecular mass enzymes), nucleic acids (free or virus-bound), lipids, phenols, and flavones (Wingender et al. 2012). Together, such compounds have been termed extracellular polymeric (or exopolymeric) substances (EPS). The environmental microbiology literature uses the term EPS widely (Decho 1990; Krembs and Deming 2008), though work in the pelagic ocean introduced the term transparent exopolymeric particles (TEP; Alldredge et al. 1993), recognized as produced by a wide range of organisms and processes. After the documentation of exopolymers in sea ice (Krembs and Engel 2001; Krembs et al. 2002), authors of that literature settled on the term EPS, as the producing organisms are clearly microbial (algae and bacteria), and EPS has long been the term for microbially produced exopolymers and specifically for exopolysaccharides (Sutherland 1972).

In this chapter, we do not use exopolysaccharides and EPS interchangeably, reserving EPS for the greater complexity of exopolymers found in natural environments. Nevertheless, our primary focus is on microbial exopolysaccharides, for two related reasons. First, despite the presence of other macromolecules, most of the material in environmental EPS is composed of polysaccharides. The dominant polysaccharides have carbon backbones of high molecular mass (up to 300 kDa; Poli et al. 2010a; Krembs et al. 2011) and major components of hexose and pentose (e.g., Underwood et al. 2010). They typically carry carboxylic acid groups in the form of uronic acids (Fazio et al. 1982; McConville et al. 1999), imparting an acidic nature, and show variable sugar compositions, including highly branched heteropolysaccharides that can contain fructose, rhamnose, mannose, D-glucose, xylose, D-glucuronic acid, galactose, and half-ester sulfate, among others (Percival et al. 1980; Mancuso-Nichols et al. 2005; Underwood et al. 2010; Carillo et al. 2015; Casillo et al. 2017).

Second, methods commonly used to quantify EPS in the environment, even when a wide array of compounds are present, have targeted the polysaccharide components, and usually the largest size fraction, captured on a 0.4 μm filter, termed particulate EPS (pEPS; Fig. 12.1a). Recently, methods have evolved to address the size fraction called dissolved EPS (dEPS; typically < 0.2 μm by filtration) and to analyze the sugar components specifically (Fig. 12.1). This evolving methodology means that more recently available data sets from cold habitats (since Krembs and Deming 2008) address exopolysaccharides of one size class or another but not all of the complex polymers that comprise EPS. Moving forward, more compound-specific chemical analyses and new analytical tools are beginning to be applied to cold habitats, including those used in metabolomics,

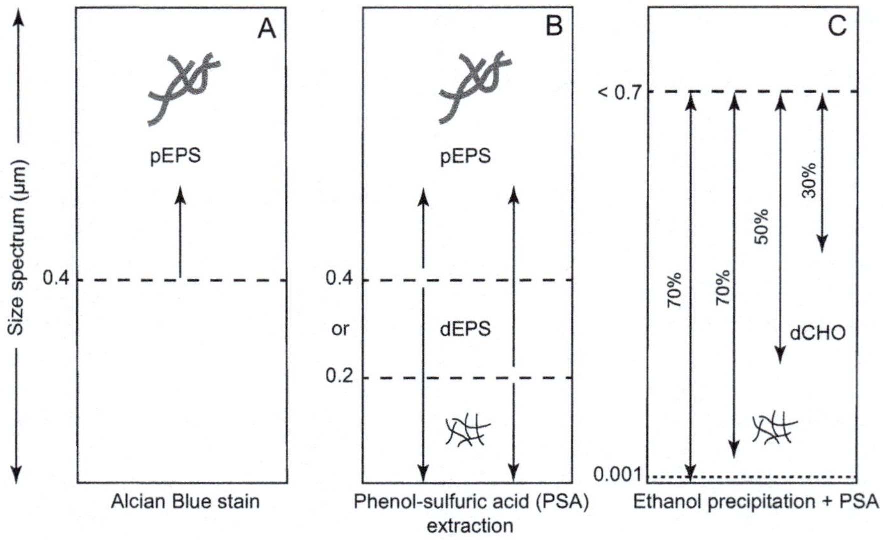

Fig. 12.1 Schematic depiction of the method-specific size spectrum of exopolysaccharides measured in sea ice. The Alcian blue staining method (**a**) uses a colorimetric stain for acidic polysaccharides to measure particulate EPS (pEPS), captured by filtration (> 0.4 µm) and calibration against the exopolysaccharide xanthan gum (Krembs et al. 2002, 2011). The phenol-sulfuric acid (PSA) extraction method (**b**) uses a colorimetric assay for neutral sugars that can quantify both pEPS and dissolved EPS (dEPS) in terms of glucose equivalents (Ewert et al. 2013; Colangelo-Lillis et al. 2016). In (**b**), the ratio pEPS/dEPS depends on choice of filter pore size. The size spectrum for exopolysaccharides within a dEPS fraction can be obtained by an ethanol precipitation approach (**c**), where dissolved carbohydrates (dCHO) are precipitated using a range of ethanol solutions, then extracted by PSA, and converted to glucose equivalents (Underwood et al. 2010, 2013). In (**c**), the dCHO fraction is first constrained in size by filtration (< 0.7 µm, as in Underwood et al. 2010, and possibly < 0.2 µm, according to Underwood et al. 2013) and by dialysis (> 10 kDa, or an estimated 0.001 µm size, in a step required to desalinate the sample; Underwood et al. 2010)

where unique chemical fingerprints of microbial activity can be evaluated in great detail.

Although difficult to apply broadly in the environment, more sophisticated analytical tools have been used to evaluate exopolysaccharides, including imaging by atomic force microscopy and transmission electron microscopy (Santchi et al. 1998). These and various experimental approaches have revealed an important feature of exopolysaccharides: they can form complex architectural structures and hydrated gels, to variable degree depending on environmental conditions of pH, salinity, and temperature (Verdugo et al. 2004; Verdugo and Santschi 2010; Verdugo 2012). In the hydrated gels of natural biofilms, microbially produced mucopolysaccharides (long unbranched polysaccharides with repeating sugar units) establish networks of largely insoluble polymers, often strengthened by cross-linking between ionizable functional groups on the polymers and ambient calcium or magnesium ions (Koerstgens et al. 2001). The strength of this architecture changes with the nature and especially size of the polymers, as has become evident in sea ice and other subfreezing habitats where even the relatively simple, size-based ratio of pEPS to dEPS can be informative. For example, in the dark subfreezing habitat of a cryopeg (liquid brine within permafrost), the pEPS/dEPS ratio is high (and filtered material highly gelatinous) compared to that in icy habitats exposed to sunlight and thus photolysis or UV-based degradation of the polymers (Krembs et al. 2011; Colangelo-Lillis et al. 2016).

As form determines function, the choice of methodology for evaluating exopolysaccharides and EPS in cold habitats, and the details involved (filter pore size, pH, changing salt concentration, and temperature), will constrain the conclusions that can be drawn about the role of these polymers in cold adaptation. No standard method has been adopted by environmental researchers, limiting comparisons of the resulting data; each of the more commonly used methods provides a different size-based result (Fig. 12.1). Common to the methods depicted, however, is the focus on exopolysaccharides.

12.2.2 General and Cold-Specific Occurrence

Exopolysaccharides co-occur inseparably with microbial assemblages in both terrestrial and aquatic environments of wide-ranging conditions, underscoring their importance and diverse functions in microbial ecology (Passow 2000; Lasa 2006; Poli et al. 2010a). Their relatively recalcitrant character and tendency to stick to each other and various surfaces can lead to accumulations that significantly alter the physical properties of the environment on a large scale, including sediments (Yallop et al. 2000) and aquifers (Battin and Sengschmitt 1999), and on a small scale, including biofilms (Costerton et al. 1995) and aquatic aggregates (Alldredge et al. 1993). Gelatinous EPS produced by aquatic microorganisms also clearly affect the biogeochemical cycles of temperate lakes and oceans (Decho 1990; Verdugo 2012), including the cold, high-latitude oceans (Schoemann et al. 2005). Exopolysaccharides produced by microorganisms cultured from these and other environments are the subject of numerous recent reviews highlighting the biotechnological potential of complex sugar compounds (Poli et al. 2010a; Freitas et al. 2011; Nwodo et al. 2012; Goff and Hartel 2013; Roca et al. 2015; Leroy and De Vuyst 2016).

Although the effects of exopolysaccharide accumulations cascade through an ecosystem, their roles at the organism level are critical in life histories and the survival of stressful conditions. Many of the advantages of producing exopolysaccharides are universal: protection against freezing (Kim et al. 2016), desiccation (Hill et al. 1997), and variations in salinity (Krembs et al. 2011); scavenging nutrients and small molecular mass compounds from solution (Decho 1990; Costerton et al. 1995), as well as free radicals (Sun et al. 2015); enhancing metal chelation (Hassler et al. 2011); and aiding cell motility and adherence (Rinta-Kanto et al. 2012). Of particular relevance to this chapter is that exopolysaccharides can provide a protective buffer zone around a cell against unfavorable shifts in the environment, including changes in ionic, osmotic, desiccation, pH, salinity, or toxic-metal conditions (Decho and Lopez 1993; Schlekat et al. 1998). Such shifts pertain to the liquid phase within the sea-ice habitat (Ewert and Deming 2013), as they do for other icy habitats like permafrost and the surface of glacial ice (Deming and Eicken 2007; Boetius et al. 2015).

Under conditions of low or fluctuating water potential, which can lead to desiccation and freeze-thaw stress in very cold habitats, microbes are stimulated to increase production of exopolysaccharides. Diatoms produce high concentrations of EPS when cultured in subfreezing brine (Aletsee and Jahnke 1992), a response that helps to explain microscopic observations of exopolysaccharides coating diatom cells in the brine pores of unmelted sea ice (Krembs et al. 2002; Krembs et al. 2011). Subfreezing temperatures (−8 °C and below) also trigger the excessive production of exopolysaccharides in cold-adapted bacteria (Marx et al. 2009), which helps to explain the small but detectable increase in bulk concentrations of exopolysaccharides in the coldest layers of Arctic sea ice during winter (where algae are absent; Collins et al. 2008). Although sea-ice bacteria can have different responses to temperature-driven fluctuations in brine salinity (Ewert and Deming 2014; Firth et al. 2016), the production of exopolysaccharides repeatedly appears in their survival strategies (Ewert and Deming 2013).

Our planet hosts a wide range of frozen environments in which the microbial inhabitants must cope with seasonal and annual fluctuations in temperature, water availability, and solute concentrations (Deming and Eicken 2007; Ewert and Deming 2013; Boetius et al. 2015). As importantly, individual microbes must prevent intracellular (and extracellular) ice-crystal growth to avoid cell damage. Although the role of exopolysaccharides in mitigating damage to cells in freshwater (glacial) ice is still virtually unexplored, the body of relevant information for sea ice has been growing (e.g., Krembs and Engel 2001; Meiners et al. 2004; Riedel et al. 2007; Ewert and Deming 2013): exopolysaccharides both protect individual microbes and favorably alter the sea-ice environment for all inhabitants (Krembs et al. 2002, 2011; Collins et al. 2008; Marx et al. 2009), affirming the importance of these extracellular components to cold adaptation.

12.3 Exopolysaccharides in Sea Ice

12.3.1 Physical Considerations: Modification of Pore Space and Contents

The processes involved in the formation and growth of the multiphase sea-ice system are complex and dynamic (Petrich and Eicken 2017). The formation stage involves brine rejection, downward into the ocean, though some brine also expels onto the surface of the new ice. Once sea ice is formed, however, temperature, brine salinity, and solute concentrations, including exopolysaccharides, become inseparably linked within the sea-ice system. As the temperature drops in winter, the porosity of the ice decreases, with the salinity of the liquid phase rising as the brine pore volume shrinks; concentrations of solutes (and particulate "impurities," including microbes and viruses) increase in parallel with the brine salinity. Wintertime in the Arctic presents the most severe shifts in temperature for sea ice, such that adequate inhabitable (fluid-filled) pore space and restricted fluid flow between pores become survival issues (Deming 2007). At the same time, the temperature-driven brine-concentration effect leads to very high contact rates between microbes and viruses (Wells and Deming 2006a; Fig. 12.2), increasing the potential for interactions, including viral infection and cell lysis (if conditions allow for viral reproduction; Wells and Deming 2006b) or horizontal gene transfer (Collins and Deming 2011a, b, 2013; Boetius et al. 2015; Deming and Collins 2017). Critical to the microbial experience within sea ice is the physical interplay between the ice matrix and the entrained solutes as the temperature changes seasonally.

Although small molecular mass solutes may have little effect on the physics of the sea-ice system, exopolysaccharides influence virtually every stage of the life cycle of sea ice. During ice growth, when exopolysaccharides accumulate at the phase boundary between ice and brine, they interfere with solute diffusion and

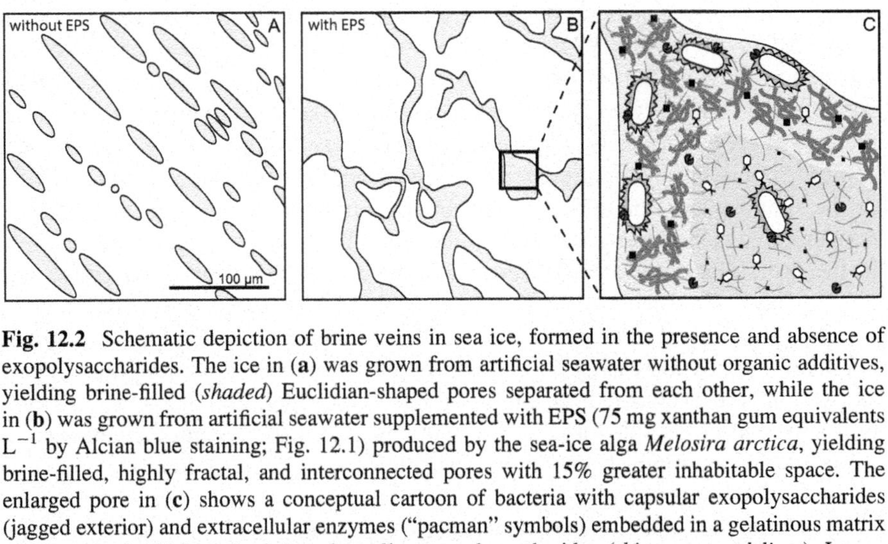

Fig. 12.2 Schematic depiction of brine veins in sea ice, formed in the presence and absence of exopolysaccharides. The ice in (**a**) was grown from artificial seawater without organic additives, yielding brine-filled (*shaded*) Euclidian-shaped pores separated from each other, while the ice in (**b**) was grown from artificial seawater supplemented with EPS (75 mg xanthan gum equivalents L^{-1} by Alcian blue staining; Fig. 12.1) produced by the sea-ice alga *Melosira arctica*, yielding brine-filled, highly fractal, and interconnected pores with 15% greater inhabitable space. The enlarged pore in (**c**) shows a conceptual cartoon of bacteria with capsular exopolysaccharides (jagged exterior) and extracellular enzymes ("pacman" symbols) embedded in a gelatinous matrix of large-sized (*thick squiggles*) and smaller exopolysaccharides (*thinner curved lines*). Larger polysaccharides bind particulate forms of Fe (*large black squares*), while smaller ones yield colloidal Fe (*small black squares*). Degree of gel hydration establishes microscale salinity differences within the pore (*dark* versus *light shading*); microbes not well protected by the hydrated gel matrix experience higher brine salinity, resulting in greater osmotic stress and higher contact rates with viruses. The overall gel matrix, by inhibiting ice-crystal nucleation and growth, provides a physical buffer zone not breached by ice crystals yet responsive to seasonal changes in temperature. Panels (**a**) and (**b**) drawn to approximate scale (scale bar of 100 µm, based on Krembs et al. 2011); features in panel (**c**) not drawn to scale

reduce the salt flux from the advancing ice surface, constraining the thermodynamic evolution of the system (Dash et al. 2006; Krembs and Deming 2008). During ice melt, similar effects must pertain, as the presence of exopolysaccharides slows the melting rate (Krembs et al. 2011). In high concentrations, exopolysaccharides clearly reduce the loss of brine from the system, translating to more inhabitable space within the ice, for microbes inhabit the liquid phase of the sea-ice matrix (Junge et al. 2001). Exopolysaccharides do more than enhance brine retention, however; they also alter the shape and connectivity of the ice pores, improving the habitat for surface interactions and fluid flow (Krembs et al. 2011; Fig. 12.2).

12.3.2 Biological Considerations: Cryoprotection, Osmoprotection, and Viral Defense

With more (ice) surface area within sea-ice pores, and the accumulation of exopolysaccharides along those surfaces (Fig. 12.2c), microbes have greater access to the various forms of protection offered by concentrations of exopolysaccharides, from cryoprotection and osmoprotection to possible defense against viral attack. The commercial field of cryobiology provides ample evidence that sugars and synthetic polymers (such as ethylene glycol) function well as cryoprotectant additives (Allegretto et al. 1992), effectively decreasing the freezing point of the solution of interest (Sutton 1991). At extremely low (eutectic) temperatures, their addition even inhibits the crystallization of salt from solution (Izutsu et al. 1995); in the case of very cold sea ice, ion precipitation as salt crystals could be extended to a colder temperature by the presence of exopolysaccharides, retaining ionic balance in the liquid phase for microbes (an untested hypothesis). Across the range of temperatures that characterize sea ice, the naturally produced organic gels present in the brine phase are also understood to depress the freezing point below that achieved by salt concentrations (Krembs et al. 2002). Such gels achieve this state by lowering the amount of available free energy for ice nucleation inside the hydrated gel matrix and by increasing the salt concentration in the surrounding liquid phase as salt ions are excluded from the gel (Hart et al. 1999, 2001; Fig. 12.2c). Any limitation on ice-crystal formation constitutes a form of cryoprotection (against ice-crystal penetration of membranes) beyond that of keeping liquid available to enable metabolic activity.

The hydrated state of exopolysaccharide gels and resulting increase in the salinity of surrounding fluids also constitute a form of osmoprotection for microbes embedded in such gels within sea ice (Krembs and Deming 2008; Fig. 12.2c). The more hydrated the gel, the greater the protective effect. The presence of low molecular mass solutes within a gel of large exopolysaccharides increases the hydration state of the gel; a microbe in sea ice may thus benefit from adjustments to the composition (large versus small molecules) of its immediate gelatinous buffers, whether by altered production schemes or enzymatic degradation of compounds already released. For those microbes surviving at the higher end of the salinity gradient created in the gel-free area of an ice pore, other forms of osmoprotection must come into play. Recent work shows that microbial communities inhabiting sea ice have efficient physiological mechanisms for avoiding an influx of damaging salt concentrations when external brine salinities are high, as well as an efflux of salt when the environment freshens during ice melt (Firth et al. 2016). These strategies involve the rapid uptake and retention of "compatible solutes" (compatible with intracellular processes, unlike damaging high salt concentrations) when the cell is exposed to high brine salinities and the rapid release or respiration of them when no longer needed for osmoprotection, as during freshening. Some of the most effective compatible solutes are small molecular mass sugars (Roberts 2005), the ultimate breakdown products of exopolysaccharides. Among the exopolymers in sea ice are cold-active extracellular hydrolytic enzymes (Huston et al. 2000). Sea-ice microbes not protected against osmotic challenges by gelatinous exopolysaccharides (of both high and low molecular mass) may well be benefitting from them indirectly by using their breakdown products as compatible solutes.

The issues of viral infection and lateral gene transfer in sea ice, and in many other very cold habitats, are the subjects of intense speculation and increasing research efforts (e.g., Deming 2002; Anesio et al. 2007; Raymond et al. 2007; Paterson and Laybourn-Parry 2012; Collins and Deming 2013; DeMaere et al. 2013; Boetius et al. 2015). The role of exopolysaccharides as a physical defense against viral attack, either by blocking viral access to cell-surface receptors or simply slowing the rate of viral diffusion toward the cell surface, has not been clearly demonstrated for very cold habitats. Evidence of such a defense is accumulating, however (Colangelo-Lillis and Deming 2013; Colangelo-Lillis et al. 2016), and would appear to merit more research, especially given broad interest in the evolutionary implications of gene flow at very low temperatures (Boetius et al. 2015).

12.3.3 Chemical Considerations: Nutrient and Metal Binding and ROS Scavenging

The complexation of metals by organic compounds (ligands) strongly influences microbial growth, by enhancing the bioavailability of essential trace metals (often limiting in the surface ocean; Hassler et al. 2011) and through chelation and subsequent detoxification of toxic heavy metals (De Philippis et al. 2011). Most research on organic ligands has focused on the class of ligands with high binding affinities (L_1, such as siderophores), but interest in understanding organic ligands with weaker binding affinities (L_2) is increasing (Gledhill and Buck 2012; Norman et al. 2015). Exopolysaccharides classify as L_2 ligands. Despite lower binding affinities in general, exopolysaccharides have the potential to outcompete the stronger class of ligands if present in high concentration, as in sea ice. Some of the exopolysaccharides in seawater EPS, however, fall into the L_1 class (Norman et al. 2015). Whether L_1 exopolysaccharides occur in sea ice is not known.

The metal-binding abilities of exopolysaccharides, especially uronic acids (which tend to dominate in sea ice), are due to their polyanionic nature, complex structures, and varying degrees of hydration (Nichols et al. 2005). The strength of their affinities for metals is highly variable, depending on the composition of functional groups and environmental conditions, such as temperature, pH, and salinity (Lamelas et al. 2006; Gutierrez et al. 2012; Causse et al. 2016). Bioavailability of metals is enhanced by increasing solubility, retarding hydrolysis and oxidation, and influencing photochemical reduction processes, thus slowing precipitation and removal (Boyd and Ellwood 2010; Hassler et al. 2011). Some protein components of EPS are also able to bind metals, stimulating research to understand the relative importance of exopolysaccharides as organic ligands.

Metal chelation by EPS is particularly important in the Southern Ocean where dissolved Fe concentrations fall in the nano- to subnanomolar range (de Baar and de Jong 2001), limiting phytoplankton growth (Boyd et al. 2000). Most (99%) of the dissolved Fe in the surface waters of this ocean is complexed by organic ligands, to which EPS (from both algae and bacteria) contribute significantly (Gledhill and van den Berg 1994; Rue and Bruland 1995). The addition of EPS to Southern Ocean phytoplankton communities clearly increases Fe bioavailability and enhances growth (Hassler et al. 2011, 2015; Gutierrez et al. 2012; Norman et al. 2015).

Exopolysaccharides produced by several cold-adapted microbes in culture, including the psychrotolerant bacterium, *Hymenobacter aerophilus*, the sea-ice bacterium *Pseudoalteromonas* sp., and the sea-ice alga *Phaeocystis antarctica*, bind not only metals but also macronutrients (NO_x, NH_3, PO_4), albeit at different affinities and with different elemental preferences (Baker et al. 2010; Norman et al. 2015). The role of naturally occurring exopolysaccharides as organic ligands within sea ice, however, remains to be elucidated. As ice forms, the solutes that concentrate within the brine phase include not only organic compounds but also nutrients and trace metals, with concentrations 1–2 orders of magnitude higher than in underlying seawater (van der Merwe et al. 2009; De Jong et al. 2015; Janssens et al. 2016). Although the sea-ice environment is thus not Fe-limited, the reliance of organisms within the ice on the bioavailability of exopolysaccharide-bound Fe is uncertain (Fig. 12.2c).

Exopolysaccharide concentration within sea ice does not correlate with Fe concentration (van der Merwe et al. 2009), suggesting that metal chelation by exopolysaccharides may take a secondary role to cryoprotection in the ice. The binding affinities of ligands, however, are known to be affected by salinity, yet the effects of high salinity sea-ice brine versus low salinity meltwater on the binding kinetics of exopolysaccharides have not been measured; much remains to be learned (Lannuzel et al. 2016). As sea ice is thought to be an important source of trace metals to the underlying ocean (Noble et al. 2013; Wang et al. 2014), the role of exopolysaccharide-bound metals within the ice is likely to be important to the biogeochemical cycles and biological productivity of polar seas.

In addition to the concentration of organic solutes, metals, and the nutrients that occur in sea ice, high concentrations of reactive oxygen species (ROS, up to 500 nmol L^{-1}) can be produced within its heterogeneous microenvironments during periods of high irradiance and primary production (Neftel et al. 1984). ROS can be highly damaging to cells, but cold-adapted organisms are well adapted for avoiding ROS damage in sea ice through high levels of non-photochemical quenching and low susceptibility to photoinactivation (Petrou et al. 2010). In addition, the exopolysaccharides of sea-ice algae, the Arctic bacterium *Polaribacter* SM1127 (Sun et al. 2015), and the marine Arctic fungi *Keissleriella* sp. YS 4108 and *Aspergillus versicolor* (Sun et al. 2004; Yan et al. 2016) all show strong ROS scavenging properties.

The mechanism by which exopolysaccharides scavenge ROS, however, is not clearly understood. While crude polysaccharide extracts have shown strong antioxidant activity, ROS scavenging by purified polysaccharides is much weaker (Wang et al. 2016). Low concentrations of proteins, peptide moieties, phenols, and flavones within an EPS matrix are thought to provide much of the ROS scavenging. The amino acid proline, the peptide glutathione, and the enzymes ascorbate peroxidase and disulfate reductase are all strong scavengers of ROS (Hayat et al. 2012), as is dimethylsulfoniopropionate (Kirst et al. 1991; Lyon et al. 2011) and its bacterial degradation product, the climate-active gas dimethyl sulfide, which are produced in high concentrations in sea ice (Sunda et al. 2002). Polysaccharides, however, that have bound metals, chemical modifications (e.g., sulfation, phosphorylation, and acetylation), large molecular masses, or high uronic acid content (as in sea ice) display enhanced ROS scavenging properties (Wang et al. 2016).

ROS are released by the irradiation and subsequent breakdown of the aromatic chromophoric fraction of dissolved organic matter (including humic substances and organic aggregates in Arctic waters). This breakdown process is enhanced in ice compared to seawater due to high concentrations of organic matter in the brines (Grannas et al. 2014). The in situ balance of ROS scavenging by EPS and ROS production through photolysis of organic matter in sea ice has yet to be determined.

12.3.4 Biogeochemical Cycling: Sea Ice to Seawater and Atmosphere

Exopolysaccharides and the hydrated gel matrix they create in sea ice play an important role in the dynamic cycling of nutrients within the ice, influencing the diffusion and bioavailability of the nutrients. The production and composition of exopolysaccharides also help to maintain the essential cellular stoichiometry of the ice inhabitants, while providing carbon sources for the heterotrophic regeneration of other nutrients. Exopolysaccharides affect sea-ice melt and organic matter aggregation, influencing the export of organic matter from the ice.

The elemental composition of exopolysaccharides is predominately carbon, with lower concentrations of other nutrients including nitrogen and sulfur. Many organisms regulate exopolysaccharide production as a means to dissipate excess organic carbon and maintain the desired cellular carbon stoichiometry, particularly as a response to carbon overconsumption (Toggweiler 1993). For example, long-term culture under conditions of high CO_2 and replete nutrients results in enhanced production of dissolved organic carbon by the sea-ice diatom *Nitzschia lecointei* (Torstensson et al. 2015). This partitioning of carbon into different particulate and dissolved organic pools can change elemental stoichiometry. In the cyanobacterium *Microcystis aeruginosa*, production of soluble exopolysaccharides correlates with a decrease in particulate C:N, whereas production of capsular exopolysaccharides elevates particulate C:N (Pannard et al. 2016). The use of exopolysaccharides to regulate and maintain elemental stoichiometry within a cell is particularly relevant to sea-ice environments because nutrient concentrations can vary dramatically due to physical concentration and biological uptake and regeneration, and organisms adapted to cold temperatures likely have high concentrations of protein and ribosomes to compensate for slow catalytic rates, altering the cellular requirement for N and P (Toseland et al. 2013).

Labile exopolysaccharides are an important carbon source for heterotrophic consumption. Autotrophically produced dissolved organic carbon (< 0.4 µm), which includes dissolved exopolysaccharides, is rapidly consumed by Arctic bacterial communities and by bacteria associated with cultures of the Antarctic diatom *Nitzschia lecointei* (Engel et al. 2013; Torstensson et al. 2015). Exopolysaccharides produced by algae can provide a carbon source that promotes the growth of only specific bacteria, implying a mutualistic relationship (Haynes et al. 2007). Utilization of exopolysaccharides as a carbon source can enhance heterotrophic activity, resulting in higher rates of ammonium regeneration within sea ice (Lavoie et al. 2005; Riedel et al. 2007). We caution, however, that these apparent high rates may also be an artifact of having melted the sea ice to measure ammonium regeneration, thus triggering the respiration of nitrogen-rich compatible solutes no longer required for osmotolerance (Firth et al. 2016).

Exopolysaccharides represent a significant source of carbon within sea ice, being enriched by 1–2 orders of magnitude over that in seawater (Meiners et al. 2004; Riedel et al. 2006). In first-year ice, most of the carbohydrates register as dissolved and small-sized and correlate strongly with algal biomass (Aslam et al. 2016). In thicker ice, a higher proportion of the carbohydrates are measured as particulates (Aslam et al. 2016), indicating heterotrophic consumption of the small-sized molecules and/or new production of large ones. Although particulate and dissolved exopolysaccharides are entrained and concentrated by physical processes (Juhl et al. 2011), and mechanical damage of cells during ice formation can contribute to these pools (Meiners et al. 2004), the predominant source of exopolysaccharides in sea ice is the in situ production by microorganisms. Their concentrations regularly correlate with algal and bacterial biomass (Meiners et al. 2008; Niemi et al. 2011; Aslam et al. 2016), as emphasized by a global-scale analysis (Underwood et al. 2013). In spring and summer, the correlation with chlorophyll is particularly strong, indicating the autotrophic community as the dominant source. In winter, a decoupling occurs, indicating contributions of exopolysaccharides by heterotrophic organisms (Niemi et al. 2011) or a change in algal physiology, as the photosynthetic apparatus (including chlorophyll content) is reduced, and large amounts of exopolysaccharides are produced for cryopreservation. Protists that produce large amounts of exopolysaccharides (e.g., pennate diatoms) are more likely to survive within the ice over winter than those that do not (e.g., dinoflagellates; Niemi et al. 2011).

As sea ice melts, the released organisms can remain in the upper ocean or be exported to depth. Key to their fate is the balance between buoyancy imparted by gelatinous exopolysaccharides and ballast created as denser particles aggregate due to the stickiness of exopolysaccharides. In the Arctic during the melting season, examples of both fates have been documented. In one study, little relationship was found between exopolysaccharide and chlorophyll concentrations, suggesting little influence on the export of algal particles, yet the export of bacteria was strongly correlated with exopolysaccharides, with the association of bacteria with diatom-bound polymers enhancing the bacterial sinking rate (Riedel et al. 2006). Even after microbial biomass is lost from sea ice, a network of exopolysaccharides can be retained within the remaining sea-ice structure (Juhl et al. 2011), as happens when brines are drained from thick ice into "sackholes" for sampling purposes. Thus the seasonal and within-ice variations in exopolysaccharide production (and composition) can influence the export of organic matter from sea ice. Once released from sea ice, however, buoyant exopolysaccharides and other components of EPS can accumulate in a surface microlayer at the air-sea interface, providing a potentially important sea-ice source of organic aerosols (Galgani et al. 2016) not yet considered in models of cloud formation at high latitudes (Orellana et al. 2011; Lohmann 2015).

What is clear is that all types of natural sea ice, regardless of source seawater, growth phase, thickness, or longevity, contain exopolysaccharides (Fig. 12.3), as well as bacteria (Deming 2010), with algae virtually always present in nutrient-bathed locations during seasons that enable photosynthesis (Zhou et al. 2013). The correlation between exopolysaccharides and chlorophyll is significant on the global scale (Underwood et al. 2013), despite differences between Arctic and Antarctic sea ice (e.g., the surface of the latter is often flooded with seawater due to freeboard depression by heavy snow cover) and relatively subtle shifts in composition as autotrophy is displaced by heterotrophy in winter (Collins et al. 2008; Underwood et al. 2010). Because chlorophyll also scales with the permeability and brine volume of sea ice (Zhou et al. 2013), we have broadly depicted the scope of exopolysaccharides in sea ice (Fig. 12.3) according to the contours of these physical parameters, as presented in Zhou et al. (2013). The many co-linked attributes of sea ice, all driven by low temperature, will continue to make sea ice an excellent natural laboratory for studying the role of exopolysaccharides in microbial adaptation to the cold.

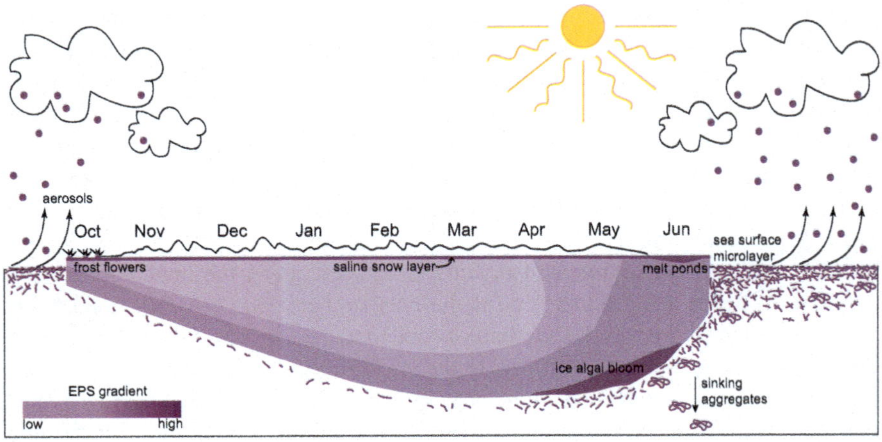

Fig. 12.3 Schematic depiction of the seasonal occurrence of EPS in Arctic landfast sea ice and surroundings. The presence and thickness of sea ice (*shaded area*) changes with season. EPS occurs throughout the ice but varies in concentration (intensity of shading) according to initial content in seawater (*squiggles*), season-based production by algae and bacteria, and the brine volume of the ice, which generally correlates with algal production (Zhou et al. 2013). As ice forms in autumn, EPS entrain into the ice matrix, concentrate in frost flowers and brine skim on the surface of the ice (Bowman and Deming 2010; Barber et al. 2014), and wick into the saline snow layer following snowfall (Ewert et al. 2013). Through winter, bacteria produce exopolysaccharides in the upper colder portions of the ice for cryoprotection, while during spring, ice algae bloom in nutrient-bathed bottom ice, and in some surface melt ponds, releasing copious amounts of exopolysaccharides. After ice melt and before freeze-up, exopolysaccharides and other exopolymers derived from the ice concentrate at the sea-surface microlayer where they are available for aerosolization, likely contributing to cloud formation (Galgani et al. 2016). Gelatinous exopolysaccharides contribute to the formation and sinking of organic-rich aggregates. Ice growth and contours based on correlation between brine volume and algal production in Zhou et al. (2013)

12.4 Organism-Specific Exopolysaccharides

12.4.1 Advances with Cold-Adapted Bacteria

A hallmark of exopolysaccharides is their role in biofilm formation on virtually any surface that occurs in nature. Although exopolysaccharides are clearly important in creating additional surface area within the sea-ice matrix and forming hydrated gels along those surfaces (Sect. 12.3; Fig. 12.2), observing their direct attachment to ice crystals has been elusive. The focus has instead been on the ice-binding properties of proteins (Chap. 11). Ice-binding proteins have a characteristic planar face that interacts with ice crystals, preventing the conversion of small ice crystals into fewer larger ones (a process called ice recrystallization or IR) that can be more damaging to the cell (Davies 2014). The ice-binding protein of the sea-ice bacterium *Marinomonas primoryensis*, for example, presents as its ice binding faces a flat, repetitive two-dimensional array of specific amino acid residues, threonine and asparagine (Dolev et al. 2016). Remarkably, and unique among ice-binding proteins, this large *Marinomonas* protein is held by the bacterium at the cell surface, and its ice-binding region serves to adhere the organism to the face of an ice crystal (Dolev et al. 2016). This discovery of a mechanism enabling bacterial attachment to ice invokes early-stage biofilm formation on a mineral surface (ice) otherwise largely ignored in the bacterial biofilm literature. Algal production of ice-binding proteins is well known, including that algae may have received the genes for these proteins via lateral gene transfer from bacteria (Raymond et al. 2007; Chap. 11). Ice-binding proteins may contribute to the ability of diatoms to colonize natural forms of ice; however, ice-active bacteria intimately associated with diatoms have been directly implicated in the successful algal colonization of ice (D'souza et al. 2013).

Although proteins are by far the best-known ice-binding compounds, recent work on EPS has uncovered some complex polysaccharides with amino acid residues that impart an IR inhibitory activity to the polysaccharide similar to that of ice-binding proteins. The possible existence of ice-binding polysaccharides was considered in earlier work with natural sea ice, the sea-ice alga *Melosira arctica*, and the model cold-adapted marine bacterium *Colwellia psychrerythraea* strain 34H, but in the absence of detailed structural information (only reactions to heat treatment), the evidence for ice binding was ascribed to glycoproteins (Ewert and Deming 2011; Krembs et al. 2011). Subsequent biochemical characterization of the exopolysaccharides produced by *C. psychrerythraea* 34H, however, has revealed two forms of IR-inhibiting polysaccharides, both of which are unique to the biochemical literature: the capsular polysaccharide, held by the cell on its surface (Carillo et al. 2015), and an extracellular polysaccharide, released into solution when the organism is grown at 4 °C (Casillo et al. 2017). The capsular polysaccharide is composed of a repetitive tetramer of sugar moieties (two galacturonic acid and two glucuronic acid residues) decorated with threonine in a manner accounting for IR inhibition, while the extracellular polysaccharide is a repetitive sugar trimer (one quinovosamine and two galacturonic acid residues) decorated with alanine (as are ice-binding proteins of fish; Davies 2014) for a similar effect (Fig. 12.4). Although both polymers inhibit ice recrystallization, the capsular exopolysaccharide, imparting immediately proximate protection for the cell, appears to be the stronger inhibitor. Given the temperature growth range of this organism, from −12 °C to 18 °C (Wells and Deming 2006b), and its superior genomic capacity for membrane transport and responding to environmental information (Huston et al. 2000; Methé et al. 2005; Techtmann et al. 2016), these two structural forms of IR-inhibiting polysaccharides may only be examples of a larger portfolio of ice-binding polymers for responding to the fluctuating and extreme environmental conditions that define sea ice (Ewert and Deming 2014). The potential ice-binding capacities of exopolysaccharides produced by other algae and other organisms remain to be explored.

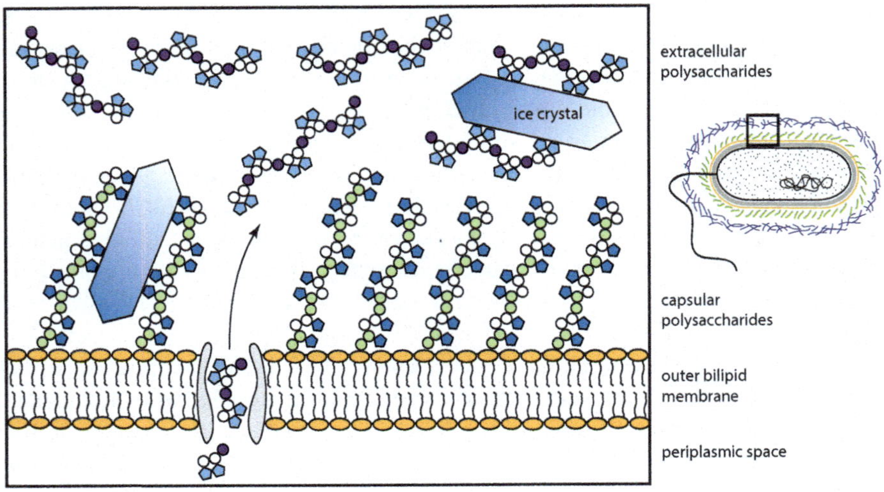

Fig. 12.4 Schematic depiction of the structure-based ice-binding function of the exopolysaccharides produced by a cold-adapted marine bacterium. *Colwellia psychrerythraea* 34H produces both capsular and extracellular polysaccharides capable of ice binding. The structure of the *capsular polysaccharide*, shown attached to the outer bilipid membrane, involves repeating tetramers of the sugars glucuronic acid (*dark circles*) and galacturonic acid (*light circles*) decorated with the amino acid threonine (*pentagons*). In contrast, the *extracellular polysaccharide*, shown assembled through the membrane and then free of the cell, involves repeating trimers of quinovosamine (*dark circles*) and galacturonic acid (*light circles*) decorated with the amino acid alanine (*pentagons*). In both cases, the amino acids impart the ability of the polysaccharide to bind to ice crystals, thus impeding ice-crystal growth and damage to the membrane. Exopolysaccharides are drawn to show the repeating sugar feature with amino acid decoration but do not convey the extensive length or size and shape of the molecules. Primary structures and impedance of ice-crystal growth based on data in Carillo et al. (2015) and Casillo et al. (2017)

12.4.2 Cyanobacteria and Algae

Cyanobacteria are generally absent from cold marine environments (Koh et al. 2012), though *Trichodesmium*-like cyanobacteria have been identified in sea-ice brines in the Arctic (Díez et al. 2012), and cyanobacterial sequences have been detected in frost flowers on the surface of new sea ice in coastal Antarctica (Bowman and Deming 2016). Cold-adapted cyanobacteria, however, are important primary producers in other polar environments, in polar lakes (Taton et al. 2006; Jungblut et al. 2016), streams (Rochera et al. 2013; Chrismas et al. 2016), sediments (de los Ríos et al. 2014), and cryoconites (Takeuchi et al. 2001; Hodson et al. 2010). Metagenomic analysis of the microbial mats they form in close association with bacteria on Arctic and Antarctic ice shelves shows that cyanobacteria are important contributors to the exopolysaccharide biosynthesis gene pool (Varin et al. 2012). Like many other cyanobacteria (and bacteria), cold-adapted cyanobacteria isolated in culture produce large amounts of exopolysaccharides that aid survival of extreme conditions, protecting against desiccation (Hill et al. 1997), high salt and heavy metals (De Philippis et al. 2011; Jittawuttipoka et al. 2013), UV radiation (Garcia-Pichel and Castenholz 1991; Song et al. 2016), and ROS (Shirkey et al. 2000). Recent advances on the genetic regulation of exopolysaccharide biosynthesis in cyanobacteria (reviewed by Kehr and Dittmann 2015) include work with an Arctic cyanobacterium *Phormidesmis preistleyi* BC1401 (Chrismas et al. 2016).

Eukaryotic algae are found across the spectrum of cold habitats: in sea ice, melt ponds, the surface ocean, cryoconite holes, glacial runoff, snow, and freshwater ponds and lakes. The cold-adapted algae encompass groups from evolutionarily distinct backgrounds but appear to be dominated by green algae and diatoms. To date, about 1000 species of unicellular eukaryotic species have been identified in Arctic sea ice, though this count includes heterotrophic protists along with microalgae (Poulin et al. 2011). Diatoms dominate sea-ice algal species (71%) and thus exopolysaccharide production in the ice, with pennate diatoms being the most prolific. Other ice algal groups include dinoflagellates, prasinophytes, silicoflagellates, prymnesiophytes, and chlorophytes and chrysophytes. Given this diversity, the biosynthesis pathways for exopolysaccharides are likely to be variable, though only two completed genomes for polar algae are available to explore the genetic aspects of exopolysaccharide production: the marine diatom *Fragilariopsis cylindrus* (Mock et al. 2017) and the green algae *Coccomyxa subellipsoidea* C-169 (Blanc et al. 2012).

Work with cultures, however, reveals that even within the same microalgal group, exopolysaccharide production and composition can vary considerably. The sea-ice diatoms *Synedropsis* sp., *Fragilariopsis cylindrus*, and *F. curta* all produce frustule-associated and colloidal exopolysaccharides, but *Synedropsis* produces small colloidal and relatively soluble compounds, while the *Fragilariopsis* species produce less soluble colloidal forms (Aslam et al. 2012). The exopolysaccharides of diatoms are often gelatinous and can be produced in a variety of forms, including stalks, tubes, apical pads, adhering films, fibrils, and cell coatings; the latter range from rigid fibrils to highly hydrated mucilaginous capsules (Hoagland et al. 1993; Gügi et al. 2015). These distinctions suggest that the specific algal composition of sea ice (and thus the specific exopolysaccharides produced within it) may be worth exploring to better understand the microscale features of the habitable pore space in sea ice (Sect. 12.3.1).

12.4.3 Other Microorganisms

Like other cold-adapted microbes, filamentous fungi and yeast also produce exopolysaccharides for protection against harsh environments (Selbmann et al. 2002) and competition (Mukhopadhyay et al. 2014). While fungi have been found in a number of cold environments (but generally not sea ice; Gunde-Cimerman et al. 2003), their high molecular mass exopolysaccharides are produced in highest concentrations at relatively warm temperatures (> 15 °C; Selbmann et al. 2002, Poli et al. 2010b; Buzzini et al. 2012). In contrast, the Archaea are known for their ability to thrive in low-energy environments, conserving energy in their cellular lifestyles. Archaea are important in selected cold habitats (e.g., Antarctic lakes; DeMaere et al. 2013), but present in (organic) energy-rich sea ice only in low numbers (Deming and Collins 2017). Their role in exopolysaccharide production in cold habitats is not evident, but they can be expected to benefit from all of the protective properties of hydrated gels produced by other microbes (Sect. 12.3.2).

12.5 Conclusions

Focusing on the dynamic sea-ice environment has allowed us to bring forward many of the fundamental aspects of exopolysaccharides and their role in microbial adaptation to very cold habitats. The sculpting of the habitable space within sea ice by exopolysaccharides highlights the power of organisms to improve their physical setting by the release of extracellular products, as well as the accrual of resulting benefits to all inhabitants, including those that do not produce their own exopolysaccharides. These benefits, all tied to the gelatinous nature of exopolysaccharides, include cryoprotection, osmoprotection, and possibly defense against viral attack. The ability of hydrated polysaccharide gels to bind and scavenge nutrients and metals also contributes to microbial success in such an extreme environment, at the same time serving as a repository of essential elements which upon release from melting ice can figure importantly in the overall productivity of the larger marine ecosystem. The high concentrations of exopolysaccharides in sea ice, produced microbially for multiple benefits in situ, eventually return to the ocean where their fate may include aerosolization and contribution to cloud formation, a complex but important component of the overall heat budget, especially in the Arctic.

In the last decade, several advances emerging from the study of environmental exopolysaccharides, especially in relation to sea ice, help to define future directions for both basic and applied research. How iron limitation is overcome in a cold ocean has brought attention to the iron-storage capacity of sea ice which, in turn, has led to the study of exopolysaccharides as organic ligands, binding nutrients and metal, and potentially reactive oxygen species, at subfreezing temperatures. The precise mechanisms and compounds involved, and their specific behaviors under the fluctuating extremes of temperature and brine salinity that characterize sea ice, remain to be determined. More sophisticated methodologies than those commonly applied to quantifying EPS in sea ice will be required in this pursuit, as will attention to potential depression of salt precipitation temperatures by exopolysaccharides. The discovery of novel ice-binding exopolysaccharides in a cold-adapted bacterium invites new ways to consider the microbial colonization of icy habitats, and ice formation itself, and suggests the promise of examining exopolysaccharides produced by other organisms for similar traits. Ice binding is no longer the sole province of proteins, opening wide the potential for biotechnological applications of ice-binding polysaccharides.

Acknowledgments We acknowledge support from NSF (ARC-1203267 to JWD), the UW Astrobiology Program, and the School of Oceanography, through the Karl M. Banse Endowed Professorship (to JWD) and start-up funds (to JNY), to develop this chapter. Numerous colleagues have influenced our thinking, but we especially thank Christopher Krembs, Marcela Ewert, Eric Collins, and Jeff Bowman for their contributions to the study of EPS in the cold and Shelly Carpenter for help with graphics and many EPS measurements over the years.

References

Aletsee L, Jahnke J (1992) Growth and productivity of the psychrophilic marine diatoms *Thalassiosira antarctica* Comber and *Nitzschia frigida* Grunow in batch cultures at temperatures below the freezing point of sea water. Polar Biol 11:643–647

Alldredge AL, Passow U, Logan BU (1993) The abundance and significance of a class of large, transparent organic particles in the ocean. Deep Sea Res 40:1131–1140

Allegretto EA, Caple G, Culbertoson LB (1992) Inhibition of ice nucleation by synthetic polymers. ABS PAP ACS 183:19

Anesio AM, Mindl B, Laybourn-Parry J, Hodson AJ, Sattler B (2007) Viral dynamics in cryoconite holes on a High Arctic glacier (Svalbard). J Geophys Res 112:G04S31

Aslam SN, Cresswell-Maynard T, Thomas DN, Underwood GJC (2012) Production and characterization of the intra- and extracellular carbohydrates and polymeric substances (EPS) of three sea-ice diatom species, and evidence for a cryoprotective role for EPS. J Phycol 48:1494–1509

Aslam SN, Michel C, Niemi A, Underwood GJC (2016) Patterns and drivers of carbohydrate budgets in ice algal assemblages from first year Arctic sea ice. LimnolOceanogr 61:919–937

Baker MG, Lalonde SV, Konhauser KO, Foght JM (2010) Role of extracellular polymeric substances in the surface chemical reactivity of *Hymenobacter aerophilus*, a psychrotolerant bacterium. Appl Environ Microbiol 76:102–109

Barber DG, Ehn JK, Pućko M, Rysgaard S, Deming JW, Bowman JS, Papakyriakou T, Galley RJ, Søgaard DH (2014) Frost flowers on young Arctic sea ice: the climatic, chemical, and microbial significance of an emerging ice type. J Geophys Res Atmos 27:11,593–11,612

Battin TJ, Sengschmitt D (1999) Linking sediment biofilms, hydrodynamics, and river bed clogging: evidence from a large river. Microb Ecol 37:185–196

Blanc G, Agarkova I, Grimwood J, Kuo A, Brueggeman A, Dunigan DD, Gurnon J, Ladunga I, Lindquist E, Lucas S, Pangilinan J, Pröschold T, Salamov A, Schmutz J, Weeks D, Yamada T, Lomsadze A, Borodovsky M, Claverie J-M, Grigoriev IV, Van Etten JL (2012) The genome of the polar eukaryotic microalga *Coccomyxa subellipsoidea* reveals traits of cold adaptation. Genome Biol 13:R39

Boetius A, Anesio AM, Deming JW, Mikucki JA, Rapp JZ (2015) Microbial ecology of the cryosphere: sea ice and glacial habitats. Nat Rev Microbiol 13:677–690. doi:10.1038/nrmicro3522

Bowman JS, Deming JW (2010) Elevated bacterial abundance and exopolymers in saline frost flowers and implications for atmospheric chemistry and microbial dispersal. J Geophys Lett 37: L13501. doi:10.1029/2010GL043020

Bowman JS, Deming JW (2016) Evidence for broad, wind-driven distribution of bacterial phylotypes in McMurdo Sound, Antarctica. Polar Biol. doi:10.1007/s00300-016-1921-2

Boyd PW, Ellwood MJ (2010) The biogeochemical cycle of iron in the ocean. Nat Geosci 3:675–682

Boyd PW, Watson AJ, Law CS, Abraham ER, Trull T, Murdoch R, Bakker DCE, Bowie AR, Buesseler KO, Chang H, Charette M, Croot P, Downing K, Frew R, Gall M, Hadfield M, Hall J, Harvey M, Jameson G, LaRoche J, Liddicoat M, Ling R, Maldonado MT, McKay RM, Nodder S, Pickmere S, Pridmore R, Rintoul S, Safi K, Sutton P, Strzepek R, Tanneberger K, Turner S, Waite A, Zeldis J (2000) A mesoscale phytoplankton bloom in the polar Southern Ocean stimulated by iron fertilization. Nature 407:695–702

Buzzini P, Branda E, Goretti M, Turchetti B (2012) Psychrophilic yeasts from worldwide glacial habitats: diversity, adaptation strategies and biotechnological potential. FEMS Microbiol Ecol 82:217–241

Carillo S, Casillo A, Pieretti G, Parrilli E, Sannino F, Bayer-Giraldi M, Cosconati S, Novellino E, Ewert M, Deming JW, Lanzetta R (2015) A unique capsular polysaccharide structure from the psychrophilic marine bacterium *Colwellia psychrerythraea* 34H that mimics antifreeze (glyco) proteins. J Am Chem Soc 137(1):179–189

Casillo A, Parrilli E, Sannino F, Mitchell DE, Gibson MI, Marino G, Lanzetta R, Parrilli M, Cosconati S, Novellino E, Randazzo A (2017) Structure-activity relationship of the exopolysaccharide from a psychrophilic bacterium: a strategy for cryoprotection. Carbohydr Polym 156:364–371

Causse B, Spadini L, Sarret G, Faure A, Travelet C, Madern D, Delolme C (2016) Xanthan exopolysaccharide: Cu2+ complexes affected from the pH-dependent conformational state: implications for environmentally relevant biopolymers. Environ Sci Technol 50:3477–3485

Chrismas NAM, Barker G, Anesio AM, Sánchez-Baracaldo P (2016) Genomic mechanisms for cold tolerance and production of exopolysaccharides in the Arctic cyanobacterium *Phormidesmis priestleyi* BC1401. BMC Genomics 17:1–14

Colangelo-Lillis JR, Deming JW (2013) Genomic analysis of cold-active *Colwelliaphage* 9A and psychrophilic phage-host interactions. Extremophiles 17:99–114. doi:10.1007/s00792-012-0497-1

Colangelo-Lillis J, Eicken H, Carpenter SD, Deming JW (2016) Evidence for marine origin and microbial-viral habitability of sub-zero hypersaline aqueous inclusions within permafrost near Barrow, Alaska. FEMS Microbiol Ecol 92(5.) fiw053. doi:10.1093/femsec/fiw053

Collins RE, Deming JW (2011a) Abundant dissolved genetic material in Arctic sea ice, Part I: Extracellular DNA. Polar Biol 34:1819–1830. doi:10.1007/s00300-011-1041-y

Collins RE, Deming JW (2011b) Abundant dissolved genetic material in Arctic sea ice, Part II: Virus dynamics during autumn freeze-up. Polar Biol 34:1831–1841. doi:10.1007/s00300-011-1008-z

Collins RE, Deming JW (2013) Identification of an inter-Order lateral gene transfer event enabling the catabolism of compatible solutes by *Colwellia* spp. Extremophiles 17(4):601–610

Collins RE, Carpenter SD, Deming JW (2008) Spatial heterogeneity and temporal dynamics of particles, bacteria, and pEPS in Arctic winter sea ice. J Mar Syst 74(3):902–917

Costerton JW, Lewandowski Z, Caldwell DE, Korber DR, Lappin-Scott HM (1995) Microbial biofilms. Annu Rev Microbiol 49:711–745

Dash JG, Rempel AW, Wettlaufer JS (2006) The physics of premelted ice and its geophysical consequences. Rev Mod Phys 78(3):695–741

Davies PL (2014) Ice-binding proteins: a remarkable diversity of structures for stopping and starting ice growth. Trends Biochem Sci 39(11):548–555

de Baar HJW, de Jong JTM (2001) Distributions, sources and sinks of iron in seawater. In: Turner DR, Hunter KA (eds) The biogeochemistry of iron in seawater. Wiley, New York, NY, pp 123–254

De Jong JTM, Stammerjohn SE, Ackley SF, Tison JL, Mattielli N, Schoemann V (2015) Sources and fluxes of dissolved iron in the Bellingshausen Sea (West Antarctica): the importance of sea ice, icebergs and the continental margin. Mar Chem 177(Part 3):518–535

de los Ríos A, Cary C, Cowan D (2014) The spatial structures of hypolithic communities in the Dry Valleys of East Antarctica. Polar Biol 37:1823–1833

De Philippis R, Colica G, Micheletti E (2011) Exopolysaccharide-producing cyanobacteria in heavy metal removal from water: molecular basis and practical applicability of the biosorption process. Appl Microbiol Biotechnol 92:697–708

Decho AW (1990) Microbial exopolymer secretions in ocean environments: their role(s) in food webs and marine processes. Oceanogr Mar Biol Annu Rev 28:73–153

Decho AW, Lopez GR (1993) Exopolymer microenvironments of microbial flora: multiple and interactive effects on trophic relationships. Limnol Oceanogr 38:1633–1645

DeMaere MZ, Williams TJ, Allen MA, Brown MV, Gibson JA, Rich J, Lauro FM, Dyall-Smith M, Davenport KW, Woyke T, Kyrpides NC, Tringe G, Cavicchioli R (2013) High level of intergenera gene exchange shapes the evolution of haloarchaea in an isolated Antarctic lake. Proc Natl Acad Sci USA 110(42):16939–16944

Deming JW (2002) Psychrophiles and polar regions. Curr Opin Microbiol 3(5):301–309

Deming JW (2007) Life in ice formations at very cold temperatures. In: Gerday C, Glansdorff N (eds) Physiology and biochemistry of extremophiles. ASM Press, Washington, DC, pp 133–145

Deming JW (2010) Sea ice bacteria and viruses. In: Thomas DN, Dieckmann GS (eds) Sea ice – an introduction to its physics, chemistry, biology and geology, 2nd edn. Blackwell, Oxford, pp 247–282

Deming JW, Collins RE (2017) Sea ice as a habitat for bacteria, archaea and viruses. In: Thomas DN (ed) Sea ice – an introduction to its physics, chemistry, biology and geology, 3rd edn. Wiley. in press

Deming JW, Eicken H (2007) Life in ice. In: Sullivan WT, Baross JA (eds) Planets and life: the emerging science of astrobiology. Cambridge University Press, Cambridge, pp 292–312

Díez B, Bergman B, Pedrós-Alió C, Antó M, Snoeijs P (2012) High cyanobacterial nifH gene diversity in Arctic seawater and sea ice brine. Environ Microbiol Rep 4:360–366

Dolev MB, Bernheim R, Guo S, Davies PL, Braslavsky I (2016) Putting life on ice: bacteria that bind to frozen water. J Roy Soc Interface 13(121):20160210

D'souza NA, Kawarasaki Y, Gantz JD, Lee RE, Beall BF, Shtarkman YM, Kocer ZA, Rogers SO, Wildschutte H, Bullerjahn GS, McKay RM (2013) Diatom assemblages promote ice formation in large lakes. ISME J 7(8):1632–1640

Engel A, Borchard C, Piontek J, Schulz K, Riebesell U, Bellerby R (2013) CO_2 increases ^{14}C primary production in an Antarctic plankton community. Biogeosciences 10:1291–1298

Ewert M, Deming JW (2011) Selective retention in saline ice of extracellular polysaccharides produced by the cold-adapted marine bacterium *Colwellia psychrerythraea* strain 34H. Ann Glaciol 52(57):111–117

Ewert M, Deming JW (2013) Sea ice microorganisms: environmental constraints and extracellular responses. Biology 2(2):603–628

Ewert M, Deming JW (2014) Bacterial responses to fluctuations and extremes in temperature and brine salinity at the surface of Arctic winter sea ice. FEMS Microbiol Ecol 89(2):476–489

Ewert M, Carpenter SD, Colangelo-Lillis J, Deming JW (2013) Bacterial and extracellular polysaccharide content of brine-wetted snow over Arctic winter first-year sea ice. J Geophys Res-Oceans 118(2):726–735

Fazio SA, Uhlinger DJ, Parker JH, White DC (1982) Estimations of uronic acids as quantitative measures of extracellular and cell wall polysaccharide polymers from environmental samples. Appl Environ Microbiol 47:135–143

Firth E, Carpenter SD, Sørensen HL, Collins RE, Deming JW (2016) Bacterial use of choline to tolerate salinity shifts in sea-ice brines. Elem Sci Anth 4(1):000120. doi:10.12952/journal.elementa.000120

Freitas F, Alves VD, Reis MA (2011) Advances in bacterial exopolysaccharides: from production to biotechnological applications. Trends Biotechnol 29(8):388–398

Galgani L, Piontek J, Engel A (2016) Biopolymers form a gelatinous microlayer at the air-sea interface when Arctic sea ice melts. Sci Rep 6:29465

Garcia-Pichel F, Castenholz RW (1991) Characterization and biological implications of scytonemin, a cyanobacterial sheath pigment. J Phycol 27:395–409

Gledhill M, Buck K (2012) The organic complexation of iron in the marine environment: a review. Front Microbiol 3:69

Gledhill M, van den Berg CMG (1994) Determination of complexation of iron(III) with natural organic complexing ligands in seawater using cathodic stripping voltammetry. Mar Chem 47:41–54

Goff HD, Hartel RW (2013) Ice cream structure. In: Ice cream. Springer, New York, NY, pp 313–352

Grannas AM, Pagano LP, Pierce BC, Bobby R, Fede A (2014) Role of dissolved organic matter in ice photochemistry. Environ Sci Technol 48:10725–10733

Gügi B, Le Costaouec T, Burel C, Lerouge P, Helbert W, Bardor M (2015) Diatom-specific oligosaccharide and polysaccharide structures help to unravel biosynthetic capabilities in diatoms. Mar Drugs 13:5993–6018

Gunde-Cimerman N, Sonjak S, Zalar P, Frisvad JC, Diderichsen B, Plemenitaš A (2003) Extremophilic fungi in arctic ice: a relationship between adaptation to low temperature and water activity. Phys Chem Earth 28:1273–1278

Gutierrez T, Biller DV, Shimmield T, Green DH (2012) Metal binding properties of the EPS produced by *Halomonas* sp. TG39 and its potential in enhancing trace element bioavailability to eukaryotic phytoplankton. Biometals 25:1185–1194

Hart TD, Chamberlain AHL, Lynch JM, Newling B, McDonald PJ (1999) A stray field magnetic resonance study of water diffusion in bacterial exopolysaccharides. Enzyme Microb Technol 24:339–347

Hart TD, Lynch JM, Chamberlain AHL (2001) Anion exclusion in microbial and soil polysaccharides. Biol Fertil Soil 34:201–209

Hassler CS, Alasonati E, Mancuso Nichols CA, Slaveykova VI (2011) Exopolysaccharides produced by bacteria isolated from the pelagic Southern Ocean – role in Fe binding, chemical reactivity, and bioavailability. Mar Chem 123:88–98

Hassler CS, Norman L, Mancuso Nichols CA, Clementson LA, Robinson C, Schoemann V, Watson RJ, Doblin MA (2015) Iron associated with exopolymeric substances is highly bioavailable to oceanic phytoplankton. Mar Chem 173:136–147

Hayat S, Hayat Q, Alyemeni MN, Wani AS, Pichtel J, Ahmad A (2012) Role of proline under changing environments: a review. Plant Signal Behav 7:1456–1466

Haynes K, Hofmann TA, Smith CJ, Ball AS, Underwood GJC, Osborn AM (2007) Diatom-derived carbohydrates as factors affecting bacterial community composition in estuarine sediments. Appl Environ Microbiol 73:6112–6124

Hill DR, Keenan TW, Helm RF, Potts M, Crowe LM, Crowe JH (1997) Extracellular polysaccharide of *Nostoc commune* (Cyanobacteria) inhibits fusion of membrane vesicles during desiccation. J Appl Phycol 9:237–248

Hoagland KD, Rosowski JR, Gretz MR, Roemer SR (1993) Diatom extracellular polymeric substances: function, fine structure, chemistry, and physiology. J Phycol 29:537–566

Hodson A, Bøggild C, Hanna E, Huybrechts P, Langford H, Cameron K, Houldsworth A (2010) The cryoconite ecosystem on the Greenland ice sheet. Ann Glaciol 51:123–129

Huston AL, Krieger-Brockett BB, Deming JW (2000) Remarkably low temperature optima for extracellular enzyme activity from Arctic bacteria and sea ice. Environ Microbiol 2 (4):383–388

Izutsu K, Yoshioka S, Kojima S (1995) Effect of cryoprotectants on the eutectic crystallization of NaCl in frozen solution studied by differential scanning calorimetry (DSC) and broad-line pulsed NMR. Chem Pharm Bull 43:1804–1806

Janssens J, Meiners KM, Tison J-L, Dieckmann GS, Delille B, Lannuzel D (2016) Incorporation of iron and organic matter into young Antarctic sea ice during its initial growth stages. Elem Sci Anth 4:123

Juhl AR, Krembs C, Meiners KM (2011) Seasonal development and differential retention of ice algae and other organic fractions in first-year Arctic sea ice. Mar Ecol Prog Ser 436:1–16. doi:10.3354/meps09277

Jittawuttipoka T, Planchon M, Spalla O, Benzerara K, Guyot F, Cassier-Chauvat C, Chauvat F (2013) Multidisciplinary evidences that *Synechocystis* PCC6803 exopolysaccharides operate in cell sedimentation and protection against salt and metal stresses. PLoS ONE 8:e55564

Jungblut AD, Hawes I, Mackey TJ, Krusor M, Doran PT, Sumner DY, Eisen JA, Hillman C, Goroncy AK (2016) Microbial mat communities along an oxygen gradient in a perennially ice-covered Antarctic lake. Appl Environ Microbiol 82:620–630

Junge K, Krembs C, Deming J, Stierle A, Eicken H (2001) A microscopic approach to investigate bacteria under in-situ conditions in sea-ice samples. Ann Glaciol 33:304–310

Kehr J-C, Dittmann E (2015) Biosynthesis and function of extracellular glycans in cyanobacteria. Life 5:164–180

Kim SJ, Kim B-G, Park HJ, Yim JH (2016) Cryoprotective properties and preliminary characterization of exopolysaccharide (P-Arcpo 15) produced by the Arctic bacterium *Pseudoalteromonas elyakovii* Arcpo 15. Prep Biochem Biotechnol 46:261–266

Kirst GO, Thiel C, Wolff H, Nothnagel J, Wanzek M, Ulmke R (1991) Dimethylsulfoniopropionate (DMSP) in ice algae and its possible biological role. Mar Chem 35:381–388

Koerstgens V, Flemming H-C, Wingender J, Borchard W (2001) Influence of calcium ions on the mechanical properties of a model biofilm of mucoid *Pseudomonas aeruginosa*. Water Sci Technol 43:49–57

Koh EY, Cowie ROM, Simpson AM, O'Toole R, Ryan KG (2012) The origin of cyanobacteria in Antarctic sea ice:marine or freshwater? Environ Microbiol Rep 4:479–483

Krembs C, Deming JW (2008) The role of exopolymers in microbial adaptation to sea ice. In: Margesin R, Schinner F, Marx J-C, Gerday C (eds) Psychrophiles: from biodiversity to biotechnology. Springer, Berlin, pp 247–264

Krembs C, Engel E (2001) Abundance and variability of microorganisms and TEP across the ice-water interface of melting first-year sea-ice in the Laptev Sea (Arctic). Mar Biol 138:173–185

Krembs C, Eicken H, Junge K, Deming JW (2002) High concentrations of exopolymeric substances in Arctic winter sea ice: implications for the polar ocean carbon cycle and cryoprotection of diatoms. Deep Sea Res 49:2163–2181

Krembs C, Eicken H, Deming JW (2011) Exopolymer alteration of physical properties of sea ice and implications for ice habitability and biogeochemistry in a warmer Arctic. Proc Natl Acad Sci U S A 108(9):3653–3658

Lamelas C, Benedetti M, Wilkinson KJ, Slaveykova VI (2006) Characterization of H+ and Cd2+ binding properties of the bacterial exopolysaccharides. Chemosphere 65:1362–1370

Lannuzel D, Vancoppenolle M, van der Merwe P, de Jong J, Meiners KM, Grotti M et al (2016) Iron in sea ice: review and new insights. Elem Sci Anth 4:130. doi:10.12952/journal.elementa.000130

Lasa I (2006) Towards the identification of the common features of bacterial biofilm development. Int Microbiol 9:21–28

Lavoie D, Denman K, Michel C (2005) Modeling ice algal growth and decline in a seasonally ice-covered region of the Arctic (Resolute Passage, Canadian Archipelago). J Geophys Res Oceans 110:C11009

Leroy F, De Vuyst L (2016) Advances in production and simplified methods for recovery and quantification of exopolysaccharides for applications in food and health. J Dairy Sci 99 (4):3229–3238

Lohmann U (2015) Aerosol-cloud interactions and their radiative forcing. In: North G, Pyle J, Zhang F (eds) Encyclopedia of atmospheric sciences, 2nd edn, vol 1. Academic Press, pp 17–22. doi: 10.1016/B978-0-12-382225-3.00052-9

Lyon BR, Lee PA, Bennett JM, DiTullio GR, Janech MG (2011) Proteomic analysis of a sea-ice diatom: Salinity acclimation provides new insight into the dimethylsulfoniopropionate production pathway. Plant Physiol 157:1926–1941

Mancuso-Nichols CA, Guezennec J, Bowman JP (2005) Bacterial exopolysaccharides from extreme marine environments with special consideration of the southern ocean, sea ice, and deep-sea hydrothermal vents: a review. Mar Biotechnol 7:253–271

Marx JG, Carpenter SD, Deming JW (2009) Production of cryoprotectant extracellular polysaccharide substances (EPS) by the marine psychrophilic bacterium *Colwellia psychrerythraea* strain 34H under extreme conditions. Can J Microbiol 55(1):63–72

McConville MJ, Wetherbee R, Bacic A (1999) Subcellular location and composition of the wall and secreted extracellular sulphated polysaccharides/proteoglycans of diatom *Stauroneis amphioxys* Gregory. Protoplasma 206:188–200

Meiners K, Brinkmeyer R, Granskog MA, Lindfors A (2004) Abundance, size distribution and bacterial colonization of exopolymer particles in Antarctic sea ice (Bellingshausen Sea). Aquat Microb Ecol 35:283–296

Meiners KM, Krembs C, Gradinger R (2008) Exopolymer particules: microbial hotspots of enhanced bacterial activity in Arctic fast ice (Chukchi Sea). Aquat Microb Ecol 52:195–207

Methé BA, Nelson KE, Deming JW, Momen B, Melamud E, Zhang X, Moult J, Madupa R, Nelson WC, Dodson RJ, Brinkac LM, Daugherty SC, Durkin AS, DeBoy RT, Kolonay JF, Sullivan SA, Zhou L, Davidsen TM, Wu M, Huston AL, Lewis M, Weaver B, Weidman JF, Khouri H, Utterback TR, Feldblyum TV, Fraser CM (2005) The psychrophilic lifestyle as revealed by the genome sequence of *Colwellia psychrerythraea* 34H through genomic and proteomic analyses. Proc Natl Acad Sci U S A 102(31):10913–10918

Mock T, Otillar RP, Strauss J, McMullan M, Paajanen P, Schmutz J, Salamov A, Sanges R, Toseland A, Ward BJ, Allen AE (2017) Evolutionary genomics of the cold-adapted diatom *Fragilariopsis cylindrus*. Nature 16:1–5

Mukhopadhyay SK, Chatterjee S, Gauri SS, Das SS, Mishra A, Patra M, Ghosh AK, Das AK, Singh SM, Dey S (2014) Isolation and characterization of extracellular polysaccharide Thelebolan produced by a newly isolated psychrophilic Antarctic fungus *Thelebolus*. Carbohydr Polym 104:204–212

Neftel A, Jacob P, Klockow D (1984) Measurements of hydrogen peroxide in polar ice samples. Nature 311:43–45

Nichols CM, Lardière SG, Bowman JP, Nichols PD, JAE G, Guézennec J (2005) Chemical characterization of exopolysaccharides from Antarctic marine bacteria. Microb Ecol 49:578–589

Niemi A, Michel C, Hille K, Poulin M (2011) Protist assemblages in winter sea ice: setting the stage for the spring ice algal bloom. Polar Biol 34:1803–1817

Noble AE, Moran DM, Allen AE, Saito MA (2013) Dissolved and particulate trace metal micronutrients under the McMurdo Sound seasonal sea ice: basal sea ice communities as a capacitor for iron. Front Chem 1:25

Norman L, Worms IAM, Angles E, Bowie AR, Nichols CM, Ninh Pham A, Slaveykova VI, Townsend AT, Waite TD, Hassler CS (2015) The role of bacterial and algal exopolymeric substances in iron chemistry. Mar Chem 173:148–161

Nwodo UU, Green E, Okoh AI (2012) Bacterial exopolysaccharides: functionality and prospects. Int J Mol Sci 11:14002–14015

Orellana MV, Matrai PA, Leck C, Rauschenberg CD, Lee AM, Coz E (2011) Marine microgels as a source of cloud condensation nuclei in the high Arctic. Proc Natl Acad Sci U S A 108 (33):13612–13617

Pannard A, Pédrono J, Bormans M, Briand E, Claquin P, Lagadeuc Y (2016) Production of exopolymers (EPS) by cyanobacteria: impact on the carbon-to-nutrient ratio of the particulate organic matter. Aquat Ecol 50:29–44

Passow U (2000) Formation of transparent exopolymer particles, TEP, from dissolved precursor material. Mar Ecol Prog Ser 192:1–11

Paterson H, Laybourn-Parry J (2012) Antarctic sea ice viral dynamics over an annual cycle. Polar Biol 35:491–497

Percival E, Rahman MA, Weigel H (1980) Chemistry of the polysaccharide of the diatom *Coscinodiscus nobilis*. Phytochemistry 19:809–811

Petrich C, Eicken H (2017) Overview of sea ice growth and properties. In: Thomas DN (ed) An introduction to its physics, chemistry, biology and geology, 2nd edn. Wiley. in press

Petrou K, Hill R, Brown CM, Campbell DA, Doblin MA, Ralph PJ (2010) Rapid photoprotection in sea-ice diatoms from the East Antarctic pack ice. Limnol Oceanogr 55:1400–1407

Poli A, Anzelmo G, Nicolaus B (2010a) Bacterial exopolysaccharides from extreme marine habitats: production, characterization and biological activities. Mar Drugs 8(6):1779–1802

Poli A, Anzelmo G, Tommonaro G, Pavlova K, Casaburi A, Nicolaus B (2010b) Production and chemical characterization of an exopolysaccharide synthesized by psychrophilic yeast strain *Sporobolomyces salmonicolor* AL1 isolated from Livingston Island, Antarctica. Folia Microbiol 55:576–581

Poulin M, Daugbjerg N, Gradinger R, Ilyash L, Ratkova T, von Quillfeldt C (2011) The pan-Arctic biodiversity of marine pelagic and sea-ice unicellular eukaryotes: a first-attempt assessment. Mar Biodivers 41:13–28

Raymond JA, Fritsen C, Shen K (2007) An ice binding protein from an Antarctic sea ice bacterium. FEMS Microbiol Ecol 61:214–221

Riedel A, Michel C, Gosselin M (2006) Seasonal study of sea-ice exopolymeric substances on the Mackenzie shelf: implications for transport of sea-ice bacteria and algae. Aquat Microb Ecol 45:195–206

Riedel A, Michel C, Gosselin M, LeBlanc B (2007) Enrichment of nutrients and organisms in newly formed sea ice on the Mackenzie shelf: significance of heterotrophic regeneration and exopolymeric substances, nutrients and EPS in newly formed sea ice. Mar Ecol Prog Ser 342:55–67

Rinta-Kanto JM, Sun S, Sharma S, Kiene RP, Moran MA (2012) Bacterial community transcription patterns during a marine phytoplankton bloom. Environ Microbiol 14:228–239

Roberts MF (2005) Organic compatible solutes of halotolerant and halophilic microorganisms. Saline Syst 1(5):1–30

Roca C, Alves VD, Freitas F, Reis MA (2015) Exopolysaccharides enriched in rare sugars: bacterial sources, production, and applications. Front Microbiol 6:288. doi:10.3389/fmicb.2015.00288

Rochera C, Fernández-Valiente E, Van de Vijver B, Rico E, Toro M, Vincent WF, Quesada A, Camacho A (2013) Community structure and photosynthetic activity of benthic biofilms from a waterfall in the maritime Antarctica. Polar Biol 36:1709–1722

Rue EL, Bruland KW (1995) Complexation of iron(III) by natural organic ligands in the Central North Pacific as determined by a new competitive ligand equilibration/adsorptive cathodic stripping voltammetric method. Mar Chem 50:117–138

Santchi PH, Banois E, Wilkinson KJ, Zhang J, Buffle J (1998) Fibrillar polysaccharides in marine macromolecular organic matter as imaged by atomic force microscopy and transmission electron microscopy. Limnol Oceanogr 43:896–908

Schlekat CE, Decho AW, Chandler GT (1998) Sorption of cadmium to bacterial extracellular polymeric sediment coatings under estuarine conditions. Environ Toxicol Chem 17:1867–1874

Schoemann V, Becquevort S, Stefels J, Rousseau V, Lancelot C (2005) *Phaeocystis* blooms in the global ocean and their controlling mechanisms: a review. J Sea Res 53(1):43–66

Selbmann L, Onofri S, Fenice M, Federici F, Petruccioli M (2002) Production and structural characterization of the exopolysaccharide of the Antarctic fungus *Phoma herbarum* CCFEE 5080. Res Microbiol 153:585–592

Shirkey B, Kovarcik DP, Wright DJ, Wilmoth G, Prickett TF, Helm RF, Gregory EM, Potts M (2000) Active Fe-containing superoxide dismutase and abundant sodF mRNA in *Nostoc commune* (Cyanobacteria) after years of desiccation. J Bacteriol 182:189–197

Song W, Zhao C, Zhang D, Mu S, Pan X (2016) Different resistance to UV-B radiation of extracellular polymeric substances of two cyanobacteria from contrasting habitats. Front Microbiol 7:1208

Sun C, Wang J-W, Fang L, Gao X-D, Tan R-X (2004) Free radical scavenging and antioxidant activities of ePS2, an expolysaccharide produced by a marine filamentous fungus *Keissleriella* sp. YS4108. Life Sci 75:1063–1073

Sun M-L, Zhao F, Shi M, Zhang X-Y, Zhou B-C, Zhang Y-Z, Chen X-L (2015) Characterization and biotechnological potential analysis of a new exopolysaccharide from the Arctic marine bacterium Polaribacter sp. SM1127. Scientific Rep 5:18435

Sunda W, Kieber DJ, Kiene RP, Huntsman S (2002) An antioxidant function for DMSP and DMS in marine algae. Nature 418:317–320

Sutherland IW (1972) Bacterial exopolysaccharides. Adv Microb Physiol 8:143–213

Sutton RL (1991) Critical cooling rates to avoid ice crystallization in aqueous cryoprotectant solutions containing polymers. Chem Soc Faraday Trans 87:3747–3751

Takeuchi N, Kohshima S, Seko K (2001) Structure, formation, and darkening process of albedo-reducing material (cryoconite) on a Himalayan glacier: a granular algal mat growing on the glacier. Arct Antarct Alp Res 33:115–122

Taton A, Grubisic S, Balthasart P, Hodgson DA, Laybourn-Parry J, Wilmotte A (2006) Biogeographical distribution and ecological ranges of benthic cyanobacteria in East Antarctic lakes. FEMS Microbiol Ecol 57:272–289

Techtmann SM, Fitzgerald KS, Stelling SC, Joyner DC, Uttukar SM, Harris AP, Alshibli NK, Brown SD, Hazen TC (2016) *Colwellia psychrerythraea* strains from distant deep sea basins show adaptation to local conditions. Front Environ Sci 4:33

Toggweiler JR (1993) Carbon overconsumption. Nature 363:210–211

Torstensson A, Hedblom M, Mattsdotter Björk M, Chierici M, Wulff A (2015) Long-term acclimation to elevated pCO_2 alters carbon metabolism and reduces growth in the Antarctic diatom Nitzschia lecointei. Proc Roy Soc B Biol Sci 282:20151513

Toseland A, Daines SJ, Clark JR, Kirkham A, Strauss J, Uhlig C, Lenton TM, Valentin K, Pearson GA, Moulton V, Mock T (2013) The impact of temperature on marine phytoplankton resource allocation and metabolism. Nat Clim Change 3:979–984

Underwood GJ, Fietz S, Papadimitriou S, Thomas DN, Dieckmann GS (2010) Distribution and composition of dissolved extracellular polymeric substances (EPS) in Antarctic sea ice. Mar Ecol Progr Ser 404:1–9

Underwood GJ, Aslam SN, Michel C, Niemi A, Norman L, Meiners KM, Laybourn-Parry J, Paterson H, Thomas DN (2013) Broad-scale predictability of carbohydrates and exopolymers in Antarctic and Arctic sea ice. Proc Natl Acad Sci U S A 110(39):15734–15739

van der Merwe P, Lannuzel D, Nichols CAM, Meiners K, Heil P, Norman L, Thomas DN, Bowie AR (2009) Biogeochemical observations during the winter–spring transition in East Antarctic sea ice: evidence of iron and exopolysaccharide controls. Mar Chem 115:163–175

Varin T, Lovejoy C, Jungblut AD, Vincent WF, Corbeil J (2012) Metagenomic analysis of stress genes in microbial mat communities from Antarctica and the High Arctic. Appl Environ Microbiol 78:549–559

Verdugo P (2012) Marine microgels. Annu Rev Mar Sci 4:375–400

Verdugo P, Santschi PH (2010) Polymer dynamics of DOC networks and gel formation in seawater. Deep Sea Res II 57(16):1486–1493

Verdugo P, Alldredge AL, Azam F, Kirchman DL, Passow U, Santschi PH (2004) The oceanic gel phase: a bridge in the DOM-POM continuum. Mar Chem 92:67–85

Wang S, Bailey D, Lindsay K, Moore JK, Holland M (2014) Impact of sea ice on the marine iron cycle and phytoplankton productivity. Biogeosciences 11:4713–4731

Wang J, Hu S, Nie S, Yu Q, Xie M (2016) Reviews on mechanisms of in vitro antioxidant activity of polysaccharides. Oxid Med Cell Longev 2016:5692852

Wells LE, Deming JW (2006a) Modeled and measured dynamics of viruses in Arctic winter sea-ice brines. Environ Microbiol 8(6):1115–1121

Wells LE, Deming JW (2006b) Characterization of a cold-active bacteriophage on two psychrophilic marine hosts. Aquat Microb Ecol 45:15–29

Wingender J, Neu TR, Flemming HC (eds) (2012) Microbial extracellular polymeric substances: characterization, structure and function. Springer Science and Business Media, Heidelberg

Yallop ML, Paterson DM, Wellsbury P (2000) Interrelationships between rates of microbial production, exopolymer production, microbial biomass, and sediment stability in biofilms of intertidal sediments. Microb Ecol 39:116–127

Yan M-X, Mao W-J, Liu X, Wang S-Y, Xia Z, Cao S-J, Li J, Qin L, Xian H-L (2016) Extracellular polysaccharide with novel structure and antioxidant property produced by the deep-sea fungus *Aspergillus versicolor* N2bc. Carbohydr Polym 147:272–281

Zhou J, Delille B, Eicken H, Vancoppenolle M, Brabant F, Carnat G, Geilfus NX, Papakyriakou T, Heinesch B, Tison JL (2013) Physical and biogeochemical properties in landfast sea ice (Barrow, Alaska): insights on brine and gas dynamics across seasons. J Geophys Res Oceans 118(6):3172–3189

Chapter 13
Molecular Structure of Lipopolysaccharides of Cold-Adapted Bacteria

Maria Michela Corsaro, Angela Casillo, Ermenegilda Parrilli, and Maria Luisa Tutino

Abstract Cold-adapted Gram-negative bacteria inhabit snow, ice water, and frozen grounds and are perfectly adapted to live at very low temperatures. They have evolved several adaptation strategies, some of which concern the cellular membrane. Bacterial outer membrane, that functions as a selective barrier allowing the influx of nutrients and confers protection to the cell, is also considered a primary sensor of the cold. It is well known that one of the main responses to cold concerning the cell membrane is the increment of fluidity and the upregulation of genes encoding for proteins and membrane transporters. In this context, also the lipopolysaccharides, macromolecules constituting approximately 75% of the outer surface, may be modified in their structure in response to the conditions prevailing in the environment. In this chapter, for the first time, the role played by lipopolysaccharides structures in response to cold adaptation is analyzed. Some structural features of the lipopolysaccharides are modified as a consequence of living at low temperatures, thus confirming that all the outer membrane components are involved in adaptation and survival molecular mechanisms.

Contents

M.M. Corsaro (✉) • A. Casillo • E. Parrilli • M.L. Tutino
Department of Chemical Sciences, University of Naples Federico II, Via Cintia 4, 80126 Naples, Italy
e-mail: corsaro@unina.it

R. Margesin (ed.), *Psychrophiles: From Biodiversity to Biotechnology*,
DOI 10.1007/978-3-319-57057-0_13

13.1 Introduction

The presence of liquid water represents one of the most important parameters for the life on Earth and is obviously related to pressure and temperature. Among these parameters, temperature is a key point (McKay 2014): for example, frozen water impairs the habitability of an environment. Life in cold habitats, such as snow, ice water, and frozen grounds, is still a challenge, as the combination of low temperatures and low liquid water availability contributes to making these regions extremely inhospitable to all forms of existence (Casanueva et al. 2010). Nevertheless, cold environments represent an unexploited reservoir of microbial population. Cold-adapted microorganisms are classified as stenothermal (true or obligate psychrophiles) or eurythermal (facultative psychrophiles) (Feller and Gerday 2003). These microorganisms have evolved a complex range of structural and functional adaptations to survive in harsh conditions. In particular, the membrane, which acts as an interface between the external and internal environment of the cell, is considered one of the primary sensors of the cold. In this chapter, the molecular structures of several lipopolysaccharides of cold-adapted Gram-negative bacteria are described, and their role in the physiological response to cold environments is investigated.

13.2 The Lipopolysaccharide Molecule

Gram-negative bacteria cells are surrounded by a cell wall, which comprises two membranes, an inner and an outer one. These two layers are separated by a thin leaflet of a peptidoglycan, the structure of which is slightly different from that of Gram-positive microorganisms. The inner membrane is mainly constituted by a phospholipid bilayer, whereas the outer membrane (OM) contains also lipopolysaccharides (LPSs).

The LPSs are the major components of the OM of almost all Gram-negative bacteria and of some cyanobacteria (Wilkinson 1977; Lüderitz et al. 1982; Westphal et al. 1986; Carillo et al. 2014), constituting approximately 75% of the outer surface. They are endotoxin heat-stable amphiphilic molecules indispensable for the viability and survival of Gram-negative bacteria, as they heavily contribute to the structural integrity of the OM and to the protection of the bacterial cell envelope (Alexander and Rietschel 2001).

The colony morphology of Gram-negative bacteria can appear as smooth or rough, as a consequence of a different structure of the LPSs, named smooth (S-LPS) or rough (R-LPS), respectively (Fig. 13.1). The structure of a S-LPS molecule can be divided into three covalently linked domains: (1) the glycolipid anchor, called lipid A; (2) the intermediate core oligosaccharide (core); and (3) the O-specific polysaccharide (O-chain) (Caroff and Karibian 2003). Instead, the R-LPSs (also called lipooligosaccharides, LOSs) are completely devoid of the O-specific

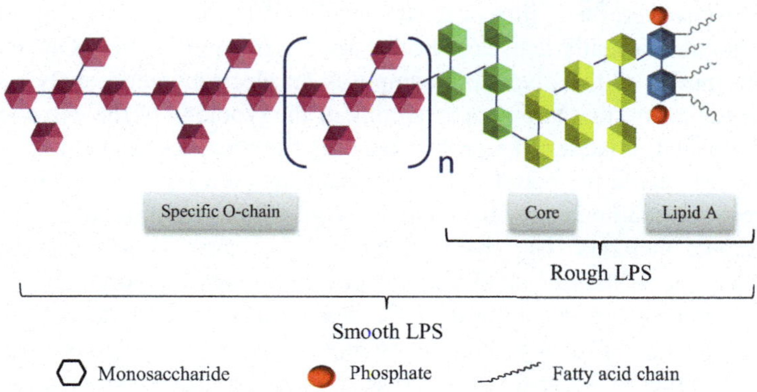

Fig. 13.1 Chemical structure of a LPS from Gram-negative bacteria

polysaccharide chain, either due to genetic mutation or to the inherent nature of bacteria (Lüderitz et al. 1966).

The LPSs of Gram-negative bacteria are better represented as a family of glycoforms, due to the structural variations that occur in each portion of the entire molecule. The LPSs are anchored in the OM through the lipid A, that is the most conserved region even among different species belonging to the same genus (Holst et al. 1996). The phosphate groups, usually located on the lipid A, contribute to maintain the integrity of the OM, through the bridging action of divalent cations, such as Ca^{2+} and Mg^{2+} (Nikaido 2003).

The importance of LPSs mainly lies in their external location, being the closest molecules to the surrounding environment. In particular, in pathogens they are involved in all the aspects of host–bacterium interactions, such as recognition, adhesion, colonization, and virulence, whereas in commensal bacteria they participate in symbiosis and tolerance. It is then reasonable to suppose that LPSs are not only involved in cooperation of surviving under harsh conditions, but also in cold-adaptation mechanisms.

13.3 Methods for Structural Investigation of Lipopolysaccharides

LPSs or LOSs can be isolated from intact bacteria cells by exploiting a variety of solvent extraction methods. Conventional procedures are used for the extraction of S- or R-type LPSs from bacteria, such as the hot phenol/water extraction procedure (Westphal et al. 1952) or the phenol/chloroform/petroleum ether (PCP) method (Galanos et al. 1969). Depending on the polarity of the glycolipid, it can be recovered in the aqueous, phenol, or organic phase. To remove phospholipid components, cell washing with chloroform is usually employed, while enzymatic treatments with nuclease and protease are required to eliminate nucleic acids and proteins. Moreover, the ability of these glycolipids to form micellar aggregates in aqueous solution usually does not allow a trouble-free purification. One method to check the purity of the glycolipid sample is an electrophoretic analysis using a denaturating agent, that allows ascertaining of the typology of the extracted material, too. In fact, a ladder-like profile suggests the presence of a S-LPS, whereas bands exclusively at the bottom of the gel indicate a R-LPS.

Current methodologies to characterize the glycolipid exploit both chemical and spectroscopic methods. The chemical methods are used to establish the monosaccharides types, mainly involving the gas chromatography-mass spectrometry (GC-MS) analysis of sugar derivatives. For qualitative and quantitative sugar analysis, a portion of the glycolipid is usually subjected to a complete degradation with acids, either in water (hydrolysis) or in methanol (methanolysis). After neutralization and concentration, the component sugars are converted into their alditol acetates or acetylated methyl glycosides. The components are separated by GC and identified by their relative retention times and electron ionization-mass spectrometry (EI-MS) fragmentation patterns (Kenne and Lindberg 1983). Methylation analysis is then performed to obtain information about the linkage positions. It starts with a complete methylation of the polysaccharide by the Hakamori method (Hakamori 1964), followed by hydrolysis with an acid to cleave the intact permethylated polysaccharide into simple partially methylated reducing monosaccharides. Following purification by gel chromatography and reduction with sodium borohydride, these partially methylated monosaccharides are converted into open chain alditols, which are subsequently acetylated to give partially methylated alditol acetates (PMAAs). The analysis of the PMAAs is always carried by GC-MS, where examination of the relative retention times and the MS fragmentation patterns unambiguously reveal the positions of attachment and the ring size of the residues in the carbohydrate chain. The absolute configurations of the sugars are normally determined by GC-MS analysis of their acetylated (*R*)-2-butyl (Gerwig et al. 1979) or (*R*)-2-octyl glycoside derivatives (Leontein et al. 1978). The anomeric configurations are usually deduced from ^{13}C nuclear magnetic resonance (NMR) chemical shifts and from the $^{1}J_{H,H}$ coupling constants of the anomeric signals in the proton spectrum of the intact molecule.

Due to the amphiphilic nature of the LPSs, it is convenient to separate the lipidic from the saccharidic portion to achieve a better solubility. Two different approaches can be used, which lie in an acid or an alkaline hydrolysis.

The mild acid treatment exploits the labile glycosidic linkage of the 3-deoxy-D-*manno*-oct-2-ulosonic acid (Kdo) residue, which links the core region to the lipid A. This kind of procedure allows the isolation of the intact lipid A and of the oligo/polysaccharidic portion (Fig. 13.2). The reaction conditions lead to dehydration of Kdo with consequent formation of artifacts (Volk et al. 1972). This increases the sample heterogeneity and can make the interpretation of NMR spectra arduous, because the artifacts can be present in a not negligible amount.

The alkaline treatment consists of a first *O*-deacylation of the LPSs, and a successive *N*-deacylation, which leads to the isolation of an oligo/polysaccharide containing the glycosidic portion of the lipid A (Holst 2000). The reaction

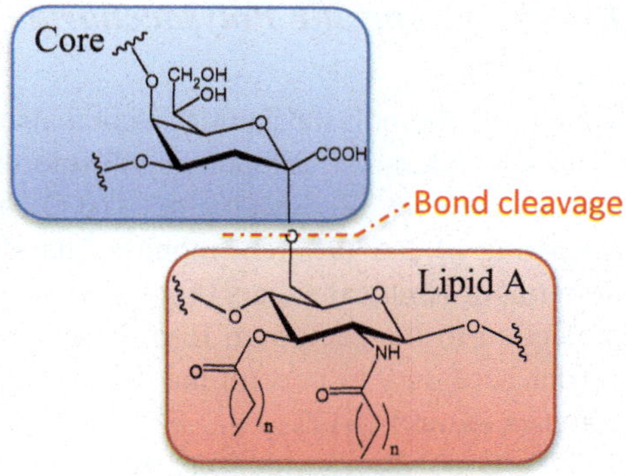

Fig. 13.2 Schematic illustration of the Kdo linkage cleavage during the mild acid treatment

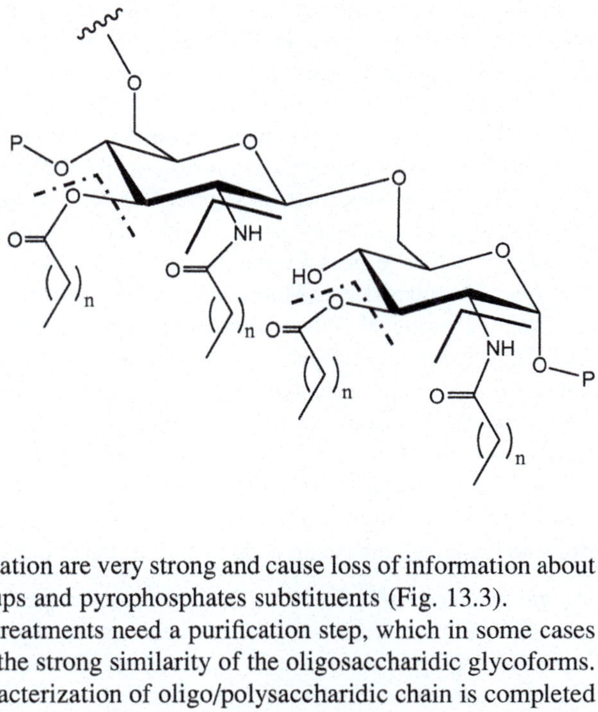

Fig. 13.3 Schematic representation of the lipid A deacylation. *Dotted* and *solid lines* indicate the linkages hydrolyzed during treatment with hydrazine or KOH, respectively

conditions for the *N*-deacylation are very strong and cause loss of information about the presence of acetyl groups and pyrophosphates substituents (Fig. 13.3).

Both acid and alkaline treatments need a purification step, which in some cases could be very hard, due to the strong similarity of the oligosaccharidic glycoforms. After purification, the characterization of oligo/polysaccharidic chain is completed by $^{1}H/^{13}C$ mono- and two-dimensional NMR spectroscopy and by Mass Spectrometry.

13.4 Cold-Adapted Lipopolysaccharides

The single components of the LPS molecules isolated from several strains of cold-adapted bacteria are described below.

13.4.1 O-Specific Polysaccharide Structures

The O-specific polysaccharide, named also O-antigen or O-polysaccharide, represents the outermost component of smooth LPSs, linked to the lipid A moiety through the core oligosaccharide. The O-chain is made of oligosaccharidic repeats consisting of 2–8 different monosaccharide residues (heteroglycans) or, in some bacteria, of identical sugars (homoglycans) (Knirel 2011). The O-antigen provides protection to the bacterium from host immune system and constitutes a physical barrier necessary to prevent the cellular dehydration and promote the adhesion of colonies (Knirel 2011).

The O-polysaccharide represents the most variable portion of the LPSs, depending on the composition, the structure, and the length. The variability is related to the occurrence of monosaccharides widely distributed in nature together with unusual sugars (Jansson 1999) and for the presence of uncommon nonsugar constituents. These decorations range from sulfate and phosphate groups to amino acids, lactic, and hydroxybutyric acids (Jansson 1999; Whitfield 2006). Sometimes these substitutions are in a non-stoichiometric amount and complicate the identification of a precise O-repeating unit (Raetz and Whitfield 2002).

The number of identified LPSs produced by cold-adapted bacteria is constantly increasing, although many of these lack the O-antigen portion. The O-antigen structures that have been currently elucidated belong to LPSs from the bacterial families of Moraxellaceae, Moritellaceae, Alteromonadaceae, and Flavobacteriaceae.

The first group of polysaccharides here reported were isolated from LPSs of bacteria belonging to the genus *Acinetobacter*. Bacteria of this genus are commonly found as soil organisms (Gerischer 2008), and recently the structures of the O-polysaccharide from the LPSs of cold-tolerant bacteria isolated from Siberian permafrost soil were elucidated. Among the genus *Acinetobacter*, the species *Acinetobacter lwoffii* EK30A, *Acinetobacter* sp. VS-15, and *A. lwoffii* EK67 were described. Interestingly, all of them show in their repeating unit acidic residues and substituents that increase the total negative charge density of the O-antigen moiety.

***Acinetobacter lwoffii* EK30A** The strain was isolated from 1.6–1.8 million-year-old Siberian permafrost subsoil sediments (Arbatsky et al. 2010) and produces a smooth lipopolysaccharide at 25 °C. After purification and mild acid hydrolysis, an O-chain with a tetrasaccharide repeating unit containing galactose (Gal), 2-amino-2-deoxy-galactose (GalN), 2-amino-2,6-dideoxy-galactose (FucN), and 4-amino-4,6-dideoxy-glucose (Qui4N) was isolated. The significance of this structure is the presence of novel derivatives of 4-amino-4,6-dideoxy-D-glucose with N-acetyl-D-homoserine and N-[(*S*)-3-hydroxybutanoyl]-D-homoserine.

→3)-β-D-Qui*p*4NAcyl-(1→6)-α-D-Gal*p*-(1→4)-α-D-Gal*p*NAc-(1→3)-α-D-Fuc*p*NAc-(1→
Acyl=N-acetyl-orN-[(*S*)-3-hydroxybutanoyl]-D-homoseryl

***Acinetobacter* sp. VS-15 and *A. lwoffii* EK67** This structure was found to be very different from those reported for O-polysaccharides of *Acinetobacter* sp. VS-15 and *A. lwoffii* EK67, both isolated from 2 to 3 million-year-old Siberian permafrost and structurally identical to each other. In fact, the O-polysaccharide of both strains displays a pentasaccharide repeating unit containing only neutral sugars, such as 6-deoxy-mannose (Rha), Gal, GalN, 2-amino-2-deoxy-glucose (GlcN), and QuiN.

$$\begin{array}{c} \rightarrow 2)\text{-}\alpha\text{-L-Rha}p\text{-}(1 \rightarrow 6)\text{-}\alpha\text{-D-Gal}p\text{-}(1 \rightarrow 4)\text{-}\alpha\text{-D-Gal}p\text{NAc-}(1 \rightarrow 3)\text{-}\alpha\text{-D-Qui}p\text{NAc-}(1 \rightarrow \\ 3 \\ \uparrow \\ 1 \\ \beta\text{-D-Glc}p\text{NAc} \end{array}$$

***Psychrobacter cryohalolentis* K5^T^** The *Psychrobacter* genus comprises psychrophilic to mesophilic, halotolerant, aerobic, nonmotile, Gram-negative coccobacilli (Bowman 2005). These bacteria live in extremely cold habitats, such as Antarctic ice, soil, and sediments, as well as in deep sea environments (Maruyama et al. 2000; Romanenko et al. 2002). Studies on *Psychrobacter* macromolecules structures and metabolic pathways may aid in research on potential extraterrestrial cryo-dwelling organisms and provide better insight into the evolution of microbes (Kondakova et al. 2012a).

Psychrobacter cryohalolentis K5^{T} was recovered from permafrost samples within the Kolyma lowland region of Siberia (Gilichinsky et al. 1992) and grown at 24 °C. On the basis of structural characterization, the polysaccharide was found to include D-Gal and L-Rha, together with several di-amino and tri-amino sugars, including: 2,4-diamino-2,4,6-trideoxy-glucose (Qui2,4N), 2,3-diamino-2,3-dideoxy-glucuronic acid (Glc2,3NA), and 2,3,4-triamino-2,3,4-trideoxy-arabinose (Ara2,3,4N), which has not been hitherto reported in natural carbohydrates.

The polysaccharide is neutral, as the constituent acidic monosaccharide Glc2,3NA occurs in the amide form (Glc2,3NAN). Another peculiar feature of this polysaccharide is the presence of a novel unique triamino sugar, 2,3,4-triamino-2,3,4-trideoxy-arabinose, which occupies the terminal position in a rather long, tetrasaccharide side chain. The terminal monosaccharides of carbohydrate antigens are most accessible to environmental factors, such as immune system and bacteriophages; their diversity provides specificity to the bacterial cell surface and is believed to be important for bacterial survival and niche adaptation (Kondakova et al. 2012a).

$$\begin{array}{r} \beta\text{-L-Ara}p2,3,4\text{NAc-}(1 \rightarrow 4)\text{-}\beta\text{-D-Glc}p2,3\text{NAcAN-}(1 \rightarrow 3)\text{-}\alpha\text{-L-Rha}p^{\text{II}}\text{-}(1 \rightarrow 2)\text{-}\alpha\text{-L-Rha}p^{\text{I}} \\ 1 \\ \downarrow \\ 4 \\ \rightarrow 3)\text{-}\beta\text{-D-Qui}p2,4\text{NAc-}(1 \rightarrow 3)\text{-}\alpha\text{-D-Gal}p\text{-}(1 \rightarrow \end{array}$$

***Psychrobacter muricolla* 2pS^T^** The strain was isolated from the same habitat as reported for the previous bacterium and produces a polysaccharide displaying a disaccharide repeating unit. The peculiarity of this structure is the presence of the amino acid glycine substituting the position C-6 of the 2-amino-2-deoxy-guluronic acid (GulNA) residue, in turn linked to a D-GlcNAc residue, resulting in an acidic polysaccharide (Kondakova et al. 2012b).

$$\rightarrow 4)\text{-}\alpha\text{-L-Gul}p\text{NAcA6Gly-}(1 \rightarrow 3)\text{-}\beta\text{-D-Glc}p\text{NAc-}(1 \rightarrow$$

***Moritella viscosa* strain M2-226** The Moritellaceaea family embraces the genera *Moritella* and *Paramoritella*. All species, isolated from marine environments, were characterized as halophilic anaerobes. In addition, the genus *Moritella* consists solely of psychrophilic species (Hidetoshi 2014). *Moritella viscosa* strain M2-226 is a psychrophilic bacterium isolated from Atlantic salmon in Norway, Iceland, and Scotland (Benediktsdottir et al. 1998) and is considered responsible for the salmon winter ulcer. The strain was grown at 12 °C and was found to produce a low amount of smooth LPSs under these conditions. However, mild acid hydrolysis afforded, after purification, an O-polysaccharide with a trisaccharide repeating unit, containing D-GlcNAc, D-glucuronic acid (D-GlcA), and L-Fuc.

$$\rightarrow 3)\text{-}\beta\text{-D-Glc}p\text{NAc-}(1 \rightarrow 4)\text{-}[\alpha\text{-D-Glc}p\text{A-}(1 \rightarrow 3)]\text{-}\alpha\text{-L-Fuc}p\text{-}(1 \rightarrow$$

***Flexibacter psychrophilum* strain 259-93** The genera belonging to the Flavobacteriaceae family have colonized diverse ecological niches (Bernardet and Nakagawa 2006), and the only genus *Flavobacterium* includes more than 106 validly described species (Verma and Rathore 2015). *Flavobacterium psychrophilum* is the etiological agent of rainbow trout fry syndrome and bacterial cold water disease, septicemic infections that can cause significant early losses in hatchery-reared salmonids (MacLean 2001). Several immunogenic cell surface molecules that may be involved in pathogenesis, including LPSs, were identified as potential vaccine candidates (Crump et al. 2001). The structure of the antigenic O-polysaccharide contained in the LPS of *Flexibacter psychrophilum* strain 259-93 was found to be an unbranched polymer of trisaccharide repeating units composed of L-Rha, L-FucNAc, and D-Qui2,4N, where R is 3*S*,5*S*-3,5-dihydroxyhexanoyl (MacLean 2001).

$$\rightarrow 4)\text{-}\alpha\text{-L-Fuc}p\text{NAc-}(1 \rightarrow 3)\text{-}\alpha\text{-D-Qui}p\text{2NAc4NR-}(1 \rightarrow 2)\text{-}\alpha\text{-L-Rha}p\text{-}(1 \rightarrow$$
$$R = (3S, 5S)\text{-}CH_3CH(OH)CH_2CH(OH)CH_2CO\text{-}$$

***Idiomarina zobellii* KMM 231^T** *Idiomarina zobellii* KMM 231^T was isolated from a seawater sample taken at a depth of 4000 m in the Pacific Ocean; this deep-sea strain was found to be halophilic and psychrotolerant. On the basis of chemical and spectroscopic analyses, it was concluded that a pentasaccharide is the repeating unit of the O-antigen from *Idiomarina zobellii* KMM 231^T lipopolysaccharide. The polysaccharide is distinguished by the presence of both uronic acids and amino sugars. In addition, another peculiarity is the absence of acyl substituents on two amino sugars of the repeating unit, thus conferring both positive and negative charges to the entire polysaccharide (Kilcoyne et al. 2004).

→3)-α-D-Qui*p*4N-(1→4)-α-D-Glc*p*A-(1→6)-α-D-Glc*p*NAc-(1→4)-α-L-Gul*p*NA-(1→3)-α-D-Fuc*p*NAc-(1→

13.4.2 Core Structures

The core moiety consists of a linear or branched oligosaccharide that joins up the lipid A to the polysaccharidic portion. In R-LPSs, the core constitutes the most external region of the entire molecule, thus playing a key role in the bacterial cell interaction with the external environment. The core structures of pathogenic bacteria typically contain 8–12 monosaccharides and in most cases are characterized by the presence of Kdo (2→6)-linked to GlcNII of lipid A (Caroff and Karibian 2003). In addition, the majority of core regions possess L-*glycero*-D-*manno*-heptose (L,D-Hep) substituting the Kdo in turn linked to the lipid A. These peculiar monosaccharides identify the inner core; the outer core, instead, is built up by common hexose monosaccharides. The currently characterized core structures from cold-adapted microorganisms belong to those of the genera *Alteromonas*, *Psychromonas*, *Colwellia*, and *Psychrobacter*. Actually, none of these glycolipids show smooth LPSs.

The genus *Alteromonas*, belonging to the family Alteromonadaceae, was established by Baumann and coworkers (Baumann et al. 1972) for marine Gram-negative heterotrophic bacteria. Later, the genus *Alteromonas* was revised in 1995 to contain only one species, *A. macleodii*, while the remaining species were reclassified as *Pseudoalteromonas* (Bosi et al. 2015).

The marine Gammaproteobacteria *Pseudoalteromonas haloplanktis* strains TAC 125 and TAB 23 (Feller et al. 1992) were isolated from Antarctic coastal sea water sample collected in the vicinity of the French Antarctic station Dumont d'Urville, Terre Adelie. The LPSs isolated from both bacteria grown at 15 °C are rough and share the same oligosaccharidic skeleton, except for nonreducing end monosaccharide, that it is a 2-amino-2-deoxy-mannose (ManN) in *Ph*TAC 125 whereas it is a galactose in *Ph*TAB 23 (Corsaro et al. 2001; Carillo et al. 2011). This is not surprising, as they are phylogenetically close (Bosi et al. 2015). In addition, both LOSs show the presence of the D-*glycero*-D-*manno*-heptose, a feature never revealed in the genus *Pseudoalteromonas* (Silipo et al. 2004). An interesting difference between the two strains is the phosphorylation pattern, as the structure from *Ph*TAB 23 shows up to five phosphate groups, whereas in *Ph*TAC 125 only three were found. This is an important feature, in consideration of the role that the phosphorylation plays in cold adaptation (Ray et al. 1994; Kumar et al. 2002).

***Pseudoalteromonas haloplanktis* TAC 125:**

β-D-Gal*p*3R^{I}(1 → 4)-α-D, D-Hep*p*-(1 → 5)-α-D-Kdo*p*4*P*, 8R-(2 →
R = H_2PO_3 or H
R^{I} = α-D-Man*p*NAc

***Pseudoalteromonas haloplanktis* TAB 23**:

α-D, D-Hep*p*3R, 6R, 4R^{I}-(1 → 5)-α-Kdo*p*4*P*-(2 →
R = H_2PO_3 or H
R^{I} = *α*-Gal*p*-(1 → 4)-β-Gal*p*-(1 → or H

Psychromonas arctica The LPSs from *Psychromonas arctica* grown at 4 °C also showed to be rough, but with a longer oligosaccharidic portion, when compared to *Pseudoalteromonas haloplanktis* strains. The occurrence in this structure of two residues of galacturonic acid, together with five phosphate groups, indicates for this structure a very high negative charge density (Corsaro et al. 2008). In addition, another intriguing difference with respect to the LOSs of *P. haloplanktis* strains is the presence of a terminal β-fructofuranose residue in *P. arctica*. The occurrence of such a monosaccharide as terminal residue is quite rare in this type of structure and has been found only in the core of LPSs from *Vibrio cholerae* strains (Vinogradov et al. 1995; Knirel et al. 1997).

R^{I}-P ─┐
3
α-D-GalA*p*3*P*(1 → 2)-α-L, D-Hep*p*-(1 → 5)-α-D-Kdo*p*4*P*-(2 → 6)-β-D-GlcN4*P*-(1 → 6)-α-D-GlcN1*P*
4
R^{I}-β-D-Glc*p* ─┘

R = α-D-GalA*p*-(1 → or H
R^{II} = β-D-Fru*f*-(2 → 4) or H

***Colwellia psychrerythraea* 34H** proved to be an obligate psychrophile (Ewert and Deming 2013). It has been isolated from enriched Arctic marine sediments at −1 °C, and it is considered a model organism for cold-adaptation mechanism (Boetius et al. 2015). The LOS produced from *Colwellia* is almost unique, as it does not show any heterogeneity relatively to the carbohydrate backbone (Carillo et al. 2013). Indeed, it consists of only one glycoform, where the more frequently found heptose linked to the inner core Kdo is substituted by a mannose. This structural feature is commonly found in the Rhizobiaceae family, but has never been found before in extremophiles. In addition, the structure displays the unusual substitution of a phosphoglycerol moiety (GroP), which has sometimes found in O-chain structures. Finally, differently from *Pseudoalteromonas* structures, *Colwellia* shows a high negative charge density.

α-L-Col*p*-(1 → 2)-α-D-Gal*p*-(1 → 2)-α-D-Man*p*-[3*P*-D-Gro]-(1 → 5)-α-D-Kdo*p*4*P*-(2 →

Psychrobacter arcticus **strain 273-4** is a Gram-negative bacterium isolated from a 20,000–30,000 years-old continuously frozen permafrost horizon in the Kolyma region in Siberia. The strain has been chosen as a model for cold-adaptation mechanism in permafrost, due to its growth at subzero temperatures and widespread prevalence (Ayala-del-Río et al. 2010). The LOS isolated from the cells of this bacterium grown at 4 °C has been characterized after acetic acid hydrolysis and removing of lipid A (Casillo et al. 2015). The great majority of the residues in the core oligosaccharides is constituted by neutral sugars, with a muramic acid, a typical component of bacterial cell-wall peptidoglycan, as terminal nonreducing residue.

```
                                                                      β-D-Glcp
                                                                         1
                                                                         ↓
                                                                         6
α-D-NAM-(1 → 3)-α-L-Rhap-(1 → 3)-β-D-Galp-(1 → 4)-β-D-Glcp-(1 → 4)-β-D-Glcp-(1 → 5)-α-D-Kdop4P-(2 →
                                                                         4                4
                                                                         ↑                ↑
                                                                         1                2
                                                                      β-D-Glcp        α-D-Kdop
```

This feature has been found up to now only in O-chain structures. Other elements quite unusual with respect to the marine cold-adapted LOSs are the substitution of the first Kdo with a second unit of Kdo, which is characteristic of enteric bacteria, and with a glucose unit instead of the heptose. The lack of this residue has been found so far in the Moraxellaceae (Masoud et al. 1994) and Rhizobiaceae families (Carlson and Krishnaiah 1992; Forsberg and Carlson 1998).

13.4.3 Lipid A

Lipid A moiety displays a quite conservative structure, even if there is a significant structural variability among Gram-negative bacteria, even within a single genus or species of the Gammaproteobacteria (Montminy et al. 2006). The lipid A, which represents the hydrophobic anchor of LPSs in the OM, is the endotoxic principle of the entire molecule; indeed, it acts as a potent elicitor of the host innate immune system *via* Toll-like receptor 4/myeloid differentiation factor 2 (TLR4/MD2) receptor complex.

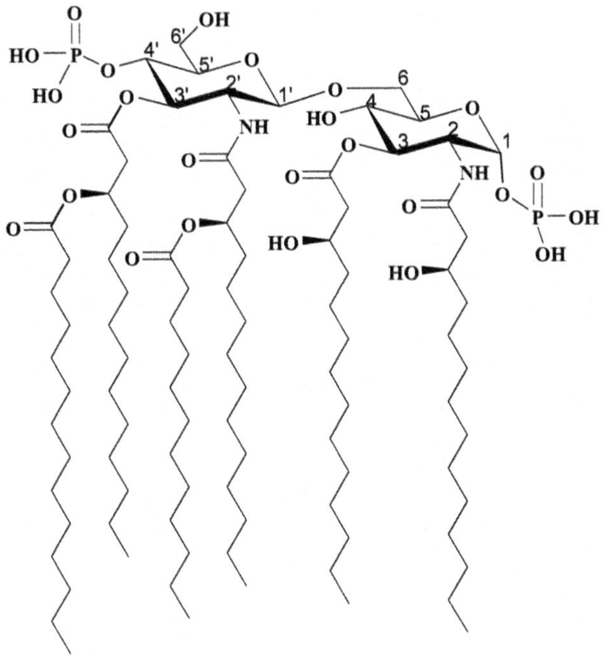

Fig. 13.4 Lipid A from *Escherichia coli*

Chemically, lipid A consists of a typical structure characterized by a β-(1→6)-linked amino sugar disaccharide backbone which, in most cases, comprises two D-Glc*p*N. Usually, lipid A is phosphorylated at position 1 of the proximal α-Glc*p*N (GlcNI) and 4′ of the distal β-Glc*p*N (GlcNII), respectively, and both amino sugar units are substituted with ester- and amide-linked fatty acids (primary fatty acids). Some of them are peculiar because they bear a β-hydroxyl function, which can be esterified in turn by other fatty acid residues (secondary fatty acids) (Fig. 13.4).

Sometimes, one of the phosphate groups may be absent or can be replaced with another substituent (Holst and Molinaro 2009). As the lipid A portion is totally embedded in the OM, it is crucial to understand the involvement of its chemical structure in the maintenance of membrane integrity. At low temperatures, in bacteria that are not naturally adapted to cold (mesophiles such as *E. coli* and *B. subtilis*), cold shock induces homeoviscous adaptation of phospholipid acyl chains, with the increased incorporation of unsaturated, short, and/or branched fatty acids. This response is required to maintain membrane fluidity as part of a set of physiological actions necessary to the survival in cold conditions (Carty et al. 1999; Graumann and Marahiel 1999). Despite a thorough biochemical study of lipid A in the Enterobacteriaceae, little is known about its detailed structure in psychrophilic species. The only lipid A structures from cold-adapted bacteria, reported up to now, belong to the *Acinetobacter*, *Pseudomonas, Psychromonas,* and *Colwellia* genera.

Psychromonas marina and *Psychrobacter arcticus* are two psychrophiles taxonomically divergent, though both were originally cultured from cold marine

Table 13.1 Distribution of acyl and 3-acyloxacyl residues over hydroxyl and amino groups of the lipid A backbone of cold-adapted bacteria in the main lipid A fraction

	GlcNII			GlcNI		
Bacteria	O-4′	O-3′	N-2′	O-3	N-2	O-1
P. cryohalolentis	P[a]	12:0 [3-O (10:0)]	12:0 [3-O (10:0)]	12:0 (3-OH)	12:0 (3-OH)	P
P. arcticus 273-4	P	12:0 [3-O (10:0)]	12:0 [3-O (10:0)]	12:0 (3-OH)	12:0 (3-OH)	P
P. marina	P	14:0 [3-O (12:0)]	14:0 [3-O (14:2)]	14:0 (3-OH)	14:0 (3-OH)	P
P. arctica	P	14:0 [3-O (12:0)[b]]	14:0 (3-OH)	14:0 [3-O (14:1)[b]]	14:0 (3-OH)	P
C. psychrerythraea 34H	P	12:0 (3-OH)	14:1 [3-O (12:0)]	12:0 [3-O(GroP (10:0)(16:0))]	12:0 (3-OH)	P
C. piezophila	P	14:0	11:0 [3-O (10:0)]	14:0	11:0 (3-OH)	P
C. hornerae	P	14:0	11:0 [3-O (12:0)]	14:0	11:0 (3-OH)	P
P. haloplanktis TAB 23	P	12:0 (3-OH)	12:0 [3-O (12:0)]	12:0 (3-OH)	12:0 (3-OH)	P
P. haloplanktis TAC 125	P	12:0 (3-OH)	12:0 [3-O (12:0)]	12:0 (3-OH)	12:0 (3-OH)	P

[a]P indicates a phosphate group; [b]The position of these acyl chains is interchangeable

environments (*P. marina* from −1 °C sea water and *P. cryohalolentis* from a cryopeg embedded in a marine region of Siberian permafrost), and both are capable of growth at temperatures below 0 °C (Kawasaki et al. 2002; Bakermans et al. 2006).

Psychrobacter cryohalolentis (Sweet et al. 2015) and *Psychrobacter arcticus* 273-4 (data not published) lipid A structures display features that are related to the increasing of membrane fluidity. The lipid A of these bacteria displays a high degree of acyl variations, including shorter acyl chains and odd-chain acyl variants. The most abundant species, a hexa-acylated lipid A, displays four 3-hydroxydodecanoyl residues [C14:0(3-OH)] as primary fatty acids and two decanoyl residues (C10:0) as secondary fatty acids at the positions 2′ and 3′ of GlcNII (Table 13.1).

P. cryohalolentis lipid A structure was compared to that of a closely related mesophile, *Acinetobacter baumannii*, to evaluate the effect due to the temperature lowering. Looking at the hexa-acylated species, the two reported structures differ significantly in both primary and secondary fatty acids substitution. The overall results suggest that the psychrotolerant *P. cryohalolentis* counteracts the temperature lowering by producing shorter acyl chains.

With the same approach, the lipid A structure of *Psychromonas marina* was compared with that of the closely related mesophile *E. coli*. The *P. marina* penta-acylated lipid A structure displays, as that of *E. coli*, four C14:0(3-OH) as primary fatty acids and one dodecanoyl residue (C12:0) as acyoxacyl at 2′ position, whereas the hexa-acylated one shows a tetradecenoic (C14:1) as additional secondary fatty acid, in contrast to a tetradecanoic (C14:0) in the mesophile (Sweet et al. 2014). The lipid A of another bacterium belonging to the *Psychromonas* genus, *Psychromonas arctica*, is considered. This psychrotolerant seawater bacterium produces a lipid A structure containing unsaturated fatty acids as secondary acyl chain, even if the exact distribution is not yet defined.

The family Alteromonadaceae embraces different genera, including *Pseudoalteromonas* and *Colwellia*, and comprises an essential part of the marine microbial population with very different habitats. Regarding the *Colwellia* genus, the lipid A structures currently elucidated belong to *C. hornerae* (Bowman et al. 1998), *C. piezophila* (Nogi et al. 2004), and *C. psychrerythraea* 34H (Methé et al. 2005) (Table 13.1).

Colwellia psychrerythraea 34H lipid A displays some features related to its ability to enhance the fluidity of the membrane (unpublished results). In particular, the acylation pattern of primary fatty acids is asymmetric, being the reducing GlcN substituted at positions 2 and 3 by two 3-hydroxydodecanoic acids [C12:0(3-OH)] while the distal GlcN carries a C12:0(3-OH) and a 3-hydroxy-tetradecenoic acid [C14:1Δ^4(3-OH)] linked at positions 3′ and 2′, respectively. In addition, this structure displays a GroP moiety on a secondary acylation site at 3-position of the reducing GlcN. To our knowledge, these characteristics were found only in the lipid A of the closely related mesophile *Vibrio fischeri* (Phillips et al. 2011). A comparison of *C. psychrerythraea* 34H lipid A with the structures of lipid A from *C. piezophila* and *C. hornerae* revealed substantial differences, of which the most important is the absence of the phosphoglycerol moiety and unsaturated fatty acids. Instead, *C. piezophila* and *C. horneraea* lipid A structures share the same pattern of primary fatty acids, that are 3-hydroxyundecanoic [C11:0(3-OH)] linked through amide bonds, while those linked through ester bonds are tetradecanoic acids (Table 13.1). The two structures differ only for the secondary acylation pattern, that is a decanoic acid and a dodecanoic ones in *P. piezophila* and *P. hornerae* species, respectively.

P. haloplanktis TAC 125 (Corsaro et al. 2002) and *P. haloplanktis* TAB 23 (Carillo et al. 2011) lipid A structures are the only structures, of this species, currently characterized. The lipid A structures of these bacteria contain the classical bisphosphorylated glucosamine disaccharide. The structures display the presence of 3-hydroxydodecanoyl residues linked both as esters and amides at 2′ and 3′ (GlcN II), and 2 and 3 positions (GlcN I) of the sugar backbone, respectively. The only difference between the main structures of *P. haloplanktis* TAB 23 and TAC 125 is the position of the secondary fatty acid, constituted by a dodecanoyl residue. Actually, in both cases it is localized on GlcNII, being acyloxyamide and acyloxyacyl linked in TAB 23 and TAC 125 structures, respectively.

13.5 The Role of Lipopolysaccharides in Cold Adaptation

Gram-negative psychrophiles can be isolated from sediments, marine water, and sea-ice, and, for this reason, they require seawater-based culture media for their optimal growth. In addition, the temperature parameter clearly influences microbial growth. Then, the best way to establish if the LPS structure is correlated with cold adaptation is to grow the bacterial cells at temperatures corresponding to those of the natural environment of the strains.

Many papers report on the relationship between chilling and maintenance of membrane integrity of Gram-negative bacteria. Indeed, membrane alterations seemed to be the principal cause of the cold shock injury during a rapid temperature variation (Cao-Hoang et al. 2010). Then, one physiological response to the cold environment is the alteration of membrane components, such as the presence of unsaturated and branched fatty acids in phospholipids that maintain membrane fluidity (Chattopadhyay 2006). When *Escherichia coli* cells are grown at 12 °C, they produce lipid A structures displaying an almost complete substitution of the C12:0 with hexadecenoic acid (C16:1) (Carty et al. 1999). This replacement of laurate with palmitoleate in lipid A may reflect the desirability of maintaining the optimal fluidity of the OM, and lowers its phase transition temperature, counteracting the effect of low temperature.

Consistent with these results are the structures of the lipid A of the obligate psychrophiles *Colwellia psychrerythraea* 34H and *Psychromonas marina*, which display unsaturated acyl chains both as primary and/or secondary fatty acids. In addition, all the cold-adapted currently studied lipid As show shortened acyl chains with respect to mesophiles. The only current case of a lipid A structure of eurythermals displaying unsaturated fatty acids chains is that of *Psychromonas arctica*. An explanation could be the fact the *P. arctica* was isolated from cold marine environments, which are usually dominated by stenothermal microorganisms.

All the core structures currently characterized belong to R-LPS and display a high negative charge density. This last feature can be related to the capability to sequester divalent cations, the concentration of which is higher in cold sea-water and in brine channels. In addition, the appearance of shortened LPS molecules could enhance the flexibility and stability of the outer membrane.

It is tempting to speculate that the growth temperature influences the production of S-LPS and/or R-LPS molecules. Indeed, all the characterized O-chains here reported were isolated from bacteria grown at temperatures above 20 °C. A proven evidence of this correlation is *Psychromonas arctica*, the LOS molecule of which is longer when the temperature shifts from 4 °C to 20 °C.

13.6 Conclusions

All the molecular components of the outer membrane of Gram-negative bacteria are entangled in the adaptation to extreme environments, including LPSs. In this chapter, the involvement of each of the three domains of this complex glycolipid has been described. The characterization of other LPSs from cold-adapted bacteria together with an analysis of their physicochemical properties can shed light on a better comprehension of cold-adaptation involvement of these membrane constituents.

References

Alexander C, Rietschel ET (2001) Bacterial lipopolysaccharides and innate immunity. J Endotoxin Res 7:167–202

Arbatsky NP, Kondakova AN, Shashkov AS et al (2010) Structure of the O-antigen of *Acinetobacter lwoffii* EK30A; identification of D-homoserine, a novel non-sugar component of bacterial polysaccharides. Org Biomol Chem 8:3571–3577

Ayala-del-Río HL, Chain PS, Grzymski JJ et al (2010) The genome sequence of *Psychrobacter arcticus* 273-4, a psychroactive Siberian permafrost bacterium, reveals mechanisms for adaptation to low-temperature growth. Appl Environ Microbiol 76:2304–2312

Bakermans C, Ayala-del-Rio HL, Ponder MA et al (2006) *Psychrobacter cryohalolentis* sp. nov. and *Psychrobacter arcticus* sp. nov., isolated from Siberian permafrost. Int J Syst Evol Microbiol 56:1285–1291

Baumann L, Baumann P, Mandel M et al (1972) Taxonomy of aerobic marine Eubacteria. J Bacteriol 110:402–429

Benediktsdottir E, Helgason S, Sigurjonsdottir H (1998) *Vibrio* spp. isolated from salmonids with shallow skin lesions and reared at low temperature. J Fish Dis 21:19–28

Bernardet JF, Nakagawa Y (2006) An introduction to the family Flavobacteriaceae. In: Dworkin M, Falkow S, Rosenberg E, Schleifer K-H, Stackebrandt E (eds) The prokaryotes, vol 7, 3rd edn. Springer, New York, NY, pp 455–480

Boetius A, Anesio AM, Deming JW et al (2015) Microbial ecology of the cryosphere: sea ice and glacial habitats. Nat Rev Microbiol 13:677–690

Bosi E, Fondi M, Maida I et al (2015) Genome-scale phylogenetic and DNA composition analyses of Antarctic *Pseudoalteromonas* bacteria reveal inconsistencies in current taxonomic affiliation. Hydrobiologia 761(1):85–95

Bowman JP (2005) The genus *Psychrobacter*. In: Dworkin M (ed) The prokaryotes: an evolving electronic resource for the microbiological community, vol 6, 3rd edn. Springer, New York, NY, pp 920–930

Bowman JP, Gosink JJ, McCammon SA et al (1998) *Colwellia demingiae* sp. nov., *Colwellia hornerae* sp. nov., *Colwellia rossensis* sp. nov. and *Colwellia psychrotropica* sp. nov.: psychrophilic Antarctic species with the ability to synthesize docosahexaenoic acid. Int J Syst Bacteriol 48:1171–1180

Cao-Hoang L, Dumont F, Marechal PA et al (2010) Inactivation of *Escherichia coli* and *Lactobacillus plantarum* in relation to membrane permeabilization due to rapid chilling followed by cold storage. Arch Microbiol 192:299–305

Carillo S, Pieretti G, Parrilli E et al (2011) Structural investigation and biological activity of the lipooligosaccharide from the psychrophilic bacterium *Pseudoalteromonas haloplanktis* TAB 23. Chem Eur J 17:7053–7060

Carillo S, Pieretti G, Lindner B et al (2013) Structural characterization of the core oligosaccharide isolated from the lipopolysaccharide of the psychrophilic bacterium *Colwellia psychrerythraea* Strain 34H. Eur J Org Chem:3771–3779

Carillo S, Pieretti G, Bedini E et al (2014) Structural investigation of the antagonist LPS from the cyanobacterium *Oscillatoria planktothrix* FP1. Carbohydr Res 388:73–80

Carlson RW, Krishnaiah B (1992) Structures of the oligosaccharides obtained from the core regions of the lipopolysaccharides of *Bradyhizobium japonicum* 61A101c and its symbiotically defective lipopolysaccharide mutant, JS314. Carbohydr Res 231:205–219

Caroff M, Karibian D (2003) Structure of bacterial lipopolysaccharides. Carbohydr Res 338:2431–2447

Carty SM, Sreekumar KR, Raetz CRH et al (1999) Effect of cold shock on lipid A biosynthesis in *Escherichia coli*. Induction At 12 degrees C of an acyltransferase specific for palmitoleoyl-acyl carrier protein. J Biol Chem 274:9677–9685

Casanueva A, Tuffin M, Cary C et al (2010) Molecular adaptations to psychrophily: the impact of 'omic' technologies. Trends Microbiol 18:374–381

Casillo A, Parrilli E, Sannino F et al (2015) Structural investigation of the oligosaccharide portion isolated from the lipooligosaccharide of the permafrost psychrophile *Psychrobacter arcticus* 273-4. Mar Drugs 13:4539–4555

Chattopadhyay K (2006) Mechanism of bacterial adaptation to low temperature. J Biosci 31 (1):157–165

Corsaro MM, Lanzetta R, Parrilli E et al (2001) Structural investigation on the lipooligosaccharide fraction of psychrophilic *Pseudoalteromonas haloplanktis* TAC 125 bacterium. Eur J Biochem 268:5092–5097

Corsaro MM, Dal Piaz F, Lanzetta R et al (2002) Lipid A structure of *Pseudoalteromonas haloplanktis* TAC 125: use of electrospray ionization tandem mass spectrometry for the determination of fatty acid distribution. J Mass Spectrom 37(5):481–488

Corsaro MM, Pieretti G, Lindner B et al (2008) Highly phosphorylated core oligosaccaride structures from cold-adapted *Psychromonas arctica*. Chem Eur J 14:9368–9376

Crump EM, Perry MB, Clouthier SC et al (2001) Antigenic characterization of the fish pathogen *Flavobacterium psychrophilum*. Appl Environ Microbiol 67:750–759

Ewert M, Deming JW (2013) Sea ice microorganisms: environmental constraints and extracellular responses. Biology 2(2):603–628

Feller G, Gerday C (2003) Psychrophilic enzymes: hot topics in cold adaptation. Nat Rev Microbiol 1:200–208

Feller G, Lonhienne T, Deroanne C et al (1992) Purification, characterization, and nucleotide sequence of the thermolabile alpha-amylase from the antarctic psychrotroph *Alteromonas haloplanctis* A23. J Biol Chem 267:5217–5221

Forsberg LS, Carlson RW (1998) The structures of the lipopolysaccharides from *Rhizobium etli* Strains CE358 and CE359. J Biol Chem 273:2747–2757

Galanos C, Luderitz O, Westphal O (1969) A new method for the extraction of R lipopolysaccharides. Eur J Biochem 9:245–249

Gerischer U (2008) *Acinetobacter* molecular biology. Int Microbiol 11:147–150

Gerwig GJ, Kamerling JP, Vliegenthart JF (1979) Determination of the absolute configuration of monosaccharides in complex carbohydrates by capillary GLC. Carbohydr Res 77:10–17

Gilichinsky D, Vorobyova EA, Erokhina LG et al (1992) Long-term preservation of microbial ecosystems in permafrost. Adv Space Res 12:255–263

Graumann PL, Marahiel MA (1999) Cold shock response in *Bacillus subtilis*. J Mol Microbiol Biotechnol 1:203–209

Hakamori S (1964) A rapid permethylation of glycolipid and polysaccharide catalysed by methyl sulfinil carbanion in dimethyl sulfoxide. J Biochem (Tokyo) 55:205–208

Hidetoshi U (2014) The family Moritellaceae. In: Rosenberg E, DeLong EF, Lory S, Stackebrandt E, Thompson F (eds) The prokaryotes: gammaproteobacteria. Springer, Berlin, pp 477–489

Holst O (2000) Deacylation of lipopolysaccharides and isolation of oligosaccharide phosphates. Methods Mol Biol 145:345–353

Holst O, Molinaro A (2009) Core oligosaccharide and lipid A components of lipopolysaccharides. In: Moran A, Brennan P, Holst O, von Itszstein M (eds) Microbial glycobiology:structures relevance and applications. Elsevier, San Diego, CA, pp 29–56

Holst O, Ulmer AJ, Brade H et al (1996) Biochemistry and cell biology of bacterial endotoxins. FEMS Immunol Med Microbiol 16:83–104

Jansson PE (1999) The chemistry of O-polysaccharide chains in bacterial lipopolysaccharides. In: Brade H, Morrison DC, Opal S, Vogel S (eds) Endotoxin in health and disease. Marcel Dekker, New York, NY, pp 155–178

Kawasaki K, Nogi Y, Hishinuma M et al (2002) *Psychromonas marina* sp. nov., a novel halophilic, facultatively psychrophilic bacterium isolated from the coast of the Okhotsk Sea. Int J Syst Evol Microbiol 52:1455–1459

Kenne L, Lindberg B (1983) Bacterial polysaccharides. In: Aspinall GO (ed) The polysaccharides. Academic Press, New York, NY, pp 287–363

Kilcoyne M, Perepelov AV, Tomshich SV et al (2004) Structure of the O-polysaccharide of *Idiomarina zobellii* KMM 231T containing two unusual amino sugars with the free amino group, 4-amino-4,6-dideoxy-D-glucose and 2-amino-2-deoxy-L-guluronic acid. Carbohydr Res 339:477–482

Knirel YA (2011) Structure of O-antigens. In: Knirel YA, Valvano MA (eds) Bacterial lipopolysaccharides. Springer, Vienna, pp 41–115

Knirel YA, Widlman G, Senchenkova SN et al (1997) Structural studies on the short-chain lipopolysaccharide of *Vibrio cholerae* O139 Bengal. Eur J Biochem 247:402–410

Kondakova M, Novototskaya-Vlasova KA, Arbatsky NP et al (2012a) Structure of the O-specific polysaccharide from the lipopolysaccharide of *Psychrobacter cryohalolentis* K5^T containing a 2,3,4-triacetamido-2,3,4-trideoxy-L-arabinose Moiety. J Nat Prod 75:2236–2240

Kondakova M, Novototskaya-Vlasova KA, Drutskaya MS et al (2012b) Structure of the O-polysaccharide chain of the lipopolysaccharide of *Psychrobacter muricolla* 2pS^T isolated from overcooled water brines within permafrost. Carbohydr Res 349:78–81

Kumar GS, Jagannadham MV, Ray MK (2002) Low-temperature-induced changes in composition and fluidity of lipopolysaccharides in the antarctic psychrotrophic bacterium *Pseudomonas syringae*. J Bacteriol 184:6746–6749

Leontein K, Lindberg B, Lönngren J (1978) Assignment of absolute configuration of sugars by g.l.c. of their acetylated glycosides from chiral alcohols. Carbohydr Res 62:359–362

Lüderitz O, Risse R, Schlecht S et al (1966) Structural relationship of *Salmonella* O and R antigens. Ann N Y Acad Sci 133:349–374

Lüderitz O, Freudenberg MA, Galanos C et al (1982) Lipopolysaccharides of Gram-negative bacteria. In: Razin S, Rottem S (eds) Current topics in membranes and transport. Academic Press, New York, NY, pp 79–51

MacLean LL (2001) The structure of the lipopolysaccharide O-antigen produced by *Flavobacterium psychrophilum* (259-93). Eur J Biochem 268:2710–2716

Maruyama A, Honda D, Yamamoto H et al (2000) Phylogenetic analysis of psychrophilic bacteria isolated from the Japan Trench, including a description of the deep-sea species *Psychrobacter pacificensis* sp. nov. Int J Syst Evol Microbiol 50:835–846

Masoud H, Perry MB, Brisson JB et al (1994) Structural elucidation of the backbone oligosaccharide from the lipopolysaccharide of *Moraxella catarrhalis* serotype A. Can J Chem 72:1466–1477

McKay CP (2014) Requirements and limits for life in the context of exoplanets. Proc Natl Acad Sci 111:12628–12633

Methé BA, Nelson KE, Deming JW et al (2005) The psychrophilic lifestyle as revealed by the genome sequence of *Colwellia psychrerythraea* 34H through genomic and proteomic analyses. Proc Natl Acad Sci U S A 102:10913–10918

Montminy SW, Khan N, McGrath S et al (2006) Virulence factors of *Yersinia pestis* are overcome by a strong lipopolysaccharide response. Nat Immunol 7:1066–1073

Nikaido H (2003) Molecular basis of bacterial outer membrane permeability revisited. Microbiol Mol Biol Rev 67:593–656

Nogi Y, Hosoya S, Kato C et al (2004) *Colwellia piezophila* sp. nov., a novel piezophilic species from deep-sea sediments of the Japan Trench. Int J Syst Evol Microbiol 54:1627–1631

Phillips N, Adin DM, Stabb EV et al (2011) The lipid A from *Vibrio fischeri* lipopolysaccharide. A unique structure bearing a phosphoglycerol mojety. J Biol Chem 286(24):21203–21219

Raetz CRH, Whitfield C (2002) Lipopolysaccharide endotoxins. Annu Rev Biochem 71:635–700

Ray MK, Kumar GS, Shivaji S (1994) Phosphorylation of lipopolysaccharides in the Antarctic psychrotroph *Pseudomonas syringae*: a possible role in temperature adaptation. J Bacteriol 176:4243–4249

Romanenko LA, Schumann P, Rohde M et al (2002) *Psychrobacter submarinus* sp. nov. and *Psychrobacter marincola* sp. nov., psychrophilic halophiles from marine environments. Int J Syst Evol Microbiol 52:1291–1297

Silipo A, Leone S, Lanzetta R et al (2004) The complete structure of the lipooligosaccharide from the halophilic bacterium *Pseudoalteromonas issachenkonii* KMM 3549^{T}. Carbohydr Res 339:1985–1993

Sweet CR, Alpuche GM, Landis CA et al (2014) Endotoxin structures in the psychrophiles *Psychromonas marina* and *Psychrobacter cryohalolentis* contain distinctive acyl features. Mar Drugs 12:4126–4147

Sweet CR, Watson RE, Landis CA et al (2015) Temperature-dependence of lipid A acyl structure in *Psychrobacter cryohalolentis* and Arctic isolates of *Colwellia hornerae* and *Colwellia piezophila*. Mar Drugs 13:4701–4720

Verma DK, Rathore G (2015) New host record of five *Flavobacterium* species associated with tropical fresh water farmed fishes from North India. Braz J Microbiol 46(4):969–976

Vinogradov EV, Bock K, Holst O et al (1995) The Structure of the lipid A-core region of the lipopolysaccharides from *Vibrio cholerae* O1 smooth strain 569B (Inaba) and rough mutant strain 95R (Ogawa). Eur J Biochem 233:152–158

Volk WA, Salomonsky NL, Hunt D (1972) Isolation of 4,7-anhydro- and 4,8-anhydro-3-deoxy-octulosonic acid followixg acid hydrolysis of *Xanthomonas sinensis* lipopolysaccharide. J Biol Chem 247(12):3881–3388

Westphal O, Luderitz O, Bister F (1952) Über die Extraktion von Bakterien mit Phenol/Wasser. Z Naturforsch 7b:148–155

Westphal O, Lüderitz O, Galanos C et al (1986) The story of bacterial endotoxin. In: Chedid L, Hadden JW, Spreafico F (eds) Advances in immunopharmacology. Pergamon Press, Oxford, pp 13–34

Whitfield C (2006) Biosynthesis and assembly of capsular polysaccharides in *Escherichia coli*. Annu Rev Biochem 75:39–68

Wilkinson SC (1977) Composition and structure of bacterial lipopolysaccharides. In: Sutherland IW (ed) Surface carbohydrates of the prokaryotic cell. Academic Press, New York, Ny, pp 97–105

Chapter 14
Polar Microalgae: Functional Genomics, Physiology, and the Environment

Amanda Hopes, David N. Thomas, and Thomas Mock

Abstract Microalgae underpin most foodwebs in polar regions as terrestrial primary production is too limited to support these complex and productive ecosystems. The success of microalgae in these extreme and highly variable ecosystems is rooted in their evolution and adaptation. The recent application of omics approaches in addition to biochemical and physiological measurements enabled a step change in our understanding of how these important organisms are adapted to their environment and how they have evolved from non-polar anchestors. This chapter is focused on diatoms and green algae as both groups of microalgae are most prevalent in polar regions. First genomes, transcriptomes, and reverse genetic tools have recently become available for representative species from both groups. They serve as important platforms to advance studies on their ecology, evolution, and adaptation. We highlight some of the key findings from these studies and link them with biochemical and physiological data to give insights into how genes and their products have shaped important microalgae in their diverse polar environments such as oceans, sea ice, permanently frozen lakes, snow and glaciers. Data from these studies will pave the way for understanding how these key organisms and their communities are going to respond to global climate change. They already provide novel genetic resources for various different biotechnological applications.

Contents

A. Hopes • T. Mock (✉)
School of Environmental Sciences, University of East Anglia, Norwich Research Park, Norwich NR4 7TJ, UK
e-mail: T.Mock@uea.ac.uk

D.N. Thomas
School of Ocean Sciences, College of Natural Sciences, Bangor University, Menai Bridge, Anglesey LL59 5AB, UK

Finnish Environment Institute (SYKE), Helsinki, Finland

R. Margesin (ed.), *Psychrophiles: From Biodiversity to Biotechnology*,
DOI 10.1007/978-3-319-57057-0_14

14.1 Introduction

Protists inhabiting polar regions have been the subject of intense interest ever since the first explorers ventured into the inhospitable seas of the Arctic and Southern Oceans (Ehrenberg 1841, 1853; Hooker 1847; Sutherland 1852). The first records of microbial biodiversity in extreme environments were made with the most basic of microscopes, and until the mid 1900s (ultimately when scientific programs in polar regions became more common) much of the work on protists remained largely descriptive and restricted to the more robust physiological experiments that could be attempted under unfavorable field conditions. Despite the fact that there has been nearly 170 years of research into algae living in the Arctic and Antarctic it is only in the last 20 years that there has been a revolution in laboratory facilities available at remote sites, and of course the technological advances that allow collection, extraction, and subsequent cultivation of organisms in home laboratories. Coupled to this we now have the sophisticated molecular tools to determine the true extent of this diversity, and in turn the molecular and physiological capabilities that permit life to continue at the extremes of low temperature. That is not to belittle the need to still look down the microscopes as works such as Scott and Marchant (2005) quite eloquently demonstrate.

This review is restricted to a discussion of microalgae found in Arctic and Antarctic regions, and most of the discussion will concentrate on diatoms and green algae (Fig. 14.1) living in sea ice, lake waters and snow because most physiological and molecular studies have been conducted with species from these two groups. However, there are other seasonally ice covered sub-polar regions, such as the Baltic Sea and Sea of Okhotsk where much of our understanding about cryogenic adaptations and microbial ecology are forwarding our understanding (see Granskog et al. 2006). Both in Polar and sub-Polar systems a huge diversity of microalgal species exist that for the purposes of this review are split into either psychrophiles, organisms with an optimal growth temperature at or below 15 °C, and a maximum growth temperature below 20 °C, or psychrotrophes, organisms with the ability to grow at temperatures below 15 °C but exhibiting maximum growth rates at temperature optima above 18 °C (Deming 2002).

Naturally permanently low temperatures combined with strong seasonality of solar irradiance are the most important environmental factors for evolution and life of polar photosynthetic organisms. Despite this a wide range of phylogenetic groups of algae have successfully adapted to these extreme environmental

Fig. 14.1 Polar marine microalgal community composed of a green alga (*Chlamydomonas* sp.) and a chain-forming diatom (*Melosira arctica*). Image courtesy by Brian Eddie, Arizona State University, USA

conditions despite the polar regions being a geologically young habitat in the Earth System. The Antarctic continent, with the formation of a permanently cold-water ocean (Southern Ocean), was formed ca. 25 Mio years ago whereas the Arctic Ocean formed ca. 6 Mio years ago (Beil and Thiede 1990; Hansom and Gordon 1998; Thomas et al. 2008). However, most of the algal groups are older than 25 Mio years (Kooistra and Medlin 1996). Thus, the formation of polar environmental conditions was a major radiation event where new species developed that were able to grow under these extreme conditions.

14.2 Environmental Conditions

14.2.1 Light

The strong seasonality of solar irradiance is the major factor that influences the availability of light for photoautotrophic organisms at high latitudes. However, snow and ice thickness very much determine how much of the light is able to penetrate to regions where photosynthetic organisms are living (Eicken 1992). It cannot be forgotten that despite high latitude regions being commonly thought of as light limited systems, in fact during summer periods irradiances on snow and ice surfaces can be extremely high with high doses of harmful ultraviolet radiation being a commonly reported stress factor (reviewed by Brierley and Thomas 2002). Therefore, a wide range of photoadaption is exhibited and is a prerequisite for photoautotrophic organisms living at high latitudes (Kirst and Wiencke 1995). Snow algae, which are mainly chlorophytes, grow on and within snow and ice surfaces and may therefore be exposed to high doses of UV radiation. This is of course true for the microalgal assemblages of glacial lakes, cryoconite holes on glacial systems, and seasonally formed melt features both in terrestrial systems and on the surfaces of pack ice (Vincent et al. 2000; Säwström et al. 2002; Hodson et al. 2005; Mindl et al. 2007).

In contrast, microalgae growing inside or under sea ice, as well as in or under permanent ice covers of the Antarctic dry valley lakes are photosynthetically active in a light environment almost without UV radiation and less than 1% of incident photosynthetically active radiation (PAR). Beyond the aquatic systems hypoliths—mostly cyanobacteria and chlorophytes—that grow on the underside of stones and rocks in periglacial systems, where they utilize irradiances far less than >0.1% of the incident light for photosynthesis (Cockell and Stokes 2004). Far beyond these habitat-specific light irradiance differences, all photoautotrophs in high latitude regions must have the physiological ability to survive several months of darkness.

14.2.2 Seawater

In high latitudes, the Polar Oceans are the major habitat of microalgae in terms of biomass, abundance, and species diversity. Most of these cold water masses are characterized by seasonal surface-freezing and strong vertical mixing due to katabatic polar winds, convection at frontal zones or deep-water formation (Cottier et al. 2017; Meredith and Brandon 2017). However, the central Arctic Ocean and the Southern Ocean are not similar regarding their physical and chemical conditions. The Arctic Ocean is a Central Ocean surrounded by landmasses with a permanent cover of multi-year sea ice in its central area around the north pole. The Southern Ocean surrounds the Antarctic continent with a series of circumpolar fronts, and different water masses with distinct physical and chemical characteristics occur between these fronts (Tomczak and Godfrey 2003).

Most of the sea ice in the Southern Ocean is seasonal with an advance and retreat of 16 million km^2 in sea ice around the continent within 1 year (Fig. 14.2). Multi-year sea ice only occurs close to the continent in inlets or bays or in major ocean gyre systems in the Weddell and Ross Seas (Stammerjohn and Maksym 2017). This results in Antarctic sea-ice having a mean thickness of about 0.55 m. In contrast, the mean thickness of Arctic sea ice is approximately 3 m, due to 50% of the Arctic pack ice being multiyear ice lasting between 2 and 11 years (Dieckmann and Hellmer 2003). The latitudinal influence of sea ice is greater in the Arctic covering a region that extends from 90°N to 44°N, whereas in the Southern Ocean the region is only 75°S to 55°S, although at the maximum extents Arctic ice covers an area of 16 million km^2 and Antarctic sea ice 19 million km^2 (Meier 2017; Stammerjohn and Maksym 2017).

Overwhelmingly, in the past decade, it has been the reduction in summer sea ice in the Arctic Ocean that has captured the imagination of policy makers, media, and the non-specialized audience. This loss in the sea ice extent and volume in the northern hemisphere has been contrasted by slight increases in overall sea ice extent in the Southern Ocean (Stroeve et al. 2005; Stoeve and Notz 2015; Meier 2017; Stammerjohn and Maksym 2017). This results in the overall global sea ice extent

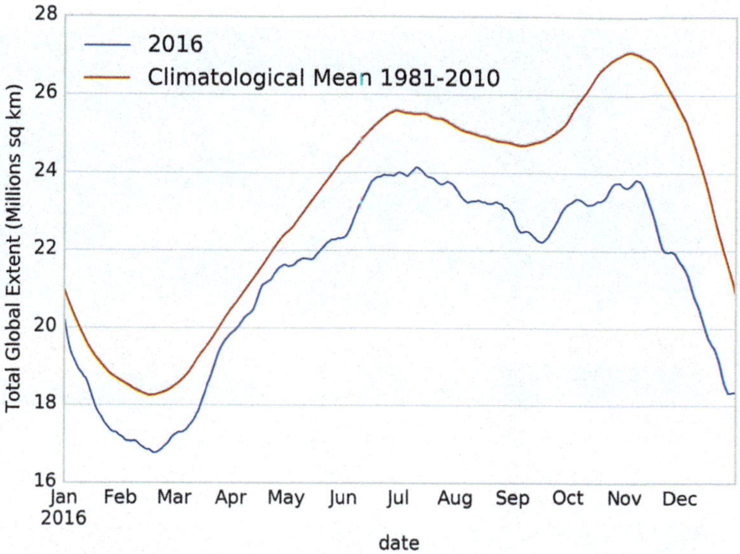

Fig. 14.2 Time series of daily global sea ice extent (Arctic plus Antarctic). It shows the global sea ice extent for 2016 tracking below the 1981–2010 average. Image provided by the National Snow and Ice Data Center, University of Colorado, Boulder, USA (http://nsidc.org/arcticseaicenews/)

varying on average between 18 and 26 million km^2, although in 2016 the range was from 17 to 24 due to the changes in the Arctic sea ice compared to the 1981–2010 climatological mean (Fig. 14.2).

The Southern Ocean has the highest inventory of unused macronutrients in the World Ocean (Fig. 14.3) and is the most important province for the export and burial of biogenic silica from diatoms (Smetacek 1998; Smetacek and Nicol 2005). The discovery of high macronutrient concentrations and relatively low phytoplankton concentrations in the Southern Ocean led to the concept of the "Antarctic Paradox" that was subsequently referred to as high nitrogen-low chlorophyll a region (HNLC). Micronutrients such as iron are considered to be the reason for this Antarctic Paradox, and several international large-scale iron fertilization experiments confirmed this hypothesis (reviewed by Boyd et al. 2007). Thus, the supply of iron to Southern Ocean phytoplankton (iron is a requirement for proteins involved in photosynthetic carbon assimilation) resulted in marked increases in both carbon fixation and nitrate utilization rates. However, it is assumed that many offshore species do have a lower requirement for iron and therefore are well adapted to these conditions.

In contrast, the Arctic Ocean is relatively rich in micronutrients such as iron because of terrigenous sources of micronutrients primarily via river runoff and also dust and sediments deposited in shallow coastal water masses. Thus, macronutrients are more important in limiting phytoplankton biomass in the Arctic Ocean compared to the Southern Ocean. However, the most important factors regulating the large-scale distribution of phytoplankton production and biomass in the Arctic

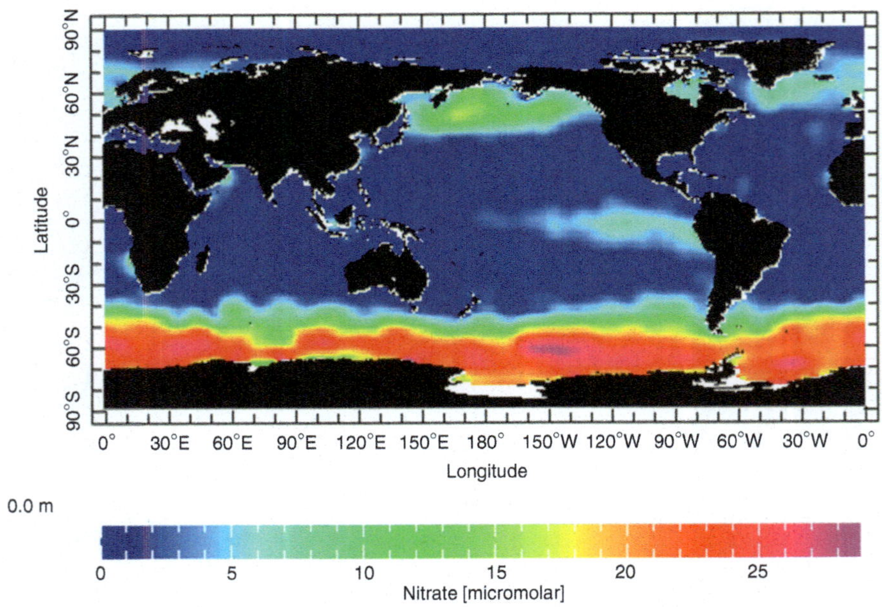

Fig. 14.3 Map of high nutrient-low chlorophyll (HNLC) regions around the world. Measurement in map is of nitrate, with the scale as a gradient of color pictured on the *bottom* (http://www.atmosphere.mpg.de/media/archive/1058.gif)

Ocean are probably the surface ice cover and the depth of the surface mixed layer, thus the availability of light (Sakshaug and Slagstad 1991). The distribution of macronutrients in the Arctic Ocean is highly heterogeneous between basins resulting in significant regional differences in primary production dynamics (Wheeler et al. 1997; Dittmar and Kattner 2003; Jones et al. 2003).

Beside the dominance of diatoms in both polar oceans, Prymesiophytes such as *Phaeocystis* spp. and *Emiliania* spp. are the second most abundant algal group. They even may form large blooms under more stable conditions and therefore outgrow bloom-forming diatoms (e.g., Smith et al. 1991; Merico et al. 2003). Dinoflagellates, Chlorophytes, Prasinophytes, and other algal groups are underrepresented in polar oceans (Kopczynska et al. 1986; Smetacek et al. 2002). However, a study by Lovejoy et al. (2006) who used 18S rRNA clone gene libraries indicated a high diversity of microbial eukaryotes in the Arctic Ocean. This is either indicative of a large number of endemic species or a high number of under-sampled taxa. Nevertheless, the dominance of diatoms in polar oceans makes this group ecologically the most important group of polar microalgae (Lizotte 2003a, b). Diatoms in general are estimated to contribute to at least 50% of the global marine primary production (Nelson et al. 1995).

Due to the presence of glaciers and permafrost, photosynthetic biomass on land in polar regions is negligible compared to that found in the ocean. Consequently, polar diatoms are of interest not only because of their important role as the main food source for the whole polar food web (terrestrial and aquatic) but also because of their ability to thrive in this extreme ecosystem.

14.2.3 Sea Ice

Sea ice being one of the most extreme and largest habitats in polar oceans is important in structuring the whole polar ecosystem (Eicken 1992; Brierley and Thomas 2002; Arrigo and Thomas 2004). At its maximum, it covers 13% of the Earth's surface (Comiso 2003). The physical characteristics pertinent to the biology living in sea ice have been reviewed by Petrich and Eicken (2017): Sea ice, in contrast to fresh water ice, is not solid but is comprised of a system of brine channels (Fig. 14.4) that provide a habitat characterized by low temperature (ca. −2 to −20 °C), high salinity (35–200), high pH (up to 11), and low irradiances that can be below 1 μmol photons m^{-2} s^{-1} despite the fact the ice is only 1–2 m from the ocean surface (Eicken 1992; Gleitz et al. 1995; Kirst and Wiencke 1995).

Sea water typically containing about 34 g of dissolved salts and ions (mostly sodium, chloride, sulfate, magnesium, calcium, and potassium) does not begin to freeze until temperatures drops below −1.86 °C. At this temperature, ice crystals begin to form and rise to the surface. These initial ice crystals (termed frazil ice) vary in shape, from plates to needles, and size, from ≤ millimeter to centimeter in length. The crystals consolidate by wind and water motion within hours to form loosely aggregated discs (termed pancakes). After a few days of growth by accumulation of more and more ice crystals that form in the upper water column, pancakes can be several meters across and up to 50 cm thick. They freeze together and after 1 or 2 days a closed ice cover has formed (termed pack ice). As temperatures continue to decrease this pack ice thickens, not necessarily by the accumulation of more ice crystals but by the growth of columnar ice at the ice–water interface. This type of ice is formed by the vertical elongation of frazil

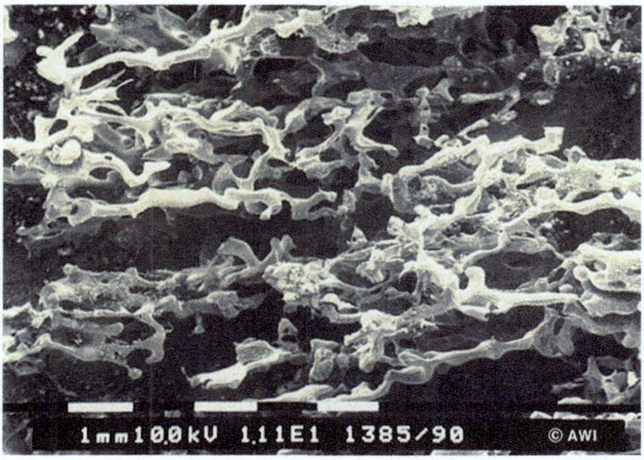

Fig. 14.4 Brine channel system in columnar sea ice made visible by filling the system with epoxy resin under a vacuum. Picture by Alfred-Wegener Institute for Polar and Marine Research, Bremerhaven, Germany, based on the work of J. Weissenberger et al. (1992)

ice crystals. The proportion of frazil ice to columnar ice depends largely on the turbulence of the water in which it was formed. The more turbulent the water the more frazil ice is usually formed. Antarctic sea ice thus contains up to 80% frazil ice as it is formed under more turbulent conditions. In the Arctic, sea ice is formed under more calm conditions containing up to 80% columnar ice (Petrich and Eicken 2017; Fig. 14.5 = ice from Arctic and Antarctic, polarization).

When ice is formed from seawater, salt ions and air in the water cannot be incorporated into the ice crystals and are therefore concentrated as salty brine either into inclusions of pockets and channels (Fig. 14.4) or released into the water below the ice. Thus, sea ice is a solid matrix penetrated by a labyrinth of channels and pores that contain highly concentrated brine and air bubbles (Fig. 14.4). Brine channels vary in size from a few micrometers through several millimeters in diameter and are the main habitat for all microorganisms in sea ice (reviewed by Brierley and Thomas 2002; Deming 2002; Lizotte 2003a; Mock and Thomas 2005). Their volume and the concentration of salt in them is directly proportional to temperature (Fig. 14.5) (Weissenberger et al. 1992; Krembs et al. 2000; Petrich and Eicken 2017). When temperatures decrease, brine volume decreases and salt content increases. Thus, the colder ice contains brine channels with highly salty brines and overall fewer, smaller, and less interconnected channels than warmer ice. Since ice at the sea-ice air interface is usually colder than ice in contact to the underlying water, a temperature gradient exists through the ice, resulting in a gradient in brine salinity and the overall volume of brine in sea ice as well. A host of protists and zooplankton have been recorded from sea ice (Horner 1985; Palmisano and Garrison 1993; Lizotte 2003a; Werner 2006; Bluhm et al. 2017),

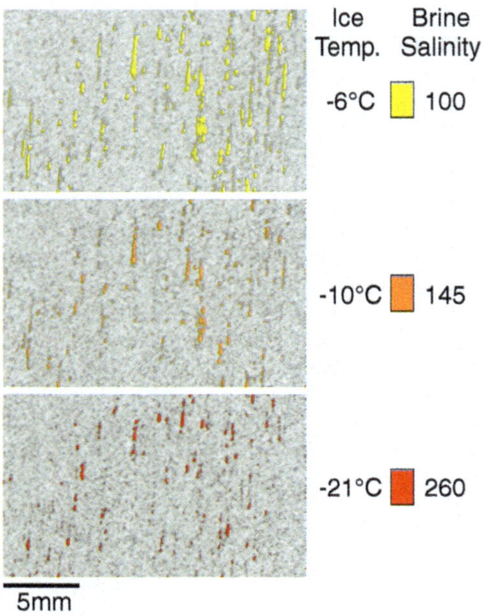

Fig. 14.5 Color-enhanced magnetic resonance images of the same piece of ice shows how the pore space and size of the brine channels and pockets reduces with decreasing temperature, with corresponding increase in salinity of brines contained within the pores. Image after Thomas and Dieckmann (2002), based on the work of Eicken et al. (2000)

Fig. 14.6 Ice floe (*upside down*, about 80 cm thick) with dense populations of pennate diatoms at the sea-ice water interface (indicated by *brown* color that is caused by their main light-harvesting pigment fucoxanthin). Image from David N. Thomas

although among the photoautotrophs the most studied are the diatoms. All organisms living within the sea ice matrix have to have plastic physiologies to cope with these ever-changing physical and chemical conditions of their environment, which are dominated by the temperature and salinity changes.

Microalgae are mainly introduced into the ice as it is forming. They get caught between ice crystals or simply stick to them as crystals rise through the water when it freezes in fall. During the formation of consolidated ice, diatoms become trapped within brine channels. Pennate diatoms are the most conspicuous organisms in sea ice along with other microalgae (e.g., Dinoflagellates, flagellates), heterotrophic protists (e.g., ciliates) and bacteria (Brierley and Thomas 2002; Thomas and Dieckmann 2002; Caron et al. 2017). These micrometer-sized algae, with their main light-harvesting pigment being fucoxanthin, can reach such concentrations in sea ice that they discolor the ice visibly brown (Fig. 14.6). The time for acclimation to the new conditions in sea ice is not very long since day light hours are continually decreasing as winter approaches. Nevertheless, diatoms, especially at the ice–water interface where conditions are most similar to the water below the ice, are often able to photoacclimate rapidly and can accumulate to high biomass even before the winter begins (Gleitz and Thomas 1993). Sea-ice diatoms are very efficient in using solar irradiance and are able to grow at irradiance levels below 1 μmol photons m^{-2} s^{-1} (Mock and Gradinger 1999). Light levels are minimal during high latitude winters, not only due to short days or complete darkness, but also due to snow cover on top of the ice that is a very efficient reflector of solar irradiance (Perovich 2017).

Sea ice is mostly an ephemeral feature since after its formation and consolidation the majority of it melts resulting in the release of all organisms within to the underlying water. Increase in solar irradiance is the most important factor that causes the ice to melt. A common feature and a sign for the beginning of the ice melt is the formation of melt ponds on the surface of the ice (Fig. 14.7). They are more common in the Arctic than Antarctic. One reason for fewer melt ponds in the Antarctic is that more heat is derived from underlying water and melting from

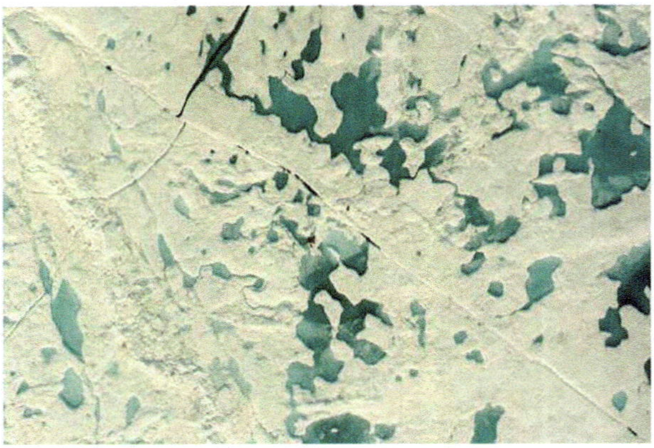

Fig. 14.7 Melt ponds on top of Arctic sea ice. This picture is 100 m across. The melt ponds have different shades of *green-blue*, which is determined by the optical properties of the ice underneath the water in the melt ponds (http://www.arcticice.org/close100m.htm)

above is less significant (Haas 2017). When melting continues due to increasing water temperatures and solar irradiance on top of the ice, the ice gets thinner and more porous. Large pores and brine channels that are filled with seawater characterize warm ice, and the ice itself has very little strength and is easily broken up. However, not all the ice that is formed in fall actually melts during next summer. If it survives the summer, refreezing occurs during the following winter that makes the ice even thicker. The longevity of the ice depends on the geographical location, on the wind, and ocean currents. Sea ice of northern Greenland and the Canadian archipelago can be up to 15 years old with an average thickness of 6–8 m (Haas 2017), although the extent of such thick multiyear ice is considerably reduced in the past 20 years. Such differences in physical properties of the ice also result in differences in the abundance, activity, and composition of the microbial communities within sea ice.

Despite the high diversity of autotrophs within sea ice, that also includes Prasinophytes, autotrophic dinoflagellates and cilliates, two small pennate diatoms, *Fragilariopsis cylindrus* (Grunow) Krieger and *F. curta* (Van Heurck) Hustedt, and the prymnesiophyte *Phaeocystis antarctica* Karsten are the dominant species in blooms in the Antarctic sea ice zone (Leventer 1998; Lizotte 2001). Gleitz et al. (1998) found that at high diatom standing stocks species diversity decreases. This has also been reported by Gleitz and Thomas (1993), who showed that as first-year sea ice grew and high algal standing stocks established, the assemblages were dominated by only a very few small diatom species. Taking into consideration the findings of other studies, Gleitz and Thomas (1993) suggested that pore and channel size was the major factor in the preferential accumulation of a few smaller species within sea ice. However, Gleitz et al. (1998) subsequently concluded that it was the physiological capacity of these species to maintain high growth rates in the spring and summer, in connection with their life history cycles, that may be the key to the prominence of so few diatom species in the ice.

Phaeocystis species are more usually found in sea-ice habitats not constrained by the brine channel systems, such as surface ponds, rotten summer sea ice, or freeboard/infiltration layers. Especially in the latter, these are situations where the constraints of salinity, temperature, and low light do not inhibit primary production as they do in interior ice assemblages, thereby enabling high standing stocks (including diatoms) to accumulate (Haas et al. 2001; Kennedy et al. 2002; Kattner et al. 2004).

Dense dinoflagellate and chrysophyte assemblages can develop in the upper sea-ice interior, and high rates of primary production have been measured at these sites, especially in spring when the upper sea ice temperature is low and brine salinities are high (Stoecker et al. 1997, 1998, 2000). These algal assemblages are often poorly defined, but they may make an important additional contribution to total sea-ice primary production.

Even psychrophilic, halotolerant *Chlamydomonas* spp. have been isolated from sea ice in both the Arctic and Antarctic (Hsiao 1983; Ikävalko and Gradinger 1997; Krembs and Engel 2001; Eddie et al. 2008). While diatoms have received most of the research attention in sea ice work, other groups of organisms such as *Chlamydomonas* species will increasingly attract effort, especially since similar organisms are routinely isolated from saline and freshwater lakes in Arctic and Antarctic sites.

14.2.4 Snow

In regions where snow persists during the summer such as in high-altitude and in the low-latitude polar regions, its color may change from white to red, pink, green, yellow, or orange. The largest patches are often red and therefore called "blood snow" or "watermelon snow" (Fig. 14.8). These macroscopic expressions are based on massive growth of unicellular psychrophilic green algae. These algae are reviewed by Hoham and Duval (2001). Most snow algae belong to the genera

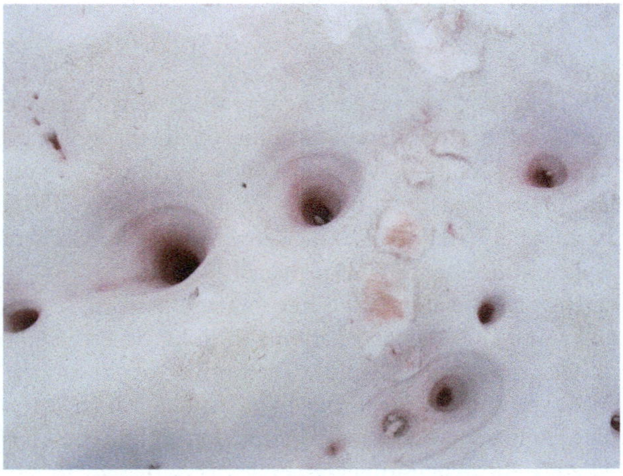

Fig. 14.8 Watermelon snow pits superimposed with an orange bootprint. The coloration is caused by *Chlamydomonas nivalis* (https://en.wikipedia.org/wiki/Watermelon_snow)

Chlamydomonas and *Chloromonas* (Chlorophyta, Volvocales), and these are most active in spring and summer. The beginning of snow melt in spring provides liquid water between ice crystals that is essential for the vegetative stages. The snow has to be neither too cold nor too dry such as freshly fallen snow. Green flagellated stages are often observed within this wet snow (Fig. 14.9a). They are able to move within the snow layer to reach optimal depths for their light and temperature requirement. They can form massive blooms and color the snow green if enough nutrients are available as is commonly found close to bird colonies or nutrient-rich streams or ponds.

During summer and fall, they have to acclimate to extreme temperature regimes, high irradiance and UV radiation, and low nutrient levels. For instance, in high-altitude regions (above 2500 m), UV radiation can be very high and spherical integrated photosynthetic active radiation (PAR) can often reach 4500 μmol photons m^{-2} s^{-1} and occasionally up to 6000 μmol photons m^{-2} s^{-1}. If less liquid water becomes available and therefore also nutrients, most flagellated stages turn into immotile hypnoblast stages (Fig. 14.9b) because this form is the most resistant to environmental changes. The transformation into hypnoblasts is characterized by a massive incorporation of reserve material, including sugars, lipids, and by formation of esterified extraplastidal secondary carotenoides. Studies have shown that the cells mainly form oxycarotenoides and in particular astaxanthin that has a red color and therefore gave the snow its name "blood snow" (Müller et al. 1998). Theses hypnocygotes and other resting cells have thick cell walls and sometimes mucilaginous envelopes (Müller et al. 1998). They can survive dry and warm periods in a dormant state and tolerate high pressure such as under thick snow. They also tolerate freezing in ice blocks at temperatures down to −35 °C during winter. However, some of these resting stages can remain photosynthetically active even under very high photon flux densities because of well-protected photosystems by secondary carotenoids (Remias et al. 2005).

14.2.5 Rock Surfaces

Most of the Antarctic continent is covered with a several kilometer thick layer of meteoric ice. However, parts of this continent are ice free such as the McMurdo Dry Valleys in southern Victoria Land (Fig. 14.10). With ca. 4500 km^{-2}, this is the largest ice-free area on this continent. Precipitation in this region is below 10 cm $year^{-1}$ that makes it one of the driest deserts on earth, and air temperature ranges from 5 to −55 °C (Priscu 1998).

Periglacial activity, the freezing and thawing of ground water, often sorts rocks and stones into defined patterns in Polar deserts, regions of permafrost, and high altitude stone rubble fields. The sorting of the stones results in a high degree of spatial heterogeneity in the light incident under different parts of the stone/rock patterning and vegetation patterns. In turn, this results in regions of the stone field where light penetrating the stones is sufficient to support photosynthetic carbon

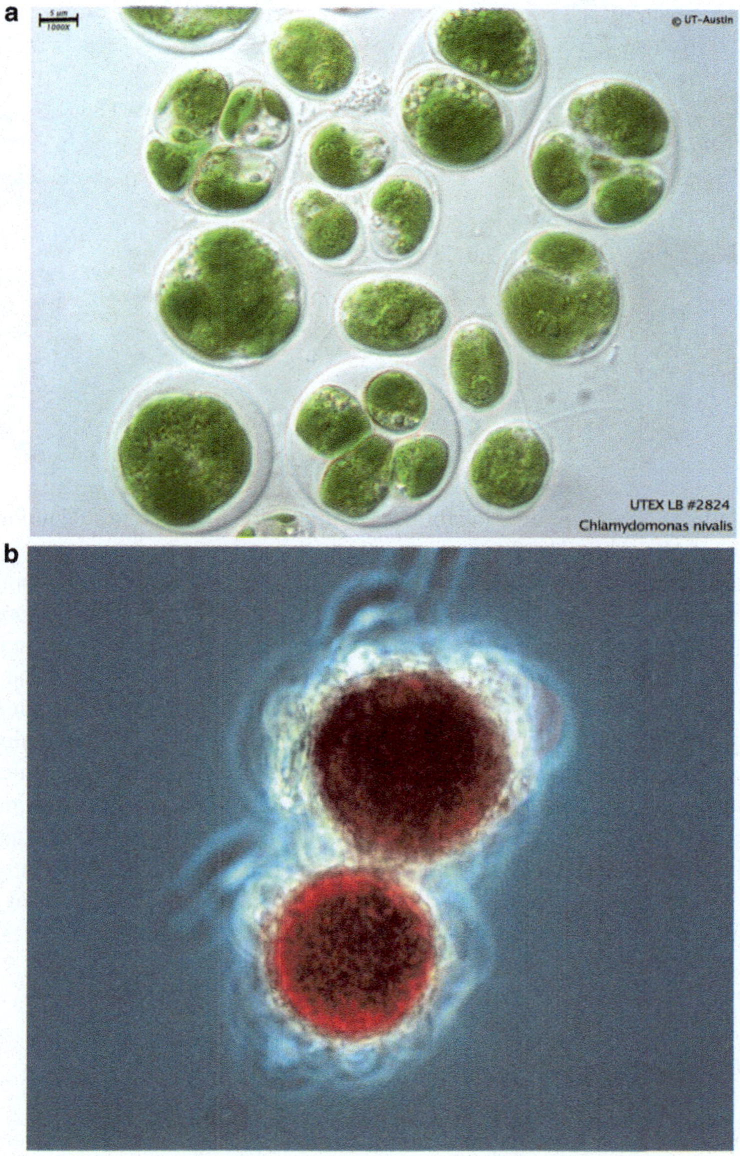

Fig. 14.9 (**a**) Vegetative stages of *Chlamydomonas nivalis*. Some of the single cells show two flagella. Picture from UTEX Image Bank: (http://www.bio.utexas.edu/research/utex/photogallery/c/Chlamydomonas %20nivalis %20LB %202824.htm); (**b**) *Chlamydomonas nivalis* aplanospores filled with the *red* pigment astaxanthin and with attached particles. Image courtesy by Brian Duval and Lynn Rothschild

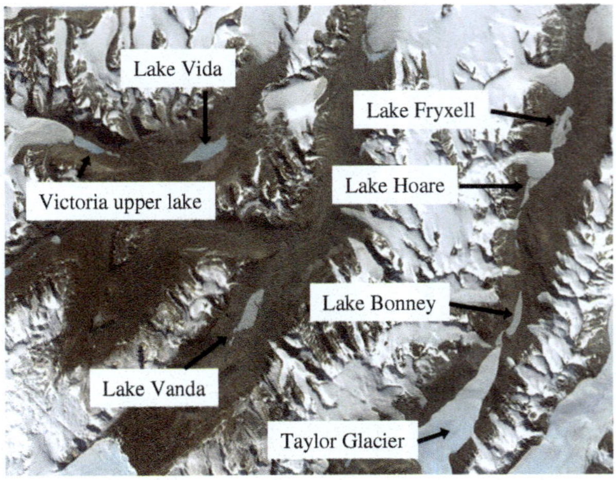

Fig. 14.10 Satellite image (NASA, USA) from McMurdo Dry Valleys. Best studied lakes are Lake Bonney, Hoare, and Fryxell that are located within the Taylor Valley

assimilation by the hypoliths inhabiting the underside of the stones (Cannone et al. 2004; Cockell and Stokes 2004).

The patterning of stone fields influenced by periglacial activity is often polygonal, and both the Arctic and Antarctic rocks at the edges of the polygons support well-developed assemblages of photosynthetic organisms, whereas in the center of the polygons colonization by hypoliths is significantly reduced (Cockell and Stokes 2004). These cyanobacteria and unicellular algae are growing, even thriving, in an extreme environment, where temperatures sink below −30 °C, water is minimal, and light conditions are reduced to virtually nothing. Likewise, there are microalgae associated with cryptoendolithic layers growing within sandstones in regions such as the Dry Valleys. Typically, these communities have layers of fungi and cyanobacteria, but microalgae such as *Hemichloris antarctica* are frequently found in the lowest bands of endolithic systems where the irradiance can be as low as 0.05 μmol photons m^{-2} s^{-1} (Johnston and Vestal 1991; Friedmann et al. 1993).

14.2.6 Permanently Ice Covered Lakes

Besides short-term glacier melt events, most of the liquid water, and therefore accumulation of organisms, is available in the perennial ice-covered lakes that are characteristic for the Dry Valleys and also regions such as the Vestford Hills. However, the source of lake waters is assumed to be glacier meltwater that penetrates the lakes without melting the surface ice cover (Priscu 1995). A sensitive balance of freeze-thaw cycles is assumed to keep the lakes permanently covered with ice but also accessible for water in- and outflow from underneath. There are numerous permanently covered lake systems in the McMurdo Dry Valleys, but the best studied are Lake Bonney, Hoare, and Fryxell that are located within the Taylor Valley (Fig. 14.10).

The major food-web components in all lake systems are unicellular eukaryotes and prokaryotes. The main groups are green algae, diatoms, ciliates, rotifers, heterotrophic nanoflagellates, bacteria, and viruses. Autotrophic phytoplankton play an essential role in functioning of the food web by production of organic carbon. However, photosynthesis is strongly limited by the availability of solar irradiance. Measured irradiance below the ice cover of Lake Bonney never exceeds 50 μmol photons m^{-2} s^{-1}, and the wavelength of maximum transmission through the water column is in the range from 480 to 520 nm with longer wavelengths (>600 nm) being diminished. This light is only available from late September through mid-March and this seasonality is the trigger for spring phytoplankton growth in these non-turbulent waters. Vertical stratification is very pronounced in most of the lakes and is sometimes accompanied with strong gradients in salinity and nutrients. Vertical stratification is also pronounced for many phytoplankton species. The water layer immediately beneath the ice cover in Lake Bonney is dominated by the crytomonad *Chroomonas* sp. and *Chlamydomonas intermedia* whereas *Chlamydomonas raudensis* is confined to the deep saline and low-irradiance layers of the photic zone (Morgan-Kiss et al. 2006).

14.3 Adaptation of Microalgae at High Latitudes

14.3.1 Diatoms (Bacillariophyceae)

Psychrophilic diatoms are one of the most abundant groups of phytoplankton in polar oceans. This is mainly due to the presence of higher silicate concentrations in these waters and to their successful adaptation to strong vertical mixing in polar waters, strong seasonality in solar irradiance, freezing temperatures, and extremes of salinity (Cota 1985; Fiala and Oriol 1990; Boyd 2002; Mock and Valentin 2004; Ryan et al. 2004; Ralph et al. 2005). Due to their importance as primary producers, many physiological studies with polar diatoms were related either to growth and its dependency on nutrients and temperature or to regulation of photosynthesis under typical polar condition. This section aims to provide a comprehensive overview of new data regarding physiological and in particular molecular adaptation for this important group of polar algae.

Maximum growth rates for many polar diatoms are in the range of 0.25–0.75 divisions per day, that is two- to threefold slower than growth at temperatures above 10 °C (Sommer 1989). Many of these diatoms are psychrophilic and not able to live at warmer temperatures (above ca. 15 °C), which is indicative of the presence of specific molecular adaptations that enable these diatoms to grow under freezing temperatures.

14.3.1.1 Functional Genomics

Approaches to uncover the gene repertoire of a polar diatom have been dominated by the genus Fragilariopsis, in particular *Fragilariopsis cylindrus*, a marine indicator species for cold water, found at both poles (von Quillfeldt 2004) and in seasonally cold waters (Hendey 1974; Hällfors 2004).

The first approaches involved constructing and sequencing two expressed sequence tag (EST) libraries, one generated under freezing temperatures (Mock et al. 2005) and the another under increased salinity (Krell 2006). 966 EST were generated from the cold stress library and 1691 from the salt stress library. There are now over 21,000 EST from *F. cylindrus* on the EST-databank at NCBI and about 200 gene-specific oligonucleotides (70mers) from the original EST libraries for functional gene-array experiments (Mock and Valentin 2004). An important addition to algal research, particularly in terms of understanding polar adaptation, is the recent publication of the *F. cylindrus* genome and RNA-sequencing data generated under a range of polar conditions (Mock et al. 2017). This is the third diatom genome to be published and the first polar diatom. There is only one other polar microalga with a published genome, the psychrotolerant freshwater green alga *Coccomyxa subellipsoidea* (Blanc et al. 2012).

All EST-sequences were compared against the genomes of *Thalassiosira pseudonana* and *Phaeodactylum tricornutum*. In addition, 11 algae and plant databanks were consulted to annotate sequences that were not found in the temperate diatom genomes. Nevertheless, over 50% of sequences showed no similarity to known sequences in these databanks and to both diatom genomes even when using a comparatively high e-value of $\leq 10^{-4}$ (Mock et al. 2005).

In the cold-stress EST library, the most abundant functional categories were related to translation, posttranslational modification of proteins, and transport of amino acids and peptides by ABC transporters. Some of these ABC transporters displayed homology to bacterial permeases and others appeared to be involved in translational or posttranslational control. However, most of them could not be assigned a function.

The presence of six different DNA/RNA helicases in the cold-stress library indicated that DNA and RNA coiling and uncoiling are important under freezing temperatures. Minimizing the likely formation of secondary structures and duplexes of mRNAs under low temperature stress is necessary to initiate translation. However, protein domains of DNA/RNA helicases are also the eighth most abundant protein domain in the genome of *T. pseudonana* (Armbrust et al. 2004), and therefore more evidence is necessary to conclude that these enzymes are essential to cope with freezing temperatures. The most abundant sequences in this library in terms of their redundancy were either sequences that were related to energy generation (e.g., fucoxanthin-chlorophyll a, c-binding proteins) or completely unknown sequences (Mock et al. 2005).

In the salt-stress library, the most abundant functional categories of sequences were related to posttranslational modification of proteins (e.g., heat-shock proteins;

hsps) and ion-transport (Krell 2006). Most of them were hsps and different ionic transporter genes reflecting the requirement to reestablish homeostasis under salt stress. Several sequences of different kinds of V-type H+-ATPases and antiporters for various ions such as sodium, potassium and calcium were found in this library. V-type H+-ATPases are of great importance in establishing an electrochemical proton gradient across the tonoplast to drive sodium sequestration into the vacuole (Shi et al. 2003).

One important organic osmolyte under salt stress in diatoms is the amino acid proline. Many genes involved in proline synthesis were found in the salt-stress-EST library indicating that this pathway was active under experimental conditions (Krell 2006). The gene coding for pyrroline-5-carboxylate reductase (P5CR, catalyzing the final step in proline synthesis) could be identified among the most abundant sequences in the salt-stress library (Krell 2006). Furthermore, seven proteins involved in the proline synthesis pathway increased in abundance in response to high salinity (Lyon et al. 2011). This indicates that proline may be important for salt stress acclimation.

One of the interesting aspects of the *F. cylindrus genome* is the high number of divergent alleles. Approximately, 25% of the diploid genome consists of alleles that are highly divergent, particularly in comparison to the temperate diatom genomes of *T. pseudonana* and *P. tricornutum* (Mock et al. 2017).

Differential expression can be seen between divergent alleles under different conditions, many of which are commonplace in the polar environment, including prolonged darkness, freezing and elevated temperatures, iron starvation, and increased CO_2 concentration (Fig. 14.11b). In addition, dN/dS analysis suggests that there may be a positive correlation between allelic differentiation and diversifying selection (Mock et al. 2017).

Copper rather than iron-binding proteins are enriched in the *F. cylindrus* genome as are plastocyanin/azurin-like domains. This may facilitate electron transport during photosynthesis while reducing iron dependence. In terms of photosynthesis, a large number of light-harvesting complex (LHC) proteins are also present including Lhcx, which is involved in stress response. There are also a larger number of methionine sulfoxide reductase (MSR) genes in the *F. cylindrus* genome compared to *T. pseudonana* or *P. tricornutum* that are linked to oxidative stress under cold temperatures (Lyon and Mock 2014).

A large number of zinc-binding proteins can be found in this genome compared to the sequenced temperate diatoms. These contain myeloid-Nervy-DEAF-1 domains (MYND) which are associated with protein–protein interactions and regulation.

Enrichment of specific gene groups can be found within the diverged alleles; these include: catalytic activity, transport, membrane proteins, and metabolic processes (Fig. 14.11a). Furthermore, divergent alleles were found to be differentially expressed under different conditions, suggesting that they may be involved in adaptation to polar conditions. Given the low sequence identity between promoters of divergent alleles and their differential regulation, it is seems likely that individual copies are under different regulatory controls. RNA-seq data focused on

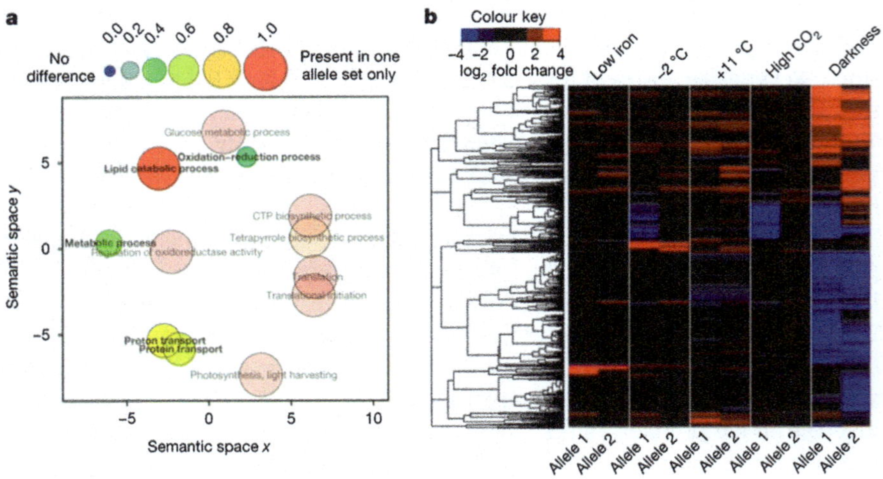

Fig. 14.11 Bi-allelic transcriptome and metatranscriptome profiling. (**a**) REViGO semantic similarity scatterplot of biological process gene ontology terms for *Fragiariopsis cylindrus*-like sequences (E-value $\leq 1 \times 10^{-10}$) in Southern Ocean metatranscriptome samples. Gene ontology terms that are overrepresented in the set of diverged alleles compared to non-diverged alleles are shown in bold. (**b**) Hierarchical clustering of 4030 differentially expressed allelic gene pairs in *F. cylindrus* (likelihood ratio test, $P < 0.001$; log2 fold change ≤ -2 or $\geq +2$) under low iron, freezing temperature (−2 °C), elevated temperature (+11 °C), elevated carbon dioxide (1000 ppm CO_2) and prolonged darkness, relative to optimal growth conditions. Each experimental treatment corresponds to two separate columns for both allelic variants and each single-haplotype gene to a single row. Image is taken from Mock et al. (2017)

changes in expression under prolonged (7 days) darkness as this condition gave rise to the highest number of up- and downregulated genes (Fig. 14.11b). Downregulated genes include those involved in photosynthesis, light harvesting, photoprotection, and translation. Genes involved in regulation of gene expression, DNA replication, signal transduction, and starch, sucrose, or lipid metabolism were upregulated (Mock et al. 2017). RNA-seq data suggests that during darkness, photosynthetic activity and supporting processes are reduced while processes such as chrysolaminarin and fatty acid storage are used instead.

Interestingly, as well as displaying the largest differential expression, growth under prolonged darkness also led to double the number of RNA-seq reads (30%) that did not map to predicted genes compared to any other condition. Alleles with the largest dN/dS ratios tended to show strong differences in expression between conditions; in addition, the majority of these alleles have no known function. As mentioned, this suggests a positive correlation between diversifying selection and allelic differentiation. It also highlights the necessity for reverse genetics in polar species to determine the function of these sequences and in turn understand how they are adapted to polar environments.

One of the most interesting discoveries in the *F. cylindrus* EST salt-stress library was a gene involved in antifreeze processes (Krell 2006; Krell et al. 2008). The presence of ice-binding protein (IBP) genes in this species was verified following sequencing of the genome (Mock et al. 2017). Shortly after, IBPs were identified and characterized in the polar diatom *Navicula glaciei* (Janech et al. 2006). Since then several papers have been produced which explore the function of IBPs in polar diatoms; this is discussed in more detail in the next section. In diatoms, ice-binding proteins have been identified in *F. cylindrus*, *Fragilariopsis curta*, *N. glaciei*, *C. neogracile*, *Attheya* sp., *Amphora* sp., and *Nitzschia stellate* (Janech et al. 2006; Krell et al. 2008; Bayer-Giraldi et al. 2010; Gwak et al. 2010; Raymond and Kim 2012).

The N-terminal sequences of the identified IBPs of *N. glaciei, F. cylindrus,* and each of the *T. ishikariensis* antifreeze isoforms are most likely signal peptides and have low probabilities of being mitochondrial- or chloroplast-targeting peptides (Janech et al. 2006; Fig. 14.12). N-terminal sequences were found in *Attheya* sp. but not *Amphora* sp. or *Nitzschia stellate* and therefore may not be secreted (Raymond and Kim 2012).

Many diatom genes show homology to bacterial or fungal genes suggesting origins from horizontal gene transfer (HGT). *N. glaciei* and *F. cylindrus* IBPs show sequence similarity to several antifreeze isoforms of the Basidomycete fungus *Typhula ishikaiensis* (Figs. 14.12 and 14.13), which is known to inhabit sea ice (Janech et al. 2006). Sorhannus (2011) also found homology between IBPs of *F. cylindrus* and *F. curta* to IBPs from basidomycetes; however, in contrast to findings from Janech et al. (2006), IBPs from *N. glaciei* are placed in a separate clade and are suggested to originate from ancestral genes along with IBPs from *C. neogracile*.

Similarities between *F. cylindrus* and *N. glaciei* IBPs to hypothetical proteins from Gram-negative bacteria such as *Cytophaga hutchinsonii* and *Shewanella denitrificans* (between 43 and 58% amino acid sequence identity) have been observed. These bacteria have frequently been isolated from Arctic and Antarctic sea ice (Junge et al. 2002), and Cytophaga–Flavobacterium–bacteroides, which include *C. hutchinsonii*, are important in well-established sea-ice algal assemblages (Bowman et al. 1997) and the coldest (wintertime) sea ice (Junge et al. 2004). Raymond and Kim (2012) found IBPs from *Attheya* sp., *Amphora* sp., and *Nitzschia stellate* to show greatest homology to bacterial IBPs. These diatom IBPs contain no introns, and furthermore, *Flavobacterium frigoris*, which produces an IBP with 47% amino acid identity to an IBP in *Nitzschia stellate*, was isolated from Antarctic sea ice in the same layer as diatoms.

Expression of IBPs have also been demonstrated in the Antarctic bacterium *Marinomonas primoryensis,* where they aid adherence to ice, allowing *M. primoryensis* to remain near the top of the water column (Guo et al. 2012) and in an Antarctic *Colwellia* sp. where they inhibit ice recrystallization (Raymond et al. 2007). In other organisms, antifreezes appear to have arisen from a variety of proteins with other functions, although some retain the original functions (Cheng 1998). Other genes with homology to bacteria found in the *F. cylindrus* genome include ABC transporters with similarities to bacterial permeases and proton-

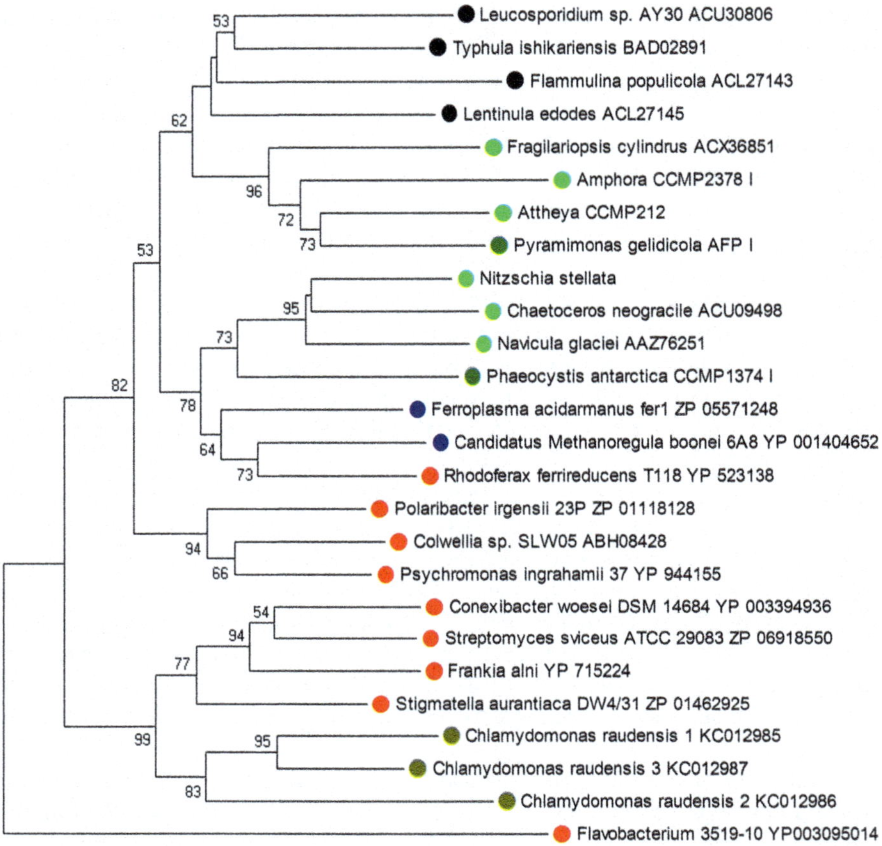

Fig. 14.12 Neighbor-joining tree constructed from amino acid sequences of selected ice-binding proteins (IBP) and IBP-like proteins. The *Chlamydomonas raudensis* IBPs (*olive*) are closest to IBP-like proteins in several bacteria and relatively distant from other algal IBPs. The tree was rooted with the Flavobacterium 3519-10 IBP. *Numbers at nodes* indicate bootstrap values for 500 replications. Values less than 50 are not shown. Colors: *black*, fungi; *light green*, diatoms; *dark green*, prasinophyte and prymnesiophyte; *blue*, archaea; *red*, bacteria; *olive*, *C. raudensis*

pumping proteorhodopsins, for trace-metal-independent ATP synthesis (Strauss et al. 2013).

14.3.1.2 Molecular Physiology

The presence of genes in a genome only indicates the potential for physiological adaptation, but knowledge of the expression and regulation of genes and their respective proteins leads to an actual understanding of how these diatoms cope with the extreme polar conditions. Expression analysis can be done by focusing on

```
Navicula           ----MMFLAKTVTLLVALVASSVAAEQ-SAVDLGTAGDFAVLSKAGVSTT 45
Fragilariopsis     MMNLNLFLISAAAMVSVASASTALPPSPPAVNLGTAEDFVILAKAGVTNV 50
Typhula            -----MFSASSLLAVIALAVSSVSAAGPSAVPLGTAGNYVILASTGVSTV 45
Cytophaga          TGIKDAAGAALPSNVTWSFTTGANASVLAVVNLRTAVNYVLLAKTAINNN 146
Ferroplasma        VSPASPVTVNGAAITVNVSFTKLAPVSISPVNLGTAGNYAILAKTGISNT 129

Navicula           GPTEVTGDIGTSPIASTALTGFALIKDSSNTFSTSSLVTGKIYAADYTAP 95
Fragilariopsis     PGGAITGDIGVSPIAASAMTGFDLVMDSSNEFSTSTEITGKAYAPDYMSP 100
Typhula            PQSVITGAVGVSPGTAASLTGFSLILSGTGTFSTSSQVTGQLTGADYGTP 95
Cytophaga          PTSAVTGAIGLSPAATSYITGFSLT-NATG-YATSSQVTGHIFAADMVSP 194
Ferroplasma        GTTSIVGNIGVSPASSTYITGLSLTMNSSGQFSTSSMVTGNVYAATYASP 179

Navicula           TPSKMTTAISDMSTAFTDAAGR-----------SDPDFLELGAGSIEGET 134
Fragilariopsis     TGTKLTTAVSDMLTAYNDAAARPVTGGPFGNSLSGETYTNLGAGEIGGLT 150
Typhula            TPSILTTAIGDMGTAYINAATR-----------SGPDFLEIYTGALGGTT 134
Cytophaga          TSSNLTTAINDMQTAYTDAAGR-----------KTPDYVELGTGNIGGKT 233
Ferroplasma        TPSTLTTAVGDMQTAYTNAAGR-----------TNPNYVNLGAGDLNGMT 218

Navicula           LVAGLYKWGTDVSFTS-SLVFDGSATDVWILQVAKDFIVGNGAQMYLTGT 183
Fragilariopsis     LTRGVYTYDINVSITSGKVTFHGGADDVFIIKTSKSVLQAANTEVVLTGG 200
Typhula            LLPGLYKWTSSVGASA-DFTISGTSTDTWIFQIDGTLDVATGKQITLVGG 183
Cytophaga          LQPGLYKWTSSVSVPS-DVTISGGANDVWIFQISGNLSLSAGAKITLSGG 282
Ferroplasma        LVPGLYKWGTGVSIST-SITLTGNSSSVWIFQISGGLTFGNGAHIILSGG 267

Navicula           AKAENIFIQVSGAVNIGTTAHVEGNILSATAIALQTGSSLNGKALSQTAI 233
Fragilariopsis     AQAKNIFWSVAQEVNVGAGAHMEGILLVKTAVKFITGSSFVGRVLSATAV 250
Typhula            AQAKNIIWVVAGAVNIEVGAKFEGTILAKTAVTFKTGSSLNGRILAQTAV 233
Cytophaga          AQAKNIFWQVAGTVTAGTTSHIEGVILSKTGITFNTGASLKGRALAQTAI 332
Ferroplasma        AQPQNIFWQVASGATIGTGATFYGTILSQTAITIATGSSMTGLALAQTAV 317

Navicula           TLDSVTIVS----------------------------------------- 242
Fragilariopsis     TLQSAAITAPATSAPTTRRGPRGLQVA----------------------- 277
Typhula            ALQSATIVEK---------------------------------------- 243
Cytophaga          ILDGNTVTQP---------------------------------------- 342
Ferroplasma        TLQSDTITAPLEPQSITAAMYGVTFTEAGLPSGTQWNVTLNGVLLSSTVP 367
```

Fig. 14.13 ClustalW alignment of ice-binding proteins from *Navicula glaciei* (Acc. no. DQ062566), *Fragilariopsis cylindrus* (CN212299), and *Typhula ishikariensis* (AB109745), and hypothetical proteins from *Cytophaga hutchinsonii* (ZP_00309837) and *Ferroplasma acidarmanus* (ZP_500309837). Predicted signal peptides are *underlined*. Gaps have been inserted to improve alignment. Conserved residues are *shaded*. The N-terminal sequence of Cytophaga protein and the N- and C-terminal sequences of Ferroplasma protein are truncated. Residue numbers are shown at *right*. Alignment is taken from Janech et al. (2006)

single genes (e.g., northern blots or quantitative PCR) or multiple genes through gene arrays or RNA sequencing. Arrays can be composed of known genes (gene-specific arrays) or the whole genome sequence (tiling arrays).

One of the most dramatic environmental changes in polar marine sea-ice habitats is the freezing of seawater and melting of the ice. The inclusion of organisms into newly formed sea ice represents a strong selective pressure. Only those organisms that are capable of acclimation to the relatively fast-changing conditions of temperature, irradiance, and salinity can survive.

Several experiments have been conducted to investigate gene expression under polar conditions including freezing temperatures, high salinity, high irradiance, and prolonged darkness. Some study multiple genes using macro-arrays (Mock and Valentin 2004) or RNA-seq (Mock et al. 2017) while others focus on specific genes such as ice-binding proteins (Bayer-Giraldi et al. 2010, 2011).

Data from EST libraries has been used to produce arrays for two polar diatom species, *F. cylindrus* (Mock and Valentin 2004) and *C. neogracile* (Hwang et al. 2008; Park et al. 2010). About 200 70mer oligonucleotides were compiled into a nylon-membrane-based macro-array to study short-, mid-, and long-term acclimation to freezing temperatures under high and low irradiance in *F. cylindrus*. 1400 *C. neogracile* transcripts were analyzed using micro-arrays to observe expression at 4 and 10 °C (Hwang et al. 2008) as well as under high, moderate, low, and changing light intensities (Park et al. 2010).

The short-term response to freezing temperatures, which simulates the incorporation into newly formed sea ice during fall, was characterized by downregulation of genes encoding proteins for photosystem II (psbA and psbC) and carbon fixation (RUBISCO large subunit, rbcL) regardless of light intensity used (3 and 35 μmol photons m^{-2} s^{-1}). However, under higher irradiance (35 μmol photons m^{-2} s^{-1}), upregulation of genes encoding chaperons (hsp 70) and genes for plastid protein synthesis and turnover (elongation factor EfTs, ribosomal rpS4 and plastidial ftsH protease) were observed (Mock and Valentin 2004).

In *Chaetoceros neogracile,* increased irradiance led to both up- and downregulation of particular LHCx proteins and fucoxanthin-chlorophyll a, c-binding proteins (FCPs) (Park et al. 2010). Several genes for cell division, transcription, and signaling were upregulated while many genes for photosynthesis (including LHC, FCPs, and PSII-associated proteins) were downregulated along with some transporter genes including members from the ABC-transporter family.

In *Fragilariopsis cylindrus*, freezing accompanied with a reduction in irradiance (from 35 to 3 μmol photons m^{-2} s^{-1}) showed a typical response to low-light acclimation by upregulation of genes encoding specific FCPs without signs of a cold stress response. FCPs are a diverse gene family composed of genes involved in light harvesting as well as dissipation of light (see Sect. 14.2; Mock and Valentin 2004). Low irradiance in this species also leads to an increase in chloroplast PUFAs which can maintain electron flow by increasing fluidity of the thylakoid membrane (Mock and Kroon 2002a).

Upregulation of stress response genes and genes for protein turnover only under higher light intensities and decreasing temperatures indicates that a decrease in temperature at such light intensities mimics a further increase in light that could be more stressful than the actual decrease in temperature was by itself (Mock and Valentin 2004). This phenomenon is probably part of a cold-shock response that is also known from temperate plants when they get exposed to lower temperatures (Allen and Ort 2001).

Entomoneis kufferathii, a sea-ice diatom, showed high catalase activity, which is linked to protection against oxidative damage, in response to high irradiance and low temperatures (Schriek 2000). Genes for glutathione metabolism, an important antioxidant, were upregulated soon after exposure to high light in *C. neogracile*, although glutathione S-transferase and superoxide dismutase (SOD), two enzymes involved in scavenging ROS, were downregulated (Park et al. 2010). A gradual increase in heat-shock proteins was observed under the same conditions over 6 h. Shifts in irradiance from either low to high or high to low light resulted in an increase in SOD in *Chaetoceros brevis* (Janknegt et al. 2008). An increase in temperature in *C. neogracile* from 4 °C to 10 °C resulted in the upregulation of several antioxidant genes including monoascorbate reductase, glutaredoxin, glutathione peroxidase, glutathione S-transferase, and alternative oxidase (Hwang et al. 2008). Polar diatoms appear to have tailored multiple resources for dealing with stress caused by the extreme polar environment.

Psychrophilic plants and diatoms are able to acclimate to higher irradiances under low temperatures (Streb et al. 1998; Mock and Hoch 2005; Ralph et al. 2005; Morgan-Kiss et al. 2006; Park et al. 2010). Long-term acclimation experiments to higher irradiances at freezing temperatures, when compared to the same light intensity but higher temperatures (+5 °C), revealed that cells kept at lower temperatures showed a typical response known from high-light acclimation: higher non-photochemical quenching, upregulation of the gene psbA, and upregulation of high-light FCPs that are involved in energy dissipation (Mock and Valentin 2004; Mock and Hoch 2005). A rapid increase in diatoxanthin (Dtx) in *C. neogracile* under high light also demonstrates energy dissipation through NPQ, along with an increase in expression of specific FCPs (Park et al. 2010).

In *F. cylindrus*, a reduction in expression of other photosynthesis-related genes (such as rbcL) was not observed after several months under freezing conditions indicating that long-term acclimation had been achieved.

Temperature effects that are less dependent on adjustments of the energy flow under freezing temperatures could also be identified by gene expression analysis (Mock and Valentin 2004). In the Mock and Valentin (2004) study, genes were selected that were either abundant in the EST libraries (e.g., ABC transporters) or were important for general acclimation to freezing temperatures (e.g., IBP, fatty-acid desaturase). Three unknown but abundant genes (in EST libraries) were also selected to see whether at least one of them is upregulated under freezing temperatures. Expression of these genes was investigated at +5 °C and 9 days after reducing the temperatures to −1.8 °C.

Upregulation of a gene encoding a delta5-desaturase under freezing temperatures indicated the necessity for production of polyunsaturated fatty acids (PUFAs) to maintain membrane fluidity at lower temperatures. Delta-5 desaturases produce omega3-fattyacids such as EPA (20:5 n−3), one of the most abundant fatty acid in diatoms and the main fatty acid in the galactolipids MGDG and DGDG. Thus, it can be assumed that more EPA is necessary under freezing temperatures to keep the thylakoid membrane fluid for electron transport or other membrane-bound processes.

Teoh et al. (2013) also found that PUFA concentration increased in *N. glaciei* with a decrease in temperature. In contrast, a delta-12 desaturase gene also known for producing PUFAs was not upregulated in temperate cyanobacteria (Nishida and Murata 1996). This indicates a different mechanism of gene regulation for this enzyme in psychrophilic diatoms.

An ABC-transporter gene was strongly upregulated at −1.8 °C in *F. cylindrus*, however, the family of ABC transporters is composed of genes with very diverse functions so it unclear of its specific function in response to freezing temperatures (Mock and Valentin 2004).

Extracellular polymeric substances (EPS) are linked to adaptation of diatoms in polar environments as both cryoprotectants and through maintenance of the cells microclimate (Underwood et al. 2010). An example of this can be seen in the sea-ice diatom *Melosira arctica*, in which EPS from the sea-ice diatom *Melosira arctica* altered the microstructure of ice-pore morphologies leading to salt retention (Krembs et al. 2011). EPS can include, but are not limited to substances such as polysaccharides (Aslam et al. 2012), uronic acid, peptides, proteins and glycoproteins (Krembs et al. 2011; Underwood et al. 2013).

Cryoprotectants can also help to maintain both the internal and external environment in polar cells and include solutes such as proline, DMSP and betaine (Lyon and Mock 2014). Proline synthesis genes were enriched in the *F. cylindrus* cold stress EST library (Mock et al. 2005).

DMSP pathway-linked protein concentrations were also increased in response to high salinity as were two protein isoforms with homology to bacterial/archaeal glycine betaine methyltransferase (Lyon et al. 2011). DMSP, which has been found in high concentrations in ice-diatom communities, has been shown to protect enzymes against denaturation in freezing conditions (DiTullio et al. 1998).

Studies on ice-binding proteins in diatoms show that they have antifreeze properties and are able to inhibit ice recrystallization (Gwak et al. 2010; Bayer-Giraldi et al. 2011; Raymond 2011). As several IBPs have similarities to bacterial or fungal sequences, it is hypothesized that they have been acquired through HGT (see Sect. 14.3.1.1) and may have allowed diatoms to colonize sea ice.

An IBP protein in *F. cylindrus* was strongly upregulated (ca. 50-fold) under freezing temperatures (Mock and Valentin 2004), while Bayer-Giraldi et al. (2011) found several isoforms in *F. cylindrus* and *F. curta* to be differentially regulated depending on temperature and salt stress. *F. cylindrus* IBPs in both of these studies have been identified in the recently published genome (Mock et al. 2017). Proteomics studies on *C. neogracile* also showed an increase in concentration of IBPs in response to freezing conditions (Gwak et al. 2010). Furthermore, isolation of IBP transcripts from Arctic and Antarctic sea ice suggests that they are found at similar levels as genes with essential metabolic processes such as photosynthesis (Uhlig et al. 2015). Within the same study, it was found that most IBP transcripts originated from diatoms, haptophytes, and crustaceans; however, many of the IBPs have not been previously characterized (Uhlig et al. 2015). These results support the hypothesis that these proteins are of great importance not only under salt stress but also under freezing temperatures to protect the cells from injury by growing ice crystals.

An important adaptation for polar photosynthetic organisms is the need to survive for periods of prolonged darkness. As discussed in Sect. 14.3.1.1, 7 day darkness in *F. cylindrus* leads to a decrease in photosynthesis and associated processes. Genes which are involved in starch, sugar, and fatty acid metabolism are upregulated (Mock et al. 2017) suggesting that *F. cylindrus* is able to use existing cellular resources in place of photosynthesis. Diatoms are able to store glucan for use in periods of extended darkness (van Oijen et al. 2003) and are able to uptake molecules such as sugar and starch (Palmisano and Garrison 1993). The urea cycle in diatoms has been suggested as a means to process inorganic carbon and nitrogen, particularly during low nitrogen availability (Allen et al. 2011). All genes for the urea cycle can be found in the *F. cylindrus* genome. Proton-pumping proteorhodopsins, for trace-metal-independent ATP synthesis (Strauss et al. 2013), were upregulated under darkness, suggesting a role in energy production. There are also ATP-independent enzymes available to *F. cylindrus* which may save chemical energy such as pyrophosphate-dependent phospho-fructo-kinase which was elevated during salinity acclimation (Lyon et al. 2011).

Information is steadily becoming available for polar diatoms. New insights are being gained into their adaptations and the importance of their roles in polar communities. Although much has been learned, there are vast numbers of genes with unknown or partially characterized functions in many of these studies. For example, many identified transcripts have no homology to existing sequences (Mock et al. 2005, 2017; Krell 2006), and different FCPs and LHC proteins are both up- and downregulated under the same conditions (Park et al. 2010). Reverse genetics is needed in order to establish the function and roles of these genes and their pathways. A transformation system for *F. cylindrus* has been successfully established—as far as we are aware, this is the first transformation system for any eukaryotic polar species (Hopes and Mock, unpublished). Furthermore, CRISPR-Cas for gene knock-out and gene silencing in the temperate diatoms *Thalassiosira pseudonana* (Hopes et al. 2016; Kirkham and Mock, unpublished) and *Phaeodactylum tricornutum* (Nymark et al. 2016; De Riso et al. 2009) have been established. Work on CRISPR-Cas in *F. cylindrus* is also currently ongoing.

With the establishment of additional, elegant molecular tools for diatoms, there is a much greater scope for potential research and therefore our understanding of these psychrophilic and psychrotolerant organisms and their environment.

14.3.2 Green Algae (Chlorophyceae)

Most polar green algae live in freshwater ecosystems such as snow, permanently ice-covered lakes, or more ephemeral habitats like creeks or melt ponds on top of snow or sea ice. Most species belong either to the genera *Chlamydomonas, Chloromonas*, or *Chlorella*, and many of them are very motile due to the presence of flagella.

Ecologically important species that are physiologically and molecularly well characterized are *Chlamydomonas raudensis, Chlamydomonas nivalis*, and *Chlamydomomas* sp. ICE-L. *C. raudensis* is an abundant species in permanently ice-covered lakes and the clone UWO241 has been studied for decades (see review by Morgan-Kiss et al. 2006). *C. nivalis* is a dominant representative of the snow-algae community and also intensively studied (Williams et al. 2003). Therefore, this discussion will mainly focus on *Chlamydomonas* sp. There is less research in this area in terms of functional genomics; however, the genome sequencing of *Coccomyxa subellipsoidea* provides some insight into polar adaptations within the Chlorophyceae as does the cold shock EST library for *Pyramimonas gelidicola*.

14.3.2.1 Functional Genomics

Coccomyxa subellipsoidea is a psychrotolerant green alga that has been isolated from dried algal peat in Antarctica, and although it can grow at low temperatures it shows optimal growth at around 20 °C (Blanc et al. 2012). Despite not being a true psychrophile, its genome has some pronounced differences to mesophilic chylorphytes and offers several insights into polar adaptation. Although the genomes of several green algae have been sequenced, this is the first genome to be published from a polar microalga. An EST library has also been generated under cold shock conditions for the psychrophilic *Pyramimonas gelidicola*, a dominant primary producer from Antarctic sea ice (Jung et al. 2012).

In comparison to other sequenced chlorophytes, *C. subellipsoidea* has a large number of mitochondrial and chloroplast sequences integrated into its nuclear genome. GC content of these organelle genomes is also comparatively high. It is important to maintain homeostasis and efficient cellular functions under the extreme conditions found in polar regions. This includes lipid metabolism and membrane fluidity. Four lipid protein families were over-represented in *C. subellipsoidea*: type-I-fatty acid synthases, FA elongases, FA ligases, and type 3 lipases. There were also three fatty acid desaturases present that were not found in temperate counterparts (Blanc et al. 2012). An increase in double bonds in membrane based lipids helps to increase fluidity at cold temperatures (Los and Murata 2004). Within the same species, there were a high number of genes involved in polysaccharide and cell wall metabolism (Blanc et al. 2012). As previously mentioned, both glycoproteins and polysaccharides can act as cryopreservants in microalgae. Two genes involved in cryoprotection with homology to late embryogenesis abundant (LEA) proteins have also been found in *C. subellipsoidea* (Liu et al. 2011).

Structural parts such as the cytoskeleton of the cell also have to be adapted to low temperatures in order to conduct mitosis, meiosis, secretion, and cell motility.

The tubulin alpha chain protein domain was the fifth most abundant in the EST library from *P. gelidicola* (Jung et al. 2012). Willem et al. (1999) showed that alpha-tubulin from two *Chloromonas* spp. had five amino acid substitutions compared to the mesophilic *Chlamydomonas reinhardtii*. Two of these substitutions occurred in the region of inter dimer contacts that could therefore positively influence microtubule assembly under low temperatures.

Translation elongation factor-1a was prominent in ESTs from *P. gelidicola* (Jung et al. 2012). Furthermore, a translation elongation factor-1a was found in the *C. subellipsoidea* genome that is able to functionally replace elongation factor like EFL found in previously sequenced chlorophytes. Upregulation of an elongation factor involved in protein synthesis has also been observed in cold shock diatoms (Mock and Valentin 2004).

Given that polar species may be exposed to freezing temperatures and high light, many adaptive strategies include proteins involved in stress response and protection against ROS. DOPA-dioxygenase which provides protection against solubilized oxygen was identified in the *C. subellipsoidea* genome, as were two genes with homologs to phospholipase D and chalcone synthase. The former is involved in stress response, while homologs of the latter are involved in metabolites for UV photoprotection and antimicrobial defense in plants (Blanc et al. 2012).

Both heat shock protein 70 (hsp70) and stress-related chlorophyll a/b binding protein were enriched in *P. gelidicola* ESTs. Heat shock protein 70 appears to be a key component involved in adaptation of several polar microalgae species (Mock and Valentin 2004; Krell 2006; Liu et al. 2010).

When comparing *C. subellipsoidea* to temperate chlorophytes, Blanc et al. (2012) found that as well as enrichment of certain gene families and gene additions there were also several key genes missing. This includes PsaN, which is involved in docking plastocyanin to the PSI complex. This leads to a drop in electron transfer from plastocyanin to PSI which may be beneficial in a polar environment as low temperatures create an excess of electrons through this system which in turn leads to an increase in ROS. As PsaN is not crucial for photosynthesis, loss of this gene may protect the cell from oxidative damage (Blanc et al. 2012). *C. subellipsoidea* also has genes for dioxygenase and FA desaturases that utilize dioxygen and therefore may provide further protection against ROS (Blanc et al. 2012).

One gene loss which could reduce cellular efficiency in *C. subellipsoidea*, however, is a pyruvate phosphate dikinase (PPDK), which produces ATP through glycolysis. Function of this gene appears to be replaced by three pyruvate kinases, which potentially produce less chemical energy (Blanc et al. 2012).

In terms of nutrient acquisition, *C. subellipsoidea* has a large number of genes for amino acid permeases and transporters which may enhance uptake of organic nutrients. It also has cobalamin-dependent methionine synthase but lacks the cobalamin-dependent version of this gene MetH (Blanc et al. 2012), suggesting that this species is not dependent on this often bacteria-associated cofactor (Croft et al. 2005) for synthesis of this important amino acid.

There is still much to discover in establishing the function and origins of many genes specific to polar species. There were a higher number of ESTs with unknown functions under freezing conditions in *P. gelidicola* compared to 4 °C (Jung et al. 2012). Furthermore, there are over 2300 genes in the *C. subellipsoidea* genome with no known homologs in sequenced mesophilic chlorophytes. The majority of these genes show homology to Streptophytes and other Eukaryotes, suggesting origins from a common ancestor to chlorophytes. Interestingly rather than displaying homology to green algae, most of the genes involved in defense, detoxification, and carbohydrate metabolism show higher sequence similarity to bacteria, suggesting possible acquisition by HGT.

As discussed in Sect. 14.3.1.1, ice-binding proteins in diatoms appear to have bacterial or fungal origins. Several IBPs have also been identified in psychrophilic or psychrotolerant green algae including *Pyramimonas gelidicola* (Jung et al. 2014), *Chlamydomonas raudensis* (Raymond and Morgan-Kiss 2013), *Chlamydomonas* sp. strain CCMP681 (Raymond et al. 2009), *Chloromonas sp.* (Jung et al. 2016) and *Chloromonas brevispina* (Raymond 2014). Raymond and Morgan-Kiss (2013) separate ice-binding proteins into two different groups: IBP I, a group of similar proteins appearing to have fungal or bacterial origins (Raymond and Morgan-Kiss 2013; Sorhannus 2011; Raymond and Kim 2012; Jung et al. 2014; Raymond 2011) and IBP II. So far all studied algal species have type I IBPs with the exception of *Chlamydomonas* sp. strain CCMP681 which has four type II isoforms isolated from ESTs (Raymond et al. 2009; Raymond and Morgan-Kiss 2013). A polyphyletic origin for IBPs has been suggested given their sequential and structural differences, as well as a lack of IBPs in temperate species (Fig. 14.13; Raymond and Morgan-Kiss 2013).

As more genomes and transcriptomes become available for polar chlorophytes, more light can be shed on their intricacies and adaptations to extreme environments. The genome of an important Antarctic sea ice chlorophyte, *Chlamydomonas* ICE-L, has been recently sequenced (personal communication with Naihao Ye, Yellow Sea Fisheries Research Institute, Chinese Academy of Fishery Sciences, Qingdao 266071, China) and Raymond and Morgan-Kiss (2013) plan to compare the transcriptome of *C. raudensis* to a temperate counterpart.

14.3.2.2 Molecular Physiology

Maximum growth rates of polar green algae are comparable to those from polar diatoms. They range from 0.2 to 0.4 day^{-1} (Tang et al. 1997). Temperatures above 18 °C are mostly lethal to these algae. *C. raudensis* has its maximum photosynthetic rates at 8 °C, which declines steadily with increasing temperatures (Morgan-Kiss et al. 2006). This indicates maximal efficiency in converting light into photosynthetic energy at low temperatures. The quality of light also plays an important role, and *C. raudensis* is not able to grow under red light (Morgan-Kiss et al. 2005). This is probably a consequence of almost never being exposed to a longer wavelength spectrum in the natural habitat of permanently ice covered lakes where the ice absorbs all longer wavelengths of solar irradiance (Fritsen and Priscu 1999; Morgan-Kiss et al. 2006). However, the majority of this light is reflected on the white surface of ice and scattered while passing through ice. Thus, the environment below the ice is characterized by low intensities enriched in blue-green wavelengths (Lizotte and Priscu 1992).

Many physiological and molecular investigations have been conducted with *C. raudensis* to find the reasons for successful photo adaptation under these extreme conditions. A comparison with the temperate *C. reinhardtii* partly uncovered the mechanisms of photo adaptation in *C. raudensis* (Morgan-Kiss et al. 2005, 2006): In contrast to the temperate *C. reinhardtii*, the psychrophile has lost its ability to

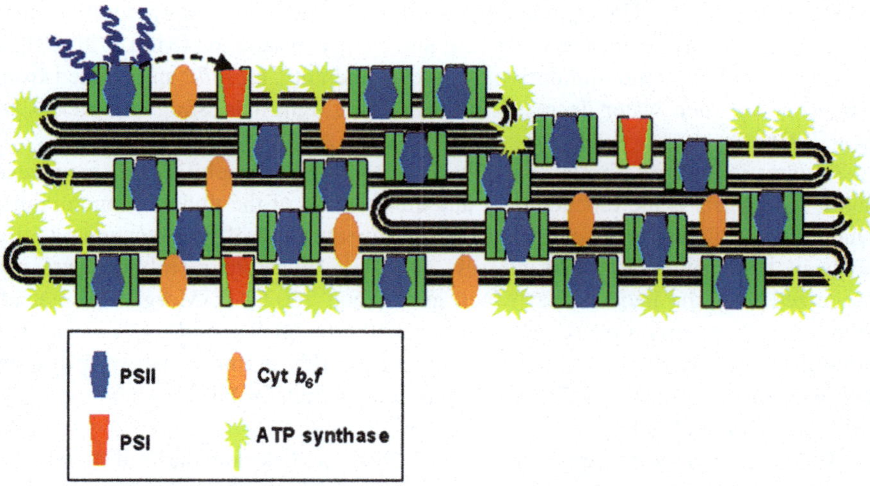

Fig. 14.14 Model for organization of thylakoid pigment–protein complexes of the electron transport chain in the psychrophilic *Chlamydomonas raudensis* UWO 241. In the natural, extremely stable light environment of extreme shade and predominantly *blue-green* wavelengths (*blue lines*), the majority of available light would be preferentially absorbed by PSII. Adaptation in *C. raudensis* to this light environment has led to an unusually high PSII/PSI stoichiometry and highly efficient energy transfer from LHCII to PSII. Conversely, PSI and associated light-harvesting complexes are both structurally and functionally downregulated. Given the severe reduction in light-harvesting capacity of PSI, it is proposed that PSI centers are largely excited via a spillover energy transfer mechanism from PSII (*dotted line*). Photosynthetic membranes may be arranged as loose stacks rather than distinct granal and stromal regions to promote energy spillover between the photosystems. Picture from Morgan-Kiss et al. (2006)

live under high light but increased its efficiency of light harvesting under low light in the blue-green spectrum. This adaptation can be seen in structural changes of the photosynthetic apparatus (Fig. 14.14). For instance, *C. raudensis* has an unusually high ratio of photosystem II to I and significantly higher levels of light-harvesting II complexes than its temperate counterpart *C. reinhardtii*. These changes are probably an adaptive advantage under constant exposure to blue light of low photon flux densities because the light-harvesting apparatus of photosystem II (PSII) utilizes chlorophyll b and short-wavelength-absorbing chlorophyll a to absorb light predominantly in the blue region. Interestingly, most marine algae (e.g., red algae, diatoms), which are also living in a blue-green light environment because of optical properties of the seawater, also show a high ratio of PSII to PSI due to chromatic regulation (Fujita 2001). However, most of them, and even psychrophilic diatoms, have the physiological ability to grow under high irradiance levels.

While the ability to dissipate excess energy through NPQ has been reduced in *C. raudensis* (Morgan-Kiss et al. 2006), other polar species in this genera have retained this ability which allows them to photosynthesize under high-light conditions. *Chlamydomonas* sp. ICE-L shows an upregulation of light-harvesting complex (LHC) genes LhcSR1 and LhcSR2, accompanied by an increase in NPQ following high light, UV-B radiation, and high salinity. This suggests that these LHC genes play a role in stress response and energy dissipation (Mou et al. 2012). In order to mitigate photoinhibition, a psychrotolerant *Chlorella* sp. isolated from Arctic glacier melt water decreases the size of its light-harvesting complex (Cao et al. 2016).

Another interesting similarity between diatoms (psychrophilic and temperate) and *C. raudensis* is the biochemistry and architecture of the thylakoid membrane. Diatoms, as well as *C. raudensis*, have high concentrations of polyunsaturated fatty acids in their thylakoid lipids classes, and their thylakoid membranes are not organized in grana and stroma (Mock and Kroon 2002a, b; Morgan-Kiss et al. 2006). This possibly means that looser membrane stacks in *C. raudensis* and homogeneously folded membranes in diatom plastids promote energy spill over between photosystems and therefore light energy transfer between photosystems (Morgan-Kiss et al. 2006).

An increase in transcripts for omega-3 fatty acid desaturase (CiFAD3) was measured in a *Chlamydomonas* sp. ICE-L under both high (12 °C) and low temperatures (0 °C) compared to a control at 6 °C, as well as at high salinity (Zhang et al. 2011; An et al. 2013). This suggests that PUFAs may also play a role in heat stress and high salinity acclimation. Consumption of PUFAs was also observed in the same species during darkness (Xu et al. 2014), indicating that PUFAs are an important aspect of adaptation to several extreme conditions found in the polar regions. As with diatoms, antioxidants also play an important role in cold-shock adaptation, as seen in an Antarctic *Chlamydomonas* sp. in which an increase in glutathione S-transferase was observed (Kan et al. 2006).

The snow alga *C. nivalis* is exposed to the full spectrum of solar irradiance (UVC to infrared) and must therefore have a completely different photosynthesis performance compared to the low light adapted *C. raudensis* (Remias et al. 2005). The most striking difference between photosynthesis of both psychrophilic green algae is that *C. nivalis* does not seem to be inhibited by high solar irradiances. Even an exposure of cells to photon flux densities of 1800 μmol photons m^{-2} s^{-1} for 40 min at 1.5 °C did not inhibit net photosynthesis (Remias et al. 2005). This extreme photosynthetic performance is only possible by a change in the life cycle. A combination of factors may trigger the formation of immotile red hypnoblast stages that are most resistant to environmental changes (Müller et al. 1998; Remias et al. 2005).

The transformation into hypnoblasts is characterized by a substantial incorporation of sugars and lipids and by the formation of esterified extraplastidal secondary carotenoids (Hoham and Duval 2001). The most important carotenoid is astaxanthin which is located in cytoplasmatic lipid globuli (Müller et al. 1998; Remias et al. 2005) and is assumed to be responsible for the high photostability and therefore the absence of photoinhibition under strong solar irradiance on top of snow (Remias et al. 2005). Mature hypnoblasts can contain about 20 times more of this pigment than chlorophyll a, where the astaxanthin is possibly acting as a filter to reduce the irradiance that would otherwise be damaging to the photosynthetic activity inside the plastids. Exposure to UV-B in *Chlamydomonas* sp. ICE-L led to an increase in expression of heat shock protein 70 (Liu et al. 2010) which suggests a role in protection against high irradiance.

High solar irradiance is not the only harsh condition on top of snow. Drought due to freezing of water is another main stress on the hypnoblast stages of *Chlamydomonas nivalis*. Like cacti in the desert, these stages have very rigid cell walls as the outer boundary to an extreme environment (Müller et al. 1998; Remias et al. 2005). Sometimes cells secrete carbohydrates to produce a visible mucilage sheet around them (Müller et al. 1998). These carbohydrates are not only attractive to bacteria that use them as a substrate but they also trap particles transported into the snow by wind. These particle-covered cells increase the absorption of solar irradiance and therefore the production of heat. This heat might cause melting of surrounding snow crystals and therefore provide liquid water to the cells (Takeuchi 2002). Such small spots of melt events around warm bodies (e.g., rock debris, cells) are called cryoconite holes (Takeuchi 2002). However, these adhering particles may also shade and thus protect *C. nivalis* against high irradiance. This is not universal and hypnoblasts from *C. nivalis*, for example, never show such attached structures.

Chemical reactions are influenced by temperature according to the relationship described by Arrhenius. In general, a 10 °C reduction in growth temperature causes biochemical reaction rates to decline 2–3 times. However, doubling times of psychrophilic algae can be comparable to mesophilic algae (Sommer 1989) which means that rates of enzyme catalyzed reactions must be optimized to low temperatures in these organisms (Feller and Gerday 2003). Studies with the enzyme nitrate reductase (NR), for instance, showed that these enzymes from psychrophilic algae possess structural modifications that make them more cold adapted, being more catalytically efficient at lower temperatures but at the same time less thermally stable, than NRs from mesophilic species (Di Martino Rigano et al. 2006). It also appears that light and salinity may influence nitrogen metabolism in *Chlamydomonas* sp. ICE-L (Wang et al. 2015).

In contrast to NR, the temperature maximum for carboxylase activity of ribulose-1-5-bisphosphatecarboxylase/oxygenase (RUBISCO), one of the most critical enzymes for inorganic carbon fixation in photoautotrophes, was not altered in some psychrophilic green algae and the specific activity at low temperatures was actually lower in the psychrophilic if compared to the mesophilic Rubisco (Devos et al. 1998). Decreased catalytic efficiency of these RUBISCOs under low temperature seems to be at least partly compensated by an increased cellular concentration of the protein. This is supported by the presence of RUBISCO as the fifth most abundant EST in *P. gelidicola* (Jung et al. 2012). An increase in ribosomal proteins seen at colder temperatures may counteract reduced efficiencies in translation (Toseland et al. 2013); alternatively, it may help to cope with upregulation of proteins due to reduced activity. Cao et al. (2016) found that a strain of arctic *Chlorella* increased both proteins and lipids at lower temperatures.

Expression and secretion of ice-binding proteins in polar Chlorophytes helps to maintain a fluid environment and reduce damage from ice crystals. Studies which look at IBPs through recombinant proteins and culture supernatant have demonstrated functions including changes in ice morphology, ice pitting, recrystallization inhibition, and the creation of smaller brine pockets which helps to maintain salinity (Raymond et al. 2009; Raymond and Kim 2012; Raymond and Morgan-Kiss 2013; Jung et al. 2014).

As with diatoms, molecular tools are constantly improving for Chlorophytes, including techniques for activating gene expression with transcription activator-like effectors (TALEs) (Gao et al. 2014) and gene editing with CRISPR-Cas (Shin et al. 2016; Wang et al. 2016). So far only temperate green algae have been selected for targeted gene knock-out, but as molecular tools such as these become available in their polar counter-parts, the potential to discover the function and role of important genes and pathways in polar adaptation drastically increases.

14.4 Conclusions

The application of omics approaches in combination with biochemical and physiological measurements has revealed unique adaptations in polar microalgae. Unsurprisingly, there is evidence that the extreme and highly variable conditions in polar ecosystems were driving those adaptations. While some of these adaptations (e.g., allelic divergence, gene duplications) were the consequence of mutations and subsequent diversification, others were based on biotic interactions that enabled transfer of genes (e.g., ice-binding) between different species and therefore the entire community to thrive under the extreme conditions of polar ecosystems. These mechanisms of adaptive evolution are not unique to polar microalgae but how they are used to produce unique phenotypes required to survive temperatures below freezing, long periods of darkness, strong seasonality, and fluctuations in nutrients and salinity is still unknown. Once we have obtained genetically tractable model species such as *Fragilariopsis cylindrus* and *Chlamydomonas* sp. ICE, we'll be able to better understand how genotypes impact phenotypes that matter to thrive under polar conditions. With these model species, we will be able to test, through experimental evolution approaches, how their populations respond to global warming, which is still largely unknown. Results from these model species can be used to inform studies on natural populations (e.g., barcoding, metatranscriptomes, and metagenomes) in terms of identifying their standing pool of genetic variation and evolutionary potential to respond to global warming. Identification of genetic diversity in these organisms not only provides new insights into their evolution and adaptation but also contributes to extend the pool of marine genetic resources, which so far is dominated by genes and their products from non-polar organisms.

References

Allen DJ, Ort DR (2001) Impacts of chilling temperatures on photosynthesis in warm-climate plants. Trends Plant Sci 6:36–42. doi:10.1016/S1360-1385(00)01808-2

Allen AE, Dupont CL, Oborník M et al (2011) Evolution and metabolic significance of the urea cycle in photosynthetic diatoms. Nature 473:203–207

An M, Mou S, Zhang X et al (2013) Expression of fatty acid desaturase genes and fatty acid accumulation in Chlamydomonas sp. ICE-L under salt stress. Bioresour Technol 149:77–83. doi:10.1016/j.biortech.2013.09.027

Armbrust EV, Berges JA, Bowler C, Green BR, Martinez D, Putnam NH, Zhou S, Allen AE, Apt KE, Bechner M, Brzezinski MA, Chaal BK, Chiovitti A, Davis AK, Demarest MS, Detter JC, Glavina T, Goodstein DS (2004) The genome of the diatom Thalassiosira pseudonana: ecology, evolution, and metabolism. Science 306:79–86

Arrigo KR, Thomas DN (2004) Large scale importance of sea ice biology in the Southern Ocean. Antarct Sci 16:471–486. doi:10.1017/S0954102004002263

Aslam SN, Cresswell-Maynard T, Thomas DN, Underwood GJC (2012) Production and characterization of the intra- and extracellular carbohydrates and polymeric substances (Eps) of three sea-ice diatom species, and evidence for a cryoprotective role for Eps. J Phycol 48:1494–1509. doi:10.1111/jpy.12004

Bayer-Giraldi M, Uhlig C, John U et al (2010) Antifreeze proteins in polar sea ice diatoms: diversity and gene expression in the genus *Fragilariopsis*. Environ Microbiol 12:1041–1052. doi:10.1111/j.1462-2920.2009.02149.x

Bayer-Giraldi M, Weikusat I, Besir H, Dieckmann G (2011) Characterization of an antifreeze protein from the polar diatom *Fragilariopsis cylindrus* and its relevance in sea ice. Cryobiology 63:210–219. doi:10.1016/j.cryobiol.2011.08.006

Beil U, Thiede J (1990) Geophysical history of polar oceans: Arctic versus Antarctic. Kluwer Academic Publishers, Netherlands

Blanc G, Agarkova I, Grimwood J et al (2012) The genome of the polar eukaryotic microalga Coccomyxa subellipsoidea reveals traits of cold adaptation. Genome Biol 13:R39. doi:10.1186/gb-2012-13-5-r39

Bluhm BA, Swadling KM, Gradinger R (2017) Sea ice as a habitat for macrograzers. In: Thomas DN (ed) Sea ice, 3rd edn. Wiley-Blackwell, Oxford, pp 394–414

Bowman JP, Sa MC, Brown MV et al (1997) Diversity and association of psychrophilic bacteri in Antarctic sea ice. Appl Environ Microbiol 63:3068–3078

Boyd PW (2002) Review of environmental factors controlling phytoplankton processes in the Southern Ocean 1. J Phycol 38:844–861. doi:10.1046/j.1529-8817.2002.t01-1-01203.x

Boyd PW, Jickells T, Law CS et al (2007) Mesoscale iron enrichment experiments 1993-2005: synthesis and future directions. Science 315:612–617. doi:10.1126/science.1131669

Brierley AS, Thomas DN (2002) Ecology of Southern Ocean pack ice. Adv Mar Biol 43:171–276

Cannone N, Guglielmin M, Gerdol R (2004) Relationships between vegetation patterns and periglacial landforms in northwestern Svalbard. Polar Biol 27:562–571. doi:10.1007/s00300-004-0622-4

Cao K, He M, Yang W et al (2016) The eurythermal adaptivity and temperature tolerance of a newly isolated psychrotolerant Arctic *Chlorella* sp. J Appl Phycol 28:877–888. doi:10.1007/s10811-015-0627-0

Caron DA, Gast RJ, Garneau M-E (2017) Sea ice for a habitat for micrograzers. Sea ice. In: Thomas DN (ed) Sea ice, 3rd edn. Wiley-Blackwell, Oxford, pp 370–393

Cheng CHC (1998) Evolution of the diverse antifreeze proteins. Curr Opin Genet Dev 8:715–720. doi:10.1016/S0959-437X(98)80042-7

Cockell CS, Stokes MD (2004) Widespread colonization by polar hypoliths H. Nature 431:414. doi:10.1038/nature03019

Comiso J (2003) Large-scale characteristics and variability of the global sea ice cover. In: Thomas DN, Dieckmann G (eds) Sea ice: an introduction to its physics, chemistry, biology and geology. Blackwell Science, Oxford, pp 112–142

Cota GF (1985) Photoadaptation of high Arctic ice algae. Nature 315:219–222

Cottier F, Steele M, Nilsen F (2017) Sea ice and Arctic Ocean oceanography. In: Thomas DN (ed) Sea ice, 3rd edn. Wiley-Blackwell, Oxford, pp 197–215

Croft MT, Lawrence AD, Raux-Deery E et al (2005) Algae acquire vitamin B12 through a symbiotic relationship with bacteria. Nature 438:90–93. doi:10.1038/nature04056

De Riso V, Raniello R, Maumus F et al (2009) Gene silencing in the marine diatom Phaeodactylum tricornutum. Nucleic Acids Res 37:e96. doi:10.1093/nar/gkp448

Deming JW (2002) Psychrophiles and polar regions. Curr Opin Microbiol 5:301–309

Devos N, Ingouff M, Loppes R, Matagne RF (1998) Rubisco adaptation to low temperatures: a comparative study in psychrophilic and mesophilic unicellular algae. J Phycol 34:655–660. doi:10.1046/j.1529-8817.1998.340655.x

Di Martino Rigano V, Vona V, Lobosco O et al (2006) Temperature dependence of nitrate reductase in the psychrophilic unicellular alga *Koliella antarctica* and the mesophilic alga *Chlorella sorokiniana*. Plant Cell Environ 29:1400–1409. doi:10.1111/j.1365-3040.2006.01523.x

Dieckmann GS, Hellmer HH (2003) The importance of sea ice: an overview. In: Thomas DN, Dieckmann GS (eds) Sea ice – an introduction to its physics, chemistry, biology and geology. Blackwell Science, Oxford, pp 1–21

Dittmar T, Kattner G (2003) The biogeochemistry of the river and shelf ecosystem of the Arctic Ocean: a review. Mar Chem 83:103–120. doi:10.1016/S0304-4203(03)00105-1

Ditullio GR, Garrison DL, Mathot S (1998) Dimethylsulfonopropionate in sea ice algae from the Ross Sea polynya. In: Lizotte M, Arrigo K (eds) Antarctic sea ice: biological processes, interaction and variability, Antarctic research series, vol 73, pp 139–146

Eddie B, Krembs C, Neuer S (2008) Characterization and growth response to temperature and salinity of psychrophilic, halotolerant *Chlamydomonas* sp. ARC isolated from Chukchi Sea ice. Mar Ecol Prog Ser 354:107–117. doi:10.3354/meps07243

Ehrenberg CG (1841) Einen Nachtrag zu dem Vortrage über Verbreitung und Einfluß des mikroskopischen Lebens in Süd- und Nordamerika. Berichte über die zur Bekanntmachung geeigneten Verhandlung der K Akad der Wissenschaften zu Berlin. Monatsberichte 1841:202–207

Ehrenberg CG (1853) Über neue Anschauungen des kleinstein nördlichen Polarlebens. Berichte über die zur Bekanntmachung geeigneten Verhandlung der Königlich-Preussischen Akad der Wissenschaften zu Berlin. Monatsberichte 1853:522–529

Eicken H (1992) The role of sea ice in structuring Antarctic ecosystems. Polar Biol 12:3–13

Eicken H, Bock C, Wittig R et al (2000) Magnetic resonance imaging of sea-ice pore fluids: methods and thermal evolution of pore microstructure. Cold Reg Sci Technol 31:207–225. doi:10.1016/S0165-232X(00)00016-1

Feller G, Gerday C (2003) Psychrophilic enzymes; hot topics in cold adaptation. Nat Rev Microbiol 1:200–208

Fiala M, Oriol L (1990) Light-temperature interactions on the growth of Antarctic diatoms. Polar Biol 10:629–636

Friedmann EI, Kappen L, Meyer MA et al (1993) Long-term productivity in the cryptoendolithic microbial community of the Ross Desert, Antarctica. Microb Ecol 25:51–69

Fritsen CH, Priscu JC (1999) Seasonal change in the optical properties of the permanent ice cover on Lake Bonney, Antarctica: consequences for lake productivity and phytoplankton dynamics. Limnol Oceanogr 44(2):447–454

Fujita Y (2001) Chromatic variation of the abundance of PSII complexes observed with the red alga Prophyridium cruentum. Plant Cell Physiol 42:1239–1244. doi:10.1093/pcp/pce164

Gao H, Wright D, Li T et al (2014) TALE activation of endogenous genes in Chlamydomonas reinhardtii. Algal Res 5:52–60. doi:10.1016/j.algal.2014.05.003

Gleitz M, Thomas DN (1993) Variation in phytoplankton standing stock, chemical composition and physiology during sea-ice formation in the southeastern Weddell Sea, Antarctica. J Exp Mar Biol Ecol 173:211–230. doi:10.1016/0022-0981(93)90054-R

Gleitz M, v.d. Loeff MR, Thomas DN et al (1995) Comparison of summer and winter inorganic carbon, oxygen and nutrient concentrations in Antarctic sea ice brine. Mar Chem 51:81–91. doi:10.1016/0304-4203(95)00053-T

Gleitz M, Bartsch A, Dieckmann GS, Eicken H (1998) Composition and succession of sea ice diatom assemblages in the eastern and southern Weddell Sea, Antarctica. Antarct Res Ser 73:107–120

Granskog M, Kaartokallio H, Kuosa H et al (2006) Sea ice in the Baltic Sea – a review. Estuar Coast Shelf Sci 70:145–160. doi:10.1016/j.ecss.2006.06.001

Guo S, Garnham CP, Whitney JC, Graham LA, Davies PL, Hensel M (2012) Re-evaluation of a bacterial antifreeze protein as an adhesin with ice-binding activity. PLoS One 7(11):e48805

Gwak IG, sic Jung W, Kim HJ et al (2010) Antifreeze protein in Antarctic marine diatom, *Chaetoceros neogracile*. Mar Biotechnol 12:630–639. doi:10.1007/s10126-009-9250-x

Haas C (2017) Sea ice thickness distribution. In: Thomas DN (ed) Sea ice, 3rd edn. Wiley-Blackwell, Oxford, pp 42–64

Haas C, Thomas DN, Bareiss J (2001) Surface properties and processes of perennial Antarctic sea ice in summer. J Glaciol 47:613–625. doi:10.3189/172756501781831864

Hällfors G (2004) Checklist of Baltic Sea phytoplankton species (including some heterotrophic protistan groups). Balt Sea Environ Proc 95:208

Hansom J, Gordon J (1998) Antarctic environments and resources – a geographical perspective. Addison Wesley Longman, Harlow

Hendey NI (1974) A revised checklist of British diatoms. J Mar Biol Assoc UK 54:277–300

Hodson AJ, Mumford PN, Kohler J, Wynn PM (2005) The High Arctic glacial ecosystem: new insights from nutrient budgets. Biogeochemistry 72:233–256. doi:10.1007/s10533-004-0362-0

Hoham RW, Duval B (2001) Microbial ecology of snow and freshwater ice with emphasis on snow algae. In: Jones HG, Pomeroy JW, Walker DA, Hoham RW (eds) Snow ecology: an interdisciplinary examination of snow-covered ecosystems. Cambridge University Press, Cambridge, pp 168–228

Hooker J (1847) The botany of the Antarctic voyage of H.M. Discovery ships Erebus and Terror in the years 1838–1843 Part 1. Flora Antarctica. Reeve Brothers, London

Hopes A, Nekrasov V, Kamoun S, Mock T (2016) Editing of the urease gene by CRISPR-Cas in the diatom *Thalassiosira pseudonana*. Plant Methods 12:49–60. doi:10.1186/s13007-016-0148-0

Horner R (1985) Sea ice biota. CRC Press, Baco Raton, FL

Hsiao S (1983) A checklist of marine phytoplankton and sea ice microalgae recorded from Arctic Canada. Nova Hedwigia 37:225–314

Hwang YS, Jung G, Jin E (2008) Transcriptome analysis of acclimatory responses to thermal stress in Antarctic algae. Biochem Biophys Res Commun 367:635–641. doi:10.1016/j.bbrc.2007.12.176

Ikävalko J, Gradinger R (1997) Flagellates and heliozoans in the Greenland Sea ice studied alive using light microscopy. Polar Biol 17:473–481. doi:10.1007/s003000050145

Janech MG, Krell A, Mock T et al (2006) Ice-binding proteins from sea ice diatoms (Bacillariophyceae). J Phycol 42:410–416. doi:10.1111/j.1529-8817.2006.00208.x

Janknegt PJ, Van De Poll WH, Visser RJW et al (2008) Oxidative stress responses in the marine antarctic diatom *Chaetoceros brevis* (Bacillariophyceae) during photoacclimation. J Phycol 44:957–966. doi:10.1111/j.1529-8817.2008.00553.x

Johnston CG, Vestal JR (1991) Photosynthetic carbon incorporation and turnover in Antarctic cryptoendolithic microbial communities: are they the slowest-growing communities on earth? Appl Environ Microbiol 57:2308–2311

Jones EP, Swift JH, Anderson LG et al (2003) Tracing Pacific water in the North Atlantic Ocean. J Geophys Res 108:1–10. doi:10.1029/2001JC001141

Jung W, Lee SG, Kang SW et al (2012) Analysis of expressed sequence tags from the Antarctic psychrophilic green algae, *Pyramimonas gelidicola*. J Microbiol Biotechnol 22:902–906. doi:10.4014/jmb.1201.01002

Jung W, Gwak Y, Davies PL et al (2014) Isolation and characterization of antifreeze proteins from the Antarctic marine microalga *Pyramimonas gelidicola*. Mar Biotechnol 16:502–512. doi:10.1007/s10126-014-9567-y

Jung W, Campbell RL, Gwak Y et al (2016) New cysteine-rich ice-binding protein secreted from antarctic microalga, chloromonas sp. PLoS ONE 11:1–26. doi:10.1371/journal.pone.0154056
Junge K, Imhoff F, Staley T, Deming JW (2002) Phylogenetic diversity of numerically important Arctic sea-ice bacteria cultured at subzero temperature. Microb Ecol 43:315–328. doi:10.1007/s00248-001-1026-4
Junge K, Eicken H, Deming JW (2004) Bacterial Activity at -2 to -20 degrees C in Arctic wintertime sea ice. Appl Environ Microbiol 70:550–557. doi:10.1128/AEM.70.1.550
Kan GF, Miao JL, Shi CJ, Li GY (2006) Proteomic alterations of antarctic ice microalga Chlamydomonas sp. under low-temperature stress. J Integr Plant Biol 48:965–970. doi:10.1111/j.1744-7909.2006.00255.x
Kattner G, Thomas DN, Haas C et al (2004) Surface ice and gap layers in Antarctic sea ice: highly productive habitats. Mar Ecol Prog Ser 277:1–12. doi:10.3354/meps277001
Kennedy H, Thomas DN, Kattner G et al (2002) Particulate organic matter in Antarctic summer sea ice: concentration and stable isotopic composition. Mar Ecol Prog Ser 238:1–13. doi:10.3354/meps238001
Kirst GO, Wiencke C (1995) Ecophysiology of polar algae. J Phycol 31(2):181–199
Kooistra WHCF, Medlin LK (1996) Evolution of the diatoms (Bacillariophyta). Mol Phylogenet Evol 6:391–407
Kopczynska EE, Weber LH, El-Sayed SZ (1986) Phytoplankton species composition and abundance in the Indian sector of the Antarctic Ocean. Polar Biol 6:161–169
Krell A (2006) Salt stress tolerance in the psychrophilic diatom *Fragilariopsis cylindrus*. Dissertation, University of Bremen, Germany
Krell A, Beszteri B, Dieckmann G et al (2008) A new class of ice-binding proteins discovered in a salt-stress-induced cDNA library of the psychrophilic diatom *Fragilariopsis cylindrus* (Bacillariophyceae). Eur J Phycol 43:423–433. doi:10.1080/09670260802348615
Krembs C, Engel A (2001) Abundance and variability of microorganisms and transparent exopolymer particles across the ice-water interface of melting first-year sea ice in the Laptev Sea (Arctic). Mar Biol 138:173–185. doi:10.1007/s002270000396
Krembs C, Gradinger R, Spindler M (2000) Implications of brine channel geometry and surface area for the interaction of sympagic organisms in Arctic sea ice. J Exp Mar Biol Ecol 243:55–80. doi:10.1016/S0022-0981(99)00111-2
Krembs C, Eicken H, Deming JW (2011) Exopolymer alteration of physical properties of sea ice and implications for ice habitability and biogeochemistry in a warmer Arctic. Proc Natl Acad Sci U S A 108:3653–3658. doi:10.1073/pnas.1100701108
Leventer A (1998) The fate of Antarctic "Sea ice diatoms" and their use as paleoenvironmental indicators. Antarct Res Ser 73:121–137
Liu S, Zhang P, Cong B et al (2010) Molecular cloning and expression analysis of a cytosolic Hsp70 gene from Antarctic ice algae *Chlamydomonas* sp. ICE-L. Extremophiles 14:329–337. doi:10.1007/s00792-010-0313-8
Liu X, Wang Y, Gao H, Xu X (2011) Identification and characterization of genes encoding two novel LEA proteins in Antarctic and temperate strains of *Chlorella vulgaris*. Gene 482:51–58. doi:10.1016/j.gene.2011.05.006
Lizotte MP (2001) The contributions of sea ice algae to Antarctic marine primary production. Am Zool 41:57–73. doi:10.1668/0003-1569(2001)041[0057:TCOSIA]2.0.CO;2
Lizotte M (2003a) Microbiology of sea ice. In: Thomas DN, Dieckmann GS (eds) Sea ice – an introduction to its physics, chemistry, biology and geology. Blackwell Science, Oxford, pp 184–210
Lizotte M (2003b) The influence of sea ice on Ross Sea biogeochemical processes. Antarct Res Ser 78:107–122
Lizotte MP, Priscu J (1992) Spectral irradiance and biooptical properties in perennial ice-covered lakes of the dry valleys (McMurdo Sound Antarctica). Antarct Res Ser 57:1–14
Los D, Murata N (2004) Membrane fluidity and its roles in the perception of environmental signals. Biochim Biophys Acta Biomembr 1666:142–157. doi:10.1016/j.bbamem.2004.08.002

Lovejoy C, Massana R, Pedro C (2006) Diversity and distribution of marine microbial eukaryotes in the Arctic Ocean and adjacent seas diversity. Appl Environ Microbiol 72:3085–3095. doi:10.1128/AEM.72.5.3085

Lyon B, Mock T (2014) Polar microalgae: new approaches towards understanding sdaptations to an extreme and changing environment. Biology (Basel) 3:56–80. doi:10.3390/biology3010056

Lyon BR, Lee P, Bennett JM et al (2011) Proteomic analysis of a sea-ice diatom: salinity acclimation provides new insight into the dimethylsulfoniopropionate production pathway. Plant Physiol 157:1926–1941. doi:10.1104/pp.111.185025

Meier W (2017) Losing Arctic sea ice: observations of th erecent decline and the long-term context. In: Thomas DN (ed) Sea ice, 3rd edn. Wiley-Blackwell, Oxford, pp 261–289

Meredith MP, Brandon M (2017) Oceanography and sea ice in the Southern Ocean. In: Thomas DN (ed) Sea ice, 3rd edn. Wiley-Blackwell, Oxford, pp 216–238

Merico A, Tyrrell T, Brown CW et al (2003) Analysis of satellite imagery for *Emiliania huxleyi* blooms in the Bering Sea before 1997. Geophys Res Lett 30:1337. doi:10.1029/2002GL016648

Mindl B, Anesio AM, Meirer K et al (2007) Factors influencing bacterial dynamics along a transect from supraglacial runoff to proglacial lakes of a high Arctic glacieri. FEMS Microbiol Ecol 59:307–317. doi:10.1111/j.1574-6941.2006.00262.x

Mock T, Gradinger R (1999) Determination of Arctic ice algal production with a new in situ incubation technique. Mar Ecol Prog Ser 177:15–26. doi:10.3354/Meps177015

Mock T, Hoch N (2005) Long-term temperature acclimation of photosynthesis in steady-state cultures of the polar diatom *Fragilariopsis cylindrus*. Photosynth Res 85:307–317. doi:10.1007/s11120-005-5668-9

Mock T, Kroon BM (2002a) Photosynthetic energy conversion under extreme conditions—I: important role of lipids as structural modulators and energy sink under N-limited growth in Antarctic sea ice diatoms. Phytochemistry 61:41–51. doi:10.1016/S0031-9422(02)00216-9

Mock T, Kroon BM (2002b) Photosynthetic energy conversion under extreme conditions—II: the significance of lipids under light limited growth in Antarctic sea ice diatoms. Phytochemistry 61:53–60. doi:10.1016/S0031-9422(02)00216-9

Mock T, Valentin K (2004) Photosynthesis and cold acclimation: molecular evidence from a polar diatom. J Phycol 40:732–741. doi:10.1111/j.1529-8817.2004.03224.x

Mock T, Thomas D (2005) Sea ice – recent advances in microbial studies. Environ Microbiol 7:605–619

Mock T, Krell A, Glöckner G et al (2005) Analysis of expressed sequence tags (ESTS) from the polar diatom *Fragilariopsis cylindrus*. J Phycol 42:78–85. doi:10.1111/j.1529-8817.2005.00164.x

Mock T, Otillar RP, Strauss J et al (2017) Evolutionary genomics of the cold-adapted diatom *Fragilariopsis cylindrus*. Nature 541:536–540. doi:10.1038/nature20803

Morgan-Kiss RM, Ivanov AG, Pocock T et al (2005) The antarctic psychrophile, *Chlamydomonas raudensis* Ettl (UWO241) (Chlorophyceae, Chlorophyta), exhibits a limited capacity to photoacclimate to red light. J Phycol 41:791–800. doi:10.1111/j.1529-8817.2005.04174.x

Morgan-Kiss RM, Priscu JC, Pocock T et al (2006) Adaptation and acclimation of photosynthetic microorganisms to permanently cold environments. Microbiol Mol Biol Rev 70:222–252. doi:10.1128/MMBR.70.1.222-252.2006

Mou S, Zhang X, Ye N et al (2012) Cloning and expression analysis of two different LhcSR genes involved in stress adaptation in an Antarctic microalga, *Chlamydomonas* sp. ICE-L. Extremophiles 16:193–203. doi:10.1007/s00792-011-0419-7

Müller T, Bleiß W, Martin CD et al (1998) Snow algae from northwest Svalbard: their identification, distribution, pigment and nutrient content. Polar Biol 20:14–32. doi:10.1007/s003000050272

Nelson DM, Tréguer P, Brzezinski MA et al (1995) Production and dissolution of biogenic silica in the ocean: revised global estimates, comparison with regional data and relationship to biogenic sedimentation. Glob Biogeochem Cycles 9:359–372

Nishida I, Murata N (1996) Chilling sensitivity in plants and cyanobacteria: the crucial contribution of membrane lipids. Annu Rev Plant Physiol Plant Mol Biol 47:541–568. doi:10.1146/annurev.arplant.47.1.541

Nymark M, Sharma AK, Sparstad T et al (2016) A CRISPR/Cas9 system adapted for gene editing in marine algae. Sci Rep 6:24951. doi:10.1038/srep24951

Palmisano AC, Garrison DL (1993) Microorganisms in Antarctic sea ice. In: Friedmann E (ed) Antarctic microbiology. Wiley-Liss, New York, NY, pp 167–218

Park S, Jung G, Hwang YS, Jin E (2010) Dynamic response of the transcriptome of a psychrophilic diatom, *Chaetoceros neogracile*, to high irradiance. Planta 231:349–360. doi:10.1007/s00425-009-1044-x

Perovich D (2017) Sea ice and sunlight. In: Thomas DN (ed) Sea ice, 3rd edn. Wiley-Blackwell, Oxford, pp 110–137

Petrich C, Eicken H (2017) Overview of sea ice growth and properties. In: Thomas DN (ed) Sea ice, 3rd edn. Wiley-Blackwell, Oxford, pp 1–41

Priscu JC (1995) Phytoplankton nutrient deficiency in lakes of the McMurdo Dry Valleys, Antarctica. Freshw Biol 34:215–227

Priscu JC (1998) Ecosystem dynamics in a Polar desert: the McMurdo Dry Valleys, Antarctica. American Geophysical Union, Washington, DC

Ralph PJ, McMinn A, Ryan KG, Ashworth C (2005) Short-term effect of temperature on the photokinetics of microalgae from the surface layers of Antarctic pack ice. J Phycol 41:763–769. doi:10.1111/j.1529-8817.2005.00106.x

Raymond J (2011) Algal ice-binding proteins change the structure of sea ice. Proc Natl Acad Sci 108:E198. doi:10.1073/pnas.1106288108

Raymond J (2014) The ice-binding proteins of a snow alga, *Chloromonas brevispina*: probable acquisition by horizontal gene transfer. Extremophiles 18:987–994. doi:10.1007/s00792-014-0668-3

Raymond J, Kim HJ (2012) Possible role of horizontal gene transfer in the colonization of sea ice by algae. PLoS ONE 7:35968. doi:10.1371/journal.pone.0035968

Raymond J, Morgan-Kiss R (2013) Separate origins of ice-binding proteins in Antarctic *Chlamydomonas* species. PLoS ONE 8:e59186. doi:10.1371/journal.pone.0059186

Raymond J, Fritsen C, Shen K (2007) An ice-binding protein from an Antarctic sea ice bacterium. FEMS Microbiol Ecol 61:214–221. doi:10.1111/j.1574-6941.2007.00345.x

Raymond J, Janech MG, Fritsen CH (2009) Novel ice-binding proteins from a psychrophilic antarctic alga (Chlamydomonadaceae, Chlorophyceae). J Phycol 45:130–136. doi:10.1111/j.1529-8817.2008.00623.x

Remias D, Lutz-Meindl U, Lutz C (2005) Photosynthesis, pigments and ultrastructure of the alpine snow alga *Chlamydomonas nivalis*. Eur J Phycol 40:259–268. doi:10.1080/09670260500202148

Robinson DH, Kolber Z, Sullivan CW (1997) Photophysiology and photoacclimation in surface sea ice algae from McMurdo Sound, Antarctica. Mar Ecol Prog Ser 147:243–256. doi:10.3354/meps147243

Ryan KG, Ralph P, McMinn A (2004) Acclimation of Antarctic bottom-ice algal communities to lowered salinities during melting. Polar Biol 27:679–686. doi:10.1007/s00300-004-0636-y

Sakshaug E, Slagstad D (1991) Light and productivity of phytoplankton in polar marine ecosystems: a physiological view. Polar Res 10:69–86. doi:10.1111/j.1751-8369.1991.tb00636.x

Säwström C, Mumford P, Marshall W et al (2002) The microbial communities and primary productivity of cryoconite holes in an Arctic glacier (Svalbard 79 degrees N). Polar Biol 25:591–596. doi:10.1007/s00300-002-0388-5

Schriek R (2000) Effects of light and temperature on the enzymatic antioxidative defense systems in the Antarctic ice diatom *Entomoneis kufferathii* Manguin. Rep Polar Res 349:1–130

Scott FJ, Marchant H (eds) (2005) Antarctic marine protists. Australian Biological Resources Study/Australian Antarctic Division, Canberra

Shi H, Lee B, Wu S-J, Zhu J-K (2003) Overexpression of a plasma membrane Na+/H+ antiporter gene improves salt tolerance in *Arabidopsis thaliana*. Nat Biotechnol 21:81–85. doi:10.1038/nbt766

Shin S-E, Lim J-M, Koh HG et al (2016) CRISPR/Cas9-induced knockout and knock-in mutations in *Chlamydomonas reinhardtii*. Sci Rep 6:27810. doi:10.1038/srep27810

Smetacek V (1998) Diatoms and the silicate factor. Nature 391:224–225. doi:10.1111/j.1600-0579.1998.tb00036.x

Smetacek V, Nicol S (2005) Polar ocean ecosystems in a changing world. Nature 437:362–368. doi:10.1038/nature04161

Smetacek V, Klaas C, Menden-Deuer S, Rynearson TA (2002) Mesoscale distribution of dominant diatom species relative to the hydrographical field along the Antarctic Polar Front. Deep Res Part II Top Stud Oceanogr 49:3835–3848. doi:10.1016/S0967-0645(02)00113-3

Smith WO Jr, Codispoti LA, Nelson DM et al (1991) Importance of Phaeocystis blooms in the high-latitude ocean carbon cycle. Nature 352:514–516

Sommer U (1989) Maximum growth rates of Antarctic phytoplankton: only weak dependence on cell size. Limnol Oceanogr 34:1109–1112

Sorhannus U (2011) Evolution of antifreeze protein genes in the diatom genus *Fragilariopsis*: evidence for horizontal gene transfer, gene duplication and episodic diversifying selection. Evol Bioinforma 2011:279–289. doi:10.4137/EBO.S8321

Stammerjohn S, Maksym T (2017) Gaining (and losing) Antarctic sea ice: variability, trends and mechanisms. In: Thomas DN (ed) Sea ice, 3rd edn. Wiley-Blackwell, Oxford, pp 290–303

Stoecker DK, Gustafson DE, Merrell JR et al (1997) Excystment and growth of chryophytes and dinoflagellates at low temperatures and high salinities in Antarctic sea-ice. J Phycol 33:585–595

Stoecker DK, Gustafson DE, Black MMD, Baier CT (1998) Population dynamics of microalgae in the upper land-fast sea ice at a snow-free location. J Phycol 34:60–69. doi:10.1046/j.1529-8817.1998.340060.x

Stoecker DK, Gustafson DE, Baier CT, Black MMD (2000) Primary production in the upper sea ice. Aquat Microb Ecol 21:275–287

Stoeve J, Notz D (2015) Insights on past and future sea-ice evolution from combining observations and models. Glob Planet Chang 135:119–132

Strauss J, Gao S, Morrissey J, et al (2013) A light-driven rhodopsin proton pump from the psychrophilic diatom *Fragilariopsis cylindrus*. In: Proceeding of EMBO workshop: the molecular life of diatoms, Paris, France, 25–28 June 2013

Streb P, Shang W, Feierabend J, Bligny R (1998) Divergent strategies of photoprotection in high-mountain plants. Planta 207:313–324. doi:10.1007/s004250050488

Stroeve JC, Serreze MC, Fetterer F et al (2005) Tracking the Arctic's shrinking ice cover: another extreme September minimum in 2004. Geophys Res Lett 32:1–4. doi:10.1029/2004GL021810

Sutherland PC (1852) Journal of a voyage in Baffin's Bay and Barrow Straits in the years 1850–51, performed by H.M. ships "Lady Franklin" and "Sophia," under the command of Mr.William Penny in search of the missing crews of H.M. ships "Erebus" and "Terror", vol 1 and 2. Longman, Brown, Green, and Longmans, London

Takeuchi N (2002) Optical characteristics of cryoconite (surface dust) on glaciers: the relationship between light absorbency and the property of organic matter contained in the cryoconite. Ann Glaciol 34:409–414

Tang EPY, Vincent WF, Proulx D et al (1997) Polar cyanobacteria versus green algae for tertiary wast-water treatment in cool climates. J Appl Phycol 9:371–381

Teoh ML, Phang SM, Chu WL (2013) Response of Antarctic, temperate, and tropical microalgae to temperature stress. J Appl Phycol 25:285–297. doi:10.1007/s10811-012-9863-8

Thomas DN, Dieckmann GS (2002) Antarctic sea ice–a habitat for extremophiles. Science 295:641–644. doi:10.1126/science.1063391

Thomas DN, Fogg GE, Convey P et al (2008) Biology of polar regions. Oxford University Press, Oxford

Tomczak M, Godfrey JS (2003) Regional oceanography: an introduction, 2nd edn. Elsevier Science, Tarrytown, NY

Toseland A, Daines SJ, Clark JR et al (2013) The impact of temperature on marine phytoplankton resource allocation and metabolism. Nat Clim Chang 3:979–984. doi:10.1038/nclimate1989

Uhlig C, Kilpert F, Frickenhaus S et al (2015) In situ expression of eukaryotic ice-binding proteins in microbial communities of Arctic and Antarctic sea ice. ISME J 9:2537–2540. doi:10.1038/ismej.2015.43

Underwood GJC, Fietz S, Papadimitriou S et al (2010) Distribution and composition of dissolved extracellular polymeric substances (EPS) in Antarctic sea ice. Mar Ecol Prog Ser 404:1–19. doi:10.3354/meps08557

Underwood GJC, Aslam SN, Michel C et al (2013) Broad-scale predictability of carbohydrates and exopolymers in Antarctic and Arctic sea ice. Proc Natl Acad Sci U S A 110:15734–15739. doi:10.1073/pnas.1302870110

Van Oijen T, van Leeuwe M, Gieskes WWC (2003) Variation of particulate carbohydrate pools over time and depth in a diatom-dominated plankton community at the Antarctic Polar Front. Polar Biol 26:195–201. doi:10.1007/s00300-002-0456-x

von Quillfeldt C (2004) The diatom *Fragilariopsis cylindrus* and its potential as an indicator species for cold water rather than for sea ice. Vie Milieu 54:137–143

Vincent WF, Ja G, Pienitz R et al (2000) Ice shelf microbial ecosystems in the high arctic and implications for life on snowball earth. Naturwissenschaften 87:137–141. doi:10.1007/s001140050692

Wang DS, Xu D, Wang YT et al (2015) Adaptation involved in nitrogen metabolism in sea ice alga Chlamydomonas sp. ICE-L to Antarctic extreme environments. J Appl Phycol 27:787–796. doi:10.1007/s10811-014-0372-9

Wang Q, Lu Y, Xin Y et al (2016) Genome editing of model oleaginous microalgae *Nannochloropsis* spp. by CRISPR/Cas9. Plant J 88:1071–1081. doi:10.1111/tpj.13307

Weissenberger J, Dieckmann G, Gradinger R, Spindler M (1992) Sea ice: a cast technique to examine and analyze brine pockets and channel structure. Limnol Oceanogr 37:179–183. doi:10.4319/lo.1992.37.1.0179

Werner I (2006) Seasonal dynamics, cryo-pelagic interactions and metabolic rates of arctic pack-ice and under-ice fauna – a review. Polarforschung 75:1–19

Wheeler P, Watkins JM, Hansing RL (1997) Nutrients, organic carbon and organic nitrogen in the upper water column of the Arctic Ocean: implications for the sources of dissolved organic carbon. Deep Sea Res Part II Top Stud Oceanogr 44:1571–1592. doi:10.1016/S0967-0645(97)00051-9

Willem S, Srahna M, Devos N et al (1999) Protein adaptation to low temperatures: a comparative study of α-tubulin sequences in mesophilic and psychrophilic algae. Extremophiles 3:221–226. doi:10.1007/s007920050119

Williams WE, Gorton HL, Vogelmann TC (2003) Surface gas-exchange processes of snow algae. Proc Natl Acad Sci U S A 100:562–566. doi:10.1073/pnas.0235560100

Xu D, Wang Y, Fan X et al (2014) Long-term experiment on physiological responses to synergetic. Environ Sci Technol 48:7738–7746

Zhang P, Liu S, Cong B et al (2011) A novel omega-3 fatty acid desaturase involved in acclimation processes of polar condition from Antarctic ice algae *Chlamydomonas* sp. ICE-L. Mar Biotechnol 13:393–401. doi:10.1007/s10126-010-9309-8

Chapter 15
Genomics of Psychrophilic Bacteria and Archaea

John P. Bowman

Abstract Genomes are available for a wide range of psychrophilic bacteria and archaea. As of early 2017, approximately 130 cold-adapted species have genome sequences. Several studies complement this data with functional studies. In this review the cold adaptation traits of psychrophilic microorganisms are explored from a genome-centric point of view including surveys of traits across genomes. A broader view of psychrophiles in terms of growth rates amongst life on Earth explaining what a psychrophile represents is presented. Trait surveys, limited to the perspective of gene gain, reveal prevalence of genes demonstratively providing better growth at low temperature including compatible solute uptake and synthesis, antifreeze proteins and polyunsaturated fatty acids and investigate their functional relevance to psychrophily. This includes revealing prevalent antifreeze DUF3494-type proteins that occur in all domains of life but is limited to cold-adapted taxa and is absent in higher-temperature adapted life.

Contents

J.P. Bowman (✉)
School of Land and Food, University of Tasmania, Hobart, TAS, 7005, Australia
e-mail: john.bowman@utas.edu.au

R. Margesin (ed.), *Psychrophiles: From Biodiversity to Biotechnology*,
DOI 10.1007/978-3-319-57057-0_15

15.1 Introduction

Psychrophilic organisms grow best at low temperature and fail to grow or grow slowly at temperatures significantly above room temperature (Morita 1975). This is distinct from organisms termed "psychrotolerant" (or psychroactive; Panikov and Sizova 2006) that typically grow from refrigeration temperatures (4–5 °C) to a temperature equivalent to that of the human body (37 °C). In most taxonomic studies, temperature growth ranges of bacteria and archaea are tested at these temperatures as they are well-known signposts. A few bacteria and even fewer archaea available in culture are bona fide extreme psychrophiles, unable to grow appreciably on agar at 20 °C or higher temperatures. This is not necessarily due to them being hard to grow, but rather environmental conditions do not select from them strongly, and consistently very cold ecosystems will contain many microorganisms that only have mild levels of cold adaptation and are fully capable of growing under standard laboratory temperature environments (22 °C to >30 °C). Given this, psychrophiles have an interesting place in biology and have representatives in all domains of life. Indeed, highly psychrophilic eukaryotic microorganisms are just as likely to be encountered in cold habitats since their inherently more complex structure and biology can be particularly fine-tuned to external temperatures (Mock and Junge 2007; Cvetkovska et al. 2017). Bacteria, on the other hand, have a greater degree of flexibility, and depending on the situation, fast growth rates are not generally selected for in ecosystems with sporadic resource distribution (Roller and Schmidt 2015); thus perfect alignment of growth optima to in situ temperatures of low temperature environments does not occur. For example, most bacteria that dwell in sea ice, which is typically −1 to −10 °C, have optimal growth at 10–25 °C (Bowman et al. 1997). In frozen permafrost, glacier, boreal forest and tundra soils, which can be lower than −10 °C, many bacteria and archaea are found to occur and still show metabolic activity at subzero temperatures (Steven et al. 2008; Drotz et al. 2010; Tuorto et al. 2013), suggesting many could be psychrophilic but usually have reasonably wide growth temperature ranges, for example, *Planococcus halocryophilus* (−15 °C to >30 °C; Mykytczuk et al. 2013) and *Psychrobacter arcticus* (< −5 °C to ~30 °C; Bergholz et al. 2009).

The ability of psychrophiles to grow is also influenced by the presence of other hurdles present in the given ecosystem, for example, pH, water activity and nutrient availability. In the example of sea ice, hurdles include nutrient availability, salinity (water activity), freezing (ice crystal formation and desiccation) and oxidative stress (O_2 supersaturation). Thus, a psychrophile often has a host of adaptations that suit its particular niche inclusive of cold adaptation traits. Typically extreme pH and saline ecosystems have few if any psychrophiles since the organisms would need to successfully develop adaptations to overcome the challenge of multiple hurdles, which have an additive effect on reducing growth rates (McMeekin et al. 2000). Temperature-dependent growth optima as the combination of the rate and biomass generation capability (growth yield) also should be considered in relation to psychrophilic growth. This is because the fastest growth rates are typically close to the maximum temperatures for growth, and additional stressors reduce the growth rate limits. Cold adaptation traits overlap with traits also required for other forms of stress. For example, compatible solutes (osmolytes), low molecular weight compounds that can be concentrated to high concentrations in the cell cytoplasm (Roberts 2005), are generally beneficial for survival at low temperature as well as salty ecosystems as they stabilise proteins, DNA and cell membranes from temperature-induced entropy in general and help maintain osmotic pressure control. Psychrophilic microorganisms in general have to make trade-offs since the available biological innovation that has evolved on Earth to low temperature appears to have had limits. The innovations are of course coded in and stored in genomes; thus, the study of genomes provides a window into what the diversity and patterns of adaptations of psychrophiles possess within our current state of understanding of biology. This review investigates the state of knowledge of psychrophilic bacteria and archaea in terms of genome-level data. Using the available data, in which approximately 127 genomes are available (compared to only a dozen 10 years ago), the prevalence of cold adaptation traits is examined. The overall consensus of many studies on cold-adapted microorganisms indicates psychrophilic microorganisms achieve efficient growth aided by specific cold adaptation traits and through elimination of unnecessary and deleterious traits. Genomes open the way to study known traits, discover novel proteins and determine the associated biological processes they carry out in greater detail.

15.2 A Broadscale View of Growth Rates and Psychrophily

Crucial to understanding psychrophiles is the concept where they fit in terms of life on Earth. Psychrophiles are determined purely via temperature and growth. In this respect biokinetic temperature/growth rate distributions have notional maximums (T_{max}) and minimums (T_{min}) which are modelled values as they are points in which growth rate is zero (Ratkowsky et al. 1982, 1983). A point in the distribution, T_{opt}, indicates where growth rate is most rapid. The T_{opt} position is usually skewed to the left on the temperature x-axis, and the temperature region between T_{opt} and T_{max} is termed the supraoptimal temperature domain. This represents a range of temperatures and, though growth permissive, causes an organism to experience thermal stress that is compensated for by energy-dependent maintenance processes (Zakhartsev et al. 2015). Since energy for maintenance is finite and entropy from heat grows as the temperature increases, the supraoptimal domain features a rapid decline in growth rate usually within a window of 5–10 °C. At temperatures $>T_{max}$, the temperature is lethal resulting in permanent inactivation (cell death). It has been found where growth is otherwise not possible—due to some other limitation of the system (pH, salt level, lack of nutrients, radiation and more)—that temperature governs the inactivation rate (Ross et al. 2008; Zhang et al. 2010). As a result psychrophiles are constrained by a temperature barrier but are compensated by traits that give them fitness advantages in their particular cold habitats. Some microorganisms by virtue of a range of stress tolerance traits can achieve high levels of ubiquity, for example, common bacteria of permafrost and Antarctic soils, *Psychrobacter* and *Exiguobacterium*, can be found worldwide in all climatic zones (Rodrigues et al. 2009) partly because the members have wide temperature ranges (<0 °C to ~35–40 °C).

On the opposite side of T_{opt}, growth rate declines as temperature declines. This response essentially follows Arrhenius kinetics (Arrhenius 1889) in that cell reactions are governed by temperature and that activation energies of enzymes within cells collectively operate over defined temperature ranges that set the organisms' temperature growth range. Psychrophiles have enzymes that have low activation energy requirements due to the greater flexibility of the structures (Feller 2013) that aids in formation of enzyme-substrate complexes. Cold temperatures can also cause proteostasis-related problems (Bednarska et al. 2013) for cells due to protein denaturation (Privalov 1990).

It was found quite some time ago that a square-root growth rate model (Ratkowsky et al. 1983) readily interprets temperature growth responses and is able to derive T_{min}, T_{max} and T_{opt} values. The Ratkowsky model forms a reliable albeit empirical interpretation of temperature/growth relationships (Heitzer et al. 1991). The exact same model can be applied to estimate temperature/growth relationships of communities (Pietikäinen et al. 2005). Attempting to bring a level of mechanistic understanding to this relationship is useful and relevant to understanding psychrophily since it provides a better working knowledge of how organisms operate in nature. The main approach is to increase the biological connectivity of models, such as the development of thermodynamic models that try to explain responses to temperature with a biochemical basis. Some recent examples include a soil-oriented model based on activation energy (Schipper et al. 2014); a second utilising the concept of 'exergy', the available energy in a cell for biochemical reactions that takes into account entropy or reduction of available energy (Desmond-Le Quéméner and Bouchez 2014); and a third on protein stability (Corkrey et al. 2012, 2014).

To explore patterns of growth rate temperature, a large number of data sets were compiled from as many sources in the scientific literature as possible to create what is referred to as the biokinetic spectrum of temperature (Corkrey et al. 2016) or the 'BKS'. This triangle-shaped relation peaks at a temperature of about 45 °C, corresponding to the rapid growth rate of *Clostridium perfringens*, and has two distinct descending curves towards increasingly colder and hotter temperatures. The BKS provides an interesting overview of temperature relationships for bacteria and archaea that makes up the bulk of the data used by Corkrey and colleagues. Cold-adapted bacteria occupy the right-hand corner of the BKS (Fig. 15.1). A close-up view shows the overlay of growth rate relationships but also shows that the data has a degree of limitation in that it is especially sparse below 0 °C (Fig. 15.1). Nevertheless, the protein stability thermodynamic model (Corkrey et al. 2014) was found to fit the BKS concept quite well though it only provides so far a partial understanding of the underpinning mechanistics of low- and high-temperature adaptation. The BKS, however, does provide a credible means of assessing the

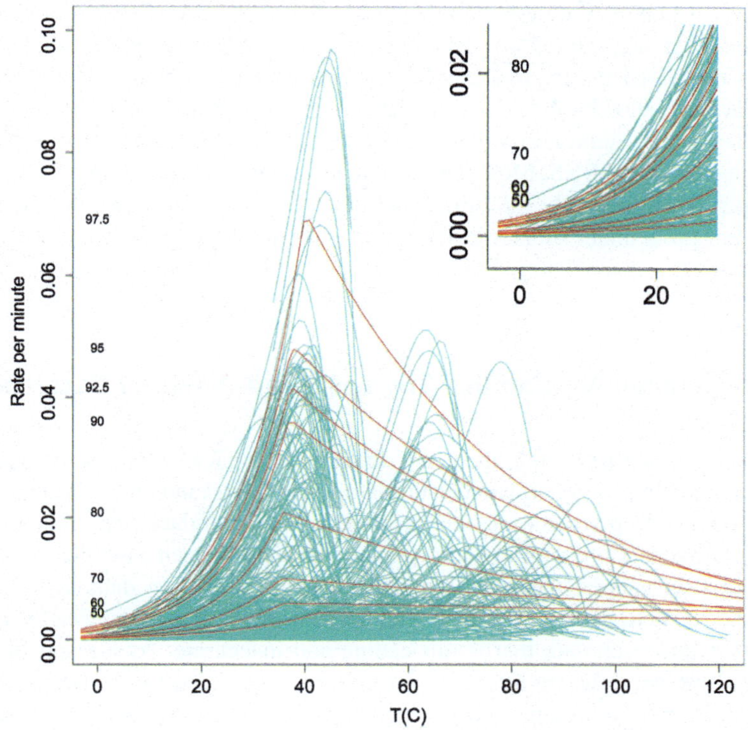

Fig. 15.1 Compilation of temperature versus growth rate *curves* forming the biokinetic spectrum of temperature (Corkrey et al. 2016). The insetted graph magnifying the *bottom right hand corner* of the BKS shows growth *curves* of psychrophilic bacteria and archaea and also shows data sparseness below 0 °C. The graph was adapted from Corkrey et al. (2017)

limits to life on Earth in terms of temperature as well as in terms of growth rate. Consequently, the BKS was used to predict maximum growth rates—essentially the largest possible rates for life on Earth across the entire temperature spectrum. The fastest rates are those performed by mesophiles and moderate thermophiles such as *Vibrio natriegens* and *E. coli* (T_{opt} 35–39 °C, generation time (GT) 10–20 min; Weinstock et al. 2016), *C. perfringens* (T_{opt} 42–46 °C, GT 10 min; Juneja et al. 2010) and *Thermobrachium celere* (T_{opt} 60 °C, GT 10–15 min; Engle et al. 1996). Cold-adapted bacteria obviously grow considerably more slowly, and at the temperature extremes of ≤ -5 °C, growth rates cannot be estimated easily due to freezing problems. Bacteria from permafrost and sea ice show growth down to −10 to −15 °C, including *Colwellia psychrerythraea* (Junge et al. 2003), *Psychrobacter* spp. (Bakemans et al. 2003; Bergholz et al. 2009), *Psychromonas ingrahamii* (Breezee et al. 2004) and *Planococcus halocryophilus* (Mykytczuk et al. 2013) with GT calculated at 10–50 days. *Planococcus halocryophilus*, which doubled roughly every 40–50 days at about −15 °C, is an exemplar of the type of microorganism that would be expected to compete well in frozen and other cold soils. The maximum growth rate predicted at −20 °C by Corkrey et al. (2017) was estimated at 250 h (95% confidence limits and when pure water is the solute). This is compared to the smallest GT estimated at 6.2–8.5 min at 45–46 °C, a difference of 1800–2400-fold. These data are only inferred from what is currently available in the literature, which as mentioned above tends to be sparse at the extreme ends of the BKS. More data would be valuable in providing more accurate concepts of what is truly the limit to life in the cryosphere and how fast microorganisms grow under subzero conditions taking into account in situ factors of the environment.

15.3 Genome Availability Amongst Cold-Adapted Bacteria

Currently, as of early 2017, there are about 91,000 bacterial genomes and 2200 archaeal genomes. A general metadata survey of bacterial and archaeal genera from polar regions, deep sea and marine locations, alpine and glacial environments and other low temperature natural and artificial ecosystems (such as chilled food) was performed. In general the state of metadata for genomes on databases (such as NCBI and PATRIC) is patchy and often of limited value as they usually include scant data and even data that is misleading and inaccurate. As a result literature surveys were necessary to largely cover the available state of knowledge as of the end of 2016 on psychrophilic bacterial and archaeal genomes. Out of 257 bacterial and archaeal genera (1335 species) with sequence data surveyed that had some connection to low-temperature ecosystems, only 59 genera contain genome-sequenced strains that could be considered highly cold adapted. A further 137 genera include a proportion of species that are psychrotolerant, while the remaining genera examined had genome-sequenced species that were exclusively mesophilic. Of the 59 genera with cold-adapted bacteria, almost all also contain psychrotolerant and/or mesophilic sister species; 23 genera contained all three types and so are especially useful for temperature-oriented genome comparisons. A total of 16 genera had only psychrophilic species, but most of these only have a few extant species and are likely incomplete in terms of the coverage of diversity. Time will tell if mesophilic species relatives are found for these genera. These genera are also denoted in Table 15.1. It is notable that at the species level strains can vary in temperature growth range. For example, one strain of *Acetobacterium bakii* grew well at 30 °C, while two other strains stopped growing at 30 °C (Kotsyurbenko et al. 1995). Thus, the distinctions of what is psychrophilic, psychrotolerant and mesophilic are fairly arbitrary. Taken together organisms listed as psychrophilic grow well at low temperatures (~5–10 °C) while mesophiles by comparison will not grow or grow slowly and with reduced biomass levels.

From the data 137 genomes were found that belonged to psychrophilic bacteria ($n = 134$) and archaea ($n = 3$) where corroborating growth data is generally available. The phylogenetic relationship of these bacteria is given in Fig. 15.2. These included strains that grow optimally at 20 °C or less (Table 15.2). Most of the

Table 15.1 List of cold-adapted bacterial and archaeal strains with complete or draft genomes

Species	Strain no. (type strain^T)	Opt. growth temp. (pressure)	Source environment (s)	Accession code	Mbp	G+C mol %	16S rRNA copies, no. of tRNA
Actinobacteria							
Arthrobacter alpinus (3)[a]	R3.8	18	Alpine soil	CP012676-7[b]	4.046	62.2	6, 51
Cryobacterium arcticum	PAMC 27867	16	Permafrost soil	CP016282–4	4.351	68.4	3, 51
Cryobacterium flavum	CGMCC 1.11215	9	Glacier ice core	FNIB00000000	4.040	64.7	–, 45
Cryobacterium levicorallinum	GMCC 1.11211	8	Glacier ice core	FOPW00000000	3.754	64.4	–, 45
Cryobacterium luteum	CGMCC 1.11210	10	Glacier ice core	FOCN00000000	3.834	65.1	–, 47
Cryobacterium psychrotolerans	CGMCC 1.5382	15	Glacial soil	FNFU00000000	3.247	68.3	–, 45
Cryobacterium roopkundense	RuG17	8	Glacial soil	JPXF00000000	4.356	65.3	–, 47
Demequina lutea	NBRC 106155	18	Permafrost soil	BBRC01000000	2.643	65.9	–, 43
Glaciibacter superstes	DSM 21135	15	Permafrost ice wedge	ATWH00000000	4.801	64.4	–, 50
Leifsonia rubra	CMS 76R	15	Polar microbial mat	ATIA00000000	2.671	59.2	–, 45
Tomitella biformata	AHU 1821	17	Permafrost ice wedge	BAVQ00000000	4.688	68.1	–, 50
Firmicutes							
Bacillus cecembensis	DSM 21993	18	Glacial soil	LMBZ00000000	4.782	36.9	–, 49
Bacillus psychrosaccharolyticus	ATCC 23296	18	Polar soil	AJTN00000000	4.590	38.8	–, 85
Brochothrix campestris	FSL F6-1037_c9	18	Chilled meat	AODH00000000	2.372	40.2	–, 73
Lactobacillus algidus	DSM 15638	18	Chilled meat	AZDI00000000	1.590	36.0	–, 46
Paenibacillus antarcticus	CECT 5836	19	Lake sediment	LVJI00000000	5.372	40.5	–, 107

(continued)

Table 15.1 (continued)

Species	Strain no. (type strain^T)	Opt. growth temp. (pressure)	Source environment (s)	Accession code	Mbp	G+C mol %	16S rRNA copies, no. of tRNA
Paenibacillus glacialis	DSM 22343	18	Glacial soil	LVJH00000000	5.709	40.7	–, 108
Paenibacillus macquariensis (2)	DSM 2	18	Polar soils	LVJF00000000	6.274	40.7	–, 100
Planococcus antarcticus	DSM 14505	18	Polar microbial mat	CP016534	3.788	43.2	9, 72
Planococcus faecalis	CECT 8759	18	Polar (ornithogenic) soil	MBMU00000000	3.534	40.8	–, 95
Planococcus halocryophilus	Or1	18	Permafrost soil	CP016537	3.425	40.0	10, 72
Planomicrobium glaciei	CGMCC 1.6846	18	Glacial soil	FNDC00000000	3.917	46.7	–, 55
Carnobacterium alterfunditum	DSM 5972	16	Polar anoxic saline lake	JQLG00000000	2.616	36.0	–, 75
Carnobacterium funditum	DSM 5970	16	Polar anoxic saline lake	JQLL00000000	2.352	34.6	–, 73
Carnobacterium iners	DSM 28070	16	Polar pond sediment	FOAH00000000	2.661	34.0	–,53
Carnobacterium pleistocenium	FTR1	16	Permafrost ice	JQLQ00000000	2.376	35.4	–, 74
Acetobacterium bakii	DSM 8239	18	Freshwater sediment	LGYO00000000	4.135	41.2	–, 47
Bacteroidetes							
Aequorivita sublithincola	DSM 14238	18	Polar sublithic biofilm	CP003280	3.520	36.2	–, 36
Bizionia algoritergicola	APA-3	16	Polar saline lake	LZRO00000000	3.528	34.1	–, 35
Bizionia argentinensis	JUB59	18	Polar seawater	AFXZ00000000	3.279	33.8	–, 36
Bizionia psychrotolerans	PB-M7	18	Marine fauna	JNGS00000000			
Cellulophaga algicola	DSM 14237	18	Sea ice	CP002453	4.888	33.8	5, 44
Chryseobacterium antarcticum	LMG 24720	16	Polar soil and moss	JPEP00000000	3.123	36.1	–, 37

Chryseobacterium frigidisoli	DSM 26000	13	Glacial forefield soil	FOQT00000000	3.021	33.3	–, 35
Crocinitomix catalasitica	ATCC 23190	18	Polar marine sand	JHXV00000000	4.623	34.1	–, 41
Flavobacterium antarcticum	DSM 19726	16	Polar soil	ATTM00000000	3.068	34.9	
Flavobacterium branchiophilum	FL-15	18	Cold freshwater fish	FQ859182–3	3.559	32.9	3, 43
Flavobacterium degerlachei	DSM 15718	18	Polar microbial mat	FNMV00000000	3.856	34.0	–, 43
Flavobacterium frigidarium	DSM 17623	14	Polar marine sediment	AUDO00000000	3.624	34.1	–, 46
Flavobacterium frigoris	DSM 15719	15	Polar microbial mat	FOFZ00000000	4.045	34.2	–, 47
Flavobacterium fryxellicola	DSM 16209	16	Polar microbial mat	LVJE00000000	3.721	34.6	–, 42
Flavobacterium gelidilacus	DSM 15343	16	Polar microbial mat	AUGN00000000	3.428	30.0	–, 37
Flavobacterium gillisiae	DSM 22376	17	Sea ice	FNRD00000000	4.377	34.3	–, 46
Flavobacterium hibernum	DSM 12611	19	Polar freshwater lake	JPRK00000000	5.283	33.2	–, 57
Flavobacterium micromati	DSM 17659	18	Polar microbial mat	FQWF00000000	3.692	33.1	–, 42
Flavobacterium noncentrifugens	CGMCC 1.10076	17	Glacial meltwater	FNEZ00000000	4.032	40.7	–, 41
Flavobacterium omnivorum	CGMCC 1.2747	8	Glacial soil	FNDB00000000	3.808	34.3	–, 43
Flavobacterium rivuli	WB3.3–2	18	Cold mineral spring	ARKJ00000000	4.484	39.6	–, 47
Flavobacterium psychrophilum (29)	DSM 3660	15	Cold freshwater fish	FMVE00000000	2.626	32.3	6, 49
Flavobacterium sinopsychrotolerans	CGMCC 1.8704	16	Glacial soil	FODN00000000	3.520	34.1	–, 44
Flavobacterium sp.	ACAM123	10	Polar saline lagoon	AJXL00000000	3.930	34.8	–, 42

(continued)

Table 15.1 (continued)

Species	Strain no. (type strain^T)	Opt. growth temp. (pressure)	Source environment (s)	Accession code	Mbp	G+C mol %	16S rRNA copies, no. of tRNA
Flavobacterium tegetincola	DSM 22377	17	Polar microbial mat	AUDN00000000	3.759	35.0	–, 39
Flavobacterium xanthum	DSM 3661	17	Polar microbial mat	FRBU00000000	3.755	34.4	–, 44
Flavobacterium xueshanense	CGMCC 1.9227	11	Glacier ice core	FONQ00000000	3.471	34.2	–, 41
Gelidibacter algens	ACAM 536	18	Sea ice, polar saline lakes, sublithic biofilms	LZRN00000000	4.506	37.3	–, 38
Gillisia limnaea	DSM 15749	18	Polar microbial mat	AHKR00000000	3.959	37.6	–, 44
Lacinutrix algicola	AKS293	18	Marine alga	LIQH01000000	3.661	31.4	–, 36
Lacinutrix himadriensis	E4-9a	13	Polar marine sediment	LIQI01000000	4.166	32.6	–, 35
Lacinutrix jangbogonensis	PAMC 27137	10	Polar marine sediment	JSWF00000000	4.017	32.2	–, 37
Lacinutrix mariniflava	AKS432	15	Marine alga	LIQG01000000	3.973	31.8	–, 37
Maribacter antarcticus	DSM 21422	10	Polar marine alga	JHZC00000000	4.853	37.4	–, 38
Polaribacter atrinae	KACC 17473	18	Marine mollusc	LVWE00000000	3.942	30.4	–, 42
Polaribacter franzmannii	ATCC 700399	6	Sea ice	ARJD00000000		32.5	
Polaribacter irgensii	23-P	6	Sea ice	AAOG00000000	2.763	34.5	–, 37
Pricia antarctica	DSM 23421	16	Polar marine sediment	FNAO00000000	4.851	43.8	–, 37
Psychroflexus gondwanensis	ACAM 44	18	Polar hypersaline lake	APLF00000000	3.306	35.8	–, 38
Psychroflexus torquis	ATCC700755	8	Sea ice	AAPR00000000	4.321	34.5	3, 37
Psychroserpens burtonensis	DSM 12212	8	Polar saline lagoon	AUDE00000000	3.961	33.4	–, 35
Psychroserpens jangbogonensis	PAMC 27130	12	Polar marine sediment	JSWG00000000	3.633	32.7	–, 34

Rhodonellum psychrophilum	GCM71	12	Fjord tufa columns	AWXR00000000	5.120	41.8	–, 37
Winogradskyella psychrotolerans	RS-3	18	Polar marine sediment	ATMR00000000	4.279	33.5	–, 40
Arcticibacter eurypsychrophilus	MJ9–5	15	Glacier ice core	MDFN00000000	4.867	38.5	–, 37
Arcticibacter svalbardensis	MN12–7	16	Polar soil	AQPN00000000	4.681	38.2	–, 39
Pedobacter arcticus	A12	15	Polar soil	AKZJ00000000	4.221	36.9	–, 39
Pedobacter hartonius	DSM 19033	18	Cold mineral spring	FNRA00000000	5.188	43.0	–, 51
Hymenobacter psychrophilus	CGMCC 1.8975	15	Alpine soil	FNOV00000000	4.758	60.8	–, 44
Alphaproteobacteria							
Devosia psychrophila	CGMCC 1.10210	13	Glacier cryoconite	FOMB00000000	4.328	61.2	–, 46
Erythrobacter sp.	QSSC1–22B	15	Polar sublithic biofilm	LZRP00000000	3.344	63.5	1, 46
Phaeobacter arcticus	DSM 23566	15	Polar marine sediment	AXBF00000000	5.049	59.3	–, 58
Octadecabacter antarcticus	307	10	Sea ice	CP003740–1	4.875	54.6	2, 42
Octadecabacter arcticus	238	10	Sea ice	CP003742–44	5.478	55.2	2, 42
Octadecabacter temperatus	SB1	17	Seawater	CP012160–1	3.264	54.7	1, 38
Robiginitomaculum antarcticum	DSM 21748	17	Polar seawater	AQOY00000000	2.768	52.5	–, 36
Deinococcus-Thermus							
Deinococcus frigens	DSM 12807	15	Polar endolith rocks	JNIW00000000	4.031	65.4	–, 52
Deinococcus marmoris	DSM 12784	15	Polar endolith rocks	JNIV00000000	4.800	64.4	–, 54

(continued)

Table 15.1 (continued)

Species	Strain no. (type strain[T])	Opt. growth temp. (pressure)	Source environment (s)	Accession code	Mbp	G+C mol %	16S rRNA copies, no. of tRNA
Betaproteobacteria							
Polaromonas glacialis	R3–9	18	Glacier cryoconite	JMDZ00000000	5.284	62.3	–, 53
Polaromonas naphthalenivorans	CJ2	18	Cold freshwater stream	CP000529–37	5.366	61.7	2, 51
Simplicispira psychrophila	DSM 11588	17	Polar moss	JHYS00000000	3.600	61.5	–, 50
Gammaproteobacteria							
Aliivibrio logei	ATCC 35077	18	Cold-water marine fish	ASAH00000000	5.288	38.9	–, 70
Aliivibrio salmonicida	LFI1238	10	Cold-water marine fish	FM178379–84	4.531	39.0	12, 91
Aliivibrio wodanis	AWOD1	15	Cold-water marine fish	LN554846–51	4.518	38.2	7, 87
Colwellia chukchiensis	CGMCC 1.9127	18	Polar seawater	FOBI00000000	4.023	41.9	–, 55
Colwellia hornerae	PAMC 20917	12	Sea ice	CP014944	4.683	37.9	6, 73
Colwellia piezophila	ATCC BAA-637	5 (60 MPa)	Deep sea fauna	ARKQ00000000	5.475	38.9	–, 67
Colwellia psychrerythraea (3)	34H	8	Marine fauna, sea ice	CP000083	5.373	38.0	9, 88
Dasania marina	DSM 21967	18	Polar marine sediment	ARDZ00000000	4.101	47.4	–, 45
Glaciecola pallidula	ACAM 615	13	Sea ice	BAEQ00000000	4.355	41.2	–, 46
Glaciecola punicea (2)	ACAM 611	14	Sea ice	BAET00000000	3.072	43.1	–, 45
Idiomarina abyssalis	KMM 227	18	Deep seawater	LGOW01000000	2.684	47.2	–, 55
Idiomarina zobellii	KMM 231	18	Deep seawater	LHSG01000000	2.618	47.1	–, 52
Marinobacter psychrophilus	20,041	15	Sea ice	CP011494	3.998	53.8	3, 49
Marinomonas ushuaiensis	DSM 15871	18	Subpolar seawater	JAMB00000000	3.332	41.1	–, 58
Moritella dasanensis	ArB 0140	10	Polar seawater	AKXQ00000000	4.882	40.8	–, 90
Moritella marina	ATCC 15381	12	Marine sediment	ALOE00000000	4.624	40.5	–, 48
Moritella viscosa		16	Cold-water fish	LN554852–4	5.093	39.4	11, 131

Neptunomonas antarctica	S3–22	15	Polar marine sediment	LJSI01000000	4.568	45.7	–, 60
Neptunomonas japonica	DSM 18939	17	Whale carcass	AUGR00000000	4.389	43.7	–, 54
Oleispira antarctica	RB-8	15	Polar seawater	FO203512	4.406	42.2	5, 50
Paraglaciecola arctica	BSs20135	16	Polar marine sediment	BAEO00000000	5.961	40.1	–, 50
Paraglaciecola chathamensis	S18 K6	18	Marine sediment	BAEM00000000	5.255	44.1	–, 49
Paraglaciecola polaris	LMG 21857	18	Sea ice, polar seawater	BAER00000000	5.225	44.1	–, 48
Paraglaciecola psychrophila	170	10	Sea ice	CP003837	5.374	40.4	6, 59
Photobacterium profundum	SS9	10 (20 MPa)	Deep sea samples	CR354531	6.403	41.7	16,169
Pseudoalteromonas arctica	A 37–1-2	16	Sea ice, polar seawater	AHBY00000000	4.621	39.0	–, 94
Pseudoalteromonas denitrificans	DSM 6059	12		FOLO00000000	6.094	34.7	–, 72
Psychrobacter arcticus	273–4	16	Permafrost	CP000082	2.650	42.8	4, 49
Psychrobacter cryohalolentis		18	Permafrost	CP000323	3.101	42.2	4, 48
Psychrobacter glacincola	BNF20	10	Sea ice	LIQB01000000	3.490	42.8	–, 47
Psychrobacter urativorans	R310.10B	16	Polar soil	CP012678	2.931	42.1	6, 52
Psychromonas arctica	DSM 14288	17	Sea ice, polar seawater	AXWP00000000	4.719	37.8	–, 65
Psychromonas hadalis	ATCC BAA-638	6 (60 MPa)	Deep-sea sediment	ATUO00000000	3.977	39.1	–, 42
Psychromonas ingrahamii	37	8	Sea ice	CP000510	4.559	40.1	10, 86
Psychromonas ossibalaenae	ATCC BAA-1528	18	Whale carcass	ARML00000000	5.200	41.9	–, 57
Shewanella benthica	KT99	7 (60 MPa)	Deep-sea samples	ABIC00000000	4.320	46.0	13, 86
Shewanella frigidimarina	NCBM4000	16	Sea ice, seawater	CP000447	4.845	41.6	9, 96
Shewanella halifaxensis	HAW-EB4	16	Deep-sea sediment	CP000931	5.226	44.6	10, 125

(continued)

Table 15.1 (continued)

Species	Strain no. (type strain^T)	Opt. growth temp. (pressure)	Source environment (s)	Accession code	Mbp	G+C mol %	16S rRNA copies, no. of tRNA
Shewanella piezotolerans	WP3	16 (20 MPa)	Deep-sea sediment	CP000472	5.396	43.3	8, 89
Shewanella sediminis	HAW-EB3	16	Deep-sea sediment	CP000821	5.517	46.1	12, 125
Shewanella violacea	DSS12	7 (60 MPa)	Deep-sea samples	AP011177	4.962	44.7	14, 126
Shewanella woodyi	ATCC 51908	16	Seawater	CP000961	5.935	43.7	10, 126
Thiomicrospira arctica	DSM 13458	11	Polar marine sediment	ARLF00000000	2.551	41.9	–, 45
Deltaproteobacteria							
Desulfotalea psychrophila	LsV54	7	Polar marine sediment	CR522870–2	3.523	46.6	7, 65
Desulfovibrio ferrireducens	DSM 16995	18	Polar marine sediment	FNGA00000000	3.872	42.8	–, 57
Desulfovibrio frigidus	DSM 17176	18	Polar marine sediment	JONL00000000	4.185	42.7	–, 56
Epsilonproteobacteria							
Sulfurimonas gotlandica	GD1	12	Hypoxic seawater	AFRZ00000000	2.952	33.6	–, 47
Archaea (Methanomicrobia)							
Methanococcoides burtonii	DSM 6242	19	Polar saline lake	CP000300.1	2.575	40.8	3, 49
Methanogenium frigidum		10	Polar saline lake		~2.2		
Methanolobus psychrophilus	R15	15	Alpine wetland sediment	CP003083.1	3.072	44.6	–, 52

[a]Number of strains sequenced
[b]Genome includes multiple elements (multiple chromosomes and/or plasmids)

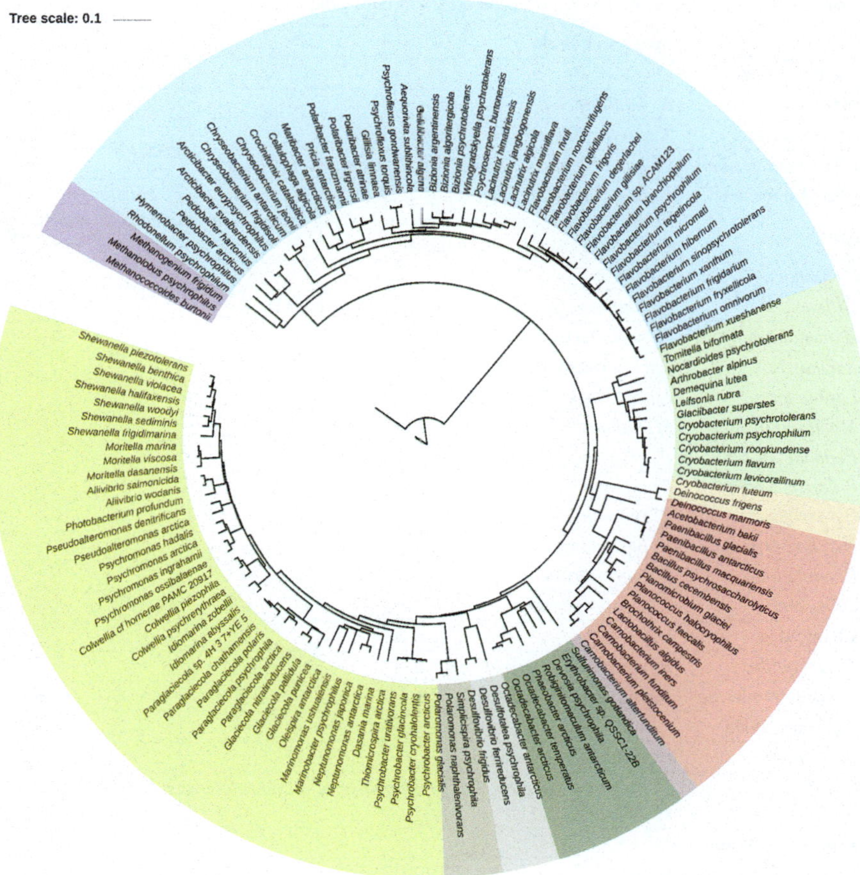

Fig. 15.2 A phylogenetic tree showing 16S rRNA gene sequences of genome-sequenced psychrophilic species. The *coloured* sections correspond to the major taxonomic groups to which the species belong (see Table 15.1). The tree was created using the Interactive Tree of Life online application (itol.embl.de/) using maximum likelihood and neighbour joining algorithms

strains are type strains. All speciated strains are taxonomically valid except for *Methanolobus psychrophilus* (Zhang et al. 2008). A temperature of 25–30 °C represents supraoptimal to lethal temperatures for these strains. Many strains that are not speciated, and potentially represent undescribed species, are also available with genomes, but a few have any available growth data and so were not considered further here except in a few cases where data was available. Genomes extracted from environmental metagenomes are also not considered for the same reason since little data is available that provides credible growth rate versus temperature data for uncultured microorganisms, let alone microorganisms of low-temperature systems. The use of amino acid distributions could be used to assess their potential for psychrophily (see next section); however, this was not done here. Overall a much

Table 15.2 Distribution of DUF3494 type AFPs in cold-adapted bacteria[a]

Taxonomic group	Species or strain
Bacteroidetes	
Aequorivita	*A. sublithincola*
Arcticibacter	*A. eurypsychrophilus*, *A. barbardensis*
Chryseobacterium	*C. antarcticum*, *C. gregarium*, *C. jeonii*, *C. luteum*
Flavobacterium	*F. degerlachei*, *F. frigoris*, *F. fryxellicola*, *F. geldilacus*, *F. gillisiae*, *F. micromati*, *F. noncentrifugens*, *F. omnivorum*, *F. urumqiense*, *F. xanthum*, *F. xinjiangense*, *F. xueshanense*
Gelidibacter	*G. algens*
Gillisia	*G. limnaea*, *Gillisa* sp. CAL575, *Gillia* sp. JM1
Hymenobacter	*Hymenobacter* sp. DG5B
Lacinutrix	*Lacinutrix* sp. Hel_I_90
Lutibacter	*Lutibacter* sp. BRH_c52
Mucilaginibacter	*M. paludis*
Pedobacter	*P. arcticus*, *P. oryzae*, *P. ruber*
Polaribacter	*P. irgensii*
Psychroflexus	*P. sediminis*, *P. torquis*
Psychroserpens	*P. burtonensis*
Rhodonellum	*R. antarcticum*
Spirosoma	*S. spitsbergense*
Firmicutes	
Paenibacillus	*P. wynnii*
Paenisporosarcina	*Paenisporosarcina* sp. TG-14
Actinobacteria	
Demequina	*D. aestuarii*, *D. lutea*
Cryobacterium	*C. flavum*, *C. luteum*, *C. roopkundense*, *C. psychrotolerans*, *Cryobacterium* sp. MLB-32
Gammaproteobacteria	
Colwellia	*Colwellia hornerae* PAMC 20917, *Colwellia* sp. SLW05
Dasania	*D. marina*
Idiomarina	*I. salinarum*, *Idiomarina* sp. A28L
Marinobacter	*Marinobacter* sp. ELB17, *M. subterrani*
Pseudomonas	*P. xinjiangensis*, *P. pelagia*
Psychromonas	*P. ingrahamii*
Saccharospirillum	*S. impatiens*
Shewanella	*S. frigidimarina*, *S. denitrificans*
Thiomicrospira	*T. arctica*
Deltaproteobacteria	
Desulfovibrio	*Desulfovibrio* sp. DV
Geospychrobacter	*G. electrodiphilus*

[a]Also detected in the following species and strains: *Cytophaga aurantiaca*, *Cytophaga hutchinsonii*, *Sulfuricaulis limicola*, *Roseivirga echinicomitans*, *Nafulsella turpanensis*, *Marivirga tractuosa*, *Pontibacter akesuensis*, *Pelosinus* sp. UW01, *Streptomyces* spp. and some other actinomycetes, *Rhodoferax ferrireducens*, *Janthinobacterium* sp. CG3, *Candidatus* 'Gallionella acididurans', *Sphaerochaeta* spp. and some other spirochaetes. Also observed in several uncultured taxa from metagenomes (e.g. *Candidatus* 'Ignavibacteria'). Archaea that contain orthologs include *Methanoregula boonei* from a cold acidic bog, *Haloterrigena* sp., *Halohasta litchfieldiae* (from Deep lake, Antarctica) and *Halovivax asiaticus* from a lake from Inner Mongolia. Eukaryotes possessing similar proteins include fungi (*Piloderma croceum*, *Fibulorhizoctonia* sp., *Typhula ishakariensis* and others) and sea-ice diatom (*Fragilariopsis cylindrus*)

improved pool of data is available for comparative analysis as compared to a decade ago. Amongst psychrophilic strain genomes, the specific statistics (Table 15.1) do not show any unusual features, genome lengths are typical and G+C mol% content is consistent with the genus where they belong and includes high and low values depending on the source phyla. The number of rRNA operons (were available) and tRNAs also highly varies between taxa but is generally consistent with higher-temperature relatives.

15.4 Protein Amino Acid Content as Markers for Psychrophily as Discerned from Genomes

Cold-active enzymes and low-temperature adjusted cell structures, such as tRNA and ribosomes, make life possible in freezing conditions since they can process biochemical reactions and mechanically function at low temperature successfully. For example, tRNA nucleotide content in *Archaea* seems to increase with increasing T_{opt} and thus alters in flexibility based on genome data (Saunders et al. 2003); however, psychrophiles cannot be readily distinguished from mesophiles on this basis. The G+C content of genomes has not been found to be related to T_{opt} when relatively large genome sets are compared (Zeldovich et al. 2007). Similarly for proteins the differences in amino acid content between psychroactive and mesophilic counterparts are subtle (Georlette et al. 2004; D'Amico et al. 2006; Siddiqui and Cavicchioli 2006; De Vendittis et al. 2008). Only specific changes are needed to affect significantly the temperature-based catalytic properties of an enzyme. Folding also effects discrete changes in protein structure that affect catalytic efficiency at different temperatures (Feller 2013). It was found that the Hsp70 (GroEL/GroES) complex from *Oleispira antarctica* (Table 15.1) when expressed in *E. coli* resulted in *E. coli* being able to grow at low temperature where before it could not (Ferrer et al. 2003). Based on their data, the growth rate of the *E. coli* transgenic train was improved 141-fold at 8 °C (close to the wild-type strain notional T_{min} value) and was threefold faster at 15 °C. Growth occurred at 4 °C, and the notional T_{min} was estimated to be −13.7 °C after using the Ratkowsky square-root model. The result importantly shows that the capability of psychrotolerant bacteria to grow at low temperature is strongly controlled by rate-limiting steps in protein folding, a process that requires a substantial investment of cell energy (Rothman and Schekman 2011). Other traits possessed by *E. coli* seem to provide fitness for low-temperature growth, in the absence of other challenges of cold ecosystems (such as freezing).

Cold-active enzymes are generally comparatively thermolabile due to a different distribution of non-polar and charged amino acid residues. This allows substantial activity at ecosystem relevant temperatures, such as $\leq$0 °C in sea ice (Huston et al. 2000). Several studies have examined genome data to determine trends in amino acid substitutions affecting low-temperature catalytic rates and/or stability. Some groups of microorganisms such as the haloarchaea seem to have an inherent limit to the flexibility of their enzymes. For example, *Halorubrum lacusprofundi*, one of the dominant life forms in Deep Lake, Antarctica, and one of the most extreme still liquid environments on Earth (annual temperature 5 °C to −21 °C, ~30% salinity), has subtle changes to proteins that help with both high-salinity and low-temperature adaptation (DasSarma et al. 2013). This is of interest when considering the limits of life on Earth. Examination of a substantial number of proteomes of psychrophilic versus mesophilic bacteria and archaea have provided further insights into amino acid preferences and the locations of amino acid changes that influence protein structure (Metpally and Reddy 2009; Ayala-del-Río et al. 2010). A number of criteria have been defined for cold-active proteins that can be useful in predicting psychrophily of overall proteomes when compared to mesophilic relatives:

1. Relative levels of arginine (R) versus arginine plus lysine (E). Arginine creates more salt bridges than lysine, promoting stability (Siddiqui et al. 2006). It should be noted higher levels of lysine also stabilise thermophilic proteins through entropic processes (Berezovsky et al. 2005).
2. Relative amount of acidic residues—glutamate (Q) and aspartate (D). Collectively reduced levels of Q and D reduce the number of salt bridge forming with R and E.
3. Amount of proline (P). Less proline enhances chain flexibility of protein secondary structure (Sakaguchi et al. 2003). Less proline also helps avoid the limitations of the rate-limiting step of proline isomerisation during protein folding, which is slowed further at low temperature (Feller 2013). Trigger factor, which acts as a prolyl *cis-trans* isomerase, is more abundant at low temperature (Piette et al. 2010), and cold-active versions have efficient prolyl isomerase activity (Godin-Roulling et al. 2015).
4. Lower levels of hydropathicity, scored as the grand average of hydropathicity (GRAVY) (Kyte and Doolittle 1982). Typically cold-adapted bacteria have less buried hydrophobic residues that would affect folding processes (Sælensminde et al. 2009; Feller 2013).
5. A fifth criterion is the measure of the aliphatic index, which provides an estimate of the protein packing volume (Ikai 1980). However, this measure appears less sensitive unless buried amino acids are specifically analysed since the link with cold-active enzymes seems to mainly relate to buried amino acid residues and overall measurements lose information (Ayala-del-Río et al. 2010).

Though there is little universality of these amino acid substitutions, they can provide valuable insight into modifications that can be applied to improve enzyme function for practical applications (Metpally and Reddy 2009). In the big picture sense, seven amino acids (I, V, Y, W, R, E, L) were found to be collectively the most influential, correlating broadly to overall bacterial and archaeal T_{opt} (Zeldovich et al. 2007). For psychrophilic bacteria the ratios of these amino acids are typically 38–39% of total amino acid residues versus 40–41% for most mesophiles. The relationship of IVYWREL content to T_{opt} is not strong enough for it to be able to accurately predict the T_{opt} of psychrophiles. In the study of the

Planococcus halocryophilus strain Or1, which has a large temperature growth range (Mykytczuk et al. 2013), only three of the five above criteria fitted. The wide temperature growth range of this organism suggests it has not undergone selection for enhanced protein flexibility and thus allows it to maintain fitness under a range of conditions. It is also recognised that low temperature effects the permeability of the lipid bilayer and could denature membrane proteins that carry out vital processes such as transport, adhesion and motility. Data suggests that exposed parts of membrane proteins seem more likely to have amino acid differences that provide cold adaptation benefits (Kahlke and Thorvaldsen 2012). This in turn suggests the lipid portion of the membrane is also crucial for cold adaptation.

15.5 Comparative and Functional Studies of Genomes of Psychrophilic Bacteria and Archaea

Comparative genomic studies provide a perspective on traits acquired and lost in microorganisms and are especially useful when empirical phenotypic, biochemical and genetic data is available to support interpretations. Furthermore, annotated genomes can provide predictions on unrealised functionality as well as provide a detailed view of evolutionary and taxonomic relationships. For psychrophilic archaea and bacteria, comparative genomics places emphasis on the basis of cold adaptation. This includes questions related to what are the cold adaptation-linked traits present; how they might contribute to temperature preferences, and how these traits collectively relate to the ecosystem the psychrophile inhabits.

Comparative and functional studies have been performed utilising sequenced complete and draft genomes of single psychrophilic strains and, where available, multiple genomes from closely related species. In several cases these studies have been coupled with transcriptomic and proteomic analysis. Comparative studies have been performed on groups of species including cold-adapted archaeal species *Methanococcoides burtonii* and *Methanogenium frigidum* (Saunders et al. 2003) and species of the bacterial genera *Shewanella* (Zhao et al. 2010), *Octadecabacter* (Vollmers et al. 2013), *Glaciecola* (Qin et al. 2014), *Psychroflexus* (Feng et al. 2014, 2015), *Paenibacillus* (Moreno Switt et al. 2014), Arctic seawater-derived *Psychrobacter* strains (Moghadam et al. 2016), cold-water fish pathogens (Touchon et al. 2011; Castillo et al. 2016; Vincent et al. 2016) and *Pseudoalteromonas* (Bosi et al. 2017). Examples of intensive genome studies on specific psychrophilic and psychrotolerant strains include *Colwellia psychrerythraea* strain 34H, the first psychrophile to be genome sequenced (Methé et al. 2005; Nunn et al. 2015); *Methanococcoides burtonii* (Allen et al. 2009; Burg et al. 2010; Campanaro et al. 2011; Williams et al. 2010a, b, 2011); *Methanolobus psychrophilus* R15 (Chen et al. 2012, 2015); the deep sea species *Luteimonas abyssi* (Zhang et al. 2015); Arctic *Mesorhizobium* strain N33 (Ghobakhlou et al. 2015); *Shewanella livingstonensis* Ac10 (Kawamoto et al. 2007; Park et al. 2012); *Psychromonas ingrahamii* strain 37 (Riley et al. 2008); *Planococcus halocryophilus* Or1 (Mykytczuk et al. 2013); *Pseudoalteromonas haloplanktis* TAC125 (Fondi et al. 2015); the type strain of *Pseudomonas extremaustralis* (Tribelli et al. 2015); and Antarctic desert-soil strain *Nesterenkonia* sp. AN1 (Aliyu et al. 2016). The genomes of these species have been investigated in terms of ecology, lifestyle, metabolic models, environmental processes, and almost universally cold adaptation strategies.

Through research and bioinformatics investigations, certain cellular traits can be discerned that are closely connected to psychrophily and cold adaptation though it is rare to find any that are exclusively connected to only cold adaptation since most traits have broad functionality and thus have a degree of temperature independence. The main traits discussed further in this review have been distilled from genomic comparisons and functional studies as listed above. In some cases relevant traits are firmly linked to cold adaptation either because they are induced at low temperature or when disabled result in poor or no growth at low temperature, for example, osmolyte uptake (Angelidis et al. 2002) and fatty acid biosynthesis (Zhu et al. 2005). Psychrophily also can represent the loss of traits that allow growth at higher temperatures. These are much more difficult to discern since the phenotype of psychrophily needs to be evaluated in the context of temperature by empirical experiments. Thus, this review focuses only on traits gained rather than lost. Furthermore, the traits considered are generally stable within genomes since they would theoretically contribute to positive and purifying selection (Bosi et al. 2017) in a given cold ecosystem.

More specifically comparisons of genomes coupled with functional studies provide insights that in general suggest that psychrophiles in their diversity have arrived at their psychrophily in a myriad of different ways though they use a limited pool of traits that are just part of a fundamental conserved framework found in psychrophilic to at least mesophilic species. For example, gene expression and proteome studies on the saline Antarctic lake-derived methanogen *Methanococcoides burtonii* have demonstrated that cell envelope, RNA and protein structure are compromised at low temperatures if not appropriately maintained and managed (Williams et al. 2010a, b, 2011; Campanaro et al. 2011). This lead to the discovery of TRAM domain proteins that seem to act as RNA chaperones analogous to cold-shock proteins (CspA family proteins) in bacteria. These proteins are found in most if not all archaea (Taha et al. 2016). The studies also demonstrated that growth rates are directly linked with metabolic and energy-generation pathway activity as inferred from transcript levels and via proteomics. The overall activity declines with temperature as would be expected with a declining growth rate with lowering temperature. In *M. burtonii* transporters that import glycine betaine, molybdate, iron and phosphate were more abundant indicating the greater need for the resources for maintaining fitness under cold conditions.

A second example is the species *Psychroflexus torquis* ATCC 700755, a strictly aerobic flavobacterial epiphyte which occurs in sea ice algal assemblages. Genome comparative analysis with the closely related Antarctic hypersaline lake species *P. gondwanensis* was performed (Feng et al. 2014) (Fig. 15.3). Genome data

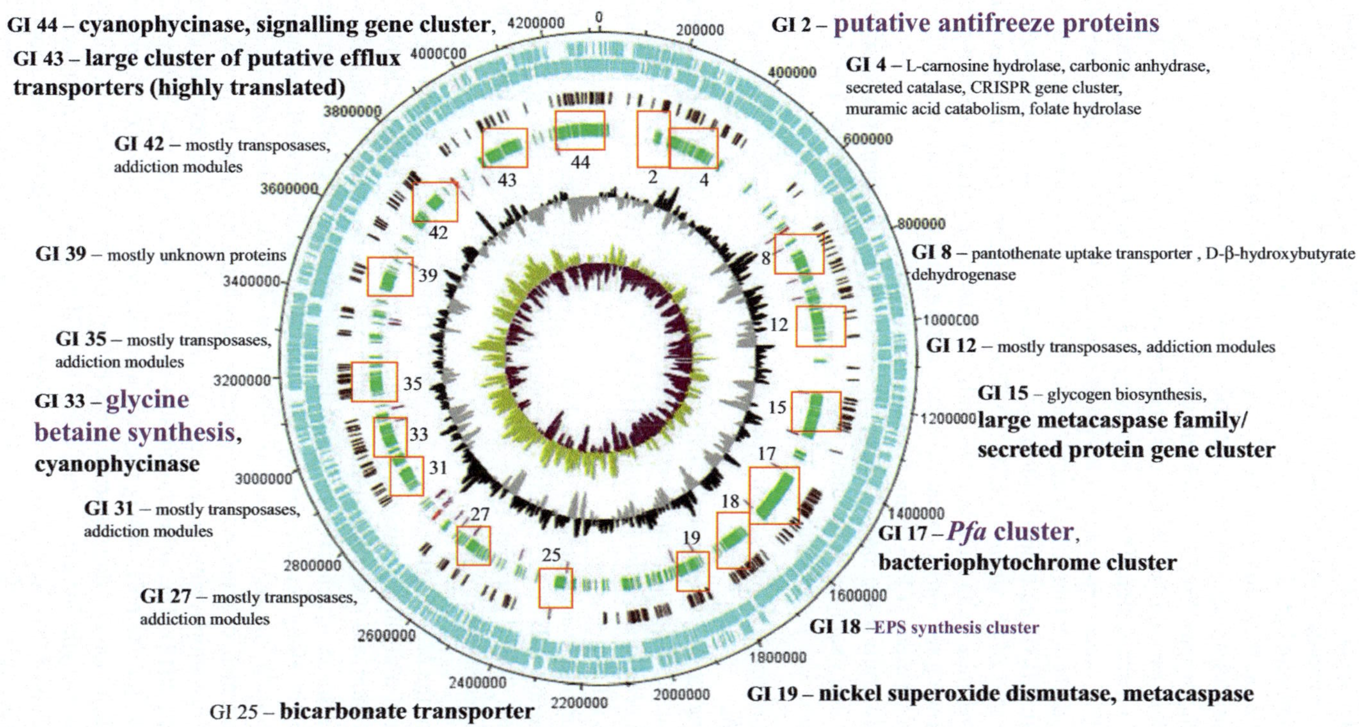

Fig. 15.3 Standout genes located on the genome of *Psychroflexus torquis* ATCC 700755 compared to close relative *Psychroflexus gondwanensis* ACAM 44 (99% 16S rRNA gene sequence similarity). Genomic islands containing genes of interest are defined from the *green ring* that denotes genes present in *Psychroflexus torquis* ATCC 700755 but absent in *Psychroflexus gondwanensis* ACAM 44. Genes that are relevant to a psychrophilic lifestyle are shown in *purple* while other genes potentially relevant to the unique lifestyle of the species (in sea ice and as an algal epiphyte) are in bold. Other features of the genome can be obtained from Feng et al. (2014)

revealed *P. torquis* has a consistently larger genome than all of its five genome-sequenced sister species (4.3 Mb vs. 2.7–3.3 Mb) (Table 15.1). This extra genetic material forms 'genomic islands' (Langille et al. 2010) that are completely absent in the other genomes and are likely acquired through horizontal gene transfer (HGT) processes based on their divergent G+C values and the presence of HGT-related genes (transposases, integrons) that might be selfish (addiction modules, group II retrons) (Zimmerly and Wu 2015) and thus could affect fitness. Likewise, the other species have their own unique genomic islands but on a smaller scale and which are not linked to cold adaptation. There seems extensive evidence of both gene gain and loss through HGT processes in *P. torquis* (Feng et al. 2014). This has been observed in other cold-adapted bacteria such as *Octadecabacter arcticus* and *O. antarcticus* (Vollmers et al. 2013) and in fish pathogen *Aeromonas salmonicida* (Vincent et al. 2016). In the case of *P. torquis* ATCC 700755, these islands contain genes that provide psychrophilic traits including the ability to synthesise polyunsaturated fatty acids (see following section), compatible solute uptake, putative antifreeze proteins and exopolysaccharides (EPS) (Feng et al. 2014).

Similar cold traits have been frequently identified in the genomes of other psychrophiles. For example, in the analysis of the *Psychromonas ingrahamii* strain 37 genome there was suggestion of:

1. Hypothetical proteins that seem cold adapted based on their amino acid composition.
2. An EPS gene cluster that formed an EPS that seems to protect cells from freezing by acting as a cryoprotectant (Breezee et al. 2004); a similar phenotype was found for *Colwellia psychrerythraea* 34H (Methé et al. 2005; Marx et al. 2009; Nunn et al. 2015).
3. The main osmolyte accumulated seemed to be glycine betaine.
4. A large number of TRAP family transporters could aid nutrient acquisition (Riley et al. 2008). Similar types of adaptations though never exactly the same are noted in other studies.

The study of the *M. burtonii* genome suggested it had high plasticity and seemed to have acquired traits that allowed it to fit into a cold niche (Allen et al. 2009). A similar observation was observed in the genomes of *O. arcticus* and *O. antarcticus* isolated from sea ice, which have very high levels of infiltration of HGT elements that have brought along features to the cell that could aid in survival, including xanthorhodopsin, a means to energise membrane electron transport via light harvesting, and also cyanophycinase, a means to obtain nitrogen from poly-L-aspartic acid (cyanophycin), a nitrogen storage polymer common in algae (Vollmers et al. 2013). The ability to combat oxidative stress was also noted as a significant trait, for example, aiding the cold adaptation and survival of *Nesterenkonia* sp. AN1, in soils of the Antarctic Ross Desert, one of the driest and coldest environments on Earth that can sustain some life. This strain relative to its warm temperature-adapted relatives (such as *Nesterenkonia alba*) had a greater proportion of cold-shock protein, DNA repair, fatty acid biosynthesis, osmoprotective and oxidative management stress genes in terms of genome gene proportion (Aliyu et al. 2016).

In the case of *Colwellia psychrerythraea* 34H, a number of cold adaptation traits are identified including those already mentioned including EPS, compatible solutes, omega-3 polyunsaturated fatty acids, oxidative stress management and enzyme flexibility (Methé et al. 2005). Proteomics performed on cells of 34H grown at −1 °C and at −10 °C in ice confirmed these observations especially in regards to EPS and osmolyte management (Nunn et al. 2015). At −10 °C more effort was emphasised in cells in terms of DNA repair, chemotaxis and sensing processes. Modulation also occurred in cell envelope, metabolism, iron and nitrogen uptake (Nunn et al. 2015). Overall, the differences were still rather subtle and tend to suggest that the strain uses a similar array of traits and works at a slowed pace at subzero temperatures and that energy is directed to cell maintenance and stress responses, which are largely compensatory in nature ensuring cell survival. This is not unlike other forms of stress where when growth permissiveness is challenged, the overall processes become increasingly temperature dependent since disorder (entropy) in the system rapidly increases, especially when moving beyond growth limits (Ross et al. 2008; Zhang et al. 2010; Corkrey et al. 2014).

The genome of *Planococcus halocryophilus* strain Or1 featured several cold adaptation traits, such as a large number of compatible transporters (Mykytczuk et al. 2013). Expression studies revealed cell envelope modification including EPS synthesis, energy metabolism (increased ATP synthesis), ion transport, transcription and translation modulation that were important for survival at ≤−10 °C. The studies of Nunn et al. (2015) and those on *M. burtonii* mentioned above parallel these findings. The overall interpretation is that psychrophily relies at least to some extent on adaptive plasticity to acquire more cold-adaptation-relevant traits. These seem to be coupled with efficient use of resources allowing generation of residual energy above and beyond what is needed to counter deleterious effects of extremely low temperature including protein denaturation, slow enzymatic reactions and associated metabolic bottlenecks and membrane permeability limitations. The combined use of genomics, gene expression and proteomic analysis has the power to develop a more systems-based understanding of psychrophily in model organisms, for example, as done with *Pseudoalteromonas haloplanktis* TAC125 in terms of metabolic models (Fondi et al. 2015). Studying strains spread over a number of phyla and from contrasting ecosystems would be illuminating for investigating the limits to life.

Metagenomic studies also clearly have their place in this area. Many psychrophiles are likely being uncovered by the ever growing number of population genomes and metagenomes created in recent times from samples collected from cold ecosystems (Simon et al. 2009; Berlemont et al. 2011; Han et al. 2012; DeMaere et al. 2013; Lay et al. 2013; Ugalde et al. 2013; Wright et al. 2013; Bowman et al. 2014; Choudhari et al. 2014; Glass et al. 2014; Klippel et al. 2014; Tveit et al. 2014; Anderson et al. 2015; Bowman and Ducklow 2015; Lee et al. 2015; Chrismas et al. 2016; Colangelo-Lillis et al. 2016; Le et al. 2016; Lopatina et al. 2016; Raymond 2016; Tschitschko et al. 2016; Goordial et al. 2017).

However, proving they are psychrophilic may require use of predictive tools such as using developing inferences from amino acid content and the presence of genes found in cultured psychrophilic and psychrotolerant species. This has yet to be done in any systematic way and could be useful, especially aligned with larger scale concepts such as the BKS.

An exemplary example of a metagenomic analysis of cold ecosystems was the study of Ace-C, a dominant bacterial taxon within meromictic, marine saline Ace Lake located in the Vestfold Hills of Eastern Antarctica (Ng et al. 2010). Ace-C, a phototrophic member of the phylum *Chlorobia* (green sulphur bacteria), is the only reported cold-adapted bacteria within this phylum to date with substantial data, including genomic, transcriptomic and proteomic information. The study by Ng et al. (2010) revealed the traits these organisms use to survive without needing to culture, though this was aided by the microbe completely dominating a particular depth of Ace Lake, at about the oxycline. Primary adaptations used by Ace-C include (1) specific bacteriochlorophyll and pigment production that likely maximises photon capture; (2) tight regulation of its sulphur metabolism in addition to syntrophic relations developed with sulphate-reducing bacteria in the anoxic layer of Ace Lake for sulphur supply; and (3) oxidative stress management that was found to be important for its survival since its biomass concentrated near the oxycline where it could be periodically exposed to cold O_2-rich waters in the upper layer of the lake.

Genomics has been useful in revealing features of cold adaptation including cold-inducible proteins that provide a perspective on how bacterial and archaeal cells manage low temperature at a fundamental level. For example, cold shocking of *Shewanella livingstonensis* Ac10 (Kawamoto et al. 2007) revealed via proteomics that it increased the abundance of proteins associated with RNA synthesis (RpoA, GreA, CspA), protein folding (including Tuf, Efp, LysU, Tig), membrane transport (OmpA and OmpC) and flagella proteins (FlgE and FlgL). Experiments can also tend to be more practical and provide underpinning knowledge in man-made systems. An example of such a study investigated enzymes produced by pasteurisation of cold-adapted *Paenibacillus* spp. at 6–7 °C (Moreno Switt et al. 2014). This is useful since most wastage of food via spoilage in modern systems is due to a 'food microbiome' consisting largely of psychrotolerant and psychrophilic bacteria including *Pseudomonas*, *Shewanella*, *Psychrobacter*, *Photobacterium*, *Carnobacterium* and other cold-adapted lactobacilli, *Brochothrix*, *Clostridium* and other spore-forming bacilli for which much genome data is available now. Genomic approaches may also pave the way to counter cold-water fish infectious diseases caused by the pathogens *Flavobacterium psychrophilum* (Castillo et al. 2016), *Aeromonas salmonicida* (Vincent et al. 2016) and *Aliivibrio salmonicida* (Kashulin et al. 2017).

It was found that by examining the evolutionary connections of cold-adapted traits between *Pseudoalteromonas* spp. (Bosi et al. 2017) and *Glaciecola* spp. (Qin et al. 2014), there is support of the concept that they are highly beneficial and would support increased fitness in appropriate cold ecosystems. Traits that have been highlighted tended to focus on central processes or comprise a particular pool of traits. The distribution of major cold adaptation traits amongst psychrophilic microorganisms as determined by examining genome data is covered in the next sections.

15.6 Membrane Phospholipid Content in Psychrophiles and Associated Gained Genes and Proteins

Membrane fluidity homeostasis is critical for growth at low temperature. Disabling the ability to synthesise certain enzymes in model bacterial species leads to mutants unable to grow at low temperature. For example, disabling branched chain fatty acid synthesis in the psychrotolerant human pathogen *Listeria monocytogenes* leads to a mutant with very poor growth at 10 °C (Zhu et al. 2005). The proportionally increased levels of fatty acids that relatively promote increased fluidity are a major tactic by bacteria to enable growth over a wide temperature range. The shift by some microorganisms to the colder end of the BKS may suggest that the fluidisation of membranes becomes 'fixed' and that the ability to apply thermotolerance to membranes becomes lost. Fatty acid profiles tend to be relatively conserved at the genus level with quantitative rather than qualitative differences predominating. Changes in the ratios of acyl chain length, proportion of unsaturated and branched chain fatty acid types, comprise a homeostatic mechanism that involves enzymes in at least some bacteria activated transiently by a thermosensory two-component histidine kinase/response regulator system. These systems are likely widespread in bacteria and archaea (Sengupta and Garrity 2013; Inda et al. 2014; Porrini et al. 2014). Thus, psychrotolerant versus mesophilic bacteria could differ in that psychrotolerant strains can better fluidise their membranes via either desaturation of lipids or changing branched chain fatty acid (BCFA) levels, for example, by increasing anteiso C15:0 levels as found in *L. monocytogenes*. Various mesophiles like *E. coli* already have some flexibility in this sense, while other bacteria, such as *Campylobacter jejuni*, seem inflexible and may reflect their econiche preferences and degree of host adaptation (Hughes et al. 2009). Various psychrophiles as well as psychrophilic piezophiles have overtly adapted to low temperature by gaining genes that allow synthesis of omega-3 polyunsaturated fatty acids (PUFA) including eicosapentaenoic acid (EPA, 20:5ω3) and docosahexaenoic acid (DHA, 22:6ω3c). Some PUFA-producing bacteria also can produce arachidonic acid (AA, 20:4ω6c). AA, EPA and DHA all increase membrane fluidity while shorter chain, less unsaturated lipids such as oleic acid (18:1), linoleic acid (18:2), and α-linolenic acid (18:3) have substantially weaker effects (Yang et al. 2011).

The synthesis of PUFA has come through the apparent HGT of a polyketide synthase (PKS) gene cluster that includes five genes (*pfaABCDE*). Three classes of bacterial PUFA synthesis PKS are known (Shulse and Allen 2011a, b) (Fig. 15.4). One class termed group A by Shulse and Allen (2011b) creates EPA, group B creates DHA, and class D creates AA and EPA depending on the species. PUFA *pfa*

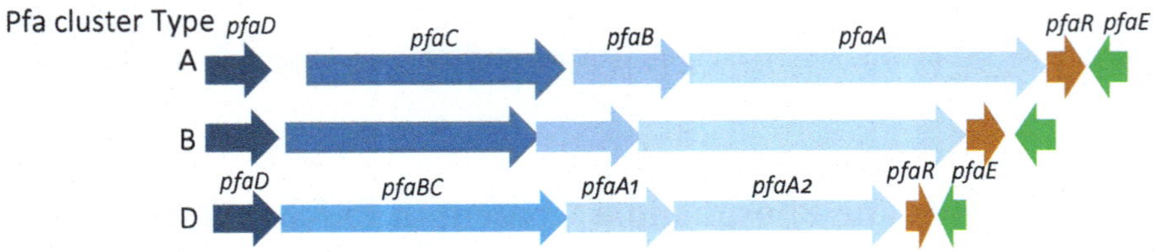

Fig. 15.4 Polyketide synthase type *pfaABCDE* gene clusters that allow synthesis of PUFA as defined by Shulse and Allen (2011a, b). Group A, B and D *pfa* gene cluster synthesise EPA, DHA and EPA and/or AA, respectively. The gene pfaR is a putative transcriptional regulator.

gene clusters are only found in marine bacteria (Shulse and Allen 2011a), so far restricted to *Gammaproteobacteria* and *Bacteroidetes*, primarily in psychrophiles. Based on surveys of the list of psychrophilic and other cold-adapted bacterial and archaeal genomes mentioned previously (Table 15.1), group A *pfa* genes are mainly possessed by members of the genus *Shewanella* as well as in the deep sea moderate piezophile *Photobacterium profundum* and include all psychrophilic species with genomes available, as well as some psychrotolerant species (such as *S. putrefaciens*). Group A *pfa* cluster genes are also found in some psychrotolerant *Vibrio* spp. such as *V. splendidus* and *V. tasmaniensis*. Group B *pfa* genes are found in most if not all *Colwellia*, *Psychromonas* and *Moritella* species, as well as in the species *Pseudoalteromonas denitrificans*, the most cold-adapted species of that genus. Group D *pfa* genes are found in members of the phylum *Bacteroidetes* and is rare, spread across several genera but typically just in specific psychrophilic species or strains including *Psychroflexus torquis*, *Maribacter antarcticus*, *Psychroserpens jangbogonensis*, *Aureispira* spp. and unclassified strains within the genera *Ulvibacter* and *Dokdonia*. *Aureispira* species, which predate other bacteria (via ixotrophy), produce large amounts of AA, and this has been shown to be made by the class D pfa/PKS system (Ujihara et al. 2014). Ujihara and colleagues also showed PUFA synthesis requires the PfaE protein, a 4-′-phosphopantetheinyl (4′-PP) transferase that post-translationally modifies apo-acyl carrier protein to holo-ACP by adding a 4′-PP moiety to an invariant serine residue. Holo-ACP is activated to allow thioesterification of the bound fatty acid acyl chain from the 4′PP distal end. In *P. torquis* the *pfa* cluster is located on a genomic island (see Fig. 15.3) and unlike *Aureispira* spp. instead makes mainly EPA and only some AA. Regulatory systems involved in controlling expression of PUFA proteins are unstudied, but a putative regulator gene is located between *pfaA* and *pfaE* in all PUFA-producing strains suggesting it plays a role in controlling expression of some or all *pfa* genes. There is possibly a link to thermosensing since in *P. torquis* the EPA/AA ratio increases as temperature drops (Nichols et al. 1997) suggesting expression is temperature dependent. At 2 °C proteomics revealed the *pfa* cluster in *P. torquis* was strongly translated (Feng et al. 2014); however, the specific thermosensory regulation if any remains to be elucidated.

15.7 Antifreeze Proteins with the DUF3494 Domain Are Prevalent Amongst Psychrophilic Bacteria

Freezing of water is potentially detrimental to microorganisms with cryodamage worsened when the freezing process is relatively slow. This is due to formation of larger ice crystals than what is found in flash freezing. Ice crystals that form in the cytoplasm act as a solute, and large amounts lead to the intake of water resulting in the increase of osmotic pressure and subsequent membrane rupture. Alternatively, external ice formation causes water loss from cells as it reduces water availability. Bacteria combat freezing in two ways: first by concentrating compatible solutes in the cytoplasm and second by utilising ice-active or antifreeze proteins (AFPs) that hinder or control the recrystallisation of water in the immediate region around the cell or within the cytoplasm itself (Lorv et al. 2014; Bar Dolev et al. 2016a, b; Cheung et al. 2017). Compatible solutes are discussed in the next section.

AFPs are a diverse set of proteins belonging to a number of different protein superfamilies, several with no classification. In general AFPs are not subgrouped owing to their diverse nature both structurally and functionally. Until more data is available, AFPs are generally only associated with the groups of organisms they are found (Lorv et al. 2014; Bar Dolev et al. 2016b; Cheung et al. 2017). AFPs have been studied extensively in bacteria and eukaryotes.

One famous group of AFPs are called ice-nucleation proteins (INPs). INPs are mainly restricted to plant-associated bacteria including *Pseudomonas* spp., *Xanthomonas* spp. and *Pantoea* spp. and can trigger ice crystal nucleation events at high-subzero temperatures (Bar Dolev et al. 2016a, b). The ice crystal formation is used as a means to disrupt plant cells to gain access to nutrients. INP-forming bacteria, such as the plant pathogen *Pseudomonas syringae*, are psychrotolerant but not psychrophilic. Screening for INPs has failed to find them in freezing ecosystems such as sea ice or polar lakes where one theory was that they could act as nucleators of ice formation. However, they are ubiquitous in precipitation worldwide (Christner et al. 2008), and thus wind- and rain-transported bacteria with INPs could act as nuclei for ice formation. INPs are large proteins of around 130 kDa with numerous repeat domains. No INP-type orthologs are found in the strain genomes listed in Table 15.1 and seem entirely restricted to pseudomonads and some other members of class *Gammaproteobacteria*.

Several diverse proteins have been found to have antifreeze properties, which allow bacteria and eukaryotes living in freezing-prevalent environments to control the size and shape of ice crystals in supercooled liquids. These proteins do this by thermal hysteresis and by controlling ice recrystallisation. Based on the literature, AFPs use both of these processes though the level of hysteresis can be quite minimal. Some AFPs can strongly depress the equilibrium freeze-temperature point (by up to 6 °C), a process called hyperactive thermal hysteresis, and simultaneously raise the melting temperature of ice (hysteresis melting), essentially heating the ice by up to around 0.1–0.2 °C. Ice crystals neither grow nor melt in the hysteresis-controlled-temperature range; however, when the temperature drops below the equilibrium point, crystal formation occurs, often explosively, in a burst pattern. The temperature range of hysteresis varies widely, and in general it was thought bacterial thermal hysteresis ranges of bacterial AFPs tend to be smaller (0.1–0.3 °C) than those of fish, plants and insects (Middleton et al. 2012; Lorv et al. 2014). Several recently studied recombinant bacterial hyperactive AFPs (see below) have hysteresis temperatures equivalent to insect and fish AFPs (2–5 °C) (Kawahara et al. 2007; Hanada et al. 2014) when at sufficient concentrations and in the presence of ligands and solutes. Bacterial extracts alone give weak results likely because the protein is not highly abundant and possibly because it requires some form of activation.

Reducing ice recrystallisation by AFPs seems a prevalent activity and prevents ice crystal size increasing in size. This is because the amount of protein required can be very low, indeed nanomolar levels (Mangiagalli et al. 2017). This is important since as subzero temperatures drop ice crystals tend to be bigger due to small ice crystals combining together. This impediment of crystal size reduces the probability of cryodamage and applies to either the cytoplasm if the AFP is cytosolic or the external environment if the AFP is secreted. AFPs have been noted in a range of bacteria, but relatively few examples have been studied in depth with functionality characterised and crystal structures of proteins also obtained. An AFP from *Pseudomonas putida* strain GR12–2 (Sun et al. 1995; Muryoi et al. 2004) was characterised that had modest thermal hysteresis activity but promoted survival of the bacterium in freezing conditions. The protein precursor of this AFP is 50 kDa in size and is unusual in that is highly glycosylated and lipidated (increasing molecular mass to 164 kDa). Based on genome comparisons performed here, this AFP is restricted to only strain Gr12–2 so far based on conserved domain structure and homology. Another AFP was isolated from an Antarctic psychrophile classified oddly as *Moraxella catarrhalis* (Yamashita et al. 2002); most likely this is a member of the genus *Psychrobacter*; however, no sequence is available to confirm this nor is there a protein sequence for the AFP. This is the case for many other reports of AFP activity in bacteria but still suggests AFP activity is widespread (Wilson and Walker 2010; Wilson et al. 2012; Wu et al. 2012).

An isolate of the species *Marinomonas primoryensis* (not the type strain) detected in a large screen for bacterial AFP activity from Antarctic lakes in the Vestfold Hills region produces an unusual large (1.5 MDa) adhesin-like AFP that has a large thermal hysteresis (2 °C) and seems to be used by the bacterium to survive freezing (Gilbert et al. 2005). This AFP is complex in having five subunits and requires Ca^{2+} for stabilisation of the complex that forms (Garnham et al. 2011). Only subunit IV interacts with ice clathrates and is about 30 kDa in size. It was shown that this definitively is able to bind to ice and that the ice interaction is not just a chance feature (Bar Dolev et al. 2016a). Most of the size of this adhesin is due to 120 tandem repeats of subunit II that forms a kind of cell extension to provide better access to surfaces. Comparison with available genomes indicates this AFP/adhesin is unique to the *M. primoryensis* strain it is from. The type strain of the species has not been sequenced yet, and so no data is available if it is prevalent in the species *M. primoryensis* or relatively unique being acquired by horizontal gene transfer.

Proteins that have been classified with functionally unknown DUF3494 domains have been shown to be AFPs. These AFPs were initially characterised from a number of polar environment-sourced bacteria including *Flavobacterium frigoris* PS1 (Raymond et al. 2007), *Flavobacterium xanthum* (Kawahara et al. 2007) and *Colwellia* sp. SWL05 (Hanada et al. 2014) (Fig. 15.5). Similar proteins have been found in metagenome surveys of Antarctic moss bacteria (Raymond 2016), and orthologs are found in yeast and other ascomycetous fungi, algae and ciliates. Crystal structures have been solved for the versions from *F. frigoris* PS1 (Do et al. 2014), *Colwellia* sp. SWL05 (Hanada et al. 2014) and the snow mould *Typhula ishikariensis* (Kondo et al. 2012) (Fig. 15.5). These AFPs have large thermal hysteresis values, lack repeat domains and can control recrystallisation and formation of ice crystals. Surveying psychrophile genomes here reveals orthologs of this AFP family are prevalent in psychrophilic bacterial species and are also found in some archaea from cold ecosystems (Table 15.2). Most importantly they rarely if at all appear in bacteria that do not have some level of cold adaptation, either proven or inferred (from their ecosystem source). Orthologs detected in archaea include a methanogen *Methanoregula boonei* (Bräuer et al. 2011) and haloarchaea, including *Halohasta litchfieldiae* from Deep Lake, Antarctica (Mou et al. 2012) (Table 15.2). Some interesting variations of DUF3494 AFPs occur including 35 kDa orthologs in the Antarctic species *Aequorivita sublithincola* that have C-terminal T9SS secretion domains. The type IX secretion system (T9SS) terminal domains represent characteristic signal peptide domains used by members of the *Bacteroidetes* for the export of many proteins (McBride and Nakane 2015), including those involved in gliding motility. Another version occurs in *Psychroflexus torquis* in which the DUF3934 domain is located at the N-terminal ends of putatively T9SS secreted proteins of 100–110 kDa. The DUF3494 domains

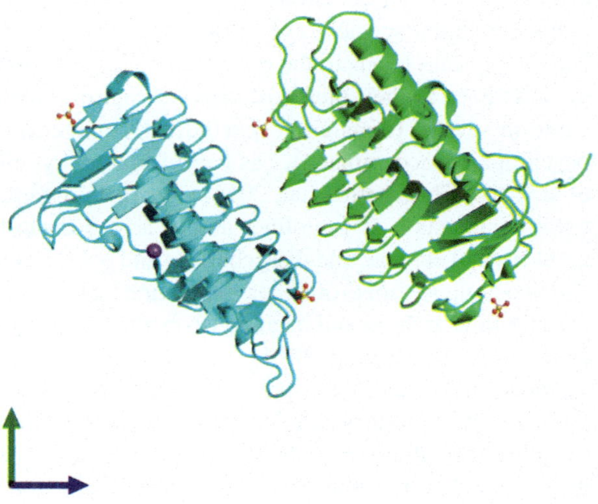

Fig. 15.5 Diagrammatic representation of the structure of the DUF3494-type antifreeze protein from *Flavobacterium frigoris* PS1 as determined by Do et al. (2014). The two chains forming the protein are separately coloured and viewed from the front. The associated single Na^+ (*purple sphere*) and four sulphate ligands are also shown. Image obtained from the EMBL-EBI Protein Data Bank

in this case are quite divergent. The aforementioned AFP from *F. xanthum* has several repeat regions accompanying the DUF3494 domain (Hanada et al. 2014). Orthologs of this protein are prevalent in *Flavobacterium* and some other members of the phylum *Bacteroidetes*. Other proteins with DUF3494 domains occur with additional domains in other taxa. Functionally it remains to be seen if these proteins also have antifreeze activity. The survey suggests AFPs are commonly distributed in bacteria and other life forms from cold regions though they require further study to verify functionality and cellular roles.

15.8 Compatible Solute Uptake Transporters and Synthesis Pathways Revealed by Genomics in Psychrophiles

Compatible solutes are low molecular mass compounds that can be accumulated to high levels in the cytoplasm without affecting protein and other cell functions. Their main role is to act as osmoprotectants and stabilise proteins and membranes; doing so provides defence against a range of stress including mainly low water activity but also low temperature, freezing, heat, starvation, desiccation and solvent stresses (Wood et al. 2001; Roberts 2005; Hoffmann and Bremer 2011, 2017; Vyrides and Stuckey 2017).

Bacteria and archaea can accumulate compatible solutes from the external environment via a range of transporters (Hoffmann and Bremer 2011, 2017). Compatible solutes can also be synthesised de novo (Roberts 2005). Most bacteria and archaea have a capacity to either accumulate or create compatible solutes, and so distinctions in terms of capability are usually not relevant when comparing psychrophiles to mesophiles. This is also complicated further by the fact that osmoprotection by accumulated solutes is also highly important for halophilic and halotolerant species. The most simple solute that is accumulated and usually held at constant levels in the cytoplasm is K^+, commonly used by haloarchaea and other archaea as their means of osmotic control (Grant 2004). Transporters for K^+ come in several forms and occur virtually universally. Thus, it is assumed most psychrophiles accumulate K^+ as part of normal cytoplasmic homeostasis (Binepal et al. 2016; Checchetto et al. 2016). The main organic compatible solutes include amino acids such as L-proline; amino acid derivatives such as glycine betaine, carnitine and ectoine; and disaccharides and polyols, in particular trehalose, sucrose and inositol. A range of other compounds are formed as compatible solutes in thermophilic archaea and bacteria (Roberts 2005; Empadinhas et al. 2007; Lamosa et al. 2013; Borges et al. 2014).

Most psychrophilic taxa contained betaine-choline-carnitine transporter (BCCT)-type uptake systems that take up a combination of glycine betaine, choline, carnitine or (hydroxy)ectoine. By comparing sequences of BCCT transporters to the genomes listed in Table 15.1, they appear to be quite prevalent in psychrophiles

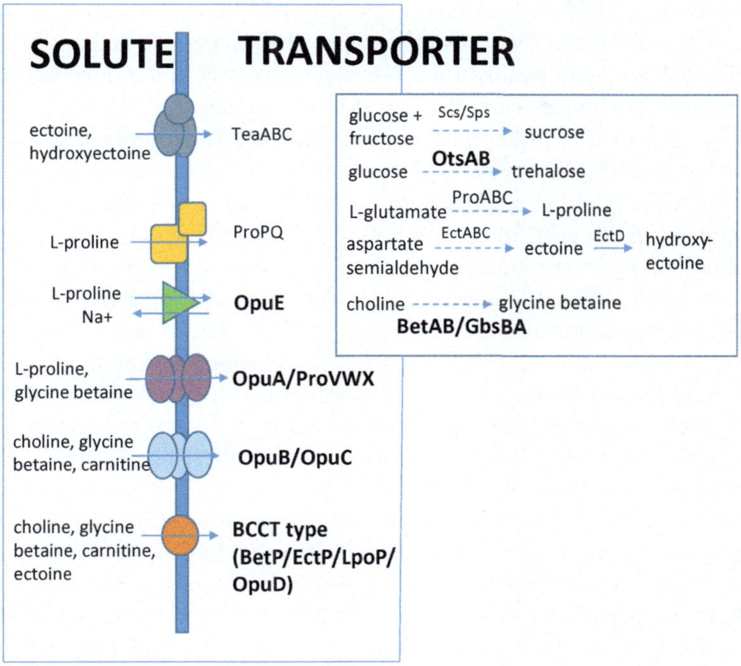

Fig. 15.6 The main types of compatible solute uptake systems and synthesis pathways present in bacterial and archaeal psychrophiles. Proteins shown in bold are the most extensively distributed

(Fig. 15.6). Not all species possess them, and the pattern seems random, not linked to halophily or halotolerance. Many species have multiple paralogs present in their genomes, for example, *Planococcus halocryophilus* contains four paralogs likely working together to allow growth at both low temperatures and high-salt levels (Mykytczuk et al. 2013). The distribution of transporters is widespread in terms of phylogenetic groups.

Proline can also be transported by the MHS-type transporter called ProP. This transporter has an osmosensor regulatory domain and is activated by the effector protein ProQ. ProP and ProQ were only detected in *Pseudomonas* spp. and *Devosia* spp. amongst the cold-adapted taxa. Virtually all taxa could in any case synthesise proline from glutamate (via proABC), but whether proline is used as a compatible solute requires direct chemical investigation. Data from studies on *B. subtilis* suggests L-proline accumulation does not provide cold tolerance (Hoffmann and Bremer 2011), which may explain the limited distribution of ProPQ. Various halophilic and halotolerant species that include many marine psychrophiles, for example, species of *Colwellia*, *Glaciecola*, *Oleispira*, *Paraglaciecola*, *Psychrobacter*, *Psychromonas*, *Shewanella*, *Planococcus*, *Psychroflexus* and *Psychroserpens*, can transport proline via Na^{+}-based symporter OpuE (Fig. 15.6). Whether L-proline only provides osmotolerance protection requires further analysis in marine psychrophiles.

Compatible solutes can also be transported via ABC-type transporters, notably the OpuA/ProVWX and OpuB/OpuC systems (Hoffmann and Bremer 2017). The OpuA/ProVWX system mainly transports glycine betaine and proline while OpuB/OpuC transports choline, carnitine and glycine betaine (Fig. 15.6). These transporters occur primarily in the more cold-adapted species, but this was not a strict relationship. It is likely that Opu-type transporters contribute equally to osmotic and cold tolerance as found in other bacteria such as *L. monocytogenes* (Angelidis et al. 2002) and *B. subtilis* (Hoffmann and Bremer 2011).

Ectoine and hydroxyectoine have been found to be transported by TRAP-type proteins called TeaABC (Grammann et al. 2002). TeaABC proteins are only found amongst *Halomonas* and its relatives including *Marinobacter* spp. and the Antarctic hypersaline lake psychrophile *Saccharospirillum impatiens* (Labrenz et al. 2003). Ectoine and hydroxyectoine are synthesised from aspartate semialdehyde by a pathway including three enzymes (EctABC) for ectoine and an additional hydroxylase EctD for hydroxyectoine (Cánovas et al. 1997; Roberts 2005). *Devosia*, *Marinobacter*, *Marinomonas*, some *Paenibacillus* spp., *Photobacterium halotolerans*, *Pseudomonas*, *Dasania marina* and the permafrost ice species *Tomitella biformata* (Katayama et al. 2010) possessed either EctABC or EctABCD pathways (Fig. 15.6). The results suggest only a small proportion of psychrophiles accumulates these compounds and most that do have both psychrotolerant and halotolerant capabilities. Ectoine, like proline, was found not to confer cold tolerance to *B. subtilis* (Hoffmann and Bremer 2011) and is likely more used for survival at low water activity.

De novo synthesis of compatible solutes is very prevalent but not universal amongst cold-adapted bacteria and occurs as frequently as in their more warm-temperature adapted relatives. The pathway for glycine betaine synthesis (BetA/GpsB and BetB/GpsB) was, however, very commonly detected in taxa listed in Table 15.1 across all major taxonomic groups. The ability to synthesise trehalose (via OtsAB) was also relatively common, especially in the more salt-tolerant taxa but was more species specific in terms of distribution. Sucrose synthase or sucrose phosphate synthase, which can potentially allow accumulation of sucrose as a compatible solute, often observed in cyanobacteria and proteobacteria with low halotolerance (Roberts 2005), was found in a few taxa including several *Flavobacterium* species and the glacial soil species *Planomicrobium glaciei* (Zhang et al. 2009).

Overall, psychrophiles have a broad capability to take up and synthesise compatible solutes of conventional sorts. The BCCT and OpuA/OpuB/OpuC transporters are the most prevalent accumulating choline, glycine betaine and carnitine. Many can make glycine betaine from choline or synthesise and accumulate trehalose. Whether there are any other forms of solutes present that are more novel in nature or known compounds not covered here requires direct chemical and biochemical analysis to be conducted.

15.9 Other Traits Relevant to but Not Directly Connected to Psychrophily

Certain stress protective traits seem universal at least in many well-known mesophiles often used as genetic models, including *E. coli*, *B. subtilis*, *P. aeruginosa* and *Streptomyces coelicolor*. All these species have wide growth temperature ranges (typically 40 °C) and are well endowed with protective systems. These species seem to have equal or more capability than many psychrophiles in some aspects of stress defence such as oxidative stress management and possession of multiple cold-shock proteins (CspA family proteins), which act as RNA chaperones, and other enzymes linked to cold temperature survival. These additional enzymes include, for example, as tRNA-dihydrouridine synthetases (Dus proteins) and ATP-dependent RNA DEAD-box helicases (CshA, CshB, RhlB, SrmB) that provide flexibility to tRNA and stabilise ribosomes against cold denaturation (Saunders et al. 2003). Chill stress and low-temperature shocks often lead to these proteins becoming more abundant as mentioned previously.

Low temperature encourages greater solubility of O_2, which can become supersaturated especially where there is active primary production (Marks 2008); thus, cold environment-associated bacteria likely need some form of oxidative stress protection. But this equally applies to any aerobic environment where high dissolved O_2 levels can accumulate or where production of reactive oxygen is used as a chemical defence. The above model species have 17–22 proteins associated with dealing with toxic oxidative reaction products including catalases (KatE, KatG, KatN), thiol peroxidases (peroxiredoxin Tpx and Bcp types), glutathione peroxidase (BtuE), cytochrome *c* and heme-type peroxidases, dye-reducing peroxidase (EfeB), vanadium-dependent haloperoxidases and superoxide dismutases (SodN, SodA/SodB, SodC). They also have multiple cold-shock proteins, Dus and RNA helicase proteins and compatible solute uptake systems and often some flexibility in terms of fatty acid synthesis. It is possible psychrophiles may possess unknown means for oxidative stress protection synthesising unusual or novel antioxidants; however, data suggests at present psychrophiles do not generally possess innovative or unusual distributions of stress protection mechanisms.

Light-harvesting proton and Na^+ pump proteins proteorhodopsin and xanthorhodopsin (Pinhassi et al. 2016) seem concentrated in psychrophilic species. These membrane proteins have a β-carotene-derived cofactor; thus, most strains produce carotenoid-pigmented colonies. *Glacieola* spp. are exceptions, and their pigmentation (pale to bright pink) is due to the accumulation of proteorhodopsin in cell membranes. Psychrophiles possessing these proteins are listed in Table 15.3. Most are marine where rhodopsins abound in marine bacteria to such an extent genes have been observed in viruses through which they are believed to have been spread via HGT (Philosof and Béjà 2013). Proteorhodopsin likely accumulates because large marine systems are cold, and in all probability most marine bacteria are thus cold-adapted to some extent. It is possible rhodopsins provide energetic advantages

Table 15.3 Psychrophilic and psychrotolerant bacterial species and strains possessing rhodopsin family proteins

Genus	Species or strain
Bacteroidetes	
Bizionia	*Bizionia* sp. PZ-13
Cellulophaga	*Cellulophaga* sp. He1_I_12
Chryseobacterium	*C. frigidisoli*, *C. humi*
Erythrobacter	*Erythrobacter* sp. QSSC1–22B[a]
Flavobacterium	*F. antarcticum*, *F. aquatile*, *F. fontis*, *F. frigidarium*, *F. frigoris*, *F. fryxellicola*, *F. gelidilacus*, *F. gillisiae*, *F. micromati*[b], *F. omnivorum*, *F. segetis*, *F. sinopsychrotolerans*, *F. urumqiense*, *F. xanthum*, *F. xinjiangense F. xueshanense*, *F. cucumis*, *Flavobacterium* sp. ACAM 123
Gelidibacter	*G. algens*
Gillisia	*G. limnaea*[b], *Gillisia* sp. JM1
Lacinutrix	*L. jangbogonensis*, *Lacinutrix* sp. He_I_90
Maribacter	*M. antarcticus*, *M. orientalis*
Pedobacter	*P. glucosidilyticus*
Polaribacter	*P. irgensii*, *P. dokdonensis*
Pricia	*P. antarctica*
Psychroflexus	*P. torquis*, *P. gondwanensis*, *P. salarius*, *P. sediminis*[b], *P. tropicus*
Psychroserpens	*P. burtonensis*
Rhodonellum	*R. psychrophilum*
Spirosoma	*S. linguale*, *S. spitsbergense*
Alphaproteobacteria	
Devosia	*D. psychrophila*
Octadecabacter	*O. antarcticus*, *O. arcticus*
Gammaproteobacteria	
Glaciecola	*G. punicea*, *G. pallidula*, *G. nitratireducens*
Marinobacter	*M. psychrophilus*[a]
Paraglaciecola	*P. psychrophila*
Photobacterium	*P. angustum*
Pseudomonas	*P. psychrotolerans*, *P. graminis*, *P. coleopterorum* (very diverged orthologs)
Actinobacteria	
Cryobacterium	*C. arcticum*, *C. flavum*, *C. levicorallinum*, *C. luteum*, *C. roopkundense*
Leifsonia	*L. rubra*
Firmicutes	
Carnobacterium	*C. iners*, *C. funditum* (diverged orthologs)
Exiguobacterium	*E. profundum*, *E. acetylicum*, *E. antarcticum*, *E. aurantiacum*, *E. enclense*, *E. indicum*, *E. sibiricum*, *E. undae*
Deinococcus-Thermus	
Deinococcus	*D. pimensis*, *D. marmoris*, *D. aquatilis*, *D. puniceus*

[a]Xanthorhodopsin present
[b]Also has a Na^+ pump rhodopsin

in marine systems such as sea ice (Koh et al. 2010; Feng et al. 2013, 2014, 2015) where several species with proteorhodopsin were isolated.

In summary, a rate-limiting step, such as protein folding (Ferrer et al. 2003), likely stops mesophiles from growing at low temperature. Otherwise many environmental mesophiles, not closely linked to host systems (such as gut microbiota), are not any different to psychrophiles in the sense of stress protection and thus have the ability to survive chill stress. Furthermore, gene content does not necessarily translate to a heightened tolerance; such phenotypes need to be empirically tested.

15.10 Conclusions and Future Prospects

Beneficial traits that seem more species (or strain) specific and potentially more concentrated in psychrophiles include traits mentioned above. They provide fitness in the context of particular econiches where psychrophiles abound. Since cold ecosystems are diverse—ranging from high mountains to the deepest oceans—the adaptations though having some general similarity have substantial variation in terms of specific details. The sum of these differences likely defines the ecophysiological profile of the psychrophile, including the temperature range for growth and their inherent growth rate. Genomes provide sources of knowledge and a start point to further explore evolved cold adaptation traits and other phenotypes, some potentially innovative and unrealised. From genome surveys and functional discovery, some traits could indeed develop greater significance, for example, proteins that provide unrealised cold stress protection but have unknown functionality. One recent example was a sequence extracted from an Antarctic desert soil metagenome that coded a protein with a WHy (water stress and hypersensitivity response) domain (Singh et al. 2005) that when expressed in *E. coli* provided chill and desiccation stress protection (Anderson et al. 2015). When WHy proteins are compared to psychrophile genomes, they can only be found in some psychrotolerant *Pseudomonas* species. The protein type seems associated with species with limited halotolerance and could act to protect bacteria against desiccation stress possibly by acting as an effector protein. There could be other proteins with analogous functions that could be found in psychrophilic bacteria that tangibly provides benefits for survival. Further research into the functional aspects of psychrophilic bacteria and archaea is now well served with the abundance of genome data currently available.

References

Aliyu H, De Maayer P, Cowan D (2016) The genome of the Antarctic polyextremophile *Nesterenkonia* sp. AN1 reveals adaptive strategies for survival under multiple stress conditions. FEMS Microbiol Ecol 92:fiw032

Allen MA, Lauro FM, Williams TJ et al (2009) The genome sequence of the psychrophilic archaeon, *Methanococcoides burtonii*: the role of genome evolution in cold adaptation. ISME J 3:1012–1035

Anderson D, Ferreras E, Trindade M et al (2015) A novel bacterial water hypersensitivity-like protein shows in vivo protection against cold and freeze damage. FEMS Microbiol Lett 362: fnv110

Angelidis AS, Smith LT, Smith GM (2002) Elevated carnitine accumulation by *Listeria monocytogenes* impaired in glycine betaine transport is insufficient to restore wild-type cryotolerance in milk whey. Int J Food Microbiol 75:1–9

Arrhenius S (1889) Uber die Reaktionsgeschwindigkeit bei der Inversion von Rohrzucker durch Säuren. Z Phys Chem 4:226–248

Ayala-del-Río HL, Chain PS, Grzymski JJ et al (2010) The genome sequence of *Psychrobacter arcticus* 273-4, a psychroactive Siberian permafrost bacterium, reveals mechanisms for adaptation to low-temperature growth. Appl Environ Microbiol 76:2304–2312

Bakermans C, Tsapin AI, Souza-Egipsy V et al (2003) Reproduction and metabolism at -10°C of bacteria isolated from Siberian permafrost. Environ Microbiol 5:321–326

Bar Dolev M, Bernheim R, Guo S et al (2016a) Putting life on ice: bacteria that bind to frozen water. J Roy Soc Interface 13:20160210

Bar Dolev M, Braslavsky I, Davies PL (2016b) Ice-binding proteins and their function. Annu Rev Biochem 85:515–542

Bednarska NG, Schymkowitz J, Rousseau F et al (2013) Protein aggregation in bacteria: the thin boundary between functionality and toxicity. Microbiology 159:1795–1806

Berezovsky IN, Chen WW, Choi PJ et al (2005) Entropic stabilization of proteins and its proteomic consequences. PLoS Comput Biol 1:e47

Bergholz PW, Bakermans C, Tiedje JM (2009) *Psychrobacter arcticus* 273-4 uses resource efficiency and molecular motion adaptations for subzero temperature growth. J Bacteriol 191: 2340–2352

Berlemont R, Pipers D, Delsaute M et al (2011) Exploring the Antarctic soil metagenome as a source of novel cold-adapted enzymes and genetic mobile elements. Rev Argent Microbiol 43: 94–103

Binepal G, Gill K, Crowley P et al (2016) Trk2 Potassium Transport System in *Streptococcus mutans* and its role in potassium homeostasis, biofilm formation, and stress tolerance. J Bacteriol 198:1087–1100

Borges N, Jorge CD, Gonçalves LG et al (2014) Mannosylglycerate: structural analysis of biosynthesis and evolutionary history. Extremophiles 18:835–852

Bosi E, Fondi M, Orlandini V et al (2017) The pangenome of (Antarctic) *Pseudoalteromonas* bacteria: evolutionary and functional insights. BMC Genomics 18:93

Bowman JS, Ducklow HW (2015) Microbial communities can be described by metabolic structure: a general framework and application to a seasonally variable, depth-stratified microbial community from the coastal west Antarctic Peninsula. PLoS One 10:e0135868

Bowman JS, Berthiaume CT, Armbrust EV et al (2014) The genetic potential for key biogeochemical processes in Arctic frost flowers and young sea ice revealed by metagenomic analysis. FEMS Microbiol Ecol 89:376–387

Bowman JP, McCammon SA, Brown MV et al (1997) Diversity and association of psychrophilic bacteria in Antarctic sea ice. Appl Environ Microbiol 63:3068–3078

Bräuer SL, Cadillo-Quiroz H, Ward RJ (2011) *Methanoregula boonei* gen. nov., sp. nov., an acidiphilic methanogen isolated from an acidic peat bog. Int J Syst Evol Microbiol 61:45–52

Breezee J, Cady N, Staley JT (2004) Subfreezing growth of the sea ice bacterium "*Psychromonas ingrahamii*". Microb Ecol 47:300–304

Burg DW, Lauro FM, Williams TJ et al (2010) Analyzing the hydrophobic proteome of the Antarctic archaeon *Methanococcoides burtonii* using differential solubility fractionation. J Proteome Res 9:664–676

Campanaro S, Williams TJ, Burg DW et al (2011) Temperature-dependent global gene expression in the Antarctic archaeon *Methanococcoides burtonii*. Environ Microbiol 13:2018–2038

Cánovas D, Vargas C, Iglesias-Guerra F et al (1997) Isolation and characterization of salt-sensitive mutants of the moderate halophile *Halomonas elongata* and cloning of the ectoine synthesis genes. J Biol Chem 272:25794–25801

Castillo D, Christiansen RH, Dalsgaard I et al (2016) Comparative genome analysis provides insights into the pathogenicity of *Flavobacterium psychrophilum*. PLoS One 11:e0152515

Checchetto V, Segalla A, Sato Y et al (2016) Involvement of potassium transport systems in the response of *Synechocystis* PCC 6803 cyanobacteria to external pH change, high-intensity light stress and heavy metal stress. Plant Cell Physiol 57:862–877

Chen Z, Yu H, Li L, Hu S et al (2012) The genome and transcriptome of a newly described psychrophilic archaeon, *Methanolobus psychrophilus* R15, reveal its cold adaptive characteristics. Environ Microbiol Rep 4:633–641

Chen Z, Feng D, Zhang B et al (2015) Proteomic insights into the temperature responses of a cold-adaptive archaeon *Methanolobus psychrophilus* R15. Extremophiles 19:249–259

Cheung RC, Ng TB, Wong JH (2017) Antifreeze proteins from diverse organisms and their applications: an overview. Curr Protein Pept Sci 18:262–283

Choudhari S, Lohia R, Grigoriev A (2014) Comparative metagenome analysis of an Alaskan glacier. J Bioinforma Comput Biol 12:1441003

Chrismas NA, Barker G, Anesio AM et al (2016) Genomic mechanisms for cold tolerance and production of exopolysaccharides in the Arctic cyanobacterium *Phormidesmis priestleyi* BC1401. BMC Genomics 17:533

Christner BC, Cai R, Morris CE et al (2008) Geographic, seasonal, and precipitation chemistry influence on the abundance and activity of biological ice nucleators in rain and snow. Proc Natl Acad Sci U S A 105:18854–18859

Colangelo-Lillis J, Eicken H, Carpenter SD et al (2016) Evidence for marine origin and microbial-viral habitability of sub-zero hypersaline aqueous inclusions within permafrost near Barrow, Alaska. FEMS Microbiol Ecol 92:fiw053

Corkrey R, Olley J, Ratkowsky D et al (2012) Universality of thermodynamic constants governing biological growth rates. PLoS One 7:e32003

Corkrey R, McMeekin TA, Bowman JP et al (2014) Protein thermodynamics can be predicted directly from biological growth rates. PLoS One 9:e96100

Corkrey R, McMeekin TA, Bowman JP et al (2016) The biokinetic spectrum for temperature. PLoS One 11:e0153343

Corkrey R, McMeekin TA, Bowman JP et al (2017) The maximum growth rate of life on Earth. Int J Astrobiol. doi:10.1017/S1473550416000501

Cvetkovska M, Hüner NPA, Smith DR (2017) Chilling out: the evolution and diversification of psychrophilic algae with a focus on *Chlamydomonadales*. Polar Biol. doi:10.1007/s00300-016-2045-4

D'Amico S, Collins T, Marx JC et al (2006) Psychrophilic microorganisms: challenges for life. EMBO Rep 7:385–389

DasSarma S, Capes MD, Karan R et al (2013) Amino acid substitutions in cold-adapted proteins from *Halorubrum lacusprofundi*, an extremely halophilic microbe from Antarctica. PLoS One 8:e887

De Vendittis E, Castellano I, Cotugno R et al (2008) Adaptation of model proteins from cold to hot environments involves continuous and small adjustments of average parameters related to amino acid composition. J Theor Biol 250:156–171

DeMaere MZ, Williams TJ, Allen MA (2013) High level of intergenera gene exchange shapes the evolution of haloarchaea in an isolated Antarctic lake. Proc Natl Acad Sci U S A 110: 16939–16944

Desmond-Le Quéméner E, Bouchez T (2014) A thermodynamic theory of microbial growth. ISME J 8:1747–1751

Do H, Kim SJ, Kim HJ, et al (2014) Structure-based characterization and antifreeze properties of a hyperactive ice-binding protein from the Antarctic bacterium *Flavobacterium frigoris* PS1 Acta Crystallogr D Biol Crystallogr 70:1061–1073

Drotz SH, Sparrman T, Nilsson MB et al (2010) Both catabolic and anabolic heterotrophic microbial activity proceed in frozen soils. Proc Natl Acad Sci U S A 107:21046–21051

Empadinhas N, Mendes V, Simões C et al (2007) Organic solutes in *Rubrobacter xylanophilus*: the first example of di-myo-inositol-phosphate in a thermophile. Extremophiles 11:667–673

Engle M, Li Y, Rainey F (1996) *Thermobrachium celere* gen. nov., sp. nov., a rapidly growing thermophilic, alkalitolerant, and proteolytic obligate anaerobe. Int J Syst Bacteriol 46: 1025–1033

Feller G (2013) Psychrophilic enzymes: from folding to function and biotechnology. Scientifica (Cairo) 2013:512840

Feng S, Powell SM, Wilson R et al (2013) Light-stimulated growth of proteorhodopsin-bearing sea-ice psychrophile *Psychroflexus torquis* is salinity dependent. ISME J 7:2206–2213

Feng S, Powell SM, Wilson R et al (2014) Extensive gene acquisition in the extremely psychrophilic bacterial species *Psychroflexus torquis* and the link to sea-ice ecosystem specialism. Genome Biol Evol 6:133–148

Feng S, Powell SM, Wilson R et al (2015) Proteomic insight into functional changes of proteorhodopsin-containing bacterial species *Psychroflexus torquis* under different illumination and salinity levels. J Proteome Res 14:3848–3858

Ferrer M, Chernikova TN, Yakimov MM et al (2003) Chaperonins govern growth of *Escherichia coli* at low temperatures. Nat Biotechnol 21:1266–1267

Fondi M, Maida I, Perrin E et al (2015) Genome-scale metabolic reconstruction and constraint-based modelling of the Antarctic bacterium *Pseudoalteromonas haloplanktis* TAC125. Environ Microbiol 17:751–766

Garnham CP, Campbell RL, Davies PL (2011) Anchored clathrate waters bind antifreeze proteins to ice. Proc Natl Acad Sci U S A 108:7363–7367

Georlette D, Blaise V, Collins T et al (2004) Some like it cold: biocatalysis at low temperatures. FEMS Microbiol Rev 28:25–42

Ghobakhlou AF, Johnston A, Harris L et al (2015) Microarray transcriptional profiling of Arctic *Mesorhizobium* strain N33 at low temperature provides insights into cold adaption strategies. BMC Genomics 16:383

Gilbert JA, Davies PL, Laybourn-Parry J (2005) A hyperactive, Ca^{2+-}dependent antifreeze protein in an Antarctic bacterium. FEMS Microbiol Lett 245:67–72

Glass JB, Yu H, Steele JA, Dawson KS et al (2014) Geochemical, metagenomic and metaproteomic insights into trace metal utilization by methane-oxidizing microbial consortia in sulphidic marine sediments. Environ Microbiol 16:1592–1611

Godin-Roulling A, Schmidpeter PA, Schmid FX et al (2015) Functional adaptations of the bacterial chaperone trigger factor to extreme environmental temperatures. Environ Microbiol 17:2407–2420

Goordial J, Davila A, Greer CW et al (2017) Comparative activity and functional ecology of permafrost soils and lithic niches in a hyper-arid polar desert. Environ Microbiol 19:443–458

Grammann K, Volke A, Kunte HJ (2002) New type of osmoregulated solute transporter identified in halophilic members of the bacteria domain: TRAP transporter TeaABC mediates uptake of ectoine and hydroxyectoine in *Halomonas elongata* DSM 2581(T). J Bacteriol 184:3078–3085

Grant WD (2004) Life at low water activity. Philos Trans R Soc Lond Ser B Biol Sci 359: 1249–1266

Han J, Jung J, Hyun S, Park H, Park W (2012) Effects of nutritional input and diesel contamination on soil enzyme activities and microbial communities in Antarctic soils. J Microbiol 50: 916–924

Hanada Y, Nishimiya Y, Miura A et al (2014) Hyperactive antifreeze protein from an Antarctic sea ice bacterium *Colwellia* sp. has a compound ice-binding site without repetitive sequences. FEBS J 281:3576–3590

Heitzer A, Kohler HP, Reichert P et al (1991) Utility of phenomenological models for describing temperature dependence of bacterial growth. Appl Environ Microbiol 57:2656–2665

Hoffmann T, Bremer E (2011) Protection of *Bacillus subtilis* against cold stress via compatible-solute acquisition. J Bacteriol 193:1552–1562

Hoffmann T, Bremer E (2017) Guardians in a stressful world: the Opu family of compatible solute transporters from *Bacillus subtilis*. Biol Chem 398:193–214

Hughes RA, Hallett K, Cogan T et al (2009) The response of *Campylobacter jejuni* to low temperature differs from that of *Escherichia coli*. Appl Environ Microbiol 75:6292–6298

Huston AL, Krieger-Brockett BB, Deming JW (2000) Remarkably low temperature optima for extracellular enzyme activity from Arctic bacteria and sea ice. Environ Microbiol 2:383–388

Ikai A (1980) Thermostability and aliphatic index of globular proteins. J Biochem 88:1895–1898

Inda ME, Vandenbranden M, Fernández A et al (2014) A lipid-mediated conformational switch modulates the thermosensing activity of DesK. Proc Natl Acad Sci U S A 111:3579–3584

Juneja VK, Porto-Fett AC, Gartner K et al (2010) Potential for growth of *Clostridium perfringens* from spores in pork scrapple during cooling. Foodborne Pathog Dis 7:153–157

Junge K, Eicken H, Deming JW (2003) Motility of *Colwellia psychrerythraea* strain 34H at subzero temperatures. Appl Environ Microbiol 69:4282–4284

Kahlke T, Thorvaldsen S (2012) Molecular characterization of cold adaptation of membrane proteins in the *Vibrionaceae* core-genome. PLoS One 7:e51761

Kashulin A, Seredkina N, Sørum H (2017) Cold-water vibriosis. The current status of knowledge. J Fish Dis 40:119–126

Katayama T, Kato T, Tanaka M et al (2010) *Tomitella biformata* gen. nov., sp. nov., a new member of the suborder *Corynebacterineae* isolated from a permafrost ice wedge. Int J Syst Evol Microbiol 60:2803–2807

Kawahara H, Iwanaka Y, Higa S et al (2007) A novel, intracellular antifreeze protein in an Antarctic bacterium, *Flavobacterium xanthum*. Cryo Lett 28:39–49

Kawamoto J, Kurihara T, Kitagawa M, Kato I, Esaki N (2007) Proteomic studies of an Antarctic cold-adapted bacterium, *Shewanella livingstonensis* Ac10, for global identification of cold-inducible proteins. Extremophiles 11:819–826

Klippel B, Sahm K, Basner A et al (2014) Carbohydrate-active enzymes identified by metagenomic analysis of deep-sea sediment bacteria. Extremophiles 18:853–863

Koh EY, Atamna-Ismaeel N, Martin A et al (2010) Proteorhodopsin-bearing bacteria in Antarctic sea ice. Appl Environ Microbiol 76:5918–5925

Kondo H, Hanada Y, Sugimoto H et al (2012) Ice-binding site of snow mold fungus antifreeze protein deviates from structural regularity and high conservation. Proc Natl Acad Sci U S A 109:9360–9365

Kotsyurbenko OR, Simankova MV, Nozhevnikova AN et al (1995) New species of psychrophilic acetogens: *Acetobacterium bakii* sp. nov., *A. paludosum* sp. nov., *A. fimetarium* sp. nov. Arch Microbiol 163:29–34

Kyte J, Doolittle RF (1982) A simple method for displaying the hydropathic character of a protein. J Mol Biol 157:105–132

Labrenz M, Lawson PA, Tindall BJ et al (2003) *Saccharospirillum impatiens* gen. nov., sp. nov., a novel gamma-Proteobacterium isolated from hypersaline Ekho Lake (East Antarctica). Int J Syst Evol Microbiol 53:653–660

Lamosa P, Rodrigues MV, Gonçalves LG et al (2013) Organic solutes in the deepest phylogenetic branches of the Bacteria: identification of α(1-6)glucosyl-α(1-2)glucosylglycerate in *Persephonella marina*. Extremophiles 17:137–146

Langille MG, Hsiao WW, Brinkman FS (2010) Detecting genomic islands using bioinformatics approaches. Nat Rev Microbiol 8:373–382

Lay CY, Mykytczuk NC, Yergeau É et al (2013) Defining the functional potential and active community members of a sediment microbial community in a high-arctic hypersaline subzero spring. Appl Environ Microbiol 79:3637–3648

Le PT, Makhalanyane TP, Guerrero LD et al (2016) Comparative metagenomic analysis reveals mechanisms for stress response in hypoliths from extreme hyperarid deserts. Genome Biol Evol 8:2737–2747

Lee DH, Choi SL, Rha E et al (2015) A novel psychrophilic alkaline phosphatase from the metagenome of tidal flat sediments. BMC Biotechnol 15:1

Lopatina A, Medvedeva S, Shmakov S et al (2016) Metagenomic analysis of bacterial communities of Antarctic surface snow. Front Microbiol 7:398

Lorv JSH, Rose DR, Glick BR (2014) Bacterial ice crystal controlling proteins. Scientifica 2014: 1–20

Mangiagalli M, Bar-Dolev M, Tedesco P et al (2017) Cryo-protective effect of an ice-binding protein derived from Antarctic bacteria. FEBS J 284:163–177

Marks R (2008) Dissolved oxygen supersaturation and its impact on bubble formation in the southern Baltic sea coastal waters. Hydrol Res 39:229–236

Marx JG, Carpenter SD, Deming JW (2009) Production of cryoprotectant extracellular polysaccharide substances (EPS) by the marine psychrophilic bacterium Colwellia psychrerythraea strain 34H under extreme conditions. Can J Microbiol 55:63–72

McBride MJ, Nakane D (2015) Flavobacterium gliding motility and the type IX secretion system. Curr Opin Microbiol 28:72–77

McMeekin TA, Presser K, Ratkowsky D et al (2000) Quantifying the hurdle concept by modelling the bacterial growth/no growth interface. Int J Food Microbiol 55:93–98

Methé BA, Nelson KE, Deming JW et al (2005) The psychrophilic lifestyle as revealed by the genome sequence of *Colwellia psychrerythraea* 34H through genomic and proteomic analyses. Proc Natl Acad Sci U S A 102:10913–10918

Metpally RP, Reddy BV (2009) Comparative proteome analysis of psychrophilic versus mesophilic bacterial species: insights into the molecular basis of cold adaptation of proteins. BMC Genomics 10:11

Middleton AJ, Marshall CB, Faucher F et al (2012) Antifreeze protein from freeze-tolerant grass has a beta-roll fold with an irregularly structured ice-binding site. J Mol Biol 416:713–724

Mock T, Junge K (2007) Psychrophilic diatoms: mechanisms for survival in freeze–thaw cycles. In: Seckbach J (ed) Algae and cyanobacteria in extreme environments. Cellular origin, life in extreme habitats and astrobiology, vol 11. Springer, New York, NY, pp 343–364

Moghadam MS, Albersmeier A, Winkler A (2016) Isolation and genome sequencing of four Arctic marine *Psychrobacter* strains exhibiting multicopper oxidase activity. BMC Genomics 17:117

Moreno Switt AI, Andrus AD, Ranieri ML et al (2014) Genomic comparison of sporeforming bacilli isolated from milk. BMC Genomics 15:26

Morita RY (1975) Psychrophilic bacteria. Bacteriol Rev 39:144–167

Mou YZ, Qiu XX, Zhao ML et al (2012) *Halohasta litorea* gen. nov. sp. nov., and *Halohasta litchfieldiae* sp. nov., isolated from the Daliang aquaculture farm, China and from Deep Lake, Antarctica, respectively. Extremophiles 16:895–901

Muryoi N, Sato M, Kaneko S et al (2004) Cloning and expression of *afpA*, a gene encoding an antifreeze protein from the arctic plant growth-promoting rhizobacterium *Pseudomonas putida* GR12-2. J Bacteriol 186:5661–5671

Mykytczuk NC, Foote SJ, Omelon CR et al (2013) Bacterial growth at −15 °C; molecular insights from the permafrost bacterium *Planococcus halocryophilus* Or1. ISME J 7:1211–1226

Ng C, DeMaere MZ, Williams TJ et al (2010) Metaproteogenomic analysis of a dominant green sulfur bacterium from Ace Lake, Antarctica. ISME J 4:1002–1019

Nichols DS, Brown JL, Nichols PD, McMeekin TA (1997) Production of eicosapentaenoic and arachidonic acids by an Antarctic bacterium: response to growth temperature. FEMS Microbiol Lett 152:349–354

Nunn BL, Slattery KV, Cameron KA et al (2015) Proteomics of *Colwellia psychrerythraea* at subzero temperatures – a life with limited movement, flexible membranes and vital DNA repair. Environ Microbiol 17:2319–2335

Panikov NS, Sizova MV (2006) Growth kinetics of microorganisms isolated from Alaskan soil and permafrost in solid media frozen down to −35°C. FEMS Microbiol Ecol 59:500–512
Park J, Kawamoto J, Esaki N, Kurihara T (2012) Identification of cold-inducible inner membrane proteins of the psychrotrophic bacterium, *Shewanella livingstonensis* Ac10, by proteomic analysis. Extremophiles 16:227–236
Philosof A, Béjà O (2013) Bacterial, archaeal and viral-like rhodopsins from the Red Sea. Environ Microbiol Rep 5:475–482
Pietikäinen J, Pettersson M, Bååth E (2005) Comparison of temperature effects on soil respiration and bacterial and fungal growth rates. FEMS Microbiol Ecol 52:49–58
Piette F, D'Amico S, Struvay C, Mazzucchelli G (2010) Proteomics of life at low temperatures: trigger factor is the primary chaperone in the Antarctic bacterium *Pseudoalteromonas haloplanktis* TAC12. Mol Microbiol 7:120–132
Pinhassi J, DeLong EF, Béjà O et al (2016) Marine bacterial and archaeal ion-pumping rhodopsins: genetic diversity, physiology, and ecology. Microbiol Mol Biol Rev 80:929–954
Porrini L, Cybulski LE, Altabe SG et al (2014) Cerulenin inhibits unsaturated fatty acids synthesis in *Bacillus subtilis* by modifying the input signal of DesK thermosensor. Microbiologyopen 3(2):213–224
Privalov PL (1990) Cold denaturation of proteins. Crit Rev Biochem Mol Biol 25:281–305
Qin QL, Xie BB, Yu Y et al (2014) Comparative genomics of the marine bacterial genus *Glaciecola* reveals the high degree of genomic diversity and genomic characteristic for cold adaptation. Environ Microbiol 16:1642–1653
Ratkowsky DA, Olley J, McMeekin TA et al (1982) Relationship between temperature and growth rate of bacterial cultures. J Bacteriol 149:1–5
Ratkowsky DA, Lowry RK, McMeekin TA et al (1983) Model for bacterial culture growth rate throughout the entire biokinetic temperature range. J Bacteriol 154:1222–1226
Raymond JA (2016) Dependence on epiphytic bacteria for freezing protection in an Antarctic moss, *Bryum argenteum*. Environ Microbiol Rep 8:14–19
Raymond JA, Fritsen C, Shen K (2007) An ice-binding protein from an Antarctic sea ice bacterium. FEMS Microbiol Ecol 61:214–221
Riley M, Staley JT, Danchin A et al (2008) Genomics of an extreme psychrophile, *Psychromonas ingrahamii*. BMC Genomics 9:210
Roberts MF (2005) Organic compatible solutes of halotolerant and halophilic microorganisms. Saline Syst 1:5
Rodrigues DF, da C Jesus E, Ayala-Del-Río HL et al (2009) Biogeography of two cold-adapted genera: *Psychrobacter* and *Exiguobacterium*. ISME J 3:658–665
Roller BK, Schmidt TM (2015) The physiology and ecological implications of efficient growth. ISME J 9:1481–1487
Ross T, Zhang D, McQuestin OJ (2008) Temperature governs the inactivation rate of vegetative bacteria under growth-preventing conditions. Int J Food Microbiol 128:129–135
Rothman JE, Schekman R (2011) Molecular mechanism of protein folding in the cell. Cell 146: 851–854
Sælensminde G, Halskau Ø Jr, Jonassen I (2009) Amino acid contacts in proteins adapted to different temperatures: hydrophobic interactions and surface charges play a key role. Extremophiles 13:11–20
Sakaguchi M, Matsuzaki M, Niimiya K (2003) Mechanisms of thermal adaptation revealed from the genomes of the Antarctic archaea *Methanogenium frigidum* and *Methanococcoides burtonii*. Genome Res 13:1580–1588
Saunders NFW, Thomas T, Curmi PM et al (2003) Mechanisms of thermal adaptation revealed from the genomes of the Antarctic archaea *Methanogenium frigidum* and *Methanococcoides burtonii*. Genome Res 13:1580–1588
Schipper LA, Hobbs JK, Rutledge S et al (2014) Thermodynamic theory explains the temperature optima of soil microbial processes and high Q10 values at low temperatures. Glob Chang Biol 20:3578–3586

Sengupta P, Garrity P (2013) Sensing temperature. Curr Biol 23:R304–R307
Shulse CN, Allen EE (2011a) Diversity and distribution of microbial long-chain fatty acid biosynthetic genes in the marine environment. Environ Microbiol 13:684–695
Shulse CN, Allen EE (2011b) Widespread occurrence of secondary lipid biosynthesis potential in microbial lineages. PLoS One 6:e20146
Siddiqui KS, Cavicchioli R (2006) Cold-adapted enzymes. Annu Rev Biochem 75:403–433
Siddiqui KS, Poljak A, Guilhaus M et al (2006) Role of lysine versus arginine in enzyme cold-adaptation: modifying lysine to homo-arginine stabilizes the cold-adapted alpha-amylase from *Pseudoalteramonas haloplanktis*. Proteins 64:486–501
Simon C, Wiezer A, Strittmatter AW et al (2009) Phylogenetic diversity and metabolic potential revealed in a glacier ice metagenome. Appl Environ Microbiol 75:7519–7526
Singh S, Cornilescu CC, Tyler RC (2005) Solution structure of a late embryogenesis abundant protein (LEA14) from *Arabidopsis thaliana*, a cellular stress-related protein. Protein Sci 14: 2601–2609
Steven B, Pollard WH, Greer CW et al (2008) Microbial diversity and activity through a permafrost/ground ice core profile from the Canadian high Arctic. Environ Microbiol 10:3388–3403
Sun X, Griffith M, Pasternak JJ et al (1995) Low temperature growth, freezing survival, and production of antifreeze protein by the plant growth promoting rhizobacterium *Pseudomonas putida* GR12-2. Can J Microbiol 41:776–784
Taha SKS, Campanaro S et al (2016) Single TRAM domain RNA-binding proteins in Archaea: functional insight from Ctr3 from the Antarctic methanogen *Methanococcoides burtonii*. Environ Microbiol 18:2810–2824
Touchon M, Barbier P, Bernardet JF et al (2011) Complete genome sequence of the fish pathogen *Flavobacterium branchiophilum*. Appl Environ Microbiol 77:7656–7662
Tribelli PM, Solar Venero EC, Ricardi MM et al (2015) Novel essential role of ethanol oxidation genes at low temperature revealed by transcriptome analysis in the Antarctic bacterium *Pseudomonas extremaustralis*. PLoS One 10:e0145353
Tschitschko B, Williams TJ, Allen MA et al (2016) Ecophysiological distinctions of haloarchaea from a hypersaline Antarctic lake as determined by metaproteomics. Appl Environ Microbiol 82:3165–3173
Tuorto SJ, Darias P, McGuinness LR et al (2013) Bacterial genome replication at subzero temperatures in permafrost. ISME J 8:139–149
Tveit AT, Urich T, Svenning MM (2014) Metatranscriptomic analysis of arctic peat soil microbiota. Appl Environ Microbiol 80:5761–5772
Ugalde JA, Gallardo MJ, Belmar C et al (2013) Microbial life in a fjord: metagenomic analysis of a microbial mat in Chilean Patagonia. PLoS One 8:e71952
Ujihara T, Nagano M, Wada H et al (2014) Identification of a novel type of polyunsaturated fatty acid synthase involved in arachidonic acid biosynthesis. FEBS Lett 588:4032–4036
Vincent AT, Trudel MV, Freschi L et al (2016) Increasing genomic diversity and evidence of constrained lifestyle evolution due to insertion sequences in *Aeromonas salmonicida*. BMC Genomics 17:44
Vollmers J, Voget S, Dietrich S et al (2013) Poles apart: Arctic and Antarctic *Octadecabacter* strains share high genome plasticity and a new type of xanthorhodopsin. PLoS One 8:e63422
Vyrides I, Stuckey DC (2017) Compatible solute addition to biological systems treating waste/wastewater to counteract osmotic and other environmental stresses: a review. Crit Rev Biotechnol 2:1–15
Weinstock MT, Hesek ED, Wilson CM et al (2016) *Vibrio natriegens* as a fast-growing host for molecular biology. Nat Methods 13:849–881

Williams TJ, Burg DW, Ertan H et al (2010a) Global proteomic analysis of the insoluble, soluble, and supernatant fractions of the psychrophilic archaeon *Methanococcoides burtonii*. Part II: The effect of different methylated growth substrates. J Proteome Res 9:653–663

Williams TJ, Burg DW, Raftery MJ et al (2010b) Global proteomic analysis of the insoluble, soluble, and supernatant fractions of the psychrophilic archaeon *Methanococcoides burtonii*. Part I: The effect of growth temperature. J Proteome Res 9:640–652

Williams TJ, Lauro FM, Ertan H et al (2011) Defining the response of a microorganism to temperatures that span its complete growth temperature range (−2°C to 28°C) using multiplex quantitative proteomics. Environ Microbiol 13:2186–2203

Wilson SL, Walker VK (2010) Selection of low-temperature resistance in bacteria and potential applications. Environ Technol 31:943–956

Wilson SL, Grogan P, Walker VK (2012) Prospecting for ice association: characterization of freeze-thaw selected enrichment cultures from latitudinally distant soils. Can J Microbiol 58: 402–412

Wood JM, Bremer E, Csonka LN, Kraemer R et al (2001) Osmosensing and osmoregulatory compatible solute accumulation by bacteria. Comp Biochem Physiol A Mol Integr Physiol 130:437–460

Wright KE, Williamson C, Grasby SE et al (2013) Metagenomic evidence for sulfur lithotrophy by *Epsilonproteobacteria* as the major energy source for primary productivity in a sub-aerial arctic glacial deposit, Borup Fiord Pass. Front Microbiol 4:63

Wu Z, Kan FW, She YM et al (2012) Biofilm, ice recrystallization inhibition and freeze-thaw protection in an epiphyte community. Prikl Biokhim Mikrobiol 48:403–410

Yamashita Y, Nakamura N, Omiya K, Nishikawa J, Kawahara H, Obata H (2002) Identification of an antifreeze lipoprotein from *Moraxella* sp. of Antarctic origin. Biosci Biotechnol Biochem 66:239–247

Yang X, Sheng W, Sun GY, Lee JCM (2011) Effects of fatty acid unsaturation numbers on membrane fluidity and α-secretase-dependent amyloid precursor protein processing. Neurochem Int 8:321–329

Zakhartsev M, Yang X, Reuss M et al (2015) Metabolic efficiency in yeast *Saccharomyces cerevisiae* in relation to temperature dependent growth and biomass yield. J Therm Biol 52: 117–129

Zeldovich KB, Berezovsky IN, Shakhnovich EI (2007) Protein and DNA sequence determinants of thermophilic adaptation. PLoS Comput Biol 3:e5

Zhang G, Jiang N, Liu X, Dong X (2008) Methanogenesis from methanol at low temperatures by a novel psychrophilic methanogen, "*Methanolobus psychrophilus*" sp. nov., prevalent in Zoige wetland of the Tibetan plateau. Appl Environ Microbiol 74:114–120

Zhang DC, Liu HC, Xin YH, Yu Y, Zhou PJ, Zhou YG (2009) *Planomicrobium glaciei* sp. nov., a psychrotolerant bacterium isolated from a glacier. Int J Syst Evol Microbiol 59:1387–1390

Zhang DL, Ross T, Bowman JP (2010) Physiological aspects of *Listeria monocytogenes* during inactivation accelerated by mild temperatures and otherwise non-growth permissive acidic and hyperosmotic conditions. Int J Food Microbiol 141:177–185

Zhang L, Wang X, Yu M et al (2015) Genomic analysis of *Luteimonas abyssi* XH031(T): insights into its adaption to the subseafloor environment of South Pacific Gyre and ecological role in biogeochemical cycle. BMC Genomics 16:1092

Zhao JS, Deng Y, Manno D et al (2010) *Shewanella* spp. genomic evolution for a cold marine lifestyle and in-situ explosive biodegradation. PLoS One 5:e9109

Zhu K, Bayles DO, Xiong A, Jayaswal RK, Wilkinson BJ (2005) Precursor and temperature modulation of fatty acid composition and growth of *Listeria monocytogenes* cold-sensitive mutants with transposon-interrupted branched-chain alpha-keto acid dehydrogenase. Microbiology 151:615–623

Zimmerly S, Wu L (2015) An unexplored diversity of reverse transcriptases in bacteria. MicrobiolSpec 3. doi:10.1128/microbiolspec.MDNA3-0058-2014

Chapter 16
Metagenomic Analysis of Low-Temperature Environments

Habibu Aliyu, Pieter De Maayer, Sara Sjöling, and Donald A. Cowan

Abstract The Earth's permanently cold biosphere is known to harbour abundant microbial biomass and represents a rich resource for the discovery of novel cold-adapted microorganisms, many of which form part of the 'microbial dark matter' which cannot be analysed using traditional culture-dependent approaches. The recent development of metagenomics and related multi-omics strategies has provided a means by which entire microbial communities can be studied directly, without the prerequisite of culturing. The advancement of the 'omic' methods is directly linked to recent progress in high-throughput sequencing, robust data processing capabilities and the application of cutting-edge analytical tools for high-throughput detection of biomolecules. The combined application of these tools and strategies has provided an unprecedented access to the structure and potential function of microbial communities in cold environments, providing increasingly comprehensive insights into the taxonomic richness and functional capacity of the indigenous microorganisms. Applications of 'omic' strategies have enhanced our understanding of psychrophilic adaptation mechanisms, revealing the versatility and adaptability of life in the 'cryosphere'. In addition to the predicted roles of psychrophiles in biogeochemical cycling, recent multi-omic studies have further emphasised the importance of the 'cryosphere' in influencing global atmospheric conditions. Finally, metagenomic bioprospecting of cold environments has yielded a variety of novel bioactive molecules including novel 'psychrozymes', with a wide range of potential industrial and biotechnological applications. Here, we have provided an overview of recent developments in metagenomic technologies and their application in the study of the cold biosphere.

H. Aliyu • D.A. Cowan (✉)
Centre for Microbial Ecology and Genomics, University of Pretoria, Pretoria 0028, South Africa
e-mail: don.cowan@up.ac.za

P. De Maayer
School of Molecular and Cell Biology, University of the Witwatersrand, Private Bag 3, Wits, 2050 Johannesburg, South Africa

S. Sjöling
School of Natural Sciences, Technology and Environmental Studies, Sodertorn University, 141 89 Huddinge, Sweden

R. Margesin (ed.), *Psychrophiles: From Biodiversity to Biotechnology*,
DOI 10.1007/978-3-319-57057-0_16

Contents

16.1 Introduction

Cold environments collectively constitute the predominant portion (ca. 80%) of the Earth's biosphere. These include the oceans, the polar terrestrial masses, layers above the troposphere and artificial environments such as refrigerators and freezers (De Maayer et al. 2014a). While long considered to be inhospitable to life, natural cold environments have been shown to harbour abundant and diverse bacterial and archaeal communities, which form a significant proportion of the Earth's microbial biomass (Rodrigues and Tiedje 2008; Casanueva et al. 2010; Margesin and Miteva 2011; Cavicchioli 2015). This biomass represents a largely untapped resource for the identification of novel psychrophilic species, biochemical pathways, genes and novel compounds with potential biotechnological applications (Margesin and Feller 2010).

It has been estimated that over 99% of the predicted 10^6 distinct prokaryotic 'species' on Earth represent the so-called microbial dark matter that are uncultivable using standard protocols (Gasc et al. 2015). This created a 'bottleneck' in prokaryotic biology for decades, as pure cultures of microorganisms were required for taxonomic characterisation and bioprospecting (Schloss and Handelsman 2005b). This bottleneck has largely been resolved in recent years through the development of the field of metagenomics, which involves the use of genomic tools to analyse the collective genomes of a microbial community in an environmental sample, without the requisite isolation and cultivation of each member of the community. Metagenomics has thus provided primary insights into the 'true' microbial diversity in a given environment, has revealed the complex functions of microbial ecosystems and has served as a tool for bioprospecting novel genes and gene products.

In its initial form, metagenomics involved the extraction of DNA directly from an environmental sample, shearing of the DNA into fragments of a suitable size and cloning of these fragments into a suitable vector. Subsequently, the cloned community DNA was sequenced using whole-genome shotgun approaches. This has been extensively reviewed elsewhere (Sjöling et al. 2006; Sjöling and Cowan 2008; De Maayer et al. 2014b). The development of next-generation sequencing technologies has, however, greatly facilitated the sequencing of metagenomes. As such, community DNA can now be sequenced directly without the need for prior cloning strategies (Forde and O'Toole 2013; Reuter et al. 2015). The classical cloning-based approach, however, still serves as the backbone for functional metagenomics studies, whereby the cloned DNA is transformed into a suitable host, and the transformants are screened for a particular function, such as antibiotic or enzyme activities.

In a previous version of this chapter (Sjöling and Cowan 2008), we provided extensive coverage of the functional and sequence-based metagenomics approaches and discussed results from studies of microbial communities associated with cold environments. Given the rapid evolution of this field, and the progressive development of novel metagenomic technologies, we here provide an update of recent developments with a particular focus on the metagenomics of psychrophiles.

16.2 Approaches in Metagenomic Analysis

Generally, metagenomic analysis starts with the extraction of nucleic acids from an environmental sample. The method of collection and processing of the samples, efficiency of the nucleic acid extraction protocol and post-extraction procedures have direct influence on subsequent downstream analyses of the metagenome data (Nayfach and Pollard 2016). For example, differences in the integrity of the cell walls of the microorganisms in a given environment may give rise to a bias in the amount and quality of the DNA recovered. The completeness, quality and quantity of DNA obtained from environmental samples have also been shown to significantly vary with the method of extraction employed (Kennedy et al. 2014; Brooks et al. 2015). The efficiency of DNA extraction from soil is also greatly affected by the presence of inhibitors such as humic compounds, which are co-extracted with the DNA (Handelsman et al. 2002; Howeler et al. 2003). Two methods, namely, direct and indirect extraction, are widely applied for the extraction of DNA. Direct extraction entails approaches in which microbial cells are lysed without separation from the environmental sample, while the indirect method involves the initial separation of the cells from the sample prior to lysis (Delmont et al. 2011). The indirect approach minimises the contamination of the extracted DNA, but the procedure is tedious and requires specialised equipment when compared with the direct method (Delmont et al. 2011).

Several techniques have also been developed to maximise the amount of DNA obtained from environments where low numbers of organisms are present. For example, large volumes of marine water are pumped through a series of selective filters (0.16–0.22 μm) that recover different particulate sizes (Gilbert et al. 2011). DNA is subsequently extracted from the retentate on the filters that are impermeable to the microorganisms contained in the sample (Tseng et al. 2015). Similarly, airborne microorganisms are collected using pumps fitted with filters that trap bioaerosols or by using liquid impingers in which the microorganisms become trapped in a liquid medium (Fahlgren et al. 2011; Gandolfi et al. 2013; Behzad et al. 2015).

16.2.1 Metagenome Sequencing

16.2.1.1 Metagenome Sequencing in the High-Throughput Sequencing Era

In classical metagenomic approaches, the nucleic acid extraction step was followed by metagenomic library construction and subsequent prospecting for desired genes/ gene products (discussed below) or shotgun metagenome sequencing using Sanger sequencing chemistry (Sanger et al. 1977). Sanger sequencing produces long (up to 1000 bp) and highly accurate reads (Shendure and Ji 2008). Major drawbacks of the Sanger chemistry, however, included low throughput and the high cost of sequencing (Perkel 2009), both of which are prohibitive when applied in metagenomic projects that typically generate gigabases or terabases of sequencing data (Sunagawa et al. 2015; White et al. 2016). In addition, the tedious nature of producing clone libraries and the inherent cloning bias (Thomas et al. 2012; De Maayer et al. 2014b) have limited the application of shotgun libraries and sequencing approaches in metagenomic projects. However, Sanger sequencing was instrumental in pioneering a number of large-scale metagenomic projects. For instance, the shotgun sequencing of the Sargasso Sea yielded 1.6 Gb of DNA sequence and the discovery of 140 novel bacterial species (Venter et al. 2004).

Current metagenomic approaches utilise an array of available high-throughput sequencing (HTS) technologies which offer massively parallel sequencing and generate several orders of magnitude greater throughput at much lower cost than Sanger sequencing (Escalona et al. 2016). For instance, a single sequencing run using the Illumina HiSeq 2000 produces ca. 600 Gb of sequences per run at a much lower cost than with Sanger sequencing chemistries (Liu et al. 2012). The major limitation of the current HTS technologies is the short read length, which ranges between 50 and 500 bases. However, this is compensated for by the depth of sequencing (Loman and Pallen 2015). The recent development and optimisation of long read sequencing technologies are significantly enhancing genomics and metagenomics by offering unprecedented read lengths (up to kilobases in length) which are less amenable to misassembly in subsequent downstream analyses (Land et al. 2015; Reuter et al. 2015). For example, the long read single molecule real-time (SMRT) chemistry of the PacBio RS II sequencing platform (Pacific Biosciences, http://www.pacificbiosciences.com/) produces reads up to 40 kb in length, while the portable MinION™ platform manufactured by Oxford Nanopore (Quick et al. 2014) yields reads that are between 6 and 60 kb in length (Ashton et al. 2014; Jain et al. 2015). The SMRT technology is, however, subject to high levels of sequencing errors (mainly indels), with up to 15% errors in reads produced using PacBio platform (Carneiro et al. 2012; Kremkow and Lee 2015) and 18% for the MinION™ technology (Jain et al. 2015). These errors could be minimised by performing multiple sequencing runs and subsequently generating consensus reads (Quick et al. 2016). The application of SMRT technologies in metagenomic studies may also be limited by the relative low throughput of the currently available platforms (White et al. 2016). Despite these challenges, long read sequencing technologies have been used recently to sequence metagenomes from clinical samples (Greninger et al. 2015; Schmidt et al. 2016) and cryoconite hole debris samples and red snow algal bloom samples from a glacier adjacent to the Foxfonna ice cap at Svalbard (Edwards et al. 2016).

16.2.1.2 Assembling Metagenomic Reads

In order to gain access to the genetic and biological information within a metagenome dataset, which is generally comprised of millions or billions of short DNA sequence fragments, metagenomic reads must be assembled; whereby overlapping sequencing reads are positioned into correctly ordered and properly oriented long contiguous sequences (contigs) (Earl et al. 2011; Edwards and Holt 2013). The ultimate objective is to reproduce a complete nucleotide sequence identical to the original DNA template from which the reads were produced (Paszkiewicz and Studholme 2010). The assembled contigs provide access to syntenous coding regions and facilitate the accurate prediction of functions (Kim et al. 2013).

Two approaches, namely, de novo and reference-based assembly, have been implemented to assemble metagenomic reads (Zhang et al. 2011; Thomas et al. 2012). The complex nature of metagenomic data, the size of the dataset, differences in microbial abundance and the occurrence of highly conserved genomic regions across different genomes all affect the process by which metagenomic reads are assembled (Logares et al. 2012; Leung et al. 2015). As such, regular genome assemblers, which were designed for the assembly of reads in the mega- to gigabase range, are limited in their applicability to terabase-sized metagenomes (Leung et al. 2015). Consequently, several assemblers have been optimised for the assembly of very large metagenomic read datasets. For instance, IDBA-UD (Peng et al. 2012), Meta-IDBA (Peng et al. 2011), MetaRay (Boisvert et al. 2012) and MetaVelvet (Namiki et al. 2012) have been optimised for reads obtained from Illumina sequencers (Leung et al. 2015).

16.2.1.3 Binning of Metagenomic Contigs

Sequences assembled from metagenomic data originate from different organisms, and the genomic composition, distribution and number of contigs reflect the diversity and abundance of members within the microbial community from which the data was obtained. On the basis of the nucleotide composition (G+C contents, codon usage and tetranucleotide frequencies) and/or sequence homology, these contigs can be sorted, or binned, into taxonomic clusters representing the organisms of their origin (Simon and Daniel 2011; Thomas et al. 2012). Some of the widely applied sequence binning tools, including PCAHIER (Zheng and Wu 2010), PhyloPythia (McHardy et al. 2007) and TETRA (Teeling et al. 2004) implement algorithms that bin the metagenomic contigs on the basis of nucleotide composition. Several homology-based tools have also been developed based on algorithms that take into account the percentage sequence similarity to sort the metagenomic contigs into distinct clusters. These include CARMA (Krause et al. 2008), MEGAN (Huson and Weber 2012) and MetaPhyler (Liu et al. 2010). Other tools, such as PhymmBL (Brady and Salzberg 2009) and RITA (MacDonald et al. 2012), incorporate both composition- and sequence alignment-based classification approaches, which could be used to improve binning accuracy.

16.2.1.4 Recovering Novel Genomes from Metagenome Data

As an extension to the metagenome binning strategies, several complete and high-quality draft genome sequences of both known and novel microorganisms have been reconstructed from metagenomes. This endeavour began with the recovery of near-complete genomes of *Leptospirillum* group II and *Ferroplasma* type II, by assigning scaffolds assembled from shotgun sequencing metagenome data into distinct species clusters on the basis of average G+C contents (Tyson et al. 2004). With the advent of HTS, binning of genome fragments into genomic bins, or metagenome assembled genomes (MAGs), and reconstructing genomes from metagenome dataset are now becoming routine (Sharon and Banfield 2013; Hugerth et al. 2015). For example, Albertsen et al. (2013) proposed a pipeline for binning high-quality genomes from metagenomes. The principle involves (1) de novo assembly and scaffolding of the metagenome sequences; (2) binning of the sequences using information about population coverage, genome composition metrics (% G+C content, tetranucleotide frequency) and essential single copy genes; (3) extraction of reads associated with the cluster (or organism) of interest and (4) reassembly of the extracted reads to recover the genome of the predicted organism (Albertsen et al. 2013). Several tools for automated binning which implement variants of this protocol have been developed, including CONCOCT (Alneberg et al. 2014), MaxBin (Wu et al. 2014), MetaBAT (Kang et al. 2015) and MyCC (Lin and Liao 2016). Since the environmental sample may contain multiple strains of the same species and closely related species, the recovered genomes may be subject to chimeric assembly or binning (Sangwan et al. 2016). To assess the quality of population genomes, Parks et al. (2015) developed the CheckM method which ascertains the levels of completeness and contamination in the recovered genomes using a set of taxa-specific genetic markers and information regarding their colocation (Parks et al. 2015).

Metagenome sequence datasets from various cold habitats have been used for the reconstruction of genomes of novel psychrophiles and the prediction of their potential survival strategies. For instance, a novel 1.9 Mb draft methanogen genome was identified from an Alaskan permafrost site. Given evidence of the abundance of these organisms, a major potential role in methane production under cold condition was predicted (Mackelprang et al. 2011). Similarly, a novel methanogen genome binned from a thawing permafrost metagenome (Candidatus *Methanoflorens stordalenmirensis*) has recently been described. (Mondav et al. 2014). Binning of metagenomes from the Alaskan permafrost active layers and bog soils yielded a novel *Acidobacterium* genome that encodes the dissimilatory iron reduction system. In addition, a novel member of the phylum *Proteobacteria* with the capacity for iron and nitrate reduction and two novel members of the phylum *Chloroflexi* (*Anaerolinea* spp.) encoding sulphate reduction and iron utilisation capabilities were recovered (Hultman et al. 2015).

16.2.1.5 Annotation of Metagenomic Contigs

Annotation involves the assignment of putative functions to genes in assembled genomes or metagenome datasets (Zhang et al. 2011; Pareja-Tobes et al. 2012) and is accomplished using bioinformatic tools which work on the basis of one of two approaches: homology-based searches or algorithm-based prediction (Logares et al. 2012). The homology-based approach relies on comparison of the genomic or metagenomic data against annotated reference genomes (Richardson and Watson 2013). The algorithm-based approach involves de novo gene prediction, using an array of different gene prediction tools which delineate genes based on a combination of criteria, including information regarding the genetic code and gene length. Thereafter the resultant coding sequences are searched against specific genomic or metagenomic databases using various computational tools (Stothard and Wishart 2006; Edwards and Holt 2013). Predicted coding sequences which differ significantly from those deposited in the databases are considered as putative novel genes (Stothard and Wishart 2006). Unlike single genome annotation, the annotation of metagenome data involves the processing of huge amounts of data, which may contain sequences from eukaryotes, prokaryotes and viruses. A number of software packages and web servers, including SEED (Overbeek et al. 2014), KEGG (Kanehisa et al. 2012), LCA (last common ancestor) (Huson et al. 2007), CAMERA (Seshadri et al. 2007; Altintas et al. 2010), EBI metagenomics (Hunter et al. 2014), IMG/M (Markowitz et al. 2014), MG-RAST (Meyer et al. 2008) and VMGAP (Lorenzi et al. 2011), have been developed specifically for the functional annotation of metagenome data.

16.2.1.6 Metagenome Sequencing from Cold Environments

Recent advances in the field of metagenomics have resulted in the accumulation of unprecedented volumes of metagenomic data from a wide range of environmental and man-made environments. Sequence reads are typically deposited in public databases, e.g. European Nucleotide Archive (ENA), the Integrated Microbial Genomes & Microbiomes (IMG/M) (Markowitz et al. 2014) at the Department of Energy Joint Genome Institute or the MG-RAST server. Currently (September 2016), 264,406 metagenomes, totalling 118.96 terabases and 917 billion sequences, have been deposited in the MG-RAST server (Meyer et al. 2008) by over 12,000 users. Of these, a total of 36,628 'metagenomes', consisting of 15,062 Gb and 128,714 sequences, are publicly available. Metadata searches against the MG-RAST database, using keywords such as 'cold', 'ice', 'marine', 'ocean', 'permafrost' and 'sea', identified 1395 'metagenome' datasets derived from cold environments. This includes 910 amplicon sequences or metataxonomic datasets (described in further detail below) and 463 metagenome sequencing datasets. The number of metagenomes from cold environments are underestimated, due to non-compliance of metagenome depositors with the requirement for minimum metadata information (Field et al. 2008; Yilmaz et al. 2011) regarding the deposited projects and also the number of metagenomic projects which have deposited sequences in other databases. Some of the pertinent cold-environment metagenomes in the MG-RAST, ENA or IMG/M databases are highlighted in Table 16.1.

Table 16.1 Metagenome studies of cold environments

Source of metagenome	Location	References
Hunt Ice and the Markham Ice Shelves	Canadian High Arctic	Varin et al. (2012)
McMurdo Ice Shelf	Ross Sea, Antarctica	Varin et al. (2012)
Marine water	Southern Ocean and Ross Sea, Antarctica	Brown et al. (2012)
Sea ice	Offshore of Pt. Barrow, Alaska, USA	Bowman et al. (2014)
University Valley permafrost	McMurdo Dry Valleys, Antarctica	Goordial et al. (2016a)
McKelvey Valley soil and rock surfaces	McMurdo Dry Valleys, Antarctica	Chan et al. (2013)
Miers Valley soil	McMurdo Dry Valleys, Antarctica	Wei et al. (2015)
Glacier ice	Northern Schneeferner, Germany	Simon et al. (2009)
Seawater	Off Palmer Station, Antarctica	Grzymski et al. (2006)
Deep sea >2000 m	North Pacific Subtropical Gyre	Konstantinidis et al. (2009)
Miers Valley hypolith	McMurdo Dry Valleys, Antarctica	Le et al. (2016)
Peatland, permafrost and thermokarst bog soil	Fairbanks, Alaska	Hultman et al. (2015)
Permafrost	Hess Creek, Alaska	Mackelprang et al. (2011)
Taylor Valley soil High-latitude coastal sediments	McMurdo Dry Valleys, Antarctica Antarctica, Patagonia, Arctic, Baltic Sea	Van Horn et al. (2014) Matos et al. (2016)
Deep seawater and sediment Lower terrace surface soil	Baltic Sea, Landsort Deep Mars Oasis, Antarctic Peninsula	Thureborn et al. (2013, 2016) Pearce et al. (2012)

Analyses of metagenomic data obtained from cold environments have contributed significantly to our understanding of the adaptations and ecological functions of various microbial communities (Boetius et al. 2015). For example, evaluation of the metagenome of the Northern Schneeferner glacier provided one of the first detailed insights into the adaptations of microorganisms living in a permanently cold environment. The data predicted a wide range of potential metabolic capacities among the microbiota from this environment, with a predominance of autotrophic lifestyles as indicated by abundance of genes encoding enzymes of both the tricarboxylic acid (TCA) and carboxylate cycles. This represents a probable adaptation to cope with the scarcity of nutrients in the cold environment (Simon et al. 2009). A large variety of genes involved in psychrophilic adaptation were also identified, including those involved in the production of unsaturated fatty acids and carotenoids (associated with maintaining membrane fluidity), antioxidants (catalases and superoxide dismutase) and cryo- and osmoprotectants (betaine, choline, glycine and glutamate) (Simon et al. 2009). Metagenomic analyses of supraglacial cryoconite sediments from Pakistan and glaciers in the Italian Alps showed that light directly supplements the energy demand of some of the glacial bacterial strains, allowing them to use organic molecules which otherwise would be respired as carbon source. In these cold environments, CO_2 could also be produced by microbiologically mediated oxidation of CO, produced by photodegradation of organic matter (Franzetti et al. 2016). With declining mountain glacier environments, these results give insights into the potential climatic effects on the global cryosphere.

Other studies have addressed molecular adaptation of psychrophiles. For example, fosmid clone sequencing of Antarctic coastal surface water bacterioplankton revealed aspects of low-temperature protein adaptation. In order to reduce conformational rigidity and enhance catalytic efficiency, psychrophilic proteins showed an under-representation of arginine, aspartic acid and glutamic acid residues (reducing salt-bridge formation) and reduced proline and hydrophobic contents (Grzymski et al. 2006).

Metagenomic surveys of microbial mat communities from both Arctic and Antarctic ice shelves have provided significant evidence on psychrophilic adaptive strategies in response to multiple stress factors (Varin et al. 2012). For instance, in addition to a retinue of known cold stress response genes encoding cold-shock proteins, cryo- and osmoprotectants and proteins linked to modulation of membranes and response to oxidative stress, which were common to both metagenomes, the data showed greater representation of genes involved in copper homeostasis, which may be due to exposure of the Arctic ice shelf to pollutants. The Antarctic microbial mat showed significant over-representation of genes encoding proteins involved in photosystems I and II, phycobilisome biosynthesis and circadian rhythms, reflecting the dominance of *Cyanobacteria* in this sample (Varin et al. 2012).

Comparisons of the metagenomes of cryptoendolithic and permafrost communities from University Valley in the Quartermain Mountains, Antarctica, revealed an abundance of genes linked to catabolism of a wide range of organic compounds in both communities. The permafrost community was enriched with genes involved in cell wall degradation, a significant nutrient recycling adaptation in this nutrient scarce environment, while the cryptoendolith community showed enrichment of genes associated with CO_2 fixation and photosystems, reflecting their versatile metabolic capacities for survival under fluctuating environmental conditions (Goordial et al. 2016a). Overall, the data highlighted two divergent strategies for coping with cold: (1) in freezing temperatures beneath the threshold for metabolism, organisms 'overwinter' in dormant forms, as indicated by the higher proportion of genes involved in dormancy and sporulation and the under-representation of stress response genes, and (2) where the temperatures are within the limits where normal metabolic activities can occur, such as is typical for cryptoendolithic conditions, the organisms utilise a wide range of adaptation genes, including genes encoding antioxidants, cold-shock proteins, cryo-and osmoprotectants, phage shock proteins and sigma B response proteins and genes linked to nutrient cycling (Goordial et al. 2016a).

Consistent with the depauperate nature of the Antarctic Dry Valleys, a recent comparison of hypolith metagenome datasets from both hot and cold arid deserts showed an under-representation of genes associated with carbohydrate metabolism in the Antarctic samples, compared to the seasonally vegetated Namib desert (Le et al. 2016). Antarctic hypolith metagenomes were dominated with genes involved in DNA replication and repair, which is linked to the capacity of the microorganisms to acquire genomic elements that confer competitive advantages under extreme conditions. There was also an enrichment of sigma factor genes, which have been linked to cold stress responses. Genes involved in the response to oxidative stress and oxygen limitation were more prevalent in the Namib desert hypolith's metagenome (Le et al. 2016).

Comparisons of 23 marine sediment metagenomes from four high-latitude regions of both hemispheres revealed that alginates, with a key role in the marine carbon cycle, exhibited a similar profile in Arctic and Antarctic marine sediment communities, whereas brackish sediment communities showed distinct structures with a higher proportion of novel genes (Matos et al. 2016). Since alginates are important in converting brown algae polysaccharides into an energy source for marine heterotrophic bacteria, this information is valuable for understanding carbon fluxes in cold coastal environments (Matos et al. 2016).

Metagenomic and metatranscriptomic analyses of communities along a steep oxygen gradient to the base of the eutrophied and polluted Baltic Sea provided an understanding of ecosystem functions in this environment with community interactions providing links between carbon, nitrogen and sulphur cycling through fermentation, sulphate reduction, methanogenesis and methane oxidation (Thureborn et al. 2013, 2016). The analyses also showed the presence of heavy metal-resistant communities with a lifestyle adapted to sinking organic material, oxygen depletion and low temperature. An over-representation of integron-integrase transcripts, compared with other metatranscriptomes globally, suggested an adaptational capacity of the sediment community despite the fact that the environment is relatively stable at 500 m of depth (Thureborn et al. 2016).

16.2.2 *Targeted Gene Discovery and Bioprospecting*

A valuable objective of metagenomic studies is the screening of community DNA extracted from environmental samples for novel biomolecules for industrial and biotechnological applications. Permanently cold environments are of special interest because the 'indigenous' microorganisms and biomolecules derived from them are adapted to function optimally at low temperatures (Cavicchioli et al. 2002; Struvay and Feller 2012). These biomolecules, particularly cold-adapted enzymes (psychrozymes), are potentially cost-effective and energy-efficient biocatalysts for some biotechnological and industrial processes, especially in the detergent, food, paper and textile industries (Cavicchioli et al. 2011; Adrio and Demain 2014; Vester et al. 2014). The metagenomic bioprospecting of environmental samples for novel biomolecules typically involves one of two approaches, namely, sequence homology or functional product-targeted screening. Sequence homology-based screening targets genes directly from environmental DNA. Product-targeted functional metagenomic strategies rely on the construction of metagenomic clone libraries and the subsequent screening for predetermined bioactive products (Thomas et al. 2012).

16.2.2.1 Sequence-Based Screening

Several sequence-dependent screening approaches have been widely applied to the identification of genes or gene clusters in metagenomes (Tuffin et al. 2009). One common approach involves the design of consensus or gene-specific primers and subsequent PCR amplification, including genome walking strategies, to obtain the full-length target gene (Kotik 2009). An alternative method involves the use of hybridisation probes to identify gene clusters in metagenomic libraries (Jogler et al. 2009; Ausec et al. 2011). A third approach is the sequencing of pooled fosmids which can be combined with HMM (hidden Markov model) screening of the data to identify clones of interest (Ausec et al. 2011, 2015). Novel gene clusters can also be identified using PCR amplification methods (Courtois et al. 2003). Gene-targeted metagenomic (GT metagenomics) methods, whereby PCR amplification with degenerate primer sets is coupled with amplicon sequencing, have been implemented to recover full-length gene sequences (Iwai et al. 2009, 2011). Large genomic regions harbouring gene clusters of interest have been explored using target capture probes combined with NG sequencing of the enriched genomic fragments (Denonfoux et al. 2013).

Although homology-directed metagenomic screening is limited to the identification of genes encoding new variants of known enzymes (Simon and Daniel 2011), heterologous expression of genes identified on the basis of homology has resulted in the discovery of large number of novel enzymes from metagenomic samples (Uria and Zilda 2016). Sequence-based screening of fosmid libraries constructed from a South China Sea metagenome has yielded a novel thermostable fumarase (FumF) showing a maximum of 43% similarity to class II fumarases (Jiang et al. 2010). Sequence-based screening of fosmid libraries constructed from the Sargasso Sea metagenome database and from a metagenome derived from the western Arctic Ocean was used to identify a new variant of an endoglucanase gene (*celM*) (Cottrell et al. 2005). Screening of a BAC metagenomic library constructed from surface water samples obtained from the South China Sea has also yielded a novel thermostable laccase (Fang et al. 2011).

Sequence data obtained from whole metagenome sequencing can also be mined for specific genes that encode novel and potentially useful enzymes using synthetic metagenomic strategies (Bayer et al. 2009). Using this approach, genes of interest, identified via homology screening of NGS metagenome data, are chemically synthesised (often codon optimised for subsequent heterologous expression), cloned and tested for protein expression (Bayer et al. 2009; Allgaier et al. 2010; Iqbal et al. 2012).

16.2.2.2 Function-Based Screening

One of the primary pitfalls of sequence-based genomics is that it targets the identification of gene or gene products on the basis of homology to existing or characterised genes or functions, thereby restricting this approach to the identification of new variants of known genes (Handelsman 2004). In addition, to recover the entire genes or the entire gene clusters which are required for the biosynthesis of some products, homology screening requires genome walking strategies which are often complex and tedious in the metagenomic context (Meiring et al. 2011). This limitation is surmounted by function-based metagenomics whereby specific gene functions or activities of interest, e.g. antibiotic or enzymatic activities (Handelsman 2004), can be detected from the metagenome. Function-based screening of metagenomes also holds an enormous potential for the recovery of unknown genes that may encode highly novel gene products (Culligan et al. 2014).

Functional metagenomics involves several steps that start with the isolation and fragmentation of environmental DNA using strategies that will enable maximum nucleic acid yield and suitable mean insert size (Mirete et al. 2016). The metagenomic DNA fragments are then cloned into a suitable vector, such as a plasmid, fosmid, cosmid, phage or bacterial artificial chromosome (BAC). The fragment-containing vectors are subsequently transformed into a suitable expression host, and phenotypic assays are performed for the screening of the desired function (Simon and Daniel 2011; Ufarté et al. 2015). The chance of a successful screening assay (i.e. hit rate) in function-based metagenomics is generally low due to problems that can arise at any point between the environmental DNA extraction and heterologous expression of the gene (Uchiyama and Miyazaki 2009; Ferrer et al. 2016). However, wide variations in hit rates have been reported, ranging from one phosphatase-positive test per 2.7 Mb library (Lämmle et al. 2007) to one clone with antifungal activity in 113,700 (ca. 3.9 Gb) clones constructed from forest soils (Chung et al. 2008). Despite these limitations, hundreds of biomolecules have been identified and characterised via metagenomic expression screening (Adrio and Demain 2014; Mirete et al. 2016).

In order to improve hit rate, Uchiyama et al. (2005) developed a substrate-induced gene expression (SIGEX) methodology (Uchiyama et al. 2005; Uchiyama and Watanabe 2008). SIGEX was designed on the premise that certain substrates induce catabolic genes by interacting with regulatory genetic elements. The strategy involves the construction of metagenomic libraries using a vector in which green fluorescent protein (GFP) gene has been incorporated (Uchiyama et al. 2005; Uchiyama and Watanabe 2008). Positive hits are detected by *gfp* expression in the presence of the inducing substrate (Uchiyama et al. 2005; Yun and Ryu 2005; Uchiyama and Watanabe 2008).

In an extended version of the reporter assay, the product-induced gene expression (PIGEX), *gfp*, is coupled with an upstream product inducible promoter which initiates the expression of *gfp* gene upon the formation of products from the activity of a specific enzyme positive clone (Uchiyama and Miyazaki 2010). A similar approach is the metabolite-regulated expression (METREX) in which a quorum-sensing promoter induces the production of GFP when the concentration of signal molecules produced by a clone reaches a certain threshold (Williamson et al. 2005; Simon and Daniel 2011).

The genetic enzyme screening system (GESS) allows the in vivo detection, and measurement of the activities of enzymes such as cellulases, esterases, lyases and phosphatases, whose activities release *p*-nitrophenol, has been reported (Choi et al. 2013). In the GESS, the promoter (P_o dmp) of the phenol degradation operon (dmp), which is induced by the DmpR transcriptional activator, is placed upstream of the *gfp* gene (Choi et al. 2013). Enzymatic hydrolysis of *p*-nitrophenyl phosphate releases *p*-nitrophenol which binds to the DmpR transcriptional activator to trigger the expression of the enhanced *gfp* gene (Lee et al. 2015).

High-throughput bioprospecting of metagenomic libraries using microfluidic-based screening strategies was recently incorporated in functional metagenomics and has proven valuable in improving both hit rates and sensitivity with which bioactive molecules could be recovered (Hosokawa et al. 2015; Ufarté et al. 2015). Such methods involve the use of microfluidic devices to produce emulsion microdroplets, each containing single metagenomic clones and a fluorogenic substrate system. The system is often coupled with other reactions such as cell lysis and PCR to enhance recovery of the targeted genes and gene products (Colin et al. 2015; Hosokawa et al. 2015). Microdroplets containing positive clones are typically identified using fluorescence sorting chips (Najah et al. 2014; Colin et al. 2015; Hosokawa et al. 2015).

Function-based metagenomic screening approaches have proven effective for the identification and subsequent characterisation of a wide array of cold-active enzymes from cold environments (Table 16.2). For instance, a cold-active and alkali-tolerant carboxymethyl cellulose hydrolysing enzyme (RBcel1) was mined by screening BAC clones obtained from an Antarctic metagenome (Berlemont et al. 2009). The first metagenomically derived low-temperature lipase was isolated via expression screening of a Baltic Sea sediment fosmid library (Hårdeman and

Table 16.2 Example of novel cold-active enzymes identified via functional metagenomics

Isolation source	Enzyme	Vector	Reference
Antarctic soil	Alkali-tolerant carboxymethyl cellulose hydrolysing enzyme (RBcel1)	BAC	Berlemont et al. (2009)
Antarctic soil	Alkaliphilic esterase	Fosmid	Heath et al. (2009) Hu et al. (2012)
Arctic soil	Esterases (EstM-N1 and EstM-N2)	Fosmid	Yu et al. (2011)
Arctic intertidal sediment	Esterase Est97	Fosmid	Fu et al. (2013)
Arctic seashore sediment	Esterases (EstAT1 and EstAT11)	Fosmid	Jeon et al. (2009)
Ikaite columns, Greenland	Carbohydrate-active enzymes:α-amylase, α-galactosidase β-galactosidase, β-mannanase β-xylanase, cellulose	BAC	Vester et al. (2013, 2014)
Tidal flat sediments, west coast of Korea	Alkaline phosphatase	Fosmid	Lee et al. (2015)
Baltic Sea sediment	Lipase h1Lip1 (DQ118648)	Fosmid	Hårdeman & Sjöling (2007)
Baltic seawater	Glycoside hydrolase	BAC	Wierzbicka-Woś et al. (2013)
South China Sea	Laccase Lac21	BAC	Fang et al. (2012)

Sjöling 2007). Screening of various fosmid libraries, constructed from Antarctic soil samples, has revealed a novel cold-adapted and alkaliphilic esterase, which is active at temperatures ranging from 5 to 54 °C (Heath et al. 2009; Hu et al. 2012). Similarly, functional screening of Arctic metagenomes has resulted in the characterisation of several novel cold-active esterases; e.g. EstM-N1 and EstM-N2 (active between 0 and 35 °C) obtained from Arctic soils (Yu et al. 2011) and Est-97 (active between 0.5 and 55 °C) isolated from an Arctic intertidal metagenomic library (Fu et al. 2013).

A combination of culture- and function-based screening strategies has been used for the isolation of novel cold-adapted carbohydrate-active enzymes, including α-amylase, α-galactosidase, β-galactosidase, β-mannanase, β-xylanase and cellulose, from cultures and metagenomic libraries from ikaite (calcium carbonate) columns in Greenland (Vester et al. 2013, 2014). In a separate approach, Lee et al. (2015) recently isolated a novel cold-adapted alkaline phosphatase (AP) from a metagenomic library produced from samples of tidal flat sediments using the GESS approach (Lee et al. 2015).

16.2.2.3 Metataxonomics

The power of metagenomics is not restricted to the insights it can provide into the functional diversity of organisms in an environment or the identification of industrially or biotechnologically relevant biomolecules. The combination of total community DNA extraction, targeted PCR amplification of specific genomic markers [e.g. 16S rRNA genes in bacteria; eukaryal internal transcribed spacers (ITS)] and high-throughput sequencing can be used to evaluate both the taxonomic diversity and abundance of organisms in an environmental sample (Fierer et al. 2007; Uyaguari-Diaz et al. 2016). This approach, now known as metataxonomics or amplicon sequencing (Marchesi and Ravel 2015), is particularly pertinent when considering microbial 'dark matter', the fraction of microbial community missed when using traditional culture-dependent techniques. Metataxonomic methodologies have provided key insights into the phylogenetic species richness and biodiversity of cold environments (Table 16.3; discussed in greater detail in Chaps. 3, 4, 5, 6, 7, 8 and 9 of this book).

An essential first step in taxonomic analyses of environmental samples is the selection of a suitable phylogenetic marker depending on the specific objectives of the study (De Maayer et al. 2014b). For example, the 16S rRNA gene is a widely used phylogenetic marker, because of its universality among the prokaryotes and the availability of a very large volume of highly curated 16S rRNA sequence data (Sunagawa et al. 2013; Aguiar-Pulido et al. 2016). Although highly conserved, the 16S rRNA gene incorporates nine highly variable regions (V1–V9) which provide sufficient resolution for the discrimination of prokaryotes in environmental samples (Claesson et al. 2010; Nguyen et al. 2016). The conserved regions flanking the hypervariable region(s) are used as templates for primer (Baker et al. 2003; Wang and Qian 2009).

Both older (Sanger's) and modern high-throughput sequencing technologies are routinely used for the sequencing of the phylogenetic markers. Amplicon reads obtained from these platforms are assessed for quality using various in silico techniques and tools such as AmpliconNoise (Quince et al. 2011) and PRINSEQ (Schmieder and Edwards 2011). After the quality control step, the sequences are clustered into taxonomic bins known as operational taxonomic units (OTUs) based on sequence and/or composition similarity thresholds (Liu et al. 2008; Schloss and Westcott 2011). The distinct OTUs are then queried for homology to known taxa against curated databases, for example, the Ribosomal Database Project (RDP), SILVA or Greengenes databases (DeSantis et al. 2006; Pruesse et al. 2007; Cole et al. 2009). Various tools have been developed for these homology searches, and phylogenetic classification has been developed, including ARB (Ludwig et al. 2004), Greengenes classifier (DeSantis et al. 2006), MEGAN (Huson et al. 2007), PPLACER (Matsen et al. 2010), Ribosomal Database Project (RDP) classifier (Cole et al. 2009) and SILVAngs (Pruesse et al. 2007).

Ecological indices can be computed to assess the distribution of the clustered OTUs (Sogin et al. 2006). Some of the most pertinent statistics include measures of

Table 16.3 The most abundant taxa identified from different cold environments using metataxomic approaches

Isolation site	Location	Genomic regions	Dominant phyla	References
Mitchell Peninsula	Antarctic	16S rRNA/fungal ITS	*Chloroflexi/Ascomycota*	Ji et al. (2016)
Russian Antarctic Progress Station	Antarctic	16S rRNA	*Proteobacteria*	Lopatina et al. (2016)
Upper dry valley permafrost	Antarctic	16S rRNA/fungal ITS	*Proteobacteria* *Halobacteria* *Dothideomycetes*	Goordial et al. (2016b)
Dry valley permafrost	Antarctic	16S rRNA	*ProteobacteriaFirmicutes*	Gilichinsky et al. (2007)
Taylor Valley permafrost	Antarctic	16S rRNA/ *rpoB*	*Acidobacteria*	Bakermans et al. (2014)
Antarctic Dry Valley soils	Antarctic	16S rRNA		Niederberger et al. (2015)
Wet soils			*Cyanobacteria*	
Arid soils			*Bacteroidetes*	
Arctic permafrost	Arctic	16S rRNA	*Actinobacteria*	Wilhelm et al. (2011)
Canadian high Arctic permafrost	Arctic		*Actinobacteria*	Steven et al. (2007)
Arctic permafrost	Arctic	16S rRNA microarray		Yergeau et al. (2010)
Surface layer 2 m depth			*Actinobacteria* *Betaproteobacteria*	
Drake Passage and Gerlache Strait (3000 m deep sea)	Antarctic Polar Front	16S rRNA	*Gammaproteobacteria*	López-Garcıa et al. (2001)
30 deep ocean sites	North and South Atlantic, Indian, North and South Pacific Oceans	16S rRNA	*Gammaproteobacteria*	Salazar et al. (2016)
Coastal waters	Oarai, Ibaraki, Japan	16S rRNA	*Alphaproteobacteria*	Haider et al. (2016)
Coastal waters	Off Plymouth, UK	16S rRNA	*Alphaproteobacteria*	Gilbert et al. (2012)

taxon diversity and richness (Whittaker et al. 2001). Alpha diversity, which measures species diversity within microbial communities, and beta diversity, which measures species diversity between communities, are among the most widely applied diversity indices (Lozupone and Knight 2008; Gotelli and Colwell 2011).

Several online and stand-alone software packages have been developed for the computation of these taxon indices. These include DOTUR (Schloss and Handelsman 2005a), EstimateS (Colwell 2013), MOTHUR (Schloss et al. 2009), Phylocom (Webb et al. 2008), QIIME (Caporaso et al. 2010) and SPECIES (Wang 2011).

16.3 Combining Metagenomics and Other Omic Technologies

Metagenomic approaches have proven to be highly successful as culture-independent means of exploring the structure and potential functional capacity of communities inhabiting different environments (Handelsman 2004; Franzosa et al. 2015). The major limitation of metagenomic methods, however, is the uncertainty about what proportion of the community DNA comes from extant life forms, from organism existing in dormant forms, or from dead cells as legacy DNA (Prosser 2015). Several other 'omic' technologies, such as metatranscriptomics, metaproteomics and meta-metabolomics, which target the community transcripts, proteins and metabolites, respectively, have been developed and applied to complement metagenomic approaches (Hultman et al. 2015; Roume et al. 2015).

Metatranscriptomics, which involves the sequencing of total mRNA from environmental samples, provides access to the gene expression profiles of microbial communities under a given set of environmental conditions and/or at a specific time point (Warnecke and Hess 2009). Early metatranscriptomic analyses (Poretsky et al. 2005) were based on DNA microarrays involving the hybridisation of cDNA to oligonucleotide probes which were designed on the bases of previously known gene sequences (Shendure 2008; Van Vliet 2010). Because these probes were based on information from sequenced organisms, detection of community gene expression will be restricted to variants of known transcripts (Shendure 2008; Siezen et al. 2010). Additionally, the application of microarrays was also limited by low transcript detection efficiency and low quantitative sensitivity (Shendure 2008; Siezen et al. 2010). These problems have largely been circumvented through the development of high-throughput RNA sequencing (RNA-seq) technologies (Van Vliet 2010; Mutz et al. 2013).

RNA-seq-based metatranscriptomic approaches involve the extraction of RNA from environmental samples, messenger RNA (mRNA) enrichment and nucleic acid sequencing (Carvalhais et al. 2012). Owing to the instability of RNA (Moran et al. 2013), several protocols have been developed that include snap freezing in liquid nitrogen or sample storage in solutions such as RNAlater (Ambion, USA) and LifeGuard™ Soil Preservation Solution (MO BIO Laboratories, CA) that ensure the integrity of extracted RNA (Warnecke and Hess 2009; Moran et al. 2013).

Various estimates have shown that mRNA comprised only about 1–5% of the total RNA in environmental samples (He et al. 2010). It is thus frequently necessary to include an mRNA extraction step prior to sequencing. The most commonly used methods include subtractive hybridisation and exonuclease digestion, incorporated in kits such as MICROBExpress Bacterial mRNA Enrichment kit (Ambion, USA) and Prokaryotic mRNA Isolation kit (EPICENTRE Biotechnologies, USA) (He et al. 2010). The enriched mRNA is either converted to cDNA prior to sequencing (Warnecke and Hess 2009) or sequenced directly using methods that circumvent the cDNA synthesis step (Ozsolak and Milos 2011).

Metatranscriptomic analyses have revealed the precise functioning of different communities in a number of marine systems and a few other cold ecosystems (Jones et al. 2016). For instance, a survey of the deep marine subsurface using samples obtained from the continental shelf of Peru has provided insights into the versatile anaerobic metabolism occurring in sea floor sediments (Orsi et al. 2013). Similar studies have revealed the functional roles of the novel taxa, *Candidatus Thiopilula* spp. and *Ca. Thiophysa* spp., in deep cold microbial communities (Jones et al. 2015). Metatranscriptome data revealed the expression of several transcripts linked to sulphur oxidation and nitrogen reduction, suggesting their potential roles in global sulphur cycling (Jones et al. 2015).

The study of transcriptome profiles allows for the identification and quantification of genes that are expressed at a given point in time and leads to inferences about their potential roles in metabolic processes (Croucher and Thomson 2010; Sorek and Cossart 2010; Cho et al. 2013). However, this approach does not account any post-transcriptional and/or post-translational changes that may ultimately determine molecular activity (Vogel and Marcotte 2012; Payne 2015). Thus, in order to elucidate the functional roles of predicted genes and their transcripts within a given environment, metaproteomic methods have been developed for the identification and quantification of the collective community protein fraction (Wilmes and Bond 2004). Several strategies have been widely applied for the study of environmental proteomes (Wang et al. 2014). These include gel-based approaches, such as 'two-dimensional polyacrylamide gel electrophoresis' (2D-PAGE) (Rabilloud et al. 2010; May et al. 2012) and mass spectrometry (MS) techniques, e.g. electrospray ionisation (ESI) and matrix-assisted laser desorption ionisation (MALDI) (Zhang et al. 2010; Catherman et al. 2014; Jungblut 2014). These MS approaches are often combined with different analysers such as ion trap, time of flight and triple quad analysers (Kelleher 2004; De Maayer et al. 2014b). Metaproteomics has provided significant insight into the mechanisms of microbial adaptions in cold environments. For example, capillary liquid chromatography (LC) coupled with mass spectrometry was used to identify the metaproteome associated with various adaption strategies in the oligotrophic Sargasso Sea (Sowell et al. 2008). Similarly, LC-MS/MS studies of metaproteomes isolated from Oregon shelf revealed abundance of methylotrophic proteins supporting previous predictions that methylotrophs were the major drivers of coastal biogeochemistry (Sowell et al. 2011).

Metabolomics refers to a set of techniques which are used to profile the total small molecule or metabolites content (metabolome or metabolic profile) of a given biological sample (Lankadurai et al. 2013; Cortassa et al. 2015). Environmental metabolomics (or meta-metabolomics) can be applied to identify the metabolic activities of natural communities under different environmental conditions (Simpson and Bearden 2007; Lankadurai et al. 2013). This approach has considerable potential for detecting subtle shifts in microbial community composition caused by changing environmental conditions (Jones et al. 2013). Analytical techniques used in profiling metabolites include various nuclear magnetic resonance (NMR) techniques (Simpson and Bearden 2007) and mass spectroscopy coupled with various techniques such as Fourier transform (FT), gas chromatography (GC), capillary electrophoresis (CE) and liquid chromatography (LC) analyses (O'Malley 2013).

Various 'omics' methods have been integrated to enhance the understanding of the functional capacity of microbial communities under cold conditions (Mackelprang et al. 2014). For example, metagenomic and metaproteomic approaches were employed to investigate the molecular determinants of adaptation in permanently cold deep (746 m) marine sediments (Stokke et al. 2012). Meta-omic data revealed that the adaptation of the sediment community was defined by a large number of well-known cold-adapted proteins and those involved in gas vesicle production (Stokke et al. 2012). Multi-'omic' approaches have also been used to study the phylogenetic diversity and functional potential of the microbial communities in permafrost, active layer and thermokarst bog samples from Alaska (Hultman et al. 2015). The catalogue of sequences obtained using targeted 16S rRNA metataxonomics, metagenomics, metatranscriptomics and metaproteomics yielded over 100 Gb of sequence data and ca. 7000 proteins (Hultman et al. 2015). The combined data was consistent with previous findings on the low taxonomic diversity and functional potential of the permafrost communities. The permafrost and active layer of metatranscriptome revealed abundant cold tolerance and other stress response proteins compared to the bog soil (Hultman et al. 2015). However, new observations were made; it was concluded that certain microorganisms were capable of active metabolism under the extreme permafrost conditions. Dissimilatory iron reduction was identified as the potential energy conserving metabolic strategy used by some permafrost organisms. The possibility that all the sequences were derived from preserved genomic materials in the permafrost cannot, however, be overruled (Hultman et al. 2015).

16.4 Concluding Remarks

Since its inception three decades ago, the field of metagenomics has developed from a concept on how to disentangle the phylogenetics of mixed bacterial cultures (Pace et al. 1986) to a formidable and powerful tool for the study of the diversity, ecology and function of complex communities in virtually every accessible part of the Earth, from the depths of the ocean floor to the remotest parts of the polar regions. The enormous progress achieved in the field is tightly linked to advances in HTS sequencing technologies and the concurrent development of bioinformatic tools specifically optimised for metagenomic analysis. Initial concerns regarding the wastage of much of the enormous data generated using these tools are now being addressed by the improvements in advanced computational strategies and the development of supercomputers and various data storage options, including cloud storage.

Metagenomic strategies have provided unprecedented access to the structure and function of the cold biosphere, revealing the detailed taxonomic richness and functional capacity of the inhabiting communities. The information has added a wealth of knowledge on the versatility and expanse of life in cold ecosystems and even rekindled the hope of finding life on other planets (Bonilla-Rosso et al. 2012; Siddiqui et al. 2013). Metagenomic analyses have also provided insights into ecosystem health and sustainability: for instance, recent surveys have shown that permafrost traps a volume of carbon that equals or even surpasses the carbon reserves in plant biomass (Mackelprang et al. 2011). Continued melting of permafrost as a consequence of anthropogenic activities and global warming can potentially alter the global carbon biogeochemical cycle (Mackelprang et al. 2011; Franzetti et al. 2016).

There is no doubt that our knowledge of cold environments has been greatly enhanced through the application of metagenomic and related omic tools. However, numerous pertinent questions still remain unanswered regarding the ecological relevance of 'omics' data, especially when considering the complex and dynamic biological interactions that occur in nature (Vilanova and Porcar 2016). This problem is even more complex in the cryosphere where nucleic acids and proteins are likely to be preserved over long periods, and logistical problems may hinder in situ tracing of metabolic signals. Furthermore, because functional annotation of biomolecules recovered from environmental samples depends on reference genomes of cultured microorganism, many derived sequences (up to 60% in some instances) (Prakash and Taylor 2012) remain uninformative. Thus, despite the current accomplishments in microbial ecology resulting from the age of meta-omics, continued and focused research is still needed to establish the complex biology of natural populations.

References

Adrio JL, Demain AL (2014) Microbial enzymes: Tools for biotechnological processes. Biomolecules 4(1):117–139

Aguiar-Pulido V, Huang W, Suarez-Ulloa V, Cickovski T, Mathee K, Narasimhan G (2016) Metagenomics, metatranscriptomics, and metabolomics approaches for microbiome analysis. Evol Bioinforma 12(Suppl 1):5–16

Albertsen M, Hugenholtz P, Skarshewski A, Nielsen KL, Tyson GW, Nielsen PH (2013) Genome sequences of rare, uncultured bacteria obtained by differential coverage binning of multiple metagenomes. Nat Biotechnol 31(6):533–538

Allgaier M, Reddy A, Park JI, Ivanova N, D'haeseleer P, Lowry S, Sapra R, Hazen TC, Simmons BA, VanderGheynst JS (2010) Targeted discovery of glycoside hydrolases from a switchgrass-adapted compost community. PLoS One 5(1):e8812

Alneberg J, Bjarnason BS, de Bruijn I, Schirmer M, Quick J, Ijaz UZ, Lahti L, Loman NJ, Andersson AF, Quince C (2014) Binning metagenomic contigs by coverage and composition. Nat Methods 11(11):1144–1146

Altintas I, Lin AW, Chen J, Churas C, Gujral M, Sun S, Li W, Manansala R, Sedova M, Grethe JS (2010) Camera 2.0: a data-centric metagenomics community infrastructure driven by scientific workflows. In: Proceedings of the 6th World Congress on Services, 2010. IEEE, pp 352–359

Ashton PM, Nair S, Dallman T, Rubino S, Rabsch W, Mwaigwisya S, Wain J, O'Grady J (2014) MinION nanopore sequencing identifies the position and structure of a bacterial antibiotic resistance island. Nat Biotechnol 33:296–300

Ausec L, Zakrzewski M, Goesmann A, Schlüter A, Mandic-Mulec I (2011) Bioinformatic analysis reveals high diversity of bacterial genes for laccase-like enzymes. PLoS One 6(10):e25724

Ausec L, Črnigoj M, Šnajder M, Ulrih NP, Mandic-Mulec I (2015) Characterization of a novel high-pH-tolerant laccase-like multicopper oxidase and its sequence diversity in *Thioalkalivibrio* sp. Appl Microbiol Biotechnol 99(23):9987–9999

Baker G, Smith JJ, Cowan DA (2003) Review and re-analysis of domain-specific 16S primers. J Microbiol Methods 55(3):541–555

Bakermans C, Skidmore ML, Douglas S, McKay CP (2014) Molecular characterization of bacteria from permafrost of the Taylor Valley, Antarctica. FEMS Microbiol Ecol 89(2):331–346

Bayer TS, Widmaier DM, Temme K, Mirsky EA, Santi DV, Voigt CA (2009) Synthesis of methyl halides from biomass using engineered microbes. J Am Chem Soc 131(18):6508–6515

Behzad H, Gojobori T, Mineta K (2015) Challenges and opportunities of airborne metagenomics. Genome Biol Evol 7(5):1216–1226

Berlemont R, Delsaute M, Pipers D, D'Amico S, Feller G, Galleni M, Power P (2009) Insights into bacterial cellulose biosynthesis by functional metagenomics on Antarctic soil samples. ISME J 3(9):1070–1081

Boetius A, Anesio AM, Deming JW, Mikucki JA, Rapp JZ (2015) Microbial ecology of the cryosphere: sea ice and glacial habitats. Nat Rev Microbiol 13:677–690

Boisvert S, Raymond F, Godzaridis É, Laviolette F, Corbeil J (2012) Ray meta: scalable de novo metagenome assembly and profiling. Genome Biol 13(12):R122

Bonilla-Rosso G, Peimbert M, Alcaraz LD, Hernández I, Eguiarte LE, Olmedo-Alvarez G, Souza V (2012) Comparative metagenomics of two microbial mats at Cuatro Ciénegas Basin II: community structure and composition in oligotrophic environments. Astrobiology 12 (7):659–673

Bowman JS, Berthiaume CT, Armbrust EV, Deming JW (2014) The genetic potential for key biogeochemical processes in Arctic frost flowers and young sea ice revealed by metagenomic analysis. FEMS Microbiol Ecol 89(2):376–387

Brady A, Salzberg SL (2009) Phymm and PhymmBL: metagenomic phylogenetic classification with interpolated Markov models. Nat Methods 6(9):673–676

Brooks JP, Edwards DJ, Harwich MD, Rivera MC, Fettweis JM, Serrano MG, Reris RA, Sheth NU, Huang B, Girerd P, Strauss JF, Jefferson KK, Buck GA (2015) The truth about metagenomics: quantifying and counteracting bias in 16S rRNA studies. BMC Microbiol 15 (1):66

Brown MV, Lauro FM, DeMaere MZ, Muir L, Wilkins D, Thomas T, Riddle MJ, Fuhrman JA, Andrews-Pfannkoch C, Hoffman JM, McQuaid JB, Allen A, Rintoul SR, Cavicchioli R (2012) Global biogeography of SAR11 marine bacteria. Mol Syst Biol 8:595

Caporaso JG, Kuczynski J, Stombaugh J, Bittinger K, Bushman FD, Costello EK, Fierer N, Pena AG, Goodrich JK, Gordon JI (2010) QIIME allows analysis of high-throughput community sequencing data. Nat Methods 7(5):335–336

Carneiro MO, Russ C, Ross MG, Gabriel SB, Nusbaum C, DePristo MA (2012) Pacific biosciences sequencing technology for genotyping and variation discovery in human data. BMC Genomics 13:375–375

Carvalhais LC, Dennis PG, Tyson GW, Schenk PM (2012) Application of metatranscriptomics to soil environments. J Microbiol Methods 91(2):246–251

Casanueva A, Tuffin M, Cary C, Cowan DA (2010) Molecular adaptations to psychrophily: the impact of 'omic' technologies. Trends Microbiol 18(8):374–381

Catherman AD, Skinner OS, Kelleher NL (2014) Top down proteomics: facts and perspectives. Biochem Biophys Res Commun 445(4):683–693

Cavicchioli R (2015) On the concept of a psychrophile. ISME J 10(4):793–795

Cavicchioli R, Charlton T, Ertan H, Mohd OS, Siddiqui KS, Williams TJ (2011) Biotechnological uses of enzymes from psychrophiles. Microb Biotechnol 4(4):449–460

Cavicchioli R, Siddiqui K, Andrews D, Sowers K (2002) Low-temperature extremophiles and their applications. Curr Opin Biotechnol 13(3):253–261

Chan Y, Van Nostrand JD, Zhou J, Pointing SB, Farrell RL (2013) Functional ecology of an Antarctic dry valley. Proc Natl Acad Sci USA 110(22):8990–8995

Cho S, Cho Y, Lee S, Kim J, Yum H, Kim SC, Cho B-K (2013) Current challenges in bacterial transcriptomics. Genomics Inform 11(2):76–82

Choi S-L, Rha E, Lee SJ, Kim H, Kwon K, Jeong Y-S, Rhee YH, Song JJ, Kim H-S, Lee S-G (2013) Toward a generalized and high-throughput enzyme screening system based on artificial genetic circuits. ACS Synth Biol 3(3):163–171

Chung EJ, Lim HK, Kim J-C, Choi GJ, Park EJ, Lee MH, Chung YR, Lee S-W (2008) Forest soil metagenome gene cluster involved in antifungal activity expression in *Escherichia coli*. Appl Environ Microbiol 74(3):723–730

Claesson MJ, Wang Q, O'Sullivan O, Greene-Diniz R, Cole JR, Ross RP, O'Toole PW (2010) Comparison of two next-generation sequencing technologies for resolving highly complex microbiota composition using tandem variable 16S rRNA gene regions. Nucleic Acids Res 38 (22):e200

Cole JR, Wang Q, Cardenas E, Fish J, Chai B, Farris RJ, Kulam-Syed-Mohideen A, McGarrell DM, Marsh T, Garrity GM (2009) The Ribosomal Database Project: improved alignments and new tools for rRNA analysis. Nucleic Acids Res 37(Suppl 1):D141–D145

Colin P-Y, Kintses B, Gielen F, Miton CM, Fischer G, Mohamed MF, Hyvönen M, Morgavi DP, Janssen DB, Hollfelder F (2015) Ultrahigh-throughput discovery of promiscuous enzymes by picodroplet functional metagenomics. Nat Commun 6:10008

Colwell R(2013) EstimateS: Statistical estimation of species richness and shared species from samples. Version 9.1.0. User's guide and application available online: http://purl.oclc.org/estimates

Cortassa S, Caceres V, Bell LN, O'Rourke B, Paolocci N, Aon MA (2015) From metabolomics to fluxomics: a computational procedure to translate metabolite profiles into metabolic fluxes. Biophys J 108(1):163–172

Cottrell MT, Yu L, Kirchman DL (2005) Sequence and expression analyses of *Cytophaga*-like hydrolases in a Western Arctic metagenomic library and the Sargasso Sea. Appl Environ Microbiol 71(12):8506–8513

Courtois S, Cappellano CM, Ball M, Francou F-X, Normand P, Helynck G, Martinez A, Kolvek SJ, Hopke J, Osburne MS, August PR, Nalin R, Guérineau M, Jeannin P, Simonet P, Pernodet J-L (2003) Recombinant environmental libraries provide access to microbial diversity for drug discovery from natural products. Appl Environ Microbiol 69(1):49–55

Croucher NJ, Thomson NR (2010) Studying bacterial transcriptomes using RNA-seq. Curr Opin Microbiol 13(5):619–624

Culligan EP, Sleator RD, Marchesi JR, Hill C (2014) Metagenomics and novel gene discovery: promise and potential for novel therapeutics. Virulence 5(3):399–412
De Maayer P, Anderson D, Cary C, Cowan DA (2014a) Some like it cold: understanding the survival strategies of psychrophiles. EMBO Rep 15(5):508–517
De Maayer P, Valverde A, Cowan DA (2014b) The current state of metagenomic analysis. In: Poptsova MS (ed) Genome analysis: current procedures and applications. Caister Academic Press, Norfolk, UK, pp 183–220
Delmont TO, Robe P, Clark I, Simonet P, Vogel TM (2011) Metagenomic comparison of direct and indirect soil DNA extraction approaches. J Microbiol Methods 86(3):397–400
Denonfoux J, Parisot N, Dugat-Bony E, Biderre-Petit C, Boucher D, Morgavi DP, Le Paslier D, Peyretaillade E, Peyret P (2013) Gene capture coupled to high-throughput sequencing as a strategy for targeted metagenome exploration. DNA Res 20(2):185–196
DeSantis TZ, Hugenholtz P, Larsen N, Rojas M, Brodie EL, Keller K, Huber T, Dalevi D, Hu P, Andersen GL (2006) Greengenes, a chimera-checked 16S rRNA gene database and workbench compatible with ARB. Appl Environ Microbiol 72(7):5069–5072
Earl D, Bradnam K, St John J, Darling A, Lin D, Fass J, Yu HO, Buffalo V, Zerbino DR, Diekhans M, Nguyen N, Ariyaratne PN, Sung WK, Ning Z, Haimel M, Simpson JT, Fonseca NA, Birol I, Docking TR, Ho IY, Rokhsar DS, Chikhi R, Lavenier D, Chapuis G, Naquin D, Maillet N, Schatz MC, Kelley DR, Phillippy AM, Koren S, Yang SP, Wu W, Chou WC, Srivastava A, Shaw TI, Ruby JG, Skewes-Cox P, Betegon M, Dimon MT, Solovyev V, Seledtsov I, Kosarev P, Vorobyev D, Ramirez-Gonzalez R, Leggett R, MacLean D, Xia F, Luo R, Li Z, Xie Y, Liu B, Gnerre S, MacCallum I, Przybylski D, Ribeiro FJ, Yin S, Sharpe T, Hall G, Kersey PJ, Durbin R, Jackman SD, Chapman JA, Huang X, DeRisi JL, Caccamo M, Li Y, Jaffe DB, Green RE, Haussler D, Korf I, Paten B (2011) Assemblathon 1: a competitive assessment of de novo short read assembly methods. Genome Res 21(12):2224–2241
Edwards A, Debbonaire AR, Sattler B, Mur LA, Hodson AJ (2016) Extreme metagenomics using nanopore DNA sequencing: a field report from Svalbard, 78 N. BioRxiv. doi:10.1101/073965
Edwards DJ, Holt KE (2013) Beginner's guide to comparative bacterial genome analysis using next-generation sequence data. Microb Inform Exp 3(1):2
Escalona M, Rocha S, Posada D (2016) A comparison of tools for the simulation of genomic next-generation sequencing data. Nat Rev Genet 17(8):459–469
Fahlgren C, Bratbak G, Sandaa R-A, Thyrhaug R, Zweifel UL (2011) Diversity of airborne bacteria in samples collected using different devices for aerosol collection. Aerobiologia 27 (2):107–120
Fang Z, Li T, Wang Q, Zhang X, Peng H, Fang W, Hong Y, Ge H, Xiao Y (2011) A bacterial laccase from marine microbial metagenome exhibiting chloride tolerance and dye decolorization ability. Appl Microbiol Biotechnol 89(4):1103–1110
Fang Z-M, Li T-L, Chang F, Zhou P, Fang W, Hong Y-Z, Zhang X-C, Peng H, Xiao Y-Z (2012) A new marine bacterial laccase with chloride-enhancing, alkaline-dependent activity and dye decolorization ability. Bioresour Technol 111:36–41
Ferrer M, Martínez-Martínez M, Bargiela R, Streit WR, Golyshina OV, Golyshin PN (2016) Estimating the success of enzyme bioprospecting through metagenomics: current status and future trends. Microb Biotechnol 9(1):22–34
Field D, Garrity G, Gray T, Morrison N, Selengut J, Sterk P, Tatusova T, Thomson N, Allen MJ, Angiuoli SV, Ashburner M, Axelrod N, Baldauf S, Ballard S, Boore J, Cochrane G, Cole J, Dawyndt P, De Vos P, dePamphilis C, Edwards R, Faruque N, Feldman R, Gilbert J, Gilna P, Glockner FO, Goldstein P, Guralnick R, Haft D, Hancock D, Hermjakob H, Hertz-Fowler C, Hugenholtz P, Joint I, Kagan L, Kane M, Kennedy J, Kowalchuk G, Kottmann R, Kolker E, Kravitz S, Kyrpides N, Leebens-Mack J, Lewis SE, Li K, Lister AL, Lord P, Maltsev N, Markowitz V, Martiny J, Methe B, Mizrachi I, Moxon R, Nelson K, Parkhill J, Proctor L, White O, Sansone S-A, Spiers A, Stevens R, Swift P, Taylor C, Tateno Y, Tett A, Turner S, Ussery D, Vaughan B, Ward N, Whetzel T, San Gil I, Wilson G, Wipat A (2008) The minimum information about a genome sequence (MIGS) specification. Nat Biotechnol 26(5):541–547

Fierer N, Breitbart M, Nulton J, Salamon P, Lozupone C, Jones R, Robeson M, Edwards RA, Felts B, Rayhawk S, Knight R, Rohwer F, Jackson RB (2007) Metagenomic and small-subunit rRNA analyses reveal the genetic diversity of bacteria, archaea, fungi, and viruses in soil. Appl Environ Microbiol 73(21):7059–7066

Forde BM, O'Toole PW (2013) Next-generation sequencing technologies and their impact on microbial genomics. Brief Funct Genomics 12(5):440–453

Franzetti A, Tagliaferri I, Gandolfi I, Bestetti G, Minora U, Mayer C, Azzoni RS, Diolaiuti G, Smiraglia C, Ambrosini R (2016) Light-dependent microbial metabolisms drive carbon fluxes on glacier surfaces. ISME J 10:2984–2988

Franzosa EA, Hsu T, Sirota-Madi A, Shafquat A, Abu-Ali G, Morgan XC, Huttenhower C (2015) Sequencing and beyond: integrating molecular 'omics' for microbial community profiling. Nat Rev Microbiol 13(6):360–372

Fu J, Leiros H-KS, de Pascale D, Johnson KA, Blencke H-M, Landfald B (2013) Functional and structural studies of a novel cold-adapted esterase from an Arctic intertidal metagenomic library. Appl Microbiol Biotechnol 97(9):3965–3978

Gandolfi I, Bertolini V, Ambrosini R, Bestetti G, Franzetti A (2013) Unravelling the bacterial diversity in the atmosphere. Appl Microbiol Biotechnol 97(11):4727–4736

Gasc C, Ribière C, Parisot N, Beugnot R, Defois C, Petit-Biderre C, Boucher D, Peyretaillade E, Peyret P (2015) Capturing prokaryotic dark matter genomes. Res Microbiol 166(10):814–830

Gilbert JA, Laverock B, Temperton B, Thomas S, Muhling M, Hughes M (2011) Metagenomics. High-throughput next generation sequencing: methods and applications. In: Kwon YM, Ricke SC (eds) Methods in molecular biology, vol 733. Springer Science+Business Media, Heidelberg, pp 173–183

Gilbert JA, Steele JA, Caporaso JG, Steinbruck L, Reeder J, Temperton B, Huse S, McHardy AC, Knight R, Joint I, Somerfield P, Fuhrman JA, Field D (2012) Defining seasonal marine microbial community dynamics. ISME J 6(2):298–308

Gilichinsky D, Wilson G, Friedmann E, McKay C, Sletten R, Rivkina E, Vishnivetskaya T, Erokhina L, Ivanushkina N, Kochkina G (2007) Microbial populations in Antarctic permafrost: biodiversity, state, age, and implication for astrobiology. Astrobiology 7(2):275–311

Goordial J, Davila A, Greer CW, Cannam R, DiRuggiero J, McKay CP, Whyte LG (2016a) Comparative activity and functional ecology of permafrost soils and lithic niches in a hyper-arid polar desert. Environ Microbiol. doi:10.1111/1462-2920.13353

Goordial J, Davila A, Lacelle D, Pollard W, Marinova MM, Greer CW, DiRuggiero J, McKay CP, Whyte LG (2016b) Nearing the cold-arid limits of microbial life in permafrost of an upper dry valley, Antarctica. ISME J 10(7):1613–1624

Gotelli NJ, Colwell RK (2011) Estimating species richness. In: Magurran AE, McGill BJ (eds) Biological diversity: frontiers in measurement and assessment, vol 12. Oxford University Press, Oxford, pp 39–54

Greninger AL, Naccache SN, Federman S, Yu G, Mbala P, Bres V, Stryke D, Bouquet J, Somasekar S, Linnen JM, Dodd R, Mulembakani P, Schneider BS, Muyembe-Tamfum J-J, Stramer SL, Chiu CY (2015) Rapid metagenomic identification of viral pathogens in clinical samples by real-time nanopore sequencing analysis. Genome Med 7(1):99

Grzymski JJ, Carter BJ, DeLong EF, Feldman RA, Ghadiri A, Murray AE (2006) Comparative genomics of DNA fragments from six Antarctic marine planktonic bacteria. Appl Environ Microbiol 72(2):1532–1541

Haider MN, Nishimura M, Kogure K (2016) Bacterial community structure and diversity of closely located coastal areas. Open J Marine Sci 6:423–439

Handelsman J (2004) Metagenomics: application of genomics to uncultured microorganisms. Microbiol Mol Biol Rev 68(4):669–685

Handelsman J, Liles M, Mann D, Riesenfeld C, Goodman RM (2002) Cloning the metagenome: culture-independent access to the diversity and functions of the uncultivated microbial world. Methods Microbiol 33:241–255

Hårdeman F, Sjöling S (2007) Metagenomic approach for the isolation of a novel low-temperature-active lipase from uncultured bacteria of marine sediment. FEMS Microbiol Ecol 59(2):524–534

He S, Wurtzel O, Singh K, Froula JL, Yilmaz S, Tringe SG, Wang Z, Chen F, Lindquist EA, Sorek R, Hugenholtz P (2010) Validation of two ribosomal RNA removal methods for microbial metatranscriptomics. Nat Methods 7(10):807–812

Heath C, Hu XP, Cary SC, Cowan D (2009) Identification of a novel alkaliphilic esterase active at low temperatures by screening a metagenomic library from Antarctic desert soil. Appl Environ Microbiol 75(13):4657–4659

Hosokawa M, Hoshino Y, Nishikawa Y, Hirose T, Yoon DH, Mori T, Sekiguchi T, Shoji S, Takeyama H (2015) Droplet-based microfluidics for high-throughput screening of a metagenomic library for isolation of microbial enzymes. Biosens Bioelectron 67:379–385

Howeler M, Ghiorse WC, Walker LP (2003) A quantitative analysis of DNA extraction and purification from compost. J Microbiol Methods 54(1):37–45

Hu XP, Heath C, Taylor MP, Tuffin M, Cowan D (2012) A novel, extremely alkaliphilic and cold-active esterase from Antarctic desert soil. Extremophiles 16(1):79–86

Hugerth LW, Larsson J, Alneberg J, Lindh MV, Legrand C, Pinhassi J, Andersson AF (2015) Metagenome-assembled genomes uncover a global brackish microbiome. Genome Biol 16:279

Hultman J, Waldrop MP, Mackelprang R, David MM, McFarland J, Blazewicz SJ, Harden J, Turetsky MR, McGuire AD, Shah MB (2015) Multi-omics of permafrost, active layer and thermokarst bog soil microbiomes. Nature 521(7551):208–212

Hunter S, Corbett M, Denise H, Fraser M, Gonzalez-Beltran A, Hunter C, Jones P, Leinonen R, McAnulla C, Maguire E (2014) EBI metagenomics—a new resource for the analysis and archiving of metagenomic data. Nucleic Acids Res 42(D1):D600–D606

Huson DH, Weber N (2012) Microbial community analysis using MEGAN. Methods Enzymol 531:465–485

Huson DH, Auch AF, Qi J, Schuster SC (2007) MEGAN analysis of metagenomic data. Genome Res 17(3):377–386

Iqbal HA, Feng Z, Brady SF (2012) Biocatalysts and small molecule products from metagenomic studies. Curr Opin Chem Biol 16(1–2):109–116

Iwai S, Chai B, Sul WJ, Cole JR, Hashsham SA, Tiedje JM (2009) Gene-targeted-metagenomics reveals extensive diversity of aromatic dioxygenase genes in the environment. ISME J 4 (2):279–285

Iwai S, Chai B, Jesus EC, Penton CR, Lee TK, Cole JR, Tiedje JM (2011) Gene-targeted metagenomics (GT Metagenomics) to explore the extensive diversity of genes of interest in microbial communities. In: de Bruijn FJ (ed) Handbook of molecular microbial ecology I. Wiley-Blackwell, pp 235–243

Jain M, Fiddes IT, Miga KH, Olsen HE, Paten B, Akeson M (2015) Improved data analysis for the MinION nanopore sequencer. Nat Methods 12:351–356

Jeon JH, Kim J-T, Kang SG, Lee J-H, Kim S-J (2009) Characterization and its potential application of two esterases derived from the arctic sediment metagenome. Mar Biotechnol 11 (3):307–316

Ji M, van Dorst J, Bissett A, Brown MV, Palmer AS, Snape I, Siciliano SD, Ferrari BC (2016) Microbial diversity at Mitchell Peninsula, Eastern Antarctica: a potential biodiversity "hotspot". Polar Biol 39(2):237–249

Jiang C, Wu L-L, Zhao G-C, Shen P-H, Jin K, Hao Z-Y, Li S-X, Ma G-F, Luo F-F, Hu G-Q, Kang W-L, Qin X-M, Bi Y-L, Tang X-L, Wu B (2010) Identification and characterization of a novel fumarase gene by metagenome expression cloning from marine microorganisms. Microb Cell Factories 9:91–91

Jogler C, Lin W, Meyerdierks A, Kube M, Katzmann E, Flies C, Pan Y, Amann R, Reinhardt R, Schüler D (2009) Toward cloning of the magnetotactic metagenome: identification of magnetosome island gene clusters in uncultivated magnetotactic bacteria from different aquatic sediments. Appl Environ Microbiol 75(12):3972–3979

Jones DS, Flood BE, Bailey JV (2015) Metatranscriptomic analysis of diminutive *Thiomargarita*-like bacteria ("Candidatus *Thiopilula*" spp.) from abyssal cold seeps of the Barbados Accretionary Prism. Appl Environ Microbiol 81(9):3142–3156

Jones DS, Flood BE, Bailey JV (2016) Metatranscriptomic insights into polyphosphate metabolism in marine sediments. ISME J 10(4):1015–1019

Jones OA, Maguire ML, Griffin JL, Dias DA, Spurgeon DJ, Svendsen C (2013) Metabolomics and its use in ecology. Austral Ecol 38(6):713–720

Jungblut PR (2014) The proteomics quantification dilemma. J Proteome 107:98–102

Kang DD, Froula J, Egan R, Wang Z (2015) MetaBAT, an efficient tool for accurately reconstructing single genomes from complex microbial communities. PeerJ 3:e1165

Kanehisa M, Goto S, Sato Y, Furumichi M, Tanabe M (2012) KEGG for integration and interpretation of large-scale molecular data sets. Nucleic Acids Res 40:D109–D114. doi:10.1093/nar/gkr988

Kelleher NL (2004) Peer reviewed: top-down proteomics. Anal Chem 76(11):196A–203A

Kennedy NA, Walker AW, Berry SH, Duncan SH, Farquarson FM, Louis P, Thomson JM, Satsangi J, Flint HJ, Parkhill J, Lees CW, Hold GL, Other members not named within the manuscript author l (2014) The impact of different DNA extraction kits and laboratories upon the assessment of human gut microbiota composition by 16S rRNA gene sequencing. PLoS One 9(2):e88982

Kim M, Lee K-H, Yoon S-W, Kim B-S, Chun J, Yi H (2013) Analytical tools and databases for metagenomics in the next-generation sequencing era. Genomics Inform 11(3):102–113

Konstantinidis KT, Braff J, Karl DM, DeLong EF (2009) Comparative metagenomic analysis of a microbial community residing at a depth of 4,000 meters at station ALOHA in the North Pacific subtropical gyre. Appl Environ Microbiol 75(16):5345–5355

Kotik M (2009) Novel genes retrieved from environmental DNA by polymerase chain reaction: current genome-walking techniques for future metagenome applications. J Biotechnol 144 (2):75–82

Krause L, Diaz NN, Goesmann A, Kelley S, Nattkemper TW, Rohwer F, Edwards RA, Stoye J (2008) Phylogenetic classification of short environmental DNA fragments. Nucleic Acids Res 36(7):2230–2239

Kremkow B, Lee K (2015) Sequencing technologies for animal cell culture research. Biotechnol Lett 37(1):55–65

Lämmle K, Zipper H, Breuer M, Hauer B, Buta C, Brunner H, Rupp S (2007) Identification of novel enzymes with different hydrolytic activities by metagenome expression cloning. J Biotechnol 127(4):575–592

Land M, Hauser L, Jun S-R, Nookaew I, Leuze MR, Ahn T-H, Karpinets T, Lund O, Kora G, Wassenaar T (2015) Insights from 20 years of bacterial genome sequencing. Funct Integr Genomics 15(2):141–161

Lankadurai BP, Nagato EG, Simpson MJ (2013) Environmental metabolomics: an emerging approach to study organism responses to environmental stressors. Environ Rev 21(3):180–205

Le PT, Makhalanyane TP, Guerrero LD, Vikram S, Van de Peer Y, Cowan DA (2016) Comparative metagenomic analysis reveals mechanisms for stress response in hypoliths from extreme hyperarid deserts. Genome Biol Evol 8(9):2737–2747

Lee D-H, Choi S-L, Rha E, Kim SJ, Yeom S-J, Moon J-H, Lee S-G (2015) A novel psychrophilic alkaline phosphatase from the metagenome of tidal flat sediments. BMC Biotechnol 15 (1):1–13

Leung HCM, Wang Y, Yiu SM, Chin FYL (2015) Next-generation sequencing for metagenomic data: assembling and binning. In: Nelson KE (ed) Encyclopedia of metagenomics: genes, genomes and metagenomes: basics, methods, databases and tools. Springer US, Boston, MA, pp 539–544

Lin H-H, Liao Y-C (2016) Accurate binning of metagenomic contigs via automated clustering sequences using information of genomic signatures and marker genes. Sci Rep 6:24175

Liu B, Gibbons T, Ghodsi M, Pop M (2010) MetaPhyler: taxonomic profiling for metagenomic sequences. In: Proceedings of the IEEE international conference on bioinformatics and biomedicine (BIBM), pp 95–100

Liu L, Li Y, Li S, Hu N, He Y, Pong R, Lin D, Lu L, Law M (2012) Comparison of next-generation sequencing systems. J Biomed Biotechnol 2012:251364

Liu Z, DeSantis TZ, Andersen GL, Knight R (2008) Accurate taxonomy assignments from 16S rRNA sequences produced by highly parallel pyrosequencers. Nucleic Acids Res 36(18):e120

Logares R, Haverkamp TH, Kumar S, Lanzén A, Nederbragt AJ, Quince C, Kauserud H (2012) Environmental microbiology through the lens of high-throughput DNA sequencing: synopsis of current platforms and bioinformatics approaches. J Microbiol Methods 91(1):106–113

Loman NJ, Pallen MJ (2015) Twenty years of bacterial genome sequencing. Nat Rev Microbiol 13 (12):787–794

Lopatina A, Medvedeva S, Shmakov S, Logacheva MD, Krylenkov V, Severinov K (2016) Metagenomic analysis of bacterial communities of Antarctic surface snow. Front Microbiol 7:398

López-Garcıa P, López-López A, Moreira D, Rodrıguez-Valera F (2001) Diversity of free-living prokaryotes from a deep-sea site at the Antarctic Polar Front. FEMS Microbiol Ecol 36 (2–3):193–202

Lorenzi HA, Hoover J, Inman J, Safford T, Murphy S, Kagan L, Williamson SJ (2011) The Viral MetaGenome Annotation Pipeline (VMGAP): an automated tool for the functional annotation of viral metagenomic shotgun sequencing data. Stand Genomic Sci 4(3):418

Lozupone CA, Knight R (2008) Species divergence and the measurement of microbial diversity. FEMS Microbiol Rev 32(4):557–578

Ludwig W, Strunk O, Westram R, Richter L, Meier H, Buchner A, Lai T, Steppi S, Jobb G, Förster W (2004) ARB: a software environment for sequence data. Nucleic Acids Res 32 (4):1363–1371

MacDonald NJ, Parks DH, Beiko RG (2012) Rapid identification of high-confidence taxonomic assignments for metagenomic data. Nucleic Acids Res 40(14):e111

Mackelprang R, Waldrop MP, DeAngelis KM, David MM, Chavarria KL, Blazewicz SJ, Rubin EM, Jansson JK (2011) Metagenomic analysis of a permafrost microbial community reveals a rapid response to thaw. Nature 480(7377):368–371

Mackelprang R, Saleska SR, Jacobsen CS, Jansson JK, Taş N (2014) Permafrost meta-omics and climate change. Annu Rev Earth Planet Sci 44:439–462

Marchesi JR, Ravel J (2015) The vocabulary of microbiome research: a proposal. Microbiome 3 (1):1–3

Margesin R, Feller G (2010) Biotechnological applications of psychrophiles. Environ Technol 31 (8–9):835–844

Margesin R, Miteva V (2011) Diversity and ecology of psychrophilic microorganisms. Res Microbiol 162(3):346–361

Markowitz VM, Chen IMA, Chu K, Szeto E, Palaniappan K, Pillay M, Ratner A, Huang J, Pagani I, Tringe S, Huntemann M, Billis K, Varghese N, Tennessen K, Mavromatis K, Pati A, Ivanova NN, Kyrpides NC (2014) IMG/M 4 version of the integrated metagenome comparative analysis system. Nucleic Acids Res 42:D568–D573

Matos MN, Lozada M, Anselmino LE, Musumeci MA, Henrissat B, Jansson JK, Mac Cormack WP, Carroll J, Sjöling S, Lundgren L (2016) Metagenomics unveils the attributes of the alginolytic guilds of sediments from four distant cold coastal environments. Environ Microbiol. doi:10.1111/1462-2920.13433

Matsen FA, Kodner RB, Armbrust EV (2010) pplacer: linear time maximum-likelihood and Bayesian phylogenetic placement of sequences onto a fixed reference tree. BMC Bioinformatics 11:538

May C, Brosseron F, Pfeiffer K, Meyer HE, Marcus K (2012) Proteome analysis with classical 2D-PAGE. In: Marcus K (ed) Quantitative methods in proteomics. Humana Press, New York, pp 37–46

McHardy AC, Martin HG, Tsirigos A, Hugenholtz P, Rigoutsos I (2007) Accurate phylogenetic classification of variable-length DNA fragments. Nat Methods 4(1):63–72
Meiring TL, Bauer R, Scheepers I, Ohloff C, Tuffin I, Cowan DA (2011) Metagenomics and beyond: current approaches and integration with complementary technologies. Metagenomics: current innovations and future trends. Caister Academic Press, Norfolk, pp 1–19
Meyer F, Paarmann D, D'Souza M, Olson R, Glass EM, Kubal M, Paczian T, Rodriguez A, Stevens R, Wilke A (2008) The metagenomics RAST server–a public resource for the automatic phylogenetic and functional analysis of metagenomes. BMC Bioinformatics 9:386
Mirete S, Morgante V, González-Pastor JE (2016) Functional metagenomics of extreme environments. Curr Opin Biotechnol 38:143–149
Mondav R, Woodcroft BJ, Kim E-H, McCalley CK, Hodgkins SB, Crill PM, Chanton J, Hurst GB, VerBerkmoes NC, Saleska SR (2014) Discovery of a novel methanogen prevalent in thawing permafrost. Nat Commun 5:3212
Moran MA, Satinsky B, Gifford SM, Luo H, Rivers A, Chan L-K, Meng J, Durham BP, Shen C, Varaljay VA (2013) Sizing up metatranscriptomics. ISME J 7(2):237–243
Mutz K-O, Heilkenbrinker A, Lönne M, Walter J-G, Stahl F (2013) Transcriptome analysis using next-generation sequencing. Curr Opin Biotechnol 24(1):22–30
Najah M, Calbrix R, Mahendra-Wijaya IP, Beneyton T, Griffiths Andrew D, Drevelle A (2014) Droplet-based microfluidics platform for ultra-high-throughput bioprospecting of cellulolytic microorganisms. Chem Biol 21(12):1722–1732
Namiki T, Hachiya T, Tanaka H, Sakakibara Y (2012) MetaVelvet: an extension of Velvet assembler to de novo metagenome assembly from short sequence reads. Nucleic Acids Res 40(20):e155
Nayfach S, Pollard KS (2016) Toward accurate and quantitative comparative metagenomics. Cell 166(5):1103–1116
Nguyen N-P, Warnow T, Pop M, White B (2016) A perspective on 16S rRNA operational taxonomic unit clustering using sequence similarity. Npj Biofilms Microbiomes 2:16004
Niederberger TD, Sohm JA, Gunderson TE, Parker AE, Tirindelli J, Capone DG, Carpenter EJ, Cary SC (2015) Microbial community composition of transiently wetted Antarctic Dry Valley soils. Front Microbiol 6:9
O'Malley MA (2013) Metametabolomics. In: Dubitzky W, Wolkenhauer O, Cho K-H, Yokota H (eds) Encyclopedia of systems biology. Springer, New York, pp 1296–1297
Orsi WD, Edgcomb VP, Christman GD, Biddle JF (2013) Gene expression in the deep biosphere. Nature 499(7457):205–208
Overbeek R, Olson R, Pusch GD, Olsen GJ, Davis JJ, Disz T, Edwards RA, Gerdes S, Parrello B, Shukla M (2014) The SEED and the Rapid Annotation of microbial genomes using Subsystems Technology (RAST). Nucleic Acids Res 42:D206–D214
Ozsolak F, Milos PM (2011) Single-molecule direct RNA sequencing without cDNA synthesis. Wiley Interdiscip Rev RNA 2(4):565–570
Pace N, Stahl D, Lane D, Olsen G (1986) The analysis of natural microbial populations by rRNA sequences. Adv Microb Ecol 9:1–55
Pareja-Tobes P, Manrique M, Pareja-Tobes E, Pareja E, Tobes R (2012) BG7: a new approach for bacterial genome annotation designed for next generation sequencing data. PLoS One 7(11): e49239
Parks DH, Imelfort M, Skennerton CT, Hugenholtz P, Tyson GW (2015) CheckM: assessing the quality of microbial genomes recovered from isolates, single cells, and metagenomes. Genome Res 25(7):1043–1055
Paszkiewicz K, Studholme DJ (2010) De novo assembly of short sequence reads. Brief Bioinform 11(5):457–472
Payne SH (2015) The utility of protein and mRNA correlation. Trends Biochem Sci 40(1):1–3
Pearce DA, Newsham KK, Thorne MA, Calvo-Bado L, Krsek M, Laskaris P, Hodson A, Wellington EM (2012) Metagenomic analysis of a southern maritime Antarctic soil. Front Microbiol 3:403

Peng Y, Leung HC, Yiu S-M, Chin FY (2011) Meta-IDBA: a de Novo assembler for metagenomic data. Bioinformatics 27(13):i94–i101

Peng Y, Leung HC, Yiu S-M, Chin FY (2012) IDBA-UD: a de novo assembler for single-cell and metagenomic sequencing data with highly uneven depth. Bioinformatics 28(11):1420–1428

Perkel JM (2009) Life science technologies: Sanger who? Sequencing the next generation. Science 324(5924):275–279

Poretsky RS, Bano N, Buchan A, LeCleir G, Kleikemper J, Pickering M, Pate WM, Moran MA, Hollibaugh JT (2005) Analysis of microbial gene transcripts in environmental samples. Appl Environ Microbiol 71(7):4121–4126

Prakash T, Taylor TD (2012) Functional assignment of metagenomic data: challenges and applications. Brief Bioinform 13(6):711–727

Prosser JI (2015) Dispersing misconceptions and identifying opportunities for the use of 'omics' in soil microbial ecology. Nat Rev Microbiol 13(7):439–446

Pruesse E, Quast C, Knittel K, Fuchs BM, Ludwig W, Peplies J, Glöckner FO (2007) SILVA: a comprehensive online resource for quality checked and aligned ribosomal RNA sequence data compatible with ARB. Nucleic Acids Res 35(21):7188–7196

Quick J, Quinlan AR, Loman NJ (2014) A reference bacterial genome dataset generated on the MinION™ portable single-molecule nanopore sequencer. Gigascience 3(1):22

Quick J, Loman NJ, Duraffour S, Simpson JT, Severi E, Cowley L, Bore JA, Koundouno R, Dudas G, Mikhail A, Ouédraogo N, Afrough B, Bah A, Baum JHJ, Becker-Ziaja B, Boettcher JP, Cabeza-Cabrerizo M, Camino-Sánchez Á, Carter LL, Doerrbecker J, Enkirch T, Dorival IG, Hetzelt N, Hinzmann J, Holm T, Kafetzopoulou LE, Koropogui M, Kosgey A, Kuisma E, Logue CH, Mazzarelli A, Meisel S, Mertens M, Michel J, Ngabo D, Nitzsche K, Pallasch E, Patrono LV, Portmann J, Repits JG, Rickett NY, Sachse A, Singethan K, Vitoriano I, Yemanaberhan RL, Zekeng EG, Racine T, Bello A, Sall AA, Faye O, Faye O, Magassouba NF, Williams CV, Amburgey V, Winona L, Davis E, Gerlach J, Washington F, Monteil V, Jourdain M, Bererd M, Camara A, Somlare H, Camara A, Gerard M, Bado G, Baillet B, Delaune D, Nebie KY, Diarra A, Savane Y, Pallawo RB, Gutierrez GJ, Milhano N, Roger I, Williams CJ, Yattara F, Lewandowski K, Taylor J, Rachwal P, Turner DJ, Pollakis G, Hiscox JA, Matthews DA, MKO S, Johnston AM, Wilson D, Hutley E, Smit E, Di Caro A, Wölfel R, Stoecker K, Fleischmann E, Gabriel M, Weller SA, Koivogui L, Diallo B, Keïta S, Rambaut A, Formenty P, Günther S, Carroll MW (2016) Real-time, portable genome sequencing for Ebola surveillance. Nature 530(7589):228–232

Quince C, Lanzen A, Davenport RJ, Turnbaugh PJ (2011) Removing noise from pyrosequencing amplicons. BMC Bioinform 12:38

Rabilloud T, Chevallet M, Luche S, Lelong C (2010) Two-dimensional gel electrophoresis in proteomics: past, present and future. J Proteome 73(11):2064–2077

Reuter JA, Spacek DV, Snyder MP (2015) High-throughput sequencing technologies. Mol Cell 58 (4):586–597

Richardson EJ, Watson M (2013) The automatic annotation of bacterial genomes. Brief Bioinform 14(1):1–12

Rodrigues DF, Tiedje JM (2008) Coping with our cold planet. Appl Environ Microbiol 74 (6):1677–1686

Roume H, Heintz-Buschart A, Muller EEL, May P, Satagopam VP, Laczny CC, Narayanasamy S, Lebrun LA, Hoopmann MR, Schupp JM, Gillece JD, Hicks ND, Engelthaler DM, Sauter T, Keim PS, Moritz RL, Wilmes P (2015) Comparative integrated omics: identification of key functionalities in microbial community-wide metabolic networks. Npj Biofilms Microbiomes 1:15007

Salazar G, Cornejo-Castillo FM, Benitez-Barrios V, Fraile-Nuez E, Alvarez-Salgado XA, Duarte CM, Gasol JM, Acinas SG (2016) Global diversity and biogeography of deep-sea pelagic prokaryotes. ISME J 10(3):596–608

Sanger F, Nicklen S, Coulson AR (1977) DNA sequencing with chain-terminating inhibitors. Proc Natl Acad Sci 74(12):5463–5467

Sangwan N, Xia F, Gilbert JA (2016) Recovering complete and draft population genomes from metagenome datasets. Microbiome 4(1):8

Schloss PD, Handelsman J (2005a) Introducing DOTUR, a computer program for defining operational taxonomic units and estimating species richness. Appl Environ Microbiol 71 (3):1501–1506

Schloss PD, Handelsman J (2005b) Metagenomics for studying unculturable microorganisms: cutting the Gordian knot. Genome Biol 6(8):1

Schloss PD, Westcott SL (2011) Assessing and improving methods used in operational taxonomic unit-based approaches for 16S rRNA gene sequence analysis. Appl Environ Microbiol 77 (10):3219–3226

Schloss PD, Westcott SL, Ryabin T, Hall JR, Hartmann M, Hollister EB, Lesniewski RA, Oakley BB, Parks DH, Robinson CJ (2009) Introducing mothur: open-source, platform-independent, community-supported software for describing and comparing microbial communities. Appl Environ Microbiol 75(23):7537–7541

Schmidt K, Mwaigwisya S, Crossman L, Doumith M, Munroe D, Pires C, Khan A, Woodford N, Saunders N, Wain J (2016) Identification of bacterial pathogens and antimicrobial resistance directly from clinical urines by nanopore-based metagenomic sequencing. J Antimicrob Chemother 72(1):dkw397. doi:10.1093/jac/dkw397

Schmieder R, Edwards R (2011) Quality control and preprocessing of metagenomic datasets. Bioinformatics 27:863–864

Seshadri R, Kravitz SA, Smarr L, Gilna P, Frazier M (2007) CAMERA: a community resource for metagenomics. PLoS Biol 5(3):e75

Sharon I, Banfield JF (2013) Genomes from metagenomics. Science 342(6162):1057–1058

Shendure J (2008) The beginning of the end for microarrays? Nat Methods 5(7):585–587

Shendure J, Ji H (2008) Next-generation DNA sequencing. Nat Biotechnol 26(10):1135–1145

Siddiqui KS, Williams TJ, Wilkins D, Yau S, Allen MA, Brown MV, Lauro FM, Cavicchioli R (2013) Psychrophiles. Annu Rev Earth Planet Sci 41:87–115

Siezen RJ, Wilson G, Todt T (2010) Prokaryotic whole transcriptome analysis: deep sequencing and tiling arrays. Microb Biotechnol 3(2):125–130

Simon C, Daniel R (2011) Metagenomic analyses: past and future trends. Appl Environ Microbiol 77(4):1153–1161

Simon C, Wiezer A, Strittmatter AW, Daniel R (2009) Phylogenetic diversity and metabolic potential revealed in a glacier ice metagenome. Appl Environ Microbiol 75(23):7519–7526

Simpson MJ, Bearden DW (2007) Environmental metabolomics: NMR techniques. eMagRes 2 (4):549–560

Sjöling S, Cowan DA (2008) Metagenomics: microbial community genomes revealed. In: Margesin R, Schinner F, Marx J-C, Gerday C (eds) Psychrophiles: from biodiversity to biotechnology. Springer, Berlin, pp 313–332

Sjöling S, Stafford W, Cowan D (2006) Soil metagenomics: exploring and exploiting the microbial gene pool. In: van Elsas JD, Jansson JK, Trevors JT (eds) Modern soil microbiology. CRC Press, Boca Raton, pp 409–434

Sogin ML, Morrison HG, Huber JA, Mark Welch D, Huse SM, Neal PR (2006) Microbial diversity in the deep sea and the underexplored "rare biosphere". Proc Natl Acad Sci USA 103 (32):12115–12120

Sorek R, Cossart P (2010) Prokaryotic transcriptomics: a new view on regulation, physiology and pathogenicity. Nat Rev Genet 11(1):9–16

Sowell SM, Wilhelm LJ, Norbeck AD, Lipton MS, Nicora CD, Barofsky DF, Carlson CA, Smith RD, Giovanonni SJ (2008) Transport functions dominate the SAR11 metaproteome at low-nutrient extremes in the Sargasso Sea. ISME J 3(1):93–105

Sowell SM, Abraham PE, Shah M, Verberkmoes NC, Smith DP, Barofsky DF, Giovannoni SJ (2011) Environmental proteomics of microbial plankton in a highly productive coastal upwelling system. ISME J 5(5):856–865

Steven B, Briggs G, McKay CP, Pollard WH, Greer CW, Whyte LG (2007) Characterization of the microbial diversity in a permafrost sample from the Canadian high Arctic using culture-dependent and culture-independent methods. FEMS Microbiol Ecol 59(2):513–523

Stokke R, Roalkvam I, Lanzen A, Haflidason H, Steen IH (2012) Integrated metagenomic and metaproteomic analyses of an ANME-1-dominated community in marine cold seep sediments. Environ Microbiol 14(5):1333–1346

Stothard P, Wishart DS (2006) Automated bacterial genome analysis and annotation. Curr Opin Microbiol 9(5):505–510

Struvay C, Feller G (2012) Optimization to low temperature activity in psychrophilic enzymes. Int J Mol Sci 13(9):11643–11665

Sunagawa S, Mende DR, Zeller G, Izquierdo-Carrasco F, Berger SA, Kultima JR, Coelho LP, Arumugam M, Tap J, Nielsen HB, Rasmussen S, Brunak S, Pedersen O, Guarner F, de Vos WM, Wang J, Li J, Doré J, Ehrlich SD, Stamatakis A, Bork P (2013) Metagenomic species profiling using universal phylogenetic marker genes. Nat Methods 10(12):1196–1199

Sunagawa S, Coelho LP, Chaffron S, Kultima JR, Labadie K, Salazar G, Djahanschiri B, Zeller G, Mende DR, Alberti A (2015) Structure and function of the global ocean microbiome. Science 348(6237):1261359

Teeling H, Waldmann J, Lombardot T, Bauer M, Glöckner FO (2004) TETRA: a web-service and a stand-alone program for the analysis and comparison of tetranucleotide usage patterns in DNA sequences. BMC Bioinformatics 5:163

Thomas T, Gilbert J, Meyer F (2012) Metagenomics-a guide from sampling to data analysis. Microb Inform Exp 2:3

Thureborn P, Lundin D, Plathan J, Poole AM, Sjöberg B-M, Sjöling S (2013) A metagenomics transect into the deepest point of the Baltic Sea reveals clear stratification of microbial functional capacities. PLoS One 8(9):e74983

Thureborn P, Franzetti A, Lundin D, Sjöling S (2016) Reconstructing ecosystem functions of the active microbial community of the Baltic Sea oxygen depleted sediments. PeerJ 4:e1593

Tseng C-H, Chiang P-W, Lai H-C, Shiah F-K, Hsu T-C, Chen Y-L, Wen L-S, Tseng C-M, Shieh W-Y, Saeed I, Halgamuge S, Tang S-L (2015) Prokaryotic assemblages and metagenomes in pelagic zones of the South China Sea. BMC Genomics 16(1):1–16

Tuffin M, Anderson D, Heath C, Cowan DA (2009) Metagenomic gene discovery: how far have we moved into novel sequence space? Biotechnol J 4(12):1671–1683

Tyson GW, Chapman J, Hugenholtz P, Allen EE, Ram RJ, Richardson PM, Solovyev VV, Rubin EM, Rokhsar DS, Banfield JF (2004) Community structure and metabolism through reconstruction of microbial genomes from the environment. Nature 428(6978):37–43

Uchiyama T, Miyazaki K (2009) Functional metagenomics for enzyme discovery: challenges to efficient screening. Curr Opin Biotechnol 20(6):616–622

Uchiyama T, Miyazaki K (2010) Product-induced gene expression, a product-responsive reporter assay used to screen metagenomic libraries for enzyme-encoding genes. Appl Environ Microbiol 76(21):7029–7035

Uchiyama T, Watanabe K (2008) Substrate-induced gene expression (SIGEX) screening of metagenome libraries. Nat Protoc 3(7):1202–1212

Uchiyama T, Abe T, Ikemura T, Watanabe K (2005) Substrate-induced gene-expression screening of environmental metagenome libraries for isolation of catabolic genes. Nat Biotechnol 23 (1):88–93

Ufarté L, Potocki-Veronese G, Laville É (2015) Discovery of new protein families and functions: new challenges in functional metagenomics for biotechnologies and microbial ecology. Front Microbiol 6:563

Uria AR, Zilda DS (2016) Metagenomics-guided mining of commercially useful biocatalysts from marine microorganisms. In: Se-Kwon K, Fidel T (eds) Advances in food and nutrition research, vol 78. Academic Press, New York, pp 1–26

Uyaguari-Diaz MI, Chan M, Chaban BL, Croxen MA, Finke JF, Hill JE, Peabody MA, Van Rossum T, Suttle CA, Brinkman FSL, Isaac-Renton J, Prystajecky NA, Tang P (2016) A comprehensive method for amplicon-based and metagenomic characterization of viruses, bacteria, and eukaryotes in freshwater samples. Microbiome 4(1):20

Van Horn DJ, Okie JG, Buelow HN, Gooseff MN, Barrett JE, Takacs-Vesbach CD (2014) Soil microbial responses to increased moisture and organic resources along a salinity gradient in a polar desert. Appl Environ Microbiol 80(10):3034–3043

Van Vliet AH (2010) Next generation sequencing of microbial transcriptomes: challenges and opportunities. FEMS Microbiol Lett 302(1):1–7

Varin T, Lovejoy C, Jungblut AD, Vincent WF, Corbeil J (2012) Metagenomic analysis of stress genes in microbial mat communities from Antarctica and the High Arctic. Appl Environ Microbiol 78(2):549–559

Venter JC, Remington K, Heidelberg JF, Halpern AL, Rusch D, Eisen JA, Wu D, Paulsen I, Nelson KE, Nelson W, Fouts DE, Levy S, Knap AH, Lomas MW, Nealson K, White O, Peterson J, Hoffman J, Parsons R, Baden-Tillson H, Pfannkoch C, Rogers Y-H, Smith HO (2004) Environmental genome Shotgun Sequencing of the Sargasso Sea. Science 304(5667):66–74

Vester JK, Lylloff JE, Glaring MA, Stougaard P (2013) Microbial diversity and enzymes in ikaite columns; a cold and alkaline environment in Greenland. In: Seckbach J, Oren A, Stan-Lotter H (eds) Cellular origin, life in extreme habitats and astrobiology. Springer, Netherlands, pp 365–380

Vester JK, Glaring MA, Stougaard P (2014) Discovery of novel enzymes with industrial potential from a cold and alkaline environment by a combination of functional metagenomics and culturing. Microb Cell Factories 13(1):1–14

Vilanova C, Porcar M (2016) Are multi-omics enough? Nat Microbiol 1:16101

Vogel C, Marcotte EM (2012) Insights into the regulation of protein abundance from proteomic and transcriptomic analyses. Nat Rev Genet 13(4):227–232

Wang D-Z, Xie Z-X, Zhang S-F (2014) Marine metaproteomics: current status and future directions. J Proteome 97:27–35

Wang J-P (2011) SPECIES: an R package for species richness estimation. J Stat Softw 40(9):1–15

Wang Y, Qian P-Y (2009) Conservative fragments in bacterial 16S rRNA genes and primer design for 16S ribosomal DNA amplicons in metagenomic studies. PLoS One 4(10):e7401

Warnecke F, Hess M (2009) A perspective: metatranscriptomics as a tool for the discovery of novel biocatalysts. J Biotechnol 142(1):91–95

Webb CO, Ackerly DD, Kembel SW (2008) Phylocom: software for the analysis of phylogenetic community structure and trait evolution. Bioinformatics 24(18):2098–2100

Wei STS, Fernandez-Martinez M-A, Chan Y, Van Nostrand JD, de los Rios-Murillo A, Chiu JMY, Ganeshram AM, Cary SC, Zhou J, Pointing SB (2015) Diverse metabolic and stress-tolerance pathways in chasmoendolithic and soil communities of Miers Valley, McMurdo Dry Valleys, Antarctica. Polar Biol 38(4):433–443

White RA III, Callister SJ, Moore RJ, Baker ES, Jansson JK (2016) The past, present and future of microbiome analyses. Nat Protoc 11(11):2049–2053

Whittaker RJ, Willis KJ, Field R (2001) Scale and species richness: towards a general, hierarchical theory of species diversity. J Biogeogr 28(4):453–470

Wierzbicka-Woś A, Bartasun P, Cieśliński H, Kur J (2013) Cloning and characterization of a novel cold-active glycoside hydrolase family 1 enzyme with β-glucosidase, β-fucosidase and β-galactosidase activities. BMC Biotechnol 13(1):22

Wilhelm RC, Niederberger TD, Greer C, Whyte LG (2011) Microbial diversity of active layer and permafrost in an acidic wetland from the Canadian High Arctic. Can J Microbiol 57 (4):303–315

Williamson LL, Borlee BR, Schloss PD, Guan C, Allen HK, Handelsman J (2005) Intracellular screen to identify metagenomic clones that induce or inhibit a quorum-sensing biosensor. Appl Environ Microbiol 71(10):6335–6344

Wilmes P, Bond PL (2004) The application of two-dimensional polyacrylamide gel electrophoresis and downstream analyses to a mixed community of prokaryotic microorganisms. Environ Microbiol 6(9):911–920

Wu Y-W, Tang Y-H, Tringe SG, Simmons BA, Singer SW (2014) MaxBin: an automated binning method to recover individual genomes from metagenomes using an expectation-maximization algorithm. Microbiome 2(1):26

Yergeau E, Hogues H, Whyte LG, Greer CW (2010) The functional potential of high Arctic permafrost revealed by metagenomic sequencing, qPCR and microarray analyses. ISME J 4 (9):1206–1214

Yilmaz P, Kottmann R, Field D, Knight R, Cole JR, Amaral-Zettler L, Gilbert JA, Karsch-Mizrachi I, Johnston A, Cochrane G, Vaughan R, Hunter C, Park J, Morrison N, Rocca-Serra P, Sterk P, Arumugam M, Bailey M, Baumgartner L, Birren BW, Blaser MJ, Bonazzi V, Booth T, Bork P, Bushman FD, Buttigieg PL, Chain PSG, Charlson E, Costello EK, Huot-Creasy H, Dawyndt P, DeSantis T, Fierer N, Fuhrman JA, Gallery RE, Gevers D, Gibbs RA, Gil IS, Gonzalez A, Gordon JI, Guralnick R, Hankeln W, Highlander S, Hugenholtz P, Jansson J, Kau AL, Kelley ST, Kennedy J, Knights D, Koren O, Kuczynski J, Kyrpides N, Larsen R, Lauber CL, Legg T, Ley RE, Lozupone CA, Ludwig W, Lyons D, Maguire E, Methe BA, Meyer F, Muegge B, Nakielny S, Nelson KE, Nemergut D, Neufeld JD, Newbold LK, Oliver AE, Pace NR, Palanisamy G, Peplies J, Petrosino J, Proctor L, Pruesse E, Quast C, Raes J, Ratnasingham S, Ravel J, Relman DA, Assunta-Sansone S, Schloss PD, Schriml L, Sinha R, Smith MI, Sodergren E, Spor A, Stombaugh J, Tiedje JM, Ward DV, Weinstock GM, Wendel D, White O, Whiteley A, Wilke A, Wortman JR, Yatsunenko T, Glockner FO (2011) Minimum information about a marker gene sequence (MIMARKS) and minimum information about any (x) sequence (MIxS) specifications. Nat Biotechnol 29(5):415–420

Yu EY, Kwon M-A, Lee M, Oh JY, Choi J-E, Lee JY, Song B-K, Hahm D-H, Song JK (2011) Isolation and characterization of cold-active family VIII esterases from an arctic soil metagenome. Appl Microbiol Biotechnol 90(2):573–581

Yun J, Ryu S (2005) Screening for novel enzymes from metagenome and SIGEX, as a way to improve it. Microb Cell Factories 4(1):8

Zhang J, Chiodini R, Badr A, Zhang G (2011) The impact of next-generation sequencing on genomics. J Genet Genomics 38(3):95–109

Zhang W, Li F, Nie L (2010) Integrating multiple 'omics' analysis for microbial biology: application and methodologies. Microbiology 156(2):287–301

Zheng H, Wu H (2010) Short prokaryotic DNA fragment binning using a hierarchical classifier based on linear discriminant analysis and principal component analysis. J Bioinforma Comput Biol 8(06):995–1011

Chapter 17
Proteomic Insights of Psychrophiles

Jun Kawamoto, Tatsuo Kurihara, and Nobuyoshi Esaki

Abstract Proteomics is a comprehensive quantification and identification of proteins synthesized in the target of interests and, based on their whole genome sequences, provides a deep understanding of molecular basis for various biological processes that occur in the target. In the field of microbiology, proteomic studies have uncovered the molecular basis for metabolic systems, infection to host cells, stress adaptation, and so on. As for studies on psychrophiles, proteomics is also a powerful approach to quantitatively identify the proteins involved in their cold adaptation. In this chapter, we review various isolates from cold environments and research on their proteomics to understand how psychrophiles adapt to their cold habitats widely spread over the surface of the Earth. In order to highlight the growing of the proteomic studies on psychrophiles, we also discuss the proteomics of the next era, which has been developed by the combination of next-generation DNA sequencing technologies and high-throughput mass spectrometry.

Contents

J. Kawamoto • T. Kurihara (✉) • N. Esaki
Institute for Chemical Research, Kyoto University, Gokasho, Uji, Kyoto 611-0011, Japan
e-mail: kurihara@mbc.kuicr.kyoto-u.ac.jp

R. Margesin (ed.), *Psychrophiles: From Biodiversity to Biotechnology*,
DOI 10.1007/978-3-319-57057-0_17

17.1 Introduction

The rise of next-generation DNA sequencing technologies allows the accurate, rapid, and cost-effective genome sequencing and has been providing an opportunity to obtain whole genome information of isolates from various environments and also environmental genome samples. As for psychrophilic organisms [in this chapter, psychrophilic and psychrotolerant (or psychrotrophic) microorganisms are not distinguished, and the term "psychrophilic microorganisms" is used throughout], about 190 complete and draft genome sequences have been registered in the Genomes OnLine Database (as of September 2016, http://www.genomesonline.org/), and the number of available genome sequences of psychrophilic organisms has been increased more than 10 times compared to those of a decade ago (Pagani et al. 2012; De Maayer et al. 2014). The availability of a large number of genome information on psychrophiles and related mesophilic strains has significantly enhanced the comparative genomic analyses and also the proteomic identifications based on the vast microbial genomic database. As well as the whole genome sequencing, the workflow for the proteomic researches has become more comprehensive by using two-dimensional electrophoresis (2DE) and one-dimensional (1D) or two-dimensional liquid chromatography (2D-LC) coupled with MS (tandem mass spectrometry MS/MS) analysis, which has provided a deep understanding of the microbial cold adaptation. In this chapter, proteomic studies of psychrophilic microorganisms, *Methanococcoides burtonii, Bacillus psychrosaccharolyticus, Psychrobacter cryohalolentis* K5, *Psychrobacter arcticus* 273-4, and *Shewanella livingstonensis* Ac10, as well as those of an unsequenced strain, *Pedobacter cryoconitis* A37^{T}, are reviewed.

17.2 Global Identification of Proteins Produced at Low Temperatures

Living organisms, including psychrophilic microorganisms, produce different sets of proteins depending on environmental temperatures. Because proteins that are inducibly produced at low temperatures are supposed to play important roles at low temperatures, many studies have been carried out to identify cold-inducible proteins to gain insight into cold-adaptation mechanisms of psychrophilic microorganisms.

17.2.1 Proteomics of a Nonsequenced Cold-Adapted Bacterium, Pedobacter cryoconitis *A37^{T}*

Identification of cold-inducible proteins is greatly facilitated by whole genome sequence data. However, comprehensive identification of cold-inducible proteins is possible without whole genome sequence data by a recently developed method. Pereira-Medrano and coworkers reported a high-throughput proteomic workflow by using *Pedobacter cryoconitis* A37^T as a model bacterium without whole genome information (Pereira-Medrano et al. 2012). Although the lack of whole genome information and a comprehensive bioinformatics database of the targeted strain has critical disadvantages in identification, quantification, and validation of the cold-inducible proteins comprehensively, the authors developed a rapid and confident proteome profiling of the unsequenced strain by the combination of ^{15}N-metabolic labeling, 2DE, de novo sequencing of peptide by MS/MS, and informatics tools.

P. cryoconitis A37^T was isolated from alpine glacier cryoconite and is a Gram-negative, rod-shaped, non-flagellated, motile by gliding, and non-spore forming bacterium. This strain can grow at a temperature range of 1–25 °C. The cells were grown with ^{14}N- or ^{15}N-labeled ammonium sulfate as a sole nitrogen source at 1 and 20 °C. The soluble protein extracts of two different cell mixtures at 1:1 (v/v) were applied to 2DE. Protein spots having similar normalized volume reproducibly (size, shape, and intensity) between biological replicates, as well as having the highest normalized volume, were analyzed by LC-coupled Q-TOF-MS/MS (quadrupole time-of-flight mass spectrometer) system. As the result, 13 proteins were identified as cold-inducible proteins of *P. cryoconitis* A37^T (ratio 1/20 °C: >1.5-fold), which are involved in carbohydrate metabolism, translation, protein metabolism and modification, and oxidative stress response. Among them, superoxide dismutase might function as antioxidation defense and reduce the intracellular level of the reactive oxygen species (ROS) at low temperatures where oxygen solubility is increased.

Although proteomic identification of unsequenced bacteria still demands considerable data analysis compared to that by using sequenced strains, rapid and confident proteome analysis of unsequenced strains remains required and likely to increase in the future, especially in the field of environmental proteomics and systems biology.

17.2.2 Methanococcoides burtonii

M. burtonii is a flagellated, motile methanogen isolated from permanently cold (1–2 °C), methane-saturated waters from the bottom of Ace Lake, Antarctica. Cold-inducible proteins that are predicted to be important for cold adaptation of this archaeon were examined by comparing 2DE profiles for cells grown to late log phase at low temperature (4 °C) and at the optimum temperature for growth (23 °C) (Goodchild et al. 2004). Cold-inducible proteins were also identified by isotope-coded affinity tag chromatography and LC-MS (Goodchild et al. 2005). Proteins identified in these studies are listed in Table 17.1.

These studies revealed that key aspects of cold adaptation of this archaeon relate to transcription, protein folding, and metabolism. In particular, specific roles for RNA polymerase subunit E, a response regulator, and peptidyl-prolyl *cis-trans* isomerase were suggested. RNA polymerase subunit E was unique to growth at 4 °C. This protein may fulfill a specific role in regulating the transcription of genes involved in low-temperature growth or in facilitating transcription at low

Table 17.1 Proteins of psychrophilic microorganisms inducibly produced at low temperatures[a]

Microorganism	*M. burtonii*[b]	*B. psychrosaccharolyticus*	*P. cryohalolentis* K5	*P. **arcticus*** 273-4	*S. livingstonensis* Ac10
Temperature	4/23 °C[c]	0 or 15/30 °C[d]	−4/16 or −4/4 °C[e]	4/22 or 4/22 °C-S[f]	4/18 °C[g]
Transcription	RNA polymerase subunit E (>1.8), Response regulator (>2.7), TATA-box-binding protein (1.7)	Transcription elongation factor	Antitermination factor (NusA) (3.37[h])	DNA-directed RNA polymerase, 30–40 kDa subunit: RNA polymerase α chain, bacterial and organelle (2.6)	DNA-directed RNA polymerase α subunit (RpoA) (26), Transcription elongation factor (GreA) (86)
Translation	SSU ribosomal protein (2.0), Peptidyl-prolyl *cis-trans* isomerase (3.1)	SSU ribosomal protein S30P, Chaperonin HSP10	Ctc form of ribosomal protein L25 (RplY) (1.33), Ribosomal protein S2 (RpsB) (2.18[h]), Cold shock protein (CspA) (>25), GTP-binding protein or elongation factor Tu (TypA) (2.00 and >6.6[i]), Elongation factor Ts (Tsf) (>8.9)	Ribosomal protein S15 (3.2[j]), tRNA synthetases, class II (G, H, P, and S): seryl-tRNA synthetase class IIa: aminoacyl-tRNA synthetase, class II (2.0), Translation elongation factor Tu (EF-Tu) (3.8), Putative chaperonin HSP60 family (2.8), Peptidyl-prolyl *cis-trans* isomerase, cyclophilin type (2.2), Putative chaperonin HSP10 (unique)	Cold shock protein (CspA) (577), GTPase-translation elongation factor (TufB) (33), Translation elongation factor P/translation initiation factor (Efp) (39), FKBP-type peptidyl-prolyl *cis-trans* isomerase (trigger factor) (Tig) (5.6), Lysyl-tRNA synthase class II (LysU) (120), Protease Do (DegP) (3.9), Phage shock protein A (PspA) (2.3), Peptidase S9 prolyl oligopeptidase (Dpp4) (5.1), PpiC-type peptidyl-prolyl *cis-trans* isomerase (SurA) (2.1)

Metabolism	Methyl coenzyme M reductase, α subunit (2.3), Methyl coenzyme M reductase, ß subunit (2.0), Methyl coenzyme M reductase, γ subunit (2.0), Trimethylamine methyltransferase (>3.3 & 4.2[i]), $F_{420}H_2$ dehydrogenase 40 kDa subunit (3.6), Methylcobalamine: CoM methyltransferase (2.0), Monomethylamine methyltransferase (2.0), Pyridoxine biosynthesis protein (2.6), Riboflavin biosynthesis protein (2.5), Dihydroorotate dehydrogenase (3.7), 3-Isopropylmalate dehydratase (>2.1), Adenosylhomocysteinase (>6.7), Glutamate dehydrogenase (5.9), Isopropylmalate synthase (2.1), Glyceraldehyde 3-phosphate dehydrogenase (2.4)	Glyceraldehyde 3-phosphate dehydrogenase, Vegetative catalase 1, Catalase, Electron transfer flavoprotein α subunit, ATP synthase (subunit ß), Pyrimidine nucleoside phosphorylase, Deoxyribose-phosphate aldolase, Succinyl-CoA synthetase (α subunit)	Malate/lactate dehydrogenase (Mdh) (3.34[h]), Isocitrate lyase (AceA) (2.78), Acetate kinase (AckA) (2.11), F_1/F_0 ATP synthase ß subunit, H^+/Na^+ translocating (AtpF) (>17), NH_3-dependent (glutamine-hydrolyzing) $NAD(^+)$ synthetase (NadE) (2.70), Glycine dehydrogenase (2.37), Shikimate 5-dehydrogenase (AroE) (>10), Amino transferase (>7.1), Xanthine dehydrogenase (XdhA) (>8.4), Pyridine nucleotide transhydrogenase (2.00)	D-3-Phosphoglycerate dehydrogenase (2.1[j]), Aminotransferase (2.1[j]), Fumarate hydratase class II (fumarase) (4.5), Possible acetone carboxylase γ subunit (3.1), δ-Aminolevulinic acid dehydratase (3.4), Nucleoside diphosphate kinase (2.3), Aldehyde dehydrogenase family protein (2.7)	Inorganic pyrophosphatase/exopolyphosphatase (Ppx1) (34), Phosphoribosylamine-glycine ligase (PurD) (45), Deoxyribose-phosphate aldolase (DeoC) (62), Electron transfer flavoprotein α-subunit (FixB) (80), NADPH: flavin oxidoreductase (NemA) (52), Pyridoxal phosphate biosynthesis protein (PdxJ) (55), Predicted extracellular nuclease (202), Predicted carboxypeptidase (31), Pyruvate kinase (PykF) (3.0), Alanine dehydrogenase (Ald) (2.6), Isocitrate dehydrogenase (LeuB) (2.4), Succinate dehydrogenase/fumarate reductase (SdhA) (2.2), Dihydrolipoamide acyltransferase (SucB) (2.8), Ribose-phosphate pyrophosphokinase (2.3), Methyl-accepting

(continued)

Table 17.1 (continued)

Microorganism	*M. burtonii*[b]	*B. psychrosaccharolyticus*	*P. cryohalolentis* K5	*P. **arcticus*** 273-4	*S. livingstonensis* Ac10
					chemotaxis sensory transducer (AtoS) (>161)
Transport		ABC transporter-ATP-binding protein	ATPase of ABC transporters with duplicated ATPase domains (Uup) (2.78), ABC-type Fe^{3+} transporter, periplasmic component (AfuA) $6# (2.38[h]), TRAP-T family transporter, substrate-binding subunit (DctP) (1.41[h]), Outer membrane efflux system (TolC) (>12), ABC lipoprotein exporter, ATPase (LolD) (>8.9), Outer membrane receptor for Fe^{3+} dicitrate, TonB-dependent (FecA) (>7.5)		Porin (OmpC) (3.6), Outer membrane protein A (OmpA) (611), Porin (Sliv_c417088) (2.6)
Cell division					Cell division GTPase (FtsZ) (39), MreB (2.5), FtsY (2.0)
Motility					Flagellar basal body and hook protein (FlgE) (26), Flagellin and related hook-associated protein (FlgL) (32)

Others		General stress protein 17M	Chemotaxis protein histidine kinase (CheA) (2.04), Hydroperoxide detoxification protein (OsmC) (>14)	Putative PhoH-like protein, Predicted ATPase (3.0), Putative cold-shock protein (unique)	6Fe-6S prismane cluster-containing protein (33), Predicted outer membrane protein (51), Nucleotide-binding outer membrane protein (Tsx) (63), Transposase (143)
References	Goodchild et al. (2004, 2005)	Seo et al. (2004)	Bakermans et al. (2007)	Zheng et al. (2007)	Kawamoto et al. (2007), Park et al. (2012)

[a]Relative abundance at indicated temperatures is shown in parenthesis; the value is not available for *B. psychrosaccharolyticus* in the literature

[b]TATA-box-binding protein and monomethylamine methyltransferase were identified by using isotope-coded affinity tag chromatography and liquid chromatography-mass spectrometry (Goodchild et al. 2005); all the other proteins were identified by comparing two-dimensional electrophoresis profiles (Goodchild et al. 2004)

[c]Cells were grown to late log phase at indicated temperatures

[d]Cells were grown to mid-log phase at 30 °C and further incubated at indicated temperatures for 1 h; quantitative data indicating relative abundance of each protein are not provided in the literature

[e]Cells were grown at indicated temperatures until OD_{600} reached 0.22

[f]Cells were acclimatized to indicated temperatures by cultivation in 1/2 tryptic soy broth with or without 5% NaCl; "-S" indicates that cells were grown with 5% NaCl

[g]Cells were grown to early stationary phase at indicated temperatures

[h]Relative abundance for −4/4 °C

[i]Two spots appeared on the 2-D PAGE gel

[j]Relative abundance for 4/22 °C-S

temperatures in general. A response regulator consisting of a CheY-like receiver domain and a winged-helix DNA-binding domain was also unique to growth at 4 °C. The increased abundance of the response regulator reinforces the finding that cold adaptation involves transcriptional regulation, and that it may involve a temperature-responsive two-component regulatory system similar to those found in bacteria. Peptidyl-prolyl *cis-trans* isomerase was one of the most abundant proteins with higher (3.1-fold) spot intensity at 4 °C and probably plays an important role in protein folding at low temperatures.

Proteins with increased spot intensities at 4 °C included the 40 kDa subunit of $F_{420}H_2$ dehydrogenase involved in methanogenesis. $F_{420}H_2$ dehydrogenase is a membrane-bound proton pump, which generates a proton gradient that drives most cellular processes including ATP synthesis. In methylotrophic methanogens, ATP can also be produced by a sodium motive force. However, at low temperatures, a proton motive force is easier to maintain than a sodium motive force. Higher level of $F_{420}H_2$ dehydrogenase at 4 °C probably leads to the generation of a proton motive force, which is preferred in *M. burtonii* during low-temperature growth.

In addition to these reports, more comprehensive proteomic analysis of *M. burtonii* was carried out by using isobaric tag labeling (iTRAQ) of the N-terminus of expressed proteins and 2D-LC-MS/MS (Williams et al. 2010). To identify new proteins involved in cold adaptation, the soluble, membrane, and secreted protein fractions were collected from the cells grown at different temperatures (4 and 23 °C) with different carbon sources (trimethylamine and methanol). The results demonstrated the presence of a complex adaptive process to cold environments, in which proteins involved in the archaeal cell envelope synthesis, protein transport, cell division, transcriptional regulation, oxidative stress, translation, and protein folding might play important roles in the cold adaptation of *M. burtonii* (Williams et al. 2010).

17.2.3 Bacillus psychrosaccharolyticus

B. psychrosaccharolyticus is a facultative anaerobic sugar-digesting bacterium found in soil and lowland marshes. It grows well in glucose broth under anaerobic conditions. It grows at 0 °C, but does not grow at temperatures above 30 °C. Spores are produced at 0 °C and germinate at 0 °C. Proteomic analysis was conducted for this bacterium to identify abundantly expressed proteins and cold stress response proteins (Seo et al. 2004). The cells were grown to mid-log phase at 30 °C and further incubated at 0, 15, and 30 °C for 1 h. Proteins extracted from the cells were analyzed by 2DE. Proteins inducibly synthesized by temperature downshift are listed in Table 17.1. Cold-induced proteins in *B. psychrosaccharolyticus* were similar to the proteins expressed for the cold shock response in *B. subtilis* (Brigulla et al. 2003). Cold-induced proteins included chaperonin HSP10. It was suggested that recomposition of proteins involved in metabolic functions and stress response is a major factor for psychrophily of *B. psychrosaccharolyticus*.

17.2.4 Psychrobacter cryohalolentis *K5*

P. cryohalolentis K5 is a Gram-negative, nonmotile bacterium isolated from Siberian permafrost. It grows at temperatures from −10 to 30 °C. Patterns of protein abundance were examined for the cells grown at 16, 4, and −4 °C (Bakermans et al. 2007). It was found that growth temperature substantially reprogrammed the proteome: about 31% of the proteins at each growth temperature responded to temperature. Proteins whose relative abundance increased at −4 °C are listed in Table 17.1.

Two ribosomal proteins (S2 and the Ctc form of L25) were inducibly synthesized at −4 °C, suggesting that these proteins specifically contribute to ribosomal function at low temperatures. S2 and Ctc are among the eight ribosomal proteins (of 53 in total) that are transcribed individually in *P. cryohalolentis* K5, suggesting that they are regulated independently of the "core" of ribosomal proteins. Two elongation factors (EF-Ts and TypA) and CspA, which probably exhibit RNA chaperone activity, were also induced at low temperatures. Induction of these proteins involved in translation suggests that modulation of the translation machinery is required for the growth of this bacterium at subzero temperatures.

Six transport-related proteins, AfuA, FecA, LolD, TolC, DctP, and Uup, were upregulated at low temperatures. These proteins are probably required to counteract lower rates of diffusion and transport across the membrane at low temperatures. Induction of the lipoprotein transporter, LolD, suggests an increased need for lipoproteins for maintaining fluidity of the membrane or activity of membrane proteins at low temperatures. The observation that two of the four iron transporters of this bacterium, AfuA and FecA, were upregulated at low temperatures suggests that this bacterium employs cold-adapted alleles to ensure adequate transport of iron into the cell.

Two enzymes of the glyoxylate cycle, malate dehydrogenase and isocitrate lyase, were upregulated at −4 °C. Acetate kinase, which is involved in the introduction of acetate into the glyoxylate cycle, was also upregulated at low temperatures. Acetate was the only carbon and energy source provided in these experiments, and these enzymes probably contribute to efficient assimilation of the carbon source and generation of energy. This speculation is supported by the fact that a possible acetate transporter, DctP, is inducibly produced at −4 °C.

Another cold-inducible protein, OsmC, is supposed to detoxify organic hydroperoxides that are produced during aerobic respiration. Oxidative stress increases at low temperatures because the concentration of oxygen radicals is higher at low temperatures due to higher solubility and slower consumption of oxygen. It is likely that OsmC counteracts higher oxidative stress at low temperatures.

17.2.5 Psychrobacter arcticus *273-4*

P. arcticus 273-4 was isolated from a 20,000–40,000-year-old Siberian permafrost core, which is characterized by low temperature, low water activity, poor nutrition, and high salinity. It survives at −10 °C and shows optimal growth rate at 22 °C. To explore how it survives in the permafrost environment, proteins in four samples of cells cultured at 4 and 22 °C in media with and without 5% sodium chloride (saline medium and nonsaline medium, respectively) were profiled and comparatively studied by 2D HPLC and MS (Zheng et al. 2007). Proteins upregulated at 4 °C are summarized in Table 17.1. These proteins are supposed to increase the resistance of cells to the cold environment.

When nonsaline medium was used, putative chaperonin HSP10 and putative cold-shock proteins were detected only in the cells grown at 4 °C. In addition, HSP60 and peptidyl-prolyl *cis-trans* isomerase (cyclophilin type) were inducibly produced at 4 °C in nonsaline medium. HSP60, together with HSP10, prevents misfolding and promotes the refolding and proper assembly of unfolded polypeptides generated under stress conditions. The overexpression of HSP60 is consistent with the increase of HSP10 at 4 °C. Peptidyl-prolyl *cis-trans* isomerase (cyclophilin type) probably accelerates protein folding at 4 °C. tRNA synthetase and EF-Tu were also upregulated at 4 °C in nonsaline medium. tRNA synthetase promotes tRNA synthesis, and EF-Tu promotes the GTP-dependent binding of aminoacyl-tRNA to the A-site of ribosomes during protein biosynthesis. These results suggest that the modulation of protein expression at the translational level is an important mechanism involved in cellular response to low temperature.

Ribosomal proteins S3, S4, S6, S15, L2, L7/L12, L15, and L28 were downregulated more than twofold at 4 °C in nonsaline medium. The decrease of ribosomal proteins at low temperature suggests that protein synthesis is decreased at low temperature to conserve energy.

It is noticeable that proteins that are significantly regulated due to temperature effects in saline medium differ from those in nonsaline medium by the amount and/or the direction of regulation. For example, HSP60 and peptidyl-prolyl *cis-trans* isomerase (cyclophilin) type were downregulated more than twofold in saline medium at 4 °C, whereas they were upregulated more than twofold in nonsaline medium at 4 °C. Downregulation of these general stress-related proteins at 4 °C in saline medium suggests that a temperature of 4 °C is favored over 22 °C in saline medium by this bacterium. The phenomenon may be related to the chemical properties of salt in solution. It is known that the ionic strength and the ion activity of salt in the solution decrease with the decline of the temperature, which results in lower external osmotic pressure exposed on cells in saline media at 4 °C than at 22 °C. High salt concentration probably exposes less stress on cells at 4 °C than at 22 °C.

17.2.6 Shewanella livingstonensis *Ac10*

S. livingstonensis Ac10 is a cold-adapted bacterium isolated from Antarctic seawater. The cells grow most rapidly at 18 °C, but can also grow well at 4 °C. The doubling time of this strain is 2.5 h at 18 °C and 8.8 h at 4 °C in LB medium. To elucidate the cold-adaptation mechanism of this bacterium, we conducted proteomic analysis based on the draft genome sequence (Kawamoto et al. 2007; Park et al. 2012). The cells were grown at 4 and 18 °C, and soluble and membrane proteins were analyzed by 2DE. Moreover, Park et al. (2012) developed a method for isolating the inner-membrane proteins by sucrose density gradient ultracentrifugation and performed proteomic analysis of this fraction of the strain, which contains various proteins involved in many key cellular processes such as energy generation, energy conversion in the respiratory chain, cell division, signal transduction, and transport processes. Proteins whose relative abundance increased at 4 °C more than twofold were identified by peptide mass fingerprinting and are listed in Table 17.1.

Two transcription-related proteins, RpoA and GreA, were upregulated at 4 °C. RpoA is a subunit of RNA polymerase playing an important role in assembly of RNA polymerase and promoter recognition. GreA induces cleavage and removal of the 3′ proximal dinucleotide from the nascent RNA to control transcriptional fidelity. Since the secondary structure of mRNA is greatly influenced by temperature, different set of proteins and/or different amount of proteins are probably required to ensure efficient and accurate transcription at different temperatures. CspA, which is supposed to function as an RNA chaperone, was also inducibly synthesized at 4 °C, suggesting that this protein is required to make the secondary structure of mRNA suitable for translation. Cold-inducible Tig probably contributes to efficient translation at low temperatures. Tig is a ribosome-associated chaperone with peptidyl-prolyl *cis-trans* isomerase activity, which facilitates proper folding of newly synthesized proteins. It is composed of three domains, an N-terminal domain, which mediates association with the large ribosomal subunit, a central substrate-binding domain with homology to FKBP proteins showing the peptidyl-prolyl *cis-trans* isomerase activity, and a C-terminal domain of unknown function. Peptidyl-prolyl *cis-trans* isomerase has been identified as a cold-inducible protein of *Shewanella* sp. SIB1 (Suzuki et al. 2004). This bacterium grows most rapidly at 20 °C and can grow at a temperature as low as 0 °C. Total soluble proteins extracted from the cells grown at 4 and 20 °C were analyzed by 2DE, and it was found that the amount of an FKBP family member protein with peptidyl-prolyl *cis-trans* isomerase activity increased at 4 °C compared to that at 20 °C. The results suggest that this protein facilitates protein folding when the bacterium grows at low temperatures. Although Tig of *S. livingstonensis* Ac10 is similar to FKBP22 in that it is supposed to have the peptidyl-prolyl *cis-trans* isomerase activity, it is different from FKBP22 in that Tig has two additional domains, N-terminal and C-terminal domains, that are not found in FKBP22.

Two putative outer membrane porin homologs, OmpA and OmpC, were inducibly produced at 4 °C. Homologs of these proteins from other bacteria have been shown to form channels for hydrophilic solutes and play important roles in uptake of nutrients. Upregulation of these proteins probably counteracts low diffusion rate of solutes at low temperatures and enables efficient uptake of nutrients.

The bacterial signal recognition particle (SRP) receptor FtsY was identified as a cold-inducible inner membrane protein. FtsY, which is a key protein for membrane protein synthesis, provides the essential link between the soluble SRP-ribosome-nascent chain complexes and the membrane-bound Sec translocon (Koch et al. 2003) and interacts with the SRP protein Ffh (Parlitz et al. 2007) to dissociate SRP from the nascent-peptide chain. It is suggested that FtsY inducibly produced at low temperatures facilitates dissociation of membrane protein precursors from SRP to produce membrane proteins efficiently.

The amounts of hook-related proteins of flagella, FlgE and FlgL, were increased at low temperatures, suggesting that *S. livingstonensis* Ac10 modulates its motility by altering the expression of these proteins. Consistent with this speculation, we found that the cells are more motile at 4 °C than at 18 °C. The physiological significance of higher motility of this strain at low temperatures remains to be clarified.

17.3 Conclusions

Proteomic studies have been performed for several psychrophilic microorganisms. A set of cold-inducible proteins produced by one psychrophilic microorganism were shown to be different from those produced by another (Table 17.1). The results suggest that different psychrophilic microorganisms take a different strategy to cope with cold environments. Nevertheless, some of those strategies appear to be common in several psychrophilic microorganisms. For example, an RNA chaperone, CspA, is inducibly produced at low temperatures by several microorganisms including *P. cryohalolentis* K5 and *S. livingstonensis* Ac10. Another example is peptidyl-prolyl *cis-trans* isomerase, which facilitates protein folding. Cold induction of this protein was observed for several psychrophilic microorganisms including *Shewanella* sp. SIB1, *M. burtonii*, *P. arcticus* 273-4, and *S. livingstonensis* Ac10, although the type of peptidyl-prolyl *cis-trans* isomerase inducibly produced at low temperatures is different in different psychrophilic microorganisms: FKBP-type protein, cyclophilin-type protein, and Tig-type protein are produced by these strains. Isomerization of peptidyl-prolyl bonds for proper protein folding is probably a crucial process for many microorganisms to survive at low temperatures. Modulation of RNA polymerase and ribosome is also supposed to be important for the cells to grow at low temperatures because subunits of these complexes are upregulated at low temperatures in several psychrophilic microorganisms. More distinctive features of proteomes of psychrophilic microorganisms will be clarified by proteomic studies supported by next-generation DNA sequencing of many other psychrophilic microorganisms and high-throughput protein identification by 1D- or 2D-LC MS/MS. Proteomic studies will undoubtedly provide us with more clues to understand how psychrophilic microorganisms adapt to cold environments in the near future, and the information will also be useful for engineering of these microorganisms for biotechnology.

References

Bakermans C, Tollaksen SL, Giometti CS, Wilkerson C, Tiedje JM, Thomashow MF (2007) Proteomic analysis of *Psychrobacter cryohalolentis* K5 during growth at subzero temperatures. Extremophiles 11:343–354

Brigulla M, Hoffmann T, Krisp A, Volker A, Bremer E, Volker U (2003) Chill induction of the SigB-dependent general stress response in *Bacillus subtilis* and its contribution to low-temperature adaptation. J Bacteriol 185:4305–4314

De Maayer P, Anderson D, Cary C, Cowan DA (2014) Some like it cold: understanding the survival strategies of psychrophiles. EMBO Rep 15:508–517

Goodchild A, Saunders NF, Ertan H, Raftery M, Guilhaus M, Curmi PM, Cavicchioli R (2004) A proteomic determination of cold adaptation in the Antarctic archaeon, *Methanococcoides burtonii*. Mol Microbiol 53:309–321

Goodchild A, Raftery M, Saunders NF, Guilhaus M, Cavicchioli R (2005) Cold adaptation of the Antarctic archaeon, *Methanococcoides burtonii* assessed by proteomics using ICAT. J Proteome Res 4:473–480

Kawamoto J, Kurihara T, Kitagawa M, Kato I, Esaki N (2007) Proteomic studies of an Antarctic cold-adapted bacterium, *Shewanella livingstonensis* Ac10, for global identification of cold-inducible proteins. Extremophiles 11:819–826

Koch HG, Moser M, Muller M (2003) Signal recognition particle-dependent protein targeting, universal to all kingdoms of life. Rev Physiol Biochem Pharmacol 146:55–94

Pagani I, Liolios K, Jansson J, Chen IM, Smirnova T, Nosrat B, Markowitz VM, Kyrpides NC (2012) The genomes online database (GOLD) v. 4: status of genomic and metagenomic projects and their associated meta-data. Nucl Acids Res 40:D571–D579

Park J, Kawamoto J, Esaki N, Kurihara T (2012) Identification of cold-inducible inner membrane proteins of the psychrotrophic bacterium, *Shewanella livingstonensis* Ac10, by proteomic analysis. Extremophiles 16:227–236

Parlitz R, Eitan A, Stjepanovic G, Bahari L, Bange G, Bibi E, Sinning I (2007) *Escherichia coli* signal recognition particle receptor FtsY contains an essential and autonomous membrane-binding amphipathic helix. J Biol Chem 282:32176–32184

Pereira-Medrano AG, Margesin R, Wright PC (2012) Proteome characterization of the unsequenced psychrophile *Pedobacter cryoconitis* using ^{15}N metabolic labeling, tandem mass spectrometry, and a new bioinformatic workflow. Proteomics 12:775–789

Seo JB, Kim HS, Jung GY, Nam MH, Chung JH, Kim JY, Yoo JS, Kim CW, Kwon O (2004) Psychrophilicity of *Bacillus psychrosaccharolyticus*: a proteomic study. Proteomics 4:3654–3659

Suzuki Y, Haruki M, Takano K, Morikawa M, Kanaya S (2004) Possible involvement of an FKBP family member protein from a psychrotrophic bacterium *Shewanella* sp. SIB1 in cold-adaptation. Eur J Biochem 271:1372–1381

Williams TJ, Burg DW, Raftery MJ, Poljak A, Guilhaus M, Pilak O, Cavicchioli R (2010) Global proteomic analysis of the insoluble, soluble, and supernatant fractions of the psychrophilic archaeon *Methanococcoides burtonii*. Part I: The effect of growth temperature. J Proteome Res 9:640–652

Zheng S, Ponder MA, Shih JY, Tiedje JM, Thomashow MF, Lubman DM (2007) A proteomic analysis of *Psychrobacter articus* 273-4 adaptation to low temperature and salinity using a 2-D liquid mapping approach. Electrophoresis 28:467–488

Chapter 18
From Transcriptomes to Metatranscriptomes: Cold Adaptation and Active Metabolisms of Psychrophiles from Cold Environments

Isabelle Raymond-Bouchard and Lyle G. Whyte

Abstract Transcriptomic and metatranscriptomic studies have yielded important insights into the mechanisms and metabolisms that allow psychrophiles to remain active in cold environments. Psychrophiles undergo numerous changes to their transcriptional profiles at colder temperatures and change the regulation of genes involved in most cellular processes, including primary metabolism and biosynthetic pathways, cell wall and peptidoglycan biosynthesis, lipid biosynthesis and cell membrane composition, translation/transcription/replication processes, protein turnover and chaperone functions, and stress responses. Increases in the expression of cold shock proteins, RNA/DNA helicases, protein chaperones, osmoprotectants, and proteins involved in the oxidative stress response are common features of cold adaptation in psychrophiles, as are changes to the cell membrane to increase fluidity and thickening of the cell wall. Metatranscriptome studies from permafrost and marine environments have begun to lay the groundwork for our understanding of the active metabolisms in these ecosystems and their potential impact on greater global processes such as biogeochemical cycles and greenhouse gas emissions. Community-specific microbial interactions, changes in temperature, degree of thaw, and nutrient, water, and organic matter availability, are all important drivers and regulators of microbial activity and metabolism, and changes in any of these factors can have significant impacts on microbial community function.

Contents

I. Raymond-Bouchard • L.G. Whyte (✉)
Department of Natural Resource Sciences, McGill University, Macdonald Campus, 21,111 Lakeshore Road, Ste-Anne-de-Bellevue, QC, Canada, H9X 3V9
e-mail: lyle.whyte@mcgill.ca

R. Margesin (ed.), *Psychrophiles: From Biodiversity to Biotechnology*,
DOI 10.1007/978-3-319-57057-0_18

18.1 Introduction

Since the discovery of active, diverse, and unique microbial communities in polar environments, there has been significant interest in understanding how these organisms are capable of sustaining their growth and survival in such harsh environments and the adaptations, whether physiological, morphological, metabolic, and/or -omic, that underline this ability. Our knowledge of psychrophiles has grown significantly in the last 20 years, with numerous cold-adapted organisms isolated and cultured. To date over 50 psychrophilic genomes have been sequenced, primarily bacterial and archaeal, providing us with important clues regarding their unique capabilities (see Chap. 15 and Bakermans et al. 2012). The number of available metagenomic analyses from cold environments is rapidly increasing, leading to the genomic characterization of novel un-culturable organisms, as well as potential microbial function and adaptions to their natural habitats (see Chap. 15 for a discussion of metagenomes). However, genomic and metagenomic analyses are limited in scope, able to focus only on gene presence and thus functional potential. It is through expression and proteomic analyses looking at active transcription of genes and presence of proteins, that we will ultimately understand the mechanisms and metabolisms that allow psychrophiles to live in cold environments. Proteomics of psychrophiles are discussed in depth in Chap. 17; this chapter will focus exclusively on current transcriptomic and metatranscriptomic studies of psychrophiles and their habitats. This first half of the chapter will aim to look at current transcriptome studies of isolates, while the latter half will look at the use of metatranscriptomics to elucidate the active members and the metabolisms of cold microbial communities.

18.2 Transcriptomics of Psychrophiles

Over 20 transcriptomes have been described for eurypsychrophiles (formerly psychrotrophs; $T_{max} > 20$ °C) and stenopsychrophiles (formerly "true psychrophiles," $T_{max} < 20$ °C), including bacteria, archaea, algae, ciliates, and diatoms, isolated from polar and cold environments (Table 18.1). Here, we will use the term psychrophile to include both eury- and steno-psychrophiles. These organisms have been isolated from diverse cold environments, including permafrost, active layer, marine systems, ice, and lakes. In almost all cases, the aim of current transcriptomic studies has been to gain insight into the metabolisms and adaptations that are important for growth at colder (<10 °C) temperatures. Currently, only five

Table 18.1 Summary of current transcriptomes from psychrophiles

Organism	Location	Taxa	Temp. (°C)	Key findings	Reference
Nesterenkonia sp. AN1	Antarctic dry permafrost	Bacteria	5 and 21	Upregulation of oxidative response, universal stress proteins, chaperone functions, protein turnover, and cell membrane biogenesis and induction of glyoxylate cycle at 5 °C	Aliyu et al. (2016)
Psychrobacter arcticus 273-4	Siberian permafrost	Bacteria	−6, 0, 17 and 22	Transcription, translation, energy production, most biosynthesis pathways decreased at lower temps, but AA synthesis, RNases, and peptidases increased. Evidence for isozyme exchange and shift to resource efficiency at lower temperatures	Bergholz et al. (2009)
Exiguobacterium antarcticum B7	Antarctic lake	Bacteria	0 and 37	Cold shock proteins most highly expressed transcripts at 0 °C. Transcripts involved in translation, transcription, RNA helicases higher at 0 °C	Dall'Agnol et al. (2014)
Shewanella oneidensis MR-1		Bacteria	8, 15, and 30	Lipid fluidity and catabolic processes for NADH and NADPH enhanced at lower temps. Differential regulation of cold shock proteins observed. Increase in translation, chemotaxis, chaperones, and DNA metabolism at lower temperatures	Gao et al. (2006)
Psychrobacter sp. PAMC 21119	Antarctic permafrost	Bacteria	−5 and 20	Translation, ribosome structure and biogenesis, protein folding, and membrane fluidity upregulated at −5 °C. Decrease in	Koh et al. (2017)

(continued)

Table 18.1 (continued)

Organism	Location	Taxa	Temp. (°C)	Key findings	Reference
				lipid transport, energy production, and metabolism. Isozyme exchange of cold shock proteins	
Planococcus halocryophilus Or1	Arctic active layer	Bacteria	−15 and 25	Cell envelope formation, membrane remodeling, and increased protein flexibility at −15 °C. Isozyme exchange of genes involved in cell division, fatty acid synthesis, solute binding, oxidative stress response, and transcriptional regulation	Mykytczuk et al. (2013)
Exiguobacterium sibiricum 255-15	Siberian permafrost	Bacteria	−2.5, 10 and 39	Constitutive adaptation to cold. Decrease in saturation and length of fatty acids in membrane, thickening of peptidoglycan at −2.5 °C. Shift in carbon source utilization and isozyme exchange at different temperatures	Rodrigues et al. (2008)
Pseudomonas extremaustralis	Antarctic pond	Bacteria	8 and 30	Transcription regulation, signal transduction overexpressed at 8 °C, but primary and AA metabolism repressed. Ethanol oxidation pathway induced and important for growth at low temperatures	Tribelli et al. (2015)
Methanolobus psychrophilus R15	Tibetan wetland	Archaea	4 and 18	Methanogenesis, biosynthesis, and protein synthesis downregulated at 4 °C, but translation, protein turnover and chaperones, and signal transduction increased	Chen et al. (2012)

(continued)

Table 18.1 (continued)

Organism	Location	Taxa	Temp. (°C)	Key findings	Reference
Methanococcoides burtonii	Antarctic Ace Lake	Archaea	4 and 23	Transcriptional regulation, not translation, largely responsible for controlling gene expression. Increase of cell surface and secretory proteins, protein turnover, and maintenance of translation and initiation at 4 °C. Energy generation and methanogenesis higher at 23 °C	Campanaro et al. (2011)
Chlamydomonas sp. ICE-L	Antarctic algal mat	Algae	−20, −10, 0, 5, 10 and 15	Upregulation of 11 DEAD-box RNA helicase genes after 36 h cold stress. Fatty acid desaturases increased at low temps with higher proportion of polyunsaturated fatty acids	An et al. (2013), Liu and Huang (2015)
Chlorella UMACC 234	Antarctic snow	Algae	4, 20 and 30	Higher expression of photosystem II reaction center at 4 °C	Chong et al. (2011)
Chaetoceros neogracile KOPRI AnM0002	Antarctic ocean	Algae	4 and 10	Metabolism and photosynthesis highest at 4 °C. Photosynthesis genes made up 15% of downregulated genes at 10 °C	Hwang et al. (2008)
Euplotes focardii and *Euplotes nobilii*	Antarctic pore waters	Ciliate	4 and 20	Heat shock protein not induced in response to heat shock; low constitutive expression at 4 °C	La Terza et al. (2001)
Chlamydomonas sp. ArF0006	Norway pond	Algae	4	Cold response and photosynthesis prominent at 4 °C, including heat shock proteins, transport functions, lipid biosynthesis,	Kim et al. (2013)

(continued)

Table 18.1 (continued)

Organism	Location	Taxa	Temp. (°C)	Key findings	Reference
				fatty acid desaturase, ribosomal proteins, antifreeze protein	
Chlorella vulgaris NJ-7	Antarctica	Algae	4 and 20	Abundant expression of antifreeze proteins, highest at 4 °C, and higher than *C. vulgaris* UTEX259. Two novel cryoprotectants identified and highly expressed in NJ-7	Li et al. (2009), Liu et al. (2011)
Fragilariopsis cylindrus Grunow	Antarctic sea ice	Diatom	−1, −1.8, 5 and 7	Translation, ribosomal structure, biogenesis, post-translation modifications dominated transcripts. Expression of six different DNA/RNA helicases. Increase in numerous photosynthesis genes at −1 °C	Mock and Hoch (2005), Mock et al. (2005)

transcriptomes exist for growth at temperatures below zero, in the bacteria *Exiguobacterium sibiricum* (−2.5 °C) (Rodrigues et al. 2008), *Psychrobacter arcticus* (−6 °C) (Bergholz et al. 2009), *Psychrobacter* sp. PAMC 21119 (−5 °C) (Koh et al. 2017), *Planococcus halocryophilus* (−15 °C) (Mykytczuk et al. 2013), and in the psychrophilic diatom *Fragilariopsis cylindrus* (−1 °C) (Mock et al. 2005). Interestingly, while these organisms are capable of growth at some of the coldest temperatures recorded to date, all, except *Fragilariopsis cylindrus*, are considered eurypsychrophiles rather than "true" psychrophiles and grow best at temperatures higher than 15 °C (Rodrigues et al. 2008; Bergholz et al. 2009; Mykytczuk et al. 2013; Koh et al. 2017). The remaining psychrophilic transcriptomes use above zero temperatures (4–5 °C) as their cold representative temperature.

18.2.1 Cold Shock Proteins and Chaperones

Cold shock proteins (CSP), which bind to nucleic acids and function as regulators of various cellular functions including translation, transcription, protein folding, and RNA degradation (De Maayer et al. 2014), and other predicted chaperones, such as heat shock proteins (HSP; protein chaperones) and DEAD-box and associated RNA helicases, are amongst the most studied features of cold shock and adaptation in microorganisms, owing in large part due to their significant importance in aiding transcription and translation processes. Almost all transcriptome studies conducted on psychrophiles to date have found increased transcript levels of at least one of the CSPs, HSPs, DEAD-box, and associated helicases and other chaperones at colder temperatures (La Terza et al. 2001; Mock et al. 2005; Gao et al. 2006; Bergholz et al. 2009; Campanaro et al. 2011; Chen et al. 2012; Mazzon et al. 2012; Kim et al. 2013; Dall'Agnol et al. 2014; Liu and Huang 2015; Koh et al. 2017). The psychrophilic alga, *Chlamydomonas* sp. ICE-L, possesses 39 DEAD-box RNA helicase genes, over 10 of which are strongly upregulated at lower temperatures (Liu and Huang 2015). Four of the six CSPs found in the genome of *Exiguobacterium antarcticum* B7 are expressed at 0 °C (Dall'Agnol et al. 2014). In the bacteria *Psychrobacter arcticus* (Bergholz et al. 2009) and *Shewanella oneidensis* (Gao et al. 2006), two isozymes of a DEAD-box RNA helicase and a CSP, respectively, are differentially expressed at colder and warmer temperatures. *Psychrobacter* sp. PAMC 21119 also makes use of differentially expressed CSP isozymes at −5 and 20 °C, with the −5 °C isozyme showing signs of possible cold adaptation with increased protein flexibility (Koh et al. 2017). CSPs and chaperones are not always upregulated at low temperatures. They may be constitutively expressed across all temperatures in some organisms, suggesting perhaps a readiness of these organisms to react quickly to changing temperatures. A CSP from *P. arcticus* was found constitutively expressed at high levels across all temperatures (22, 17, 0 and −6 °C) (Bergholz et al. 2009). *Euplotes focardii*, an Antarctic stenopsychrophilic ciliate from benthic coastal sediments, grows optimally at 4–5 °C and, while it possesses many copies of hsp70, expression of these proteins changes only minimally over a large change in temperature, contradictory to its relative, the eurypsychrotroph *Euplotes nobilii*; this indicates that *E. focardii* may have lost thermal adaptation mechanisms as a result of long-term influence of the thermally (~ −2 °C) stable waters from which it was isolated (La Terza et al. 2001).

18.2.2 Cell Wall and Cell Membrane

Modifications to fatty acid biosynthesis and cell wall structure allow organisms to overcome decreased membrane fluidity at lower temperatures and perhaps increase protection from environmental challenges such as ice crystals. As a result, elevated transcription of genes involved in cell wall membrane and fatty acid biosynthesis is observed (Gao et al. 2006; Rodrigues et al. 2008; Bergholz et al. 2009; An et al. 2013; Kim et al. 2013; Mykytczuk et al. 2013). For example, the bacteria *Exiguobacterium sibiricum*, isolated from Siberian permafrost, increases expression of fatty acid desaturase and peptidoglycan biosynthesis genes, leading to a decrease in saturation and length of fatty acids in its membrane and a thickening of the cell wall, respectively, at subzero temperatures (Rodrigues et al. 2008). The extreme cryophile (capable of subzero growth) *Planococcus halocryophilus*, capable of growth down to −15 °C, undergoes extensive modifications in its cell wall and membrane at subzero temperatures leading to the formation of a unique cellular envelope, initially observed through microscopy analyses, and confirmed by transcriptomic analysis showing significant increase in cell wall biosynthesis transcripts at −15 °C (Mykytczuk et al. 2013).

18.2.3 Universal and Oxidative Stress Responses

Universal stress response genes and genes involved in the oxidative stress response have also been established as important for cold adaptation (Chen et al. 2012; De Maayer et al. 2014). Universal stress proteins are small cytoplasmic proteins often transcribed in response to multiple cell stressors including nutrient starvation, heat shock, cell growth inhibition, presence of reactive oxygen species (ROS) (Kvint et al. 2003), and to cold conditions (Gao et al. 2006; Aliyu et al. 2016). An association between cold stress and oxidative stress has been observed, and ROS concentrations are known to increase at lower temperatures, likely the result of increased gas solubility and higher rates of enzyme activity to compensate for reduced reaction rates (Chattopadhyay et al. 2011; De Maayer et al. 2014). As a result the oxidative stress response, expression of genes encoding antioxidative enzymes, is often induced at lower temperatures by many psychrophiles (Bergholz et al. 2009; Chen et al. 2012; Mykytczuk et al. 2013; Tripathy et al. 2014; Aliyu et al. 2016). Interestingly, in *Pseudomonas extremaustralis*, genes involved in the oxidative stress response are repressed at colder temperatures, although iron-related proteins, including uptake and iron-binding proteins, were upregulated, which is theorized to play a redundant role in the reduction of oxidative stress (Tribelli et al. 2015).

18.2.4 Compatible Solutes and Antifreeze Proteins

Compatible solutes, small water soluble organic compounds such as glycine betaine, ectoine, and trehalose, are used by bacteria to maintain turgor pressure during growth. In cold-adapted bacteria, compatible solutes play an important role in allowing psychrophiles to resist osmotic pressure and prevent water loss caused by extracellular ice formation and increases in salinity often associated with cold environments (Doyle et al. 2012). Compatible solutes also play a role as cryoprotectants and increase the stability of macromolecules, membranes, and proteins, as well as enhancing folding and ligand binding in the latter (Thomas et al. 2001; Yancey 2005). Genes for the biosynthesis or transport of compatible solutes are widespread in psychrophilic microbes and increase in expression at cold temperatures in several instances (Rodrigues et al. 2008; Bergholz et al. 2009; Campanaro et al. 2011). Ice nucleating proteins or antifreeze proteins also function as cryoprotectants; they bind directly to the ice surface and prevent the growth of ice crystals (Davies et al. 2002). Higher expression of these proteins at cold temperatures is described in a number of psychrophilic organisms, including the Antarctic green algae *Chlorella vulgaris* NJ-7 and Arctic *Chlamydomonas* sp. ArF0006 (Li et al. 2009; Liu et al. 2011; Kim et al. 2013).

18.2.5 Metabolisms and Energy/Nutrient Acquisition

Temperature can have a significant impact on expression of genes involved in metabolism and energy and nutrient acquisition. While psychrophiles are able to grow at low temperatures, their optimum growth temperatures usually remain above 10 °C, and in the case of eurypsychrophiles, maximum productivity is usually in the 20–30 °C range (Lauro et al. 2011; Siddiqui et al. 2013). Given that eurypsychrophiles are generally psychrotolerant rather than truly psychrophilic and tend to have growth rates that are highest above 20 °C, it is not surprising that at lower temperatures transcripts for metabolic pathways and energy production will often be downregulated. Energy production and most biosynthetic pathways are downregulated in *Psychrobacter arcticus* 273-4 (Bergholz et al. 2009), while in *Pseudomonas extremaustralis*, primary metabolism, the TCA cycle, and amino acid metabolisms are repressed (Tribelli et al. 2015). Energy metabolism is also repressed in *Planococcus halocryophilus* at −15 °C (Mykytczuk et al. 2013). Transcripts for methanogenesis and energy production in the methanogens *Methanococcoides burtonii* and *Methanolobus psychrophilus* R15 are more abundant at higher temperatures, as was core carbon and nitrogen metabolism in *M. burtonii* (Campanaro et al. 2011) and biosynthesis pathways in *M. psycrhophilus* R15 (Chen et al. 2012). This is not the case for stenopsychrophiles, however, historically recognized as true psychrophiles, which may grow optimally at fairly low (<10 °C) temperatures. In these organisms, the opposite trend can be observed. The stenopsychrophilic algae *Chaetoceros neogracile* is very well adapted to its natural Antarctic environment, growing better at 4 °C than at higher temperatures (10 °C). This is in part due to an increase in photosynthesis and photosynthesis efficiency at lower temperatures (Hwang et al. 2008). More than 15% of all downregulated genes at 10 °C belonged to part of the photosynthesis pathway. Photosynthesis systems were >3-fold higher at 4 °C in the Antarctic algae *Chlorella* sp. UMACC 234 (Chong et al. 2011). Similar results were seen for *Fragilariopsis cylindrus*, a stenopsychrophilic diatom that increases photosynthesis and abundance of some photosynthesis transcripts at −1 °C compared to 7 °C (Mock and Hoch 2005; Mock et al. 2005).

Transcriptome studies of psychrophiles have yielded incredibly useful clues with regard to their survival and growth in extreme cold environments. However, in vitro studies occur in synthetic media and cannot replace the natural environments of these organisms and therefore cannot answer the more complex question of how these organisms function and interact in their native ecosystems. This question is next explored.

18.3 Metatranscriptomics of Cryoenvironments

While metagenomics has proven incredibly useful to increase our knowledge of the diversity, abundance, and functional potential of cold ecosystems, metatranscriptomics allows us to go one step further and gain insight into the active members of these communities and the metabolisms that fuel their growth, adaptation, and survival in situ within their habitats. Indeed, often, metatranscriptomic data reveals significant differences in the relative abundance and function of active members as compared to metagenomic data (Mackelprang et al. 2016). RNA-based studies from numerous cryoenvironments, including microarrays from high Arctic and Antarctic soils (Yergeau et al. 2007, 2009), rRNA transcript analyses from thawing permafrost (Crevecoeur et al. 2015), glacial ice (Hamilton et al. 2013; Cameron et al. 2016), and subzero saline springs (Lay et al. 2013; Lamarche-Gagnon et al. 2014), and metatranscriptomics of permafrost and permafrost-affected soils (Coolen and Orsi 2015; Hultman et al. 2015; Lau et al. 2015; Tveit et al. 2015; Buelow et al. 2016) as well as polar seawater and sea ice (Toseland et al. 2013; Bertrand et al. 2015; Pearson et al. 2015) have all added to the mounting evidence that active and cold-adapted microorganisms exist in situ in cryoenvironments and are providing novel insights into their ecosystem function. Since this chapter aims to highlight current transcriptomic research, we will focus almost exclusively on metatranscriptome studies (Table 18.2).

18.3.1 Permafrost

Concerns about climate change and the potential for the large amounts of frozen organic carbon in thawing permafrost to become a significant contributor to greenhouse gas emissions, predominantly methane and carbon dioxide, have directed much of the interest in current metatranscriptome permafrost research (Davidson et al. 2006; Olefeldt et al. 2013).

Overall, permafrost microbial communities exhibit high plasticity and adapt quickly to changing temperatures (Tveit et al. 2015). In Alaskan permafrost, the transcriptional response is dominated primarily by amino acid transport and metabolism, energy production, and stress-specific mechanisms, focusing on survival and long-term cellular maintenance (biofilm formation, pilus assembly, virulence, horizontal gene transfer, DNA repair, and the general SOS response) (Coolen and Orsi 2015). In contrast, the thawed Alaskan soils have increased microbial activity as indicated by higher levels of transcripts involved in translation, ribosomal structure,

Table 18.2 Summary of metatranscriptome studies from permafrost, permafrost-affected, and polar marine environments

Location	Sample type	Key active microbial members	Key microbial metabolic and functional activity	Reference
McMurdo Dry Valleys, Antarctica	Top soil (~12 cm)	*Bacteria*: (largely dominant): Actinobacteria, Firmicutes, and Proteobacteria; *Fungi*: Ascomycetes and Basidiomycetes	Primarily chemoorganoheterotrophic bacteria, but also photoautotrophic cyanobacteria/algae; transcripts abundant for carbohydrate/nitrogen metabolism: serine-glyoxylate cycle, maltose/maltodextrin metabolism, and nitrate/nitrite ammonification	Buelow et al. (2016)
	Amended top soil		Loss of functional diversity. Transcript loss assigned to specialized pathways in bacteria and cellular function in eukaryotes. Under-expressed: glycolysis and maltose/maltodextrin utilization. Over-expressed: pentose phosphate pathway, sulfur metabolism, and transport systems	
Kuparuk river, LTER field station, Alaska	Permafrost, acidic tundra	*Bacteria*: Proteobacteria, Firmicutes, Acidobacteria, Actinobacteria; *Archaea*: Euryarchaeota; *Fungi*: Ascomycetes	Higher stress-specific response: biofilm formation, DNA repair, SOS response, pilus assembly	Coolen and Orsi (2015)
	Thawed permafrost	*Bacteria*: Firmicutes and Bacteroidetes; *Archaea*: Euryarchaeota; *Fungi*: Ascomycetes	Higher activity: abundance of transcripts for translation, ribosomes, biogenesis, carbohydrate degradation and transport, peptidases (Euryarchaeota, Bacteriodetes), hydrolases (fungi, Firmicutes), and methanogenesis (Euryarchaeota)	
Fairbanks, Alaska	Permafrost	*Bacteria*: Proteobacteria, Acidobacteria, and Firmicutes	Metabolic processes lower than active layer, but abundance of certain transcripts, i.e., methane oxidation, similar.	Hultman et al. (2015)

(continued)

Table 18.2 (continued)

Location	Sample type	Key active microbial members	Key microbial metabolic and functional activity	Reference
			Dissimilatory ferric iron reduction potentially important process	
	Active layer	*Bacteria*: Proteobacteria, Acidobacteria, and Actinobacteria; *Archaea*: Crenarchaeota	Increase in functional activity and diversity compared to permafrost. Transcripts involved in a variety of metabolic processes, including methane oxidation, nitrate reduction, denitrification, and iron reduction	
	Thermokarst bog	*Bacteria*: Proteobacteria and Firmicutes; *Archaea*: Euryarchaeota	Higher abundance of transcripts for methanogenesis (*Methanosarcina*)	
Svalbard, Knudsenheia, Norway	Active layer peat soil	*Bacteria*: Firmicutes (Clostridiales); *Archaea*: Euryarchaeota (Methanobacteriales, Methanosarcinaceae)	Syntrophic relationship between propionate oxidation (Peptococcaceae) and methanogenesis (Methanobacteriales). Transcripts for hydrogenotrophic (Methanobacteriales) and acetotrophic (Methanosarcinaceae) methanogenesis	Tveit et al. (2015)
	Warming peat soil	*Bacteria*: Bacteriodetes; *Archaea*: Euryarchaeota (Methanomicrobiales, Methanosaetaceae)	Decrease in Peptococcaceae (Clostridiales) for Bacteriodetes. Increase in CH_4 production and shift in active methanogen community to Methanomicrobiales (hydrogenotrophic) and Methanosaetaceae (acetotrophic). Increase in abundance of transcripts for methanogenesis from methylamines	
Axel Heiberg Island, Nunavut, Canadian high Arctic	Ice-wedge polygon	*Bacteria:* Atmospheric methane-oxidizing bacteria	Active atmospheric methane sink; detection of transcripts for atmospheric CH_4-oxidizers (atmMOB). 100-fold higher atmMOB transcript reads in polygon trough compared to polygon interior. Rate	Lau et al. (2015)

			of CH_4 uptake increases with temperature; theorized to increase by factor of 5–30 as Arctic warms	
Southern Ocean (Wilkins ice shelf, Bransfield Straight, western Weddell Sea)	Ice	Heterokont algea: Dictyochophyceae	Carbohydrate and energy metabolism lowest in ice, concurrent active stress response: transcripts for antifreeze proteins and cold shock transcription factors increased. Nitrogen acquisition contrasted: transport of ammonia and urea high in Bransfield community, but absent in post-bloom Weddell sample. Cold temperature pressure on protein synthesis: Decreasing temperatures linked to increase in transcripts for translation, transcription, and ribosome processes	Pearson et al. (2015)
	Seawater	Diatoms: Bacillariophyta		
McMurdo Sound, Antarctica	Coastal seawater	*Diatoms*: *Fragilariopsis* and *Pseudonitzchia*; *Bacteria*: Gammaproteobacteria (*Oceanospirillaceae* and *Methylophaga*) and Bacteriodetes	Iron and cobalamin levels impact phytoplankton growth and are meditated by intricate bacterial-phytoplankton interactions. Transcript levels indicate diatoms dependent on cobalamin synthesis by bacteria (*Oceanospirillaceae*) for primary production, while bacteria dependent on diatoms for organic compounds	Bertrand et al. (2015)
Southern and Arctic Oceans	Surface seawater	Diatoms: Bacillariophyta	Significant impact of temperature on phytoplankton metabolism. Transcripts show translation processes most impacted by temperature. Translation may be rate-limiting step for protein synthesis at lower temperatures	Toseland et al. (2013)

(continued)

Table 18.2 (continued)

Location	Sample type	Key active microbial members	Key microbial metabolic and functional activity	Reference
Barbados accretionary prism	Deep cold seep water	*Bacteria*: Epsilon- (*Sulfurovum*), Delta- (*Desulfo*. families) and Gamma-proteobacteria (*Vibrionales*), and methanotrophs (*Methylococcales*); *Archaea*: ANME-II Methanosarcinales	Primary productivity dependent on sulfide production primarily from AOM (ANME-II and Deltaproteobacteria). Supports sulfur oxidation and nitrate reduction by Ca. *Thiopilula* population and *Sulfuvorum* and *Sulfurimonas* spp.; Organoheterotrophy very active, especially *Photobacterium* spp, with carbon metabolism and transport functions	Jones et al. (2015)

biogenesis, extracellular protein degradation, and carbohydrate transport and degradation. Active microorganisms from spore forming phyla Bacteroidetes and Firmicutes increased in abundance in thawed samples, compared to Proteobacteria, Acidobacteria, and Actinobacteria which were more dominant active members in the frozen Alaskan permafrost (Coolen and Orsi 2015). Increases in metabolic activities, as determined by the ratio of functional gene transcripts to genes, was also seen in thawed soils sequenced as part of a comparison study of three soils (permafrost, active layer, and thermokarst) in different states of thaw from Fairbanks, Alaska (Hultman et al. 2015). Indeed, the active layer community exhibited more functional diversity than the permafrost and thermokarst bog samples. Many transcripts involved in specific biogeochemical cycles including methane oxidation, nitrate reduction, denitrification, and iron reduction were detected in the active layer samples. Fewer functional processes could be assigned to the permafrost, although dissimilatory ferric iron reduction was noted as a potentially important metabolic process.

In artificially warmed (1–30 °C gradient) Arctic peat soils, CH_4 production increased rapidly, with an accompanying shift in the active methanogens (Tveit et al. 2015). Active hydrogenotrophic methanogens shifted from *Methanobacteriales* to *Methanomicrobiales*, and acetotrophic methanogens *Methanosarcinaceae* were replaced by *Methanosaetaceae*. In addition, *Methanosarcinaceae* showed a high degree of flexibility during thermal adaptation and changed their methanogenic metabolism from acetotrophic to methylotrophic, likely due to increased substrate availability, over the temperature increase. Thermokarst bog soil from Alaska, which experiences higher temperatures than neighboring permafrost soil (~6 to −2 °C for bog vs ~ −0.05 to −2 °C for permafrost), has a high relative abundance of transcripts involved in methanogenesis, with a correspondingly very high measured rate of methane production, compared to permafrost. This was attributed partly to *Methanosarcina*, which comprised 6.8–10.5% of all 16s rRNA gene sequences (Hultman et al. 2015). Similarly, transcripts for heterotrophic methanogenesis, using acetate, methanol, and methylamine, were predominant in thawed Alaskan permafrost compared to the frozen sample, while acetogenesis was exclusive to the thawed samples (Coolen and Orsi 2015). It has been suggested that organisms within these complex communities have adapted the capacity to shift their metabolisms, and consequently community metabolic contributions, across a temperature gradient, allowing them to remain competitive for available substrates (Hall et al. 2008; Tveit et al. 2015). The relationship between permafrost and net methane flux on a global scale remains unclear however. While thawed or warmed permafrost from the above studies shows increased transcripts for methanogenesis and acetogenesis, correlated with higher rates of methane production, organic poor Arctic permafrost in the Canadian high Arctic acts as a methane sink, with CH_4 uptake increasing with rising temperatures, potentially due to atmospheric methane oxidizing bacteria identified as active through transcriptomics (Lau et al. 2015).

In addition to thawing permafrost, the melting of glaciers and buried ice are also expected to be important contributors to changes in permafrost environments through the increase of water availability and soluble nutrients. In the oligotrophic soils of the lower elevation McMurdo Dry Valleys (MDV), surface soils have active heterotrophic organisms as well as active photoautotrophic cyanobacteria and algae (Buelow et al. 2016). Transcripts indicate the use of simple carbon sources including maltose, maltodextrin, and methylotrophy (serine-glyoxylate cycle), as would be expected in these low nutrient soils. These soils contain high concentrations of inorganic nitrate, and this was reflected in the transcripts of the nitrogen metabolism subsystem, 31% of which could be assigned to nitrate and nitrite ammonification and 48% to ammonia assimilation. When water and nutrients (organic matter) were added to MDV surface soils, losses occurred in both taxonomic and functional diversity. There was a large decline in abundance of eukaryotes, and significant losses of transcripts could be assigned to specialized pathways in bacteria and cellular function in eukaryotes. Pathways for glycolysis, solely assigned to the Actinobacteria, and maltose/maltodextrin utilization, mapping mostly to Actinobacteria (67%) and Gammaproteobacteria (30%), were more often underexpressed in the amended soils. Positive transcriptional responses were only seen in the OM amended soils, specifically the pentose phosphate pathway (mostly Actinobacteria and Firmicutes), sulfur metabolism (Actinobacteria), and transport systems (Firmicutes and Proteobacteria), perhaps indicative of the increased nutrient availability in the amended soils.

18.3.2 Sea Ice, Seawater, and Sediments

Phytoplankton are the dominant primary producers in marine ecosystems and are important contributors to global biogeochemical cycles, especially carbon cycling systems (Field et al. 1998). Differential transcript expression by similar microbial populations across different temperature regimes in the environment, highlighting the importance of temperature on metabolic activity, can be illustrated by eukaryotic phytoplankton from polar, temperate, and tropical temperature zones of the ocean (Toseland et al. 2013). Translation of proteins is significantly affected by temperature, with ribosomes and transcripts of key enzymes involved in purine and pyrimidine metabolism highest in polar waters, though the rate of protein synthesis decreases at lower temperatures, indicating that translation may be the rate-limiting step for phytoplankton protein synthesis at lower temperatures. Evidence for cold adaption is seen in colder waters with heat shock proteins, DEAD-box RNA helicases, and fatty acid desaturases abundantly transcribed.

In order to gain a better understanding of ecological diversity and function in diatom rich communities from the Southern Ocean, Pearson et al. (2015) performed comparative metatranscriptomic analyses of three sites, the Wilkins Ice Shelf, the Bransfield Straight, and western Weddell Sea. While a core set of expressed genes are shared across the three communities, a significant proportion (40%) of identified transcripts are community specific. Organisms in the sea ice sample are clearly in the more stressful environment, with the transcriptome showing lower levels of carbohydrate and energy metabolism compared to the pelagic communities. These communities show increased levels of ice-binding (antifreeze) proteins and cold shock transcription factors. Similarly to the work described in the above paragraph, temperature was found to have a strong impact on transcript expression, especially transcription, translation, and ribosome processes which decreased with increasing temperature, highlighting the potential pressure that cold sea environments place on efficient protein synthesis.

Phytoplankton growth in coastal Antarctic seawaters near the sea ice edge is highly susceptible to nutrient availability. Iron and cobalamin availability can greatly influence late-summer phytoplankton growth, which in turn influences biogeochemical processes (Bertrand et al. 2015). In these waters, cobalamin and iron limitation is mediated by intricate phytoplankton and bacterial interactions. Diatoms, the dominant primary producers in these waters, are dependent on cobalamin synthesis by bacteria and archaea, most importantly by members of the Gammaproteobacteria *Oceanospirillaceae* ASP10-02a. In a metatranscriptome analysis, the *Oceanospirillaceae* ASP10-02a population largely dominated expression (70%) of cobalamin biosynthesis transcripts. *Oceanospirillaceae* ASP10-02a are in turn dependent on phytoplankton to provide organic compounds that fuel cobalamin production, as shown by high levels of transcripts involved in organic matter acquisition and cell surface attachment. However, phytoplankton must also compete with other bacterial groups for these nutrients, including *Methylophaga*, who not only rely on phytoplankton for carbon and energy but are also important iron and cobalamin consumers. (Bertrand et al. 2015). This study highlights the intricate and delicate association between microbes and phytoplankton in Antarctic coastal ecosystems and helps lay the groundwork to better understand what controls primary productivity in these marine systems.

18.4 Conclusions

Transcriptomics and metatranscriptomics have begun to bridge the gap between microbial activity and function in polar habitats and their impact on the environment and on larger global processes including biogeochemical cycles and global warming. However, there are still few metatranscriptomes available from psychrophilic habitats and more will be needed if we are to gain a more complete understanding of the interconnections between microbial activity and ecosystem function.

Some important caveats should be kept in mind, however, as we move forward in this field. While the presence of mRNA transcripts tells us which genes are being transcribed, this is not always representative of the final proteome. In fact, there can be significant differences between the abundance of transcripts and those that are translated to proteins (Budde et al. 2006; Bergholz et al. 2009). Furthermore, with regard to metatranscriptomics, the abundance of rRNA transcripts, in relation to the abundance of rDNA, is not always indicative of activity, as the relationship between growth and activity is complex (Blazewicz et al. 2013). For these reasons, care must be taken when attempting to interpret transcriptomic and metatranscriptomic data and linking these to microbial adaptations, function, and activity.

Finally, it will be important to look more closely at the relationship and interplay between the growth ranges of psychrophiles, their metabolisms, their differing environments, and the range of temperatures and conditions they are likely to encounter in these habitats, when we look to make larger statements about cold adaptation and ecosystem function. Each of these will, individually, impact active transcription of genes in psychrophiles and may explain some of the differences and contradictions observed in transcriptome and metatranscriptome studies of psychrophiles and their environment.

References

Aliyu H, De Maayer P, Cowan D (2016) The genome of the Antarctic polyextremophile *Nesterenkonia* sp. AN1 reveals adaptive strategies for survival under multiple stress conditions. FEMS Microbiol Ecol 92:fiw032. doi:10.1093/femsec/fiw032

An M, Mou S, Zhang X et al (2013) Temperature regulates fatty acid desaturases at a transcriptional level and modulates the fatty acid profile in the Antarctic microalga *Chlamydomonas* sp. ICE-L. Bioresour Technol 134:151–157. doi:10.1016/j.biortech.2013.01.142

Bakermans C, Bergholz PW, Rodrigues DF et al (2012) Genomic and expression analyses of cold-adapted microorganisms. In: Miller RV, Whyte LG (eds) Polar microbiology: life in a deep freeze. ASM Press, Washington, DC, pp 126–155

Bergholz PW, Bakermans C, Tiedje JM (2009) *Psychrobacter arcticus* 273-4 uses resource efficiency and molecular motion adaptations for subzero temperature growth. J Bacteriol 191:2340–2352. doi:10.1128/JB.01377-08

Bertrand EM, McCrow JP, Moustafa A et al (2015) Phytoplankton-bacterial interactions mediate micronutrient colimitation at the coastal Antarctic sea ice edge. Proc Natl Acad Sci U S A 112:9938–9943. doi:10.1073/pnas.1501615112

Blazewicz SJ, Barnard RL, Daly RA, Firestone MK (2013) Evaluating rRNA as an indicator of microbial activity in environmental communities: limitations and uses. ISME J 7:2061–2068. doi:10.1038/ismej.2013.102

Budde I, Steil L, Scharf C et al (2006) Adaptation of *Bacillus subtilis* to growth at low temperature: a combined transcriptomic and proteomic appraisal. Microbiology 152:831–853. doi:10.1099/mic.0.28530-0

Buelow HN, Kooser AS, Van Horn DJ et al (2016) Microbial community responses to increased water and organic matter in the arid soils of the McMurdo Dry Valleys, Antarctica. Front Microbiol 7:1040. doi:10.3389/fmicb.2016.01040

Cameron KA, Stibal M, Zarsky JD et al (2016) Supraglacial bacterial community structures vary across the Greenland ice sheet. FEMS Microbiol Ecol 92:fiv164

Campanaro S, Williams TJ, Burg DW et al (2011) Temperature-dependent global gene expression in the Antarctic archaeon Methanococcoides burtonii. Environ Microbiol 13:2018–2038. doi:10.1111/j.1462-2920.2010.02367.x

Chattopadhyay MK, Raghu G, Sharma YVRK et al (2011) Increase in oxidative stress at low temperature in an antarctic bacterium. Curr Microbiol 62:544–546. doi:10.1007/s00284-010-9742-y

Chen Z, Yu H, Li L et al (2012) The genome and transcriptome of a newly described psychrophilic archaeon, *Methanolobus psychrophilus* R15, reveal its cold adaptive characteristics. Environ Microbiol Rep 4:633–641. doi:10.1111/j.1758-2229.2012.00389.x

Chong G-L, Chu W-L, Othman RY, Phang S-M (2011) Differential gene expression of an Antarctic *Chlorella* in response to temperature stress. Polar Biol 34:637–645. doi:10.1007/s00300-010-0918-5

Coolen MJL, Orsi WD (2015) The transcriptional response of microbial communities in thawing Alaskan permafrost soils. Front Microbiol 6:197. doi:10.3389/fmicb.2015.00197

Crevecoeur S, Vincent WF, Comte J, Lovejoy C (2015) Bacterial community structure across environmental gradients in permafrost thaw ponds: methanotroph-rich ecosystems. Front Microbiol 6:192. doi:10.3389/fmicb.2015.00192

Dall'Agnol HPMB, Baraúna RA, de Sá PHCG et al (2014) Omics profiles used to evaluate the gene expression of *Exiguobacterium* antarcticum B7 during cold adaptation. BMC Genomics 15:986. doi:10.1186/1471-2164-15-986

Davidson EA, Janssens IA, Marks D et al (2006) Temperature sensitivity of soil carbon decomposition and feedbacks to climate change. Nature 440:165–173. doi:10.1038/nature04514

Davies PL, Baardsnes J, Kuiper MJ, Walker VK (2002) Structure and function of antifreeze proteins. Philos Trans R Soc Lond B Biol Sci 357:927–935. doi:10.1098/rstb.2002.1081

De Maayer P, Anderson D, Cary C, Cowan D (2014) Some like it cold: understanding the survival strategies of psychrophiles. EMBO Rep 15:508–517. doi:10.1002/embr.201338170

Doyle S, Dieser M, Broemsen E, Christner B (2012) General characteristics of cold-adapted microorganisms. In: Miller RV, Whyte LG (eds) Polar microbiology: life in a deep freeze. ASM Press, Washington, DC, pp 103–125

Field CB, Behrenfeld MJ, Randerson JT, Falkowski P (1998) Primary production of the biosphere: integrating terrestrial and oceanic components. Science 281:237–240. doi:10.1126/science.281.5374.237

Gao H, Yang ZK, Wu L et al (2006) Global transcriptome analysis of the cold shock response of *Shewanella oneidensis* MR-1 and mutational analysis of its classical cold shock proteins. J Bacteriol 188:4560–4569. doi:10.1128/JB.01908-05

Hall EK, Neuhauser C, Cotner JB (2008) Toward a mechanistic understanding of how natural bacterial communities respond to changes in temperature in aquatic ecosystems. ISME J 2:471–481

Hamilton TL, Peters JW, Skidmore ML, Boyd ES (2013) Molecular evidence for an active endogenous microbiome beneath glacial ice. ISME J 7:1402–1412. doi:10.1038/ismej.2013.31

Hultman J, Waldrop MP, Mackelprang R et al (2015) Multi-omics of permafrost, active layer and thermokarst bog soil microbiomes. Nature 521:208–212. doi:10.1038/nature14238

Hwang Y, Jung G, Jin E (2008) Transcriptome analysis of acclimatory responses to thermal stress in Antarctic algae. Biochem Biophys Res Commun 367:635–641. doi:10.1016/j.bbrc.2007.12.176

Jones DS, Flood BE, Bailey JV (2015) Metatranscriptomic analysis of diminutive *Thiomargarita*-like bacteria ("Candidatus Thiopilula" spp.) from abyssal cold seeps of the barbados accretionary prism. Appl Environ Microbiol 81:3142–3156. doi:10.1128/AEM.00039-15

Kim S, Kim MJ, Jung MG et al (2013) De novo transcriptome analysis of an Arctic microalga, *Chlamydomonas* sp. Genes Genomics 35:215–223. doi:10.1007/s13258-013-0085-5

Koh HY, Park H, Lee JH et al (2017) Proteomic and transcriptomic investigations on cold-responsive properties of the psychrophilic Antarctic bacterium *Psychrobacter* sp. PAMC 21119 at subzero temperatures. Environ Microbiol 19:628–644. doi:10.1111/1462-2920.13578

Kvint K, Nachin L, Diez A, Nyström T (2003) The bacterial universal stress protein: function and regulation. Curr Opin Microbiol 6:140–145. doi:10.1016/S1369-5274(03)00025-0

La Terza A, Papa G, Miceli C, Luporini P (2001) Divergence between two Antarctic species of the ciliate Euplotes, *E. focardii* and *E. nobilii*, in the expression of heat-shock protein 70 genes. Mol Ecol 10:1061–1067. doi:10.1046/j.1365-294X.2001.01242.x

Lamarche-Gagnon G, Comery R, Greer CW, Whyte LG (2014) Evidence of in situ microbial activity and sulphidogenesis in perennially sub-0°C and hypersaline sediments of a high Arctic permafrost spring. Extremophiles 19:1–15. doi:10.1007/s00792-014-0703-4

Lau MC, Stackhouse BT, Layton AC et al (2015) An active atmospheric methane sink in high Arctic mineral cryosols. ISME J 9:1880–1891. doi:10.1038/ismej.2015.13

Lauro FM, Allen MA, Wilkins D et al (2011) Psychrophiles: genetics, genomics, evolution. In: Horikoshi K (ed) Extremophiles handbook. Springer, New York, NY, pp 1118–1130

Lay C-Y, Mykytczuk NCS, Yergeau É et al (2013) Defining the functional potential and active community members of a sediment microbial community in a high-arctic hypersaline subzero spring. Appl Environ Microbiol 79:3637–3648. doi:10.1128/AEM.00153-13

Li H, Liu X, Wang Y et al (2009) Enhanced expression of antifreeze protein genes drives the development of freeze tolerance in an Antarctica isolate of *Chlorella vulgaris*. Prog Nat Sci 19:1059–1062. doi:10.1016/j.pnsc.2008.10.012

Liu C, Huang X (2015) Transcriptome-wide analysis of DEAD-box RNA helicase gene family in an Antarctic psychrophilic alga *Chlamydomonas* sp. ICE-L. Extremophiles 19:921–931. doi:10.1007/s00792-015-0768-8

Liu X, Wang Y, Gao H, Xu X (2011) Identification and characterization of genes encoding two novel LEA proteins in Antarctic and temperate strains of *Chlorella vulgaris*. Gene 482:51–58. doi:10.1016/j.gene.2011.05.006

Mackelprang R, Saleska SR, Jacobsen CS et al (2016) Permafrost meta-omics and climate change. Annu Rev Earth Planet Sci 44:439–462. doi:10.1146/annurev-earth-060614-105126

Mazzon RR, Lang EAS, Silva CAPT, Marques MV (2012) Cold shock genes CspA and CspB from caulobacter crescentus are post transcriptionally regulated and important for cold adaptation. J Bacteriol 194:6507–6517. doi:10.1128/JB.01422-12

Mock T, Hoch N (2005) Long-term temperature acclimation of photosynthesis in steady-state cultures of the polar diatom *Fragilariopsis cylindrus*. Photosynth Res 85:307–317. doi:10.1007/s11120-005-5668-9

Mock T, Krell A, Valentin K et al (2005) Analysis of expressed sequence tags (ESTs) from the polar diatom *Fragilariopsis cylindrus*. J Phycol 42:78–85. doi:10.1111/j.1529-8817.2005.00164.x

Mykytczuk NC, Foote SJ, Omelon CR et al (2013) Bacterial growth at -15 °C; molecular insights from the permafrost bacterium *Planococcus halocryophilus* Or1. ISME J 7:1211–1226. doi:10.1038/ismej.2013.8

Olefeldt D, Turetsky MR, Crill PM, Mcguire AD (2013) Environmental and physical controls on northern terrestrial methane emissions across permafrost zones. Glob Chang Biol 19:589–603. doi:10.1111/gcb.12071

Pearson GA, Lago-Leston A, Canovas F et al (2015) Metatranscriptomes reveal functional variation in diatom communities from the Antarctic Peninsula. ISME J 9:2275–2289. doi:10.1038/ismej.2015.40

Rodrigues DF, Ivanova N, He Z et al (2008) Architecture of thermal adaptation in an *Exiguobacterium sibiricum* strain isolated from 3 million year old permafrost: a genome and transcriptome approach. BMC Genomics 9:547. doi:10.1186/1471-2164-9-547

Siddiqui KS, Williams TJ, Wilkins D et al (2013) Psychrophiles. Annu Rev Earth Planet Sci 41:87–115. doi:10.1146/annurev-earth-040610-133514

Thomas T, Kumar N, Cavicchioli R (2001) Effects of ribosomes and intracellular solutes on activities and stabilities of elongation factor 2 proteins from psychrotolerant and thermophilic methanogens. J Bacteriol 183:1974–1982. doi:10.1128/JB.183.6.1974

Toseland A, Daines SJ, Clark JR et al (2013) The impact of temperature on marine phytoplankton resource allocation and metabolism. Nat Clim Chang 3:979–984. doi:10.1038/nclimate1989

Tribelli PM, Venero ECS, Ricardi MM et al (2015) Novel essential role of ethanol oxidation genes at low temperature revealed by transcriptome analysis in the Antarctic bacterium *Pseudomonas extremaustralis*. PLoS One. doi:10.1371/journal.pone.0145353

Tripathy S, Sen R, Padhi SK et al (2014) Upregulation of transcripts for metabolism in diverse environments is a shared response associated with survival and adaptation of Klebsiella pneumoniae in response to temperature extremes. Funct Integr Genomics 14:591–601. doi:10.1007/s10142-014-0382-3

Tveit AT, Urich T, Frenzel P, Svenning MM (2015) Metabolic and trophic interactions modulate methane production by Arctic peat microbiota in response to warming. Proc Natl Acad Sci U S A 112:E2507–E2516. doi:10.1073/pnas.1420797112

Yancey PH (2005) Organic osmolytes as compatible, metabolic and counteracting cytoprotectants in high osmolarity and other stresses. J Exp Biol 208:2819–2830. doi:10.1242/jeb.01730

Yergeau E, Kang S, He Z et al (2007) Functional microarray analysis of nitrogen and carbon cycling genes across an Antarctic latitudinal transect. ISME J 1:163–179. doi:10.1038/ismej.2007.24

Yergeau E, Arbour M, Brousseau R et al (2009) Microarray and real-time PCR analyses of the responses of high-arctic soil bacteria to hydrocarbon pollution and bioremediation treatments. Appl Environ Microbiol 75:6258–6267. doi:10.1128/AEM.01029-09

Part IV
Biotechnological Perspectives

Chapter 19
Biotechnological Aspects of Cold-Active Enzymes

Mário Barroca, Gustavo Santos, Charles Gerday, and Tony Collins

Abstract Cold-adapted enzymes produced by organisms inhabiting permanently low temperature environments are typically characterized by a high activity at low to moderate temperatures and a poor thermal stability. Such characteristics make these enzymes highly attractive for various applications where they can enable more efficient, cost-effective, and environmentally friendlier processes than higher temperature-adapted enzymes. In this chapter, the biotechnological aspects of cold-adapted enzymes and their application in industry are reviewed and discussed with a focus on cleaning/detergents, food and beverages, molecular biology, biomedicine, pharmaceuticals, cosmetics, textiles, biofuels, and materials applications.

Contents

19.1 Introduction

Enzymes are highly specific biological catalysts that accelerate the rate of chemical reactions in the cells of living organisms. These natural catalysts are biodegradable, fast, efficient, and selective, and produce low amounts of by-products while also

M. Barroca • G. Santos • T. Collins (✉)
Department of Biology, Centre of Molecular and Environmental Biology (CBMA), University of Minho, Campus de Gualtar, 4710-057 Braga, Portugal
e-mail: tcollins@bio.uminho.pt

C. Gerday
Laboratory of Biochemistry, Institute of Chemistry B6, University of Liège, 4000 Liège-Sart Tilman, Belgium

R. Margesin (ed.), *Psychrophiles: From Biodiversity to Biotechnology*,
DOI 10.1007/978-3-319-57057-0_19

being less demanding with respect to process energy, raw materials and toxic components than many traditional chemical catalysts. Their exquisite catalytic power, specificity of action and reduced environmental footprint makes them seemingly ideal tools for numerous biotechnological applications. Indeed, they have already found application in many diverse industries including food, beverages, pharmaceuticals, detergents, textiles, leather, chemical, biofuels, animal feed, personal care, pulp and paper, diagnostics, and therapy, and are continuously developing into new areas. They are employed to enable improved and/or more economic and eco-friendly end-products and bioprocesses and can even facilitate the development of novel products and processes. The current market value of industrial enzymes is estimated at almost USD 5 billion (BCC Research 2017), up from USD 3.3 billion in 2010 (Blamey et al. 2017), and is expected to reach almost USD 6.3 billion by 2021 (BCC Research 2017).

The International Union of Biochemistry and Molecular Biology (IUBMB) and the International Union of Pure and Applied Chemistry (IUPAC) have classified enzymes into six main classes (Enzyme Commission, EC, numbers 1–6) based on the types of reactions catalyzed. Enzymes belonging to all six classes have found application, but the hydrolases (EC number 3) are the most widespread used, with lipases, proteases, glycosidases, and other hydrolytic enzymes accounting for almost 90% of the total industrial enzymes market share (Blamey et al. 2017). These are mainly used as high volume, commodity, or bulk enzymes in the manufacture of food, beverage, cleaning agents, textiles, biofuels, and pulp and paper. On the other hand, specialty enzymes, for use in high value, low volume products such as in diagnostics, research and development, and as biocatalysts, currently have a small market share but have shown strong growth in recent years. Improved health care and the adaption of a "personalized" medicine approach drive this growth in diagnostics. Similarly, the development of "Green chemistry," making use of the high substrate-, regio-, and enantio- selectivity of enzymes for the sustainable production, modification, and/or functionalization of pharmaceuticals, fine chemicals, flavors, fragrances, etc., drives growth in biocatalysis. Indeed, the current societal shift towards greener technologies and a more sustainable low carbon resource-efficient economy as well as the expansion of biotechnology into new fields previously dominated by petroleum-based chemicals will continue to drive growth of enzymes in the future. In fact, currently only about 11% of all chemicals are made from renewable raw materials, with the remaining being obtained from crude oil, natural gas, and coal (Blamey et al. 2017), and it is believed that enzymes will play a major role in the shift towards a greater utilization of renewables in the future and in the development of efficient, improved, environmentally friendly, and sustainable bio-based processes.

For successful integration into a particular application, the ideal enzyme needs to appropriately catalyze the desired reaction with the desired specificity under the conditions used. That is to say, the enzyme needs to have the required specificity and selectivity as well as a high activity and stability in the process environment. Unfortunately, and due mainly to the harsh conditions frequently used in industrial processes, few naturally occurring enzymes fulfill all of these requirements

(Bommarius and Paye 2013; Sarmiento et al. 2015). High temperatures, as used, for example, in pulp and paper manufacture and in bioconversion, are often required in processes to enable a better breakdown and improved solubility of substrates and products as well as reduced viscosity, higher mass transfer rates, easier separation of volatile products, reduced contamination, and a shift in the equilibrium of endothermic reactions towards products (Siddiqui 2015). Conversely, low temperatures are more suited for processes involving heat-sensitive or/and volatile components or/and where undesirable chemical side-reactions occurring at high temperatures and contamination problems are to be avoided, such as in the manufacture of many foods, beverages, fine chemicals, and pharmaceuticals. Furthermore, in addition to extremes of temperature, many industrial processes are also carried out under extremes of pH, pressure, salinity, and/or in the presence of detergents, non-aqueous solvents, etc. (Bommarius and Paye 2013; Sarmiento et al. 2015; Barroca et al. 2017). Hence, standard, naturally occurring enzymes, which are typically stable and active over a narrow range near moderate physical and chemical conditions, are unsuited for use in numerous industrial processes in which "unnatural" conditions are used. In an attempt to overcome this, scientists have turned to extremophilic enzymes (Liszka et al. 2012; Elleuche et al. 2014; Littlechild 2015; Siddiqui 2015) and protein engineering (Liszka et al. 2012; Bommarius and Paye 2013). The use of rational design or directed evolution to fine-tune or engineer the properties of a protein for a particular application has achieved some success, but the use of extremophilic enzymes already adapted to extreme conditions offers a more direct route. Indeed, extremophilic enzymes, produced by organisms inhabiting and adapted to extreme environments, have already been shown to be valuable tools for processes in which extreme conditions prevail. A large number of extremophilic enzymes are already commonly used in various diverse applications with the vast majority being thermophiles, these being active and stable at high temperatures and frequently also in the presence of harsh chemicals and detergents. In contrast, use of enzymes from other extreme environments appears much less developed. As an example, a search of patenting databases for patents on enzymes with the keywords "cold-active," "cold-adapted," "cold-resistant," or "psychrophilic" in the title or abstract identified 53 patents whereas a similar search using the keywords "thermostable," "heat-stable," "heat-resistant," and "thermophilic" gave over 60-times more hits. Interestingly, while currently being much less employed than thermophilic enzymes, cold-adapted enzymes have an enormous potential as highly valuable tools for various biotechnological applications, and this review will focus on these enzymes and their biotechnological aspects.

19.2 Cold-Adapted Enzymes

As discussed in Chap. 10 of this book and in numerous previous review papers on the subject (Santiago et al. 2016; Fields et al. 2015; Siddiqui 2015; Gerday 2013, 2014; Collins et al. 2002a, 2007, 2008; Huston 2008; D'Amico et al. 2006), cold-adapted or psychrophilic enzymes, produced by organisms inhabiting permanently low temperature environments, are typically characterized by a high activity at low temperatures and a reduced stability as compared to their mesophilic and thermophilic homologs. They are believed to have overcome the low temperature challenge and maintained high activities at low temperatures by increasing the flexibility of specific regions of their molecular edifice. This increased flexibility enables a continued mobility of those regions important for enzyme activity, even in the low energy environment characteristic of low temperatures, and is achieved via a reduction in the number and/or strength of stabilizing interactions in the protein structure which in turn leads to the observed reduced structural stability of cold-adapted enzymes. Importantly, this balancing of flexibility, activity, and stability is believed to be key in enzyme adaptation to temperature, with thermophilic enzymes, in contrast to those from psychrophiles, being generally characterized by a high stability, low flexibility, and reduced low temperature activity.

The intrinsic attributes of a high activity at low temperatures and reduced stability of cold-adapted enzymes offer many advantages for use in a variety of commercial applications. Low temperature processes are common in the food and beverages industries and cold-adapted enzymes, highly active under these conditions, offer obvious benefits for such processes. Also, it is important to note that cold-adapted enzymes are not only more highly active than their mesophilic and thermophilic homologs at low temperatures, but frequently also show a higher activity at moderate temperatures. Hence, processes can be carried out at ambient temperatures without the need for energy input (for heating or cooling) and with lower quantities of cold-adapted enzyme being required as compared to enzymes adapted to higher temperatures. That is to say that these enzymes can be instrumental in developing processes with an improved economic and environmental impact. Furthermore, the thermolability of these enzymes offers solutions for those processes where a greater control is required and where a simple selective inactivation of the enzyme can be achieved by mild heat treatment (Margesin et al. 2003). Such a characteristic may prove beneficial for preserving product quality in the food and beverages industry and in biocatalysis or for sequential multi-step processes such as those used in molecular biology (Huston 2008). In fact, cold-adapted enzymes have already found application in these industries and their now commonplace application in cleaning/detergents applications as well as development into new markets underscores the potential and market value of these enzymes. In effect, market research by the Freedonia Group (The Freedonia Group 2016) has indicated that, in the mature commodity enzymes markets where competition among enzyme makers is intense, novel enzymes presenting efficient performance at lower temperatures will play an important role in allowing for market expansion in the future. In addition, while little explored at present, it is believed that in the future, psychrophilic enzymes will offer valuable tools as specialty enzymes in the preparation of temperature-sensitive pharmaceutical ingredients, fine-chemicals, flavors, and fragrances. In the following sections, the application of cold-adapted enzymes in various industrial enzymes markets will be discussed.

19.3 Cleaning/Detergents

This is probably the best developed market for cold-adapted enzymes with a large number of different hydrolytic enzymes (EC 3) being commercialized for this application by various companies throughout the world. See Sarmiento et al. (2015) for an in-depth review.

Enzymes are used as cleaning agents in household and industrial scale laundry and dishwashing, as well as for cleaning-in-place in the food, dairy, and brewing industries and even in the cleaning of buildings, carpets, contact lenses, etc. (Cavicchioli et al. 2002; Damhus et al. 2013; Sarmiento et al. 2015; Siddiqui 2015). This market currently accounts for approximately 20–25% of total industrial enzymes sales (BCC Research 2017). Enzymes break down stains, soiling, and deposits into more soluble products, thereby allowing for improved cleaning performance in, for example, laundry and dishwashing as well as for the deblocking and cleaning of filters and equipment in the food and beverages industries. Lipases hydrolyze lipids as found, for example, in grease, butter, oil, sauces, tears, molds, and biofilms; proteases break down proteins common in grass, blood, egg, milk, cheese, yoghurt, sweat, tears, molds, etc., and amylases are used for breakdown of starch soiling from cereals, pasta, potatoes, molds, biofilms, etc. Cellulases have also been used and act on oat products, such as cereals and snack bars. Furthermore, in laundry detergents, cellulases have an added benefit of contributing to fabric care; they degrade accessible broken cotton fibers (known as fuzz or pills) and thereby remove any captured dirt but at the same time reduce fuzz build up and hence also increase the softness and color brightness of cotton fabrics. Recently, attention has also been turned to mannanases and pectinases for use in cleaning detergents for removal of difficult stains due to gum, fruit products, juices, mayonnaise, tomato sauce, salad dressing, body lotions, personal care products, etc. Pullulanases have also been shown to have potential for biofilm removal (Antranikian et al. 2004). The action of enzymes allows for improved cleaning performance and in turn this improved effectiveness enables a reduction in the use of other more hazardous components, e.g., detergents, surfactants, polymers, alkaline builders (in laundry), phosphates (in dishwashing), and organic solvents (for cleaning-in-place), as well as reduced water consumption (Damhus et al. 2013; Sarmiento et al. 2015). Indeed, up to 25–50% reduction in a laundry detergent surfactant system has even been demonstrated upon use of enzymes (Damhus et al. 2013; Siddiqui 2015). Such modifications obviously lead to environmentally friendlier wash water wastes and more sustainable wash processes.

The major benefit of using cold-adapted enzymes in cleaning/detergents is that the process temperature and hence energy input can be reduced and thus enables an improved economic and environmental impact. It has been estimated that a reduction in wash temperature from 40 to 30 °C allows for a 30% saving in energy, corresponding to 100–300 g of CO_2 per wash (Cavicchioli et al. 2011; Siddiqui 2015), and already over 50% of laundry machine washes are carried out at low temperatures (Sarmiento et al. 2015). Furthermore, in laundry washing, low temperatures also extend garment life by being less aggressive, reducing garment degradation, and lessening shrinkage and/or dye bleeding. Currently, the focus is on further reducing temperatures to approx. 20 °C and leading to further savings, and a continued growth in the use of cold-adapted enzymes in this application is thus forecast. In cleaning in-place applications in the food and beverages industries, the use of cold-adapted enzymes would avoid the need for warm cleaning washes and the poor stability of these enzymes would give a greater assurance of complete enzyme inactivation following heat treatment, a desirable characteristic for the cleaning of food, dairy, and brewing industry equipment. Additionally, cold-adapted enzymes could offer advantages in the cleaning of large immovable objects where heating is not viable, and indeed glucose oxidases, proteases, amylases, and lipases have already been shown to be effective in mold and biofilm removal from building surfaces (Webster and May 2006; Valentini et al. 2010). These were shown to enable effective building cleaning and conservation while also reducing the use of more aggressive cleaning agents. Similarly, enzymes have been shown to be effective in contact lens cleaning, and a thermolabile fish waste isolate protease has been shown to have potential as an efficient, non-hazardous cleaning agent for the removal of tear films and proteinaceous deposits on contact lens (Pawar et al. 2009).

19.4 Food and Beverages

Industrial enzymes for use in the food and beverages industries represent a relatively well developed market with current sales of nearly USD 1.5 billion (BCC Research 2017). They are used as food additives and processing aids in such diverse areas as the more traditional processes of cheese manufacturing, wine making, brewing, and bread making to the more recent applications in the preparation of functional foods and nutraceuticals (Chandrasekaran 2015). Enzymes are used in the manufacture, processing, preparation, and treatment of foods and beverages. They can enable improved process efficiency, reduced processing costs, and reduced environmental impact and, frequently also, enhance the flavor, nutritional value, appeal, digestibility, texture, and/or shelf life of the final product. Growing consumer preference for more natural, healthier, and flavorsome foods as well as an improved awareness of environmental issues and food safety is driving continued growth in the use of enzymes in this sector, and a compound annual growth rate of 4.7% through 2021 has been forecasted (BCC Research 2017).

Cold-adapted enzymes are particularly attractive in food and beverages preparation due to their high catalytic activity at temperatures that minimize spoilage, alterations in taste, and loss of nutritional value as well as their ease of inactivation (Huston 2008). They have found application in the dairy, baking, beverages, meat, and fish processing industries and in the production of functional foods.

In the dairy industry, a number of cold-adapted β-galactosidases, or lactases, have been developed for the production of lactose free milk and treatment of the waste by-product whey. Approximately 65% of the human population has a reduced ability to digest lactose after infancy, with Asian and African populations being the most affected. β-Galactosidases hydrolyze lactose to glucose and galactose and hence can be used to remove lactose from dairy products and improve product digestibility while also enhancing sweetness. A variety of β-galactosidases are currently being marketed, but cold-adapted β-galactosidases offer the advantage of efficient hydrolysis at refrigeration temperatures which minimize problems associated with contamination and alteration of product organoleptic properties (Hoyoux et al. 2001; Ghosh et al. 2012; Stougaard and Schmidt 2012; Pawlak-Szukalska et al. 2014). In the valorization of whey, a by-product of cheese production and a waste disposal problem, the glucose and galactose produced by β-galactosidase treatment can be used as sweeteners in soft drinks and confectionary products, in the production of functional foods/nutraceuticals (Van de Voorde et al. 2014), and in biofuel production (Huston 2008). A recent study showed the potential of using a cold-adapted β-galactosidase for the initial steps of preparation of tagatose, a novel, low-calorie sweetener (Van de Voorde et al. 2014). Similarly, cold-adapted β-galactosidases have also been shown to be useful in the preparation of other low calorie sweeteners, namely, galactooligosaccharides (Karasova-Lipovova et al. 2003; Nakagawa et al. 2007; Schmidt and Stougaard 2010; Pawlak-Szukalska et al. 2014). In effect, in addition to hydrolysis, many β-galactosidases also display transglycosylation activities where monosaccharides are transferred to oligosaccharides with the production of di-, tri-, tetra-, and pentasaccharides. These can be produced in the milk or from whey, and in addition to functioning as low calorie sweeteners, they have also been found to be effective prebiotics, selectively stimulating the growth of beneficial intestinal microorganisms (Pawlak-Szukalska et al. 2014). Proteases are another family of enzymes which play an important role in the dairy industry and in particular during clotting and ripening in cheese making. Lipases and phospholipases are also used and due to the low temperatures employed, cold-adapted variants have been suggested for this (Huston 2008).

Amylases, xylanases, oxidases, asparaginases, and lipases are all commonly used in baking applications so as to improve product quality (dough machinability, bread texture, and shelf life), reduce the use of chemical additives (e.g., potassium bromate, emulsifiers, etc.), and reduce the production of acrylamide (in baked or fried products including biscuits, crisps, crackers, etc.). As yet, the majority of enzymes used appear to be of mesophilic or thermophilic origin whereas dough preparation and proofing is typically carried out at moderate temperatures at which cold-adapted enzymes offer considerable cost and efficiency advantages. Nevertheless, it appears that currently only one cold-adapted enzyme has found application in baking, namely, a cold-adapted xylanase (Collins et al. 2002b, 2006, 2012; Dornez et al. 2011; Dutron et al. 2012). Xylanases improve dough machinability, giving rise to a more flexible, easier-to-handle dough, larger loaf size, and improved crumb structure, and the cold-adapted xylanase was shown to be more effective than the other commercial xylanases studied (Collins et al. 2006, 2012; Dutron et al. 2012).

In the brewing, wine, and fruit and vegetable processing industries, pectinases (polygalacturonases, pectin lyases, and pectin methylesterase) and hemicellulases such as xylanases are used to increase extraction yield, improve clarification, reduce viscosity, and enhance color and flavor (Tu et al. 2013; Adapa et al. 2014). Rhamnogalacturonases, galactanases, and arabinanases have also been recently developed for these applications. Apparently no cold-adapted enzymes have been commercialized in this sector as yet but low temperature active enzymes are available, e.g., Lallzyme EX (Lallemand) is active between 5 and 20 °C (Sarmiento et al. 2015). Pectin esterases can also be used in the manufacture of fruit preparations composed of intact fruit pieces, and a cold-active pectin methylesterase was found to increase gelling and enhance fruit integrity during processing (Pan et al. 2014). Finally, pectinases, in addition to other glycoside hydrolases (EC 3.2.1), lipases, and proteases, can likewise be used for the treatment of food and beverage industry wastes with cold-adapted enzymes allowing for a more effective ambient temperature waste management (Margesin et al. 2005; Naga Padma et al. 2011; Tsuji et al. 2013).

In meat and fish processing, cold-active proteases can be used for tenderization and taste enhancement as well as improving the nutritional and functional properties of refrigerated products (He et al. 2004; Bjarnason and Benediktsson 2010; Venugopal 2016). They can be employed in the preparation of soluble protein hydrolysates for use as flavor enhancers, meat extracts, emulsifiers, and foaming agents and which have also been shown to exert health benefits such as antihypertensive, antioxidant, and immunoregulatory activity (Cazarin et al. 2015). A study by He et al. (2004) showed how a cold-adapted protease released more taste amino acids and essential amino acids from meat than a mesophilic protease during cold storage. Similarly, a marine protease was shown to be effective in the preparation of protein hydrolysates for use as flavor enhancers in foods for human consumption and animal feed (Bjarnason and Benediktsson 2010). Finally, the use of cold-adapted enzymes (proteases, lipases, chitinase etc.) in seafood processing (fish descaling, skin removal and degreasing, waste treatment, oil extraction, etc.) has also been discussed (Shahidi and Janak Kamil 2001; Junpei et al. 2016; Venugopal 2016).

19.5 Molecular Biology

Cold-adapted alkaline phosphatases (Kobori et al. 1984; Sullivan et al. 1988; Rina et al. 2000; Nilsen et al. 2008; Muller-Greven et al. 2012), both single and double stranded nucleases (Awazu et al. 2011) and uracil-DNA *N*-glycosylases (Lanes et al. 2002), are currently being commercialized as molecular biology tools by various companies (New England Biolabs Inc., ArcticZymes, Takara-Clontech, Affymetrix, Inc.). Alkaline phosphatases are most commonly used in the dephosphorylation of the 5′ end of DNA or RNA during cloning and end-labeling procedures. Nucleases, depending on their specificity, degrade single and/or double stranded DNA and/or RNA and are used, e.g., in removing contaminating

DNA/RNA from RNA preparations, PCR master mixes, and protein solutions. Uracil-DNA *N*-glycosylases are used in PCR, RT-PCR, site-directed mutagenesis, and SNP genotyping procedures to release free uracil from uracil-containing DNA (Sarmiento et al. 2015). In all these cases, in addition to a high activity at low temperatures being beneficial, the instability of cold-adapted enzymes is a determining factor in their successful application. This latter characteristic enables for simplified enzyme inactivation by moderate heat treatment as opposed to the time consuming chemical treatments or column purifications required with mesophilic or thermophilic variants which often also lead to sample loss and downstream contamination problems.

Recently, a cold-adapted polymerase has been commercialized by Arcticzymes for use in gene sequencing, molecular diagnostics, and other markets. Furthermore, cold-adapted ligases, recombinases, and proteinase k have been called for (Huston 2008; Sarmiento et al. 2015).

19.6 Biomedicine, Pharmaceuticals, and Cosmetics

Many pharmaceuticals, active pharmaceutical ingredients, fine chemicals, flavors, and fragrances are heat sensitive or/and volatile and hence must be synthesized at low temperatures at which cold-adapted enzymes are most active. In addition, it has been proposed that as a result of their proposed high structural flexibility, cold-adapted enzymes can operate at low water activity, such as in the aqueous/organic and non-aqueous solvent systems frequently used during organic synthesis of complex molecules (Huston 2008; Karan et al. 2012). In this market sector, hydrolases, oxidoreductases, lyases, transferases, reductases, carboxylases, etc. are becoming more commonly used but only a few cold-adapted enzymes have been investigated and below an overview of these is given.

The most widely used cold-adapted enzymes in this sector are lipases and esterases for the synthesis of optically pure intermediate compounds of synthetic value. In fact, lipases (mainly CALB) from *Candida antarctica* are among the most extensively and diversely used enzymes in organic synthesis. They are used in a broad range of surprisingly diverse applications, including the modification of sugars and sugar-related compounds, desymmetrization of complex prochiral drug intermediates, and resolution of racemic alcohols and amines (Huston 2008; Kirk and Christensen 2002; Suen et al. 2004) during the synthesis of various pharmaceuticals (e.g., calcium antagonists as antihypertensive drugs, NK1/NK2 antagonist for asthma treatment), cosmetics (e.g., iso-propyl myristate, a skin emollient), flavors, and fragrance esters.

As discussed above (Sect. 19.4), cold-adapted ß-galactosidases are suited to the production of tagatose (an antihyperglycemic agent) and galactooligosaccharides (prebiotics). Moreover, these have also been shown to catalyze the synthesis of heterooligosaccharides such as lactulose (for treatment of constipation and hepatic encephalopathy, use as a prebiotic, and use in diagnostics), galactosyl-xylose (use in diagnostics), and alkyl glycosides (foaming agents) as well as glycosylated salicin (antiinflammatory agent) from lactose (Pawlak-Szukalska et al. 2014). Also, as discussed previously, cold-adapted proteases can be used for the preparation of bioactive peptides for use as antihypertensive, antioxidant, and immunoregulatory agents (Cazarin et al. 2015).

Cold-adapted proteases are currently being marketed as therapeutic agents against bacterial (biofilm breakdown) and viral (virus infectivity reduction) infections and in oral health care (plaque removal) and cosmetics (frown line reduction and dead or dried skin removal) (Fornbacke and Clarsund 2013).

Other cold-adapted enzymes with potential in biomedical applications include a marine α-galactosidases which was shown to be capable of converting B red blood cells into the universal blood type O cells for use in transfusion therapy (Balabanova et al. 2010) and a cold-active nitroreductase as a cancer prodrug activating enzyme using low temperature therapy for activation (Çelik and Yetiş 2012).

19.7 Other Applications

In the textiles industry, cold-adapted amylases, cellulases, and laccases have been developed for the rapid desizing, or starch removal, of woven fabrics, bio-finishing of cellulosic fabrics, and less abrasive enzymatic stonewashing and bleaching of denim (Sarmiento et al. 2015).

Cold-adapted enzymes, namely, cell wall degrading enzymes, amylases, laccases, lipases, and phospholipases, have been suggested for improving the energy efficiency and costs of biofuel (bioethanol, biodiesel, and biogas) production processes. In particular, they should find application in cold-cook or no-cook processes, simultaneous saccharification and fermentation (Festersen et al. 2005; Huston 2008; Gohel and Duan 2012; Ji et al. 2014; Wen et al. 2015), and low temperature biogas production (Akila and Chandra 2010).

Marine silicatein enzymes have been shown to be central in biomineralization and in the synthesis of biosilicates found in marine diatoms, radiolaria, and sponges (Shimizu et al. 1998; Wang et al. 2012). This has important implications in materials science and indicates the potential of cold-adapted enzymes for the synthesis of a variety of nanostructured mineral/organic composite materials under low temperature and mild chemical conditions. Examples of such materials include silica and siloxane polymers, bimetallic alloy nanoparticles, bimetallic perovskite-like materials, zirconia nanoparticles, spinel gallium oxide, etc. See Huston (2008) for a review.

19.8 Conclusions

The initial development of cold-adapted enzymes for use in industry was somewhat delayed as compared to mesophilic and thermophilic enzymes. Nowadays, however, they have found application in almost all sectors of the industrial enzymes markets. A continued growth in their use is envisaged in the future as novel cold-adapted enzymes with unique properties are isolated from the vast and varied cold environments in the world and as techniques for their isolation (Cavicchioli et al. 2011), efficient production (Cavicchioli et al. 2011), purification, engineering (Liszka et al. 2012; Bommarius and Paye 2013), stabilization, and immobilization (Mateo et al. 2007) are further developed. In addition, the current shift towards a more environmentally friendly and sustainable economy and the development of enzymes for biocatalysis will enable expansion into new application areas and drive growth further.

Acknowledgments T.C. is supported by the Fundação para a Ciência e a Tecnologia (FCT), the European Social Fund, the Programa Operacional Potencial Humano and the Investigador FCT Programme (IF/01635/2014). M.B. acknowledges the FCT for grant PD/BD/113810/2015 within the Doctoral Program in Applied and Environmental Microbiology. This work was funded by the European Regional Development Fund (ERDF) through project EcoAgriFood (NORTE-01-0145-FEDER-000009) via the North Portugal Regional Operational Programme (NORTE 2020) under the PORTUGAL 2020 Partnership Agreement. The FCT is thanked for their funding through EngXyl (EXPL/BBB-BIO/1772/2013-FCOMP-01-0124-FEDER-041595) and the strategic program UID/BIA/04050/2013 (POCI-01-0145-FEDER-007569). All the technical staff at the CBMA are thanked for their skillful technical assistance.

References

Adapa V, Ramya LN, Pulicherla KK, Sambasiva Rao KRS (2014) Cold active pectinases: advancing the food industry to the next generation. Appl Biochem Biotechnol 172 (5):2324–2337. doi:10.1007/s12010-013-0685-1

Akila G, Chandra TS (2010) Stimulation of biomethanation by *Clostridium* sp. PXYL1 in coculture with a *Methanosarcina* strain PMET1 at psychrophilic temperatures. J Appl Microbiol 108(1):204–213. doi:10.1111/j.1365-2672.2009.04412.x

Antranikian G, Breves R, Janßen F, Qoura FM (2004) Pullulanases from psychrophilic organisms. Patent Application WO2006032477A1, DE102004046116A1

Awazu N, Shodai T, Takakura H, Kitagawa M, Mukai H, Kato I (2011) Microorganism-derived psychrophilic endonuclease. Granted Patent US8034597B2

Balabanova LA, Bakunina IY, Nedashkovskaya OI, Makarenkova ID, Zaporozhets TS, Besednova NN, Zvyagintseva TN, Rasskazov VA (2010) Molecular characterization and therapeutic potential of a marine bacterium *Pseudoalteromonas* sp. KMM 701 α-galactosidase. Mar Biotechnol 12(1):111–120. doi:10.1007/s10126-009-9205-2

Barroca M, Santos G, Johansson B, Gillotin F, Feller G, Collins T (2017) Deciphering the factors defining the pH-dependence of a commercial glycoside hydrolase family 8 enzyme. Enzym Microb Technol 96:163–169. doi:10.1016/j.enzmictec.2016.10.011

BCC Research (2017) BCC research report. Global markets for enymes in industrial applications. BIO030J, Jan 2017. BCC Research LLC, USA

Bjarnason JB, Benediktsson B (2010) Protein hydrolysates produced with the use of marine proteases. Granted Patents CA2421058C, DE60007655D1/T2, EP1227736B1, US7070953B1

Blamey JM, Fischer F, Meyer H-P, Sarmiento F, Zinn M (2017) Enzymatic biocatalysis in chemical transformations: a promising and emerging field in green chemistry practice. In: Brahmachari G, Demain AL, Adrio JL (eds) Biotechnology of microbial enzymes, production, biocatalysis and industrial applications. Academic, New York, pp 347–403. doi:10.1016/B978-0-12-803725-6.00014-5

Bommarius AS, Paye MF (2013) Stabilizing biocatalysts. Chem Soc Rev 42:6534–6565. doi:10.1039/c3cs60137d

Cavicchioli R, Charlton T, Ertan H, Mohd Omar S, Siddiqui KS, Williams TJ (2011) Biotechnological uses of enzymes from psychrophiles. Microb Biotechnol 4(4):449–460. doi:10.1111/j.1751-7915.2011.00258.x

Cavicchioli R, Siddiqui KS, Andrews D, Sowers KR (2002) Low-temperature extremophiles and their applications. Curr Opin Biotechnol 13(3):253–261

Cazarin C, Lima G, da Silva J, Maróstica M (2015) Enzymes in meat processing. In: Chandrasekaran M (ed) Enzymes in food and beverage processing. CRC, Boca Raton, pp 337–351. doi:10.1201/b19408-17

Çelik A, Yetiş G (2012) An unusually cold active nitroreductase for prodrug activations. Bioorg Med Chem 20(11):3540-3550. doi:10.1016/j.bmc.2012.04.004

Chandrasekaran M (2015) Enzymes in food and beverage processing. CRC, Boca Raton

Collins T, Claverie P, D'Amico S, Georlette D, Gratia E, Hoyoux A, Meuwis MA, Poncin J, Sonan G, Feller G, Gerday C (2002a) Life in the cold: psychrophilic enzymes. In: Pandalai SG (ed) Recent research developments in proteins, vol 1. Transworld Research Network, Trivandrum, pp 13–26

Collins T, Meuwis MA, Stals I, Claeyssens M, Feller G, Gerday C (2002b) A novel family 8 xylanase, functional and physicochemical characterization. J Biol Chem 277 (38):35133–35139. doi:10.1074/jbc.M204517200

Collins T, Hoyoux A, Dutron A, Georis J, Genot B, Dauvrin T, Arnaut F, Gerday C, Feller G (2006) Use of glycoside hydrolase family 8 xylanases in baking. JCS 43:79–84

Collins T, D'Amico S, Marx J, Feller G, Gerday C (2007) Cold-adapted enzymes. In: Gerday C, Glansdorff N (eds) Physiology and biochemistry of extremophiles. ASM Press, Washington, DC, pp 165–179. doi:10.1128/9781555815813.ch13

Collins T, Feller G, Gerday C, Meuwis MA (2012) Family 8 enzymes with xylanolytic activity. Granted Patent US8309336B2

Collins T, Roulling F, Florence P, Marx JC, Feller G, Gerday C, D'Amico S (2008) Fundamentals of cold-adapted enzymes. In: Margesin R, Schinner F, Marx JC, Gerday C (eds) Psychrophiles: from biodiversity to biotechnology. Springer, Berlin, pp 211–227

D'Amico S, Collins T, Marx JC, Feller G, Gerday C (2006) Psychrophilic microorganisms: challenges for life. EMBO Rep 7(4):385–389

Damhus T, Kaasgaard S, Olsen HS (2013) Enzymes at work, 4th edn. Novozymes A/S, Denmark

Dornez E, Verjans P, Arnaut F, Delcour JA, Courtin CM (2011) Use of psychrophilic xylanases provides insight into the xylanase functionality in bread making. J Agric Food Chem 59 (17):9553–9562. doi:10.1021/jf201752g

Dutron A, Georis J, Genot B, Dauvrin T, Collins T, Hoyoux A, Feller G (2012) Use of family 8 enzymes with xylanolytic activity in baking. Granted Patents US8192772 (2012), EP1549147B1 (2011), CN1681392B (2010), DE60336153 D1 (2011), CA 2498014C (2011), ES2360942 (2011), DE60336153D1 (2011)

Elleuche S, Schroder C, Sahm K, Antranikian G (2014) Extremozymes—biocatalysts with unique properties from extremophilic microorganisms. Curr Opin Biotechnol 29:116–123. doi:10.1016/j.copbio.2014.04.003

Festersen RM, Olsen HS, Pedersen S (2005) Alcohol product processes. Granted Patents DE04718914T1, EP1604019B1, CN1788083B, US8772001B2

Fields PA, Dong Y, Meng X, Somero GN (2015) Adaptations of protein structure and function to temperature: there is more than one way to 'skin a cat'. J Exp Biol 218(12):1801–1811. doi:10.1242/jeb.114298

Fornbacke M, Clarsund M (2013) Cold-adapted proteases as an emerging class of therapeutics. Infect Dis Ther 2(1):15–26. doi:10.1007/s40121-013-0002-x

Gerday C (2013) Psychrophily and catalysis. Biology 2(2):719–741. doi:10.3390/biology2020719

Gerday C (2014) Fundamentals of cold-active enzymes. In: Buzzini P, Margesin R (eds) Cold-adapted yeasts: biodiversity, adaptation strategies and biotechnological significance. Springer, Berlin, pp 325–350. doi:10.1007/978-3-642-39681-6_15

Ghosh M, Pulicherla KK, Rekha VP, Raja PK, Sambasiva Rao KR (2012) Cold active beta-galactosidase from *Thalassospira* sp. 3SC-21 to use in milk lactose hydrolysis: a novel source from deep waters of Bay-of-Bengal. World J Microbiol Biotechnol 28(9):2859–2869. doi:10.1007/s11274-012-1097-z

Gohel V, Duan G (2012) No-cook process for ethanol production using indian broken rice and pearl millet. Int J Microbiol 2012:680232. doi:10.1155/2012/680232

He H, Chen X, Li J, Zhang Y, Gao P (2004) Taste improvement of refrigerated meat treated with cold-adapted protease. Food Chem 84(2):307–311. doi:10.1016/S0308-8146(03)00242-5

Hoyoux A, Jennes I, Dubois P, Genicot S, Dubail F, Francois JM, Baise E, Feller G, Gerday C (2001) Cold-adapted beta-galactosidase from the Antarctic psychrophile *Pseudoalteromonas haloplanktis*. Appl Environ Microbiol 67(4):1529–1535

Huston AL (2008) Biotechnological aspects of cold-adapted enzymes. In: Margesin R, Schinner F, Marx JC, Gerday C (eds) Psychrophiles: from biodiversity to biotechnology. Springer, Berlin, pp 347–364

Ji L, Yang J, Fan H, Yang Y, Li B, Yu X, Zhu N, Yuan H (2014) Synergy of crude enzyme cocktail from cold-adapted *Cladosporium cladosporioides* Ch2-2 with commercial xylanase achieving high sugars yield at low cost. Biotechnol Biofuels 7(1):130. doi:10.1186/s13068-014-0130-x

Junpei Z, Rui Z, Zunxi H, Zhifeng S, Xianghua T, Junjun L, Qian W (2016) Low-temperature salt-tolerant product-inhibition-resistant beta-N-acetyl glucosamine enzyme JB10NagA. Patent Application CN105483101A

Karan R, Capes MD, Dassarma S (2012) Function and biotechnology of extremophilic enzymes in low water activity. Aquat Biol 8(1):4. doi:10.1186/2046-9063-8-4

Karasova-Lipovova P, Strnad H, Spiwok V, Mala S, Kralova B, Russell NJ (2003) The cloning, purification and characterisation of a cold-active [beta]-galactosidase from the psychrotolerant Antarctic bacterium *Arthrobacter* sp. C2-2. Enzym Microb Technol 33(6):836–844

Kirk O, Christensen MW (2002) Lipases from *Candida antarctica*: unique biocatalysts from a unique origin. Org Process Res Dev 6(4):446–451. doi:10.1021/op0200165

Kobori H, Sullivan CW, Shizuya H (1984) Heat-labile alkaline phosphatase from Antarctic bacteria: rapid 5' end labelling of nucleic acids. Proc Natl Acad Sci USA 81:6691–6695

Lanes O, Leiros I, Smalas AO, Willassen NP (2002) Identification, cloning, and expression of uracil-DNA glycosylase from Atlantic cod (*Gadus morhua*): characterization and homology modeling of the cold-active catalytic domain. Extremophiles 6(1):73–86

Liszka MJ, Clark ME, Schneider E, Clark DS (2012) Nature versus nurture: developing enzymes that function under extreme conditions. Annu Rev Chem Biomol Eng 3:77–102. doi:10.1146/annurev-chembioeng-061010-114239

Littlechild JA (2015) Enzymes from extreme environments and their industrial applications. Front Bioeng Biotechnol 3:161. doi:10.3389/fbioe.2015.00161

Margesin R, Fauster V, Fonteyne PA (2005) Characterization of cold-active pectate lyases from psychrophilic *Mrakia frigida*. Lett Appl Microbiol 40(6):453–459

Margesin R, Feller G, Gerday C, Russell NJ (2003) Cold-adapted microorganisms: adaptation strategies and biotechnological potential. In: Britton G (ed) Encyclopedia of environmental microbiology. Wiley, New York. doi:10.1002/0471263397.env150

Mateo C, Palomo JM, Fernandez-Lorente G, Guisan JM, Fernandez-Lafuente R (2007) Improvement of enzyme activity, stability and selectivity via immobilization techniques. Enzyme Microb Technol 40(6):1451–1463. doi:10.1016/j.enzmictec.2007.01.018

Muller-Greven JC, Post MA, Kubu CJ (2012) Recombinant *Colwellia psychrerythraea* alkaline phosphatase and uses thereof. Granted Patents US8486665B2, US8129168B2

Naga Padma P, Anuradha K, Reddy G (2011) Pectinolytic yeast isolates for cold-active polygalacturonase production. Innov Food Sci Emerg Technol. 12(2):178–181. doi:10.1016/j.ifset.2011.02.001

Nakagawa T, Ikehata R, Myoda T, Miyaji T, Tomizuka N (2007) Overexpression and functional analysis of cold-active β-galactosidase from *Arthrobacter psychrolactophilus* strain F2. Protein Expr Purif 54(2):295–299. doi:10.1016/j.pep.2007.03.010

Nilsen I, Overbö K, Lanes O (2008) Shrimp alkaline phosphatase. Granted Patents US 7323325B2, DE60130567D1, DE60130567T2, EP1326890B1

Pan X, Tu T, Wang L, Luo H, Ma R, Shi P, Meng K, Yao B (2014) A novel low-temperature-active pectin methylesterase from *Penicillium chrysogenum* F46 with high efficiency in fruit firming. Food Chem 162:229–234. doi:10.1016/j.foodchem.2014.04.069

Pawar R, Vasudeo Z, Siddhivinayak B, Govind P (2009) Application of protease isolated from Bacillus sp. 158 in enzymatic cleansing of contact lenses. Biotechnology 8(2):276–280. doi:10.3923/biotech.2009.276.280

Pawlak-Szukalska A, Wanarska M, Popinigis AT, Kur J (2014) A novel cold-active β-d-galactosidase with transglycosylation activity from the Antarctic *Arthrobacter* sp. 32cB—Gene cloning, purification and characterization. Process Biochem 49(12):2122–2133. doi:10.1016/j.procbio.2014.09.018

Rina M, Pozidis C, Mavromatis K, Tzanodaskalaki M, Kokkinidis M, Bouriotis V (2000) Alkaline phosphatase from the Antarctic strain TAB5. Properties and psychrophilic adaptations. Eur J Biochem 267(4):1230–1238

Santiago M, Ramírez-Sarmiento CA, Zamora RA, Parra LP (2016) Discovery, molecular mechanisms, and industrial applications of cold-active enzymes. Front Microbiol 7:1408. doi:10.3389/fmicb.2016.01408

Sarmiento F, Peralta R, Blamey JM (2015) Cold and hot extremozymes: industrial relevance and current trends. Front Bioeng Biotechnol 3:148. doi:10.3389/fbioe.2015.00148

Schmidt M, Stougaard P (2010) Identification, cloning and expression of a cold-active beta-galactosidase from a novel Arctic bacterium, *Alkalilactibacillus ikkense*. Environ Technol 31 (10):1107–1114. doi:10.1080/09593331003677872

Shahidi F, Janak Kamil YVA (2001) Enzymes from fish and aquatic invertebrates and their application in the food industry. Trends Food Sci Tech 12(12):435–464. doi:10.1016/S0924-2244(02)00021-3

Shimizu K, Cha J, Stucky GD, Morse DE (1998) Silicatein α: cathepsin L-like protein in sponge biosilica. Proc Natl Acad Sci USA 95(11):6234–6238

Siddiqui KS (2015) Some like it hot, some like it cold: Temperature dependent biotechnological applications and improvements in extremophilic enzymes. Biotechnol Adv 33(8):1912–1922. doi:10.1016/j.biotechadv.2015.11.001

Stougaard P, Schmidt M (2012) Cold-active beta-galactosidase, a method of producing same and use of such enzyme. Granted Patents US8288143, EP2396403B1, EP2396403B8, CN102361974B

Suen W-C, Zhang N, Xiao L, Madison V, Zaks A (2004) Improved activity and thermostability of *Candida antarctica* lipase B by DNA family shuffling. Protein Eng Des Sel 17(2):133–140. doi:10.1093/protein/gzh017

Sullivan CW, Shizuya H, Kobori H (1988) Heat sensitive bacterial alkaline phosphatase. Granted Patent US4720458

The Freedonia Group (2016) World enzymes industry study with forecasts for 2020 & 2025, Study #3417. https://www.freedoniagroup.com/brochure/34xx/3417smwe.pdf

Tsuji M, Yokota Y, Shimohara K, Kudoh S, Hoshino T (2013) An application of wastewater treatment in a cold environment and stable lipase production of Antarctic basidiomycetous yeast *Mrakia blollopis*. PLoS One 8(3):e59376. doi:10.1371/journal.pone.0059376

Tu T, Meng K, Bai Y, Shi P, Luo H, Wang Y, Yang P, Zhang Y, Zhang W, Yao B (2013) High-yield production of a low-temperature-active polygalacturonase for papaya juice clarification. Food Chem 141(3):2974–2981. doi:10.1016/j.foodchem.2013.05.132

Valentini F, Diamanti A, Palleschi G (2010) New bio-cleaning strategies on porous building materials affected by biodeterioration event. Appl Surf Sci 256(22):6550–6563. doi:10.1016/j.apsusc.2010.04.046

Van de Voorde I, Goiris K, Syryn E, Van den Bussche C, Aerts G (2014) Evaluation of the cold-active *Pseudoalteromonas haloplanktis* β-galactosidase enzyme for lactose hydrolysis in whey permeate as primary step of D-tagatose production. Process Biochem 49(12):2134–2140. doi:10.1016/j.procbio.2014.09.010

Venugopal V (2016) Enzymes from seafood processing waste and their applications in seafood processing. In: Se-Kwon K, Fidel T (eds) Advances in food and nutrition research, vol 78. Academic, New York, pp 47–69. doi:10.1016/bs.afnr.2016.06.004

Wang X, Schloßmacher U, Wiens M, Batel R, Schröder HC, Müller WEG (2012) Silicateins, silicatein interactors and cellular interplay in sponge skeletogenesis: formation of glass fiber-like spicules. FEBS J 279(10):1721–1736. doi:10.1111/j.1742-4658.2012.08533.x

Webster A, May E (2006) Bioremediation of weathered-building stone surfaces. Trends Biotechnol 24(6):255–260. doi:10.1016/j.tibtech.2006.04.005

Wen J, Ren C, Huang N, Liu Y, Zeng R (2015) Draft genome of bagasse-degrading bacteria *Bacillus aryabhattai* GZ03 from deep sea water. Mar Genomics 19:13–14. doi:10.1016/j.margen.2014.11.004

Chapter 20
Biotechnological Improvements of Cold-Adapted Enzymes: Commercialization via an Integrated Approach

Khawar Sohail Siddiqui, Ahsan Mushir Shemsi, Gea Guerriero, Tahria Najnin, Taha, and Haluk Ertan

Abstract Cold-adapted (psychrophilic) enzymes have intrinsically high activities at the expense of low stabilities due to their flexible structures. Their higher thermolability limits their applications under numerous industrial conditions that require the process to be carried out at higher temperatures for efficient catalysis. Therefore, for effective utilization, cold-adapted enzymes need to be improved in such a way that enhances their stability with an increase or retention of their activity. This chapter discusses the thermodynamic aspects of improvement of catalytic properties and presents a unified strategy that aims at simultaneously improving the activity and stability of cold-adapted enzymes by employing not a single but a combination of approaches that include genetic and chemical

K.S. Siddiqui (✉)
Life Sciences Department, King Fahd University of Petroleum and Minerals (KFUPM), Dhahran Blvd, Dhahran 31261, Saudi Arabia
e-mail: ksiddiqui@kfupm.edu.sa; sohailsiddiqui1995@yahoo.com

A.M. Shemsi
Center for Environment and Water, Research Institute, King Fahd University of Petroleum and Minerals (KFUPM), Dhahran Blvd, Dhahran 31261, Saudi Arabia

G. Guerriero
Environmental Research and Innovation (ERIN), Luxembourg Institute of Science and Technology (LIST), 5 Avenue des Hauts-Fourneaux, Esch/Alzette 4362, Luxembourg

T. Najnin
Institute for Glycomics, Building G26, Griffith University (Gold Coast Campus), Parklands Drive, Southport, QLD 4215, Australia

Taha
School of Biotechnology and Biomolecular Sciences, The University of New South Wales, Sydney, NSW 2052, Australia

H. Ertan
School of Biotechnology and Biomolecular Sciences, The University of New South Wales, Sydney, NSW 2052, Australia

Department of Molecular Biology and Genetics, Istanbul University, 34134 Vezneciler, Istanbul, Turkey

R. Margesin (ed.), *Psychrophiles: From Biodiversity to Biotechnology*,
DOI 10.1007/978-3-319-57057-0_20

modifications, immobilization, nonaqueous solvents, and additives. This concept aims to take cold-adapted enzymes a step further from current potential to cost-effective tangible commercial applications.

Contents

20.1 Introduction

Life on Earth and probably beyond is constrained by temperature, and the majority of our biosphere is permanently cold, reaching as far low as −89 °C in Antarctica (Siddiqui et al. 2013), where complex life (amphipods and fishes) has recently been found in a subglacial lake around −2 °C (Fox 2015). Cold-adapted (psychrophilic) organisms have been found living at these low temperatures due to numerous adaptations at the molecular level that include catalytic proteins (enzymes). In fact life is made possible by enzymes, which dramatically increase the rate of reaction compared to uncatalyzed reactions due to a decrease in their activation energy (Wolfenden and Snider 2001). Generally, cold-adapted enzymes further decrease the activation energy as compared to the mesophilic and thermophilic homologues, with consequent increase in the activity (Siddiqui and Cavicchioli 2006; Feller 2008, 2013; Santiago et al. 2016).

Cold-adapted enzymes have a huge biotechnological potential due to their high intrinsic activity at low temperatures and thermolability. This is especially relevant in pharmaceutical and food-based industries where rapid inactivation is required after completion of the reaction (Cavicchioli et al. 2011; Siddiqui 2015). However, under high temperature conditions, this potential has yet to be commercially realized due to activity-stability trade-off (Siddiqui 2017), implying that an increase in activity is associated with a decrease in the stability, which limits the application of cold-adapted enzymes. It is therefore inescapable that, in order for cold-adapted enzymes to be commercially exploited, their catalytic properties (such as activity, stability, inhibition) have to be improved to suit various reaction conditions (such as pH, temperature, solvents, additives) normally encountered under industrial applications (Siddiqui 2015, 2017; Guerriero et al. 2016).

In this chapter we critically discuss the thermodynamic basis for the improvements in the activity and stability of cold-adapted enzymes employing genetic engineering (computational design, site-directed mutagenesis, directed evolution, truncation and fusion, etc.) and chemical modifications, additives/formulations, and immobilization on non-magnetic materials and magnetic nanoparticles in both aqueous and nonaqueous media. In addition, improvements in cold-adapted enzymes related to the production of biofuel from lignocellulosic biomass and waste oil are emphasized. The chapter will conclude by highlighting the future directions with a view to converting potential applications into real commercial processes and/or patents.

20.2 Thermodynamic Aspects of Improvement in Activity and Stability

There is a free energy ($\Delta G^{\#}$) barrier between the ground-state substrate or the folded protein and the transition state ($TS^{\#}$) with an equilibrium constant (K). This barrier must be overcome by supplying a minimum amount of activation energy (E_a) before the $TS^{\#}$ can proceed to the formation of products with a rate constant (k_{cat}) in case of reaction, or into the denatured form in case of protein unfolding with a rate constant, k_{inact} (Eq. 20.1).

$$\text{Reactants or folded protein} \overset{K}{\rightleftharpoons} TS^{\#} \begin{cases} \xrightarrow{k_{cat}} \text{Reaction product} \\ \xrightarrow{k_{inact}} \text{Denatured protein} \end{cases} \tag{20.1}$$

Enzymes increase the rate of reaction (k_{cat}) by decreasing $\Delta G^{\#}$, whereas the higher thermolability (increased k_{inact}) of the cold-adapted protein is also attained with a decrease in $\Delta G^{\#}$ (Siddiqui 2017). Free energy ($\Delta G^{\#}$) is composed of two components, enthalpy ($\Delta H^{\#}$) and entropy ($\Delta S^{\#}$), that are related to each other (Eq. 20.2).

$$\Delta G^{\#} = \Delta H^{\#} - T\Delta S^{\#} \tag{20.2}$$

It follows from Eq. (20.2) that in order to decrease $\Delta G^{\#}$ (k_{cat} to increase), either $\Delta H^{\#}$ has to decrease or $\Delta S^{\#}$ has to increase (more disordered $TS^{\#}$). Studies involving thermally adapted sets of homologous enzymes have revealed that cold-adapted enzymes generally increase their activity (V_{max} or k_{cat}) by decreasing their $\Delta H^{\#}$ that is structurally attained by reducing the number and magnitude of enthalpy-related interactions within the active site that are needed to be broken during the formation of transition state from the ground-state substrate (Lonhienne et al. 2000; Siddiqui and Cavicchioli 2006; Siddiqui 2017; Santiago et al. 2016).

However, the enhancement in activity as a result of increased $\Delta S^{\#}$ has not been explored in detail, probably because the majority of cold-adapted enzymes have been found to attain higher activity as a result of reduced $\Delta H^{\#}$. Theoretical calculations suggest that in the absence of enthalpy-entropy compensation, a change of only 20 kJ mol^{-1} in $\Delta H^{\#}$ can result in an enormous (~3000-fold) increase in activity, provided $\Delta S^{\#}$ does not decrease concurrently (Lonhienne et al. 2000). Even further increase in activity can be attained if the decrease in $\Delta H^{\#}$ is accompanied by an increase in $\Delta S^{\#}$, but in reality this does not happen because a decrease in $\Delta H^{\#}$ is accompanied by an equivalent decrease in $\Delta S^{\#}$ (Siddiqui 2017). However, in the soil fungus *Scopulariopsis* sp., which is not a cold-adapted organism, the less acidic xylanase (LAX) isoenzyme showed ~tenfold higher activity than the more acidic (HAX) isoenzyme. Interestingly, the increase in activity was not due to reduced $\Delta H^{\#}$ (LAX, 66; HAX, 20 kJ mol^{-1}), but due to an increase in $\Delta S^{\#}$ (LAX, 28; HAX, −138 J mol^{-1} K^{-1}) (Afzal et al. 2005).

The structural origin of activity at low temperatures (near freezing) in cold-adapted enzymes still remains debatable; is it caused by global or local (active-site) flexibility (Fields and Somero 1998; Siddiqui and Cavicchioli 2006; Isaksen et al. 2016)? Using computer simulations, it has been recently proposed that in thermally adapted trypsins, reduction in $\Delta H^{\#}$ and negative $\Delta S^{\#}$ (entropy-enthalpy compensation) are correlated with higher surface mobility rather than active-site flexibility (Isaksen et al. 2016). Recently, active-site mutants (D308A and Y309A) of cold-adapted esterase from *Pseudomonas mandelii* have revealed the trade-off between the activity and the thermal stability of the catalytic pocket (Truongvan et al. 2016). This finding is consistent with the activation thermodynamics of hydrolytic activity, which showed that an increase in the activity of genetic variants was entropically driven (increase in $\Delta S^{\#}$) compared to the wild-type enzyme (Table 20.1). The increased disorder of TS$^{\#}$ has been interpreted in view of the enhanced local flexibility of the active sites of variants, consistent with higher Stern-Volmer constants determined by dynamic fluorescence quenching (DFQ) using acrylamide as a quencher (Truongvan et al. 2016). Similarly, deletion of the linker region (LR) between the catalytic (CM) and carbohydrate-binding (CBM) modules and of both LR and CBM in cold-adapted cellulase resulted in decreased activity and increased stability (Table 20.1). Thermodynamic activation parameters showed the lower activity in both deletion mutants to be enthalpy-driven (less ordered TS$^{\#}$), and DFQ indicated lower flexibility implying a role for the linker in cold adaptation (Sonan et al. 2007).

An attractive approach to further enhance the activity of cold enzymes is to increase $\Delta S^{\#}$ by displacing more water molecules on the formation of the TS$^{\#}$ from the ground-state substrate molecule (Wolfenden and Snider 2001; Siddiqui et al. 2010; Siddiqui 2017). An increase in ~8 kJ mol^{-1} in $\Delta S^{\#}$ has been estimated upon removal of one structured water molecule from a hydrophobic cavity of a protein molecule to the bulk water (Dunitz 1994; Kotting et al. 2008); however, exclusion of a water molecule coupled to polar residues can be thermodynamically unfavorable (enthalpic penalty) due to breaking of H-bonds (Huggins 2015). Although enhancement in the activity of a cold-adapted enzyme due to an increase in $\Delta S^{\#}$ as a result of water displacement has not yet been experimentally shown, a pioneering study was carried out on GTPase-activating protein (GAP). Structural simulations showed that the Arg finger moves into the active-site cleft upon TS$^{\#}$ formation that displaces five structured water molecules into the bulk solvent resulting in positive

Table 20.1 Changes in thermodynamic parameters of cold-adapted enzymes after genetic or chemical modification

Enzyme/organism	Modification/additive	Activity	Activity		Stability	Stability		Reference
		k_{cat} (min^{-1}) V_{max}	$\Delta H^{\#}$	$T\Delta S^{\#}$	$t_{1/2}$ (min), T_{opt} (°C), T_m (°C)	$\Delta H^{\#}$	$T\Delta S^{\#}$	
α-Amylase (AHA) *Pseudoalteromonas haloplanktis*	Native	17,640 (10 °C)	35	−81	$t_{1/2}$: 0.23 (43 °C)[a]	721	635	D'Amico et al. (2003)
		41,820 (25 °C)	–	–	T_{opt}: 28, T_m: 44[F], 61[CI]	–	–	D'Amico et al. (2003)
AHA-Mutant (Q/C, A/C)	SDM	20,400 (25 °C)	23	−34	$t_{1/2}$: 2.5 (45 °C)	nd		Siddiqui et al. (2006)
AHA	Unmodified	50,400 (3 °C)	26	−93	$[\text{Urea}]_{1/2}$: 1.6 M	ΔG, 8.7		
AHA (K106/hR)	Chemically modified	7200 (3 °C)	32	−94	$[\text{Urea}]_{1/2}$: 2.5 M	ΔG, 9.5		
Alkaline phosphatase *Pandalus borealis*		*p*NPP (37 °C):			50 °C:			Siddiqui et al. (2004)
	Unmodified	48,740	35	−10	$t_{1/2}$: 12	210	348	
	Chemically modified	97,363	28	−25	$t_{1/2}$: 58	210	334	
Metalloprotease *Pseudomonas* sp.	± Ca^{2+}	AC (25 °C):			48 °C:			Ertan et al. (2015)
	Ca^{2+}-depleted apoenzyme	2.2	49.0	−0.11	$t_{1/2}$: 1.2	191	176	
	5 mM Ca^{2+}-treated enzyme	3.3	47.5	−0.11	$t_{1/2}$: 7.5	310	249	
Subtilisin protease Psychrophilic S41		Casein (10 °C):			60 °C:	nd		Miyazaki et al. (2000)
	Directed evolution (error-prone PCR) WT	1038	27	−125	$t_{1/2}$: 1, T_{opt}: 55			
	S145I/S175 T/K211P/R212A/K221E/N291I/S295 T	3090	25	−123	$t_{1/2}$: 450, T_{opt}: 70			
Laccase *Pseudomonas putida* (Himalayan glacier)	WT	11 U mg^{-1} (4 °C)	25	nd	$t_{1/2}$: 7 (4 °C)	−107	−685	Mukhopadhyay et al. (2015)
		19 U mg^{-1} (80 °C)	36	nd	$t_{1/2}$: 7 (80 °C)	−54	−453	

(continued)

Table 20.1 (continued)

Enzyme/organism	Modification/additive	Activity k_{cat} (min^{-1}) V_{max}	Activity $\Delta H^{\#}$	Activity $T\Delta S^{\#}$	Stability $t_{1/2}$ (min), T_{opt} (°C), T_m (°C)	Stability $\Delta H^{\#}$	Stability $T\Delta S^{\#}$	Reference
	Cu_2O-NP trapped in SWCNT	55 U mg^{-1} (4 °C)	1	nd	$t_{1/2}$: 73 (4 °C)	88	1	
		64 U mg^{-1} (80 °C)	3	nd	$t_{1/2}$: 436 (80 °C)	72	−128	
Cellulase (Cel5G) *P. haloplanktis*		CMC (10 °C)						Sonan et al. (2007)
	WT	720	10.6	−52.7	T_m, 45.3, 47.9	264^{cal}, 220^{cal}	nd	
	-37 AA linker	540	12.6	−51.4	T_m, 45.6, 48.8	269^{cal}, 229^{cal}	nd	
	-Linker, -CBM	360	17.6	−47.4	T_m, 50.2, 53.8	278^{cal}, 295^{cal}	nd	
Esterase *Pseudomonas mandelii*	WT	0.5 ± 0.05	20	−43.3	T_m, 50.3; T_{opt}: 57	nd		Truongvan et al. (2016)
	W208Y	1.63 ± 0.08	22	−38.4	T_m, 44.7; T_{opt}: 52.4	nd		
	D308A	1.1 ± 0.001	28.5	−33.1	T_m, 42.5; T_{opt}: 50.0	nd		
	Y309A	1.05 ± 0.02	25.6	−36.1	T_m, 44.4; T_{opt}: 50.1	nd		
Protease (Deseasin MCP-01) *Pseudoalteromonas* sp. SM9913	No TMAO	26,000 (0 °C)	18.9	−124	$t_{1/2}$: 57 (40 °C)	104	83	He et al. (2009)
	+ 1 M TMAO	50,000 (0 °C)	19.6	−116	$t_{1/2}$: 330 (40 °C)	148	174	

WT wild-type, *nd* not determined, $t_{1/2}$ half-life of inactivation, T_{opt} optimal temperature of activity, T_m melting temperature, $\Delta H^{\#}$ (kJ mol^{-1}), $\Delta S^{\#}$ (J mol^{-1} K^{-1}), *cal* calorimetric, *F* free enzyme, *CI* in the presence of competitive inhibitor, *hR* homoarginine, $Urea_{1/2}$ and ΔG (conformational stability) using transverse urea gradient-PAGE (TUG-PAGE), *CBM* cellulose-binding module, *p*NPP *p*-nitrophenylphosphate, *AC* azocasein, *CMC* carboxymethylcellulose, *AA* amino acid, *NP* nanoparticle, *SWCNT* single-wall carbon nanotube, *TMAO* trimethylamine *N*-oxide, *U* unit, amount of enzyme that can convert 1 μmol substrate per min

[a]Data for irreversibly unfolded N12R mutant were included in lieu of data for reversibly unfolded wild type

$T\Delta S^{\#}$ (39 kJ mol^{-1}) that increased the reaction rate by 10^5 times (Kotting et al. 2008). Readers are referred to a recent review (Siddiqui 2017) for a detailed account of the role of entropy in enhancing the activity of enzymes.

In contrast, to increase $\Delta G^{\#}$ and hence the stability, either proteins increase $\Delta H^{\#}$ (more electrostatic interactions in the native protein) or decrease $\Delta S^{\#}$ (more ordered $TS^{\#}$ state). In chemically modified α-amylases (Siddiqui et al. 2010) and DNA ligases, the higher thermostability has been attained by highly ordered $TS^{\#}$ state relative to the folded state (reduced $\Delta S^{\#}$), whereas in mutants of a neutral protease (Asghari et al. 2010), higher stability was achieved by increasing the number of interactions needed to be broken to reach $TS^{\#}$ (increased $\Delta H^{\#}$). The majority of biotechnologically important enzymes are large multi-domain proteins that unfold irreversibly, although exceptions do exist such as in the case of multi-domain cold-adapted α-amylase from *Pseudoalteromonas haloplanktis* that unfolds reversibly (Siddiqui et al. 2005; Feller 2008, 2013; Siddiqui and Cavicchioli 2006).

20.3 Improvements in Properties by Genetic Modifications

Modifications involving changes in gene sequence have frequently been used to enhance the stability of the cold-adapted enzymes with the retention or increase in the activity. Conversely, the low-temperature activity of mesophilic and thermostable enzymes has also been enhanced with the retention of stability (Siddiqui 2015; Table 20.2). Generally, genetic modifications (GM) can be subdivided into site-directed mutagenesis (SDM), directed evolution (DE), iterative site-saturation mutagenesis (ISM), consensus-guided mutagenesis (CGM), truncation and fusion [Cavicchioli et al. (2006), Yang et al. (2014), Siddiqui (2015) and references therein]. To date, SDM and DE alone or in combination are the most commonly used methods for the improvement of activity and/or stability of thermally adapted enzymes (Table 20.2). Whereas rational design using SDM requires precise structural information that is not frequently available (Wang et al. 2014), DE has the advantage that it involves random mutagenesis via error-prone PCR (ep-PCR) followed by selection of the desired mutant (Wintrode et al. 2001; Zhang et al. 2003; Koutsioulis et al. 2008; Struvay and Feller 2012). However, DE has drawbacks that include low efficiency (improved mutants are rare) and a necessity to screen a huge number of mutants. Recently, GM methods based on computational analysis and design are receiving attention as these are economical, reduce the number of libraries to be screened, and, more importantly, significantly boost the chance of finding an improved (highly active and/or thermostable) mutant (Bae et al. 2008; Wijma et al. 2013; Siddiqui 2015). Although numerous GMs result in activity-stability trade-off such as the esterase from *P. mandelii* (Table 20.1), Cm-IDH (isocitrate dehydrogenases), and thermophilic subtilase (Table 20.2), DE has shown extreme improvements in both activity and stability (Table 20.2) in case

Table 20.2 Improvement in activity and/or stability of thermally adapted enzymes after genetic modification

Enzyme/organism	Modification	Activity k_{cat} (min^{-1})	Stability $t_{1/2}$ (min), T_{opt} (°C), T_m (°C)	Reference
Lignin peroxidase[P/M] *Phanerochaete chrysosporium*	Ancestral amino acid substitution:	25 °C	37 °C, 30 min:	Semba et al. (2015)
	WT	696 (VA), 966 (H_2O_2)	Lost 83% activity, T_{opt}:30 °C	
	H239F/T240 L/I241L	2382 (VA), 2052 (H_2O_2)	Lost 14% activity, T_{opt}:40 °C	
Cocaine esterase[P] *Rhodococcus* sp.	Computational design +SDM:	(37 °C) Cocaine		Narasimhan et al. (2010) Fang et al. (2014)
	L169 K	3084→4800	$t_{1/2}$ (37 °C):12 min → 570 min	
	T172R/G173Q	3084→3204	$t_{1/2}$ (37 °C):12 min → 370 min	
	T172R/G173Q/L196C/I301C (2 inter-subunit S-S bridges)	3084→3450	$t_{1/2}$ (37 °C):12 min → >10 d	
			T_m: 3 → 41 °C	
Endo-1,5-α-L-arabinanase[P] *Paenibacillus polymyxa* Z6	SDM:	(40 °C) Arabinan	T_{opt}: 40 → 40 °C	Wang et al. (2014)
	WT	4122 ±498	Retained 50% activity (20 °C, 4 h)	
	H218D	5982 ± 360	Retained 60% activity (20 °C, 4 h)	
Monomeric isocitrate dehydrogenases (IDH)[P] *Colwellia maris* (*Cm*)	SDM:	(20 °C) Isocitrate	25 °C: Mutations in the *Cm*IDH led to decrease in structural stability	Kobayashi and Takada (2014)
	Cm L693F/L724Q	5520 → 960		
	Cm L693F/F735 L	5520 → 3522		
	Cm L724Q/F735 L	5520 → 768		
	Cm L693F/L724Q/F735 L	5520 → 786	Retained activity (25 °C, 10 min):	
Colwellia psychrerythraea (*Cp*)	*Cp* F693 L/Q724L	252 → 3000	6 → >40%, T_{opt}: 25→25 °C	
	Cp F693 L/L735F	252 → 3066	6 → >40%, T_{opt}: 25→30 °C	
	Cp Q724L/L735F	252 → 4032	6 → >40%, T_{opt}: 25→35 °C	
	Cp F693 L/Q724L/L735F	252 → 3084	6 → 100%, T_{opt}: 25→35 °C	

Alkaline phosphatase[P] Antarctic strain TAB5	DE:	(15 °C) *p*-NPP	55 °C, 5 min:		Koutsioulis et al. (2008)
	WT	14,160	Retained 15% activity, T_m: 45 °C		
	H135E	18,180	Retained 5.8% activity, T_m: 50 °C		
	H135E/G149D	21,660	Retained 24.5% activity, T_m: 51 °C		
	S86A	12,180	Retained 69.2% activity, T_m: 54 °C		
Subtilisin-like serine proteinase[P] Psychrotrophic *Vibrio* species	SDM:	(25 °C) sAAPF-NA	$T_{50\%}$ (°C)	T_m (°C)	Sigurdardóttir et al. (2009)
	WT	4476 ± 300	56.0 ± 0.5	63.6 ± 0.3	
	N15D	8220 ± 1020	58.8 ± 0.5	66.3 ± 0.1	
	N15D/K257R	6360 ± 516	56.2 ± 0.8	65.1 ± 0.4	
	VPRΔC	4092 ± 1020	56.4 ± 0.1	65.1 ± 0.2	
	N15DΔC	3894 ± 240	59.6 ± 0.2	68.2 ± 0.2	
Subtilisin protease[T] W146F	DE + SDM:	(25 °C)			Zhong et al. (2009)
	WT	3060 ± 300 (sAAPF-NA) 100% (casein)	T_{opt}:85 °C, $t_{1/2}$: (80 °C) 60 min		
	T7I	4200 ± 180 (sAAPF-NA) 85% (casein)	T_{opt}: 80 °C, $t_{1/2}$: (80 °C) 15 min		
	D114G/M137T/T153A/S246N	306 ± 60 (sAAPF-NA) 600% (casein)	T_{opt}: 80 °C, $t_{1/2}$: (80 °C) 4 min		

(continued)

Table 20.2 (continued)

Enzyme/organism	Modification	Activity k_{cat} (min^{-1})	Stability $t_{1/2}$ (min), T_{opt} (°C), T_m (°C)		Reference
Subtilisin S41 protease[P]	DE, ep-PCR + StEP recombination:	sAAPF-NA	$t_{1/2}$: (min, 60 °C; 1 mM $CaCl_2$)		Wintrode et al. (2001)
	WT	1380 (10 °C)	–		
		15,840 (60 °C)	0.9		
	S145I/S175T/K211P/R212A/ K221E/N291I/S295T	2880 (10 °C)	–		
		33,180 (60 °C)	449		
	N16D/Q69H/S78T/S109T/S145I/S175T/ K211P/R212A/D216E/K221E/S252A/ N291I/S295T	2460 (10 °C)	–		
		38,700 (60 °C)	1082		
Lipase[P] *Candida antarctica* (CALB)	DE:	(25 °C) pNPB	$t_{1/2}$ (70 °C)	T_m (CD)	Zhang et al. (2003)
	WT	730	8	57.7	
	V210I/A281E	2900	211	52.2	
	V210I/A281E/V221D	2500	232	50.8	
Lipase[M] *Bacillus*	DE (ep-PCR):	(37 °C) p-NPL			Goomber et al. (2016)
	WT		0 % activity (5 °C) 20% activity (10 °C) 80–90% activity (40 °C) $t_{1/2}$ (30 °C): 180–210 min		
	F19L	1.3 fold increase in k_{cat}	50% activity (5 °C) 100% activity (10 °C) 0% activity (40 °C) $t_{1/2}$ (30 °C): 45–60 min		

Xylose isomerase[T] *Thermoanaerobacterium saccharolyticum*	SDM:	(80 °C) glucose		Xu et al. (2014)
	W139F	881 → 2979	T_{opt}: 90 → 90 °C, $t_{1/2}$: (80 °C) 12 → 18 h	
Xylanase[T] *Paenibacillus campinasensis*	DE (ep-PCR) + SDM:	(70 °C) Birchwood Xylan		Zheng et al. (2014)
	V90R,/P172H/T84C/T182C/D16Y	71,790 → 109,440	T_{opt}: 60→70 °C, lost 100% activity →lost 75% activity at 90 °C (120 min)	
3-Isopropylmalate dehydrogenase (IPMDH)[T] Lactate dehydrogenase (LDH)[T] *Thermus thermophilus*	DE (ep-PCR):		No difference in stability between WT and mutants	Hayashi et al. (2011)
	IPMDH,			
	WT	34.2 ± 0.6 (25 °C) 3060 ± 60 (70 °C)		
	R85C	324 ± 12 (30 °C)		
	L254V	180 ± 6 (25 °C) 3120 ± 120 (70 °C)		
	I279V	90 ± 0.0 (25 °C) 4920 ± 60 (70 °C)		
	SDM: LDH,			
	WT	12 ± 0.6 (25 °C) 378 ± 24 (40 °C)		
	A75G	22.8 ± 1.2 (25 °C) 1020 ± 60 (40 °C)		

nd not determined, *P* psychrophilic, *M* mesophilic, *T* thermophilic. → change in properties from WT to modified enzyme, $t_{1/2}$ half-life of inactivation, T_{opt} optimal temperature of activity, T_m melting temperature, *WT* wild-type, *SDM* site-directed mutagenesis, *DE* directed evolution, *ep-PCR* Error-prone PCR, *VA* veratryl alcohol, *P-NPP* *p*-nitrophenyl phosphate, *pNPB* *p*-nitrophenylbutyrate, *sAAPF-NA* *N*-succinyl-L-Ala-L-Ala-L-Pro-L-Phe-*p*-nitroanilide, *p-NPL* *p*-nitrophenyl laurate, *StEP* staggered extension process

of cold-adapted lipase (Zhang et al. 2003) and protease (Wintrode et al. 2001). For example, lipase B from *Candida antarctica* (CALB) variant showed simultaneous increase in both activity and half-life of inactivation ($t_{1/2}$) with only two substitutions (Table 20.2). Interestingly, the increase in $t_{1/2}$ was accompanied by a reduction in melting temperature (T_m) which was attributed to a lower tendency of the variants to aggregate upon unfolding.

GM have shown that combining many single mutations into multiple amino acid substitutions sometimes has additive effects, but normally exerts nonadditive effects regarding activity and/or stability (Kobayashi and Takada 2014). In fine chemical synthesis, pharmaceuticals, and food/feed-related applications, enhanced activity at low temperature with concomitant high thermolability (for rapid enzyme inactivation) is desirable to avoid changes in the taste, texture, and composition of the products (Cavicchioli et al. 2011). Mutagenesis to convert thermophilic homologues into cold-adapted homologues has shown that all mutations (S100P, G108S, D114G, M137 T, T153A, and S246 N) substituted in the low-temperature subtilisin variants were found within or near the substrate-binding region. Importantly, none of these changes correspond to the sites found in WT psychrophilic and other GM cold-adapted subtilases, indicating there are several routes to cold adaptation (Zhong et al. 2009). Similarly, a mesophilic lipase was converted to a psychrophilic homologue by a single (F19 L) substitution that showed extreme low-temperature activity and thermolability with decreased substrate affinity (Table 20.2). Structural modeling predicted disruption of inter-secondary hydrophobic core with concomitant enhanced local flexibility (Goomber et al. 2016).

Cold-adapted enzymes are characterized by regions of higher structural flexibility (higher B-factor) from where thermal unfolding is originated, thus offering targets for rigidifying floppy regions by mutagenesis (Wijma et al. 2013; Yu and Huang 2014). Based on flexibility, a computational tool (constraint network analysis) that can envisage floppy regions and transition points upon thermal unfolding has been introduced. The method is based on ISM where each residue is replaced with each of the other 19 amino acids at the floppy site identified by computational analysis, generating small libraries. This strategy can be employed on the improved gene template for further rounds of mutagenesis on another floppy site, thus iteratively targeting numerous positions by ISM (Pfleger et al. 2013; Siddiqui 2015). Bioinformatics tools such as HoTMuSiC based on free energy of unfolding ($\Delta\Delta G$) between the WT and mutant can predict ΔT_m and therefore can be used to design thermostable cold-adapted enzymes (Pucci et al. 2016). Additional tools like Prethermut and SNAP2 (screening for nonacceptable polymorphisms) can be utilized to bypass choosing those amino acids that can decrease activity. This can be achieved by selecting residues that show more than 50% solvent accessibility and are at least 6 Å away from the active site (Tian et al. 2013) or based on biophysical properties, sequence, binding residues, evolutionary information, and structural features of proteins (Hecht et al. 2015). Although ISM can be very useful for increasing the thermostability, to the best of our knowledge, it has not yet been applied on any psychrophilic enzyme. An interesting computational method for enhancing the thermostability of enzymes is using the ancient sequence of the last universal common ancestor (LUCA) that is based on the replacement of residues with the amino acids at these sites from ancient organisms that are considered more thermostable. For example, the stability of lignin peroxidase was improved with three substitutions that were predicted by the phylogenetic tree to be 403 (H329F/T240 L) and 270 (I241L) million years old with concurrent increase in thermostability (Table 20.2; Semba et al. 2015). A bioinformatics tool (configurational entropy, ICE) based on sequence alignment information was used to increase the thermostability of both psychrophilic and mesophilic adenylate kinases with the retention of activity. The alignment displays conserved residues that were left unchanged and a set of permissible replacements in variable amino acids that do not disturb the overall protein fold or function (Bae et al. 2008).

Chimeric enzymes can be created by fusing various parts (mostly domains) from similar enzymes (such as thermally adapted homologues) or enzymes with different catalytic activities (e.g., an endocellulase and a β-glucosidase). Homochimeric enzymes are made to increase activity, stability, and/or protein yield, whereas heterochimeric enzymes are made to achieve synergistic activation (Siddiqui (2015) and references therein). The catalytic properties of these hybrid enzymes are mostly intermediate between the two original parents and strictly depend on the order in which various parts are put together, such as in the case of psychrophilic (*Colwellia maris,* Cm) and mesophilic (*Azotobacter vinelandii,* Av) IDH (Watanabe et al. 2005). While the fusion of parts from thermophilic into psychrophilic homologues has a potential to create highly stable and active enzymes (Campbell et al. 2013), surprisingly Cm-AV IDH is one of the rare examples of a hybrid involving a cold-adapted enzyme. Although the thermostability (T_{opt} and $t_{1/2}$) of the chimeric IDH was intermediate between two parents, the k_{cat} were 97 and 139 s^{-1} for Cm-Av-Cm and Av-Av-Cm hybrids, respectively, compared to 73 and 57.5 s^{-1} for WT-Av and WT-Cm (Watanabe et al. 2005). There are only few examples of truncated cold-adapted enzymes; the most notable is an esterase from *Pseudoalteromonas arctica* (Al-Khudary et al. 2010). The removal of linker and C-terminal domain (~18 kDa) resulted in a truncated esterase that showed similar T_{opt} (25 °C), activity, and substrate specificity, but a dramatic increase in the thermostability ($t_{1/2}$ at 60 and 90 °C, 42 and 2.5 h, respectively) relative to WT enzyme ($t_{1/2}$ at 40 °C: 5 h) (Al-Khudary et al. 2010). Similarly, removal of a negatively charged flexible loop from the cold-adapted protease improved the T_m by 1.5 °C. Further increase in T_m by 3 °C was attained as a result of a salt bridge (N15D) that locked the mobile arm with concurrent increase in the activity (Sigurdardóttir et al. 2009). In addition to improving activity and stability, GM have also been used to decrease the pH optimum (from 6 to 5.5) of arabinase to suit its application for the extraction of pectin or juice clarification by substituting basic by an acidic (H218D) residue around the catalytic domain (Wang et al. 2014).

20.4 Improvements in the Properties by Chemical Modifications and Immobilization

In addition to investigating the structure-function-stability relationship (Siddiqui et al. 2005), chemical modifications (CM) are routinely employed to improve the properties of enzymes and can provide an alternative (Siddiqui et al. 2004, 2006, 2009, 2010; Afzal et al. 2007; Sangeetha and Abraham 2008; Jia et al. 2013; Fresco-Taboada et al. 2014) or complementary method to GM, as in combined mutagenesis and modification (CMM) (Davis 2003). As obtaining regulatory approval for GM products is becoming harder, CM products may have a certain advantage for enzymes intended for commercial utilization (Cavicchioli et al. 2006). Unlike GM which rely on only 20 different amino acids, the variety of modifiers that can be covalently linked to the amino (Siddiqui et al. 2004, 2006, 2009), carboxyl (Siddiqui et al. 1997, 2010; Rashid and Siddiqui 1998), and/or relatively unreactive methylene (Siddiqui et al. 2004) groups is immense. CM can target a single group (amino, carboxyl, or methylene), or more than one group (such as $-NH_2$ and –COOH) can be modified simultaneously (Siddiqui et al. 1999) employing either native or GM enzymes. Although information about the X-ray structure of an enzyme is not required as in the case of SDM, the improvements in properties due to CM cannot be predicted a priori until a modification is actually carried out. Chemical modification can also be combined with immobilization to further improve the catalytic properties as in the case of CALB (Forde et al. 2010). Although CM is a simple, rapid, and inexpensive method to improve the catalytic properties, it suffers from lack of precision in targeting specific amino acids and the formation of heterogeneously modified enzymes. CMM overcomes this drawback where specific residues are mutated into cysteines, which are subsequently modified by thioalkylation using a variety of reagents (Table 20.2; Davis 2003). However, this powerful method has not yet been applied on any cold-adapted enzyme.

Cold-adapted enzymes have extrinsically high activity but suffer from low stability due to their flexible structure, whereas thermophilic homologues have low activity but enhanced stability due to their rigid structure in accordance with the activity-stability trade-off (Siddiqui 2017). As the amino acids inside or near the active site are well adapted for function, the improvement in activity is more challenging to attain, whereas the increase in stability is more practicable since the solvent-exposed amino acids are situated away from the active site on the flexible regions from which the thermal unfolding is originated (Siddiqui 2015). This implies that cold-adapted enzymes are an ideal group for stability improvement with retention of their intrinsic activity, as CM and immobilization can easily target exposed amino acids compared to the buried amino acids. Surprisingly, compared to mesophilic homologues, a relatively limited number of psychrophilic enzymes have been chemically modified or immobilized, and CALB is the most frequently modified cold-adapted enzyme.

Table 20.2 gives examples of a range of chemical modification and immobilization approaches that can be applied alone or in combination to improve the activity, productivity, stability (storage and operational), and reusability of enzymes, including cold-adapted homologues. It is noteworthy that there are numerous examples where activity-stability trade-off has been defied (both activity and stability have been enhanced concurrently, or stabilization has been achieved without sacrificing activity). Readers are referred to Tables 20.1 and 20.2 and recent reviews and references therein (Siddiqui 2015, 2017) for further details. Whereas CM enzymes remain free in solution, immobilized enzymes are non-covalently absorbed or covalently linked to a solid support, which renders their structure more rigid thus making them more resistant to harsh reaction conditions (high temperature, organic solvents, etc.) beside multiple reutilizations (Table 20.2; Homaei et al. 2013). Recycling immobilized enzymes offers economic benefits by reducing the cost of buying expensive enzymes. For example, recently β-glucosidase (~100 times more expensive than cellulases) immobilized on a nonwoven polypropylene fabric using PEGylation and cross-linking was reused 15 times (Zhu et al. 2016), whereas cross-linked magnetic CALB was recycled 10 times (Cruz-Izquierdo et al. 2014) without losing any activity (Table 20.2). Psychrophilic laccase (T_{opt}, 10 °C) supplemented with Cu_2O nanoparticles and trapped in a single-wall carbon nanotube (SWCNT) showed simultaneous enhancement in both activity and stability (Table 20.2). The remarkable aspect of this finding is that the increase in activity, as well as stability, is attained at both low (4 °C) and high (80 °C) temperatures (Mukhopadhyay et al. 2015).

Amidation is a very versatile method where a vast number of modifiers containing an $–NH_2$ group can be linked to the –COOH groups in an enzyme, whereas in acylation any anhydride can be attached to an amino group in an enzyme. Furthermore, employing both amidation and acylation to give a doubly modified endocellulase that was extremely stable in water-miscible organic solvents has also been examined (Siddiqui et al. 1999; Table 20.2). It has been shown that Arg confers structural stability due to guanidine side chain's ability to form multiple electrostatic interactions (Siddiqui and Cavicchioli 2006); therefore, guanidination of Lys residues to form homo-Arg can be used to increase the enzyme stability (Table 20.2; Siddiqui et al. 2006). It is noteworthy that methine/methylene/methyl groups are normally not targeted for modification due to their chemical inertness; however, covalent linkage of photoactivated benzophenone derivatives has been employed to increase the stability of psychrophilic alkaline phosphatase. It is interesting to point out that the addition of benzophenonetetracarboxylic acid to $–CH_3/CH_2/CH$ (photo-modification) or $–NH_2$ (acylation) groups increases thermostability in both cases; however, the activity was decreased in the former, whereas it increased in the latter case. This shows that the change in catalytic properties (activity and stability) not only depends on the modifier but also on the side chain (Table 20.2; Siddiqui et al. 2004).

Ionic liquids (IL) are not only used as an alternative nonaqueous solvent for enzyme reactions (Sect. 20.6) but are also grafted to enzymes due to their desirable chemical and physical properties. The modification of CALB with imidazolium-based IL resulted in an increase in both activity and stability including water-miscible organic solvent. Structural studies showed that IL-modified CALB had increased α-helix and β-turn content at the expense of random coil (Table 20.2; Jia et al. 2013). Glycosylation using dialdehyde polysaccharides (DAP) has routinely been used to increase the stability of enzymes including CALB, with a simultaneous increase in both activity and stability (Table 20.2; Siddiqui and Cavicchioli 2005). In a novel approach, a commercially important mesophilic protease (savinase) was transformed into a psychrophilic-like enzyme (increased activity/productivity at low temperatures and higher thermolability) by reduction in uncompetitive substrate inhibition. The kinetic and structural data indicated that the cold-adapted features of the modified protease were attained not due to higher flexibility and reduced $\Delta H^{\#}$, but as a result of bulky DAP polysaccharide modifier sterically blocking the allosteric site and relieving inhibition (Siddiqui et al. 2009).

The synthesis of cross-linked enzyme aggregates (CLEAs) has been regarded as an efficient carrier-free immobilization method that is considered intermediate between CM and carrier-bound immobilization. The method is simple, rapid, and inexpensive as it utilizes the precipitated crude enzyme preparation and possibility of cross-linking different enzymes in a same milieu. The CLEAs normally display high catalytic rates and productivities, improved storage and operational stability, ability to recycle, and product separation (Table 20.2; Sangeetha and Abraham 2008; Cui and Jia 2013). This method has potential to combine with other immobilization methods such as silica or magnetic nanoparticles (MNP) for further improvement in properties (Table 20.2; Makowski et al. 2007; Forde et al. 2010; Cruz-Izquierdo et al. 2014). In addition, CLEAs are more active in nonpolar organic solvents of lower dielectric constant (Table 20.2). The basis for operational stabilization of immobilized enzymes is prevention of subunit dissociation, aggregation and proteolysis, generation of favorable microenvironments, and rigidification of the enzyme structure via multipoint attachment (Cui and Jia 2013). One of the most promising methods for improving the catalytic properties is immobilization on iron-based MNP that has potential to increase stability against harsh conditions (chemicals, temperature, pH, etc.), relieve product inhibition, and retain or increase activity in addition to ease of recycling enzymes using magnetic field (Cruz-Izquierdo et al. 2014; Rahman et al. 2016). Surprisingly, however, only a handful of cold-adapted enzymes (esterase and lipase) have so far been immobilized on MNP (Table 20.2).

20.5 Improvement of Properties in Additives and Formulations

Compatible solutes (osmolytes) are stored inside the cell to regulate water content and consist of small inorganic and organic molecules that have also been found to act as extrinsic factors to modulate the activity and stability of enzymes at high concentrations (~0.5–2 M). These cosolutes and their analogues include sugars (trehalose, myoinositol, sorbitol, sucrose, etc.), amino acids (glycine, sarcosine, proline, taurine, arginine), amines (trimethylamine *N*-oxide (TMAO), betaine, spermidine, sulfobetaine), and salts (Shiraki et al. 2015; Yancey and Siebenaller 2015). Furthermore, the addition of macromolecules such as polyethylene glycols and proteins can also improve the activity and/or stability of enzymes (Costa et al. 2002). Some additives (Arg, sulfobetaine) prevent enzyme inactivation due to physical processes such as reduced aggregation, whereas amine-based molecules (glycinamide, NH_4Cl) are also known to prevent chemical degradation such as deamidation of Asn and Gln and disulfide exchange at high temperatures due to suppression of chemical modification of proteins (Collins et al. 2006; Shiraki et al. 2015).

The exact mechanism/s by which these osmolytes confer stability to proteins is still unclear and is subject to current investigations. Although a variety of mechanisms have been proposed to explain protein stabilization by cosolutes, generally, these processes fall into two main classes that involve the direct interaction of an additive with the protein or act indirectly via enhancing water structure by the cosolute (Canchi and García 2013). It is now believed that a combination of both mechanisms contributes toward the overall protein stabilization. In the direct mechanism, the protein stability is governed by the subtle balance between favorable and unfavorable interactions of the folded and unfolded states with the cosolutes. For instance, cosolutes are excluded from the surface of the backbone provided that intramolecular interactions in a protein are maximized and are stronger than cosolute-water interactions (osmophobic effect). Cosolutes, such as TMAO, trehalose, sarcosine, and proline, etc., have been shown to induce the formation of α-helix from random coil and result in shrinkage of the denatured state due to entropic stabilization. A relatively compact denatured state will minimize the exposed hydrophobic surface area to water, leading to a reduction in the heat capacity. On the other hand, these cosolutes are also strongly excluded from the nonpolar regions, thus stabilizing the native state of a protein against high temperature due to reduction in entropy of the denatured state (Kaushik and Bhat 2003; Cho et al. 2011).

Some cosolutes act as crowding agents that favor a more compact native state via excluded volume (volume inaccessible to the protein molecules due to crowders) effect, a mechanism similar to that found inside the cells where macromolecules occupy a large volume, thus reducing the volume of solvent (water) available to other large molecules. The surface of protein molecules in 1 M TMAO is devoid of water mimicking the crowded environment; thus crowding promotes folding since compact protein conformation occupies less volume than an unfolded molecule (Cho et al. 2011).

In addition, these cosolutes also increase the protein stability indirectly by enhancing the surface tension of water as a result of extensive and shortened H-bonds between the solute and the solvent. More energy is needed to form a cavity in the solvent to accommodate a large-sized protein upon unfolding. In this respect trehalose has been found to display exceptional properties to reduce the entropy of denatured state as well as increase the surface tension of water. For example, the T_m of RNase has been increased by as much as 18 °C in the presence of 2 M trehalose in a pH range of 2.5–7.0. Importantly, trehalose being uncharged does not interact with the enzyme structure thus minimally influencing its activity (Kaushik and Bhat 2003; Bruździak et al. 2013). In another example, a betaine analogue [2-(*N*,*N*,*N*-tri-*n*-butylammonium)acetate] enhanced the yield (ninefold of the product catalyzed by α-glucosidase) after 24 h reaction due to activation and an increase in the stability of the enzyme (Sehata et al. 2016).

Although formulations containing cosolutes have been shown to improve catalytic properties in mesophilic and thermophilic enzymes, unexpectedly very few cold-adapted enzymes have been tested. For example, in the presence of 1 M TMAO, the activity and the stability of a cold-adapted protease have been simultaneously increased (Table 20.1; He et al. 2009), whereas trehalose has been found to significantly reduce the autoproteolysis (Pan et al. 2005). Like trehalose, TMAO does not interact with the protease structure so that the enzyme can retain its flexibility at low temperature (0 °C) and increases the stability (ΔT_m, 10.5 °C) via enhancement of water structure, exclusion of cosolute from the apolar structure upon protein unfolding and promotion of crowding to favor the more compact native state via excluded volume effects (Canchi and García 2013).

20.6 Improvements of Properties in Nonaqueous Media

The properties of enzymes including cold-adapted enzymes can be improved by nonaqueous media such as organic solvents, ionic liquids, and supercritical CO_2. As discussed in Sect. 20.2, enthalpy-entropy compensation limits cold-adapted enzymes to attain higher intrinsic activity in aqueous solvent (Siddiqui 2017); however, in nonaqueous media, such as organic solvents (such as 3-hexanol) and supercritical CO_2 ($scCO_2$), it has been experimentally shown, with cold-adapted CALB-catalyzed reactions, that the enthalpy-entropy compensation relationship can be avoided (Ottosson et al. 2002) with both negative $\Delta H^{\#}$ and positive $\Delta S^{\#}$ (Ottosson et al. 2001). Other advantages of using enzymes in organic solvents compared to aqueous media include condensation reactions that are unfeasible in water due to equilibrium and product insolubility, in addition to improving stability and substrate, stereo-, regio-, and chemoselectivity. Water acts as a molecular lubricant which, when removed, makes enzymes very rigid in anhydrous organic solvents (Klibanov 2001). Although a reduced flexibility makes enzymes more stable, this also leads to a decrease in activity, in accordance with the activity-stability trade-off (Siddiqui 2017). In many cases the activity of enzymes in anhydrous solvents can be enhanced by adding small quantities of water that replace water molecules vital for catalysis (Karan et al. 2012). An intriguing phenomenon of *molecular memory* that takes advantage of the rigid structure of an enzyme in organic solvents arises. For example, a ligand added to an aqueous solution of an enzyme will form an imprint due to a conformational change, which persists after lyophilization (freeze-drying). If the ligand is then removed by extraction using an organic solvent, the imprint (memory) remains due to the enzyme's rigid structure resulting in 1–2 orders of higher activity. This effect decreases with the addition of water due to increased flexibility. A further benefit is that enzymes in such media also retain *pH memory* as their activity mirrors the pH of the last aqueous solution to which they were exposed (Klibanov 2001).

Cold-adapted enzymes show advantages over mesophilic and thermophilic homologues due to their higher intrinsic flexibility that arises by forming numerous hydrogen bonds between the surface residues and solvent molecules (Siddiqui and Cavicchioli 2006). This also counterbalances the effect of reduced viscosity at low temperatures (Siddiqui et al. 2004). Additionally, a tight hydration shell even under low-water conditions is maintained due to higher surface hydrophilicity and hydrophobicity (Siddiqui and Cavicchioli 2006), thereby promoting catalysis in organic solvents. Readers are referred to excellent reviews on this subject by Owusu Apenten (1999) and Karan et al. (2012). Cold-adapted CALB, because of its low cost and commercial availability, has been examined as a model enzyme in various nonaqueous media such as organic solvents, ionic liquids, and $scCO_2$ (Monhemi and Housaindokht (2016) and references therein) for chiral resolution and the production of fine chemicals, polymers, biodiesel, and pharmaceuticals (Sen and Puskas (2015) and references therein).

An ionic liquid (IL) is defined as a salt in the liquid state and is made of ions and short-lived ion pairs. Several are powerful solvents at or below room temperature and have been dubbed as *designer solvents of the future* that are environmentally friendly. These have many applications in biocatalysis as ILs do not easily inactivate enzymes and show higher selectivity, wide liquid range, and good dissolution power toward many substrates and high reaction rates combined with high thermal and chemical stability. The unique physical properties (such as polarity, hydrophobicity, hydrogen-bonding capability, and viscosity) can be fine-tuned through the appropriate choice of cations and anions that makes IL highly desirable as a solvent compared to organic solvents. However, IL still suffers from control of water activity, viscosity, and pH (Park and Kazlauskas 2003; Zhao 2010). Most of the work on water-immiscible ILs is centered around 1-alkyl-3-methylimidazolium cations because of their stability and possibility to adjust their properties by varying either the anion (BF_4^-, PF_6^-) or the alkyl groups on the cation, whereas the enzyme routinely employed is CALB. The transesterification reaction for the production of biodiesel by CALB in BF_4^--based IL was found to be two- and four-fold higher as compared to that in butanol and hexane, respectively, in addition to three-fold higher half-life at 50 °C. Interestingly, in a continuous reaction process, the half-life in ILs was enhanced by a factor of 300 probably due to the formation of a strong ionic medium that prevents the coating of IL on the protein surface unlike other polar solvents that inactivate the proteins by forming a solvation shell (Lozano et al. 2001). Later studies found that the higher stability of CALB in ILs was accompanied by the conversion of α-helix into β-sheet structures resulting in a more compact enzyme conformation (De Diego et al. 2005), whereas higher enzyme activities depend on the solvent-induced size (narrow, low activity; wide, higher activity) of the active-site cavity (Kim et al. 2014). Lately choline-/glycerol-based ILs that are cheaper, highly biodegradable, and of lower viscosity have been employed for the efficient production of biodiesel by commercially available immobilized CALB (Novozym 435). Miglyol oil (C8–C10 triglyceride from coconut oil) and methanol were transesterified into 97% biodiesel within 3 h reaction (Zhao et al. 2011). The further improvement in biodiesel production (100% yield) was obtained in 8 h at 60 °C due to the exceptional stabilization of CALB in hydrophobic IL with a half-life of 1370 days at the process temperature (Lozano et al. 2013). Although lots of enzymes have been studied in water-miscible (chloroperoxidase, formate dehydrogenase, β-galactosidase, laccase, cellulose, etc.) and water-immiscible (proteases, lipases) IL (Lozano et al. 2015), however apart from CALB, no fundamental studies on other cold-adapted enzymes in ILs have been reported. This is surprising because the high intrinsic activity of cold-adapted enzymes can be retained or enhanced while stabilizing protein structure with concurrent enhancement in their productivity (Ertan et al. 2015; Jayawardena et al. 2017).

A compound or a mixture acts like a supercritical fluid (SCF) below the pressure needed to convert it into a solid, but above its critical pressure and temperature. At a mild temperature and pressure (31.3 °C and 7.3 mPa, respectively), CO_2 behaves as an SCF ($scCO_2$) whose parameters (temperature and pressure) can be tuned to suit process conditions. Being inexpensive and because of its nontoxicity, nonflammability, chemical inertness under most conditions, good solvent properties for non-polar solutes, and environmentally friendly properties, $scCO_2$ is becoming a popular alternative for organic solvent-based enzymatic reactions. Additionally, because CO_2 is a gas at atmospheric pressure, a simple depressurization is sufficient to separate products from $scCO_2$ (Bernal et al. 2012). This technology has been applied with great success for the efficient production of biodiesel using CALB. As an example, for the continuous production of biodiesel in the presence of $scCO_2$ (45 °C and 18 mPa) and tert-butanol, immobilized CALB yielded 85–95% biodiesel (methyl oleate) from triolein and methanol with an operational stability extended to 45 cycles of 4–8 h each [Lozano et al. (2012), Bernal et al. (2012) and references therein]. Apart from biodiesel, CALB has also been exploited for the esterification of cellulose for the production of thermoplastics such as fibers, plastics, films, membranes, modern coatings, cosmetics, and drugs. Carrying out the reaction in $scCO_2$ at 40 °C and 18 mPa swells crystalline cellulose, thereby making it easier for the enzymes to attack the substrate in an efficient way (Gremos et al. 2012). Although CALB has commonly been studied in $scCO_2$, other cold-adapted enzymes such as proteases and glucanases have great potential. It is interesting to point out that cellulases can be utilized in $scCO_2$ for the efficient destruction of lignocellulosic biomass [reviewed in Guerriero et al. (2015)].

Recently, posttranslational modifications (PTMs) have been introduced by in silico design into CALB involving side chain of lysine residues (Monhemi and Housaindokht 2016). As shown by molecular dynamics (MD) simulations based on reduction in root mean square deviation (rmsd), methylation and carboxylation stabilized CALB in $scCO_2$, whereas the acetylated enzyme was stable in aqueous solvent. The MD simulations suggested that the native enzyme is destabilized in $scCO_2$ due to the formation of unstable carbamate between the amino groups and

CO_2 and that the blockage of amino group by methylation and carboxylation leads to stabilization.

The combination of both ILs and $scCO_2$ opened new avenues for synergistic improvements in enzyme properties (activation and stabilization) in nonaqueous media. The advantage of a biphasic system is that the substrates are transferred by $scCO_2$ to the IL phase containing the enzyme, and subsequently the product(s) are removed from the IL phase by simple decompression. Since the enzyme functions in the IL phase and the product is continuously transported and separated in the $scCO_2$ phase, product inhibition can be avoided (Lozano et al. 2015). A case in point is the transesterification reaction between 1-alkanols and vinyl esters and the kinetic resolution of rac-1-phenylethanol by CALB under wide range of pressures (100–150 bars) and temperatures (40–100 °C). In the presence of biphasic solvent system, the cold-adapted lipase showed remarkable improvement in enantioselectivity (99.9%), operational stability (enzyme lost only 15% activity after 11 cycles of 4 h), and productivity (Lozano et al. (2015) and references therein).

20.7 Efficient and Cost-Effective Lignocellulosic Degradation: An Unsolved Challenge

Lignocellulosic biomass is an abundant renewable feedstock, as it is a source of chemicals for various applications and biofuel (Guerriero et al. 2014, 2015, 2016). The use of more efficient enzymes for lignocellulosic biomass deconstruction is certainly a top priority in the current bioeconomy, since improved catalysts help in the better exploitation of this natural bioresource and hence lower our dependence on petroleum-based products. The chief lignocellulosic biomass-degrading enzymes are cellulases, xylanases, and ligninases (Ertan et al. 2012; Siddiqui et al. 2014; Guerriero et al. 2015).

Several psychrophilic cellulases have been isolated from organisms living in cold environments (Kasana and Gulati 2011). The majority of these hydrolytic psychrozymes show an optimum activity at acidic to neutral pH and between 35 and 40 °C (Kasana and Gulati 2011). A xylanase cloned directly from the environmental DNA of goat rumen retained 10% of its activity even at 0 °C, thereby demonstrating the catalytic efficiency of psychrophilic enzymes (Wang et al. 2011). Psychrozymes show higher intrinsic activity due to enhanced flexibility; therefore, they can serve as ideal biocatalysts for further improvement in stability with concurrent retention or increase in the activity (Siddiqui 2015, 2017). Surprisingly, few psychrophilic lignocellulose-degrading enzymes that have been subjected to improvement by GM were able to avoid activity-stability trade-off (Akcapinar et al. 2015). Similarly, the X-ray structure of the psychrophilic cellulase from *P. haloplanktis* revealed the presence of a linker between the catalytic domain (CD) and the cellulose-binding domain (CBD). The progressive deletion of the linker decreased the structural flexibility of the truncated cellulase with concomitant decrease in activity and an increase in thermostability (Table 20.2; Sonan et al. 2007). It is noteworthy that modifications (genetic and chemical) or media engineering (see Sects. 20.5 and 20.6) applied on other thermally adapted enzymes are similarly applicable to psychrophilic cellulases. Although cellulolytic enzymes have been employed in nonaqueous solvents (Sect. 20.6), surprisingly, this group of enzymes has not been tested in the presence of osmolytes (Sect. 20.5). Therefore, as a proof of principle, in addition to psychrophilic homologues, we have also included few examples of lignocellulose-degrading enzymes from mesophilic hosts.

Based on the structural features of cellulases, it is evident that a data-driven protein engineering approach can be envisaged, i.e., to increase the flexibility (hence activity) of linkers in these enzymes. Protocols are available to screen for increased thermostability via semi-rational approaches (Anbar and Bayer 2012), and a similar strategy can be designed to create enzyme variants that are highly active at lower temperatures. In this respect it is interesting to describe the recent rational protein engineering strategy applied to a xylanase from the anaerobic chytrid *Neocallimastix patriciarum* that led to active-site mutants displaying higher activities at temperatures lower than 60 °C (Cheng et al. 2015). A rarer case of circumventing activity-stability trade-off has been demonstrated in cold-adapted β-glucosidase employing directed evolution (Table 20.2; Gonzalez-Blasco et al. 2000).

Besides the protein engineering, another approach to improve the performance of lignocellulose-degrading enzymes is CM (Sect. 20.4). Among these strategies, it is noteworthy to mention rational charge engineering, an approach that aims at modifying the enzyme surface charge. This approach was recently carried out on succinylated cellulase (increased negative charge) that showed tolerance to nonaqueous solvent and led to an increase in activity in the presence of ILs in addition to a reduced lignin inhibition (Nordwald et al. 2014). Cellulases more stable in active detergents (for brighter garments and to reduce fuzz and pill) have been created after modification of Lys side chains with maleic anhydride and *N*-bromosuccinimide (Bund and Singhal 2002). The latter modification increases the rigidity of the protein structure, thereby retarding its denaturation in the detergent, while maleic anhydride screens the hydrophobic methylene groups from contact with water. In one instance, the double modification of both amino and carboxylic groups to remove surface charges led to a cellulase being more thermostable in water-miscible organic solvent mixture (Siddiqui et al. 1999); however, the modified enzyme has not been tested in other nonaqueous solvents (ILs and $scCO_2$). Chemical modifications have also been applied successfully to lignocellulose oxidative enzymes, notably lignin peroxidases (Yoshida et al. 1996; Wang et al. 1999), where polyethylene gylcol and aliphatic acid derivatives were used to increase the tolerance of the enzyme against organic solvents.

Biotechnological applications of cellulases are limited by their low activity, high cost, and product inhibition, whereas cold-adapted enzymes show high activity but low stability in accordance with the activity-stability trade-off (Guerriero et al.

2015, 2016; Siddiqui 2017). Immobilization on various media has potential to overcome these drawbacks where cold-adapted cellulases can be stabilized, product inhibition can be relieved, and enzymes are easily separated from the reaction mixture that can be recycled to save the production cost (Guerriero et al. 2016). In addition, immobilized enzymes can avoid activity-stability trade-off in the presence of nonaqueous solvents and can further display enhanced activity and stability (see Sect. 20.6). For example, an immobilized (on NaY zeolite) mesophilic cellulase from *Trichoderma longibrachiatum* treated with $scCO_2$ at a flow rate of 10 g min^{-1} CO_2, 54 °C, and 180 bar showed 48% higher activity than the untreated enzyme, and the enzyme could be reused ten times (Senyay-Oncel and Yesil-Celiktas 2015).

Similarly, a cellulase immobilized on MNP was applied to treat hemp hurds, and it could be shown that the enzyme retained 50% of its activity for up to five cycles and was more thermostable than the free enzyme (Abraham et al. 2014). More recently, a cold-active cellulase from *Pseudoalteromonas* sp. NJ64 immobilized via the sodium alginate-glutaraldehyde cross-linking entrapment method was used to ferment kelp cellulose (Wang et al. 2015). The immobilized psychrozyme was shown to display higher tolerance to temperature shifts and pH changes and, notably, retained ca. 58% of its activity after seven cycles.

Enzyme entrapment is an ideal method to increase the stability of cold-adapted cellulases as it enables confinement in a microenvironment providing protection against potential inhibitors and high temperature. However, to date, this method has only been applied to a mesophilic cellulase from *Trichoderma reesei* that was immobilized via sol-gel entrapment employing binary or ternary tetramethoxysilane mixtures and alkyl- or aryl-substituted trimethoxysilanes as precursors. The trapped enzyme showed increased pH and temperature stabilities with 40% activity retention upon reutilization (Ungurean et al. 2013). Entrapment in calcium alginate beads is a widespread technique to immobilize enzymes of interest; however, it suffers from enzyme leaching, as the trapped catalysts are much smaller than the pores of the beads (ca. 200 nm). To overcome this problem, immobilization was carried out on cellulase CLEA with a uniform size of 300 nm (Nguyen et al. 2016). The cross-linked immobilized cellulose retained 67% of the activity after ten cycles, while the free enzymes trapped in the beads were completely lost.

Protein-based block copolymer is a method enabling the incorporation of enzymes in nanostructures. This approach results in covalently bound proteins, which are embedded in ordered nanostructures with high stability. Enzyme-based block copolymer was recently used to manufacture LiP nanoparticles of 90 nm average size (Tay et al. 2016). Ruthenium-based amino acid monomers were used to cross-link the ligninase. The nano-LiP displayed 90% and 80% relative activity at 50 °C and 60 °C, respectively (higher than those of the free enzyme) and was stable even after 30 cycles of usage. Psychrozyme nanoparticles could therefore be manufactured using a similar approach to improve the reusability and stability of lignocellulose-degrading enzymes.

A very attractive and innovative method to modify the activity of biocatalysts is molecular imprinting that can prevent the adoption of unfavorable microconformations (Sect. 20.6; Rich and Dordick 2001; Karan et al. 2012). Molecular imprinting was recently applied to cellulase immobilization: super magnetic nanoparticles were imprinted with a mesophilic fungal cellulase as template to immobilize the enzyme (Li et al. 2014). Compared to the free enzyme, the immobilized cellulase showed 3.3-fold higher activity at 70 °C, twofold higher affinity to carboxymethylcellulose, increased thermostability ($t_{1/2}$ of 777 vs. 223 min at 70 °C), and higher T_{opt} (60 vs. 50 °C).

The use of ILs has also attracted much interest in lignocellulosic biomass degradation because they not only improve the catalytic properties (Sect. 20.6; Zhang et al. 2016) but also the solubility of crystalline cellulose for efficient deconstruction. Although the use of imidazolium-based IL is limited by its high cost, recent advances in the synthesis of ILs from lignin and hemicellulose have rendered biomass deconstruction economically feasible with comparable efficiency (Socha et al. 2014). Despite the interesting applications of ILs, it should be noted that commercial formulations of lignocellulosic biomass-degrading enzymes are inhibited by most of these nonaqueous media. Much interest is now devoted to finding IL-tolerant cellulases, which could improve biomass conversion (Gladden et al. 2014; Jaeger et al. 2015; Sriariyanun et al. 2016). Recently, the stability against 1-butyl-3-methylimidazolium chloride of a commercial cellulase immobilized on Amberlite XAD4 was improved by coating it with hydrophobic ILs (Lozano et al. 2011).

20.8 Commercialization of Cold-Adapted Enzymes and Future Avenues

Barring CALB and a few proteases, most work on the improvement of catalytic properties has been carried out on mesophilic and thermophilic homologues; however, this information can guide our future research involving cold-adapted enzymes to suit biotechnological applications. In fact, activity-stability trade-off is only valid under natural conditions because a highly active and thermostable enzyme that is unable to bestow any benefit to the organism is not likely to be sustained over the course of evolution (Siddiqui 2017). However, the present review unequivocally demonstrates that activity-stability trade-off can be surmounted by creating nonnatural enzymes (Tables 20.1, 20.2 and 20.3). Surprisingly, only few native psychrophilic enzymes have been commercialized (Cavicchioli et al. 2011; Sarmiento et al. 2015; Siddiqui 2015; Santiago et al. 2016), and even fewer modified psychrophilic enzymes have been patented (Chakraborty et al. 2015; Dasgupta et al. 2015). One way to expand the commercialization of cold-adapted enzymes is to take an integrated approach by employing multiple ways to improve their catalytic properties (Fig. 20.1). One of the rare

Table 20.3 Various chemical modification and immobilization strategies to improve the properties of biocatalysts including cold-adapted enzymes

Modification/ immobilization	General principles of modification reactions	Modifiers/immobilization supports employed with good effects	Enzyme/conditions	Improvement k_{cat} (min), $t_{1/2}$ (min), T_m (°C) and/or T_{opt}	Reference
Carboxymethylation	S – S disulphide bridge ↓ DTT, BME, TCEP SH + SH + IA → **carboxymethylated enzyme**	IA-modification is mostly used to study structure-function-stability relationship in enzymes containing disulfide bridges	α-Amylase[P]	Stability unaffected, whereas k_{cat} and T_{opt} are reduced	Siddiqui et al. (2005)
Combined mutagenesis and modification	Any amino acid in an enzyme ↓ SDM Cysteine → **thioalkylated**	CH_3–$S(O_2)$–S–R, where R, sugar, methyl, ethyl, ethylsulphonate, ethylamine, etc.	Subtilisin[M] Glycosylation	Broader substrate specificity including D-amino acids	Davis (2003)
Amidation	Enz-COOH (activation) ↓ CDI Enz-O-acylisourea ↓ + X-NH_2 Enz-X	Compounds with a (–NH_2) such as arginine methyl ester, methylamine, ethylenediamine, glucosamine, aniline	Taka α-amylase[M] + Arginine methyl ester	Productivity, 1→2 folds $t_{1/2}$, 12→340 (50 °C) T_m (DSC), 56→64 k_{cat}, 3747→2240	Siddiqui et al. (2010)
Acylation	An anhydride forms an amide linkage with NH_2 in an enzyme	Anhydrides such as acetic anhydride, BPTC-dianhydride, succinic anhydride	Alkaline phosphatase[P] BPTC-dianhydride 50 °C	k_{cat}, 48,740 ↓(37 °C) 86,593 $t_{1/2}$,15→77	Siddiqui et al. (2004)
Dual modification (COOH + NH_2) (PHM + COOH) (PHM + NH_2)	Any combination of modifications of more than one group such as amidation and acylation	COOH → dimethylamine · HCl + NH_2 → Acetic anhydride	CMCase[M], UM, aq UM, wmos DM, aq DM, wmos	$t_{1/2}$, 218 (70 °C) $t_{1/2}$, 1 (34 °C) $t_{1/2}$, 34 (70 °C) $t_{1/2}$, 147 (70 °C)	Siddiqui et al. (1999) Siddiqui et al. (2004)
Guanidination	Conversion of a Lys to homo-Arg containing a guanidine group	3,5-dimethylpyrazole-1-carboxamidine nitrate O-methylisourea	α-amylase[P] 37 °C	k_{cat}, 50,400→7200 [Urea, M], 1.6→2.5	Siddiqui et al. (2006)

(continued)

Table 20.3 (continued)

Modification/ immobilization	General principles of modification reactions	Modifiers/immobilization supports employed with good effects	Enzyme/conditions	Improvement k_{cat} (min), $t_{1/2}$ (min), T_m (°C) and/or T_{opt}	Reference
Photoactivated hydrophobic modification (PHM)	Benzophenone–C = O ↓ $h\nu$350 nm (photoactivation) Benzophenone–C = O : radical ↓ electron abstraction $^{\bullet}CH_3$, $^{\bullet}CH_2$or$^{\bullet}$CH (enzyme) ↓ radical––radical coupling **Benzophenone–Enz**	Benzophenonetetracarboxylic acid	Alkaline phosphatase[P] *Pandalus borealis* 50 °C	k_{cat}, 9341 ↓(10 °C) 880 $t_{1/2}$, 15→182	Siddiqui et al. (2004)
Ionic liquids (IL)	COOH (IL) ↓ activation by CDI → coupling to NH_2 group in an enzyme (amidation)	IL [HOOCMMIm][PF_6] with chaotropic anion and cation modified 58% of amino groups	CALB[P]	k_{cat}/K_m, 1→1.5 $t_{1/2}$, 3→34 (70 °C) V_{max} (30% methanol), 50→80%	Jia et al. (2013)
Polymer glycosylation	Sugar polymer (vicinal OH) ↓ $NaIO_4$ Oxidized DAP (vicinal HC = O) ↓ Enz – NH_2 Enz-DAP (Schiff's base)[a] Other DAPs : Ficoll, chitosan, inulin	DAP-Dextran: PEI-agarose enzyme linked to DAP-dextran	CALB[P] (*p*NPP) NDT-BP[P], 37 °C	Act,100→175% $t_{1/2}$, 18→124 (70 °C) T_{opt}, 50 → 60 °C, 25% loss in activity after 30 cycles	Siddiqui and Cavicchioli (2005) Fresco-Taboada et al. (2014)
Cross-linked enzyme aggregate (CLEA) Cross-linking/ immobilization with chitosan	Enzyme precipitation ↓ $(NH_4)_2SO_4$ or organic solvent Glutaraldehyde (GA) links two-NH_2 groups linked to two K(Lys).	Enz-NH_2 + H_2N-Enz (CLEA)	Subtilisin[M] 11 cycles (60 °C) Activity loss (60 °C) β-galactosidase[P] (*Pseudoalteromonas* sp.)	k_{cat}, 9→523 T_{opt}, 55→70 °C Retain 68% activity 65→10% T_{opt}, 40→50 °C Activity, 100→33% Inhibited→ not inhibited (glucose)	Sangeetha and Abraham (2008) Makowski et al. (2007)
	K1-NH_2+H_2N-K2 ↓ O = CH-$(CH_2)_3$-HC = O **K1 – N = CH$(CH_2)_3$** **–HC = N – K2**	Enz-NH_2 +H_2N-Chitosan			

Polyethylene glycolyation (PEGylation)	PEG $[H\text{-}(O\text{-}CH_2\text{-}CH_2)_n\text{-}OH]$ ↓ *p*-nitrophenylchloroformate PEG-nitrophenylcarbonate ↓ activated PEG + Enz-NH_2 → **PEG-Enz**	Net-cloth layer on polypropylene nonwoven fabric + diacrylate activated PEG + enzyme + GA→ polymerized cross-linked enzyme loaded fabric with only 0.4% leakage	β-glucosidase[M] (almond), using filter paper as substrate at 50 °C	Reused for 15 cycles (24 h/cycle) without losing any activity 40% increase in cellulosic biomass conversion	Zhu et al. (2016)
Cross-linking + adsorption on silica	Crosslinking of NH_2 groups using EGNHS → adsorption on mesoporous silicates	CLEA + nitrile-modified silica Enzyme protected inside the matrix channels	CALB[P] 70 °C	$t_{1/2}$, 2 → 119	Forde et al. (2010)
Cross-linked MNP	[b]Co-precipitation of Fe^{2+}/Fe^{3+} salts (1:2, [c]NH_4OH, pH 10–12) ↓ MNP-NH_2 (1-100 nm) ↙*route 1* ↘*route 2* -SL +SL ↓ ↓ MNP-NH_2 MNP-SL-NH_2 ↙+ Enz-COOH + CDI↘ **Enz-MNP** **Enz-SL-MNP** MNP-NH_2 MNP-SL-NH_2 *route 3* +GA *route 4* MNP-CHO MNP-SL-CHO + H_2N-Enz → **Enz-MNP** **Enz-SL-MNP**	CLEA + Fe_3O_4 based MNP ↓ Magnetic CLEA (75 × 100 nm) Following *route* 4 with silane linker (*SL*).	CALB[P] for making biodiesel from unrefined vegetable oils and waste cooking oil	No loss in activity after 10 consecutive 24 h cycles at 40 °C Retained more than 60% activity after 10 cycles at 60 °C. The biodiesel yield was 92% after 72 h at 30 °C	Cruz-Izquierdo et al. (2014)
Immobilization on cellulose nanogel-MNP complex	Enzyme + cellulose nanogel-MNP composite were mixed together at 4 °C for 12 h to give non-covalent immobilization	Cellulose-Fe_3O_4-based MNP composite	Esterase[P] *Zunongwangia* sp.	k_{cat}, 22,000→31,000 T_{opt}, 30→35 °C $t_{1/2}$,18→32 h (35 °C) Retained 40% activity after 10 cycles at 35 °C	Rahman et al. (2016)

(continued)

Table 20.3 (continued)

Modification/ immobilization	General principles of modification reactions	Modifiers/immobilization supports employed with good effects	Enzyme/conditions	Improvement k_{cat} (min), $t_{1/2}$ (min), T_m (°C) and/or T_{opt}	Reference
Entrapment/immobilization of Cu_2O NP in single-wall carbon nanotube (SWNT)	SWNT synthesized by catalytic chemical vapor deposition. ↓ functionalized with lipid Pseudomicelles + enzyme ↓ disruption by sonication Reassembly with trapped enzyme inside SWNT	Cu_2O nanoparticle-enzyme trapped inside a lipid functionalized SWNT	Laccase^P, 4 °C *Pseudomonas putida* (Himalaya), 80 °C	V_{max}, 11→55 $t_{1/2}$, 7→73 V_{max}, 19 → 64 $t_{1/2}$, 7 → 436	Bhattacharyya et al. (2013) Mukhopadhyay et al. (2015)

Enz enzyme, *UM* unmodified, *CM* chemical modification, *NP* nanoparticle, *MNP* magnetic nanoparticles, *P* psychrophilic, *M* mesophilic, $-NH_2$*N*-terminal or side chain (Lys) deprotonated nucleophilic amino group, *BPTC-dianhydride* benzophenonetetracarboxyl dianhydride, *PEG* polyethyleneglycol, *DAP* Dialdehyde polysaccharide, *SDM* site-directed mutagenesis, *IA* iodoacetic acid or iodoacetamide, *DTT* 1,4-dithiothreitol, *ME* 2-mercaptoethanol, *TCEP* (*tris*(2-carboxyethyl)phosphine), *aq* aqueous, *wmos* water-miscible organic solvent (50% v/v dioxan), *DM* dual modified, *pNPP p*-nitrophenyl palmitate, *SL* silane-linker (3-Aminopropyltriethoxysilane, APTS), GA glutaraldehyde, CALB *Candida antarctica* lipase B, CMCase carboxymethylcellulase, *NDT-BP* Nucleoside 2′-Deoxyribosyltransferase from *Bacillus psychrosaccharolyticus*, *PEI* Polyethyleneimine, *EGNHS* Ethylene glycol *bis*(succinimidyl succinate), *CDI* carbodiimide or carbonyldiimidazole, *CHO* aldehyde, *PHM* Photoactivated hydrophobic modification, → change in properties from WT to modified enzyme

[a]Enzyme-sugar polymer Schiff base can be further reductively aminated by $NaBH_4$, $NaCNBH_3$, or borane-pyridine complex

[b]General approaches for making iron-based MNP

[c]Coprecipitation in the presence of NH_4OH confers NH_2 functionalities on MNP, whereas NaOH treatment retains OH groups

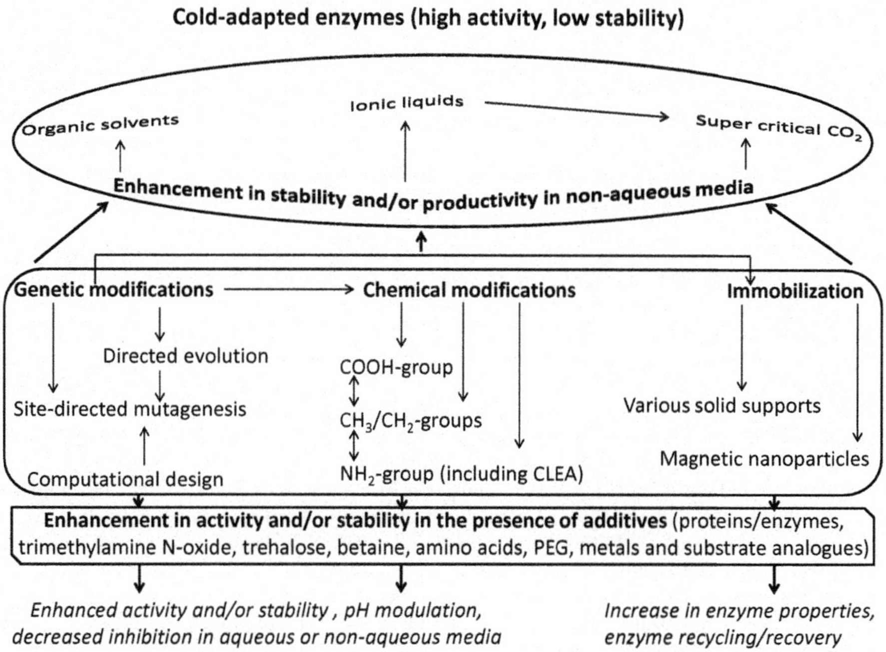

Fig. 20.1 A unified approach to improve the catalytic properties of cold-adapted enzymes in aqueous and nonaqueous media. *Curved box*, genetic and chemical modifications and immobilization; *oval*, reaction in nonaqueous media; *corner snipped box*, additives. The outcome of these approaches is shown in *italics* at the *bottom* of the flow chart. *Arrows* denote that various approaches can be combined to get optimal improvements in enzymatic properties. *Double arrows*, methods can be used in any direction or sequence. *CLEA* cross-linked enzyme aggregates

successes involving this strategy is the highly active and thermostable genetically modified cocaine esterase (Table 20.2) that was further PEGylated to enhance the enzyme's residence time in the blood for up to 3 days. The enzyme is now in clinical trial stage for the treatment of cocaine lethal dose (Fang et al. 2014). Crucially integrated approaches can also lead to the economical production of biofuels from biomass and waste oil. As an example, the production of biodiesel can be made highly efficient and cost-effective if instead of native enzyme, genetically modified CALB (Sect. 20.3; Table 20.2; Zhang et al. 2003) is used as a starting material that can be further improved and recycled by immobilization on MNP (Sect. 20.4; Cruz-Izquierdo et al. 2014) in addition to media engineering (reaction carried out in the presence of an appropriate cosolute (Sect. 20.5) and/or nonaqueous solvent (Sect. 20.6); Monhemi and Housaindokht 2016).

References

Abraham RE, Verma ML, Barrow CJ, Puri M (2014) Suitability of magnetic nanoparticle immobilized cellulases in enhancing enzymatic saccharification of pretreated hemp biomass. Bitechnol Biofuels 11(7):90

Afzal AJ, Ali S, Latif F, Rajoka MI, Siddiqui KS (2005) Innovative kinetic and thermodynamic analysis of a purified superactive xylanase from *Scopulariopsis* sp. Appl Biochem Biotechnol 120:51–70

Afzal AJ, Bokhari SA, Siddiqui KS (2007) Kinetic and thermodynamic study of a chemically modified highly active xylanase from *Scopulariopsis* sp: existence of an essential amino group. Appl Biochem Biotechnol 141:273–297

Akcapinar GB, Venturini A, Martelli PL, Casadio R, Sezerman UO (2015) Modulating the thermostability of endoglucanase I from *Trichoderma reesei* using computational approaches. Protein Eng Des Sel 28:127–135

Al-Khudary R, Venkatachalam R, Katzer M, Elleuche S, Antranikian G (2010) A cold-adapted esterase of a novel marine isolate, *Pseudoalteromonas arctica*: gene cloning, enzyme purification and characterization. Extremophiles 14:273–285

Anbar M, Bayer EA (2012) Chapter 14: Approaches for improving thermostability characteristics in cellulases. In: Gilbert HJ (ed) Methods in enzymology. Academic, London, pp 261–271

Asghari SM, Pazhang M, Ehtesham S, Karbalaei-Heidari HR, Taghdir M, Sadeghizadeh M, Naderi-Manesh H, Khajeh K (2010) Remarkable improvements of a neutral protease activity and stability share the same structural origins. Protein Eng Des Sel 23:599–606

Bae E, Bannen RM, Phillips GN Jr (2008) Bioinformatic method for protein thermal stabilization by structural entropy optimization. Proc Natl Acad Sci USA 105:9594–9597

Bernal JM, Lozano P, García-Verdugo E et al (2012) Supercritical synthesis of biodiesel. Molecules 17:8696–8719

Bhattacharyya T, Samaddar S, Dasgupta AK (2013) Reusable glucose sensing using carbon nanotube-based self-assembly. Nanoscale 5:9231–9237

Bruździak P, Panuszko A, Stangret J (2013) Influence of osmolytes on protein and water structure: a step to understanding the mechanism of protein stabilization. J Phys Chem B 117:11502–11508

Bund RK, Singhal RS (2002) An alkali stable cellulase by chemical modification using maleic anhydride. Carbohydr Polym 47:137–141

Campbell E, Chuang S, Banta S (2013) Modular exchange of substrate-binding loops alters both substrate and cofactor specificity in a member of the aldo–keto reductase superfamily. Protein Eng Des Sel 26:181–186

Canchi DR, García AE (2013) Cosolvent effects on protein stability. Annu Rev Phys Chem 64:273–293

Cavicchioli R, Curmi PMG, Siddiqui KS, Thomas T (2006) Proteins from psychrophiles. In: Rainey FA, Oren A (eds) Extremophiles—methods in microbiology. Academic, London, pp 395–436

Cavicchioli R, Charlton T, Ertan H, Omar SM, Siddiqui KS, Williams TJ (2011) Biotechnological uses of enzymes from psychrophiles. J Microbial Biotechnol 4:449–460

Chakraborty K, Mukhopadhyay A, Dutta N, Dasgupta AK (2015) Psychrophilic enzymes compositions and methods for making and using same. US patent 2015/0044731 A1

Cheng YS, Chen CC, Huang JW et al (2015) Improving the catalytic performance of a GH11 xylanase by rational protein engineering. Appl Microbiol Biotechnol 99:9503–9510

Cho SS, Reddy G, Straub JE, Thirumalai D (2011) Entropic stabilization of proteins by TMAO. J Phys Chem B 115:13401–13407

Collins T, D'Amico S, Georlette D et al (2006) A nondetergent sulfobetaine prevents protein aggregation in microcalorimetric studies. Anal Biochem 352:299–301

Costa SA, Tzanov T, Carneiro AF et al (2002) Studies of stabilization of native catalase using additives. Enzyme Microb Technol 30:387–391

Cruz-Izquierdo Á, Picó EA, López C, Serra JL, Llama MJ (2014) Magnetic cross-linked enzyme aggregates (mCLEAs) of Candida antarctica lipase: an efficient and stable biocatalyst for bio-diesel synthesis. PLoS One 9:e115202

Cui JD, Jia SR (2013) Optimization protocols and improved strategies of cross-linked enzyme aggregates technology: current development and future challenges. Crit Rev Biotechnol 35:15–28

D'Amico S, Marx JC, Gerday C, Feller G (2003) Activity-stability relationships in extremophilic enzymes. J Biol Chem 278:7891–7896

Dasgupta AK, Bhattacharyya T, Mukhopadhyay A, Dutta N, Chakraborty K (2015) Enzyme stabilization by carbon nanotubes. US patent 2015/079656A1

Davis BG (2003) Chemical modification of biocatalysts. Curr Opin Chem Biol 14:379–383

De Diego T, Lozano P, Gmouh S, Vaultier M, Iborra JL (2005) Understanding structure-stability relationships of *Candida antartica* lipase B in ionic liquids. Biomacromolecules 6:1457–1464

Dunitz JD (1994) The entropic cost of bound water in crystals and biomolecules. Science 264:670

Ertan H, Siddiqui KS, Muenchhoff J, Charlton T, Cavicchioli R (2012) Kinetic and thermodynamic characterization of the functional properties of a hybrid versatile peroxidase using isothermal titration calorimetry: insight into manganese peroxidase activation and lignin peroxidase inhibition. Biochimie 94:1221–1231

Ertan H, Cassel C, Verma A et al (2015) A new broad specificity alkaline metalloprotease from a Pseudomonas sp. Isolated from refrigerated milk: role of calcium in improving enzyme productivity. J Mol Catal B: Enzym 113:1–8

Fang L, Chow KM, Hou S et al (2014) Rational design, preparation and characterization of a therapeutic enzyme mutant with improved stability and function for cocaine detoxification. ACS Chem Biol 9:1764–1772

Feller G (2008) Enzyme function at low temperatures in psychrophiles. In: Siddiqui KS, Thomas T (eds) Protein adaptation in extremophiles. Nova Science, New York, pp 35–69

Feller G (2013) Psychrophilic enzymes: from folding to function and biotechnology. Scientifica 2013:512–840

Fields PA, Somero GN (1998) Hot spots in cold adaptation: localized increases in conformational flexibility in lactate dehydrogenase A4 orthologs of Antarctic notothenioid fishes. Proc Natl Acad Sci USA 95:11476–11481

Forde J, Vakurov A, Gibson TD et al (2010) Chemical modification and immobilisation of lipase B from *Candida antarctica* onto mesoporous silicates. J Mol Catal B: Enzym 66:203–209

Fox D (2015) Life at hell's gate. Sci Am 313:34–34

Fresco-Taboada A, Serra I, Fernández-Lucas J et al (2014) Nucleoside 2′-deoxyribosyltransferase from psychrophilic bacterium *Bacillus psychrosaccharolyticus*-preparation of an immobilized biocatalyst for the enzymatic synthesis of therapeutic nucleosides. Molecules 19:11231–11249

Gladden JM, Park JI, Bergmann J et al (2014) Discovery and characterization of ionic liquid-tolerant thermophilic cellulases from a switchgrass-adapted microbial community. Biotechnol Biofuels 7:15

Gonzalez-Blasco G, Sanz-Aparicio J, Gonzalez B, Hermoso JA, Polaina J (2000) Directed evolution of beta-glucosidase A from *Paenibacillus polymyxa* to thermal resistance. J Biol Chem 275:13708–13712

Goomber S, Kumar A, Kaur J (2016) Disruption of N terminus long range non covalent interactions shifted temp. opt 25° C to cold: evolution of point mutant *Bacillus lipase* by error prone PCR. Gene 576:237–243

Gremos S, Kekos D, Kolisis F (2012) Supercritical carbon dioxide biocatalysis as a novel and green methodology for the enzymatic acylation of fibrous cellulose in one step. Bioresour Technol 115:96–101

Guerriero G, Sergeant K, Hausman JF (2014) Wood biosynthesis and typologies: a molecular rhapsody. Tree Physiol 34:839–855

Guerriero G, Hausman JF, Strauss J et al (2015) Destructuring plant biomass: focus on fungal and extremophilic cell wall hydrolases. Plant Sci 234:180–193

Guerriero G, Hausman JF, Strauss J, Ertan H, Siddiqui KS (2016) Lignocellulosic biomass: biosynthesis, degradation and industrial utilization. Eng Life Sci 16:1–16

Hayashi S, Akanuma S, Onuki W, Tokunaga C, Yamagishi A (2011) Substitutions of coenzyme-binding, nonpolar residues improve the low-temperature activity of thermophilic dehydrogenases. Biochemistry 50:8583–8593

He HL, Chen XL, Zhang XY, Sun CY, Zou BC, Zhang YZ (2009) Novel use for the osmolyte trimethylamine N-oxide: retaining the psychrophilic characters of cold-adapted protease deseasin MCP-01 and simultaneously improving its thermostability. Marine Biotechnol 11:710–716

Hecht M, Bromberg Y, Rost B (2015) Better prediction of functional effects for sequence variants. BMC Genomics. 2015;16 Suppl 8:S1

Homaei AA, Sariri R, Vianello F, Stevanato R (2013) Enzyme immobilization: an update. J Chem Biol 6:185–205

Huggins DJ (2015) Quantifying the entropy of binding for water molecules in protein cavities by computing correlations. Biophys J 108:928–936

Isaksen GV, Aqvist J, Brandsdal BO (2016) Enzyme surface rigidity tunes the temperature dependence of catalytic rates. Proc Natl Acad Sci USA 113:7822–7827

Jaeger V, Burney P, Pfaendtner J (2015) Comparison of three ionic liquid-tolerant cellulases by molecular dynamics. Biophys J 108:880–892

Jayawardena MB, Yee LH, Poljak A, Cavicchioli R, Kjelleberg SJ, Siddiqui KS (2017) Enhancement of lipase stability and productivity through chemical modification and its application to latex-based polymer emulsions. Process Biochem. doi:10.1016/j.procbio.2017.03.014

Jia R, Hu Y, Liu L, Jiang L, Huang H (2013) Chemical modification for improving activity and stability of lipase B from *Candida antarctica* with imidazolium-functional ionic liquids. Org Biomol Chem 11:7192–7198

Karan R, Capes MD, DasSarma S (2012) Function and biotechnology of extremophilic enzymes in low water activity. Aquat Biol 8:1

Kasana RC, Gulati A (2011) Cellulases from psychrophilic microorganisms: a review. J Basic Microbiol 51:572–579

Kaushik JK, Bhat R (2003) Why is trehalose an exceptional protein stabilizer? An analysis of the thermal stability of proteins in the presence of the compatible osmolyte trehalose. J Biol Chem 278:26458–26465

Kim HS, Ha SH, Sethaphong L, Koo YM, Yingling YG (2014) The relationship between enhanced enzyme activity and structural dynamics in ionic liquids: a combined computational and experimental study. Phys Chem Chem Phys 16:2944–2953

Klibanov AM (2001) Improving enzymes by using them in organic solvents. Nature 409:241–246

Kobayashi M, Takada Y (2014) Effects of the combined substitutions of amino acid residues on thermal properties of cold-adapted monomeric isocitrate dehydrogenases from psychrophilic bacteria. Extremophiles 18:755–762

Kotting C, Kallenbach A, Suveyzdis Y et al (2008) The GAP arginine finger movement into the catalytic site of Ras increases the activation entropy. Proc Natl Acad Sci USA 105:6260–6265

Koutsioulis D, Wang E, Tzanodaskalaki M et al (2008) Directed evolution on the cold adapted properties of TAB5 alkaline phosphatase. Protein Eng Des Sel 21:319–327

Li Y, Wang XY, Zhang RZ et al (2014) Molecular imprinting and immobilization of cellulase onto magnetic Fe_3O_4@SiO_2 nanoparticles. J Nanosci Nanotechnol 14:2931–2936

Lonhienne T, Gerday C, Feller G (2000) Psychrophilic enzymes: revisiting the thermodynamic parameters of activation may explain local flexibility. Biochim Biophys Acta 1543:1–10

Lozano P, de Diego T, Carrie D, Vaultier M, Iborra JL (2001) Overstabilization of *Candida antarctica* lipase B by ionic liquids in ester synthesis. Biotechnol Lett 23:1529–1533

Lozano P, Bernal B, Bernal JM, Pucheault M, Vaultier M (2011) Stabilizing immobilized cellulase by ionic liquids for saccharification of cellulose solutions in 1-butyl-3-methylimidazolium chloride. Green Chem 13:1406–1410

Lozano P, García-Verdugo E, Bernal JM et al (2012) Immobilized lipase onto structured supports containing covalently attached ionic liquids for continuous synthesis of biodiesel in $scCO_2$. ChemSusChem 5:790–798
Lozano P, Bernal JM, Sanchez-Gomez G, Lopez-Lopez G, Vaultier M (2013) How to produce biodiesel easily using a green biocatalytic approach in sponge-like ionic liquids. Energ Environ Sci 6:1328–1338
Lozano P, Bernal JM, Nieto S, Gomez C, Garcia-Verdugo E, Luis SV (2015) Active biopolymers in green non-conventional media: a sustainable tool for developing clean chemical processes. Chem Commun 51:17361–17374
Makowski K, Białkowska A, Szczesna-Antczak M et al (2007) Immobilized preparation of cold-adapted and halotolerant Antarctic beta-galactosidase as a highly stable catalyst in lactose hydrolysis. FEMS Microbiol Ecol 59:535–542
Miyazaki K, Wintrode PL, Grayling RA, Rubingh DN, Arnold FH (2000) Directed evolution study of temperature adaptation in a psychrophilic enzyme. J Mol Biol 297:1015–1026
Monhemi H, Housaindokht MR (2016) Chemical modification of biocatalyst for function in supercritical CO_2: in silico redesign of stable lipase. J Supercrit Fluids 117:147–163
Mukhopadhyay A, Dasgupta AK, Chakrabarti K (2015) Enhanced functionality and stabilization of a cold active laccase using nanotechnology based activation-immobilization. Bioresour Technol 179:573–584
Narasimhan D, Nance MR, Gao D et al (2010) Structural analysis of thermostabilizing mutations of cocaine esterase. Protein Eng Des Sel 23:537–547
Nguyen LT, Lau YS, Yang KL (2016) Entrapment of cross-linked cellulase colloids in alginate beads for hydrolysis of cellulose. Colloids Surf B Biointerfaces 145:862–869
Nordwald EM, Brunecky R, Himmel ME et al (2014) Charge engineering of cellulases improves ionic liquid tolerance and reduces lignin inhibition. Biotechnol Bioeng 111:1541–1549
Ottosson J, Rotticci-Mulder JC, Rotticci D, Hult K (2001) Rational design of enantioselective enzymes requires considerations of entropy. Protein Sci 10:1769–1774
Ottosson J, Fransson L, Hult K (2002) Substrate entropy in enzyme enantioselectivity: an experimental and molecular modeling study of a lipase. Protein Sci 11:1462–1471
Owusu Apenten RK (1999) Low temperature organic phase biocatalysis using cold adapted enzymes. In: Margesin R, Schinner F (eds) Biotechnological applications of cold-adapted organisms. Springer, Berlin, pp 35–48
Pan J, Chen XL, Shun CY, He HL, Zhang YZ (2005) Stabilization of cold-adapted protease MCP-01 promoted by trehalose: prevention of the autolysis. Protein Pept Lett 12:375–378
Park S, Kazlauskas RJ (2003) Biocatalysis in ionic liquids—advantages beyond green technology. Curr Opin Biotechnol 14:432–437
Pfleger C, Rathi PC, Klein DL, Radestock S, Gohlke H (2013) Constraint Network Analysis (CNA): a Python software package for efficiently linking biomacromolecular structure, flexibility, (thermo-) stability and function. J Chem Inf Model 53:1007–1015
Pucci F, Bourgeas R, Rooman M (2016) Predicting protein thermal stability changes upon point mutations using statistical potentials: introducing HoTMuSiC. Sci Rep 6:23257
Rahman MA, Culsum U, Kumar A, Gao H, Hu N (2016) Immobilization of a novel cold active esterase onto $Fe_3O_4\sim$ cellulose nano-composite enhances catalytic properties. Int J Biol Macromol 87:488–497
Rashid MH, Siddiqui KS (1998) Carboxy-group modification: high-temperature activation of charge-neutralized and charge-reversed beta-glucosidases from *Aspergillus niger*. Biotechnol Appl Biochem 27:231–237
Rich J, Dordick J (2001) Imprinting Enzymes for Use in Organic Media. In: Vulfson E, Halling P, Holland H (eds) Enzymes in nonaqueous dolvents. Humana Press, New York, pp 13–17
Sangeetha K, Abraham TE (2008) Preparation and characterization of cross-linked enzyme aggregates (CLEA) of Subtilisin for controlled release applications. Int J Biol Macromol 43:314–319

Santiago M, Ramírez-Sarmiento CA, Zamora RA, Parra LP (2016) Discovery, molecular mechanisms and industrial applications of cold-active enzymes. Front Microbiol 7:1408

Sarmiento F, Peralta R, Blamey JM (2015) Cold and hot extremozymes: industrial relevance and current trends. Front Bioeng Biotechnol 3:148

Sehata S, Nakagawa Y, Genjima R, Koumoto K (2016) Quick activation/stabilization of a α-glucosidase-catalyzed hydrolysis reaction by addition of a betaine-type metabolite analogue. Chem Lett. doi:10.1246/cl.160567

Semba Y, Ishida M, Yokobori S, Yamagishi A (2015) Ancestral amino acid substitution improves the thermal stability of recombinant lignin-peroxidase from white-rot fungi, *Phanerochaete chrysosporium* strain UAMH 3641. Protein Eng Des Sel 28:221–230

Sen S, Puskas JE (2015) Green polymer chemistry: enzyme catalysis for polymer functionalization. Molecules 20:9358–9379

Senyay-Oncel D, Yesil-Celiktas O (2015) Characterization, immobilization, and activity enhancement of cellulase treated with supercritical CO_2. Cellulose 22:3619–3631

Shiraki K, Tomita S, Inoue N (2015) Small amine molecules: solvent design toward facile improvement of protein stability against aggregation and inactivation. Curr Pharm Biotechnol 17:116–125

Siddiqui KS (2015) Some like it hot, some like it cold: temperature dependent biotechnological applications and improvements in extremophilic enzymes. Biotechnol Adv 33:1912–1922

Siddiqui KS (2017) Defying activity-stability trade-off in enzymes: taking advantage of entropy to enhance activity and thermostability. Crit Rev Biotechnol. doi:10.3109/07388551.2016.1144045

Siddiqui KS, Cavicchioli R (2005) Improved thermal stability and activity in the cold-adapted lipase B from *Candida antarctica* following chemical modification with oxidized polysaccharides. Extremophiles 9:471–476

Siddiqui KS, Cavicchioli R (2006) Cold adapted enzymes. Annu Rev Biochem 75:403–433

Siddiqui KS, Saqib AAN, Rashid MH, Rajoka MI (1997) Thermostabilization of carboxymethylcellulase from *Aspergillus niger* by carboxyl group modification. Biotechnol Lett 19:325–330

Siddiqui KS, Shemsi AM, Anwar MA, Rashid MH, Rajoka MI (1999) Partial and complete alteration of surface charges of carboxymethylcellulase by chemical modification: thermostabilization in water miscible organic solvent. Enzyme Microb Technol 24:599–608

Siddiqui KS, Poljak A, Cavicchioli R (2004) Improved activity and stability of alkaline phosphatases from psychrophilic and mesophilic organisms by chemically modifying aliphatic or amino groups using tetracarboxy-benzophenone derivatives. Cell Mol Biol 50:657–667

Siddiqui KS, Feller G, D'Amico S, Gerday C, Giaquinto L, Cavicchioli R (2005) The active site is the least stable structure in the unfolding pathway of a multi-domain cold-adapted α-amylase. J Bacteriol 187:6197–6205

Siddiqui KS, Poljak A, Guilhaus M et al (2006) Role of lysine versus arginine in enzyme cold-adaptation: modifying lysine to homo-arginine stabilizes the cold-adapted alpha-amylase from *Pseudoalteramonas haloplanktis*. Proteins 64:486–501

Siddiqui KS, Parkin DM, Curmi PM et al (2009) A novel approach for enhancing the catalytic efficiency of a protease at low temperature: reduction in substrate inhibition by chemical modification. Biotechnol Bioeng 103:676–686

Siddiqui KS, Poljak A, De Francisci D et al (2010) A chemically modified alpha-amylase with a molten-globule state has entropically driven enhanced thermal stability. Protein Eng Des Sel 23:769–780

Siddiqui KS, Williams TJ, Wilkins D et al (2013) Psychrophiles. Annu Rev Earth Planet Sci 41:87–115

Siddiqui KS, Ertan H, Charlton T et al (2014) Versatile peroxidase degradation of humic substances: use of isothermal titration calorimetry to assess kinetics, and applications to industrial wastes. J Biotechnol 178:1–11

Sigurdardóttir AG, Arnórsdóttir J, Thorbjarnardóttir SH, Eggertsson G, Suhre K, Kristjánsson MM (2009) Characteristics of mutants designed to incorporate a new ion pair into the structure of a cold adapted subtilisin-like serine proteinase. Biochim Biophys Acta 1794:512–518

Socha AM, Parthasarathi R, Shi J et al (2014) Efficient biomass pretreatment using ionic liquids derived from lignin and hemicellulose. Proc Natl Acad Sci USA 111:3587–3595

Sonan GK, Receveur-Brechot V, Duez C et al (2007) The linker region plays a key role in the adaptation to cold of the cellulase from an Antarctic bacterium. Biochem J 407:293–302

Sriariyanun M, Tantayotai P, Yasurin P et al (2016) Production, purification and characterization of an ionic liquid tolerant cellulase from *Bacillus* sp. isolated from rice paddy field soil. Electron J Biotechnol 19:23–28

Struvay C, Feller G (2012) Optimization to low temperature activity in psychrophilic enzymes. Int J Mol Sci 13:11643–11665

Tay T, Köse E, Keçili R, Say R (2016) Design and preparation of nano-lignin peroxidase (nanoLiP) by protein block copolymerization approach. Polymer 8:223

Tian J, Wang P, Huang L, Chu X, Wu N, Fan Y (2013) Improving the thermostability of methyl parathion hydrolase from Ochrobactrum sp. M231 using a computationally aided method. Appl Microbiol Biotechnol 97:2997–3006

Truongvan N, Jang SH, Lee C (2016) Flexibility and stability trade-off in active site of cold-adapted *Pseudomonas mandelii* esterase EstK. Biochemistry 55:3542–3549

Ungurean M, Paul C, Peter F (2013) Cellulase immobilized by sol-gel entrapment for efficient hydrolysis of cellulose. Bioprocess Biosyst Eng 36:1327–1338

Wang G, Luo H, Wang Y et al (2011) A novel cold-active xylanase gene from the environmental DNA of goat rumen contents: direct cloning, expression and enzyme characterization. Bioresour Technol 102:3330–3336

Wang P, Woodward CA, Kaufman EN (1999) Poly (ethylene glycol)-modified ligninase enhances pentachlorophenol biodegradation in water–solvent mixtures. Biotechnol Bioeng 64:290–297

Wang S, Yang Y, Yang R et al (2014) Cloning and characterization of a cold-adapted endo-1,5-α-L-arabinanase from *Paenibacillus polymyxa* and rational design for acidic applicability. J Agric Food Chem 62:8460–8469

Wang YB, Gao C, Zheng Z et al (2015) Immobilization of cold-active cellulase from Antarctic bacterium and its use for kelp cellulose ethanol fermentation. BioResources 10:1757–1772

Watanabe S, Yasutake Y, Tanaka I, Takada Y (2005) Elucidation of stability determinants of cold-adapted monomeric isocitrate dehydrogenase from a psychrophilic bacterium, *Colwellia maris*, by construction of chimeric enzymes. Microbiology 151:1083–1094

Wijma HJ, Floor RJ, Janssen DB (2013) Structure and sequence analysis inspired engineering of proteins for enhanced thermostability. Curr Opin Struct Biol 23:588–594

Wintrode PL, Miyazaki K, Arnold FH (2001) Patterns of adaptation in a laboratory evolved thermophilic enzyme. Biochim Biophys Acta (BBA) Protein Struct Mol Enzymol 1549:1–8

Wolfenden R, Snider MJ (2001) The depth of chemical time and the power of enzymes as catalysts. Acc Chem Res 34:938–945

Xu H, Shen D, XQ W, Liu ZW, Yang QH (2014) Characterization of a mutant glucose isomerase from *Thermoanaerobacterium saccharolyticum*. J Ind Microbiol Biotechnol 41:1581–1589

Yancey PH, Siebenaller JF (2015) Co-evolution of proteins and solutions: protein adaptation versus cytoprotective micromolecules and their roles in marine organisms. J Exp Biol 218:1880–1896

Yang H, Li J, Shin HD, Du G, Liu L, Chen J (2014) Molecular engineering of industrial enzymes: recent advances and future prospects. Appl Microbiol Biotechnol 98:23–29

Yoshida S, Watanabe T, Honda Y, Kuwahara M (1996) Reaction of chemically modified lignin peroxidase of *Phanerochaete chrysosporium* in water-miscible organic solvents. Biosci Biotechnol Biochem 60:1805–1809

Yu H, Huang H (2014) Engineering proteins for thermostability through rigidifying flexible sites. Biotechnol Adv 32:308–315

Zhang DH, Xu HX, Chen N, Che WC (2016) The application of ionic liquids in enzyme immobilization and enzyme modification. Austin J Biotechnol Bioeng 3:1060

Zhang N, Suen WC, Windsor W, Xiao L, Madison V, Zaks A (2003) Improving tolerance of *Candida antarctica* lipase B towards irreversible thermal inactivation through directed evolution. Protein Eng 16:599–605

Zhao H (2010) Methods for stabilizing and activating enzymes in ionic liquids—a review. J Chem Technol Biotechnol 85:891–907

Zhao H, Baker GA, Holmes S (2011) New eutectic ionic liquids for lipase activation and enzymatic preparation of biodiesel. Org Biomol Chem 9:1908–1916

Zheng H, Liu Y, Sun M et al (2014) Improvement of alkali stability and thermostability of *Paenibacillus campinasensis* family-11 xylanase by directed evolution and site-directed mutagenesis. J Ind Microbiol Biotechnol 41:153–162

Zhong CQ, Song S, Fang N (2009) Improvement of low-temperature caseinolytic activity of a thermophilic subtilase by directed evolution and site-directed mutagenesis. Biotechnol Bioeng 104:862–870

Zhu X, He B, Zhao C et al (2016) Net-immobilization of β-glucosidase on nonwoven fabrics to lower the cost of "cellulosic ethanol" and increase cellulose conversions. Sci Rep 6:23437

Chapter 21
Heterologous Protein Expression in *Pseudoalteromonas haloplanktis* TAC125

Ermenegilda Parrilli and Maria Luisa Tutino

Abstract The Antarctic strain *Pseudoalteromonas haloplanktis* TAC125 is considered one of the model organisms of cold-adapted bacteria, and during last years, it has been exploited as an alternative expression system for recombinant protein production. *P. haloplanktis* TAC125 was the first Antarctic bacterium in which an efficient gene-expression technology was set up, and several generations of cold-adapted gene-expression vectors allow the production of recombinant proteins either by constitutive or inducible systems and to address the product toward any cell compartment or to the extracellular medium. Moreover, the development of synthetic media and efficient fermentation schemes, to upscale the recombinant protein production in automatic bioreactors, makes the industrial application of *P. haloplanktis* TAC125 more achievable and concrete. The cellular physicochemical conditions and folding processes in *P. haloplanktis* TAC125 are quite different from those observed in canonical mesophilic hosts and allowed the production of several difficult-to-express protein products, some of which are of human origin. The recently reported possibility to produce proteins within a range of temperature from 15 to −2.5 °C enhances the chances to improve the conformational quality and solubility of recombinant proteins. This chapter outlines main features and potentiality of this unconventional protein production platform.

Contents

E. Parrilli • M.L. Tutino (✉)
Department of Chemical Sciences, University of Naples Federico II, Via Cintia 4, 80126 Naples, Italy
e-mail: tutino@unina.it

R. Margesin (ed.), *Psychrophiles: From Biodiversity to Biotechnology*,
DOI 10.1007/978-3-319-57057-0_21

21.1 Introduction

The recombinant protein expression systems have greatly contributed to structural and functional analysis of proteins and to their biotechnological application. Among all the available recombinant expression systems, *Escherichia coli* facilitates gene expression by its relative plainness, inexpensive, and fast cultivation; well-known genetics; and the large number of tools available for its biotechnological application.

However, recombinant expression in *E. coli* is not always a straightforward procedure, and major obstacles are encountered when producing many eukaryotic proteins and especially membrane proteins, linked to protein proteolysis and aggregation. Incorrect folding of the nascent polypeptide chains is one of the main problems occurring during heterologous protein production in bacteria. A key role in this process is played by intermolecular hydrophobic interactions taking place among partly folded intermediates that cause protein molecules to stick together thus driving them away from the productive folding pathway (Georgiou and Valax 1996). Since formation of inclusion bodies often impairs the recombinant production of valuable proteins, many experimental approaches have been explored to minimize this undesirable effect, including expression of chimerical proteins (Mitra et al. 2005) and co-expression with chaperonines (Luo and Hua 1998) Expression of "difficult" proteins has also been carried out by lowering the temperature at the physiological limit allowed for the growth of mesophilic host organisms (between 15 °C and 18 °C for *E. coli*). Lowering the temperature, in fact, has a pleiotropic effect on the folding process, destabilizing the hydrophobic interactions occurring during intermediates aggregation (Jeon et al. 1995). Although in some cases this approach has been reported to increase yields of soluble and active recombinant protein products, the exploitation of an industrial process performed at suboptimal growth condition of the expression host might hardly be considered. A rational alternative to mesophilic organisms is the use of naturally cold-adapted bacteria as hosts for protein production at low temperature (even at around 0 °C). *Pseudoalteromonas haloplanktis* TAC125 is a Gram-negative bacterium isolated from an Antarctic coastal seawater sample collected in the vicinity of the French Antarctic station Dumont D'Urville, Terre Adélie. It can be classified as a eurypsychrophile, i.e., a bacterium growing over a wide range of low temperatures (Atlas and Bartha 1993). *P. haloplanktis* TAC125 was the first Antarctic Gram-negative bacterium from which the genome was fully sequenced and carefully annotated (Medigue et al. 2005). Genomic and metabolic features of this bacterium, accounting for its remarkable versatility and fast growth compared with other bacteria from aqueous environments, were discovered by combining genome sequencing and further *in silico* and *in vivo* analyses. Several traits make *P. haloplanktis* TAC125 an attractive host as cell factory for proteins:

1. The bacterium still duplicates at fast speed even at sub-zero temperatures and, when provided with sufficient nutrients and aeration, it grows to very high cell density (Sannino et al. 2016).

2. It is well adapted to protection against reactive oxygen species (ROS) under cold conditions by the development of a novel anti-ROS and anti-RNS (reactive nitrogen species) strategy using a 2-on-2 hemoglobin (Parrilli et al. 2010). Moreover it produces a relevant number of classical oxidative stress protecting enzymes (such as the superoxide dismutase SodB, the thioredoxin reductase TrxB, the thioredoxin-dependent peroxide reductase AhpCB, and catalase) (Medigue et al. 2005).
3. *P. haloplanktis* TAC125 *in silico* proteome composition revealed a specific bias that provides a way to resist to the protein aging features involving asparagine cyclization and deamidation (Medigue et al. 2005).
4. The Antarctic bacterium genome is characterized by a quite high number of rRNA genes and tRNA genes which justify its relevant capacity for translation at low temperatures (Medigue et al. 2005).
5. Moreover, to create genetically engineered bacterial strains with improved features as protein factories, an efficient genetic scheme for the construction of genome insertion/deletion mutants was set up (Parrilli et al. 2008a; Giuliani et al. 2012).
6. *P. haloplanktis* TAC125 is considered as one of the few model organisms of cold-adapted bacteria (Feller et al. 2003; Medigue et al. 2005), and the increasing interest in this bacterium has led to the accumulation of extended data concerning genetics and physiology of this microorganism and to a construction of a genome scale metabolic model (Fondi et al. 2015).
7. *P. haloplanktis* TAC125 was also the first Antarctic bacterium in which an efficient gene-expression technology was set up (Parrilli et al. 2008b).

21.2 Genetic Systems in *P. haloplanktis* TAC125

During the last decade, the number of reliable genetic systems for the recombinant gene expression in *P. haloplanktis* TAC125 was significantly enhanced. In this section, the genetic tools developed to achieve recombinant protein production will be described. Features of psychrophilic gene-expression vectors set up will be reported in some detail.

21.2.1 Constitutive Expression System

By combining mesophilic and psychrophilic genetic signals, a collection of psychrophilic gene-expression vectors was set up to produce recombinant proteins in *P. haloplanktis* TAC125. The mesophilic signals consist of the pUC18-derived origin of replication (OriC), a selection marker gene (a ß-lactamase-encoding gene), and the OriT sequence, the conjugational DNA transfer origin from the broad host range plasmid pJB3 (Blatny et al. 1997). This sequence is recognized by the *mob*- and *tra*-encoded functions of RK2, contained in several *E. coli λpir* strains, such as the S17-1(*λpir*) (Tascon et al. 1993), and is responsible for the plasmid mobilization from the donor to the psychrophilic recipient cells (Duilio et al. 2004a), while the psychrophilic origin of replication (OriR) from the *P. haloplanktis* TAC125 endogenous plasmid pMtBL (Tutino et al. 2001) is the psychrophilic signal. This system has made it possible to isolate constitutive psychrophilic promoters (Duilio et al. 2004b). The structural/functional characterization of *P. haloplanktis* TAC125 promoters was carried out by random cloning of genomic DNA fragments and identification of promoter sequences by evaluating their capability to express a promoter-less reporter gene (Duilio et al. 2004b). By this promoter-trap strategy, a collection of constitutive psychrophilic promoters showing different strengths at different temperatures was identified. The implementation of the above described psychrophilic promoters in the pMtBL-derived shuttle vectors resulted in the set up of cold-adapted gene-expression systems, characterized by the constitutive production of the recombinant protein. Several proteins were produced using the described constitutive expression system (Parrilli et al. 2008b).

21.2.2 Regulated Expression System

A regulative two-component system previously identified in *P. haloplanktis* TAC125 (Papa et al. 2006) was used to construct an inducible expression system that is under the control of L-malate (Papa et al. 2007). A regulative genomic region from *P. haloplanktis* TAC125 (encompassing the genes PSHAb0361–PSHAb0362) is involved in the transcriptional regulation of the gene coding for an outer membrane porin (PSHAb0363) that is strongly induced by the presence of L-malate in the growth medium. The regulative region was used to construct an inducible expression vector—named pUCRP—under the control of L-malate. Performances of the inducible system were tested for both psychrophilic and mesophilic protein production using two "difficult" proteins as control. The results obtained demonstrated that both psychrophilic β-galactosidase and yeast α-glucosidase are produced in a fully soluble and catalytically competent form. Optimal conditions for protein production, including growth temperature, growth medium, and L-malate concentration, were also reported (Papa et al. 2007; Rippa et al. 2012). Although pUCRP resulted to be highly upregulated in presence of L-malate, it was not totally repressed in absence of the inducer (Papa et al. 2007); therefore, it could not be suitable for recombinant production of protein toxic for the host cells.

21.2.3 Inducible Expression System

Aiming at the identification of new regulated promoters, the analysis of *P. haloplanktis* TAC125 genome was carried out in order to identify genes and/or operons whose transcription is expected to be induced in the presence of particular carbon sources. This kind of analysis was not easy as bacterial genes involved in primary metabolism processes are generally finely regulated in response to nutrients availability, but in environments where excess of several easily metabolized carbon sources are unlikely present simultaneously, as in Antarctic sea water, the catabolite repression is rare. The attention was captured by genes involved in D-galactose catabolism since unpublished results clearly indicate that *P. haloplanktis* TAC125 is able to grow on D-galactose, thus indicating the presence of a functional Leloir pathway (Frey 1996). The analysis indicated the presence in *P. haloplanktis* TAC125 genome of a *gal* gene cluster with a different structural organization compared to best studied *E. coli* one. In the mesophilic bacterium, the transcription regulation of *gal* operon is complex (Lewis and Adhya 2015) and involves cAMP, its receptor protein CRP, and GalS, the latter two components that are not present in *P. haloplanktis* TAC125 genome. However, the Antarctic bacterium possesses a gene coding for a *E. coli* GalR homologue, *Phgal*R (PSHAa1771), which is located upstream the *Phgal*T but in the opposite direction.

In order to obtain a finely regulated expression system, the DNA region upstream to the *Phgal* operon, i.e., the sequence containing both *Phgal*TK promoter and the gene coding for *Ph*GalR (named MAV), was suitably PCR-amplified and used to construct the psychrophilic pMAV vector (Sannino et al. 2016).

The new expression system resulted to be repressed in absence and induced in presence of D-galactose, and therefore it could be used for the production of proteins toxic for the host cells (Sannino et al. 2016).

21.2.4 Molecular Signals for Protein Addressing

Though the production of recombinant protein in the host cytoplasm is the favorite strategy in many processes due to higher production yields, this approach cannot be followed when the wanted product requires the correct formation of disulfide bonds to attain its catalytic competent conformation. Protein translocation from the cytoplasm to periplasmic space can be attained by three different routes: the Sec pathway (Matlack et al. 1998) which is a post-translational export; the Srp pathway (Luirink and Sinning 2004), which is a co-translational export and shares with the previous one some components; and the TAT pathway (Lee et al. 2006), which diverges from the others since it is able to translocate fully folded passengers. From the genome analysis, it turned out that *P. haloplanktis* TAC125 contains all the gene-encoding components of above-mentioned export machineries (Medigue et al. 2005). Therefore, a family of psychrophilic gene-expression systems for the recombinant protein production in the periplasmic space was set up. Gene fragments encoding two signal peptides (Sec- or SRP-dependent) from psychrophilic secreted proteins have been cloned under the control of constitutive or regulated promoters to generate several chimerical proteins, where the signal peptide was fused to the N-terminal of recombinant disulfide bond-containing proteins, and their efficiency as periplasm addressing tag was tested (Parrilli et al. 2008b). Furthermore, a cold-adapted secretion system, which makes use of a psychrophilic α-amylase as translocation carrier, was set up to secrete in the extracellular medium the produced recombinant proteins (Cusano et al. 2006; Parrilli et al. 2008b).

21.3 Toward *P. haloplanktis* TAC125 Implementation in Industrial Processes

The understanding of the growth physiology and the optimization of cultivation strategies are essential factors to achieve the goal of high protein production by recombinant organisms. Indeed, when the product is cell-associated, the productivity is correlated to biomass level. In this case, the optimization of the protein production is strictly depending on media composition and, in turn, closely related to the choice of the more suitable cultivation strategy to be used.

In order to widen and deeply explore the biotechnological capabilities of *P. haloplanktis* TAC125, the attention was addressed toward the development of synthetic media and the optimization of cultivation strategies.

21.3.1 Culture Media

The definition of a low cost synthetic medium is the requirement for scale-up of the *P. haloplanktis* TAC125 growth in automatic bioreactors.

LIV Medium In several studies, it has been reported that the psychrophilic bacterium of our interest shows nutritional preferences for amino acids as carbon and nitrogen source. For this reason, bacterial growth parameters were estimated in synthetic media composed of a mineral base (SCHATZ salts) with the addition of several amino acids (Giuliani et al. 2011). Obtained results suggested the use of L-leucine, in combination with L-isoleucine and L-valine, as carbon and nitrogen source and led to the formulation of LIV medium, which was then used to perform recombinant protein production both in batch and in chemostat cultivations. In detail, LIV medium contains the three selected amino acids L-leucine, L-isoleucine, and L-valine in a 1:1:2 mole ratio. The growth of the psychrophilic bacterium in LIV medium is characterized by a brief lag phase and a modest but significant increase of specific growth rate and biomass yield, compared to the growth in the medium with only L-leucine (Giuliani et al. 2011). In LIV medium, the pUCRP expression system resulted to be able to increase a reporter enzyme production yield more than seven-fold and the biomass yield three-fold, with respect to the previously optimized conditions (Giuliani et al. 2011). However, LIV medium did not allow bioprocesses at 4 °C due to the very low specific growth rate at this temperature (Giuliani et al. 2011).

GG Medium The recently published genome-scale metabolic reconstruction of *P. haloplanktis* TAC125 (Fondi et al. 2015) was used to *in silico* simulate the growth of the bacterium on 20 different minimal media. Results of this analysis revealed that predicted growth rates were much higher for the following four amino acids: L-leucine, L-proline, L-glutamate, and L-glutamine. It has previously been shown that L-leucine, when used as the sole carbon source for *P. haloplanktis* growth, at a concentration higher than 5 g L^{-1} actually inhibits the bacterial growth because of the saturation of the L-leucine uptake system (Giuliani et al. 2011). Among the other three *in silico* tested amino acids, glutamate has been already suggested in independent experimental tests to be a significant carbon and nitrogen source for *P. haloplanktis* TAC125 (Wilmes et al. 2010; Giuliani et al. 2011). For these reasons, the attention was focused on L-glutamate (Sannino et al. 2016) and on the other physiological parameters influencing the bacterial growth on this substrate. It resulted that the specific growth rate of *P. haloplanktis* TAC125, when grown in L-glutamate, depends on NaCl concentration used in growth medium.

The analysis of the genome of *P. haloplanktis* TAC125 highlighted the occurrence of the Entner–Doudoroff (ED) pathway (Medigue et al. 2005) as central catabolic pathway for carbon metabolism. *In silico* analysis revealed that genes for all enzymatic functions required in the ED pathway are present in the genome of *P. haloplanktis* TAC125. Moreover, the bacterium resulted to be able to grow in a mineral medium containing D-gluconate as the sole carbon and energy source (Sannino et al. 2016). Therefore, aiming at formulating a synthetic medium in which the Antarctic bacterium displays a balanced growth, it was decided to explore the combination of L-glutamate and D-gluconate in a mineral base. Therefore, numerous attempts have been carried out looking for the best combination of L-glutamate and D-gluconate, and the medium showing the best growth performance was named GG (10 g L-glutamate L^{-1} and 10 g D-gluconate L^{-1}). The biomass yields in GG medium resulted to be higher than those observed in media containing either L-glutamate or D-gluconate alone (Sannino et al. 2016). This improvement was more remarkable at 4 °C, and the growth kinetic parameters of the bacterium grown in GG medium at 4 °C demonstrated that this medium is optimized for *P. haloplanktis* TAC125 growth at low temperature (Sannino et al. 2016).

In detail, L-glutamate and D-gluconate furnish the bulk of cellular energy and various building blocks for cellular functions through tricarboxylic acid cycle (TCA), in the case of glutamate, and through ED and pentose phosphate pathways, in the case of gluconate. Furthermore, the L-gluconate metabolism provides a reductive environment to nullify the ROS (Singh et al. 2007).

In GG medium, the Antarctic bacterium was able to grow at 4 °C with the best biomass yield ever reported (0.44), and to efficiently grow at subzero temperatures (0 °C and −2.5 °C), displaying a doubling time of about 23 h at −2.5 °C. These growth performances are quite notable if compared to the other reported data (Mykytczuk et al. 2013; Nunn et al. 2015), where bacteria growing at subzero temperature in rich medium displayed a doubling time of several weeks. Data reported do support the choice of *P. haloplanktis* TAC125 as one of the best-adapted microorganisms to freezing lifestyle, to be pursued toward the description of the microbial metabolism at a temperature as low as −2.5 °C. Furthermore, the discovery of such a medium will be important in further studies aimed at the definition of molecular strategies adopted by microorganisms to face such extreme conditions.

21.3.2 Cultivation Strategies

The use of *P. haloplanktis* TAC125 as host for recombinant protein production for industrial purposes requires the possibility to employ different fermentation strategies.

A fed-batch cultivation strategy for this auspicious alternative expression host was reported by Wilmes and co-authors (Wilmes et al. 2010). They reported that the growth of *P. haloplanktis* TAC125 to higher cell densities on complex medium is possible by a fed-batch cultivation using casamino acids, which have a good solubility and allow a good growth control (Wilmes et al. 2010). The fed-batch process was established at 16 °C by feeding of casamino acids with a constant rate resulting in a cell dry mass of about 11 g L^{-1} which is an eight-fold increase of the cell density obtained in standard shake flask cultures.

Giuliani and co-authors established a chemostat culture of *P. haloplanktis* TAC125 at 15 °C where a feeding medium containing L-leucine as sole carbon source was used after batch cultivation in LIV medium (Giuliani et al. 2011). The L-leucine containing medium was fed to the culture at a dilution rate of 0.05 h^{-1}; the steady state was maintained for at least five resident times, in which both the cell density and the product titer remained constant. Reported data demonstrated that a continuous culture operation with *P. haloplanktis* TAC125 is feasible, and, therefore, it allows for further use of such strategy as a tool for physiological studies and process optimization for soluble production of recombinant proteins at low temperatures. A chemostat culture of *P. haloplanktis* TAC125 was established using also a rich medium (Schatz medium with tryptone and vitamins) at 4 °C at two dilution rates: 0.03 and 0.05 h^{-1} (Vigentini et al. 2006).

21.4 Examples of Heterologous Protein Produced in *P. haloplanktis* TAC125

Over the last decade, the genetic systems for the recombinant gene expression in *P. haloplanktis* TAC125 allowed the production of several recombinant proteins in diverse media using different cultivation strategies and in a wide range of temperatures from 15 °C to −2.5 °C (Table 21.1). Indeed in a recently published paper (Sannino et al. 2016), the production of a recombinant protein at subzero temperature was reported, for the first time, thus providing an innovative strategy for the recombinant production of "difficult" proteins.

Beneficial effects in using this cold-adapted protein production platform was validated by the successful production of difficult proteins (Vigentini et al. 2006; Papa et al. 2007) and biopharmaceuticals (Dragosits et al. 2011; Giuliani et al. 2011; Corchero et al. 2013; Unzueta et al. 2015). The recombinant proteins, when required, were successfully secreted in the periplasmic space and in the extracellular medium allowing the right folding of disulfide bond-containing proteins.

The successful production in a soluble and biologically competent form of aggregation-prone proteins of biopharmaceutical interest, such as antibody fragments (Table 21.1), is related to the amplification of the number of genes coding for peptidyl-prolyl *cis-trans* isomerases (PPIases), whose enzymatic activity is required to attain their correct three-dimensional conformation. Indeed, the *P. haloplanktis* TAC125 genome contains 15 genes encoding an enlarged set of PPIases (Giuliani et al. 2014). Since it is widely accepted that scFv and Fab antibody fragments folding rely on the activity of PPIases (Feige et al. 2010), *P. haloplanktis* TAC125 resulted to be a naturally optimized host for the recombinant production of antibody fragments. Moreover, it is interesting to note that insoluble aggregates of recombinant protein have never been found in *P. haloplanktis* TAC125 suggesting that its cellular physicochemical conditions and/or folding processes are rather different from those observed in mesophilic bacteria.

21.5 Conclusions

The production of recombinant proteins in *P. haloplanktis* TAC125 is not only a mature and reliable technology, but it is also a successful strategy to overcome the production and/or folding issues sometimes occurring in conventional systems. The main fall-out deriving from the development of this innovative production platform is the publication of several recent papers reporting the use of other cold-adapted bacteria as host for recombinant protein production. For example, a low-temperature expression system using an Antarctic cold-adapted bacterium, *Shewanella sp.* strain Ac10, was successfully used to produce putative peptidases, PepF, LAP, and PepQ, and a putative glucosidase, BglA, from *Desulfotalea*

Table 21.1 Heterologous proteins produced in *P. haloplanktis* TAC125

Recombinant protein	Promoter	Cell compartment	Temperature	Medium	Cultivation strategy	References
β-Galactosidase from *P. haloplanktis* TAE79	Regulated (pUCRP)	Cytoplasm	15 °C, 4 °C	MM + casamino acids	Batch	Papa et al. (2007)
	Regulated (pUCRP)	Cytoplasm	15 °C	LIV	Batch	Giuliani et al. (2011)
	Inducible (pMAV)	Cytoplasm	−2.5 °C, 0 °C, 4 °C, 15 °C	GG	Batch	Sannino et al. (2016)
α-Glucosidase from *Saccharo-myces cerevisiae*	Regulated (pUCRP)	Cytoplasm	15 °C, 4 °C	MM + casamino acids	Batch	Papa et al. (2007)
Human nerve growth factor hβ-NGF	Constitutive	Periplasmic space	4 °C	MM + tryptone and vitamins	Batch, continuous cultivation	Vigentini et al. (2006)
AmyΔCt-DsbA chimera	Constitutive	Extracellular medium	4 °C	TYP	Batch	Cusano et al. (2006)
AmyΔCt-IGPS chimera	Constitutive	Extracellular medium	4 °C	TYP	Batch	Cusano et al. (2006)
AmyΔCt-BlaM	Constitutive	Extracellular medium	4 °C	TYP	Batch	Cusano et al. (2006)
ScFv anti-oxazolone single chain antibody	Regulated (pUCRP)	Periplasmic space	15 °C	LIV	Batch, continuous cultivation	Giuliani et al. (2014)
Camelid VHHD6.1 fragment	Regulated (pUCRP)	Periplasmic space	15 °C	LIV	Batch	Giuliani et al. (2015)
3H6 Fab fragment	Regulated (pUCRP)	Periplasmic space	15 °C	LIV	Batch, continuous cultivation	Giuliani et al. (2011)
Human α-galactosidase A	Constitutive	Periplasmic space	15 °C, 4 °C	TYP	Batch	Unzueta et al. (2015)

MM, marine mix; TYP, complex medium containing bacto-tryptone, yeast extract, and NaCl; LIV, synthetic medium containing leucine, isoleucine, and valine in Schatz mineral base; GG, synthetic medium containing gluconate and glutamate in Schatz mineral base (see cited literature and text for the exact composition)

psychrophila DSM12343 (Miyake et al. 2007). Additionally, using a recently developed expression system in psychrophilic *Pseudoalteromonas sp.*, a cold-adapted protease, pseudoalterin, which cannot be maturely expressed in *E. coli*, was successfully expressed as an active extracellular enzyme (Yu et al. 2015). Furthermore, a low-temperature-inducible protein expression vector (pSW2) based on a filamentous phage (SW1) of the deep-sea bacterium *Shewanella piezotolerans* WP3 was employed to produce a patatin-like protein (PLP) that tends to form inclusion bodies in *E. coli* (Yang et al. 2015).

In conclusion, data reported confirm the value of *P. haloplanktis* TAC125 as cell factory and endorse the potential of cold-adapted bacteria as alternative hosts for recombinant protein production.

References

Atlas RM, Bartha R (1993) Microbial ecology: fundamentals and applications, 3rd edn. Benjamin/Cummings, Redwood City, CA

Blatny JM, Brautaset T, Winther-Larsen HC et al (1997) Construction and use of a versatile set of broad-host-range cloning and expression vectors based on the RK2 replicon. Appl Environ Microbiol 63:370–379

Corchero JL, Gasser B, Resina D, Smith W, Parrilli E, Vázquez F, Abasolo I, Giuliani M, Jäntti J, Ferrer P, Saloheimo M, Mattanovich D, Schwartz S Jr, Tutino ML, Villaverde A (2013) Unconventional microbial systems for the cost-efficient production of high-quality protein therapeutics. Biotechnol Adv 31(2):140–153

Cusano AM, Parrilli E, Marino G et al (2006) A novel genetic system for recombinant protein secretion in the Antarctic *Pseudoalteromonas haloplanktis* TAC125. Microb Cell Fact 5:40

Dragosits M, Frascotti G, Bernard-Granger L, Vázquez F, Giuliani M, Baumann K, Rodríguez-Carmona E, Tokkanen J, Parrilli E, Wiebe MG, Kunert R, Maurer M, Gasser B, Sauer M, Branduardi P, Pakula T, Saloheimo M, Penttilä M, Ferrer P, Luisa Tutino M, Villaverde A, Porro D, Mattanovich D (2011) Influence of growth temperature on the production of antibody Fab fragments in different microbes: a host comparative analysis. Biotechnol Prog 27(1):38–46

Duilio A, Madonna S, Tutino ML et al (2004b) Promoters from a cold-adapted bacterium: definition of a consensus motif and molecular characterization of UP regulative elements. Extremophiles 8:125–132

Duilio A, Tutino ML, Marino G (2004a) Recombinant protein production in Antarctic Gram-negative bacteria. Methods Mol Biol 267:225–237

Feige MJ, Hendershot LM, Buchner J (2010) How antibodies fold. Trends Biochem Sci 35:189–198

Fondi M, Maida I, Perrin E et al (2015) Genome-scale metabolic reconstruction and constraint-based modelling of the Antarctic bacterium *Pseudoalteromonas haloplanktis* TAC125. Environ Microbiol 17(3):751–766

Frey P (1996) The Leloir pathway: a mechanistic imperative for three enzymes to change the stereochemical configuration of a single carbon in galactose. FASEB J 10(4):461–470

Georgiou G, Valax P (1996) Expression of correctly folded proteins in *Escherichia coli*. Curr Opin Biotechnol 7:190–197

Giuliani M, Parrilli E, Ferrer P et al (2011) Process optimization for recombinant protein production in the psychrophilic bacterium *Pseudoalteromonas haloplanktis*. Process Biochem 46(4):953–959

Giuliani M, Parrilli E, Sannino F et al (2014) Recombinant production of a single-chain antibody fragment in *Pseudoalteromonas haloplanktis* TAC125. Appl Microbiol Biotechnol 98:4887–4895

Giuliani M, Parrilli E, Sannino F et al (2015) Soluble recombinant protein production in *Pseudoalteromonas haloplanktis* TAC125. Methods Mol Biol 1258:243–257

Giuliani M, Parrilli E, Pezzella C,, et al. (2012) A novel strategy for the construction of genomic mutants of the Antarctic bacterium *Pseudoalteromonas haloplanktis* TAC125. Methods Mol Biol 824:219-233

Lee PA, Tullman-Ercek D, Georgiou G (2006) The bacterial twin-arginine translocation pathway. Annu Rev Microbiol 60:373–395, Review

Jeon YH, Negishi T, Shirakawa M et al (1995) Solution structure of the activator contact domain of the RNA polymerase alpha subunit. Science 270:1495–1497

Lewis DE, Adhya S (2015) Molecular mechanisms of transcription initiation at gal promoters and their multi-level regulation by GalR, CRP and DNA loop. Biomolecules 5(4):2782–2807

Luirink J, Sinning I (2004) SRP-mediated protein targeting: structure and function revisited. Biochim Biophys Acta 1694(1–3):17–35

Luo ZH, Hua ZC (1998) Increased solubility of glutathione S-transferase-P16 (GST-p16) fusion protein by co-expression of chaperones GroES and GroEL in *Escherichia coli*. Biochem Mol Biol Int 46:471–477

Matlack KE, Mothes W, Rapoport TA (1998) Protein translocation: tunnel vision. Cell 92 (3):381–390

Medigue C, Krin E, Pascal G et al (2005) Coping with cold: the genome of the versatile marine Antarctica bacterium *Pseudoalteromonas haloplanktis* TAC125. Genome Res 15:1325–1335

Mitra A, Chakrabarti KS, Shahul Hameed MS et al (2005) High level expression of peptides and proteins using cytochrome b5 as a fusion host. Protein Expr Purif 41:84–97

Miyake R, Kawamoto J, Wei JL et al (2007) Construction of a low-temperature protein expression system using a cold-adapted bacterium, *Shewanella* sp. strain Ac10, as the host. Appl Environ Microbiol 73(15):4849–4856

Mykytczuk NC, Foote SJ, Omelon CR et al (2013) Bacterial growth at -15 °C; molecular insights from the permafrost bacterium *Planococcus halocryophilus* Or1. ISME J 7(6):1211–1226

Nunn BL, Slattery KV, Cameron KA et al (2015) Proteomics of *Colwellia psychrerythraea* at subzero temperatures—a life with limited movement, flexible membranes and vital DNA repair. Environ Microbiol 17(7):2319–2335

Papa R, Glagla S, Danchin A et al (2006) Proteomic identification of a two-component regulatory system in *Pseudoalteromonas haloplanktis* TAC125. Extremophiles 10(6):483–491

Papa R, Rippa V, Sannia G et al (2007) An effective cold inducible expression system developed in *Pseudoalteromonas haloplanktis* TAC125. J Biotechnol 127(2):199–210

Parrilli E, De Vizio D, Cirulli C et al (2008a) Development of an improved *Pseudoalteromonas haloplanktis* TAC125 strain for recombinant protein secretion at low temperature. Microb Cell Fact 7:2

Parrilli E, Duilio A, Tutino ML (2008b) Heterologous protein expression in psychrophilic hosts. In: Margesin R, Schinner F, Marx JC, Gerday C (eds) Psychrophiles: from biodiversity to biotechnology. Berlin Heidelberg, Springer, pp 365–379

Parrilli E, Giuliani M, Giordano D (2010) The role of a 2-on-2 haemoglobin in oxidative and nitrosative stress resistance of Antarctic *Pseudoalteromonas haloplanktis* TAC125. Biochimie 92(8):1003–1009

Rippa V, Papa R, Giuliani M et al (2012) Regulated recombinant protein production in the Antarctic bacterium *Pseudoalteromonas haloplanktis* TAC125. Methods Mol Biol 824:203–218

Sannino F, Giuliani M, Salvatore U et al (2016) A novel synthetic medium and expression system for subzero growth and recombinant protein production in *Pseudoalteromonas haloplanktis* TAC125. Appl Microbiol Biotechnol. doi:10.1007/s00253-016-7942-5

Singh R, Mailloux RJ, Puiseux-Dao S, Appanna VD (2007) Oxidative stress evokes a metabolic adaptation that favors increased NADPH synthesis and decreased NADH production in *Pseudomonas fluorescens*. Bacteriology 189(18):6665–6675

Tascon RI, Rodriguez-Ferri EF, Gutierrez-Martin CB et al (1993) Transposon mutagenesis in *Actinobacillus pleuropneumoniae* with a Tn10 derivative. J Bacteriol 175:5717–5722

Tutino ML, Duilio A, Parrilli E et al (2001) A novel replication element from an Antarctic plasmid as a tool for the expression of proteins at low temperature. Extremophiles 5:257–264

Unzueta U, Vázquez F, Accardi G et al (2015) Strategies for the production of difficult-to-express full-length eukaryotic proteins using microbial cell factories: production of human alpha-galactosidase A. Appl Microbiol Biotechnol 99(14):5863–5874

Vigentini I, Merico A, Tutino ML, Compagno C, Marino G (2006) Optimization of recombinant human nerve growth factor production in the psychrophilic *Pseudoalteromonas haloplanktis*. J Biotechnol 127(1):141–150

Wilmes B, Hartung A, Lalk M et al (2010) Fed-batch process for the psychrotolerant marine bacterium *Pseudoalteromonas haloplanktis*. Microb Cell Fact 21:9–72

Yang X-W, Jian H-H, Wang F-P et al (2015) pSW2, a novel low-temperature inducible gene expression vector based on a filamentous phage of the deep-sea bacterium *Shewanella piezotolerans* WP3. Appl Environ Microbiol 81:5519–5526

Yu Z-C, Tang B-L, al ZD-L (2015) Development of a cold-adapted *Pseudoalteromonas* expression system for the *Pseudoalteromonas* proteins intractable for the *Escherichia coli* system. PLoS One 10(9):e0137384

Chapter 22
Psychrophiles as a Source of Novel Antimicrobials

Erik Borchert, Stephen A. Jackson, Fergal O'Gara, and Alan D.W. Dobson

Abstract Cold environments such as Arctic and Antarctic regions and the deep sea are richly populated by microbes which encounter the same selective pressures and/or even more than their counterparts from moderate or warm environments. Microbes from moderate and warm environments have been extensively studied for their ability to produce antimicrobial compounds, and new results are scarce. Taking into account the rapid emergence of antimicrobial resistance, we need to look for new sources of antimicrobials. Here in this review, we summarise the knowledge of new antimicrobials or antimicrobial activity from microorganisms from different cold environments to date. A special focus due to their richness in bioactive compounds is given to marine environments; nonetheless, compounds from other sources are mentioned. The later-described compounds all display some kind of beneficial antimicrobial activity and are part of many different chemical classes of antimicrobial compounds (lantibiotics, spirotetronates, microcins, synoxazolidinones, indole alkaloids, monanchocidins, etc.). Furthermore, studies which identified antimicrobial producers but did not identify the actual produced compound are mentioned, and promising sources for future bioprospecting are discussed.

E. Borchert • S.A. Jackson
School of Microbiology, University College Cork, National University of Ireland, Cork T12YN60, Ireland

F. O'Gara
School of Microbiology, University College Cork, National University of Ireland, Cork T12YN60, Ireland

Biomerit Research Centre, University College Cork, National University of Ireland, Cork T12YN60, Ireland

School of Biomedical Sciences, Curtin Health Innovation Research Institute, Curtin University, Perth, WA, Australia

A.D.W. Dobson (✉)
School of Microbiology, University College Cork, National University of Ireland, Cork T12YN60, Ireland

Environmental Research Institute, University College Cork, National University of Ireland, Cork T23XE10, Ireland
e-mail: A.Dobson@ucc.ie

R. Margesin (ed.), *Psychrophiles: From Biodiversity to Biotechnology*,
DOI 10.1007/978-3-319-57057-0_22

Contents

22.1 Why Do We Need Novel Antimicrobials?

Before the 'antibiotic era', infectious diseases such as tuberculosis, syphilis, cholera, smallpox, plague, mumps and many others had been major causes of death amongst the human population. The discovery of penicillin by Alexander Fleming in 1928 (Fleming 2001) together with other antibiotics leads to major increases in life expectancy and life quality. These discoveries together with advances in health care (vaccination, antisepsis, public health measures and sanitation), in the early 1950s and, thereafter, in the so-called Golden era of antibiotic discovery, resulted in infectious diseases becoming less of a problem as a major cause of morbidity in the general population, relative to other diseases such as cardiovascular diseases, cancer and stroke.

Soon after the discovery of penicillin, concerns were raised about the possible development of antibacterial resistance to the antibiotic (even from Alexander Fleming himself) (Levy 2002). Unfortunately, these fears have been realised, not only with respect to penicillin but to a wide variety of other antibiotics (Brown and Wright 2016). Nowadays, it is widely believed that we live in a 'post antibiotic era' (Alanis 2005), where an increasing number of pathogenic bacteria are resistant to many commonly used antibiotics, resulting in infectious diseases becoming an ever more common cause of death in humans. Bacteria have acquired resistance to many antibiotics through a variety of mechanisms primarily involving mutation, conjugation, transformation and transduction, resulting in the acquisition of resistances to all classes of antibiotics currently in use (Table 22.1). Therefore, the rapid spread of resistance amongst opportunistic human pathogens to antimicrobials is a huge threat to the healthcare system and future development of the human population in general, which appears unable to keep pace with the ever-evolving microbial pathogens. It can in fact be argued that we are even speeding up the process of the evolution of antibiotic resistance ourselves by misusing antimicrobials, particularly the widespread and quite indiscriminate use of antibiotics in the agricultural/aquaculture areas (Davies and Davies 2010). To date most of the antimicrobials in use have been isolated from microorganisms from terrestrial, temperate or tropic environments. In order to find novel bioactive compounds with new modes of

Table 22.1 Antibiotic families, mechanism of action and resistance mechanism [adapted from Alanis (2005) and Davies and Davies (2010)]

Antibiotic family	Mechanism of action	Resistance mechanism
Beta-lactams	Inhibition of cell wall synthesis	Beta-lactamases, efflux, altered target
Glycopeptides	Inhibition of cell wall synthesis	Reprogramming peptidoglycan biosynthesis
Cyclic lipopeptides	Inhibition of cell wall synthesis	Altered target
Tetracyclines	Inhibition of protein synthesis	Monooxygenation, efflux, altered target
Aminoglycosides	Inhibition of protein synthesis	Phosphorylation, acetylation, nucleotidylation, efflux, altered target
Streptogramins	Inhibition of protein synthesis	C–O lyase, acetylation, efflux, altered target
Oxazolidinones	Inhibition of protein synthesis	Efflux, altered target
Macrolides	Inhibition of protein synthesis	Hydrolysis, glycosylation, phosphorylation, efflux, altered target
Lincosamides	Inhibition of protein synthesis	Nucleotidylation, efflux, altered target
Fluoroquinolones	Inhibition of DNA synthesis	Acetylation, efflux, altered target
Rifamycins	Inhibition of RNA synthesis	ADP-ribosylation, efflux, altered target
Sulfonamides	C_1 metabolism	Efflux, altered target
Polymyxins	Membrane-disorganizing agents	Efflux, altered target
Nitroimidazole	Others	Altered target
Phenicols	Inhibition of protein synthesis	Acetylation, efflux, altered target
Pyrimidines	C_1 metabolism	Efflux, altered target

action, it is widely believed that microorganisms from different environmental ecosystems, such as those from marine environments, need to be targeted.

22.2 Known Compounds from Cold Environments

While cold environments have to date been mostly overlooked in the search for new antimicrobials, nonetheless there are several studies focusing on the microbial diversity of cold environments which indicates a high level of diversity within these environments. It is widely believed that high levels of microbial diversity are indicative of high levels of potential antibiotic-producing microbes, given that these compounds are likely to be advantageous for the producing organism, particularly in an environment where there is competition for resources. Furthermore, given that marine microorganisms have survived under extremes of temperature, salinity and pressure over many millions of years, they are likely to have evolved to adapt to these extreme conditions and therefore potentially possess novel biochemistry. Thus, due to the environmental differences between cold marine environments and temperate or tropic marine environments coupled with adaptive evolution, it can be assumed that the bioactive compounds produced by microorganisms from these cold environments are likely to be quite different from many of the classes of antimicrobials currently in use. To date, mostly large-scale and rather unspecific antimicrobial screens of microorganism retrieved from, for example, alpine sites, benthic mats from Antarctic lakes and sponges from deep-sea and Arctic environments have been quarried out. Therefore, more emphasis should be given to find new antimicrobials from these sources coupled to a more in-depth analysis of the compounds/activities found, because to date only a few studies concentrated on targeting antimicrobial activity in these environments; nonetheless, most of them show promising results.

A comprehensive review on cold-water marine natural products was published in 2007 by Lebar and co-workers, covering most of the compounds identified up until 2005 from cold marine environments (Lebar et al. 2007). In this review, natural products from microbes, bacteria, fungi, microalgae, macroalgae, sponges, corals, bryozoans, molluscs, tunicates and echinoderms living in cold marine environments were described. Furthermore, Abbas and co-workers subsequently published a review on Arctic and Antarctic sponge secondary metabolites (Abbas et al. 2011). Thus, we will therefore only review more recent advances but recommend the interested reader to refer to the aforementioned publication if required. Furthermore, a general article on marine natural products is typically published on an annual basis (Blunt et al. 2016), but typically no more than 3% of the compounds which are described are retrieved from cold or deep-sea sources and even less display antimicrobial activity.

22.2.1 *Synoxazolidinones*

Synoxazolidinones A and B (Fig. 22.1) have been isolated from the ascidian *Synoicum pulmonaria* collected from the Norwegian coast in 2010 (Tadesse et al. 2010). These compounds constitute a novel family of brominated guanidinium oxazolidinones with activities against a range of Gram-positive bacteria, especially against methicillin-resistant *Staphylococcus aureus* (MRSA). Besides the synoxazolidinones A and B, the ascidian also produces synoxazolidinones C and pulmonarins, which are also brominated compounds, all of which display some

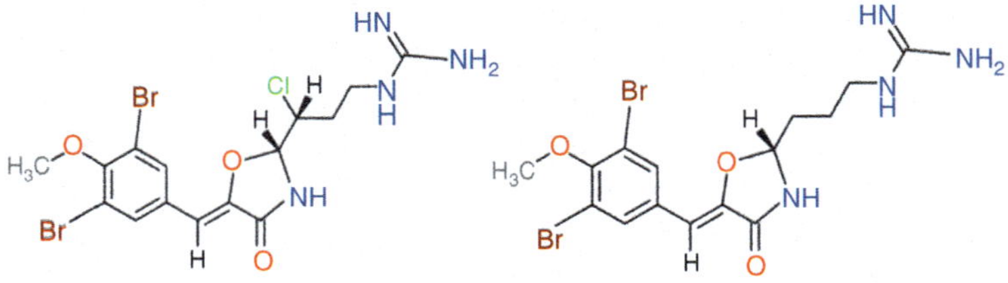

Fig. 22.1 Synoxazolidinones A and B [adapted from Tadesse et al. (2010) and visualised with 2D Sketcher (https://web.chemdoodle.com/demos/sketcher/)]

kind of antimicrobial activity especially against micro- and macrofouling organisms in the water column and are therefore of industrial interest (Trepos et al. 2014). However, one of the major bottlenecks in the use of bioactive compounds from natural resources for biopharmaceutical application is the quantities produced by the native strains, which are often quite low. Therefore, the large-scale production of these compounds would need the harvest of huge amounts of, for example, ascidians to fulfil the required demands. Therefore, the possibility of total chemical synthesis of, for example, synoxazolidinones, would be required to allow large-scale productions, thereby minimising any potential detrimental environmental impact of harvesting ascidians (Shymanska et al. 2014).

22.2.2 Microcins

In 2010, a bacteriocin-like compound named serraticin A produced by a psychrophilic microorganism closely related to *Serratia proteamaculans* was isolated from a soil sample from Isla de los Estados, Argentina, and was termed to be the first cold-active compound with antimicrobial activity from *S. proteamaculans* (Sánchez et al. 2010). The compound showed activity against an *Escherichia coli* and a *Salmonella enterica* strain but did not result in haemolysis. The mode of action was proposed to involve either blocking DNA replication or inhibition of the septation process.

22.2.3 Lantibiotics

Subtilomycin is a class I bacteriocin and was purified from a *Bacillus subtilis* strain isolated from the marine sponge *Haliclona simulans* collected on the west coast of Ireland (Phelan et al. 2013). The peptide shows very good activity against *Clostridium sporogenes*, good activity against *Bacillus cereus*, *Bacillus megaterium*, *Listeria monocytogenes* and *Listeria innocua* and also some activity against *Staphylococcus aureus*, a methicillin-resistant *S. aureus* strain and a vancomycin-resistant *S. aureus* strain. Class I lantibiotics usually interfere with the cell membranes of its target, either by inhibiting membrane biosynthesis or pore forming (McAuliffe et al. 2001).

22.2.4 Spirotetronate Antibiotics

Lobophorins are classified as medium-sized spirotetronates with a central ring system comprising of 13 carbon atoms (Vieweg et al. 2014). In 2013 Pan and co-workers identified two new groups of compounds, namely, lobophorins H and I from a *Streptomyces* sp. isolated from a South China deep-sea sediment sample (Pan et al. 2013). Lobophorin H in particular showed potent activity against *Bacillus subtilis* (Fig. 22.2), which was comparable to the activity of ampicillin; unfortunately, the mode of action for lobophorins has not as yet been described. Furthermore, the lobophorins seem to be exclusively active against Gram-positive bacteria and not against either Gram-negative bacteria or fungi, but some of them do display antitumor activities (Cruz et al. 2015).

In 2013 Wang and co-workers screened a large marine-derived library comprising of 4024 bacterial and 533 fungal isolates for growth inhibition of an attenuated strain of the bovine tuberculosis bacillus *Mycobacterium bovis* (Wang et al. 2013). Twenty-seven of the screened abstracts (0.6%) showed antibacterial activity. One of the active extracts was from a South China deep-sea sediment-derived actinomycete, *Verrucosispora* sp., which was retrieved from 2733 m below sea level. Structural elucidation of the active fractions revealed the presence of three new abyssomicin polyketides as well as four known abyssomicins, which were formerly also isolated from *Verrucosispora* sp. isolates. The newly isolated abyssomicin J (Fig. 22.3) is a dimeric thioester in contrast to other members of the abyssomicin family, which are

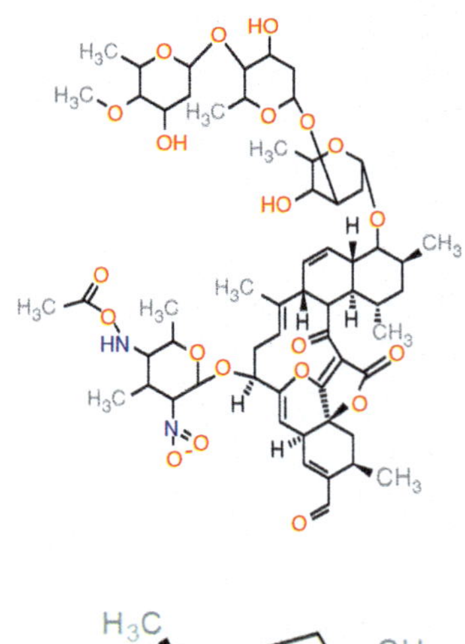

Fig. 22.2 Lobophorin H (adapted from Pan et al. (2013) and visualised with 2D Sketcher)

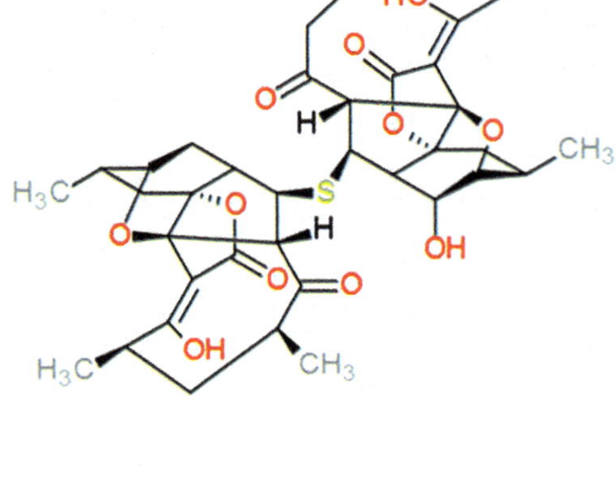

Fig. 22.3 Abyssomicin J (adapted from Wang et al. (2013) and visualised with 2D Sketcher)

typically monomeric small spirotetronates (central ring system $C = 11$). Abyssomicins are of particular interest as novel antibiotics as they target the *p*-aminobenzoic acid biosynthetic pathway, which is involved in the synthesis of tetrahydrofolate, a pathway unique to multiple microorganisms, but not found in humans.

22.2.5 Other Compounds

The actinomycete genus *Serinicoccus* which was firstly discovered in 2004 (Xiao et al. 2011) and which currently contains only three species, all of whom were isolated from marine habitats, was recently reported to produce secondary metabolites and new indole alkaloids (Fig. 22.4) with weak antimicrobial and cytotoxic activities (Yang et al. 2013).

Monanchocidins B–E (Fig. 22.5) are unusual polycyclic guanidine alkaloids isolated from the marine sponge *Monanchora pulchra* collected near Urup Island by dredging, which displayed potent antileukemic activities (Makarieva et al.

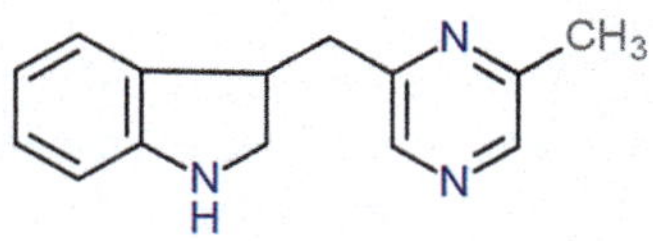

Fig. 22.4 New indole alkaloid (adapted from Yang et al. (2013) and visualised with 2D Sketcher)

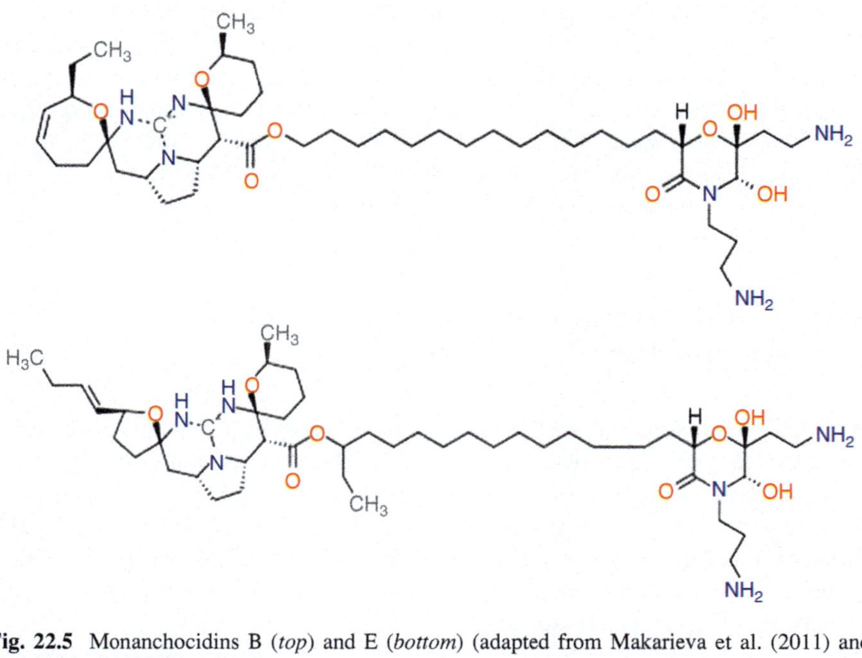

Fig. 22.5 Monanchocidins B (*top*) and E (*bottom*) (adapted from Makarieva et al. (2011) and visualised with 2D Sketcher)

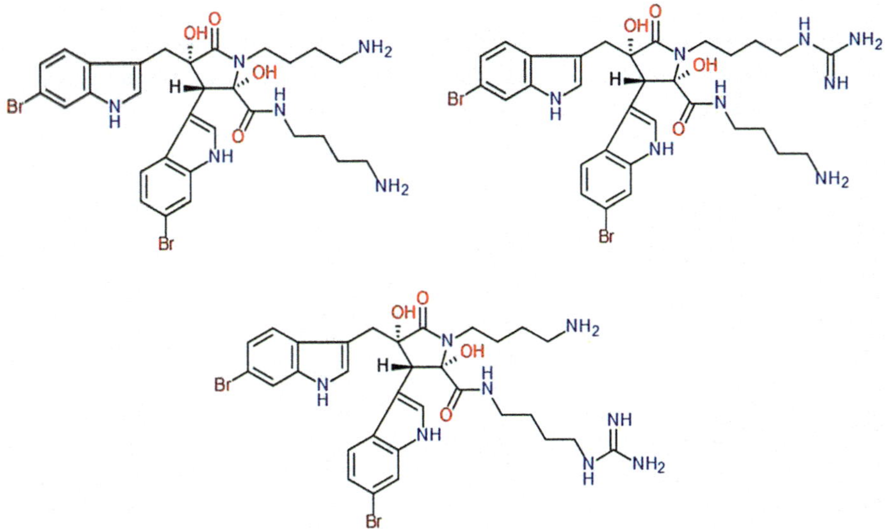

Fig. 22.6 Eusynstyelamides D, E and F (clockwise direction) (adapted from Tadesse et al. (2011) and visualised with 2D Sketcher)

2011). The monanchocidins are part of the well-known group of pentacyclic guanidine alkaloids with the first representative being ptilomycin A (Ohizumi et al. 1996). Metabolites of this compound displayed a broad range of biological activities including antifungal, antimicrobial, antimalarial and many other properties.

New antibacterial compounds, namely, ent-eusynstyelamides D, E and F (Fig. 22.6), were isolated from the arctic bryozoan *Tegella* cf. *spitzbergensis* and represented the first report of compounds with antimicrobial activity from this organism (Tadesse et al. 2011). These Eusynstyelamides are brominated tryptophan derivatives which were first isolated from the Australian ascidian *Eusynstyela latericius* (Tapiolas et al. 2009), which display activities against *E. coli*, *S. aureus*, *P. aeruginosa* and *C. glutamicum* strains.

22.3 Uncharacterised Compounds from Cold Environments

Kim and co-workers isolated bioactive microorganisms from an Arctic lichen collected in Spitsbergen (Kim et al. 2014). Lichens are a composite, symbiotic organism comprising of an algae or cyanobacteria and a filamentous fungi. They isolated five bacteria with antibacterial activity, which were closely related to either *Sphingomonas* sp. or *Burkholderia* sp. The isolates were active against Gram-positive (*S. aureus*, *B. subtilis*, *M. luteus*) as well as Gram-negative (*E. coli*, *P. aeruginosa*, *E. cloacae*) indicator strains.

Lo Giudice and colleagues studied 580 bacterial isolates retrieved from various Antarctic marine sources, such as seawater, sediment and Antarctic fish intestine collected during four oceanographic campaigns, for their antibacterial activities against Gram-positive and Gram-negative strains as well as antifungal activity against the eukaryotic yeast *Candida albicans*. Twenty-two of the isolates showed varying degree of antibacterial activity against *E. coli*, *P. mirabilis*, *M. luteus* and *B. subtilis*. The active microbial isolates were identified as belonging to two main phylogenetic groups, one being *Actinobacteria* (*Arthrobacter*, *Janibacter*, *Nesterenkonia* and *Rhodococcus* sp.) and the other being *γ-Proteobacteria* (*Pseudoalteromonas* and *Pseudomonas* sp.) by 16S rRNA gene sequencing analysis (Lo Giudice et al. 2007).

In another study, 132 bacterial isolates, retrieved from three Antarctic sponges (*Haliclonissa verrucosa*, *Anoxycalyx joubini* and *Lissodendoryx nobilis*), were screened for antimicrobial activity against more than 70 opportunistic pathogens. Most of these isolates exhibited an ability to inhibit the growth of *Burkholderia cepacia* complex bacteria, but not other pathogenic bacteria, which indicates a very specific action against these types of bacteria. The retrieved bacteria belonged mostly to the *Arthrobacter*, *Pseudoalteromonas*, *Psychrobacter*, *Shewanella* and *Roseobacter* genera. The cause of action was believed to be due to the production of an array of volatile organic compounds (VOCs) produced by these isolates rather than by bioactive secondary metabolites. Interestingly the array of VOCs produced differed from isolate to isolate and corresponded to the range of observed antimicrobial activities (Papaleo et al. 2012). The aforementioned work leads to the sequencing and comparative analysis of three *Arthrobacter* (Orlandini et al. 2014) and three *Psychrobacter* strains (Fondi et al. 2014) which displayed good antibacterial activity against *Burkholderia cepacia* complex bacteria, but unfortunately none of the studies was able to give further insight into the cause of antibacterial activity besides excluding known secondary metabolite gene clusters, suggesting an unknown type of action and/or compound.

A halophilic Antarctic *Nocardioides* sp. retrieved from Antarctic soil has been investigated for the production of enzymes and antimicrobial properties following growth on different carbon sources (Gesheva and Vasileva-Tonkova 2012). The bacterium displayed differential expression of hydrolytic enzymes and antimicrobial compounds in respect to the available carbon source, which highlights the importance of varying the growth condition in the laboratory to help unlock the 'hidden' potential from environmental isolates. The isolate displayed antimicrobial activity against Gram-positive and Gram-negative bacteria, with the highest activity against *S. aureus* and *Xanthomonas oryzae*. Further analysis suggested that glycolipids and/or lipopeptides could be responsible for the antimicrobial phenotype, depending on the carbon source on which the isolate was cultured.

Besides Arctic, Antarctic and deep-sea environments, there are also examples of where Alpine environments harbour interesting psychrophiles with promising activities. This is highlighted by a recent study which investigated the bacterial community of a hydrocarbon-contaminated industrial site in South Tyrol, Italy (Hemala et al. 2014). Forty-seven bacterial strains were isolated and around two thirds of them showed a cold-active (10 °C) antimicrobial activity against at least one of the test strains (*E. coli*, *Shigella flexneri*, *Salmonella enterica*, *P. aeruginosa*, *S. aureus*, *C. albicans* and *Cryptococcus neoformans*). The five most active isolates belonged to the *Gammaproteobacteria* class and were members of the genera *Pseudomonas* and *Serratia*.

22.4 Hidden Potential in Deep-Sea Sponge Microbiomes

As outlined in the previous sections, psychrophiles from cold environments clearly possess the ability to produce interesting compounds that display promising activities. One of the main restrictions in studying the antimicrobial activity of any given environmental sample is usually the level of cultivable microorganisms that can be generated from that sample. Metagenomic-based approaches are becoming increasingly popular in an effort to help overcome this particular limitation. Studies which specifically target the secondary metabolite production potential, by targeting secondary metabolite gene clusters within a given sample, can help bridge the gap between the antimicrobial potential of cultivable and uncultivable microorganisms. In a previous study from our group, we investigated the presence of gene fragments from secondary metabolite gene clusters in the metagenome of the three different deep-sea sponge species *Stelletta normani*, *Inflatella pellicula* and *Poecillastra compressa* which had been retrieved from depths ranging 760–2900 m below sea level (Borchert et al. 2016). Ketosynthase (KS) domains of polyketide synthases (PKS) and adenylation domains (AD) of nonribosomal peptide synthetases (NRPS), which are gene clusters known to produce bioactive compounds, were specifically targeted. These gene fragments were PCR-amplified from the sponge metagenomes and sequenced on a 454 GS-FLX+ pyrosequencing platform. After several quality-control and dereplication steps, a unique subset of 80 AD and 176 KS sequences was retrieved from all samples and further analysed. Analysis of the AD gene fragments showed considerable similarities to a wide variety of genes known to be involved in the synthesis of bioactive compounds such as thuggacin, putisolvin, gramicidin, microcystin, yersiniabactin, curacin, hapalosin, tyrocidine and many others. Sequence similarity analysis of the KS gene fragments also indicated a wide variety of these types of biosynthetic genes, to an even greater extent than those of the AD gene fragments. Notable similarities were observed to genes which are known to be involved in the biosynthesis of compounds such as rifampicin, jamaicamide, myxalamid, epothilone D, nostophycin, jerangolid, stigmatellin A, and hectochlorin, amongst others. Thus, this study highlights the huge secondary metabolite production potential in the metagenome of these deep-sea sponges. Furthermore phylogenetic trees of the obtained sequences and of other database-deposited sequences were generated, and clusters of only distantly related genes were observed for both AD and KS sequences (Fig. 22.7), adding further proof to the high probability of finding new or considerably different bioactive compounds in this special and as yet largely underexplored psychrophilic marine environment. Following further analysis of

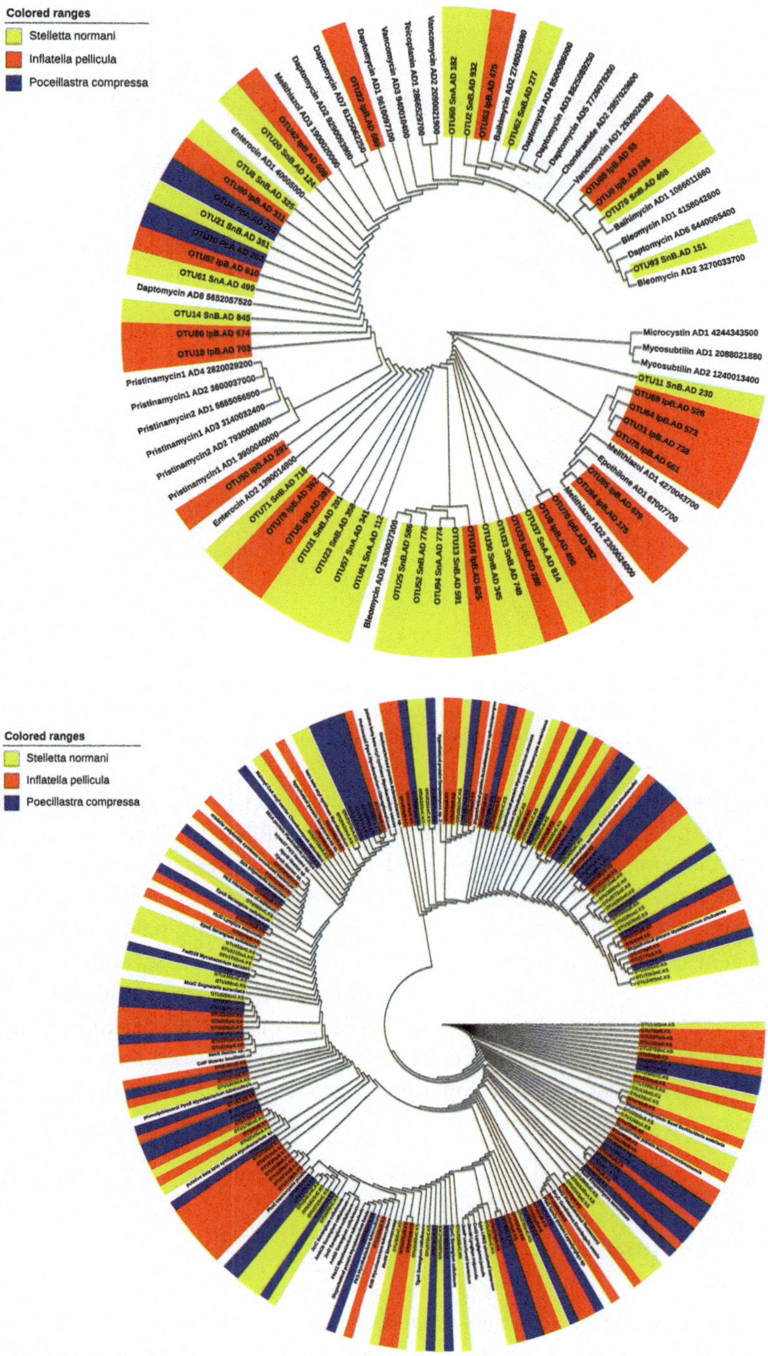

Fig. 22.7 Phylogenetic comparison of AD (*left*) and KS (*right*) sequences with references sequences. The sequences from *P. compressa* are *blue* coloured, from *I. pellicula red* coloured and from *S. normani yellow* coloured

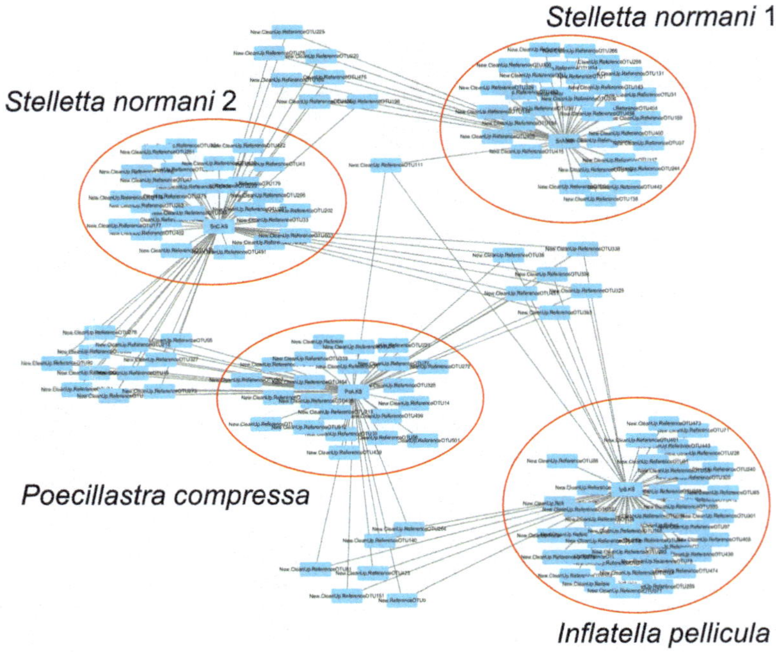

Fig. 22.8 OTU network of representative ketosynthase sequences from individual sponge samples. The *red circle* indicates the 'core/unique' ketosynthase sequences of a given sponge sample. Sequences represented outside the circles are shared amongst the different sponges

an OTU network generated from the KS sequences of some individual sponge samples (Fig. 22.8), remarkably only a small number of sequences are shared amongst the sponges and even amongst the same sponge species. This indicates the uniqueness of each sponge sample taken as it appears that each sponge has a large unique set of secondary metabolite genes, while only having a small set of shared genes. If this is a general trend in sponge microbiome secondary metabolite production, then this needs to be further evaluated. It would be imaginable that the observed differences may be as a result of depth- or location-dependent effects as these sponge samples have been collected from different locations and depths. In any case, these trends further highlight the potential of these deep-sea sponge metagenomes as a good source of novel bioactive compounds.

22.5 Conclusions

Psychrophiles are an as yet largely untapped source of novel or considerably different antimicrobial compounds. To date a sustained level of research has not focused on psychrophiles as a source of novel bioactive compounds, but with the ever-growing need for new antibiotics due to the ongoing threat of antimicrobial resistance, this will undoubtedly change in the future. Promising and encouraging advances in the field have been highlighted with synoxazolidinones A and B, lobophorin H and abyssomicin J providing some recent examples. Furthermore there are more undescribed compounds (Sect. 22.3) that require detailed identification and characterisation in the future, which will undoubtedly add to the list of new bioactive compounds. Metagenomic-based approaches may also prove useful in trying to exploit the unculturable microbial populations associated with organisms living in cold environments, such as sponges, bryozoans or ascidians. This has the capacity to bypass the bottlenecks inherently present in cultivation-dependent approaches. The deep-sea sponge microbiome in particular may prove a useful source of new compounds (Sect. 22.4), but their identification and subsequent exploitation may be hampered to a large extent by issues surrounding access to the deep cold ecosystems.

References

Abbas S, Kelly M, Bowling J, Sims J, Waters A, Hamann M (2011) Advancement into the Arctic region for bioactive sponge secondary metabolites. Mar Drugs 9(11):2423–2437

Alanis AJ (2005) Resistance to antibiotics: are we in the post-antibiotic era? Arch Med Res 36 (6):697–705

Blunt JW, Copp BR, Keyzers RA, Munro MH, Prinsep MP (2016) Marine natural products. Nat Prod Rep 33:382–431

Borchert E, Jackson SA, O'Gara F, Dobson AD (2016) Diversity of natural product biosynthetic genes in the microbiome of the deep sea sponges *Inflatella pellicula*, *Poecillastra compressa*, and *Stelletta normani*. Front Microbiol 7:1027

Brown ED, Wright GD (2016) Antibacterial drug discovery in the resistance era. Nature 529 (7586):336–343

Cruz PG, Fribley AM, Miller JR, Larsen MJ, Schultz PJ, Jacob RT, Tamayo-Castillo G, Kaufman RJ, Sherman DH (2015) Novel lobophorins inhibit oral cancer cell growth and induce Atf4- and Chop-dependent cell death in murine fibroblasts. ACS Med Chem Lett 6(8):877–881

Davies J, Davies D (2010) Origins and evolution of antibiotic resistance. Microbiol Mol Biol Rev 74(3):417–433

Fleming A (2001) On the antibacterial action of cultures of a penicillium, with special reference to their use in the isolation of *B. influenzae*. Bull World Health Org 79(8):780–790

Fondi M, Orlandini V, Perrin E, Maida I, Bosi E, Papaleo MC, Michaud L, Lo Giudice A, de Pascale D, Tutino ML, Liò P, Fani R (2014) Draft genomes of three Antarctic *Psychrobacter* strains producing antimicrobial compounds against *Burkholderia cepacia* complex, opportunistic human pathogens. Mar Genomics 13:37–38

Gesheva V, Vasileva-Tonkova E (2012) Production of enzymes and antimicrobial compounds by halophilic Antarctic *Nocardioides* sp. grown on different carbon sources. World J Microbiol Biotechnol 28(5):2069–2076

Hemala L, Zhang D, Margesin R (2014) Cold-active antibacterial and antifungal activities and antibiotic resistance of bacteria isolated from an alpine hydrocarbon-contaminated industrial site. Res Microbiol 165(6):447–456

Kim MK, Park H, Oh TJ (2014) Antibacterial and antioxidant capacity of polar microorganisms isolated from Arctic lichen *Ochrolechia* sp. Pol J Microbiol 63(3):317–322

Lebar MD, Heimbegner JL, Baker BJ (2007) Cold-water marine natural products. Nat Prod Rep 24 (4):774–797

Levy SB (2002) From tragedy the antibiotic era is born. In: Levy SB (ed) The antibiotic paradox: how the misuse of antibiotics destroys their curative powers. Perseus, Cambridge, pp 1–14

Lo Giudice A, Bruni V, Michaud L (2007) Characterization of Antarctic psychrotrophic bacteria with antibacterial activities against terrestrial microorganisms. J Basic Microbiol 47(6):496–505

Makarieva TN, Tabakmaher KM, Guzii AG, Denisenko VA, Dmitrenok PS, Shubina LK, Kuzmich AS, Lee HS, Stonik VA (2011) Monanchocidins B-E: polycyclic guanidine alkaloids with potent antileukemic activities from the sponge *Monanchora pulchra*. J Nat Prod 74 (9):1952–1988

McAuliffe O, Ross RP, Hill C (2001) Lantibiotics: structure, biosynthesis and mode of action. FEMS Microbiol Rev 25(3):285–308

Ohizumi Y, Sasaki S, Kusumi T, Ohtani II (1996) Ptilomycalin A, a novel Na+, K(+)- or Ca2(+)-ATPase inhibitor, competitively interacts with ATP at its binding site. Eur J Pharmacol 310 (1):95–98

Orlandini V, Maida I, Fondi M, Perrin E, Papaleo MC, Bosi E, de Pascale D, Tutino ML, Michaud L, Lo Giudice A, Fani R (2014) Genomic analysis of three sponge-associated *Arthrobacter* Antarctic strains, inhibiting the growth of *Burkholderia cepacia* complex bacteria by synthesizing volatile organic compounds. Microbiol Res 169(7–8):593–601

Pan HQ, Zhang SY, Wang N, Li ZL, Hua HM, Hu JC, Wang SJ (2013) New spirotetronate antibiotics, lobophorins H and I, from a South China Sea-derived *Streptomyces* sp. 12A35. Mar Drugs 11(10):3891–3901

Papaleo MC, Fondi M, Maida I, Perrin E, Lo Giudice A, Michaud L, Mangano S, Bartolucci G, Romoli R, Fani R (2012) Sponge-associated microbial Antarctic communities exhibiting antimicrobial activity against *Burkholderia cepacia* complex bacteria. Biotechnol Adv 30 (1):272–293

Phelan RW, Barret M, Cotter PD, O'Connor PM, Chen R, Morrissey JP, Dobson AD, O'Gara F, Barbosa TM (2013) Subtilomycin: a new lantibiotic from *Bacillus subtilis* strain MMA7 isolated from the marine sponge *Haliclona simulans*. Mar Drugs 11(6):1878–1898

Sánchez LA, Hedström M, Delgado MA, Delgado OD (2010) Production, purification and characterization of serraticin A, a novel cold-active antimicrobial produced by *Serratia proteamaculans* 136. J Appl Microbiol 109(3):936–945

Shymanska NV, An IH, Pierce JG (2014) A rapid synthesis of 4-oxazolidinones: total synthesis of synoxazolidinones A and B. Angew Chem Int Ed Engl 53(21):5401–5404

Tadesse M, Strøm MB, Svenson J, Jaspars M, Milne BF, Tørfoss V, Andersen JH, Hansen E, Stensvåg K, Haug T (2010) Synoxazolidinones A and B: novel bioactive alkaloids from the ascidian *Synoicum pulmonaria*. Org Lett 12(21):4752–4755

Tadesse M, Tabudravu JN, Jaspars M, Strøm MB, Hansen E, Andersen JH, Kristiansen PE, Haug T (2011) The antibacterial ent-eusynstyelamide B and eusynstyelamides D, E, and F from the Arctic bryozoan *Tegella* cf. *spitzbergensis*. J Nat Prod 74(4):837–841

Tapiolas DM, Bowden BF, Abou-Mansour E, Willis RH, Doyle JR, Muirhead AN, Liptrot C, Llewellyn LE, Wolff CW, Wright AD, Motti CA (2009) Eusynstyelamides A, B, and C, nNOS inhibitors, from the ascidian *Eusynstyela latericius*. J Nat Prod 72(6):1115–1120

Trepos R, Cervin G, Hellio C, Pavia H, Stensen W, Stensvåg K, Svendsen JS, Haug T, Svenson J (2014) Antifouling compounds from the sub-arctic ascidian *Synoicum pulmonaria*: synoxazolidinones A and C, pulmonarins A and B, and synthetic analogues. J Nat Prod 77 (9):2105–2113

Vieweg L, Reichau S, Schobert R, Leadlay PF, Süssmuth RD (2014) Recent advances in the field of bioactive tetronates. Nat Prod Rep 31(11):1554–1584

Wang Q, Song F, Xiao X, Huang P, Li L, Monte A, Abdel-Mageed WM, Wang J, Guo H, He W, Xie F, Dai H, Liu M, Chen C, Xu H, Piggott AM, Liu X, Capon RJ, Zhang L (2013) Abyssomicins from the South China Sea deep-sea sediment *Verrucosispora* sp.: natural thioether Michael addition adducts as antitubercular prodrugs. Angew Chem Int Ed Engl 52 (4):1231–1234

Xiao J, Luo Y, Xie S, Xu J (2011) *Serinicoccus profundi* sp. nov., an actinomycete isolated from deep-sea sediment, and emended description of the genus *Serinicoccus*. Int J Syst Evol Microbiol 61(Pt 1):16–19

Yang XW, Zhang GY, Ying JX, Yang B, Zhou XF, Steinmetz A, Liu YH, Wang N (2013) Isolation, characterization, and bioactivity evaluation of 3-((6-methylpyrazin-2-yl)methyl)-1H-indole, a new alkaloid from a deep-sea-derived actinomycete *Serinicoccus profundi* sp. nov. Mar Drugs 11(1):33–39

Chapter 23
Permafrost Bacteria in Biotechnology: Biomedical Applications

Anatoli Brouchkov, Vladimir Melnikov, Ludmila Kalenova, Oksana Fursova, Gennady Pogorelko, Vasiliy Potapov, Nadezhda Fursova, Sergei Ignatov, Evgeny Brenner, Vladislav Bezrukov, and Khachik Muradian

Abstract Mechanisms of survival and growth capacities of microorganisms trapped in permafrost are still under discussion, but the very fact of their existence for thousands of years is evidence of their phenomenal viability. One of these nonpathogenic bacteria, identified as *Bacillus cereus* strain BF, was found in Yakutia and was capable of enhancing longevity and immunity in *Drosophila* and mice and showed probiotic activity on the mice *Salmonella enterica* model. In developing *Drosophila melanogaster*, the application of *Bacillus cereus* strain BF resulted in a dose-dependent increase of the growth rate of *Drosophila* and in a

A. Brouchkov (✉)
Tyumen State University, 625000 Tyumen, Volodarskogo str., 6, Russia

Lomonosov Moscow State University, GSP-1, Leninskie Gory, Moscow 119991, Russia
e-mail: brouchkov@geol.msu.ru

V. Melnikov • L. Kalenova
Tyumen State University, 625000 Tyumen, Volodarskogo str., 6, Russia

Tyumen Scientific Center, Siberian Branch, Russian Academy of Science, 625026 Tyumen, P/O box 1230, Russia

O. Fursova
Lomonosov Moscow State University, GSP-1, Leninskie Gory, Moscow 119991, Russia

G. Pogorelko
Iowa State University, Ames, IA 50011, USA

V. Potapov • N. Fursova • S. Ignatov
State Research Center for Applied Microbiology and Biotechnology, Obolensk, Moscow Region 142279, Russia

E. Brenner
Institute of Chemical Biology and Fundamental Medicine of the Siberian Branch of the Russian Academy of Science, Novosibirsk 630090, Russia

V. Bezrukov • K. Muradian
State Institute of Gerontology of National Academy of Medical Sciences of Ukraine, 67, Vyshgorodskaya Str., 04114 Kiev, Ukraine

R. Margesin (ed.), *Psychrophiles: From Biodiversity to Biotechnology*,
DOI 10.1007/978-3-319-57057-0_23

decrease of larval mortality, whereas in adult imagoes gaseous exchange (VO_2 and VCO_2), spontaneous motor activity, and stress resistance were enhanced.

Contents

23.1 Introduction

Permafrost occupies a vast territory; it takes about 65% of area of the Russian Federation and 25% of the Northern Hemisphere. Its temperatures are mainly about −2 to 8 °C, and its age in some places is millions of years (Ershov 1998). The evidence of vitality of microorganisms in permafrost appeared long ago (Friedmann 1994). Bacteria, fungi, diatoms, and other microorganisms were found at the Vostok Antarctic station (Abyzov et al. 1979). Bacterial metabolism in permafrost was observed at about −10 °C (Bakermans et al. 2003). Other references also demonstrated the vitality of bacteria at temperatures close (Margesin and Miteva 2011) and below 0 °C (Clein and Schimel 1995). Microorganisms are resistant to freezing; many of them endure it easily (Lozina-Lozinsky 1972). It is known that part of water in the materials (more than 10%) remains unfrozen at temperatures below −20 °C (Brouchkov et al. 1995). Without denying the probability of development of microorganisms in permafrost soils, we should note that their growth is hampered. As we know, the growth of aging cultures stops even under laboratory conditions. Crystallization of the water and cessation of metabolism diminish the ability for growth. The width of nonfreezing water interlayers at −2 and −4 °C is about 0.01–0.10 μm (Brouchkov et al. 1995) and much less than the sizes of microorganisms. These conduction paths are actually unavailable for life support, and any noticeable transfer of cells in such material is out of question. Therefore, we can be sure that the bacteria in permafrost deposits are fossil, relict organisms (Repin et al. 2007). Their age is confirmed by geological conditions, history of formation of frozen strata, ^{14}C dating, results of the study of optical isomers of amino acids, and, indirectly, biodiversity (Katayama et al. 2007).

The nature of longevity of microorganisms in ancient permafrost cannot be exhaustively explained. It is believed that no chemical and biological reactions take place in an organism in a state of anabiosis (Lozina-Lozinsky 1972). However, we do know that most proteins in a living cell are unstable (Jaenicke 1996), being alive for minutes and more rarely for days. Genetic structures are liable to mutations, and reparations are not as effective to prevent accumulation of damages

(Cairns et al. 1994). Cell structures are affected by free radicals and radiation. Thermal motion of atoms and molecules in water solution is an independent destructive factor: permafrost temperatures are far from absolute zero (Aleksandrov 1975). The cytoplasm of cells does not freeze at such temperatures (Ashcroft 2007). It is obvious that an organism in anabiosis is also liable to degradation and decay. The data on the heat stability of nucleotides (Levy and Miller 1998) show that, due to the instability of cytosine, the period of time when the DNA chain remains functional is hardly longer than several hundreds of years. Taking a certain average mutation rate into account (Cairns et al. 1994), nearly all genes in a bacterial chromosome will be modified in 1000 years, and no traces of the genetic apparatus will remain in a million years, even subject to the functioning of repair systems. In the absence of nutrients, cells can rely only on the electrical charge of the cytoplasmic membrane, but it will suffice to synthesize only 100–150 ATP molecules. Evidently, this energy is insufficient for long-term preservation of integrity of cell structures. Ancient DNA of mummies, mammoths, insects in amber, and other organisms were found to be destroyed (Willerslev and Cooper 2005). Calculations show that even minor DNA fragments (100–500 nucleotides) can be preserved for no more than 10,000 years in the usual climate and up to 100,000 years at the most in cold regions (Willerslev and Cooper 2005). Thus, it is unclear how bacteria survive in millennial permafrost. One should suppose the existence of mechanisms preventing the accumulation of damages. We believed it would be worthwhile to study the influence of these microorganisms on higher organisms, with special regard to the immune responses of the latter.

23.2 Samples and Methods

Samples for the study of microorganisms in frozen deposits were taken from outcrop located on the left bank of the Aldan River, 325 km upstream to its confluence with the Lena River on the Mamontova Gora. Samples were taken 0.9–1.0 m deeper than the layer of seasonal thawing out. The outcrop was destroyed by the river (more than 1 m per year); therefore, the sampled sediments were obviously in a state of permafrost. At the same time, there is the annual spring failure washing-off that prevents obstructions and the mixture of soils. The latter are fine-grained sands, and their age corresponds to the Middle Miocene: 10–12 million years (Baranova et al. 1976). The fall of temperature and frost penetration into sediments began here in the end of Pliocene, about 3.0–3.5 million years ago (Ershov 1998). Later on, the sediments did not thaw out because of the cold climate of Yakutia. According to data from paleoclimatic reconstructions of the region, the average annual temperatures in Pleistocene were −12 to −32 °C in winter and +12 to +16 °C in summer (Bakulina and Spektor 2000). Thus, the age of permafrost on the Mamontova Gora may be up to 3.5 million years.

Moreover, samples were taken from the younger ice wedge of Yakutia from the underground of the Melnikov Institute of Geocryology, Siberian Branch of RAS

(Yakutsk), and from underground ice in the Fox Permafrost Tunnel (Alaska). Frozen soil samples were taken with maximum possible precautions for the field conditions, using metal instruments (bores, forceps, and scalpels) sterilized with alcohol and burned in flame. For surface sterilization, a sample of about 50 g was placed into a glass with 96% ethanol solution, then put into a burner flame, and packed into a sterile test tube.

Additionally, 4–5 kg monoliths of frozen soils were taken. Soils were stored at −5 °C (i.e. close to the natural conditions).

Samples were transported in a frozen state in thermally controlled containers with cooling agents. Samples of different sterile dilutions were added to Petri dishes containing YPD, MRS, and nutrient agar media. Samples were also added into a liquid meat–peptone broth under anaerobic and aerobic conditions. DNA was extracted using the FastDNA kit for soil. The 16S rRNA gene fragments were amplified by polymerase chain reaction with bacterial primers. The reaction was performed in a 20 mL volume by the GeneAmp PCR System 2700 (Applied Biosystems); the amplicons were exposed to electrophoresis and purified by the Wizard SV Gel and PCR Clean-Up System (Promega). Purified amplicons were cloned using the pCR2.1 vector, *Escherichia coli* culture, and the TA cloning kit (Invitrogen) in accordance with the manufacturer's instructions. Plasmid DNA carrying 165 rDNA was obtained from 24 h cultures using the Spin Miniprep Kit (Qiagen). Purified plasmid DNAs were sequenced in an ABI PRISM 3100 Genetic Analyzer using the Big Dye Terminator Cycle and a sequencing kit (Applied Biosystems). The sequence length was 1488 bp. The sequence obtained was compared to other sequences using BLAST (Basic Local Alignment Search Tool). The phylogenetic tree was constructed using the CLUSTAL W software package (Thompson et al. 1994).

23.3 Isolated Bacteria and Full Genome Sequencing

A cultivated bacterium (strain F) capable of aerobic and anaerobic growth was found in frozen Miocene sediments on the Mamontova Gora; its optimal growth temperature was defined as +37 °C. The cells are comparatively large rods (1.0–1.5 × 3–6 μm), which are joined into chains in the culture and can form spores. The microorganism is immobile, Gram-positive, and belongs to the species *Bacillus cereus*. The 16S rRNA nucleotide sequence of strain *Bacillus cereus* strain BF was deposited in the DDBJ/EMBL/GenBank (accession number AB178889; identification number 20040510203204.24251).

Also a draft sequence of the bacterial genome has been made (Brenner et al. 2013). The sequencing was performed with the use of a hybrid approach. At the first stage, a rapid fragment library was sequenced on a Roche FLX instrument following the titanium protocol, resulting in approximately 170 Mb of raw data with a mean read length of 340 nucleotides. A 2× 50 bp mate-paired library with a mean insertion size of 1 kb was additionally sequenced by using ABI SOLiD v.3.5

Workflow. The combination of different sequencing platforms and library types made it possible to perform de novo assembly from Roche FLX reads, followed by scaffolding by SOLiD reads pairing. The draft genome sequence includes nearly 5.26 Mb of nucleotide sequences and is characterized by 186 contigs. The genome obtained was annotated using the RAST annotation server. As many as 5602 coding sequences were predicted. This whole-genome shotgun project has been deposited at DDBJ/EMBL/GenBank under the accession number AHHI00000000. The version described here is the first version, AHHI01000000. Believed to be as ancient as the permafrost sample from which this strain was isolated, *Bacillus cereus* strain BF exhibits a surprisingly high level of homology with modern *Bacillus cereus* strains, particularly with *Bacillus cereus* strain ATCC 10987. The difference in chromosomal nucleotide sequences between these two strains does not exceed 1.5%, which is comparable to or even less than the difference between available chromosomal sequences of other *Bacillus cereus* strains. These observations may reflect the adaptability of *Bacillus cereus* for long-term survival and the evolution strategies of this organism in the permafrost environment (Friedmann 1994).

The growth of bacilli at low temperatures was observed previously (Lozina-Lozinsky 1972). Scanning electron microscopy has not shown colonies or even groups of cells inside of the permafrost sample, just single cells. This is additional evidence of the age of the bacterial cells. It is known that *Bacillus anthracis* easily endures freezing (Repin et al. 2007). The optimal growth temperature of *Bacillus cereus* strain BF is rather high. Bacillary spores are known to be most resistant (Nicholson et al. 2000). Therefore, the finding of a living *Bacillus* in the ancient permafrost of Mamontova Gora is not altogether surprising. Spore formation is a widespread mechanism of survival, probably involving the horizontal transfer of genetic information.

Tens of microbial species were isolated by the foregoing method from the much younger ice wedge of Yakutia and Alaska aged about 25,000–40,000 years (Katayama et al. 2007). Most of the isolated bacteria were Gram-positive and phylogenetically related to *Arthrobacter* and *Micrococcus* sp., and the fungi were close to *Geomyces* sp.

23.4 Tests on Mice

An experiment on mice was carried out with a 24 h culture of *Bacillus cereus* strain BF grown in a meat–peptone broth, the strain was frozen and thawed before injection (Brouchkov et al. 2009). Experiments were carried out in F1 CBA/Black-6 mice (15 animals per group on average). In the first series of experiments, the effects of culture doses on immune system parameters were studied in young animals (aged 3–4 months). Two groups of animals were used: one was used as a control, and animals in the other group received a saline solution. The bacterial culture was introduced once intraperitoneally: 5000; 50,000; 500,000; 5,000,000; and 50,000,000 microbial objects (m.o.) per animal. In the second series of experiments, the effect of the bacterial culture on physiological and behavioral reactions was estimated in "senior" mice (aged 17 months); the bacterial culture was introduced once intraperitoneally at a dose of 5000 m.o. The control group consisted of animals of the same age. The animals were euthanized by dislocation of the cervical vertebrae.

The following standard methods were used for estimations: index of the organ (the percentual ratio of organ mass to body mass) for morphophysiological activity of the thymus and spleen; levels of phagocytic (PC, %) and metabolic (NST, %) activities of spleen macrophages for the activity of nonspecific immune resistance factors; the reaction of a delayed-type hypersensitivity (DTH) in vivo for cell-mediated immunity; the number of antibody-forming cells in 1 million nucleus-containing cells in the spleen for the activity of humoral immunity; the load-lifting test for the muscle strength of animals; and the "open field" test for behavioral responses; the lifespan was estimated using standard methods.

It was shown that *Bacillus cereus* strain BF at a dose of 5000 m.o. favored an increase of the thymus and spleen indices by 60–80%. The bacillary culture stimulated the phagocytic activity of spleen macrophages at a small dose (5000 m.o.) and inhibited it at average doses (500,000 and 5,000,000 m.o.). The culture enhanced at almost all doses the activity of humoral immunity; the dose of 5000 m.o. favored the increase in functional activity of both cell-mediated and humoral immunity.

In this context, the dose of 5000 m.o. was chosen for studying the culture influence on lifespan. The minimal and maximal lifespans of mice from the control group were 589 days and 833 days, respectively. The minimal and maximal lifespans of mice from the experimental group were 836 and 897 days, respectively (Fig. 23.1). Two months after the bacterial culture was introduced, animals from the experimental group showed a higher body mass compared to animals from the control group. The muscle strength of the experimental animals increased by about 60% compared to their coevals from the control. The enhanced spatial intellect and exploratory activity of animals was evidenced by their more frequent visits to internal sectors of the experimental field, an increase in the number of upright

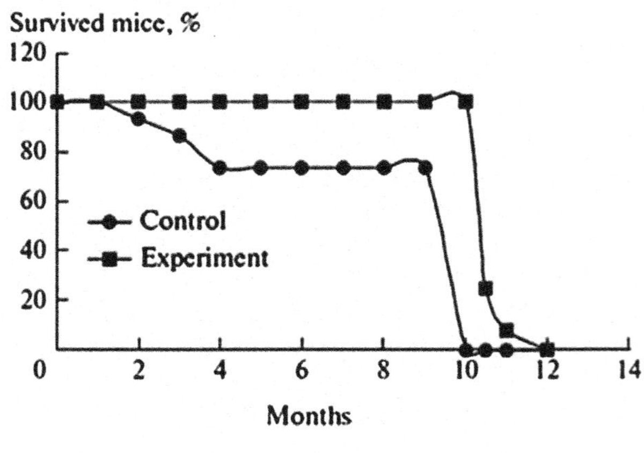

Fig. 23.1 Influence of *Bacillus cereus* strain BF, during parenteral introduction of 5000 cells, on the lifespan of laboratory mice aged 17 months

postures, and visits to burrows. It seems that intraperitoneal introduction of the bacterial culture stimulates the immune system and improves the emotional stability of laboratory animals. The enhanced lifespan of mice is evidence of the possible presence of geroprotectors in the culture of *Bacillus cereus* strain BF. We emphasize that investigations of the properties of relict microorganisms should be considered as very preliminary in respect to both their scope and performance.

23.5 Tests on *Drosophila*

Bacillus cereus strain BF was also tested in developing *Drosophila* fruit flies as follows: a culture of the bacterial strain was added to the nutrient medium of *Drosophila melanogaster*, and a set of indices characterizing growth rate and mortality at the larval and pupal stages was investigated. The level of gas exchange (VO_2 and VCO_2), body mass, and stress resistance were investigated in imagoes hatched under these conditions. *Bacillus cereus* strain BF induced dose-dependent growth acceleration and decrease in larval mortality. An increase in spontaneous motor activity, VO_2 and VCO_2, and body mass were demonstrated for imagoes. To establish optimal and toxic doses, a wide range of bacterial concentrations was tested (1–500 million cells mL^{-1} of the flies' feeding medium). Surprisingly, no toxic effects of *Bacillus cereus* strain BF could be registered even on such a "sensitive" model as the developing larvae. In fact, the rate of development, survival, and body mass gradually increased with elevation of the bacterial concentration. The gain of higher body mass within shorter periods of development could indicate enhanced anabolic and/or declined catabolic effects of *Bacillus cereus* strain BF. Higher motor activity and gaseous exchange rates were observed in imagoes developed on the media containing *Bacillus cereus* strain BF. Survival of these flies after heat shock (30 min at 38 °C) and ultraviolet irradiation (60 min, 50 W UV lamp) was increased, indicating elevated stress resistance, apparently due to stimulation of DNA repair and chaperone-mediated protection of macromolecules (Brouchkov et al. 2012).

23.6 Probiotic Effects

The search of novel probiotics sources against pathogenic bacteria is a very important direction in probiotics research. From this point of view, it is very important to analyze and protect the activity of bacteria stored in permafrost during a long period of time. We investigated the probiotic effect of *Bacillus cereus* strain BF (introduced per os to mice) against *Salmonella enterica* var. *enteritidis* when orally introduced to mice (Fursova et al. 2012). *Salmonella enterica* var. *enteritidis* and *Bacillus cereus* var. *toyoi* were received from the SRCAMB Culture Collection (#B-10130). All strains were cultivated routinely on Luria medium at 37 °C and stored after lyophilization at −70 °C in Luria broth containing 20% glycerol. Bacterial suspensions for inoculation were prepared by growing microorganisms for 4 h in nutrient broth at 37 °C. The concentration of microorganisms was determined by spectrophotometry, and appropriate dilutions were prepared in sterile 50 mM phosphate buffer solution (PBS), pH 7.0. The viable numbers of bacteria in the inocula were confirmed by plating onto nutrient agar. Six-week-old female Webster outbred mice were used in all experiments. The mice were kept in cages of up to 10 mice per cage, and food and water were given ad libitum. All experiments were approved by the Animal Care and Ethics Committee of the SRCAMB.

Webster outbred mice were inoculated per os with *S. enterica* var. *enteritidis* cells (5×10^6 CFU per mouse) for the modeling of infection. The dynamics of infection in mice was determined on the 3rd, 4th, 5th, 6th, and 7th day after infection using CO_2 euthanasia and postmortem examination. The following characteristics were used for the estimation of affection of mice: (1) changing mass of internal organs that were increased in mass 1.5–2.0-fold; (2) pathological changes in organs 10–14 days after infection, expressed as cirrhosis and/or the formation of necrosis regions in the spleen and on the border of the liver; and (3) number of *Salmonella* cells in the liver and spleen that increased to up to 10^8–10^9 CFU per mouse on the 7th day after infection. Life duration of the infected mice was 7–10 days after infection on average. Death specificity was confirmed by the isolation of *Salmonella*, growing on the selective medium SS agar (SRCAMB, Russia) from the homogenate of parenchymatous organs.

Livers and spleens of mice were removed aseptically on specific days after infection. Organs were weighed, and Mueller–Hinton broth was added to yield a 10% suspension (w/v) after homogenization in a Stomacher 80 (Thermo). Serial tenfold dilutions were plated in duplicate on Mueller–Hinton agar. After incubation for 18 h at 37 °C, colonies were counted and the number of CFU per organ was calculated. Isolates from animal organs were identified as *S. enterica* var. *enteritidis* by slide agglutination with the use of commercially available antisera to somatic (O) and flagellum (H) antigens (Immunoteks, Russia).

The bacterial pathogenicity of *Bacillus cereus* var. *toyoi* and *Bacillus cereus* strain BF was studied using Webster outbred mice (mass 20−24 g, age 10−12 weeks) by intravenous, intraperitoneal, and intragastrical inoculation of a 24 h culture at doses of 10^7, 10^8, and 10^9 CFU per mouse. The number of mice in each group was 12. The observation time after inoculation was 21 days. The control group of mice was inoculated with PBS on the same regime. Pathogenicity was estimated using the following parameters: (1) survival number of mice 21 days after infection; (2) pathomorphology of organs of euthanized mice on the 21th day after infection; and (3) presence of *Bacillus cereus* strain BF cells in the blood, liver, and spleen of the infected mice. All animal experiments and mice euthanasia was done according to the "Guide for Care and Use of Laboratory Animals."

Bacillus cereus var. *toyoi* and *Bacillus cereus* strain BF were grown in 10 mL of Luria broth (Difco) at 37 °C overnight; the cells were transferred to 50 mL of the same broth and were further incubated at 37 °C. When cell growth reached the log phase, the cells were harvested by centrifugation and resuspended in PBS. *Bacillus cereus* strain BF, at the dose of 1×10^9 CFU per mouse, was administered orally for 7 consecutive days, while the mice in the control group were treated by PBS only. One day after the last application, the group of ten mice received orally 5×10^6 CFU of *S. enterica* var. *enteritidis* that was equivalent to the LD50 dose. The number of the dead mice was recorded daily for 21 days.

Bacillus cereus strain BF was sensitive to ampicillin with a minimal inhibitory concentration (MIC) of 100 mg L^{-1}, tetracycline (MIC 25 mg L^{-1}), rifampicin (MIC 50 mg L^{-1}), gentamicin (MIC 25 mg L^{-1}), and kanamycin (MIC 50 mg L^{-1}) and resistant to spectinomycin (MIC 100 mg L^{-1}) and hygromycin (MIC 50 mg L^{-1}).

Experimental estimation of the pathogenicity of *Bacillus cereus* strain BF was done in comparison with the *Bacillus cereus* var. *toyoi* commercial strain. *Bacillus cereus* var. *toyoi* was pathogenic for mice with LD50 = $(8 \pm 2.5) \times 10^6$ CFU at intravenous injection and with LD50 = $(1 \pm 0.4) \times 10^7$ CFU at intraperitoneal injection. Moreover, single instances of mice death were detected after intragastrical injection of *Bacillus cereus* var. *toyoi*. Under the same conditions, intravenous, intraperitoneal, or intragastrical injection of *Bacillus cereus* strain BF was not the cause of the mice death (Fig. 23.2).

The intravenous and intraperitoneal injection of both *Bacillus cereus* var. *toyoi* and *Bacillus cereus* strain BF was not the cause of cutaneous edema and hyperemia one day after treatment. The absence of cells of *Bacillus cereus* var. *toyoi* and *Bacillus cereus* strain BF in the blood, liver, and spleen of the survived mice was shown on the 21st day of observation, and pathomorphological changes of organs of euthanized mice on that day were not detected. So, on the base of the presented tests, *Bacillus cereus* var. *toyoi* strain was determined as low pathogenic bacterial strain and *Bacillus cereus* strain BF as nonpathogenic bacterial strain.

Low pathogenicity of the commercial probiotic strain *Bacillus cereus* var. *toyoi* (Bactisubtil) may be explained by the following facts: its cells were found to carry

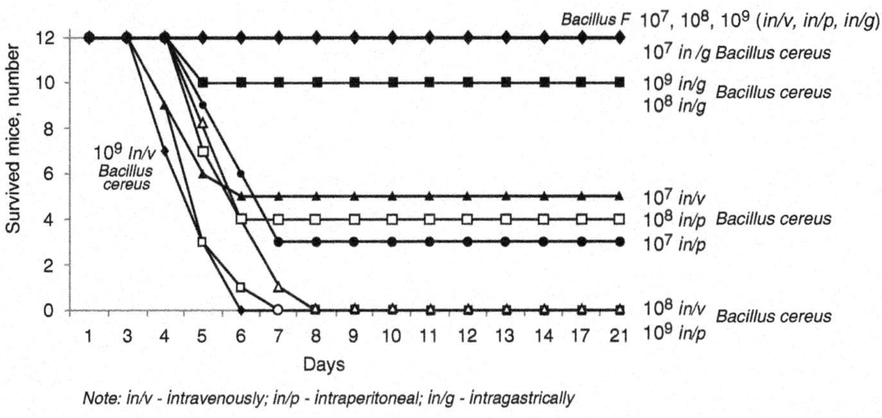

Fig. 23.2 Mice surveillance after intravenous, intraperitoneal, and intragastrical injection (10^7, 10^8, and 10^8 CFU per mouse) of *Bacillus cereus* var. *toyoi* and *Bacillus cereus* strain BF

three genes, hblA, hblC, and hblD, that encode hemolysin BL (Hbl), the primary virulence factor in *B. cereus* diarrhea, and the bceT gene, which encodes the single-component enterotoxin T. Moreover, the cells of this probiotic strain produce lecithinase (Granum and Lund 1997). We propose that the nonpathogenic properties of *Bacillus cereus* strain BF are based on the absence of such pathogenic factors (data not shown).

Experimental murine salmonellosis is a widely used experimental model for acute systemic salmonellosis in humans (Ohl and Miller 2001). Strain *S. enterica* serotype *typhi* was used for the mice model (Silva et al. 1999). Here, the *S. enterica* model was described as a variant to cause a chronic carrier state in mice after oral inoculation, as a model for a human carrier state. The maximum amount of pathogenic bacteria in organs (spleen and livers) could be detected after 7–10 days of infection (Sukupolvi et al. 1997). In our experiment, bacterial cultures from liver and spleen samples were determined after 3, 4, 5, 6, and 7 days after infection. After 7 days, there was no difference in numbers of *S. enterica* in organs obtained from control mice and mice treated with *Bacillus cereus* var. *toyoi*. But the treatment with *Bacillus cereus* strain BF decreased bacterial numbers in organs at least 10–100 times. This decrease was very important for mice survival (Fig. 23.3).

Therapy was started 4 h after per os infection with *S. enterica* with an inoculum corresponding to LD50 (5×10^6 CFU per mouse). *Bacillus cereus* strain BF was administered to groups of ten mice per os for 7 days. The treatment of mice with this strain reduced their mortality compared with that of untreated animals and animals treated with *B. cereus* var. *toyoi*. *Salmonella* cells were present in the livers and spleens of dead mice.

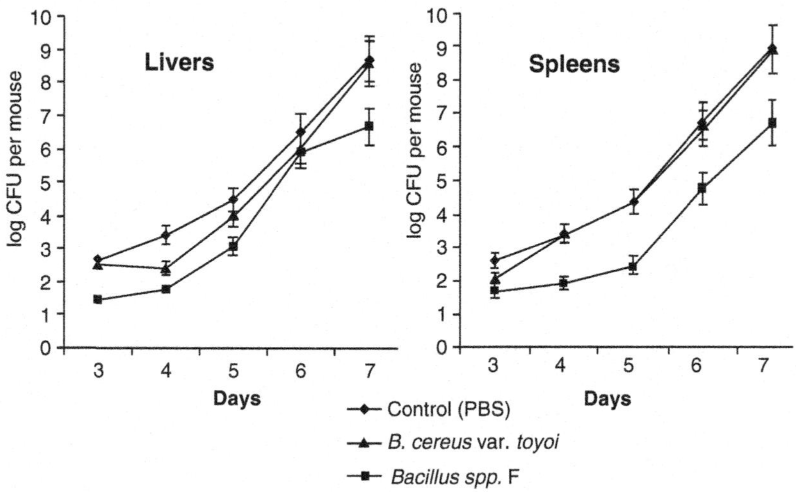

Fig. 23.3 Isolation of *S. enterica* from livers and spleens after intravenous, intraperitoneal, and intragastrical injection of *Bacillus cereus* var. *toyoi* and *Bacillus cereus* strain BF. Bars represent mean values (± standard deviation, $n = 5$)

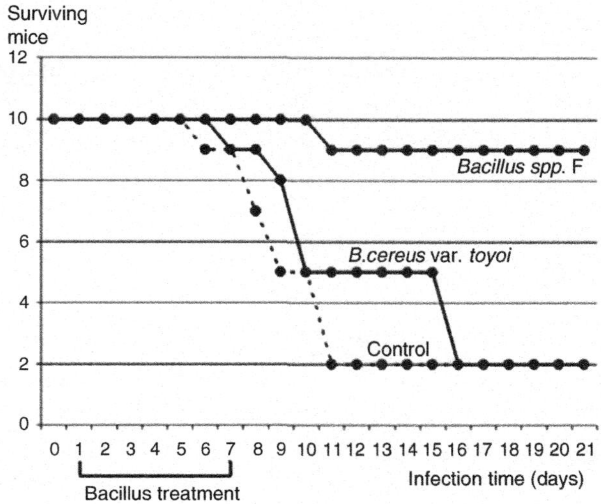

Fig. 23.4 Effect of treatment with *Bacillus cereus* var. *toyoi* and *Bacillus cereus* strain BF on mortality in outbred mice after oral inoculation with *S. enterica* var. *enteritidis* (5×10^6 CFU per mouse on day 0; ten mice in each group)

The mouse model for *S. enterica* var. *enteritidis* infection, as described here, was shown to be useful in the determination of the effect of *Bacillus cereus* strain BF on the host of the infection. The use of inbred mice with a stable genetic background ensured that the resistance to *S. enterica* var. *enteritidis* could be clearly defined in all groups of animals. Ten mice each were used for treatment with *Bacillus cereus* strain BF, treatment with *Bacillus cereus* var. *toyoi*, and for the control group. The mice were treated orally once daily with *Bacillus cereus* strain BF and *Bacillus cereus* var. *toyoi*, up to 7 days after infection. The follow-up observation was performed for 21 days after infection. Ninety percent of the mice treated with *Bacillus cereus* strain BF survived, compared to 20% of the untreated mice or mice treated with *Bacillus cereus* var. *toyoi* (Fig. 23.4).

The following three basic mechanisms could be proposed for how orally ingested nonindigenous bacteria can have a probiotic effect in a host: (1) immunomodulation, that is, stimulation of the gut-associated lymphoid tissue, e.g., induction of cytokines; (2) competitive exclusion of gastrointestinal pathogens, e.g., competition for adhesion sites; and (3) secretion of antimicrobial compounds which suppress the growth of harmful bacteria (Duc et al. 2004). We propose that the notable probiotic properties of *Bacillus cereus* strain BF as compared with *Bacillus cereus* var. *toyoi* can be explained by the production of unknown bacteriocin-like inhibitory substances (unpublished data). We described the *S. enterica* var. *enteritidis* model as a variant to cause a chronic carrier state in mice after oral inoculation, as a model for a human carrier state. The pathogenic mechanism of *S. enterica* var. *enteritidis* action was connected with the gastrointestinal tract. Therefore, we used oral infection and treatment of mice. To protect mice from *Salmonella* infection, antibiotics or bacterial extracts can be used (Deng et al. 2007). Viable microorganisms such as yeast and bacterial species have been used to protect mice from *Salmonella* infection (Szabó et al. 2009). In our case,

Bacillus cereus strain BF protected mice from *Salmonella* infection. Probiotic bacteria reduced colonization by pathogens and decreased host defense mechanisms. Preliminary results for *Bacillus cereus* strain BF showed an increase of the humoral and cell immunity of mice (Brouchkov et al. 2009). Thus, the possibility of oral treatment of mice infected with *S. enterica* var. *enteritidis* with probiotic *Bacillus cereus* strain BF, obtained from relict permafrost, was clearly demonstrated.

23.7 Conclusions

Relict microorganisms preserved in permafrost of geological periods possess a unique resistance to unfavorable factors of the internal and external environment. Their metabolic activity, DNA repair, and growth capacities are still under discussion, but the very fact of their existence in permafrost during many thousands of years is evidence of their phenomenal viability. One of these bacteria, identified as *Bacillus cereus* strain BF, has been found in Yakutia and was capable of enhancing longevity and immunity in *Drosophila* and mice and was analyzed for probiotic activity on the mice *Salmonella enterica* model. *Bacillus cereus* strain BF was assed as nonpathogenic bacterium by evaluation of pathogenicity. Probiotic activity of this strain has been shown against *Salmonella* infection in mice. The application of the strain to developing *Drosophila* fruit flies showed that levels of gas exchange (VO_2 and VCO_2), spontaneous motor activity, body mass, and stress resistance were increased in imagoes of *Drosophila melanogaster* and dose-dependent growth acceleration and decrease in larval mortality were induced. Permafrost microorganisms are possibly a perspective object for the search of new probiotics.

References

Abyzov SS, Bobin NE, Kudryashov BB (1979) Microbiological studies of glacier in the Central Antarctica. Izv AN SSSR Biol Ser 6:828–836

Aleksandrov VY (1975) Kletki, makromolekuly i temperature (Cells, macromolecules, and temperature). Nauka, Leningrad (in Russian)

Ashcroft F (2007) Life at the extremes. Harper and Collins, New York

Bakermans C, Tsapin AI, Souza-Egipsy V, Gilichinsky DA, Nealson H (2003) Reproduction and metabolism at 10°C of bacteria isolated from Siberian permafrost. Environ Microbiol 5:321–326

Bakulina NT, Spektor VB (2000) Reconstruction of climatic parameters of neogene in Yakutia by palynology data. In: Maksimov GN, Fedorov AN (eds) Klimat i merzlota (Climate and Permafrost). Institut Merzlotovedeniya: Yakutsk, pp 21–32 (in Russian)

Baranova YP, Il'inskaya IA, Nikitin VP et al (1976) Miocene of the Mamontova Gora. Nauka, Moscow

Brenner EV, Brouchkov AV, Kurilshikov AM et al (2013) Draft genome sequence of *Bacillus cereus* strain F, isolated from ancient permafrost. Genome Announc 1(4):e00561–e00513. doi:10.1128/genomeA.00561-13

Brouchkov AV, Vlasov AN, Merzlyakov VP, Talonov AV (1995) The effect of local phase transitions on deformation of frozen low plasticity soils. Geoekologiya 5:71–77

Brouchkov AV, Mel'nikov VP, Sukhovei IG, Griva GI, Repin VE, Kalenova LF, Brenner EV, Subbotin AM, Trofimova IB, Tanaka M, Kataiama T, Utsumi M (2009) Relict microorganisms of cryolithozone as possible objects of gerontology. Adv Gerontol 22:253–258

Brouchkov AV, Bezrukov VV, Griva GI, Myradyan HK (2012) The effects of the relict microorganism B. sp. on development, gas exchange, spontaneous motor activity, stress resistance, and survival of *Drosophila melanogaster*. Adv Gerontol 2:19–26

Cairns J, Overbaugh J, Miller S (1994) The origin of mutations. Nature 335:142–145

Clein JS, Schimel JP (1995) Microbial activity of tundra and taiga soils at subzero temperatures. Soil Biol Biochem 27:1231–1234

Deng SX, Cheng AC, Wang MS, Cao P (2007) Gastrointestinal tract distribution of *Salmonella enteritidis* in orally infected mice with a species-specific fluorescent quantitative polymerase chain reaction. World J Gastroenterol 13:6568–6574

Duc LH, Hong HA, Barbosa TM, Henriques AO, Cutting SM (2004) Characterization of *Bacillus* probiotics available for human use. Appl Environm Microbiol 70(4):2161–2171

Ershov ED (1998) General geocryology. Cambridge University Press, Cambridge

Friedmann EI (1994) Permafrost as microbial habitat. In: Tiedje JM, Gilichinsky D (eds) Viable microorganisms in permafrost. Russian Academy of Sciences, Pushchino, pp 21–26

Fursova O, Potapov V, Brouchkov A, Pogorelko G, Griva G, Fursova N, Ignatov S (2012) Probiotic activity of a bacterial strain isolated from ancient permafrost against salmonella infection in mice. Probiotics Antimicrob Proteins 4:145–153

Granum PE, Lund T (1997) *Bacillus cereus* and its food poisoning toxins. FEMS Microbiol Lett 157:223–228

Jaenicke R (1996) Stability and folding of ultrastable proteins: eye lens crystallins and enzymes from thermophiles. FASEB 10:84–92

Katayama T, Tanaka M, Moriizumi J, Nakamura T, Brouchkov A, Douglas T, Fukuda M, Tomita M, Asano K (2007) Phylogenetic analysis of bacteria preserved in a permafrost ice wedge for 25,000 years. Appl Environ Microbiol 73:2360–2363

Levy M, Miller SL (1998) The stability of the RNA bases: implications for the origin of life. Biochemistry 95(14):7933–7938

Lozina-Lozinsky LK (1972) Outlines of cryobiology. Nauka, Leningrad

Margesin R, Miteva V (2011) Diversity and ecology of psychrophilic microorganisms. Res Microbiol 162:346–361

Nicholson WL, Munakata N, Horneck G, Melosh HJ, Setlow P (2000) Resistance of *Bacillus* endospores to extreme terrestrial and extraterrestrial environments. Microbiol Mol Biol Rev 64:548–572

Ohl ME, Miller SI (2001) Salmonella: a model for bacterial pathogenesis. Annu Rev Med 52:259–274

Repin VE, Pugachev VG, Taranov OS et al (2007) Potential hazard of microorganisms – comers from the past. In: Boyeskorov GG, Tikhonov AN, Suzuki N (eds) Yukagir Mammoth. St Peterburg State University, Moscow, pp 183–190

Silva AM, Bambirra EA, Oliveira AL, Souza PP, Gomes DA, Vieira EC, Nicoli JR (1999) Protective effect of bifidus milk on the experimental infection with *Salmonella enteritidis* subsp. *typhimurium* in conventional and gnotobiotic mice. J Appl Microbiol 86:331–336

Sukupolvi S, Edelstein A, Rhen M, Normark SJ, Pfeifer JD (1997) Development of a murine model of chronic *Salmonella* infection. Infect Immun 65:838–842

Szabó I, Wieler LH, Tedin K, Scharek-Tedin L, Taras D, Hensel A, Appel B, Nöckler K (2009) Influence of probiotic strain of *Enterococcus faecium* on *Salmonella enterica* serovar

Typhimurium DT104 infection in a porcine animal infection mode. Appl Environ Microbiol 75:2621–2628
Thompson JD, Higgins DG, Gibson TJ (1994) CLUSTAL W: improving the sensitivity of progressive multiple sequence alignment through sequence weighting, positions-specific gap penalties and weight matrix choice. Nucleic Acids Res 22:4673–4680
Willerslev E, Cooper A (2005) Ancient DNA. Proc R Soc B 272:3–16

Chapter 24
Cold-Adapted Basidiomycetous Yeasts as a Source of Biochemicals

Giorgia Tasselli, Sara Filippucci, Ciro Sannino, Benedetta Turchetti, and Pietro Buzzini

Abstract Yeasts play a relevant role as starter cultures in traditional foods and beverages, as well as in innumerable biotechnological applications for obtaining high-value bulk and fine biochemicals. Despite a considerable number of studies on yeasts have been performed by using almost exclusively the species *Saccharomyces cerevisiae* (otherwise labeled as baker's yeast), the number of yeast species described so far accounts for more than 1600, belonging to over 130 ascomycetous and basidiomycetous genera. This huge yeast diversity includes many non-*Saccharomyces* species possessing useful, and sometimes uncommon, metabolic features potentially interesting for both food and nonfood industries. Like other organisms, cold-adapted yeasts include species able to survive and grow in cold environments. They are usually labeled as psychrophiles or psychrotolerants on the basis of their cardinal growth temperatures. Among them, yeasts belonging to the phylum *Basidiomycota* apparently exhibit a superior adaptation to cold. This apparent superiority, which could be the result of some metabolic strategies implemented for adapting life to different thermal conditions in order to overcome the adverse effect of cold, can be considered worthwhile for implementing their biotechnological application at low temperatures. Accordingly, cold-adapted basidiomycetous yeasts have attracted considerable attention for their biotechnological potential, because they have developed the ability to synthesize cold-active enzymes, as well as other important biochemicals, namely, cryoprotectant compounds, polymers, lipids, and other miscellaneous compounds.

Contents

G. Tasselli • S. Filippucci • C. Sannino • B. Turchetti • P. Buzzini (✉)
Department of Agricultural, Food and Environmental Sciences, University of Perugia, Borgo XX Giugno 74, 06121 Perugia, Italy
e-mail: pietro.buzzini@unipg.it

R. Margesin (ed.), *Psychrophiles: From Biodiversity to Biotechnology*,
DOI 10.1007/978-3-319-57057-0_24

24.1 Introduction

Yeasts are defined as those fungi whose asexual growth predominantly results from budding (fission occurs only in species of the genus *Schizosaccharomyces*) and which do not form their sexual states within or upon a fruiting body. Like filamentous fungi, they are distributed between the *Ascomycota* and *Basidiomycota* phyla. The discovery that a number of yeast taxa belong to *Basidiomycota* has expanded the common opinion on the nature of yeasts (Kurtzman et al. 2011b).

Yeasts are referred to as one of the most relevant players of a variety of technological processes. They are used for millennia as starter cultures in the production of traditional foods and beverages: the production of wine, beer, bread, and many other fermented products which are well-known examples of the universal importance of yeasts in human nutrition and culture (Johnson and Echavarri-Erasun 2011; Buzzini et al. 2017). Apart from their significance in traditional technologies, the study of the physiology and metabolic diversity of yeasts has revealed that they could be used in innumerable promising biotechnological applications for producing high-value bulk and fine biochemicals.

Most of the research on yeasts concerned the model eukaryotic organism *Saccharomyces cerevisiae* (otherwise labeled as baker's yeast), which is traditionally used for biological studies, for producing fermented foods and beverages and, more recently, biofuels and heterologous compounds (Johnson and Echavarri-Erasun 2011; Buzzini et al. 2017). Indeed, due to the vast array of engineering techniques available, the recombinant DNA technology has been extensively used for obtaining genetically modified yeasts for the overexpression of a set of commercially important molecules for biomedical applications (Hou et al. 2012). However, an alternative approach is offered by yeast biodiversity (still far to be fully discovered) as source of biochemicals. An increasing part of yeast researchers consider that *Saccharomyces* species represent only a microscopic portion of the yeast diversity (Sibirny and Scheffers 2002). Really, the number of recognized yeast species has increased spectacularly in the last decades and accounts for more than 1600 species belonging to over 130 ascomycetous and basidiomycetous genera (Kurtzman et al. 2011a; Liu et al. 2015; Wang et al. 2015a, b). This trend leads to extrapolate that the potential number of yeast species sharing the ecosystems on Earth could be attested to around 150,000 (Hawksworth 2004; Lachance 2006;

Kurtzman et al. 2011a; Starmer and Lachance 2011). Undoubtedly, this vast yeast diversity includes many species possessing useful, and sometimes uncommon, metabolic features potentially interesting for both food and nonfood industries. In addition, the ever-changing consumer preferences, the need to increase productivity at both precompetitive and competitive scales by appropriate scaling-up steps, as well as the possibility to use raw substrate as carbon and nitrogen sources can lead to increasing efforts in enhancing the number of strains expressing novel and attractive properties (Buzzini and Vaughan-Martini 2006; Johnson and Echavarri-Erasun 2011).

A few authors suggested that yeasts might be better adapted to low temperatures than bacteria (Turkiewicz et al. 2003; Shivaji and Prasad 2009; Buzzini and Margesin 2014a). Like other organisms, also cold-adapted yeasts include species that successfully survive and even grow in cold environments; they are usually labeled as psychrophiles or psychrotolerant on the basis of their minimum, maximum, and optimum growth temperatures (Buzzini and Margesin 2014b). Among them, yeast species belonging to the phylum of *Basidiomycota* apparently exhibit a superior adaptation to low temperatures than *Ascomycota* (Butinar et al. 2007; de García et al. 2007, 2012; Connell et al. 2008; Turchetti et al. 2008, 2013; Branda et al. 2010; Brandão et al. 2011; Buzzini et al. 2012). It has been suggested that this apparent superiority could be the result of a few specific metabolic strategies implemented for the adaptation of life to different thermal conditions in order to overcome the adverse effect of the cold. Their distinctive ability to grow and metabolize at low temperatures can be considered extremely interesting because of the implementation of their potential biotechnological application at low temperatures (Buzzini and Margesin 2014a).

As a result, cold-adapted basidiomycetous yeasts have attracted attention from both academia and industry for their biotechnological potential. Among them, the production of cold-active enzymes can be considered to date their most important biotechnological exploitation. All organisms including yeasts which thrive frequently or permanently at near-zero temperatures have developed the ability to synthesize cold-active enzymes to sustain their cell cycle and metabolism. However, the biotechnology of cold-adapted basidiomycetous yeasts is not merely confined to the production of cold-active enzymes: a number of recent studies postulated their upcoming exploitation as promising sources of other important biochemicals, namely, cryoprotectant compounds, polymers (e.g., polysaccharides), lipids (e.g., triacylglycerols), and other minor compounds (Buzzini et al. 2012; Buzzini and Margesin 2014a). However, despite the plethora of studies so far published, only a few biochemicals obtained from cold-adapted basidiomycetous yeasts are currently on the market, while a number of other possible products are still confined to the laboratory scale (Margesin and Feller 2010; Buzzini and Margesin 2014a).

The present chapter will provide an overview of the potential of cold-adapted basidiomycetous yeasts as a source of biochemicals.

24.2 Production of Cold-Active Enzymes

Although cold-adapted basidiomycetous yeasts are generally considered less attractive than bacteria and filamentous fungi as enzyme producers, they have been studied as sources of cold-active enzymes for eco-friendly and mild technologies for food and nonfood industries. Such enzymes may provide significant economical and technical advantages due to energy savings, use in heat-sensitive manufacturing, and easy and fast inactivation at moderate temperatures after catalysis (Feller and Gerday 1997, 2003; Gerday et al. 1997, 2000; Gerday 2014; Santiago et al. 2016; Javed and Qazi 2016).

A number of cold-adapted basidiomycetous yeast species have been found to synthesize cell-bound, extracellular, or intracellular cold-active hydrolases, namely, lipases, amylases, xylanases, proteases, pectinases, galactosidases, and cellulases (Shivaji and Prasad 2009; Buzzini et al. 2012; Carrasco et al. 2012; Buzzini and Margesin 2014a). However, most of those studies were usually performed by checking strains by qualitative/semiquantitative tests carried out in Petri dishes. Quantitative determinations of enzyme activity in cell-free extracts have been performed less often, and only a limited number of studies concerned the purification and characterization of cold-active enzymes.

24.2.1 Cold-Active Lipases

Based on the NC-IUBMB enzyme nomenclature, lipases belong to the first subclass of hydrolases (EC 3.1) and include a set of enzymes acting on ester bonds: triacylglycerol lipase (EC 3.1.1.3), lysophospholipase (EC 3.1.1.5), acylglycerol lipase (EC 3.1.1.23), galactolipase (EC 3.1.1.26), phospholipase A (EC 3.1.1.32), lipoprotein lipase (EC 3.1.1.34), hormone-sensitive lipase (EC 3.1.1.79), phospholipase D (EC 3.1.4.4), EC phosphoinositide phospholipase C (EC 3.1.4.11), glycosylphosphatidylinositol phospholipase D (EC 3.1.4.50), and *N*-acetylphosphatidylethanolamine-hydrolyzing phospholipase D (EC 3.1.4.54). Triacylglycerol lipases catalyze the hydrolysis of triacylglycerols (TAGs) to free fatty acids and glycerol and are referred to as the key biocatalysts for the food and chemical industries, biomedical sciences, detergent production, and environmental management. It was recently estimated that about 40% of industrial biocatalysis is realized by using triacylglycerol lipases, which account for 21% of the global enzyme market, whose value is about USD2.9 billion (Szczesna-Antczak et al. 2014).

The production of cold-active lipases obtained from cold-adapted yeasts is considered a rapidly growing biotechnology. These enzymes have recently attracted a growing attention because of their powerful and flexible catalysis at low temperatures (if compared with their mesophilic and thermophilic homologues), which is a favorable property for working with relatively heat-sensitive molecules (Joseph et al. 2008). The ability to produce cold-active lipases was found broadly distributed in yeasts inhabiting worldwide cold environments (Table 24.1).

Table 24.1 Production of cold-active enzymes from cold-adapted basidiomycetous yeasts

Cold-active enzymes	Biotechnological application	Genus and species (synonyms)	References
Lipases	Processing of foods and beverages, production of chemicals and detergents, environmental applications	*Cystobasidium*: *C. laryngis* (*Rhodotorula laryngis*); *C. macerans* (*Cryptococcus macerans*) *Dioszegia*: *D. crocea*; *D. fristigensis* *Glaciozyma*: *G. antarctica* (*Leucosporidium antarcticum*) *Goffeauzyma*: *G. gilvescens* (*Cryptococcus gilvescens*); *G. gastrica* (*Cryptococcus gastricus*) *Leucosporidium*: *L. creatinivorum* (*Leucosporidiella creatinivora*); *L. drummii*; *L. fragarium* (*Leucosporidiella fragaria*) *Moesziomyces*: *M. antarcticus* (*Pseudozyma antarctica*; *Candida antarctica*) *Mrakia*: *M. blollopis*; *M. gelida*; *M. robertii* *Papiliotrema*: *P. laurentii* (*Cryptococcus laurentii*) *Phenoliferia*: *P. glacialis* (*Rhodotorula glacialis*) *Rhodosporidiobolus*: *R. colostri* (*Rhodotorula colostri*) *Rhodotorula*: *R. mucilaginosa* *Solicoccozyma*: *S. terricola* (*Cryptococcus terricola*) *Sporobolomyces*: *S. ruberrimus*; *S. salmonicolor* (*Sporidiobolus*	Patkar et al. (1993); Dominguez de Maria et al. (2005); Brizzio et al. (2007); Joseph et al. (2008); Juhl et al. (2010); Margesin and Feller (2010); Rashid et al. (2010); Carrasco et al. (2012); Zaliha et al. (2012); Tsuji et al. (2013a)

(continued)

Table 24.1 (continued)

Cold-active enzymes	Biotechnological application	Genus and species (synonyms)	References
		salmonicolor) *Vishniacozyma*: *V. victoriae* (*Cryptococcus victoriae*)	
Amylases	Processing of foods and beverages, production of chemicals, detergents and pharmaceuticals, paper processing	*Cystobasidium*: *C. laryngis* (*Rhodotorula laryngis*); *C. macerans* (*Cryptococcus macerans*) *Dioszegia*: *D. crocea*; *D. fristingensis* *Goffeauzyma*: *G. gilvescens* (*Cryptococcus gilvescens*); *G. gastrica* (*Cryptococcus gastricus*) *Holtermanniella*: *H. wattica* (*Cryptococcus watticus*) *Leucosporidium*: *L. creatinivorum* (*Leucosporidiella creatinivora*) *Moesziomyces*: *M. antarcticus* (*Pseudozyma antarctica*; *Candida antarctica*) *Mrakia*: *M. blollopis*; *M. gelida*; *M. robertii* *Papiliotrema*: *P. laurentii* (*Cryptococcus laurentii*) *Phenoliferia*: *P. glacialis* (*Rhodotorula glacialis*) *Rhodotorula*: *R. svalbardensis* *Saitozyma*: *S. flava* (*Cryptococcus flavus*) *Solicoccozyma*: *S. terricola* (*Cryptococcus terricola*) *Sporobolomyces*: *S. ruberrimus*	De Mot and Verachtert (1987); Brizzio et al. (2007); Galdino et al. (2011); Carrasco et al. (2012); de García et al. (2012); Singh et al. (2014); Carrasco et al. (2016)
Xylanases	Production of biofuels, solvents, paper processing, waste degradation	*Dioszegia*: *D. fristigensis* *Naganishia*: *N. adeliensis*	Amoresano et al. (2000); Gomes et al. (2000); Petrescu et al. (2000); Scorzetti et al.

(continued)

Table 24.1 (continued)

Cold-active enzymes	Biotechnological application	Genus and species (synonyms)	References
		(*Cryptococcus adeliensis*); *N. albida* (*Cryptococcus albidus*) *Papiliotrema*: *P. laurentii* (*Cryptococcus laurentii*) *Tausonia*: *T. pullulans* (*Guehomyces pullulans*)	(2000); Duarte et al. (2013)
Proteases	Processing of foods and beverages	*Cystobasidium*: *C. laryngis* (*Rhodotorula laryngis*); *C. macerans* (*Cryptococcus macerans*) *Glaciozyma*: *G. antarctica* *Goffeauzyma*: *G. gilvescens* (*Cryptococcus gilvescens*) *Itersonilia*: *I. pannonicus* (*Udeniomyces pannonicus*) *Leucosporidium*: *L. creatinivorum* (*Leucosporidiella creatinivora*); *L. fragarium* (*Leucosporidiella fragaria*) *Mrakia*: *M. frigida* (*Leucosporidium frigidum*); *M. gelida* *Naganishia*: *N. adeliensis* (*Cryptococcus adeliensis*) *Papiliotrema*: *P. laurentii* (*Cryptococcus laurentii*) *Rhodotorula*: *R. mucilaginosa*; *R. toruloides* (*Rhodosporidium toruloides*) *Solicoccozyma*: *S. terricola* (*Cryptococcus terricola*) *Sporobolomyces*: *S. salmonicolor* (*Sporidiobolus*	Ray et al. (1992); Ogrydziak (1993); Pazgier et al. (2003); Turkiewicz et al. (2003); Brizzio et al. (2007); Huston (2008); Carrasco et al. (2012); de García et al. (2012); Duarte et al. (2013); Alias et al. (2014); Lario et al. (2015); Chaud et al. (2016)

(continued)

Table 24.1 (continued)

Cold-active enzymes	Biotechnological application	Genus and species (synonyms)	References
		salmonicolor) *Udeniomyces*: *U. megalosporus*; *U. pyricola* *Vanrija*: *V. humicola* (*Candida humicola*)	
Pectinases	Fruit and vegetable processing	*Cystobasidium*: *C. laryngis* (*Rhodotorula laryngis*); *C. macerans* (*Cryptococcus macerans*) *Cystofilobasidium*: *C. capitatum* (*Rhodosporidium capitatum*); *C. larimarini* *Dioszegia*: *D. crocea*; *D. fristigensis* *Leucosporidium*: *L. drummii* *Mrakia*: *M. aquatica* (*Mrakiella aquatica*); *M. frigida* (*Leucosporidium frigidum*); *M. psychrophila*; *M. robertii* *Phenoliferia*: *P. glacialis* (*Rhodotorula glacialis*) *Rhodotorula*: *R. mucilaginosa* *Sporobolomyces*: *S. salmonicolor* (*Sporidiobolus salmonicolor*)	Nakagawa et al. (2002, 2005a, b); Birgisson et al. (2003); Margesin et al. (2005); Carrasco et al. (2012); de García et al. (2012); Sahay et al. (2013)
Galactosidases	Fruit and vegetable processing	*Cystofilobasidium*: *C. capitatum* (*Rhodosporidium capitatum*) *Naganishia*: *N. albida* (*Cryptococcus albidus*) *Tausonia*: *T. pullulans* (*Guehomyces pullulans*)	Pavlova et al. (2002); Nakagawa et al. (2006); Song et al. (2010); Hamid et al. (2013)
Cellulases	Foods, detergents, and biofuel, processing of textile fibers	*Dioszegia*: *D. fristigensis* *Glaciozyma*: *G. antarctica*	Pavlova et al. (2002); Turkiewicz et al. (2005); Carrasco et al. (2012); Carrasco et al. (2016)

(continued)

Table 24.1 (continued)

Cold-active enzymes	Biotechnological application	Genus and species (synonyms)	References
		(*Leucosporidium antarcticum*) *Goffeauzyma*: *G. gastrica* (*Cryptococcus gastricus*) *Holtermanniella*: *H. wattica* (*Cryptococcus watticus*) *Leucosporidium*: *L. drummii* *Mrakia*: *M. blollopis*; *M. gelida*; *M. psychrophila*; *M. robertii* *Naganishia*: *N. albida* (*Cryptococcus albidus*) *Papiliotrema*: *P. laurentii* (*Cryptococcus laurentii*) *Vishniacozyma*: *V. victoriae* (*Cryptococcus victoriae*)	
Chitinase	Degradation of chitin-rich wastes, control of pathogens and microbial spoilage	*Dioszegia fristigensis* *Glaciozyma antarctica* *Mrakia psychrophila* *Sporobolomyces salmonicolor* (*Sporidiobolus salmonicolor*)	Ramli et al. (2011, 2012); Carrasco et al. (2012)
Ligninolytic enzymes	Second-generation biofuels	*Cryptococcus* sp. *Cylindrobasidium* sp. *Cystobasidium*: *C. laryngis* (*Rhodotorula laryngis*); *C. pallidum* (*Rhodotorula pallida*) *Dioszegia*: *D. hungarica* (*Bullera armeniaca*) *Goffeauzyma*: *G. gilvescens* (*Cryptococcus gilvescens*) *Leucosporidium*: *L. creatinivorum* (*Leucosporidiella creatinivora*) *Mrakia* sp. *Rhodotorula*: *R. creatinivora*;	Rovati et al. (2013)

(continued)

Table 24.1 (continued)

Cold-active enzymes	Biotechnological application	Genus and species (synonyms)	References
		R. glutinis; *R. mucilaginosa* *Solicoccozyma*: *S. terricola* (*Cryptococcus terricola*) *Vishniacozyma*: *V. victoriae* (*Cryptococcus victoriae*)	
Phytases	Feed processing	*Papiliotrema*: *P. laurentii* (*Cryptococcus laurentii*) *Rhodotorula*: *R. mucilaginosa*	Yu et al. (2015)
β-Glucanases	Processing of foods and beverages	*Glaciozyma*: *G. antarctica*	Parvizpour et al. (2015, 2016); Mohammadi et al. (2016)

Cold-active lipases A and B (labeled as CALA and CALB, respectively) were isolated and purified from *Candida antarctica* (now *Moesziomyces antarcticus*, syn. *Pseudozyma antarctica*) since 1990s (Patkar et al. 1993; Dominguez de Maria et al. 2005; Szczesna-Antczak et al. 2014). CALA and CALB are the first examples of patented cold-active enzymes from yeasts: their structure and catalytic activity (both in free and immobilized form) have been extensively studied since early 2000s (Kirk and Christensen 2002; Krishna et al. 2002; Blank et al. 2006; Chen et al. 2008; Ericsson et al. 2008; Miletić and Loos 2009; Emond et al. 2010; Forde et al. 2010; Gutarra et al. 2011; Habeych et al. 2011; Gruber and Pleiss 2012; Kahveci and Xu 2012). CALA exhibits a high stability up to 90 °C (Faber 2004; Dominguez de Maria et al. 2005; Gotor-Fernandéz et al. 2006; Liese et al. 2000; Patel 2004; Szczesna-Antczak et al. 2014). CALB shows also other interesting commercial characteristics and is at present produced by Novozymes (Denmark) using a recombinant strain of *Aspergillus niger* under the commercial names Lipozyme® CALB L (in soluble form) and Novozym® 435 (in immobilized form) (Margesin and Feller 2010; Szczesna-Antczak et al. 2014).

Other cold-active lipases have been studied. Rashid et al. (2010) found a cold-active lipase in *Leucosporidium antarcticum* (now classified as *Glaciozyma antarctica*), while Zaliha et al. (2012) patented a bifunctional lipase from the same species. This lipase exhibited the highest activity at 20 °C and retained 60% of its maximum activity at 5 °C. More recently, a cold-active lipase from a strain of *Mrakia blollopis* isolated from Antarctica has been isolated, purified, and characterized (Tsuji et al. 2013a).

The production of cold-active lipases from basidiomycetous yeasts isolated from cold ecosystems has been recently reviewed (Buzzini et al. 2012;

Szczesna-Antczak et al. 2014). Cold-adapted basidiomycetous yeasts expressing lipase activity at low temperatures are summarized in Table 24.1.

24.2.2 Cold-Active Amylases

Amylases belong to the first subclass of hydrolases (EC 3.2) and include enzymes hydrolyzing O- and S-glycosyl compounds: α-amylase (EC 3.2.1.1), β-amylase (EC 3.2.1.2), and isoamylase (EC 3.2.1.68). Cold-adapted amylases have emerged as one of the leading biocatalysts for a vast array of industrial applications in food (production of sweet syrups, maltose, and dextrins from polysaccharides, improvement of fruit juice clearness, delay of bread staling, and removal of starch residues), textile (processing denim textiles), detergent (cold washing), paper (decrease of starch viscosity for improving paper coating), and pharmaceutical industry (dietary supplements to aid digestive processes) (Kuddus et al. 2011).

Similarly to lipases, the scientific and analytic level of the studies on cold-active amylases was extremely variable, ranging from screening of amylase activity on Petri dishes to checking crude activity of cell-free extracts, while the use of fine purification and characterization protocols was less frequent. The first study on the purification to homogeneity and characterization of two cold-active amylases produced by *C. antarctica* (now *M. antarcticus*, syn. *P. antarctica*) dates back to 1980s. These enzymes exhibited a high debranching activity at optimum temperatures of 57 and 62 °C and the ability to hydrolyze soluble starch (De Mot and Verachtert 1987). More recently, a gene encoding an extracellular α-amylase (Amy1) from *Cryptococcus flavus* (now *Saitozyma flava*) was expressed in *S. cerevisiae* cells: the enzyme was subsequently purified to homogeneity (molecular mass: 67 KDa). Significant biochemical and structural discrepancies between wild-type and recombinant α-amylase with respect to K_m values, enzyme specificity, and secondary structure content were found (Galdino et al. 2011). Other cold-adapted basidiomycetous yeasts expressing important starch-degrading activity at low temperatures are shown in Table 24.1.

24.2.3 Cold-Active Xylanases

Xylanases belong to the subclass EC 3.2 of hydrolases and include endo-1,4-β-xylanase (EC 3.2.1.8), glucuronoarabinoxylan endo-1,4-b-xylanase (EC 3.2.1.136), and oligosaccharide reducing-end xylanase (EC 3.2.1.156). They catalyze the hydrolysis of hemicellulose (i.e., β-1,4-xylan) into xylose. The hydrolysis of this polysaccharide through xylanases is used in biotechnological processes, in particular in biofuel and paper industries (Amoresano et al. 2000). Cold-active xylanases can be applied in low-temperature digestion of farm, industrial, and sewage wastes (Shivaji and Prasad 2009).

A few cold-active xylanases have been purified to homogeneity and characterized. A glycosylated xylanase (338 amino acids) produced by an Antarctic strain of *Cryptococcus albidus* (now classified as *Naganishia albida*) was structurally elucidated. Mass spectra (MS) analysis revealed the occurrence of *N*-glycosylation only at Asn254, which was modified by high-mannose structure (Amoresano et al. 2000). Likewise, Petrescu et al. (2000) studied a cold-active glycosylated xylanase produced by an Antarctic strain of *Cryptococcus adeliensis* (now *Naganishia adeliensis*). The xylanase exhibited 84% identity and lower thermostability than its homologue synthesized by *C. albidus*. The cold-active xylanase exhibited a higher catalytic efficiency between 0 and 20 °C. Computerized analysis of molecular models indicated that the cold adaptation consists of discrete changes in the 3D structure. The ability of cold-adapted basidiomycetous yeasts to produce cold-active xylanases has been broadly reviewed (Table 24.1).

24.2.4 Cold-Active Proteases

Proteases (including a high number of peptidases) belong to the subclass EC 3.2 of hydrolases which act on peptide bonds. Cold-active proteases have a wide range of industrial and technological applications. In particular, they are used in food industry: (1) in low-temperature cheeses ripening as an alternative to rennet, (2) in seafood processing for improving the removal of scales and skin and for ameliorating the extraction of carotene-protein from shellfish, and (3) in meat processing for imparting tenderness to refrigerated products and for improving their flavor (Huston 2008).

The first discovery of an extracellular cold-active protease (molecular mass: 36 KDa) produced by an Antarctic strain of *Cryptococcus humicola* (now *Vanrija humicola*), active between 0 and 45°C and with an optimum activity at 37 °C, dates back to the early 1990s. This protease retained about 15% of its maximum activity at 0 °C (Ray et al. 1992).

The species *G. antarctica* was shown to be an active protease producer. This species was the source of the first purified and characterized cold-active subtilase belonging to the proteinase K subfamily (subtilisin family). The resulting protein exhibited a very high specific activity (85 U mg^{-1} protein) (Białkowska and Turkiewicz 2014). Another subtilase isolated from the same species exhibited wide substrate specificity, with the greatest affinity to substrates characteristic of chymotrypsin and chymotrypsin-like enzymes (Ogrydziak 1993; Białkowska and Turkiewicz 2014). It should be stressed that such high specific activity exhibited by *G. antarctica* subtilase is similar to that found in most other microbial serine proteases produced by the mesophilic species *Yarrowia lipolytica* (Tobe et al. 1976) and *Aureobasidium pullulans* (Ogrydziak 1993).

An extracellular serine proteinase synthesized by an Antarctic strain of *L. antarcticum* (now *G. antarctica*) was also characterized. The sequence of its 35 N-terminal amino acid residues showed 31% homology to that of proteinase K

(Pazgier et al. 2003). Likewise, Turkiewicz et al. (2003) characterized a glycosylated serine proteinase (molecular mass: 34.4 KDa) secreted by a strain of the same species. This enzyme exhibited low optimum temperature for activity (25 °C), poor thermal stability, and high catalytic efficiency from 0 to 25 °C.

The isolation of the PI12 protease gene from genomic and mRNA sequences of *G. antarctica* allowed the determination of 19 exons and 18 introns. Full-length cDNA of PI12 protease gene coding for 963 amino acids was amplified by rapid amplification of cDNA ends strategy. PI12 protease was purified to homogeneity (molecular mass: 99 KDa) and showed 42% identity with the subtilisin-like protease produced by the species *Rhodotorula toruloides* (formerly *Rhodosporidium toruloides*); no homology with other cold-active proteases was found. High recombinant protease production (28.3 U mL^{-1}) was obtained using a *Komagataella pastoris* (former *Pichia pastoris*) expression system at 20 °C (Alias et al. 2014).

Besides, the isolation, purification to homogeneity, and characterization of an extracellular protease from a strain of *Rhodotorula mucilaginosa* isolated from an Antarctic marine alga were also studied. The purified protease presented optimum pH and temperature activity at pH 5.0 and 50 °C (about 25% of activity were retained below 20 °C), respectively, and high stability in the presence of high concentrations of NaCl (Lario et al. 2015). The ability of cold-adapted basidiomycetous yeasts to produce cold-active proteases is reported in Table 24.1.

24.2.5 Cold-Active Pectinases

Pectinases are a complex of enzymes that hydrolyze pectin. They are commonly referred to as pectic enzymes and include pectinesterase, otherwise labeled as pectin methyl esterases (EC 3.1.1.11); pectin lyase, also known as pectolyase (EC 4.2.2.10); pectate lyase (EC 4.2.2.2); and polygalacturonase (EC 3.2.1.15). Cold-adapted yeasts are a good source of pectic enzymes. A number of yeast species produce several types of pectinases. Cold-active pectinases may be applied in the fruit-processing industry for reducing juice viscosity at low temperature and for improving the clarity of the final product (Alimardani-Theuil et al. 2011).

The first cold-active pectinase has been obtained from *Cystofilobasidium capitatum* (Nakagawa et al. 2002): this yeast was found to synthesize a wide range of cold-active pectic enzymes, pectin methyl esterases, polygalacturonases, and pectin lyases, but not pectate lyases. A polygalacturonase from this species was characterized: it exhibited an optimum activity of 458 U mg^{-1} protein at 45 °C (and retained at 0 °C 20% of maximum activity) and pH 4.4. The enzyme was found to be stable in the pH range of 2.5–7.0 and up to a temperature of about 50 °C (Nakagawa et al. 2005a). Furthermore, a pectin lyase from *C. capitatum* has been purified to homogeneity (molecular mass: 42 KDa) and characterized: it exhibited a high specific activity (1230 U mg^{-1} protein), at 40 °C and pH 8.0, and retained about 5–50% of its maximum activity in the range of 0–20 °C (Nakagawa et al.

2005b). A number of other cold-active pectic enzymes produced by cold-adapted basidiomycetous yeasts are listed in Table 24.1.

24.2.6 Cold-Active Galactosidases

Galactosidases include a few hydrolases (EC 3.2), namely, α-galactosidase (EC 3.2.1.22) and β-galactosidase (EC 3.2.1.23), which are able to hydrolyze O- and S-glycosyl compounds. From the technological viewpoint, cold-active β-galactosidases are applied in low-temperature hydrolysis of lactose for producing lactose-free milk and milk products (Sheridan and Brenchley 2000). Cold-active β-galactosidases also possess trans-glycosylation activity, whereby lactose hydrolysis takes place with simultaneous transfer of monosaccharides to higher oligosaccharides to form tri- and tetrasaccharides in probiotic foodstuffs (Karasová-Lipovová et al. 2003; Benesova et al. 2005; Huston 2008).

Although these enzymes exhibit sufficiently high activity at temperatures below 20 °C and, consequently, they would be a perfect alternative to mesophilic homologues due to energy saving (Białkowska et al. 2009), only a few cold-active β-galactosidases from cold-adapted yeasts have been described to date. The first β-galactosidase was found in 2000s in *Guehomyces pullulans* (now *Tausonia pullulans*) (Nakagawa et al. 2006). Crude enzymatic preparations exhibited a relatively high optimum temperature for lactose hydrolysis (50 °C), although at 0 °C about 20% of maximum activity were retained. The highest activity was found at pH 4.0, with stability in the pH range of 3–7. Another yeast β-galactosidase was found to be produced by a strain of the same species isolated from sea sediment in Antarctica. The optimum temperature and pH of the crude protein preparation are 50 °C and 4.0, respectively (Song et al. 2010). Cold-active galactosidases produced by cold-adapted basidiomycetous yeasts are reported in Table 24.1.

24.2.7 Cold-Active Cellulases

Cellulases (EC 3.2.1.4) are hydrolases that catalyze total hydrolysis of cellulose. It is generally accepted that effective hydrolysis of cellulose into glucose requires the synergistic actions of a complex of three enzymes including (1) endo-b-1,4-glucanase (EC 3.2.1.4), which randomly cleaves internal linkages; (2) cellobiohydrolase (EC 3.2.1.91), which specifically hydrolyzes cellobiose units from nonreducing ends; and (3) b-D-glucosidase (EC 3.2.1.21), which hydrolyzes glucosyl units from cello-oligosaccharides (Perez et al. 2002).

Cellulases have great potential for application in industries such as food, detergent, laundry, textile, and biofuels. The common requirement in these fields to reduce the temperatures of the processes has led to the continuous research of microorganisms able to secrete cold-active cellulases. Despite the fact that cold-adapted yeasts could be good candidates as producers of cold-active cellulases for cellulose hydrolysis at low temperature, all studies have been so far limited to check crude cellulase activity of yeast isolates or cell-free extracts (Kasana and Gulati 2011). The ability of cold-adapted basidiomycetous yeasts to produce cold-active cellulases is summarized in Table 24.1.

24.2.8 Miscellaneous Cold-Active Enzymes

A few additional cold-active enzymes produced by cold-adapted basidiomycetous yeasts have been studied (Table 24.1). Chitinases (EC 3.2.1.14) are hydrolytic enzymes that break down glycosidic bonds in chitin. The gene of a cold-active chitinase from *G. antarctica* was expressed in the mesophilic host *P. pastoris* (now *K. pastoris*). The recombinant chitinase was purified to homogeneity and characterized: two fractions corresponding to a first non-glycosylated protein (molecular mass: 39 KDa) and to a second glycosylated protein (50 KDa) were found. The enzyme exhibited high activity to efficiently hydrolyze chitin at 15 °C; at 5 °C the enzyme retained over 20% of its maximum activity. A 3D model of this recombinant chitinase was also built by first using a threading approach to search for a suitable template and to generate an optimum target-template alignment (Ramli et al. 2011, 2012).

Ligninolytic enzymes, namely, laccases (E.C. 1.10.3.2), manganese peroxidases (EC 1.11.1.13), and lignin peroxidases (EC 1.11.1.14), have been recently applied in producing second-generation biofuels (Plácido and Capareda 2015). Rovati et al. (2013) found that some Antarctic basidiomycetous yeasts exhibited extracellular ligninolytic activity (Table 24.1).

Phytases include any type of phosphatase enzymes that catalyze the hydrolysis of phytic acid (myoinositol hexakisphosphate): 3-phytase (EC 3.1.3.8), 4-phytase (EC 3.1.3.26), and 5-phytase (EC 3.1.3.72). Phytic acid is an indigestible, organic form of phosphorus occurring in grains and oil seeds; its hydrolysis releases a bioavailable form of inorganic phosphorus. An extracellular cold-active phytase produced by a strain of *R. mucilaginosa* isolated from Antarctic deep-sea sediments was purified to homogeneity (molecular mass: 63 KDa), and a specific activity of 31.6 U mg^{-1} protein was found. Its optimum temperature and pH were 50 and 5.0, respectively. It retained 85% and 20% of maximum activity at 37 and 10 °C, respectively, and showed good stability in the pH range 3.0–7.0 (Yu et al. 2015).

β-Glucanases include exo- (EC 3.2.1.58) and endo-β-1,3-glucanases (EC 3.2.1.6), which hydrolyze a few β-1,3-glucans such as callose or curdlan, and β-1,6 glucanase (EC 3.2.1.75), which hydrolyzes β-1,6-glucans. A cold-active exo-β-1, 3-glucanase (GaExg55) isolated from *G. antarctica* has been structurally modeled and characterized. The comparison of its primary sequence with those of other mesophilic, thermophilic, and hyperthermophilic homologues revealed the presence of several amino acid substitutions on the surface of cold-adapted enzyme that totally increased its structural flexibility for efficient catalytic reactions at low temperatures. Besides, its catalytic activity and structural flexibility at low temperatures were attained through a reduced amount of hydrogen bonds and salt bridges, as well as an increased exposure of the hydrophobic side chains to the solvent. A molecular dynamics simulation performed to evaluate the stability of the GaExg55 structure at varying low temperatures confirmed the above findings (Parvizpour et al. 2015, 2016; Mohammadi et al. 2016).

24.3 Production of Cryoprotectant Macromolecules

It is well known that a low freezing rate can damage the cytoplasm membrane (Kawahara 2008). Therefore, the synthesis of cryoprotectant (antifreeze) molecules is considered essential to reduce the risk of cell damage that occurs in natural habitats, where freezing rates are quite low. Among them, the ability to increase the intracellular trehalose content is broadly distributed in nature as cell-protecting mechanism against osmotic stress (Kandror et al. 2002). The high cytoplasm concentrations of trehalose observed in *Mrakia frigida* and *Leucosporidium fellii* probably contribute to the reduction of the freezing point of the intracellular fluid and to the stabilization of cold-active protease by preventing cell autolysis (Deegenaars and Watson 1997, 1998; Pan et al. 2005).

However, along with the intracellular accumulation of trehalose, other antifreeze compounds are produced by cold-adapted yeasts as adaptative strategies to overcome the adverse effects of low temperature on cell functionality. Among them, antifreeze proteins (AFPs) play an important role in many cold-adapted organisms. AFPs are generally characterized by two activities: (1) thermal hysteresis and (2) inhibition of ice recrystallization. They were found in some cold-adapted yeasts isolated from glacial environments to act in a non-colligative way by decreasing the cytoplasm intracellular freezing point and by inhibiting ice recrystallization at sub-zero temperatures (Lee et al. 2012; Hashim et al. 2013; Park et al. 2012; Singh et al. 2014).

A recombinant AFP from the Arctic yeast *Leucosporidium* sp. was obtained from a *Pichia* expression system. It exhibited the ability to reduce the damage induced to red blood cells by freeze thawing (Lee et al. 2012). More recently, the cell-free extract from *G. antarctica* demonstrated a high antifreeze activity given by both thermal hysteresis and ice recrystallization inhibition properties. The AFP gene of *G. antarctica* was cloned into an *Escherichia coli* expression system. Recombinant AFP exhibited the presence of inclusion bodies that were subsequently denatured by treatment with urea and allowed to refold in vitro (Hashim et al. 2013).

Lee et al. (2010) reported the ability of an Arctic strain of *Leucosporidium* sp. to secrete a glycosylated ice-binding protein (IBP) exhibiting both thermal hysteresis and recrystallization inhibition properties. The full-length cDNA for IBP synthesis was found to encode a 261 amino acid protein (molecular mass: 26.8 KDa) that includes an N-terminal signal peptide and one potential *N*-glycosylation site. The deduced protein showed high sequence identity with other IBP homologues from fungi, diatoms, and bacteria, clustering with a class of ice-active proteins (Lee et al. 2010). More recently, the same authors characterized an IBP from the same *Leucosporidium* sp. strain as belonging to a class of IBPs with no significant similarity in primary structure to other known AFPs; glycosylation did not cause structural changes and functionality (Park et al. 2012). The production of AFPs and IBPs by cold-adapted yeasts is summarized in Table 24.2.

24.4 Production of Exopolysaccharides

Exopolysaccharides (EPS) produced by microorganisms represent an industrially unexploited market. Some microorganisms can produce and excrete over 40 g L^{-1} of EPS in simple but sometimes expensive processes. The main function of these polymers is to increase the protection against environmental biotic and abiotic stresses, including cold conditions. Heteropolysaccharides and some homopolysaccharides are synthesized in microbial cells and then secreted into the extracellular environment. Physical or chemical extraction methods have been studied to improve the yield of EPS at an industrial level. Cold-adapted yeasts were found to produce EPS from monosaccharides, mainly from mannose and glucose. The physicochemical and rheological properties of the EPS make them suitable for using as emulsifiers and stabilizers in cosmetics (Donot et al. 2012; Pavlova 2014).

A few strains of the species *Cryptococcus flavus* (now *Saitozyma flava*) isolated from Antarctica were found to produce high amounts (5.75 g L^{-1}) of EPS. The composition of the polymers (molecular mass: 1.01 MDa) was established: 55.1% mannose, 26.1% glucose, 9.6% xylose, and 1.9% galactose (Pavlova et al. 2009). Likewise, Poli et al. (2010) studied the Antarctic yeast *Sporobolomyces salmonicolor* as a source of high-molecular-mass mannan-like EPS: the maximum yield (5.64 g L^{-1}) was achieved at 22 °C. The same species was also used to produce EPS in a stirred bioreactor with a yield of 5.5 g L^{-1} (Vlaev et al. 2013). Finally, an EPS (molecular mass: 8.0 KDa) was also produced by a strain of *Cryptococcus laurentii* (now *Papiliotrema laurentii*) isolated in Antarctica: this heteropolysaccharide was found to be composed of 45.2% xylose, 33.6% mannose, and 18.4% glucose (Rusinova-Videva et al. 2011; Kuncheva et al. 2013). The production of EPS by cold-adapted yeasts is reported in Table 24.2.

24.5 Production of Lipids

The number of reports on the ability of some microorganisms to accumulate high intracellular amounts of lipids (in particular TAGs) rapidly increased in recent years because of their important role as building blocks for obtaining oleochemical

Table 24.2 Production of biochemicals from cold-adapted basidiomycetous yeasts

Biochemicals	Biotechnological application	Genus and species (synonyms)	References
Antifreeze proteins	Processing of food	*Mrakia*: *M. frigida* (*Cryptococcus curiosus*); *M. gelida* (*Mrakia stokesii*) *Glaciozyma*: *G. antarctica* (*Leucosporidium antarcticum*) *Leucosporidium*: *L. fellii*; *Leucosporidium scottii* *Rhodotorula*: *R. svalbardensis*	Deegenaars and Watson (1998); Lee et al. (2012); Hashim et al. (2013); Singh et al. (2014)
Ice-binding proteins	Processing of food	*Glaciozyma*: *G. antarctica* (*Leucosporidium antarcticum*) *Leucosporidium* sp.	Lee et al. (2010, 2012); Park et al. (2012)
Exopolysaccharides	Processing of food and textile fibers, production of cosmetics, pharmaceuticals and detergents	*Cystobasidium*: *C. minutum* (*Rhodotorula minuta*) *Moesziomyces*: *M. antarcticus* (*Pseudozyma antarctica*; *Candida antarctica*) *Papiliotrema*: *P. laurentii* (*Cryptococcus laurentii*) *Rhodotorula*: *R. glutinis* *Saitozyma*: *S. flava* (*Cryptococcus flavus*) *Sporobolomyces*: *S. roseus*; *S. salmonicolor* (*Sporidiobolus salmonicolor*)	Pavlova et.al. (2009); Poli et al. (2010); Rusinova-Videva et al. (2011); Vlaev et al. (2013); Kuncheva et al. (2013)
Lipids	Production of biofuels, lubrificants, adhesives, solvents, biosurfactants, cosmetics, and polymers	*Apiotrichum*: *A. porosum* (*Trichosporon porosum*) *Cutaneotrichosporon*: *C. curvatus* (*Cryptococcus curvatus*) *Cystobasidium*: *C. laryngis* (*Rhodotorula laryngis*) *Cystofilobasidium*:	Rossi et al. (2009); Amaretti et al. (2010); Morita et al. (2013); Pereyra et al. (2014); Schulze et al. (2014); Garay et al. (2016); Jiru et al. (2016)

(continued)

Table 24.2 (continued)

Biochemicals	Biotechnological application	Genus and species (synonyms)	References
		C. infirmominiatum *Goffeauzyma*: *G. gilvescens* (*Cryptococcus gilvescens*) *Leucosporidium*: *L. creatinivorum* (*Leucosporidiella creatinivora*); *L. scottii* *Phenoliferia*: *P. glacialis* (*Rhodotorula glacialis*) *Pseudozyma* sp. *Rhodosporidiobolus*: *R. ruineniae* (*Sporidiobolus ruineniae*) *Rhodotorula*: *R. araucariae*; *R. dairenensis*; *R. graminis*; *R. kratochvilovae*; *R. mucilaginosa* *Saitozyma*: *S. podzolica* (*Cryptococcus podzolicus*) *Solicoccozyma*: *S. phenolicus* (*Cryptococcus phenolicus*) *Sporobolomyces*: *S. pararoseus* (*Sporidiobolus pararoseus*)	
γ-Decalactone	Processing of food and beverages	*Buckleyzyma*: *B. aurantiaca* (*Rhodotorula aurantiaca*)	Alchihab et al. (2009)
Biosurfactants	Processing of food, crude oil recovery, biomedical applications	*Kalmanozyma*: *K. fusiformata* (*Pseudozyma fusiformata*)	Cameotra and Makkar (2004)
cis-9-Heptadecenoic acid	Processing of food and beverages	*Anthracocystis*: *A. flocculosa* (*Pseudozyma flocculosa*)	Avis and Bélanger (2001)
Itaconic acid	Production of synthetic resins and coatings	*Moesziomyces*: *M. antarcticus* (*Pseudozyma antarctica*; *Candida antarctica*)	Levinson et al. (2006)

(continued)

Table 24.2 (continued)

Biochemicals	Biotechnological application	Genus and species (synonyms)	References
Heterologous proteins	Production of therapeutic compounds	*Anthracocystis*: *A. flocculosa* (*Pseudozyma flocculosa*) *Moesziomyces*: *M. antarcticus* (*Pseudozyma antarctica*; *Candida antarctica*)	Avis et al. (2005)
Ethanol	Production of biofuels	*Mrakia*: *M. blollopis*	Tsuji et al. (2013b, 2014)
Carotenoids	Processing of food and feed, production of pharmaceuticals and cosmetics	*Naganishia*: *N. albida* (*Cryptococcus albidus*) *Papiliotrema*: *P. laurentii* (*Cryptococcus laurentii*) *Phaffia*: *P. rhodozyma* *Rhodotorula*: *R. bacarum*; *R. diobovata* (*Rhodosporidium diobovatum*); *R. glutinis*; *R. graminis*; *R. mucilaginosa*; *R. toruloides* (*Rhodosporidium toruloides*) *Sporobolomyces*: *S. johnsonii* (*Sporidiobolus johnsonii*); *S. metaroseus* (*S. metaroseus*); *S. pararoseus* (*Sporidiobolus pararoseus*); *S. patagonicus*; *S. salmonicolor* (*Sporidiobolus salmonicolor*)	Dimitrova et al. (2010, 2013); Barahona et al. (2016)

products, in particular biofuels, lubricants, adhesives, solvents, biosurfactants, cosmetics, and degradable polymers (Beopoulos et al. 2009; Behr and Pérez Gomes 2011; Biermann et al. 2011). A few lipid overproducing yeasts (i.e., the ascomycetous species *Lipomyces starkeyi* and *Y. lipolytica*) have been obtained using recombinant DNA technology (Beopoulos et al. 2009; Johnson and Echavarri-Erasun 2011). However, of 1600 known species of yeasts, only about 70 are so far referred to as oleaginous (i.e., characterized by the ability to accumulate over 20% intracellular lipids), and this number is continuously increasing. Among them, some cold-adapted basidiomycetous yeasts were found to be good producers of TAGs, with intracellular lipid concentrations of over 60% lipid by dry mass. The major fatty acids recorded were oleic acid, linoleic acid, linolenic acid, palmitic acid, stearic acid, myristic acid, and pentadecanoic acid. An increase of the unsaturation degree of TAGs was found to be highly correlated with the decrease of the growth temperature (Rossi et al. 2009; Amaretti et al. 2010; Morita et al. 2013; Gunde-Cimerman et al. 2014; Pereyra et al. 2014; Schulze et al. 2014; Sitepu et al. 2014; Garay et al. 2016; Jiru et al. 2016). The production of lipids by cold-adapted basidiomycetous yeasts is summarized in Table 24.2.

24.6 Production of Miscellaneous Compounds

A number of recent studies carried out at the laboratory scale have demonstrated that cold-adapted basidiomycetous yeasts are able to produce other miscellaneous compounds of industrial interest.

Alchihab et al. (2009) studied the low-temperature synthesis of γ-decalactone by an Antarctic strain of *Rhodotorula aurantiaca* (now *Buckleyzyma aurantiaca*). Highest production (6.6 g L^{-1}) was obtained at 14 °C by using a medium containing 20 g L^{-1} castor oil. Also surfactant proteins (Cameotra and Makkar 2004), antimycotics (Avis and Bélanger 2001), itaconic acid (Levinson et al. 2006), and heterologous proteins (Avis et al. 2005) were produced by cold-adapted basidiomycetous yeasts (Table 24.2).

The largest part of literature reporting the development of fermentation processes for producing ethanol at low temperatures is related to the use of immobilized *S. cerevisiae* cells (Kourkoutas et al. 2004). However, a strain of the species *M. blollopis* isolated from Antarctica was studied for its ability to ferment various carbohydrates at low temperatures and for direct ethanol fermentation from cellulosic biomass. The addition of nonionic surfactants affected ethanol production: 48.2 g L^{-1} ethanol was formed from 12% (w/v) glucose, whereas 7.2–12.5 g L^{-1} ethanol was obtained from different cellulosic raw sources (Tsuji et al. 2013b, 2014).

Although the production of pigments is not restricted to cold-adapted yeasts, some basidiomycetous yeasts isolated from cold habitats exhibited a superior ability to accumulate high contents of intracellular carotenoids. Due to their biological properties, carotenoids are widely used in phytomedicine and in the chemical, pharmaceutical, cosmetic, food, and feed industries. Accordingly, their global market is continuously growing, and it is expected to reach about USD1.4 billion in 2018 (Mannazzu et al. 2015). Irradiation was found to significantly stimulate carotenoid biosynthesis (Dimitrova et al. 2010, 2013). A number of pigmented strains isolated from Antarctic sedimentary rocks and belonging to the yeast species *S. salmonicolor* and *Sporobolomyces metaroseus*, together with the fungal genus *Collophora*, produced several carotenoid pigments, namely, 2,3-dihydroxy-γ-carotene, β-carotene, 4-ketotorulene, torulene, β-cryptoxanthin, and spirilloxanthin (Barahona et al. 2016).

24.7 Conclusions

Although cold areas constitute one of the largest biospheres on Earth, prokaryotic and eukaryotic organisms harboring such extreme habitats are still poorly studied. In this framework, biological and metabolic diversity of yeasts living in permanently cold ecosystems is being studied not only to merely establish their species richness but also because they have developed many adaptation strategies that allow them to successfully survive and grow under such extreme conditions.

The synthesis of enzymes active at low and moderate temperatures is generally considered a key aspect of this adaptation. However, some additional mechanisms (e.g., the synthesis of cryoprotectant macromolecules, EPS, intracellular lipids, etc.) are applied for the same purpose. Such complex adaptive physiological changes have undoubtedly a noteworthy impact on biotechnological potentials expressed by microorganisms living at near-zero temperatures. Among them, cold-adapted yeasts may constitute a valuable resource for the development of microbial biotechnologies. The ever-increasing availability of genomes sequences of cold-adapted yeasts and their comparison with those of their mesophilic counterparts could give in the next future new interesting insights into the genetic basis of their adaptation to cold, thus improving the knowledge of the ecological significance and biotechnological impact of this important feature.

From a biotechnological point of view, the screening and selection of novel cold-adapted yeasts indisputably represents an indispensable step for discovering new overproducing strains. However, the fundamental bottleneck for their commercial exploitation is represented by the scaling-up from the laboratory to the industrial scale. The limited number of patented cold-active enzymes so far deposited may be the result of the high risks and costs that should be supported by pharmaceutical, chemical, and food industries in pursuing this largely unexplored but attractive field.

Some relatively new recombinant DNA technologies could certainly make possible the production of an increasing number of recombinant biochemicals in suitable expression systems, e.g., cheaper cold-active enzymes, proteins, etc., with novel or modified activity. The commercial production of proteins in heterologous hosts and their modification by chemical or protein engineering tools could be a way for obtaining higher catalytic activity and tailored catalytic selectivity, which could led to noteworthy implications in various fields of biotechnology.

References

Alchihab M, Destain J, Aguedo M, Majad L, Ghalfi H, Wathelet JP, Thonart P (2009) Production of γ-decalactone by a psychrophilic and a mesophilic strain of the yeast *Rhodotorula aurantiaca*. Appl Biochem Biotechnol 158:41–50

Alias N, Ahmad Mazian M, Salleh AB, Basri M, Rahman RN (2014) Molecular cloning and optimization for high level expression of cold-adapted serine protease from Antarctic yeast *Glaciozyma antarctica* PI12. Enzyme Res 2014:1–20

Alimardani-Theuil P, Gainvors-Claise A, Duchiron F (2011) Yeasts: an attractive source of pectinases – from gene expression to potential applications: a review. Proc Biochem 46:1525–1537

Amaretti A, Raimondi S, Sala M, Roncaglia L, De Lucia M, Leonardi A, Rossi M (2010) Single cell oils of the cold- adapted oleaginous yeast *Rhodotorula glacialis* DBVPG 4785. Microb Cell Fact 9:73–78

Amoresano A, Andolfo A, Corsaro MM, Zocchi I, Petrescu I, Gerday C, Marino G (2000) Structural characterization of a xylanase from psychrophilic yeast by mass spectrometry. Glycobiology 10:451–458

Avis TJ, Bélanger RR (2001) Specificity and mode of action of the antifungal fattyacid cis-9-heptadecenoic acid produced by *Pseudozyma flocculosa*. Appl Environ Microbiol 67:956–960

Avis TJ, Cheng YL, Zhao YY, Bolduc S, Neveu B, Anguenot R, Labbé C, Belzile F, Bélanger RR (2005) The potential of *Pseudozyma* yeast like epiphytes for the production of heterologous recombinant proteins. Appl Microbiol Biotechnol 69:304–311

Barahona S, Yuivar Y, Socias G, Alcaíno J, Cifuentes V, Baeza M (2016) Identification and characterization of yeasts isolated from sedimentary rocks of Union Glacier at the Antarctica. Extremophiles 20:479–491

Behr A, Pérez Gomes J (2011) The cross-metathesis of methyl oleate with cis-2-butene-1,4-diyl diacetate and the influence of protecting groups. Beilstein Org Chem 7:1–8

Benesova E, Markova M, Kralova B (2005) a-Glucosidase and b-glucosidase from psychrotrophic strain *Arthrobacter* sp. C2–2. Czech J Food Sci 23:116–120

Beopoulos A, Cescut J, Haddouche R, Uribelarrea JL, Molina-Jouve C, Nicaud JM (2009) *Yarrowia lipolytica* as a model for bio-oil production. Prog Lipid Res 48:375–387

Białkowska AM, Turkiewicz M (2014) Miscellaneous cold-active yeast enzymes of industrial importance. In: Buzzini P, Margesin R (eds) Cold-adapted yeasts: biodiversity, adaptation strategies and biotechnological significance. Springer, Heidelberg, pp 377–396

Białkowska AM, Cieśliński H, Nowakowska KM, Kur J, Turkiewicz M (2009) A new β-galactosidase with a low temperature optimum isolated from the Antarctic *Arthrobacter* sp. 20B: gene cloning, purification and characterization. Arch Microbiol 191:825–835

Biermann U, Bornscheuer U, Meier MAR, Metzger JO, Schäfer JH (2011) Oils and fats as renewable raw materials in chemistry. Angew Chem Int Ed 50:3854–3871

Birgisson HK, Delgado O, Arroyo LG, Hatti-Kaul R, Mattiasson B (2003) Cold-adapted yeasts as producers of cold-active polygalacturonases. Extremophiles 7:185–193

Blank K, Morfill J, Gumpp H, Gaub HE (2006) Functional expression of *Candida antarctica* lipase B in *Escherichia coli*. J Biotechnol 125:474–483

Branda E, Turchetti B, Diolaiuti G, Pecci M, Smiraglia C, Buzzini P (2010) Yeast and yeast-like diversity in the southernmost glacier of Europe (Calderone glacier, Apennines, Italy). FEMS Microbiol Ecol 72:354–369

Brandão LR, Libkind D, Vaz AB, Espírito Santo LC, Moliné M, de García V, van Broock M, Rosa CA (2011) Yeasts from an oligotrophic lake in Patagonia (Argentina): diversity, distribution and synthesis of photoprotective compounds and extracellular enzymes. FEMS Microbiol Ecol 76:1–13

Brizzio S, Turchetti B, de Garcìa V, Libkind D, Buzzini P, van Broock M (2007) Extracellular enzymatic activities (EEA) of basidiomycetous yeasts isolated from glacial and subglacial waters of northwest Patagonia (Argentina). Can J Microbiol 53:519–525

Butinar L, Spencer-Martins I, Gunde-Cimerman N (2007) Yeasts in high Arctic glaciers: the discovery of a new habitat for eukaryotic microorganisms. A van Leeuwenhoek 91:277–289

Buzzini P, Margesin R (2014a) Cold-adapted yeasts: biodiversity, adaptation strategies and biotechnological significance. Springer, Heidelberg

Buzzini P, Margesin R (2014b) Cold-adapted yeasts: a lesson from the cold and a challenge for the XXI century. In: Buzzini P, Margesin R (eds) Cold-adapted yeasts: biodiversity, adaptation strategies and biotechnological significance. Springer, Heidelberg, pp 3–22

Buzzini P, Vaughan-Martini A (2006) Yeast biodiversity and biotechnology. In: Rosa CA, Peter G (eds) Biodiversity and ecophysiology of yeasts. Springer, Berlin, pp 533–559

Buzzini P, Branda E, Goretti M, Turchetti B (2012) Psychrophilic yeasts from worldwide glacial habitats: diversity, adaptation strategies and biotechnological potential. FEMS Microbiol Ecol 82:217–241

Buzzini P, Di Mauro S, Turchetti B (2017) Yeasts as starter cultures. In: Speranza B, Bevilacqua B, Corbo MR, Sinigaglia M (eds) Starter cultures in food production. Wiley, New York, pp 16–49

Cameotra SS, Makkar RS (2004) Recent applications of biosurfactants as biological and immunological molecules. Curr Opin Microbiol 7:262–266

Carrasco M, Rozas JM, Barahona S, Alcaíno J, Cifuentes V, Baeza M (2012) Diversity and extracellular enzymatic activities of yeasts isolated from King George Island, the sub-Antarctic region. BMC Microbiol 12:251

Carrasco M, Villarreal P, Barahona S, Alcaíno J, Cifuentes V, Baeza M (2016) Screening and characterization of amylase and cellulase activities in psychrotolerant yeasts. BMC Microbiol 19:16–21

Chaud LC, Lario LD, Bonugli-Santos RC, Sette LD, Pessoa Junior A, Felipe MD (2016) Improvement in extracellular protease production by the marine Antarctic yeast *Rhodotorula mucilaginosa* L7. Nat Biotechnol 33:807–814

Chen B, Hu J, Miller EM, Xie W, Cai M, Gross RA (2008) *Candida antarctica* lipase B chemically immobilized on epoxy-activated micro- and nanobeads: catalysts for polyester synthesis. Biomacromolecules 9:463–471

Connell LB, Redman R, Craig S, Scorzetti G, Iszaard M, Rodriguez R (2008) Diversity of soil yeasts isolated from South Victoria Land, Antarctica. Microb Ecol 56:448–459

De Mot R, Verachtert H (1987) Purification and characterization of extracellular a-amylase and glucoamylase from the yeast *Candida antarctica* CBS 6678. Eur J Biochem 164:643–654

Deegenaars ML, Watson K (1997) Stress proteins and stress tolerance in an Antarctic, psychrophilic yeast, *Candida psychrophila*. FEMS Microbiol Lett 151:191–196

Deegenaars ML, Watson K (1998) Heat-shock response in psychrophilic and psychrotrophic yeast from Antarctica. Extremophiles 2:41–49

Dimitrova S, Pavlova K, Lukanov L, Zagorchev P (2010) Synthesis of coenzyme Q10 and beta-carotene by yeasts isolated from Antarctic soil and lichen in response to ultraviolet and visible radiations. Appl Biochem Biotechnol 162:795–804

Dimitrova S, Pavlova K, Lukanov L, Korotkova E, Petrova E, Zagorchev P, Kuncheva M (2013) Production of metabolites with antioxidant and emulsifying properties by Antarctic strain *Sporobolomyces salmonicolor* AL_1. Appl Biochem Biotechnol 169:301–311

Dominguez de Maria P, Carboni-Oerlemans C, Tuin B, Bergeman G, Meer A, Gemert R (2005) Biotechnological applications of *Candida antarctica* lipase A: state-of-the-art. J Mol Catal B 37:36–46

Donot F, Fontana A, Baccou J, Schorr-Galindo S (2012) Microbial exopolysaccharides: main examples of synthesis, excretion, genetics and extraction. Carbohydr Polym 87:951–962

Duarte AW, Dayo-Owoyemi I, Nobre FS, Pagnocca FC, Chaud LC, Pessoa A, Felipe MG, Sette LD (2013) Taxonomic assessment and enzymes production by yeasts isolated from marine and terrestrial Antarctic samples. Extremophiles 17:1023–1035

Emond S, Montanier C, Nicaud JM, Marty A, Monsan P, Andre I, Remaud-Simeon M (2010) New efficient recombinant expression system to engineer *Candida antarctica* lipase B. Appl Environ Microbiol 76:2684–2687

Ericsson DJ, Kasrayan A, Johansson P, Berqfors T, Sandstrom AG, Bäckvall JE, Mowbray SL (2008) X-ray structure of *Candida antarctica* lipase B shows a novel lid structure and a likely mode of interfacial activation. J Mol Biol 376:109–119

Faber K (2004) Biotransformations in organic chemistry. A textbook, 5th edn. Springer, Heidelberg

Feller G, Gerday C (1997) Psychrophilic enzymes: molecular basis of cold adaptation. Cell Mol Life Sci 53:830–841

Feller G, Gerday C (2003) Psychrophilic enzymes: hot topics in cold adaptation. Nat Rev Microbiol 1:200–208

Forde J, Vakurov A, Gibson TD, Millner P, Whelehan M, Marisin IW, Ó'Fágáin C (2010) Chemical modification and immobilisation of lipase B from *Candida antarctica* onto mesoporous silicates. J Mol Cat B 66:203–209

Galdino AS, Silva RN, Lottermann MT, Alvares AC, de Moraes LM, Torres FA, de Freitas SM, Ulhoa CJ (2011) Biochemical and structural characterization of amy1: an alpha-amylase from *Cryptococcus flavus* expressed in *Saccharomyces cerevisiae*. Enzyme Res 30:157294

Garay LA, Sitepu IR, Cajka T, Chandra I, Shi S, Lin T, German JB, Fiehn O, Boundy-Mills KL (2016) Eighteen new oleaginous yeast species. J Ind Microbiol Biotechnol 43:887–900

de García V, Brizzio S, Libkind D, Buzzini P, van Broock M (2007) Biodiversity of cold-adapted yeasts from glacial meltwater rivers in Patagonia, Argentina. FEMS Microbiol Ecol 59:331–341

de García V, Brizzio S, van Broock M (2012) Yeasts from glacial ice of Patagonian Andes, Argentina. FEMS Microbiol Ecol 82:540–550

Gerday C (2014) Fundamentals of cold-adapted yeasts. In: Buzzini P, Margesin R (eds) Cold-adapted yeasts: biodiversity, adaptation strategies and biotechnological significance. Springer, Heidelberg

Gerday C, Aittaleb M, Arpigny JL, Baise E, Chessa JP, Garsoux G, Petrescu I, Feller G (1997) Psychrophilic enzymes: a thermodynamic challenge. Biochim Biophys Acta 1342:119–131

Gerday C, Aittaleb M, Bentahir M, Chessa JP, Claverie P, Collins T, D'Amico S, Dumont J, Garsoux G, Georlette D, Hoyoux A, Lonhienne T, Meuwis MA, Feller G (2000) Cold-adapted enzymes: from fundamentals to biotechnology. Trend Biotechnol 18:103–107

Gomes J, Gomes I, Steiner W (2000) Thermolabile xylanase of the Antarctic yeast *Cryptococcus adeliae*: production and properties. Extremophiles 4:227–235

Gotor-Fernandéz V, Busto E, Gotor V (2006) *Candida antarctica* lipase B: an ideal biocatalyst for the preparation of nitrogenated organic compounds. Adv Synth Catal 348:797–812

Gruber CC, Pleiss J (2012) Lipase B from *Candida antarctica* binds to hydrophobic substrate–water interfaces via hydrophobic anchors surrounding the active site entrance. J Mol Catal B 84:48–54

Gunde-Cimerman N, Plemenitaš A, Buzzini P (2014) Changes in lipids composition and fluidity of yeast plasma membrane as response to cold. In: Buzzini P, Margesin R (eds) Cold-adapted yeasts: biodiversity, adaptation strategies and biotechnological significance. Springer, Heidelberg, pp 3–22

Gutarra MLE, Romero O, Abian O, Torres FAG, Freire DMG, Castro AM, Guisan JM, Palomo JM (2011) Enzyme surface glycosylation in the solid phase: improved activity and selectivity of *Candida antarctica* lipase B. Chem Cat Chem 3:1902–1910

Habeych DI, Juhl PB, Pleiss J, Venegas D, Eggink G, Boeriu CG (2011) Biocatalytic synthesis of polyesters from sugar-based building blocks using immobilized *Candida antarctica* lipase B. J Mol Catal B 71:1–9

Hamid B, Singh P, Mohiddin FA, Sahay S (2013) Partial characterization of cold-active β-galactosidase activity produced by *Cystophallobaidium capatitum* SPY11 and *Rodotorella musloganosa* PT1. Endocytobiosis Cell Res 24:23–26

Hashim NH, Bharudin I, Nguong DL, Higa S, Bakar FD, Nathan S, Rabu A, Kawahara H, Illias RM, Najimudin N, Mahadi NM, Murad AM (2013) Characterization of Afp1, an antifreeze protein from the psychrophilic yeast *Glaciozyma antarctica* PI12. Extremophiles 17:63–73
Hawksworth DL (2004) Fungal diversity and its implications for genetic resource collections. Stud Mycol 50:9–18
Hou J, Tyo KE, Liu Z, Petranovic D, Nielsen J (2012) Metabolic engineering of recombinant protein secretion by *Saccharomyces cerevisiae*. FEMS Yeast Res 12:491–510
Huston AL (2008) Biotechnological aspects of cold-adapted enzymes. In: Margesin R, Schinner F, Marx JC, Gerday C (eds) Psychrophiles: from biodiversity to biotechnology. Springer, Berlin, pp 347–363
Javed A, Qazi JI (2016) Psychrophilic microbial enzymes implications in coming biotechnological processes. ASRJETS 23:2313–4410
Jiru TM, Abate D, Kiggundu N, Pohl C, Groenewald M (2016) Oleaginous yeasts from Ethiopia. AMB Express 6:78
Johnson EA, Echavarri-Erasun C (2011) Yeast biotechnology. In: Kurtzman CP, Fell JW, Boekhout T (eds) The yeasts. A taxonomy study, vol 1. Elsevier, New York, pp 21–44
Joseph B, Ramteke PW, Thomas G (2008) Cold active microbial lipases: some hot issues and recent developments. Biotechnol Adv 26:457–470
Juhl PB, Doderer K, Hollmann F, Thum O, Pleiss J (2010) Engineering of *Candida antarctica* lipase B for hydrolysis of bulky carboxylic acid esters. J Biotechnol 150:474–480
Kahveci D, Xu X (2012) Bioimprinted immobilization of *Candida antarctica* lipase A for concentration of omega-3 polyunsaturated fatty acids. J Am Oil Chem Soc 89:1839–1845
Kandror O, DeLeon A, Goldberg AL (2002) Trehalose synthesis is induced upon exposure of *Escherichia coli* to cold and is essential for viability at low temperatures. Proc Natl Acad Sci U S A 99:9727–9732
Karasová-Lipovová P, Strnad H, Spiwok V, Spiwok V, Malá S, Královà B, Russell NJ (2003) The cloning, purification and characterisation of a cold-active b-galactosidase from the psychrotolerant Antarctic bacterium *Arthrobacter* sp. C2–2. Enzyme Microb Technol 33:836–844
Kasana RC, Gulati A (2011) Cellulases from psychrophilic microorganisms: a review. J Basic Microbiol 51:572–579
Kawahara H (2008) Cryoprotectants and ice-binding proteins. In: Margesin R, Schinner F, Marx JC, Gerday C (eds) Psychrophiles: from biodiversity to biotechnology. Springer, Berlin, pp 229–246
Kirk O, Christensen MW (2002) Lipases from *Candida antarctica*: unique biocatalysts from a unique origin. Org Process Res Dev 6:446–451
Kourkoutas Y, Bekatorou A, Banat IM, Marchant R, Koutinas AA (2004) Immobilization technologies and support materials suitable in alcohol beverages production: a review. Food Microbiol 21:377–397
Krishna H, Persson MM, Bornscheuer UT (2002) Enantioselective transesterification of a tertiary alcohol by lipase A from *Candida antarctica*. Tetrahedron Asymmetry 13:2693–2696
Kuddus M, Roohi Arif JM, Ramteke PW (2011) An overview of cold-active microbial a-amylase: adaptation strategies and biotechnological potentials. Biotechnology 10:246–258
Kuncheva M, Panchev I, Pavlova K, Rusinova-Videva S, Georgieva K, Dimitrova S (2013) Production and characterisation of exopolysaccharide by Antarctic yeast strain *Cryptococcus laurentii* AL62. Biotechnol Biotechnol Equip 27:4098–4102
Kurtzman CP, Fell JW, Boekhout T (2011a) The yeasts. A taxonomy study, vol 1–3. Elsevier, New York
Kurtzman CP, Fell JW, Boekhout T (2011b) Definition, classification and nomenclature of the yeasts. In: Kurtzman CP, Fell JW, Boekhout T (eds) The yeasts. A taxonomy study, vol 1. Elsevier, New York, pp 3–5
Lachance MA (2006) Yeast biodiversity: how many and how much? In: Rosa CA, Peter G (eds) Biodiversity and ecophysiology of yeasts. Springer, Berlin, pp 1–9

Lario LD, Chaud L, Md A, Converti A, Sette LD, Pessoa A Jr (2015) Production, purification, and characterization of an extracellular acid protease from the marine Antarctic yeast *Rhodotorula mucilaginosa* L7. Fungal Biol 119:1129–1136

Lee JK, Park KS, Park S, Park H, Song YH, Kang SH, Kim HJ (2010) An extracellular ice-binding glycoprotein from an Arctic psychrophilic yeast. Cryobiology 60:222–228

Lee SG, Koh HY, Lee JH, Kang SH, Kim HJ (2012) Cryopreservative effects of the recombinant ice-binding protein from the Arctic yeast *Leucosporidium* sp. on red blood cells. Appl Biochem Biotechnol 167:824–834

Levinson WE, Kurtzman CP, Kuo TM (2006) Production of itaconic acid by *Pseudozyma antarctica* NRRL Y-7808 under nitrogen-limited growth conditions. Enzym Microb Technol 39:824–827

Liese A, Weelbach K, Wandrey C (2000) Industrial biotransformations, 2nd edn. Wiley, Weinheim

Liu XZ, Wang QM, Göker M, Groenewald M, Kachalkin AV, Lumbsch HT, Millanes AM, Wedin M, Yurkov A, Boekhout T, Bai FY (2015) Towards an integrated phylogenetic classification of the Tremellomycetes. Stud Mycol 81:85–147

Mannazzu I, Landolfo S, Lopes da Silva T, Buzzini P (2015) Red yeasts and carotenoid production: outlining a future for non-conventional yeasts of biotechnological interest. World J Microbiol Biotechnol 31:1665–1673

Margesin R, Feller G (2010) Biotechnological applications of psychrophiles. Environ Technol 31:835–844

Margesin R, Fauster V, Fonteyne PA (2005) Characterization of cold-active pectate lyases from psychrophilic *Mrakia frigida*. Lett Appl Microbiol 40:453–459

Miletić N, Loos K (2009) Over-stabilization of chemically modified and cross-linked *Candida antarctica* lipase B using various epoxides and diepoxides. Aust J Chem 62:799–805

Mohammadi S, Parvizpour S, Razmara J, Abu Bakar FD, Illias RM, Mahadi NM, Murad AM (2016) Structure prediction of a novel exo-β-1,3-glucanase: insights into the cold adaptation of psychrophilic yeast *Glaciozyma antarctica* PI12. Interdiscip Sci Comput Life Sci. doi:10.1007/s12539-016-0180-9

Morita T, Fukuoka T, Imura T, Kitamoto D (2013) Production of mannosylerythritol lipids and their application in cosmetics. Appl Microbiol Biotechnol 97:4691–4700

Nakagawa T, Yamada K, Miyaji T, Tomizuka N (2002) Cold-active pectinolytic activity of psychrophilic basidiomycetous yeast *Cystofilobasidium capitatum* strain PPY-1. J Biosci Bioeng 94:175–177

Nakagawa T, Nagaoka T, Miyaji T, Tomizuka N (2005a) Cold-active polygalacturonase from psychrophilic basidiomycetous yeast *Cystofilobasidium capitatum* strain PPY-1. Biosci Biotechnol Biochem 69:419–421

Nakagawa T, Nagaoka T, Miyaji T, Tomizuka N (2005b) A cold-active pectin lyase from the psychrophilic and basidiomycetous yeast *Cystofilobasidium capitatum* strain PPY-1. Biotechnol Appl Biochem 42:193–196

Nakagawa T, Ikehata R, Uchino M (2006) Cold-active acid β-galactosidase activity of isolated psychrophilic-basidiomycetous yeast *Guehomyces pullulans*. Microbiol Res 161:75–79

Ogrydziak DM (1993) Yeast extracellular proteases. Crit Rev Biotechnol 13:1–55

Pan J, Chen XL, Shun CY, He HL, Zhang YZ (2005) Stabilization of cold-adapted protease MCP-01 promoted by trehalose: prevention of the autolysis. Protein Pept Lett 12:375–378

Park KS, Do H, Lee JH, Park SI, Ej K, Kim SJ, Kang SH, Kim HJ (2012) Characterization of the ice-binding protein from Arctic yeast *Leucosporidium* sp. AY30. Cryobiology 64:286–296

Parvizpour S, Razmara J, Jomah AF, Shamsir MS, Illias RM (2015) Structural prediction of a novel laminarinase from the psychrophilic *Glaciozyma Antarctica* PI12 and its temperature adaptation analysis. J Mol Model 21:63

Parvizpour S, Razmara J, Shamsir MS, Illias RM, Abdul Murad AM (2016) The role of alternative salt bridges in cold adaptation of a novel psychrophilic laminarinase. J Biomol Struct Dyn 5:1–8

Patel RN (2004) Biocatalytic synthesis of chiral pharmaceutical intermediates. Food Technol Biotechnol 42:305–325

Patkar SA, Björkling F, Zundel M, Schulein M, Svendsen A, Heldt-Hansen HP, Gormsen E (1993) Purification of two lipases from *Candida antarctica* and their inhibition by various inhibitors. Ind J Chem Sect B 32:76–80

Pavlova K (2014) Production of polymers and other compounds of industrial importance by cold-adapted yeasts. In: Buzzini P, Margesin R (eds) Cold-adapted yeasts: biodiversity, adaptation strategies and biotechnological significance. Springer, Heidelberg, pp 397–416

Pavlova K, Angelova G, Savova I, Grigorova D, Kupenov L (2002) Studies of Antarctic yeast for b-glucosidase production. World J Microbiol Biotechnol 18:569–573

Pavlova K, Panchev I, Krachanova M, Gocheva M (2009) Production of an exopolysaccharide by Antarctic yeast. Folia Microbiol 54:343–348

Pazgier M, Turkiewicz M, Kalinowska H, Bielecki S (2003) The unique cold-adapted extracellular subtilase from psychrophilic yeast *Leucosporidium antarcticum*. J Mol Catal B Enzym 21:39–42

Pereyra V, Martinez A, Rufo C, Vero S (2014) Oleaginous yeasts form Uruguay and Antarctica as renewable raw material for biodiesel production. Am J Biosci 2:251–257

Perez J, Munoz-Dorado J, Rubia T, Martinez J (2002) Bio-degradation and biological treatments of cellulose, hemicellulose and lignin: an overview. Int Microbiol 5:53–63

Petrescu I, Lamotte-Brasseur J, Chessa J-P, Ntarima P, Clayessens M, Devreese B, Marino G, Gerday C (2000) Xylanase from the psychrophilic yeast *Cryptococcus adeliae*. Extremophiles 4:137–144

Plácido J, Capareda S (2015) Ligninolytic enzymes: a biotechnological alternative for bioethanol production. Bioresour Bioprocess 2:1–12

Poli A, Anzelmo G, Tommonaro G, Pavlova K, Casaburi A, Nicolaus B (2010) Production and chemical characterization of an exopolysaccharide synthesized by psychrophilic yeast strain *Sporobolomyces salmonicolor* AL1 isolated from Livingston Island, Antarctica. Folia Microbiol 55:576–581

Ramli AN, Mahadi NM, Rabu A, Murad AM, Bakar FD, Illias RM (2011) Molecular cloning, expression and biochemical characterisation of a cold-adapted novel recombinant chitinase from *Glaciozyma antarctica* PI12. Microb Cell Fact 4:10–94

Ramli AN, Mahadi NM, Shamsir MS, Rabu A, Joyce-Tan KH, Murad AM, Illias RM (2012) Structural prediction of a novel chitinase from the psychrophilic *Glaciozyma antarctica* PI12 and an analysis of its structural properties and function. J Comput Aided Mol Des 26:947–961

Rashid FAA, Rahim RA, Ibrahim D (2010) Identification of lipase-producing psychrophilic yeast, *Leucosporidium* sp. Internet J Microbiol 9(1). doi:10.5580/1215

Ray MK, Devi KU, Kumar GS, Shivaji S (1992) Extracellular protease from the Antarctic yeast *Candida humicola*. Appl Environ Microbiol 58:1918–1923

Rossi M, Buzzini P, Cordisco L, Amaretti A, Sala M, Raimondi S, Ponzoni C, Pagnoni UM, Matteuzzi D (2009) Growth, lipid accumulation and fatty acid composition in obligate psychrophilic, facultative psychrophilic, and mesophilic yeasts. FEMS Microbiol Ecol 69:363–372

Rovati JI, Pajot HF, Ruberto L, Mac Cormack W, Figueroa LI (2013) Polyphenolic substrates and dyes degradation by yeasts from 25 de Mayo/King George Island (Antarctica). Yeast 30:459–470

Rusinova-Videva S, Pavlova K, Georgieva K (2011) Effect of different carbon sources on biosynthesis of exopolysaccharide from Antarctic strain *Cryptococcus*. Biotechnol Biotechnol Equip 23:888–891

Sahay S, Hamid B, Singh P, Ranjan K, Chauhan D, Rana RS, Chaurse VK (2013) Evaluation of pectinolytic activities for oenological uses from psychrotrophic yeasts. Lett Appl Microbiol 57:115–121

Santiago M, Ramìrez-Sarmiento CA, Zamora RA, Parra LP (2016) Discovery, molecular mechanisms, and industrial applications of cold-active enzymes. Front Microbiol 7:1408. doi:10.3389/fmicb.2016.01408

Schulze I, Hansen S, Großhans S, Rudszuck T, Ochsenreither K, Syldatk C, Neumann A (2014) Characterization of newly isolated oleaginous yeasts – *Cryptococcus podzolicus*, *Trichosporon porosum* and *Pichia segobiensis*. AMB Express 4:24

Scorzetti G, Petrescu I, Yarrow D, Fell JW (2000) *Cryptococcus adeliensis sp. nov.*, a xylanase producing basidiomycetous yeast from Antarctica. Antonie Van Leeuwenhoek 77:153–157

Sheridan PP, Brenchley JE (2000) Characterization of a salt-tolerant family 42 b galactosidase from a psychrophilic Antarctic *Planococcus* isolate. Appl Environ Microbiol 66:2438–2444

Shivaji S, Prasad GS (2009) Antarctic yeasts: biodiversity and potential applications. In: Satyanarayana T, Kunze G (eds) Yeast biotechnology: diversity and applications. Springer, Berlin, pp 3–16

Sibirny AA, Scheffers L (2002) Thematic section biochemistry, genetics, biotechnology and ecology of non-conventional yeasts. FEMS Yeast Res 2:293

Singh P, Singh MV, Tsuji M, Prasad GS, Hoshino T (2014) *Rhodotorula svalbardensis* sp. nov., a novel yeast species isolated from cryoconite holes of Ny-Ålesund, Arctic. Cryobiology 68:122–128

Sitepu IR, Garay LA, Sestric R, Levin D, Block DE, German JB, Boundy-Mills KL (2014) Oleaginous yeasts for biodiesel: current and future trends in biology and production. Biotechnol Adv 32:1336–1360

Song C, Chi Z, Li J, Wang X (2010) β-Galactosidase production by the psychrotolerant yeast *Guehomyces pullulans* 17–1 isolated from sea sediment in Antarctica and lactose hydrolysis. Bioprocess Biosyst Eng 33:1025–1031

Starmer WT, Lachance MA (2011) Yeast ecology. In: Kurtzman CP, Fell JW, Boekhout T (eds) The yeasts. A taxonomy study, vol 1. Elsevier, New York, pp 65–83

Szczesna-Antczak M, Kaminska J, Florczak T, Turkiewicz M (2014) Cold-active yeast lipases: recent issues and future prospects. In: Buzzini P, Margesin R (eds) Cold-adapted yeasts: biodiversity, adaptation strategies and biotechnological significance. Springer, Heidelberg

Tobe S, Takami T, Ikeda S, Mitsugi K (1976) Production and some enzymatic properties of alkaline proteinase by *Candida lipolytica*. Argic Biol Chem 40:1087–1092

Tsuji M, Yokota Y, Shimohara K, Kudoh S, Hoshino T (2013a) An application of wastewater treatment in a cold environment and stable lipase production of Antarctic basidiomycetous yeast *Mrakia blollopis*. PLoS One 8:e59376

Tsuji M, Goshima T, Matsushika A, Kudoh S, Hoshino T (2013b) Direct ethanol fermentation from lignocellulosic biomass by Antarctic basidiomycetous yeast *Mrakia blollopis* under a low temperature condition. Cryobiology 67:241–243

Tsuji M, Yokota Y, Kudoh S, Hoshino T (2014) Improvement of direct ethanol fermentation from woody biomasses by the Antarctic basidiomycetous yeast, *Mrakia blollopis*, under a low temperature condition. Cryobiology 68:303–305

Turchetti B, Buzzini P, Goretti M, Branda E, Diolaiuti G, D'Agata C, Smiraglia C, Vaughan-Martini A (2008) Psychrophilic yeasts in glacial environments of Alpine glaciers. FEMS Microbiol Ecol 63:73–83

Turchetti B, Goretti M, Branda E, Diolaiuti G, D'Agata C, Smiraglia C, Onofri A, Buzzini P (2013) Influence of abiotic variables on culturable yeast diversity in two distinct Alpine glaciers. FEMS Microbiol Ecol 86:327–340

Turkiewicz M, Pazgier M, Kalinowska H, Bielecki S (2003) A cold-adapted extracellular serine proteinase of the yeast *Leucosporidium antarcticum*. Extremophiles 7:435–442

Turkiewicz M, Pazgier M, Donachie SP, Kalinowska H (2005) Invertase and a-glucosidase production by the endemic Antarctic marine yeast *Leucosporidium antarcticum*. Pol Polar Res 26:125–136

Vlaev S, Rusinova-Videva S, Pavlova K, Kuncheva M, Panchev I, Dobreva S (2013) Submerged culture process for biomass and exopolysaccharide production by Antarctic yeast: some engineering considerations. Appl Microbiol Biotechnol 97:5303–5313

Wang QM, Groenewald M, Takashima M, Theelen B, Han PJ, Liu T, Boekhout T, Bai FY (2015a) Phylogeny of yeasts and related filamentous fungi within *Pucciniomycotina* determined from multigene sequence analyses. Stud Mycol 81:27–53

Wang QM, Yurkov A, Göker M, Lumbsch HT, Leavitt SD, Groenewald M, Theelen B, Liu XZ, Boekhout T, Bai FY (2015b) Phylogenetic classification of yeasts and related taxa within *Pucciniomycotina*. Stud Mycol 81:149–289

Yu P, Wang XT, Liu JW (2015) Purification and characterization of a novel cold-adapted phytase from *Rhodotorula mucilaginosa* strain JMUY14 isolated from Antarctic. J Basic Microbiol 55:1029–1039

Zaliha RN, Salleh AB, Basri M, Mohamad Ali MSB (2012) Cold active enzyme and method thereof. US Patent 2012/0058514 A1

Chapter 25
Perspectives of Low-Temperature Biomass Production of Polar Microalgae and Biotechnology Expansion into High Latitudes

Jana Kvíderová, Satya P. Shukla, Benjamin Pushparaj, and Josef Elster

Abstract The adaptation mechanisms of polar microalgae (including cyanobacteria and eukaryotic microalgae) evolved to withstand the harsh polar environment characterized by low temperature, freeze-thaw cycles, desiccation, salinity, and high and variable photosynthetically active and ultraviolet radiation. Hence, polar microalgae developed ecological, physiological, and molecular defensive and adaptive strategies, which include the synthesis of a tremendous diversity of compounds originating from different metabolic pathways which protect them against the abovementioned stresses. Production of different biological compounds and various biotechnological applications, for instance, water treatment technology in low-temperature environments and many others, are the perspectives for humans, which widely explore the polar regions. In this review, the nonmarine environmental conditions in polar environments and microalgal adaptations are summarized with respect to possible biotechnological applications. The review also provides a survey of the possible compounds to be exploited from polar microalgae. Possible

J. Kvíderová
Centre for Polar Ecology, Faculty of Science, University of South Bohemia, Na Zlaté Stoce 3, České Budějovice, Czech Republic

S.P. Shukla
Aquatic Environment Management Division, Central Institute of Fisheries Education, 11, Mumbai, India

B. Pushparaj
ISE-CNR Istituto per lo Studio degli Ecosistemi, via Madonna del Piano 10, 50019 Sesto Fiorentino, Italy

J. Elster (✉)
Centre for Polar Ecology, Faculty of Science, University of South Bohemia, Na Zlaté Stoce 3, České Budějovice, Czech Republic

Institute of Botany, Academy of Sciences of the Czech Republic, Dukelská, 135 Třeboň, Czech Republic
e-mail: Josef.Elster@ibot.cas.cz

R. Margesin (ed.), *Psychrophiles: From Biodiversity to Biotechnology*,
DOI 10.1007/978-3-319-57057-0_25

constructions of photobioreactors for mass cultivation of microalgae are proposed for operations in polar regions.

Contents

25.1 Introduction

It is well known that bioprospecting is a complex issue that encapsulates scientific and commercial interests, environmental concerns, ethics and equity, and considerations relating to international law and policy, including respect for national and international regulations in the Arctic and Antarctic. The search for novel bioproducts and development of improved technological processes, based on biological resources, has necessarily resulted in bioprospecting of organisms in diverse habitats of the world (Bull et al. 1992; Feller and Gerday 1997). In the last few decades, bioprospecting for the development of biotechnological applications and discovering molecules valuable for pharmacology mainly relied on compounds derived from those found in plants and heterotrophic microorganisms (Glazer and Nikaido 2007). Screening of novel natural products has become an important prerequisite for the discovery of novel bioactive and therapeutic products.

Microalgae—for the purpose of this review the term "microalgae" includes both prokaryotic cyanobacteria and eukaryotic microalgae—play a key role as primary producers in all ecosystems on Earth, since they are able to adapt to various adverse conditions. Microalgae are considered as important resources for biotechnology, since they can be grown easily under autotrophic conditions in suspension for large-scale biomass production and synthesis of desired compounds. Microalgal biomass could provide valuable biologically active compounds that could be prospectively useful for medicine and nutrition (Richmond 2004). At present, the main attention is focused on antibiotic, fungicide, and even cytostatic effects of microalgal extracts (Singh et al. 2005) which may lead to novel therapeutic methods in human and veterinary medicines. Carotenoids, polyunsaturated fatty acids (PUFAs), polysaccharides, peptides, and antioxidants are essential parts of food supplements for humans or animals (Astorg 1997). Oil-producing microalgae could not only be a prospective source of PUFAs for nutrition but also for biofuel production (Lukavský 2012). Despite there being thousands of microalgal species, only a few of them are exploited in biotechnological applications, such as *Arthrospira*

(*Spirulina*) for the production of proteins and phycocyanin (Vonshak 1997), *Chlorella* for chlorophylls and lipids (Lv et al. 2010), *Haematococcus* for astaxanthin (Del Campo et al. 2007), *Dunaliella* for carotenoids and glycerol (Ben-Amotz and Avron 1990), and *Nannochloropsis* sp., *Scenedesmus obliquus*, *Botryococcus* sp., and *Trachydiscus minutus* for oils (Lukavský 2012).

Almost all used strains originate from the tropic and temperate regions. Although there is an increasing amount of data on microalgal diversity in the polar regions (e.g., van den Vijver et al. 2010; Komárek et al. 2012; Kopalová et al. 2013; Strunecký et al. 2010, 2012; etc.), their prospective value for biotechnology remains unrevealed. Large-scale screening studies are necessary to assess the physiological and biochemical diversity of polar microalgae since they have developed many adaptation mechanisms to the polar environment which could be applied in low-temperature biotechnology.

Microalgal mass cultivation in polar regions seems to be difficult due to the harsh climate; however, it could be beneficial to local settlements. Microalgal mass cultivation should contribute to development of the local economies as well as to long-term environmental exploitation and protection. Microalgal biomass could serve as a food supplement or bioproduct for industry. The microalgae may be used for waste water treatment for removing nutrients, namely, nitrogen and phosphorus, thus reducing the risk of eutrophication of polar oligotrophic freshwater systems (Chevalier et al. 2000) and remediation of polluted areas. Especially, the last issue would become important in the near future due to expected oil production/mining activities resulting from the discovery of huge oil deposits. The development of mass cultivation methods and their implementation in polar regions must follow the standards of national and international regulations and other suggested guidelines (Nichols et al. 2002; Elster and Benson 2004). Despite the first pioneer experiments using open systems (Elster et al. 2001; Shukla et al. 2013), no other experiments focused on microalgal mass cultivation have been performed in the Arctic or Antarctic to evaluate the technical requirements and restrictions, as well as cost-and-benefit studies, on state-of-art photobioreactor operations.

This review summarizes the various aspects which are relevant to bioprospecting of polar resources with a special reference to microalgal diversity. The first part is focused on the unique features of polar environments and the adaptations of polar oxyphototrophic microorganisms to thrive in polar habitats. This is followed by a review of the present status of the potentials of polar microalgae for various biotechnological applications.

25.2 The Polar Environment

Climatologically and biogeographically, Antarctica can be divided into three zones: (1) continental Antarctica consisting of mainly the Antarctic continent, (2) maritime Antarctica composed of the Antarctic Peninsula and associated island archipelagos, and (3) sub-Antarctica consisting of the islands that lie in and around the Antarctic polar frontal zone (Holdgate 1970). Similarly, the Arctic regions can be divided into (1) the low Arctic of Scandinavia, Iceland, Continental Canada, Alaska, Russian continental tundra and (2) the high Arctic including the Canadian Islands, Greenland, and the Russian archipelagos in the Northern Oceans (Aleksandrova 1980). Maritime Antarctica, sub-Antarctica, and the Arctic can provide suitable habitats during the summer period when considerable areas in these regions are free of ice and snow with adequate quantities of water originating from snow/ice melt (Elster and Komárek 2003; Vincent and Laybourn-Parry 2008).

In both polar regions, two main environmental variables, the availability of liquid water and temperature, are crucial for the growth of microalgal communities. The microalgae thriving in the terrestrial habitats of Antarctica experience different ranges of temperature and water availability fluctuations during the summer season (Elster 2002). Water availability determines (1) lacustrine (lake) ecosystems where liquid water is always available, (2) hydro-terrestrial ecosystems where water is available for a major part of the vegetative season, and (3) terrestrial ecosystems where water is available only for a short period of time (Elster 2002). Depending upon the range of fluctuations of the ambient temperature, the Antarctic thermal environment can be classified as (1) stable low-temperature environments, such as glacial ice, permafrost, subglacial systems, glacial melting surfaces, and temporary snow fields, and (2) unstable low-temperature environments, e.g., rocky substrates, hydro-terrestrial and lacustrine environments, etc. (Elster and Benson 2004). By combining these two variables, six major types of habitats can be distinguished (Table 25.1).

Irradiance is the other factor limiting primary productivity in polar regions. Although there is enough microclimate data on global solar radiation in polar regions, the data on photosynthetically active radiation (PAR, 400–700 nm) are rare (Table 25.2).

Table 25.1 The six major types of polar habitats defined by temperature variation and water availability

	Liquid water		
	Always	Often	Rare
Temperature	Lacustrine/ marine	Hydro-terrestrial	Terrestrial
Stable	Sea	Thermal springs	Permafrost
Stenothermal	Lakes (frozen)	Subglacial systems Glacial surfaces Glacial streams	Temporal snowfields
Variable	Lakes (melting)	Seepage	Soil
Euthermal		Minor shallow habitats (freshwater and brackish)	Soil crust Aerophytic habitats Lithophytic habitats

Table 25.2 PAR (photosynthetically active radiation) in polar regions

Locality	PAR ($\mu mol\ m^{-2}\ s^{-1}$)		Reference
Antarctic			
Ross Desert	100–1325	Partly cloudy days	Friedmann et al. (1987)
	1100–1250 (up 1860)	Daily maxima	McKay et al. (1993)
James Ross Island	~80–900	Mean diel PAR	Komárek and Elster (2008)
King George Island	~160–800	Mean diel PAR	Krezel and Pecherzewski (1981)
Arctic			
Brooks Range, USA	1500	Maximum	Tenhunen et al. (1992)
Hornsund, Svalbard	~1600	Maximum	Hodson et al. (1998)
Petuniabukta, Svalbard	~900	Maximum	Láska et al. (2012)
King Christian Island, Canada	~335	Average	Addison and Bliss (1980)
Ellesmere Island, Canada	~130–850	Mean diel PAR	Labine (1994)
	~480–560	Mean diel PAR	Henry et al. (1994)
Devon Island, Canada	~450	Average	Courtin and Labine (1977)

Original global solar radiation data were recalculated to PAR. For recalculations, we used estimations of PAR as 50% of global solar radiation (Szeicz 1974) and 1 W m^{-2} = 4.57 $\mu mol\ m^{-2}\ s^{-1}$ (Thimijan and Heins 1983)

Compared to polar deserts, the mean PAR values are lower in humid areas due to higher cloudiness. PAR can be decreased to half of the average summer values during mixed rain and snow precipitation and cloud cover of 98% at King Christian Island, Canada (Addison and Bliss 1980). The local radiation input may be different even at nearby sites due to local weather conditions and site topography. For instance, Alexandra Fjord received more energy than Eureka, although both sites are located on Ellesmere Island (Labine 1994).

Local microclimates may be quite different from the prevailing conditions and be very variable (Fig. 25.1). For instance, depending upon location, seasonality, habitat characteristics, and other factors, the temperature in Antarctic habitats may vary from sub-zero to >10°C, such as that recorded on February 3, 1996, when the temperature reached 12.2°C in Schirmacher Oasis, near the Indian station "Maitri" (Chaturvedi and Kaul 1999).

If biotechnological applications and industry will be introduced to polar regions, the facilities will be built predominantly in euthermal terrestrial ecosystems.

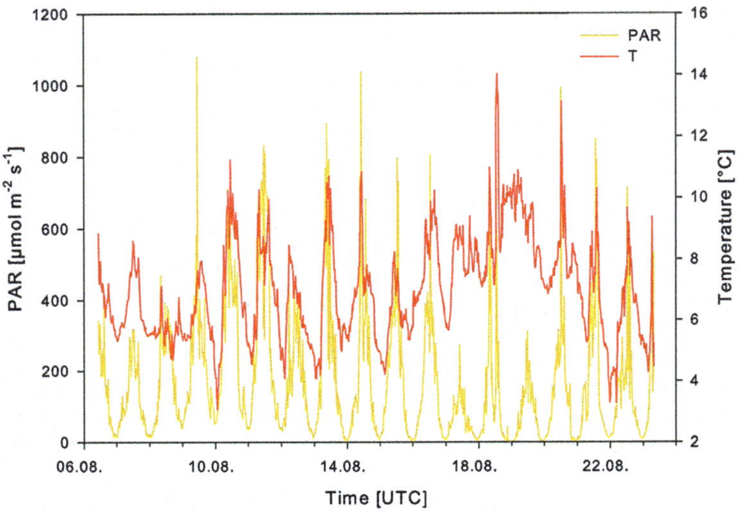

Fig. 25.1 The diel course of PAR and temperature in late summer in Longyearbyen, Svalbard (79°N)

25.3 Microalgae in Polar Ecosystems

Algae are the main primary producers in polar regions (Vincent and Laybourn-Parry 2008; Zakhia et al. 2008). Microalgae of terrestrial and hydro-terrestrial ecosystems can be found in Antarctica up to 86°29′S (Broady 1996). In the Arctic, their distribution is restricted by the presence of islands; they have been found as far north as 79°50′N (Müller et al. 1998). Under the milder conditions of sub-Antarctica, maritime Antarctica, and the Arctic, microalgae are widely distributed in lakes and streams, moist soil, on snow banks, and melting ice surfaces. In the harsher conditions of polar deserts, microalgae thrive as soil crust and endolithic communities between the grains of porous sandstone rocks and translucent quartz rocks where adequate moisture and light are available for growth (McKay et al. 1998). An examination of the value of metabolites from these microalgal taxa can provide important information about the biopotentialities of these organisms for various biotechnological applications (Vargas et al. 2002; Pushparaj et al. 2008).

The diversity of polar microalgae is controlled by many environmental variables of which the presence of liquid water and temperature are the most important (Elster 2002; Elster and Benson 2004). Although low temperatures and nutrients are slightly less important than liquid water availability, they can still limit microalgal growth (Tang et al. 1997; Wiencke et al. 2007; Shukla et al. 2011). Recently, increased UV radiation (UVR), namely, UV-B due to ozone depletion over Antarctica (Vincent and Quesada 1994; Callaghan et al. 2004) and even more recently in the Arctic (Callaghan et al. 2004; Cockell and Stokes 2006), represents an additional stress on polar microalgae, especially to those that are exposed to direct sunlight during the spring and summer periods (Vincent and Quesada 1994;

Quesada and Vincent 1997; Roos and Vincent 1998; Vincent et al. 1998; Wulff et al. 2008; Tanabe et al. 2010). Other factors may also influence the species composition and abundance of microalgae; however, they may be specific to particular polar habitats like desiccation or freezing in hydro-terrestrial and terrestrial communities (Šabacká and Elster 2006; Kvíderová et al. 2011).

25.3.1 Adaptations in Polar Microalgae

A wide array of stress conditions (low temperature, freezing, desiccation, high salinity, PAR, UVR) and their fluctuations and combinations result in a diversity of adaptive responses in polar microalgae (Elster and Benson 2004; Vincent and Laybourn-Parry 2008). Studies on the adaptive mechanisms in polar microalgae have shown that the availability of liquid water is a more limiting factor than temperature (Elster 2002; Kvíderová et al. 2011; Elster et al. 2012). Therefore, ecological, physiological, and molecular adaptations in polar microalgae often refer to freezing, desiccation, and salt stresses (Šabacká and Elster 2006; Kappen and Valladares 2007; Kvíderová et al. 2011).

Two major strategies for thriving in the polar environment include (1) stress avoidance (passive stress response) and (2) stress tolerance (active stress response). Both of them involve several adaptation mechanisms (Fig. 25.2).

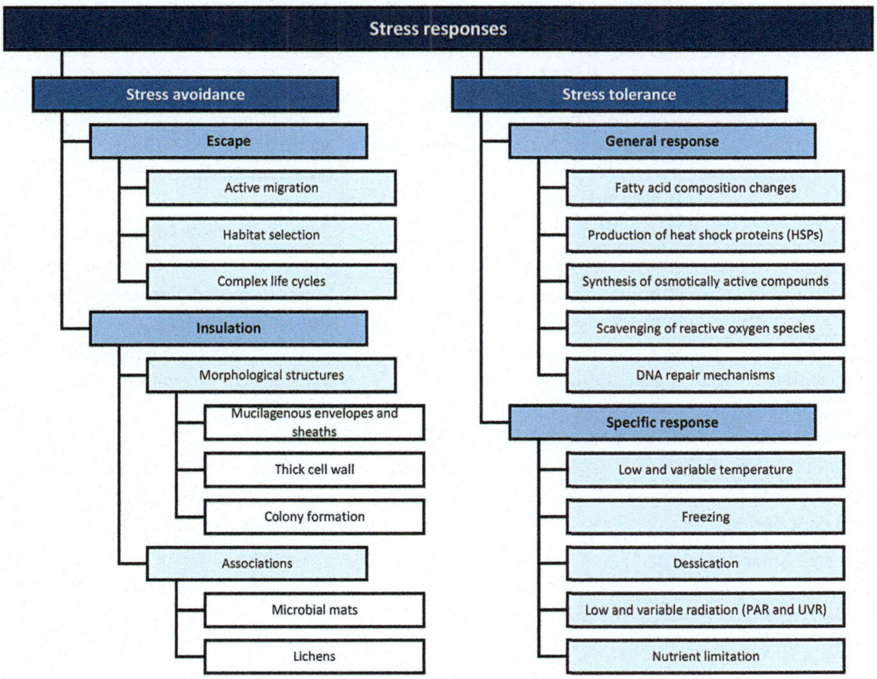

Fig. 25.2 Overview of the adaptation mechanisms in polar microalgae

25.3.2 Biotechnological Application of Microalgal Adaptations

Many of the microalgal adaptation mechanisms may provide a valuable basis for biotechnological applications (Table 25.3). Although the genomics and proteomics of polar microalgae could facilitate screening of strains according to individual requirements, such work is only beginning. In bacteria, genomic and proteomic studies have unraveled the molecular mechanisms involved in cold adaptations (Methé et al. 2005; Piette et al. 2011). Such an approach could be applicable to microalgal low-temperature biotechnology.

Especially in the case of cyanobacteria, proteomic and genomic information about various bacterial species can facilitate the investigations since their cellular properties are much similar. A variety of valuable compounds, such as PUFAs (Singh et al. 2002; Hu et al. 2008; Řezanka et al. 2008, 2009), carotenoids (Pichrtová et al. 2013), and UV-absorbing compounds (Garcia-Pichel and Castenholz 1993; Pichrtová et al. 2013), are expected to be synthesized de novo or are overproduced by polar microalgae. The results of specified screening of polar microalgal strains may be utilized in transgenic techniques for strain improvement.

Table 25.3 Adaptive features of polar microalgae and their possible applications through biotechnological and molecular biological techniques

Stress factor	Adaptations	Applications
Low temperature and freezing	Antifreeze proteins Higher superoxide dismutase activity High lipid content	Cryobiology Medicine, space biology Nutrition, biofuel
	Desiccation tolerance	Improved strains of crop plants
	Nutrient uptake at low temperature	Bioremediation at low temperature
	Nitrogen fixation at low temperature	Superior biofertilizers for temperate and alpine regions
Localized hypersaline conditions[a]	Salt-tolerant genes/compounds Salt sequestration/exclusion mechanisms	Salt-tolerant crop/microalgae Salt-tolerant crop/microalgae
High UV radiation	Mycosporine-like amino acids and other UV-protective compounds	Development of transgenic crops with higher UV resistance, nutrition
Prolonged light and dark periods	Higher photosynthetic efficiency	Improved varieties of crop plants with higher yield

[a]Coastal Antarctic lakes

25.4 Prospective Polar Microalgae

Although the mass cultivation of temperate and tropical microalgae is common (Borowitzka 2005) and they are used as food or bioproduct sources, the exploitation of polar microalgae for biotechnological applications is rare. A study on nonmarine microalgae indicates that about 700 microalgal taxa in Antarctica (Broady 1996) and about 800 microalgal taxa in Svalbard (Skulberg 1996) have the potential for producing one or more valuable metabolites; however, large screening experiments on polar microalgae are scarce (Pushparaj et al. 2008; Hrouzek et al. 2012). A number of investigations are underway to assess this possibility (Dhargalkar and Verlecar 2009), and examples on potential areas of application of polar microalgae have been presented:

- Nutrition—PUFAs: A detail investigation on fatty acid composition of 31 strains of filamentous cyanobacteria of Antarctic origin revealed that fatty acid composition varied in a strain-specific manner with relatively higher quantities of polyunsaturated fatty acids. Two strains of the cyanobacterium *Phormidium pseudopristleyi* exhibited the presence of arachidonic acid at 24% and 32% of the total fatty acid content (Pushparaj et al. 2008).
- Drugs—anticancer compounds: Secondary metabolites from cyanobacteria isolated from different climatic regions and habitats revealed significant differences in cytotoxic frequency when tested against mammalian cell lines. Further, the authors observed that (1) 60% of the isolated strains belonged to the symbiotic *Nostoc* genus and (2) even though a lower occurrence of cytotoxicity in polar isolates (36.4% frequency) was noticed, a strong inhibition was found in the extract of Antarctic *Nostoc* sp. They concluded that a proper selection of sites and strains could offer a potential resource for pharmacological and biotechnological applications (Hrouzek et al. 2012).
- Natural products—antifreeze compounds: The low-temperature environment in the Antarctic and sub-Antarctic habitats has led to the evolution of unique adaptive features in seaweeds involving the synthesis of novel compounds. Diatoms have the capacity to react to sea ice freezing by producing a "natural antifreeze," an ice-binding protein, which could have utility in biotechnological applications (Janech et al. 2006).

The hunt for Antarctic microorganisms that could be commercially exploited began in the last decade. Identification of novel strains could offer the potential for biotechnology and provide a better understanding of the polar ecosystem. The search for special cellular constituents and characteristics, especially non-catalytic proteins, pigments for surviving under extreme cold, salt, and pressure shocks, seems particularly promising (Deming 2002).

Biotechnology-based studies on Arctic genetic resources cover key areas like enzymes (used in life science research and industrial applications), antifreeze proteins, bioremediation, pharmaceuticals, nutraceuticals and dietary supplements, cosmetics, and other healthcare applications. A variety of compounds of potential

Table 25.4 Examples of patents granted for compounds isolated from polar algae

Species	Active ingredient	Properties/applications	Patent no., country
Durvillea sp.	Hyaluronic acid	Anti-wrinkle cosmetic products	JP09176036, Japan
Various marine Antarctic algae	Mycosporine-like amino acids	Sunscreen cosmetic products	US6787147, USA
Prasiola crispa	Carotenoids	Sunscreen products	WO/2002/038121
Porosira sp.	Antifreeze proteins	Industrial applications	KR20070074024, Korea
Fragilariopsis sp.	Antifreeze protein	Chemical processing	KR20060018282, Korea

Source: Bioprospecting Information Resource, United Nations University-Institute of Advanced Studies

for industrial and biotechnological applications have been characterized, and few national and international patents on the compounds isolated from polar microalgae have been granted (Table 25.4). Nonetheless, established research and development programs and considerable commercial activity by the international community need further attention when considering the implications of the emergence of bioprospecting as a new commercial activity in Antarctica.

It is worth remembering that the IPY (International Polar Year) program and its scientific collaboration were conducted by different nations, in the Arctic and Antarctica, involving more than 250 scientific projects examining a wide range of physical and biological impacts, as well as impacts on human and social systems in the circumpolar regions. Moreover, the results of IPY projects are increasing knowledge in many areas, including climate and weather, sea ice, permafrost, vegetation, and wildlife. The International Joint Committee considered the "POLARPROD" proposal on photobioreactor utilization to be a prominent and valued part of the IPY programs, as the proposal submitted includes very strong scientific education and outreach components and demonstrates a high level of adherence to IPY themes and goals.

25.5 Photobioreactors for Polar Regions

A prospective microalgal strain for polar biotechnology should be characterized by (1) high growth rates at low temperatures ranging from 0 to +10 °C, (2) at least short-term tolerance of high temperatures above 20 °C, (3) tolerance to freezing, (4) minimum requirements for nutrient additions, and (5) high photosynthetic rates across a broad temperature range. Shukla et al. (2013) proposed the unicellular green alga *Chlorella mirabilis* to be used in low-temperature biotechnology.

Microalgal mass cultivation in polar regions would require development of new types of photobioreactors to provide suitable and controlled conditions for

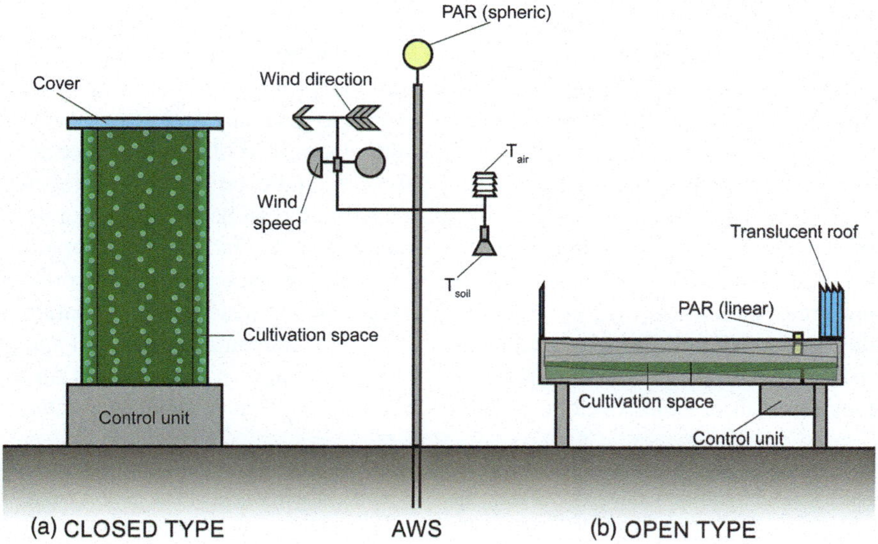

Fig. 25.3 Two technical solutions for polar photobioreactors. (**a**) Closed type, annular column, aerated by air or air + CO_2 mixture; (**b**) open type, small ponds with water movement

microalgal growth and/or biologically active compound production. The technical demands on polar photobioreactors and their operation would be stricter than on those in the tropic and temperate regions due to the harsher environmental conditions. The hardware and control systems would have to be more resistant to low temperatures and freeze-thaw cycles. The medium storage and waste treatment procedures must not contaminate the environment.

The first open photobioreactors were operated in Svalbard (Elster et al. 2001) and under similar conditions of early spring in the Czech Republic (Shukla et al. 2013).

A proposed small-scale polar photobioreactor (Fig. 25.3) for initial experiments would consist of individual subunits, allowing various technical modifications according to actual requirements (cultivation vessel, control unit, heating/cooling unit, medium and CO_2 supply, aeration, light source). The photobioreactor would be connected to an automatic weather station (AWS) which would record climatic variables/parameters (PAR, UVR, wind speed, soil and air temperatures, temperatures and PAR/UVR inside the photobioreactor at defined points) at 15–30 min intervals. The photobioreactor would be also connected to a fluorometer for monitoring the physiological status of the microalgal culture, using variable chlorophyll fluorescence measurements (Malapascua et al. 2014), and of microalgal growth using turbidity measurements. As with the meteorological data, the recording intervals should be set at 30 min. The microalgal biomass would be harvested either by centrifugation or filtration.

25.6 Conclusions

Mainly the Arctic, at the top of the Earth, warms the most rapidly and substantial ice losses are projected for the next 20–30 years due to feedback processes involving the loss of sea ice and the storage of more warmth in the Arctic Ocean. A temperature increase in the Arctic of roughly 4 °C and even up to 5 °C, together with other manifestations of climate changes across the globe, would be followed by other major physical, biological, and societal changes. Industrial explorations, e.g., industrial fishing, open seas for marine transportation, exploration of mineral resources, and others, will bring a higher-density human residential occupation in the Arctic. With all these explorations, new forms of human activities will spread into the Arctic. High technology, including biotechnology, is the only tool for the long-term sustainable development of selected polar regions such as the Arctic.

Acknowledgments The research was supported by project LM2015078—CzechPolar2—Czech Polar Research Infrastructure (MŠMT) and the long-term research development project no. RVO 67985939 (IB). We would like to thank Assoc. Prof. R. K. Edwards for language correction.

References

Addison PA, Bliss LC (1980) Summer climate, microclimate, and energy budget of a polar semidesert on King Christian Island, NWT, Canada. Arct Alp Res 12(2):161–170

Aleksandrova VD (1980) The Arctic and Antarctic: their division into geobotanical areas. Cambridge University Press, Cambridge, p 247

Astorg P (1997) Food carotenoids and cancer prevention: an overview of current research. Trends Food Sci Technol 8(12):406–413

Ben-Amotz A, Avron M (1990) The biotechnology of cultivating the halotolerant alga *Dunaliella*. Trends Biotechnol 8(5):121–126

Borowitzka MA (2005) Culturing microalgae in outdoor ponds. In: Andersen RA (ed) Algal culturing techniques. Elsevier, Amsterdam, pp 205–218

Broady PA (1996) Biodiversity, distribution and dispersal of Antarctic algae. Biodivers Conserv 5:1307–1335

Bull AT, Goodfellow M, Slater JH (1992) Biodiversity as a source of innovation in biotechnology. Annu Rev Microbiol 46(1):219–246

Callaghan TV, Björn LO, Chernov Y, Chapin T, Christensen TR, Huntley B, Ims RA, Johansson M, Jolly D, Jonasson S, Matveyeva N, Panikov N, Oechel W, Shaver G, Elster J, Jonsdottir IS, Laine K, Taulavuori K, Taulavuori E, Zöckler C (2004) Responses to projected changes in climate and UV-B at the species level. Ambio 33(7):418–435

Chaturvedi A, Kaul SMK (eds) (1999) Scientific report of fifteenth Indian expedition to Antarctica. Technical Publication no 13. Department of Ocean Development, New Delhi

Chevalier P, Proulx D, Lessard P, Vincent W, De la Noüe J (2000) Nitrogen and phosphorus removal by high latitude mat-forming cyanobacteria for potential use in tertiary wastewater treatment. J Appl Physiol 12(2):105–112

Cockell CS, Stokes MD (2006) Hypolithic colonization of opaque rocks in the Arctic and Antarctic polar desert. Arct Antarct Alp Res 38(3):335–342

Courtin GM, Labine CL (1977) Microclimatological studies of the Truelove Lowland. In: Bliss LC (ed) Truelove Lowland, Devon Island, Canada: a high arctic ecosystem. University of Alberta Press, Edmonton, pp 73–106

Del Campo JA, García-González M, Guerrero MG (2007) Outdoor cultivation of microalgae for carotenoid production: current state and perspectives. Appl Microbiol Biotechnol 74 (6):1163–1174

Deming JW (2002) Psychrophiles and polar regions. Curr Opin Microbiol 5:301–309

Dhargalkar V, Verlecar X (2009) Southern Ocean seaweeds: a resource for exploration in food and drugs. Aquaculture 287(3):229–242

Elster J (2002) Ecological classification of terrestrial algal communities in polar environments. In: Beyer L, Bötler M (eds) Geoecology of Antarctic ice-free coastal landscapes. Springer, Berlin, pp 303–326

Elster J, Benson EE (2004) Life in the polar terrestrial environment with a focus on algae and cyanobacteria. In: Fuller BJ, Lane N, Benson EE (eds) Life in the frozen state. CRC Press, Boca Raton, FL, pp 111–150

Elster J, Komárek O (2003) Ecology of periphyton in a meltwater stream ecosystem in the maritime Antarctica. Antarct Sci 15(2):189–201

Elster J, Svoboda J, Kanda H (2001) Controlled environmental platform used in temperature manipulation study of a stream periphyton in the Ny-Ålesund, Svalbard. In: Elster J, Seckbach J, Vincent WF, Lhotský O (eds) Algae and extreme environments, Nova Hedvigia Beiheft, vol 123. Cramer, Stuttgart, pp 63–75

Elster J, Kvíderová J, Hájek T, Láska K, Šimek M (2012) Impact of warming on *Nostoc* colonies (Cyanobacteria) in a wet hummock meadow, Spitzbergen. Pol Polar Res 33(4):395–420

Feller G, Gerday C (1997) Psychrophilic enzymes: molecular basis of cold adaptation. Cell Mol Life Sci 53(10):830–841

Friedmann EI, McKay CP, Nienow JA (1987) The cryptoendolithic microbial environment in the Ross Desert of Antarctica: satellite-transmitted continuous nanoclimate data, 1984 to 1986. Polar Biol 7:273–287

Garcia-Pichel F, Castenholz RW (1993) Occurrence of UV-absorbing, mycosporine-like compounds among cyanobacterial isolates and an estimate of their screening capacity. Appl Environ Microbiol 59(1):163–169

Glazer AN, Nikaido H (2007) Microbial biotechnology: fundamentals of applied microbiology. Cambridge University Press, Cambridge

Henry GHR, Svoboda J, Freedman B (1994) Standing crop and net production of non-grazed sedge meadow of a polar desert oasis. In: Svoboda J, Freedman B (eds) Ecology of a pola oasis. Alexandra Fiord, Ellesmere Island, Canada. Captus University Publications, Toronto, pp 85–95

Hodson AJ, Gurnell AM, Washington R, Tranter M, Clark MJ, Hagen JO (1998) Meteorological and runoff time-series characteristics in a small, high-Arctic glaciated basin, Svalbard. Hydrol Process 12(3):509–526

Holdgate MW (1970) Antarctic ecology. Academic Press, London

Hrouzek P, Tomek P, Lukešová A, Urban J, Voloshko L, Pushparaj B, Ventura S, Lukavský J, Štys D, Kopecký J (2012) Cytotoxicity and secondary metabolites production in terrestrial *Nostoc* strains, originating from different climatic/geographic regions and habitats: is their cytotoxicity environmentally dependent? Environ Toxicol 26(4):345–358

Hu HH, Li HY, Xu XD (2008) Alternative cold response modes in *Chlorella* (Chlorophyta, Trebouxiophyceae) from Antarctica. Phycologia 47(1):28–34

Janech MG, Krell A, Mock T, Kang J-S, Raymond JA (2006) Ice-binding proteins from sea ice diatoms (Bacillariophyceae). J Phycol 42(2):410–416

Kappen L, Valladares F (2007) Opportunistic growth and desiccation tolerance: the ecological success of poikilohydrous autotrophs. In: Pugnaire F, Valladares F (eds) Functional plant ecology. Taylor and Francis, New York, NY, pp 7–66

Komárek J, Elster J (2008) Ecological background of cyanobacterial assemblages of the northern part of James Ross Island, Antarctica. Pol Polar Res 29:17–32
Komárek J, Kováčik L, Elster J, Komárek O (2012) Cyanobacterial diversity of Petuniabukta, Billefjorden, central Spitzbergen. Pol Polar Res 33(4):347–368
Kopalová K, Nedbalová L, Nývlt D, Elster J, Van de Vijver B (2013) Diversity, ecology and biogeography of the freshwater diatom communities from Ulu Peninsula (James Ross Island, NE Antarctic Peninsula). Polar Biol 36(7):933–948
Krezel A, Pecherzewski K (1981) Preliminary data on total radiation in the region of Arctowski Station (King George Island, South Shetland Islands). Pol Polar Res 2:47–54
Kvíderová J, Elster J, Šimek M (2011) In situ response of *Nostoc commune* s.l. colonies to desiccation in Central Svalbard, Norwegian High Arctic. Fottea 11(1):87–97
Labine CL (1994) Meteorology and climatology of the Alexandra Fiord lowland. In: Svoboda J, Freedman B (eds) Ecology of a pola oasis. Alexandra Fiord, Ellesmere Island, Canada. Captus University Publications, Toronto, pp 23–39
Láska K, Witoszová D, Prošek P (2012) Weather pattern of the coastal zone of Petuniabukta, central Spitzbergen in the period 2008–2010. Pol Polar Res 33(4):297–318
Lukavský J (2012) *Trachydiscus minutus*: a new algal EPA producer. In: Krueger D, Meyer H (eds) Algae: ecology, economy uses and environmental impact. Nova Science, New York, NY, pp 77–104
Lv J-M, Cheng L-H, Xu X-H, Zhang L, Chen H-L (2010) Enhanced lipid production of *Chlorella vulgaris* by adjustment of cultivation conditions. Bioresour Technol 101(17):6797–6804
Malapascua JR, Jerez CG, Sergejevová M, Figueroa FL, Masojídek J (2014) Photosynthesis monitoring to optimize growth of microalgal mass cultures: application of chlorophyll fluorescence techniques. Aquat Biol 22:123–140
McKay CP, Nienow JA, Meyer MA, Friedmann EI (1993) Continuous nanoclimate data (1985–1988) from the Ross Desert (McMurdo Dry Valleys) cryptoendolithic microbial ecosystem. In: Bromwich DH, Stearns CR (eds) Antarctic meteorology and climatology: studies based on automatic weather stations, Antarctic research series, vol 61. American Geophysical Union, Washington, DC, pp 201–207
McKay CP, Nienow JA, Meyer MA, Friedmann EI (1998) Continuous nanoclimate data (1985–1988) from the Ross Desert (McMundo Dry Valleys) cryptoendolithic microbial ecosystem. Antarct Res Ser 61:201–207
Methé BA, Nelson KE, Deming JW, Momen B, Melamud E, Zhang X, Moult J, Madupu R, Nelson WC, Dodson RJ (2005) The psychrophilic lifestyle as revealed by the genome sequence of *Colwellia psychrerythraea* 34H through genomic and proteomic analyses. Proc Natl Acad Sci U S A 102(31):10913–10918
Müller T, Bleiss W, Martin C-D, Rogaschewski S, Fuhr G (1998) Snow algae from northwest Svalbard: their identification, distribution, pigment and nutrient content. Polar Biol 20:14–32
Nichols DS, Sanderson K, Buia A, Van De Kamp J, Holloway P, Bowman JP, Smith M, Mancuso Nichols C, Nichols P, McMeekin TA (2002) Bioprospecting and biotechnology in Antarctica. In: Jabour-Green J, Haward M (eds) The Antarctic: past, present and future. Antarctic CRC Research Report no 28, Hobart, pp 85–105
Pichrtová M, Remias D, Lewis LA, Holzinger A (2013) Changes in phenolic compounds and cellular ultrastructure of Arctic and Antarctic strains of *Zygnema* (Zygnematophyceae, Streptophyta) after exposure to experimentally enhanced UV to PAR ratio. Microb Ecol 65 (1):68–83
Piette F, D'Amico S, Mazzucchelli G, Danchin A, Leprince P, Feller G (2011) Life in the cold: a proteomic study of cold-repressed proteins in the Antarctic bacterium *Pseudoalteromonas haloplanktis* TAC125. Appl Environ Microbiol 77(11):3881–3883
Pushparaj B, Buccioni A, Paperi R, Piccardi R, Ena A, Carlozzi P, Sili C (2008) Fatty acid composition of Antarctic cyanobacteria. Phycologia 47:430–434
Quesada A, Vincent WF (1997) Strategies of adaptation by Antarctic cyanobacteria to ultraviolet radiation. Eur J Phycol 32:335–342

Řezanka T, Nedbalová L, Sigler K (2008) Unusual medium chain-chain polyunsaturated fatty acids from snow alga *Chloromonas brevispina*. Microbiol Res 163(4):373–379

Řezanka T, Nedbalová L, Elster J, Cajthaml T, Sigler K (2009) Very-long-chain *iso* and *anteiso* branched fatty acids in N-acylphosphatidylethanolamines from a natural cyanobacterial mat of *Calothrix* sp. Phytochemistry 70(5):655–663

Richmond A (2004) Handbook of microalgal culture: biotechnology and applied phycology. Wiley-Blackwell, Oxford

Roos JC, Vincent WF (1998) Temperature dependence of UV radiation effects on Antarctic cyanobacteria. J Phycol 34:118–125

Šabacká M, Elster J (2006) Response of cyanobacteria and algae from Antarctic wetland habitats to freezing and desiccation stress. Polar Biol 30(1):31–37

Shukla SP, Kvíderová J, Elster J (2011) Nutrient requirements of polar *Chlorella*-like species. Czech Polar Rep 1:1–10

Shukla SP, Kvíderová J, Tříska J, Elster J (2013) *Chlorella mirabilis* as a potential species for biomass production in low-temperature environment. Front Microbiol 4. doi:10.3389/fmicb.2013.00097

Singh SC, Sinha RP, Häder D-P (2002) Role of lipids and fatty acids synthesis in stress tolerance in cyanobacteria. Acta Protozool 41:297–308

Singh S, Kate BN, Banerjee U (2005) Bioactive compounds from cyanobacteria and microalgae: an overview. Crit Rev Biotechnol 25(3):73–95

Skulberg OM (1996) Terrestrial and limnic algae and cyanobacteria. In: Elvebakk A, Prestrud P (eds) A catalogue of Svalbard plants, fungi, algae and cyanobacteria. Norsk Polarinstitutt Skrifter, Tromsø, pp 383–395

Strunecký O, Elster J, Komarek J (2010) Phylogenetic relationships between geographically separate *Phormidium* cyanobacteria: is there a link between north and south polar regions? Polar Biol 33(10):1419–1428

Strunecký O, Komárek J, Elster J (2012) Biogeography of *Phormidium autumnale* (Oscillatoriales, Cyanobacteria) in western and central Spitsbergen. Pol Polar Res 33(4):369–382

Szeicz G (1974) Solar radiation for plant growth. J Appl Ecol 11(2):617–636

Tanabe Y, Ohtani S, Kasamatsu N, Fukuchi M, Kudoh S (2010) Photophysiological responses of phytobenthic communities to the strong light and UV in Antarctic shallow lakes. Polar Biol 33 (1):85–100

Tang EPY, Tremblay R, Vincent WF (1997) Cyanobaterial dominance of polar freshwater ecosystems: are high-latitude mat-formers adapted to low temperature? J Phycol 33:171–181

Tenhunen JD, Lange OL, Hahn S, Siegwolf R, Oberbauer SF (1992) The ecosystem role of poikilohydric tundra plants. In: Chapin FSI, Jefferies RL, Reynolds JF, Shaver GR, Svoboda J (eds) Arctic ecosystem in a changing climate. An ecological perspective. Academic Press, San Diego, CA, pp 213–237

Thimijan RW, Heins RD (1983) Photometric, radiometric and quantum light units of measure: a review of procedures for interconversion. Hortic Sci 18(6):818–822

van den Vijver B, Sterken M, Vyverman W, Mataloni G, Nedbalová L, Kopalová K, Elster J, Verleyen E, Sabbe K (2010) Four new non-marine diatom taxa from the subantarctic and arctic regions. Diatom Res 25:431–443

Vargas M, Rodriguez H, Moreno J, Olivares H, Campo JD, Rivas J, Guerrero M (2002) Biochemical composition and fatty acid content of filamentous nitrogen-fixing cyanobacteria. J Phycol 34(5):812–817

Vincent WF, Laybourn-Parry J (2008) Polar lakes and rivers: limnology of Arctic and Antarctic ecosystems. Oxford University Press, Oxford

Vincent WF, Quesada A (1994) Ultraviolet radiation effects on cyanobacteria: implications for Antarctic microbial ecosystems. In: Weiler CS, Penhale PA (eds) Ultraviolet radiation in Antarctica: measurements and biological effects, Antarctic research series, vol 62. AGU, Washington, DC, pp 111–124

Vincent WF, Rae R, Laurion I, Howard-Williams C, Priscu JC (1998) Transparency of Antarctic ice-covered lakes to solar UV radiation. Limnol Oceanogr 43(4):618–624

Vonshak A (1997) *Spirulina platensis* (*Arthrospira*): physiology, cell-biology and biotechnology. Taylor & Francis, London

Wiencke C, Clayton MN, Gómez I, Iken K, Lüder UH, Amsler CD, Karsten U, Hanelt D, Bischof K, Dunton K (2007) Life strategy, ecophysiology and ecology of seaweeds in polar waters. In: Amils R, Ellis-Evans C, Hinghofer-Szalkay H (eds) Life in extreme environments. Springer, Dordrecht, pp 213–244

Wulff A, Zacher K, Hanelt D, Al-Handal A, Wiencke C (2008) UV radiation—a thread to Antarctic benthic marine diatoms? Antarct Sci 20(1):13–20

Zakhia F, Jungblut A-D, Taton A, Vincent WF, Wilmotte A (2008) Cyanobacteria in cold ecosystems. In: Margesin R, Schinner F, Marx J-C, Gerday C (eds) Psychrophiles: from biodiversity to biotechnology. Springer, Berlin, pp 121–135

Chapter 26
Nitrification at Low Temperature for Purification of Used Water

M.J. Dempsey

Abstract Prokaryotes that can oxidize ammonia and/or nitrite are known as nitrifiers and are common in terrestrial, freshwater and marine environments. Where the temperature is commonly in the range 0–20 °C, psychrophilic strains or species can be isolated or identified using molecular techniques. It is therefore no surprise to also find psychrophilic nitrifiers in engineered systems used, for example, to remove ammonia from raw, used or wastewater or from contaminated air. In temperate regions, we have been using psychrophilic nitrifiers without most people realizing, and this chapter attempts to put their importance into context by comparing and contrasting their presence in natural and engineered systems. It concludes by describing a biofilm-based process technology, the expanded bed biofilm reactor, which the author has improved with several inventions that make this technology cost-effective for wider adoption.

Contents

26.1 Introduction

Purification of used water tends to be done at municipal wastewater treatment works and, in temperate countries, it normally relies on the activities of psychrophilic microorganisms, as the temperature of the wastewater rarely rises above

M.J. Dempsey (✉)
Faculty of Science and Engineering, Manchester Metropolitan University, John Dalton Building, Chester Street, Manchester M1 5GD, UK

Advanced Bioprocess Development Ltd., John Dalton Building, Chester Street, Manchester M1 5GD, UK
e-mail: m.dempsey@mmu.ac.uk

R. Margesin (ed.), *Psychrophiles: From Biodiversity to Biotechnology*,
DOI 10.1007/978-3-319-57057-0_26

20 °C. Because the purification of used water by secondary, biological processes involves the conversion of organic matter to microbial biomass, it represents the largest biotechnology application of cold-adapted microbes. For example, in the European Union (EU), if all wastewater was subjected to secondary treatment, approximately 4 million tonnes of biomass would be produced per year, much of it consisting of psychrophilic, heterotrophic microorganisms such as bacteria, protozoa and rotifers, as well as oligochaete and nematode worms.

In contrast, nitrification is conducted mostly by autotrophic ammonia- and nitrite-oxidizing bacteria (AOB and NOB, respectively), although bacteria capable of carrying out both processes have recently been reported (Daims et al. 2015; van Kessel et al. 2015) and ammonia-oxidizing archaea (AOA) have been found in wastewater plants in the USA by Park et al. (2006). The energy yield from oxidizing ammonia or nitrite is much less than from using organic matter as the electron donor and, for nitrite, it is one third that of ammonia. That is, 100 molecules of nitrite need to be oxidized to harvest enough energy to fix one CO_2 molecule, whereas it only requires oxidation of 35 ammonia molecules (Prosser 1986). Therefore, for nitrification, approximately 367 thousand tonnes of AOB and 86 thousand tonnes of NOB would be produced per year in Europe if tertiary treatment was used for nitrification of all domestic wastewater. Nitrification processes have become more common in recent years, for the protection of aquatic organisms and the restoration of good ecological status. An important driver for use of these processes in the EU has been the Urban Wastewater Treatment Directive (European Council 1991), which has been subsumed into the Water Framework Directive (European Council 2000).

Treatment of domestic wastewater consists of up to three stages: primary, secondary and tertiary. Primary treatment consists of physical separation of particulate materials, typically using screens and grit channels. This removes particles larger than a few millimetres or with a significantly higher density than water. The partially clarified wastewater ("settled sewage") is normally subjected to secondary, biological treatment using one of several process technologies. The oldest established technology is the trickling filter (TF), which was perfected in Salford, UK, in 1893. The second oldest, activated sludge (AS), was developed in Davyhulme, Manchester, UK, in 1914. Despite their age, these processes are still used worldwide. Oxidation ponds (lagoons) date from about the same time, but more recently developed processes include rotating biological contactors (RBC), moving bed biofilm reactors (MBBR) and membrane bioreactors (MBR) (Murdoch University Environmental Technology Centre 2000; Metcalf & Eddy et al. 2014). Essentially, secondary treatment is a process of growing mostly heterotrophic microbes on nutrients in wastewater, thus removing the polluting matter by settling out the biomass as sludge. This sludge is often combined with that from the primary settling tanks and subjected to anaerobic digestion, to reduce the final sludge volume, by about tenfold, and to generate methane-rich biogas as a renewable fuel. Although this process is normally operated at mesophilic or thermophilic temperatures, recent work by O'Flaherty's group at the National University of

Ireland, Galway, has established a method of low-temperature anaerobic digestion based on biomass granules (McKeown et al. 2009).

Following the Industrial Revolution and the growth of crowded cities, biological wastewater treatment was developed as a means of protecting human health, by breaking the cycle of waterborne disease. It also gave some basic protection to the aquatic environment. This work continues, with increasing protection of the aquatic environment driven by ever-tighter standards on effluent quality. One of the main pollutants remaining after secondary treatment, in terms of mass, is ammonia, and those eating a Western-style omnivorous diet produce about 8 g of ammoniacal nitrogen waste per day, approximately half of which is removed during secondary treatment. Therefore, to protect sensitive organisms from the damaging effects of ammonia in water bodies receiving treated effluent, a tertiary process is required for nitrification. Such processes are designed to reduce the residual ammoniacal nitrogen concentration so that the works' effluent is suitable for discharge into the aquatic environment.

26.2 Nitrification in the Natural Environment

Nitrification is a common microbiological process that occurs in terrestrial and aquatic environments, both freshwater and marine. Ammonia is oxidized to nitrite (nitritation) by one group of specialist bacteria or archaea and nitrite oxidized to nitrate (nitritation) by another group of specialist bacteria, phylogenetically unrelated to the AOB (Costa et al. 2006) but bacteria capable of carrying out both processes have recently been reported (Daims et al. 2015; van Kessel et al. 2015). No nitrite-oxidizing archaea have yet been identified. Ammonia and nitrite are electron donors and, therefore, the energy source for these prokaryotes. In recent years, ammonia-oxidizing archaea (AOA) have been isolated, with the discovery by Konneke et al. (2005) of a new species, from the rocky substratum of a tropical marine tank at the Seattle Aquarium (USA). Thus, AOA join the ammonia-oxidizing bacteria (AOB) and nitrite-oxidizing bacteria (NOB) as important prokaryotes in global nitrogen cycling.

Although ammonia and nitrite oxidation was always thought to occur in different species (Costa et al. 2006), several new candidate species of bacteria from the genus *Nitrospira* have been discovered that are capable of *com*plete *amm*onia *ox*idation, comammox, i.e. the sequential oxidation of ammonia to nitrite and nitrite to nitrate by a single species. However, as one species was isolated by Daims et al. (2015) from microbial biofilm growing on the walls of a hot water pipe (56 °C) at an oil exploration well in Aushiger (North Caucasus, Russia), it would not have been psychrophilic. However, the other was isolated by van Kessel et al. (2015) from a trickling filter system for carp (*Cyprinus carpio*) aquaculture, following enrichment at 23 $\pm$ 1 °C, which is closer to psychrophilic conditions. More recently, Chao et al. (2016) provided metagenomic evidence for *Nitrospira*-like *amoA* genes in activated sludge (AS) and biofilm samples from an aerobic suspended carrier biofilm reactor (otherwise known as a moving bed biofilm reactor, MBBR) at the Stanley WWTP in Hong Kong. This plant had a design temperature of 18 °C and an operational water temperature of about 22 °C for the Hybas™ nitrification process (Lau 2011) that was probably the source of the material for metagenomic analysis, so these comammox *Nitrospira* may not have been true psychrophiles either.

Nevertheless, there is growing evidence of significant nitrification at low temperatures, suggesting that psychrophilic nitrifiers are ubiquitous in cold environments. For example, Clark et al. (2009) provided evidence that nitrification proceeds in frozen soils (−6 °C for clay and −2 °C for loam) collected 5 km south of Québec City (Canada). This observation has been confirmed by Jusselme et al. (2016) for soils in the French Alps, where evidence for nitrification activity was obtained at soil surface temperatures in the range −0.6 to +0.8 °C. Furthermore, chemical evidence has been presented by Telling et al. (2014) for nitrification in ice-entombed cryoconite holes in the surface of glaciers in the McMurdo Dry Valley (Antarctica), where air temperatures rarely exceed 0 °C. Therefore, from an ecological perspective, nitrification by psychrophiles should occur in wastewater treatment processes whenever the wastewater is in the range 0–20 °C.

26.3 Nitrification in Engineered Environments

As well as AOB, AOA have been found in granular activated carbon filters for nitrification of raw water. The number of AOA was stable all year, even though the raw water temperature ranged from 17.7 to 28.6 °C in summer and 4.6 to 5.5 °C in winter, whereas AOB numbers were lower in winter (Niu et al. 2016). These observations indicate that psychrophilic AOA and AOB were present, with the AOA community being more resilient at low temperature. In contrast, during an investigation of the ammonia-oxidizing prokaryotes in a horizontal flow biofilm reactor (HFBR) for the treatment of ammonia-contaminated air at 10 °C, Gerrity et al. (2016) found that AOB were significantly more abundant than AOA. Nevertheless, both types of microbe were presumably psychrophilic. In a study of the relative abundance of AOB and AOA in biofilms of sequential discs in an RBC at Guelph Wastewater Treatment Plant (Ontario, Canada), Sauder et al. (2012) demonstrated that AOA became more dominant as the ammonia concentration was reduced along the reactor. This indicates that these AOA are k-strategists (Wett et al. 2011; Wu et al. 2016), as they had a higher affinity for ammonia than the AOB. Therefore, operating conditions to encourage the growth of AOA or k-strategist AOB are likely to result in a higher-quality effluent, i.e. one with a lower concentration of ammonia. Furthermore, as the average temperature in Guelph ranges from about −6 °C in winter to 20 °C in summer (http://en.climate-data.org/location/881/), both the AOB and AOA were most likely psychrophilic.

Biofilm reactors are especially useful when slow-growing process microbes have to be used, such as nitrifiers in a wastewater treatment process. Chao et al. (2016) found that the abundance and diversity of nitrogen cycle genes (both for nitrification and denitrification) were significantly higher in biofilm than that in AS. They considered that this difference indicated that the nitrogen-removing bacteria growing as biofilm contained more abundant or diverse functional genes, revealing that the increased nitrogen removal ability of biofilm processes might be mainly attributed to enhancement of removal, rather than the accumulation of nitrogen removal bacteria. Presumably, there was enhancement through physiological changes in the bacteria when growing as a biofilm, rather than a mere increase in biomass. It is widely reported that AMO and NOB cell clusters are found close to each other in biofilms (Maixner et al. 2006; Akhidime 2009), which is assumed to be the result of a selection pressure for the NOB to more readily access the nitrite produced by the AOB. Although this clustering was observed by Maixner et al. (2006) for *Nitrospira* sublineage I, it was not for sublineage II, which might be evidence that the latter were comammox bacteria and, therefore, clustering with AOB (or AOA) conferred no growth advantage.

The nitrification rate in biofilm has been calculated to decrease by about 4.5% per °C fall, compared to 10% in conventional AS systems (Boller et al. 1987, as reported by Bodık et al. 2003). This differential reduction in rate is likely to have been a consequence of the nitrifiers being washed out of the AS process faster than they can grow, owing to their low growth yield. This effect is not a problem for biofilm systems, where the microbes can be retained irrespective of growth rate.

A similar phenomenon was observed during pilot-scale trials of an expanded bed biofilm reactor (EBBR) designed for nitrification of secondary effluent. This trial was conducted at the United Utilities Davyhulme Wastewater Treatment Works (Manchester, UK), and during a particularly wet summer when it rained every day for 3 weeks, an increase in the AS plant effluent nitrite concentration was observed, increasing from about 1 to 10 mg L^{-1} NO_2^--N (unpublished observation). This was assumed to have been caused by the NOB being washed out of the AS process but not the ammonia oxidizers, owing to their higher growth yield. In contrast, the EBBR, which treated the AS settled effluent, continued to produce a treated effluent low in both ammonia and nitrite, indicating that the NOB were retained in this system, presumably as a result of immobilization in attached biofilm.

A similar effect is predicted when suspended cell and biofilm processes are operated during cold weather, when the AOB, AOA and NOB growth rates may be depressed. If the growth rate is less than the hydraulic residence time (HRT) of suspended cell reactors, cells will be washed out faster than they can reproduce, whereas in fixed biofilm systems, cells are retained because the reproduction rate is decoupled from HRT.

By starting up a process to remove ammonia from used water at temperatures in the psychrophilic range, there will be a selection pressure for species and strains that are cold adapted. These bacteria will therefore have higher growth rates at lower temperatures than if a process is established under mesophilic conditions and then operated under psychrophilic conditions. This is a problem with some research work on the effect of temperature shifts on nitrification (and other processes), where processes are established in the lab in the mesophilic range and then process performance is also evaluated in the psychrophilic range. It is likely that warm-adapted species dominate the microbial community that develops under mesophilic conditions, which then perform badly when the temperature is lowered into the psychrophilic range. Whereas, if the bioreactor is seeded with material from a psychrophilic environment and operated in the psychrophilic range, then good performance should be achieved under psychrophilic conditions but not necessarily if the temperature is increased into the mesophilic range. It is all a question of understanding microbial physiology and ecology, and applying it in an industrial context!

The local seasonal temperature range determines which species are available for seeding a bioreactor, and the temperature under which the process is started up and operated largely determines which will become established. Psychrophiles have an optimum temperature for growth at or below 15 °C; a maximum temperature of about 20 °C, above which they die; and a minimum temperature at or below 0 °C, below which they cannot grow. For example, a newly identified, psychrophilic nitrite-oxidizing bacterium, "*Candidatus* Nitrotoga arctica", has been reported from permafrost soil in the Siberian arctic (Alawi et al. 2007). This candidate species grows in the temperature range 4–17 °C, and bacteria with similar 16S rDNA sequences have now been isolated by cultivating at 10 °C or 17 °C activated sludge from the Hamburg (Germany) plant, which typically operates in the range 7–16 °C (Alawi et al. 2009).

Generally, a wastewater treatment process will develop a psychrophilic population if one is available locally and the process is operated in the relevant temperature range. During pilot-plant trials of the EBBR referred to above (Dempsey et al. 2006), good nitrification was observed down to 8 °C (Fig. 26.1), which was the lowest

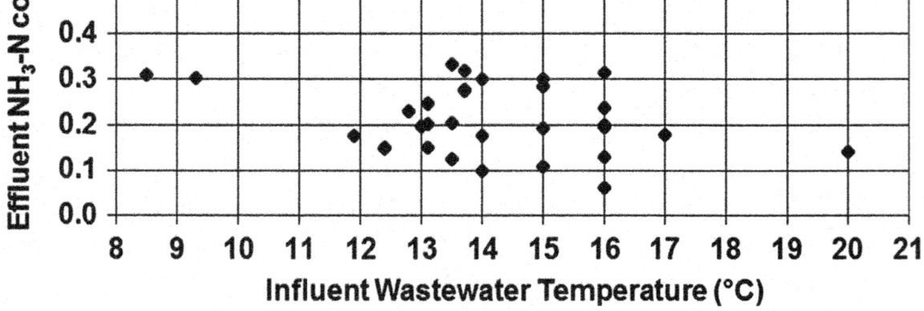

Fig. 26.1 Relationship between process temperature and effluent ammonia concentration during operation of an expanded bed biofilm reactor pilot plant (0.5 m diameter, 3 m bed depth) to achieve an effluent ammonia-nitrogen concentration below 1 mg L^{-1} when treating the activated sludge plant settled effluent at the United Utilities Davyhulme Wastewater Treatment Works (Manchester, UK)

influent wastewater temperature seen in 5 years of operation, as normally it never fell below 11 °C. The fact that this process was able to reduce the ammonia-nitrogen concentration below 1 mg L^{-1} in the temperature range 8–20 °C indicates that the active nitrifiers were psychrophilic. Currently, DNA sequencing is being conducted to determine which species of nitrifiers this process contained.

26.4 Expanded Bed Biofilm Reactor Technology

Advanced Bioprocess Development (ABD) has developed an expanded bed biofilm reactor (EBBR), which is a fixed-film, submerged-bed technology (BS-EN-12255-7 2002) where biofilms grow on small particles (0.7–1.0 mm) of porous carbon (ABDite®, Fig. 26.2; Dempsey 2001). Calculated on the basis of solid geometry, 1 mm spheres (assuming 60% packing) have a specific surface area (SSA) of 3600 m^2 m^{-3}. Therefore, ABDite® particles have a similar SSA if we ignore the material's porosity. This large SSA is almost four times greater than any other biomass support medium (Fig. 26.3). Once colonized by biofilm, the term *bioparticles* or *particulate biofilms* is used (Fig. 26.1). When expanded by 50%, the biofilm SSA is up to 2400 m^2 m^{-3} and the biomass concentration is up to 42,000 mg L^{-1} (VSS equivalent).

This large surface area of biofilm and high concentration of biomass are maintained in an active state by control of biofilm thickness, ideally in the range 100–400 μm (Akhidime and Dempsey 2009), via recycling of particles from the top of the bed to the bottom (Fig. 26.4). Bioparticles with the thickest biofilm are the least active, owing to increasing diffusional limitation as the biofilm depth increases. Fortunately, these bioparticles migrate to the top of the bed, owing to their lower fluidization velocity. Thereby, the least active biofilm is removed via

Fig. 26.2 Biomass support media particles (0.7-1.0 mm porous carbon, ABDite®). (**a**) Scanning electron micrograph (SEM), showing porous nature of these media particles (magnification ×60, pore size approx. 50 μm wide); (**b**) SEM of ABDite® colonized with thin nitrifying biofilm (bioparticle size approx. 1 mm); (**c**) light micrograph of bioparticle, showing relatively smooth nature once colonized with thick biofilm (bioparticle size 1.0–1.5 mm)

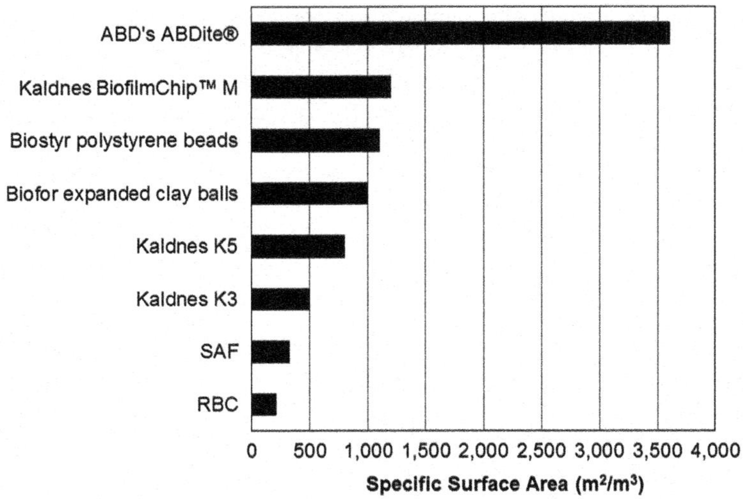

Fig. 26.3 Specific surface area of media particles used for different types of wastewater treatment technology: Suez Environment's Biofor®, Veolia Water Technologies' Biostyr®, Veolia Water Technologies' Kaldnes® moving bed biofilm reactor (MBBR), submerged aerated filter (non-proprietary), rotating biological contactor (non-proprietary) and Advanced Bioprocess Development's expanded bed biofilm reactor (EBBR)

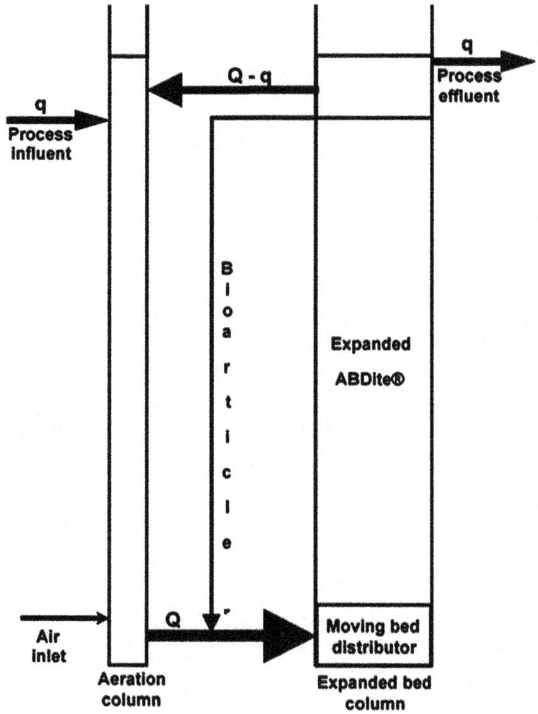

Fig. 26.4 Design of expanded bed biofilm reactor for aerobic processes

attrition during passage through regions of high turbulence, including within an injection device (Dempsey 2011), and the moving bed distributor (Dempsey 2004) at the base of the expanded bed.

The influent wastewater enters the top of the aeration column (Fig. 26.4), at flow rate q, and flows downwards. In contrast, the process air (depleted of nitrogen for highly aerobic processes, such as nitrification) flows upwards. Separation of this countercurrent flow from the immobilized biomass is unique to this technology and allows a high degree of turbulence for achieving highly efficient oxygen transfer from the gas to the liquid phase, without risking detachment of biofilm from the media particles. Aerated wastewater is pumped into the base of the expanded bed, at flow rate Q, causing particle fluidization and bed expansion. Ammonia is oxidized to nitrite by ammonia-oxidizing bacteria and the nitrite oxidized to nitrate by nitrite-oxidizing bacteria as the wastewater passes up through the bed. Treated effluent leaves at the same flow rate (q) as the influent enters, thus causing a variable recycle of partially treated wastewater back to the top of the aeration column, dependent on the works flow. Thus, diurnal variations in the flow to the works, as well as variations caused by rainfall or other precipitation, are automatically adjusted for. Bioparticles with the thickest biofilm are recycled from the top to

Fig. 26.5 Full-scale, expanded bed biofilm reactor (EBBR) prototype package plant (Technology Readiness Level 7) constructed in stainless steel by J. K. Fabrications, Newry, Northern Ireland. The expanded bed column is 1.5 m diameter and the expanded bed depth is 5 m. This size of EBBR is able to nitrify at least 8.8 kg NH_3-N per day, which is equivalent to purifying the secondary effluent from a population of about 2200 people

the bottom of the bed by another automatic system, which relies on an injector driven by the flow of wastewater induced by the fluidizing pump.

By virtue of their immobilization in the biofilm that grows on the biomass support media particles (ABDite®) and the fact that these particles are retained in the reactor, process microbes are retained despite the relatively high upflow velocity (36 m h^{-1}) of the process liquid (wastewater in this case). In contrast, planktonic cells will be washed out with the treated effluent. Thus, irrespective of their growth rate, process microbes will be available at all times, despite the fact that they will normally be reproducing more slowly than the hydraulic residence time (approx. 30–60 min for a tertiary nitrification process). This process has now been developed to Technology Readiness Level 7, with the development of a prototype package plant in collaboration with J. K. Fabrications, Newry, Northern Ireland (Fig. 26.5).

26.5 Conclusions

It is clear that nitrifying prokaryotes adapted to psychrophilic conditions (0–20 °C) are widespread in cold environments and are also found in bioprocesses operated under psychrophilic conditions. In order to establish a bioprocess, such as nitrification of raw, used or waste water, for treatment under psychrophilic conditions, it is necessary to maintain the temperature in the correct range and, ideally, to seed it with material likely to contain the psychrophilic microbes necessary for the process. In the case of nitrification, these will include ammonia-oxidizing bacteria and archaea, as well as nitrite-oxidizing bacteria and comammox bacteria; and nitrite-oxidizing archaea, should they be discovered.

Furthermore, where there is a requirement for effluent ammonia concentrations below 1 mg/L, as is becoming more frequent in Europe, the USA and other highly developed countries, it is also necessary to operate the process with nitrifiers that have a high affinity for ammonia. These nitrifiers can be enriched quite simply, by always operating the process with a low effluent ammonia concentration, which will create the conditions where k-strategists have a selective advantage over r-strategists.

References

Akhidime ID (2009) Aspects of expanded bed nitrification including treatment of oil refinery wastewaters. PhD, Manchester Metropolitan University, Manchester. http://ethos.bl.uk/OrderDetails.do?uin=uk.bl.ethos.509884

Akhidime ID, Dempsey MJ (2009) Influence of biofilm thickness on ammonia oxidation rate of bioparticles from expanded bed process for tertiary nitrification. Paper presented at the 2nd IWA Specialized Conference on Nutrient Management in Wastewater Treatment Processes, Krakow, Poland, 2009

Alawi M, Lipski A, Sanders T, Eva Maria P, Spieck E (2007) Cultivation of a novel cold-adapted nitrite oxidizing betaproteobacterium from the siberian arctic. ISME J 1(3):256–264. doi:10.1038/ismej.2007.34

Alawi M, Off S, Kaya M, Spieck E (2009) Temperature influences the population structure of nitrite-oxidizing bacteria in activated sludge. Environ Microbiol Rep 1(3):184–190. doi:10.1111/j.1758-2229.2009.00029.x

Bodık I, Kratochvıl K, Gašpariková E, Hutňan M (2003) Nitrogen removal in an anaerobic baffled filter reactor with aerobic post-treatment. Bioresour Technol 86(1):79–84. doi:10.1016/S0960-8524(02)00109-8

Boller M, Eugster J, Weber A, Gujer W (1987) Nitrification in nachgeschalteten rotierenden Tauchkörpern. Bericht über 2 Jahre Pilotuntersuchungen. EAWAG report no. 28–601, Dübendorf, Switzerland

BS-EN-12255-7 (2002) Wastewater treatment plants. Biological fixed-film reactors. British Standards Institution, London

Chao Y, Mao Y, Yu K, Zhang T (2016) Novel nitrifiers and comammox in a full-scale hybrid biofilm and activated sludge reactor revealed by metagenomic approach. Appl Microbiol Biotechnol 100(18):8225–8237. doi:10.1007/s00253-016-7655-9

Clark K, Chantigny MH, Angers DA, Rochette P, Parent L-E (2009) Nitrogen transformations in cold and frozen agricultural soils following organic amendments. Soil Biol Biochem 41 (2):348–356. doi:10.1016/j.soilbio.2008.11.009

Costa E, Perez J, Kreft JU (2006) Why is metabolic labour divided in nitrification? Trends Microbiol 14(5):213–219. doi:10.1016/j.tim.2006.03.006

Daims H, Lebedeva EV, Pjevac P, Han P, Herbold C, Albertsen M, Wagner M (2015) Complete nitrification by nitrospira bacteria. Nature 528(7583):504–509. http://www.nature.com/nature/journal/v528/n7583/abs/nature16461.html#supplementary-information. doi:10.1038/nature16461

Dempsey MJ (2001) Nitrification process. EU Patent EP 1,129,037

Dempsey MJ (2004) EU Patent No. EP 1,444,167

Dempsey MJ (2011) Improvements in and relating to fluid bed expansion and fluidisation. US Patent US 7,309,433

Dempsey MJ, Porto I, Mustafa M, Rowan AK, Brown A, Head IM (2006) The expanded bed biofilter: combined nitrification, solids destruction, and removal of bacteria. Water Sci Technol 54(8):37–46

European Council (1991) Urban wastewater treatment directive. Council Directive 91/271/EEC, L135, May 30, 1991

European Council (2000) Directive 2000/60/ec of the European parliament and of the council establishing a framework for the community action in the field of water policy, December 22, 2000

Gerrity S, Clifford E, Kennelly C, Collins G (2016) Ammonia oxidizing bacteria and archaea in horizontal flow biofilm reactors treating ammonia-contaminated air at 10°C. J Ind Microbiol Biotechnol 43(5):651–661. doi:10.1007/s10295-016-1740-z

Jusselme M-D, Saccone P, Zinger L, Faure M, Le Roux X, Guillaumaud N, Poly F (2016) Variations in snow depth modify n-related soil microbial abundances and functioning during winter in subalpine grassland. Soil Biol Biochem 92:27–37. doi:10.1016/j.soilbio.2015.09.013

Konneke M, Bernhard AE, de la Torre JR, Walker CB, Waterbury JB, Stahl DA (2005) Isolation of an autotrophic ammonia-oxidizing marine archaeon. Nature 437(7058):543–546. http://www.nature.com/nature/journal/v437/n7058/suppinfo/nature03911_S1.html

Lau FTK (2011) Application of hybrid reactor to maximize treatment capacity and nitrification with short sludge-age operation. http://www.dsd.gov.hk/EN/Files/Technical_Manual/RnD_reports/RD2038.pdf

Maixner F, Noguera DR, Anneser B, Stoecker K, Wegl G, Wagner M, Daims H (2006) Nitrite concentration influences the population structure of nitrospira-like bacteria. Environ Microbiol 8(8):1487–1495. doi:10.1111/j.1462-2920.2006.01033.x

McKeown RM, Scully C, Enright AM, Chinalia FA, Lee C, Mahony T, O'Flaherty V (2009) Psychrophilic methanogenic community development during long-term cultivation of anaerobic granular biofilms. ISME J 3(11):1231–1242. doi:10.1038/ismej.2009.67

Metcalf & Eddy Inc, Tchobanoglous G, Stensel HD, Tsuchihashi R, Burton F (2014) Wastewater engineering: treatment and resource recovery, 5th edn. WCB/McGraw-Hill, New York, NY

Murdoch University Environmental Technology Centre (2000) Environmentally sound technologies in wastewater treatment for the implementation of the unep global programme of action (gpa) "guidance on municipal wastewater". http://www.unep.or.jp/ietc/publications/freshwater/sb_summary/index.asp

Niu J, Kasuga I, Kurisu F, Furumai H, Shigeeda T, Takahashi K (2016) Abundance and diversity of ammonia-oxidizing archaea and bacteria on granular activated carbon and their fates during drinking water purification process. Appl Microbiol Biotechnol 100(2):729–742. doi:10.1007/s00253-015-6969-3

Park HD, Wells GF, Bae H, Criddle CS, Francis CA (2006) Occurrence of ammonia-oxidizing archaea in wastewater treatment plant bioreactors. Appl Microbiol Biotechnol 72 (8):5643–5647. doi:10.1128/aem.00402-06

Prosser JI (ed) (1986) Nitrification, vol 20. Published for the Society for General Microbiology by IRL, Oxford

Sauder LA, Peterse F, Schouten S, Neufeld JD (2012) Low-ammonia niche of ammonia-oxidizing archaea in rotating biological contactors of a municipal wastewater treatment plant. Environ Microbiol 14(9):2589–2600. doi:10.1111/j.1462-2920.2012.02786.x

Telling J, Anesio AM, Tranter M, Fountain AG, Nylen T, Hawkings J, Wadham JL (2014) Spring thaw ionic pulses boost nutrient availability and microbial growth in entombed antarctic dry valley cryoconite holes. Front Microbiol 5:694. doi:10.3389/fmicb.2014.00694

van Kessel MAHJ, Speth DR, Albertsen M, Nielsen PH, Op den Camp HJM, Kartal B, Lücker S (2015) Complete nitrification by a single microorganism. Nature 528(7583):555–559. doi:10.1038/nature16459

Wett B, Jimenez JA, Takacs I, Murthy S, Bratby JR, Holm NC, Ronner-Holm SGE (2011) Models for nitrification process design: one or two aob populations? Water Sci Technol 64(3):568–578. doi:10.2166/wst.2011.499

Wu J, He CD, van Loosdrecht MCM, Perez J (2016) Selection of ammonium oxidizing bacteria, (aob) over nitrite oxidizing bacteria (nob) based on conversion rates. Chem Eng J 304:953–961. doi:10.1016/j.cej.2016.07.019

Chapter 27
Biodegradation of Petroleum Oil in Cold Marine Environments

Odd Gunnar Brakstad, Synnøve Lofthus, Deni Ribicic, and Roman Netzer

Abstract The cold regions of the Earth are exposed to petroleum oil exploration, production, and transport, with risk of oil spills. Biodegradation is an essential petroleum weathering process and may remove discharged petroleum compounds completely by mineralization processes. These processes are most apparent for soluble compounds and with dispersed oil. Surface and subsurface spills will generate different situations, and in addition freezing of oil in marine ice may transport the oil over large distances. A variety of marine psychrophilic or psychrotolerant bacteria from both shallow and deepwater environments have been reported to degrade hydrocarbons in seawater or marine sediments, most of these affiliated within the phyla Proteobacteria and Bacteroidetes. Several of these may also act on hydrocarbons in sea ice, and active bacterial respiration in sea ice has been shown down to temperatures of −20 °C. The cold environments require several microbial survival and catabolism strategies, including productions of exopolysaccharides, cold-active enzymes, cold-shock, cold-acclimation and antifreeze proteins, as well as adjusting their membrane lipid composition. Oil biodegradation in cold environments is well documented by laboratory and field studies, and even oil frozen in marine ice will stimulate bacterial metabolism. Flocculation processes have also been associated with oil biodegradation, raising discussions on the fate of the oil, especially after the Deepwater Horizon blowout. Bioremediation in cold marine environments has been investigated as a labor-effective technology which generates no harmful by-products, mainly by adding fertilizers to stimulate the oil biodegradation by the indigenous bacteria (biostimulation), but also inoculation of exogenic hydrocarbonoclastic cultures (bioaugmentation) has been suggested.

O.G. Brakstad (✉) • R. Netzer
SINTEF Ocean, Marine Environmental Technology, Brattørkaia 17C, 7465 Trondheim, Norway
e-mail: odd.g.brakstad@sintef.no

S. Lofthus
Department of Biotechnology, NTNU, Sem Sælandsvei 6/8, 7491 Trondheim, Norway

D. Ribicic
Department of Cancer Research and Molecular Medicine, NTNU, Prinsesse Kristinasgt. 1, 7491 Trondheim, Norway

R. Margesin (ed.), *Psychrophiles: From Biodiversity to Biotechnology*,
DOI 10.1007/978-3-319-57057-0_27

Contents

27.1 Introduction

The cold regions represent some of the largest biotopes on the Earth, including the Arctic, Antarctic, and the permanently cold deep oceans. Oil exploration and production activities are already going on in the North American and European Arctic, and in several countries technologies are developed for offshore oil production at water depths of more than 3000 m. During the *Deepwater Horizon* (DWH) blowout in 2010 in the Gulf of Mexico (GoM), parts of the approximately 4.9 million barrels (780 million liters) of the discharged crude oil, remained as small droplets in the deep ocean (900–1300 m depth), and at water temperatures of 4–6 °C (Camilli et al. 2010; Reddy et al. 2012).

Biodegradation is an essential weathering process and important for understanding the fates and risks associated with oil discharges to marine environments. Hydrocarbon biodegradation has been the topic of several reviews (e.g., Atlas 1981, 1995; Harayama et al. 1999, 2004; Van Hamme et al. 2003; Venosa and Zhu 2003; Prince 2005, 2008; Atlas and Hazen 2011). Biodegradation of hydrocarbons in cold environments have been reviewed only by a few authors (e.g., Brakstad 2008; Margesin and Schinner 1999, 2001; Si-Zhong et al. 2009). This review will focus on the interactions between discharged oil and marine microorganisms in cold marine environments, including the use of bioremediation actions. Most of the novel data have been generated as results of field and laboratory studies of the DWH deepwater oil and gas plume, as reflected in this review.

27.2 Oil Discharges in Cold Seawater

Hydrocarbons are released to marine recipients from a variety of natural and anthropogenic sources. Natural seeps provide a continuous source of crude oil and are widely distributed in the oceans. For instance, the hydrocarbon seepage flux in the GoM region has been calculated to range between 2.2 and 18 $\times$ 10^7 tons per year, corresponding to 14–120% of the DWH discharge rate (Smith et al. 2014). Several biogenically produced hydrocarbons from marine phytoplankton and prokaryotes also have chemical structures equal to petroleum hydrocarbons. Marine microorganisms are therefore constantly exposed to hydrocarbons, and hydrocarbonoclastic prokaryotes are present in all marine systems investigated so far.

In Arctic and Antarctic regions, oil pollutions may be transported to ice-free shorelines or to the ice margins by prevailing seawater currents. In the latter case, the oil may become frozen in the ice and drift along with the ice. Investigations of the fate of stranded oil and oil frozen in sea ice are therefore important as part of the risk assessment and for the development of operational cleanup strategies, including bioremediation.

27.2.1 Oil Characteristics and Weathering

The composition of petroleum oil is extremely complex, containing thousands of different compounds, many of which never have been characterized. Crude oils contain both water-soluble and oleophilic compounds, of which the hydrocarbons are best characterized. Oil compounds are basically separated into linear or cyclic alkanes, aromatic hydrocarbons, asphalthenes, and resins, and the distributions of these compound groups vary considerably within petroleum oils. Petroleum oil also contains a variety of poorly characterized compounds containing nitrogen, sulfur, and oxygen, often referred to as "unresolved complex mixtures" (UCM), consisting of hundreds of thousands of individual compounds, many of which are recalcitrant to biodegradation (Gough and Rowland 1990; Killops and Al-Juboori 1990; Watson et al. 2002; Sutton et al. 2005).

27.2.2 Surface Oil Spills

Surface oil spills will result in several weathering processes changing the physical–chemical characteristics of the oil, which are important for the biodegradation of oil compounds. These processes are also influenced by seawater conditions like temperature, waves, and currents. Oil weathering processes include surface spreading of the oil, evaporation of volatile compounds, water-in-oil emulsions, oil-in-water dispersions, dissolution of small and charged (polar) compounds, and photo-oxidation. Also biodegradation is considered to be a weathering process (Brandvik 1997). A surface oil slick may spread by current and wind. A thin oil film (<1 μm thickness) will cover most of the surface area, while most of the oil (up to 80–90%) will be present in the slick area (Brandvik 1997). Immediate surface evaporation processes result in losses of small C_5–C_{10} alkanes and monoaromatic compounds like BTEX (benzenes, toluenes, ethylbenzenes, xylenes). Remaining compounds in the oil film may rapidly be biodegraded (Brakstad et al. 2004; Brakstad and Bonaunet 2006).

In seawater the microbes degrade soluble oil compounds and act on the oil–water interfaces of small oil droplets. Laboratory studies have shown that biodegradation of oleophilic and soluble hydrocarbons at low temperatures is considerable (Brakstad and Bonaunet 2006; Bagi et al. 2014; McFarlin et al. 2014b). Hydrocarbon evaporation is reduced in cold seawater, particularly in ice-covered waters, since ice floes will replace the oil surface. The reduced evaporation results in temporarily higher concentrations of volatile toxic compounds (e.g., BTEX) in the seawater. At high concentrations, these compounds may result in prolonged microbial lag phases and delayed onset of biodegradation due to their acute toxicity (Atlas and Bartha 1972; Hokstad et al. 1999). For wax-rich oils with high pour points, evaporation, dilution, and dispersion may be reduced in cold seawater since precipitated wax may build a matrix which limits the internal mixing of the oil and act as a diffusion barrier between the oil and the water.

The behavior of oils in freezing environments includes spreading in ice, snow, under ice, and in water with ice present. Once trapped in the ice, ocean currents can transport the oil over large distances. Oil trapped under the ice tends to move upward within the ice by density-mediated migration, even at temperatures below −15 °C, and appears essentially un-weathered at the ice surface in the summer when the ice starts to melt (Chen et al. 1974; Payne and McNabb 1991; Payne et al. 1991). Water-soluble hydrocarbons released from the ice may be transported through the brine channels (Faksness and Brandvik 2008b), thereby coming in contact with sea ice microbes in a fluid medium. The oil escapes the ice in the spring as the ice deteriorates by two general processes: (1) vertical rise of the oil through the brine channels in the ice and (2) ablation of the ice surface down to the oil lens in the ice (Fingas and Hollebone 2003).

27.2.3 Sedimentation Processes

Few oils are dense enough to sink in seawater (NRC 1985), but sinking of oil/oil residues may be caused by oil–mineral aggregation (OMA) or by incorporation of oil in aggregates of zooplankton feces or mucoid particles of phytoplankton and bacteria (Lee et al. 1996; Hazen et al. 2010; Bælum et al. 2012; Passow et al. 2012). These processes may result in dense particles with sinking properties in the marine water column. OMA is primarily a near-shore process, typically in surf zones, near river outflow, melting glaciers, and sea ice (Daly et al. 2016). Aggregates of oil and bacteria were reported in the deepwater oil plume during the DWH oil spill (Hazen et al. 2010), and it was suggested that these types of aggregates resulted in fallout of oil from the plume to the GoM seabed (Valentine et al. 2014). Sea-bed studies documented sedimented oil residues in the deepwater sediments associated with the oil spill (Stout and Payne 2016). Flocculation processes related to biodegradation of oil in cold environments will be discussed later in this chapter. Oil residues sinking to the seabed will be subject to aerobic or anaerobic biodegradation in the sediments. Biological processes like bioturbation are important for venting and oxygenation of the sediments, increasing the biodegradation.

27.2.4 Deepwater Oil Spills

A deepwater oil spill caused by a subsea blowout will result in a rising hydrocarbon plume of dispersed oil droplets and gas bubbles. The plume will entrain dense water and, eventually, the suspension will loose its buoyancy (Johansen et al. 2003). After subsurface discharges, high pressures only have limited influence on dissolution rates and dispersion (Sawamura et al. 2001), and biodegradation becomes increasingly important for depletion of toxic volatile compounds. The majority of the oil in most deepwater releases will rise to the surface within several hours. The surface slick formed once the oil reaches the surface will be thinner than that seen during a shallow-water release or a surface release. This is due in part to the fractionation of oil droplets that results in a staged arrival of the oil at the surface and in part due to diffusion or dispersion of the oil as it rises (NRC 2003). Thin oil films will subsequently be biodegraded after re-dispersion by wave actions or due to the low film thickness.

During the DWH blowout, nearly all the gas and parts of the oil remained as a plume in the deep ocean at depths of 900–1300 m and at water temperatures of 4–6 °C (Camilli et al. 2010; Reddy et al. 2012). The chemical dispersant Corexit 9500A was injected at the wellhead in early May 2010 to reduce oil surfacing with subsequent stranding along the Gulf shorelines. The use of the dispersant resulted in a deepwater plume with estimated oil droplet size of 10–50 μm in diameter which were picked up by the deepwater currents, while larger droplets <90 μm rapidly rose to the surface (Atlas and Hazen 2011; North et al. 2011). Measurements of polyaromatic hydrocarbons (PAHs) in water samples during the release showed rapid decrease with distance from the release point, being <1 ppb within 15–20 miles (24–32 km) from the release source, and the decline was partly associated with biodegradation (Atlas and Hazen 2011).

27.3 Hydrocarbon-Degrading Microorganisms

During the last decades nearly 200 bacterial, cyanobacterial, fungal, algal, or diatom genera have been shown to degrade hydrocarbons (Prince 2005). Hydrocarbon biodegradation in cold environments is mostly associated with hydrocarbonoclastic bacteria. Although hydrocarbon-degrading archaea have been reported from extreme environments, like hypersaline and high-temperature environments (Le Borgne et al. 2008; Khelifi et al. 2010), these microbes seem to be insignificant players during oil biodegradation in most cold environments.

27.3.1 Hydrocarbon Degraders in Cold Marine Environments

27.3.1.1 Arctic and Antarctic Seawater, Sediments, and Ice

Microorganisms inhabiting cold marine environments have specific adaptions making them able to metabolize at low temperatures. Characterization of microbial communities in Arctic and Antarctic seawater and ice by high-throughput methods has shown high abundances of bacteria affiliated to Alphaproteobacteria, Betaproteobacteria, Gammaproteobacteria, and Bacteroidetes (mostly Flavobacteriia) (Kirchman et al. 2010; Hatam et al. 2014; Luria et al. 2014). The changes in community composition in polar seawaters are influenced mainly by water column vertical stratification (Galand et al. 2009; Comeau et al. 2011).

Although data from previous studies showed that most bacterial and archaeal phylotypes are common between Arctic and Antarctic seawater and sea ice (Brinkmeyer et al. 2003; Bano et al. 2004), recent investigations have suggested that sea ice assemblages may differ from the assemblages in the seawater underneath the ice, the sea ice having a higher dominance of Flavobacteriia compared to seawater, where Alphaproteobacteria are abundant (Hatam et al. 2014; Boetius et al. 2015). The archaea in Arctic and Antarctic sea ice and seawater are mainly associated with the ammonia-oxidizing Thaumarchaeota and with some Eurarchaeota (Cowie et al. 2011; Luria et al. 2014). Although first-year ice and multi-year ice are dominated by the same bacterial classes, the first-year ice has higher diversity than multi-year ice (Hatam et al. 2016).

Microbial analysis of polar sediments has shown that community compositions depend on the type of substrate transported from the water column to the sediments, sometimes with higher diversity in lower than in upper sediment layers (Teske et al. 2011; Bienhold et al. 2012; Ruff et al. 2014; Learman et al. 2016).

Several of the bacterial genera described in polar seawater and marine ice include members with the ability to degrade petroleum hydrocarbons. Some specialized obligate psychrophilic hydrocarbonoclastic bacteria have been isolated, for instance *Oleispira antarctica* (Yakimov et al. 2003). Studies of psychrophilic and psychrotolerant marine bacteria have indicated that hydrocarbon degradation is mainly associated with Gammaproteobacteria, as shown in Table 27.1 (Bowman and McCuaig 2003; Yakimov et al. 2004; Deppe et al. 2005; Gerdes et al. 2005; Brakstad and Bonaunet 2006; Brakstad et al. 2008; Bagi et al. 2014; McFarlin et al. 2014a; Dong et al. 2015; Lofthus et al. 2015; Garneau et al. 2016). Other hydrocarbon-degrading bacterial genera in Arctic seawater and ice include

Table 27.1 Taxonomy of Arctic or Antarctic oil-degrading bacteria

Class	Family	Genus	Source[a]	References[b]
Alphaproteobacteria	Rhodobacteraceae	*Loktanella*	Ar, SW	1
		Sulfitobacter	Ar, SW	2
	Sphingomonadaceae	*Sphingopyxis*	Ar, SW	1
		Sphingomonas	An, SW	3
Gammaproteobacteria	Alteromonadaceae	*Alteromonas*	SW	4
		Glaciecola	Ar, SI	5
		Marinobacter	An, Ar, SI, SW	3, 6, 7
	Colwelliaceae	*Colwellia*	An, Ar, S, SI, SW	1, 2, 3, 4, 5, 8, 9, 10
		Thalassomonas	SW	4
	Moritellaceae	*Moritella*	Ar, S, SI, SW	2, 8, 9
	Pseudoalteromonadaceae	*Algicola*	Ar, SI	2
		Pseudoalteromonas	An, Ar, S, SI, SW	1, 5, 6, 8, 9, 10, 11
	Psychromonadaceae	*Psychromonas*	Ar, SW	1, 11
	Shewanellaceae	*Shewanella*	An, Ar, S, SI, SW	3, 4, 6, 7, 9, 10
	Alcanivoracaceae	*Alcanivorax*	Ar, S, SW	2, 10
	Oceanospirillaceae	*Marinomonas*	An, Ar, S, SI, SW	3, 5, 10
		Oleispira	Ar, An, SI, SW	1, 2, 3, 4, 5, 11
	Halomonadaceae	*Halomonas*	An, Ar, S, SI SW	3, 7, 10
	Moraxellaceae	*Psychrobacter*	Ar, SW	1, 6
	Pseudomonadaceae	*Pseudomonas*	An, Ar, S, SI, SW	3, 6, 7, 10
	Piscirickettsiaceae	*Cycloclasticus*	Ar, S, SW	4, 10
Epsilonproteobacteria	Campylobacteraceae	*Arcobacter*	An, Ar, SW	3, 8, 11
Bacteroidetes	Cytophagales	*Cytophagia*	An, SW	3, 11
Flavobacteriia	Flavobacteriaceae	*Ulvibacter*	Ar, SW	1
		Polaribacter	Ar, SI, SW	1, 7, 8

(continued)

Table 27.1 (continued)

Class	Family	Genus	Source[a]	References[b]
Actinobacteria	Nocardiaceae	*Rhodococcus*	An, SW	3, 12
	Microbacteriaceae	*Agreia*	Ar, SI, SW	6, 7
		Arthrobacter	An, SW	12

[a]*An* Antarctic, *Ar* Arctic, *S* sediment, *SI* sea ice, *SW* seawater

[b]1, McFarlin et al. (2014a); 2, Garneau et al. (2016); 3, Yakimov et al. (2004); 4, Lofthus et al. (2015); 5, Brakstad et al. (2008); 6, Deppe et al. (2005); 7, Gerdes et al. (2005); 8, Bagi et al. (2014); 9, Bowman and McCuaig (2003); 10, Dong et al. (2015); 11, Brakstad and Bonaunet (2006); 12, Michaud et al. (2004)

members of Alphaproteobacteria, Epsilonproteobacteria, Actinobacteria, and Bacteroidetes (Table 27.1) (Michaud et al. 2004; Yakimov et al. 2004; Deppe et al. 2005; Gerdes et al. 2005; Brakstad and Bonaunet 2006; Bagi et al. 2014; McFarlin et al. 2014a; Garneau et al. 2016). Arctic marine sediments mainly include bacteria associated with oil biodegradation, mainly Gammaproteobacteria (Table 27.1).

Previous and recent studies have shown how the microbial community dynamics changes during biodegradation of oil in cold seawater and ice. A few hydrocarbonoclastic bacteria dominate initially in the degradation, metabolizing bioavailable *n*-alkanes and volatile compounds, including Alteromonadales and Oceanospirillales (e.g., *Oleiphilus* and *Oleispira*) (Yakimov et al. 2003; Head et al. 2006; Hazen et al. 2010; Dubinsky et al. 2013; Lofthus et al. 2015; Ribicic et al. 2015). The alkane degraders are succeeded by more diverse consortia of bacteria degrading PAHs and more complex hydrocarbons, including members of the families Piscirickettsiaceae (e.g., *Cycloclasticus*), Alteromonadaceae (e.g., *Marinobacter*), Pseudoalteromonadaceae, and Shewanellaceae (Dubinsky et al. 2013; Brakstad et al. 2015a; Ribicic et al. 2015). In addition, some bacteria have complete pathways for both aliphatic and aromatic hydrocarbon degradation, like members of the genus *Colwellia*, known to be abundant in cold environments (Brakstad et al. 2004; Methé et al. 2005; Dubinsky et al. 2013; Mason et al. 2014).

27.3.1.2 Deepwater

Until recently, only a few studies have described hydrocarbon-degrading microorganisms in deepwater. Already in 1974, microorganisms collected from sediments at a depth of 4940 m in the Atlantic were reported to be capable of utilizing hydrocarbons (hexadecane) under both ambient and *in situ* pressures (Schwarz et al. 1974). Several organic-solvent-tolerant bacterial strains isolated from deep-sea seabed environments (1100–2000 m depth) were capable of degrading *n*-alkanes (*n*C7–*n*C16) from crude oil, PAHs, or cholesterol (Moriya and Horikoshi 1993). In later studies of sediments from the deep-sea Pacific (2682 m), consortia of 72 bacteria belonging to 22 genera, including *Flavobacterium*, *Cycloclasticus*, *Novosphingobium*, *Halomonas*, *Achromobacter*, *Roseovarius*, and *Alcanivorax*, proved to degrade several compounds of PAHs, including naphthalene, phenanthrene, and pyrene. Interestingly, an isolate of *Cycloclasticus* could utilize pyrene as the sole carbon source through a pathway that had not been described before (Wang et al. 2008). These researchers also reported two hydrocarbon-degrading consortia sampled from the Mid-Atlantic ridge at 2.2 m under the bottom surface, at a water depth of 3542 m. These consortia exhibited degradation potential of crude oil and PAHs (Cui et al. 2008). Isolates included *Pseudoalteromonas*, *Halomonas*, *Marinobacter*, *Thalassospira*, and *Tistrella*, while genera detected by polymerase chain reaction denaturing gradient gel electrophoresis (PCR-DGGE) were closely related to *Cycloclasticus*, *Alteromonas*, *Thalassospira*, *Alcanivorax*, and

Rhodospirillaceae. *Cycloclasticus* was identified as a key player in degradation of phenanthrene, while *Alteromonas* played a significant role in degradation of naphthalene (Cui et al. 2008).

Investigations of natural oil-seep sediments at moderate or deep-sea depths (300–1500 m) on the northwest shelf of Australia showed that 74% of isolated strains displayed preferential growth on hydrocarbon-selective media (Johnson and Hill 2003). Further studies from cold hydrocarbon seeps in deep-sea sediments (Japan trench, depth 5300 m) showed dominances of Deltaproteobacteria and Epsilonproteobacteria, the Deltaproteobacteria mostly affiliated to the sulfate-reducing prokaryotes of the genus *Desulfosarcina* (Inagaki et al. 2002).

The DWH oil spill and the rapid advances of Next Generation Sequencing (NGS) techniques offered a great opportunity to study deep-sea hydrocarbon degraders at work *in situ* exhaustively (Hazen et al. 2010; Valentine et al. 2010; Bælum et al. 2012; Lu et al. 2012; Dubinsky et al. 2013; Mason et al. 2014; King et al. 2015). The deep sea plume included small oil droplets, dissolved oil compounds, and gases, and deep sea microbes were suggested to be involved in the metabolization of both oil and gas. Throughout different studies related to the DWH oil spill, the same taxa were accentuated as the key players in biodegradation of oil compounds from the deep sea plume, most of them belonging to the Gammaproteobacteria class. It was suggested that respiration of gas compounds in the plume (mainly propane) jump-started oil biodegradation (Valentine et al. 2010). The cold environment of the deep plume (700–1300 m) was dominated by *Cycloclasticus*, *Colwellia*, and members of Oceanospirillaceae, but these were not abundant in the surface oil slicks from DWH (Hazen et al. 2010; Valentine et al. 2010; Redmond and Valentine 2012). Studies of bacterial successions in the deep sea plume showed that the early communities where dominated by *n*-alkane and cycloalkane degraders like Oceanospirillaceae and *Pseudomonas*. As the fraction of aromatic hydrocarbons increased, *Colwellia*, *Cycloclasticus*, and *Pseudoalteromonas* increased in abundance (Dubinsky et al. 2013). The gas compounds included primarily methane, ethane, and propane dissolved in the plume. Methane was the most abundant hydrocarbon released during the DWH spill (Kessler et al. 2011), and primary methane oxidation was associated by abundancies of Methylococcaceae, followed by secondary consumers of C1-compounds, including *Methylophaga* and Methylophilaceae (Redmond and Valentine 2012; Dubinsky et al. 2013). It was claimed that these methylotrophs efficiently respired most of the released methane within 120 days after the onset of the spill (Kessler et al. 2011). However, methane respiration was slower than microbial propane and ethane oxidation, as a result of the difference in initial cell abundance and growth rates of the oxidizing microorganisms (Kessler et al. 2011).

Another aspect related to the deep sea is the pressure and how this will affect biodegradation of oil. Hydrocarbon-degrading strains may react differently to increased pressure (Schedler et al. 2014), but biodegradation studies with deep sea samples at atmospheric pressure showed similar *n*-alkane biodegradation half-lives as laboratory studies with seawater collected from surface-near depth (Hazen et al. 2010). Typically, the abundant microbes in the successions of hydrocarbon-degrading bacteria from the DWH deep sea plume followed patterns similar to observations from biodegradation studies in cold shallow sea water (Dubinsky et al. 2013; Brakstad et al. 2015a; Wang et al. 2016). However, the impacts of pressure on oil biodegradation will need further considerations.

27.3.2 Hydrocarbon Metabolism at Low Temperatures

27.3.2.1 Aerobic Degradation

Aerobic hydrocarbon biodegradation can be divided into degradation of alkanes and aromatic hydrocarbons. Cell uptake is the initial step in the catabolic cascade of *n*-alkane degradation. While small water-soluble alkanes are transported directly into the cell, medium- and long-chain alkanes may be accessible for microorganisms by adhering to hydrocarbon droplets and using mechanisms based on surface-active compounds (biosurfactants) (Mohanty and Mukherji 2008; Rojo 2009). Biosurfactants have been reported to enhance direct interfacial uptake and metabolization of *n*-alkanes, but the detailed mechanistic processes are still not fully understood (Mohanty and Mukherji 2008). For *Pseudomonas putida*, it is postulated that the *alkL* gene is involved in the import of *n*-alkanes into the cell (Hearn et al. 2009). Furthermore, it is suggested that long-chain fatty acid transporter proteins (FadL) are participating in the transportation of long-chain hydrocarbons in many bacteria (van den Berg 2005).

Aerobic degradation of *n*-alkanes is typically initiated by terminal and subterminal oxidation catalyzed by monooxygenases, resulting in primary or secondary alcohols, respectively. Further degradation results in conversion to fatty acids from aldehydes by terminal oxidation or ketones and esters by subterminal oxidation (Kotani et al. 2007; Wentzel et al. 2007).

While methanotrophs oxidize methane by means of a methane monooxygenase, short-chain alkane (C2–C4) degrading microorganisms use monoxygenases similar to the methane monooxygenases (Dubbels et al. 2007). Medium-chain alkane (C5–C17) degraders typically possess integral membrane non-heme iron monooxygenases with high similarity to the well-characterized *Pseudomonas putida* GPo1 (formerly *P. oleovorans*) alkane hydroxylase encoded by *alkB* (Rojo 2009; Austin and Groves 2011). In *P. putida* GPo1, the genes involved in catabolization of *n*-alkanes include two operons localized on the OCT plasmid, *alkBFGHJKL* and *alkST*, and two genes located on the chromosome, *alkA* and *aldA* (Grund et al. 1975; van Beilen et al. 1994). The initial step in degradation of medium chain length alkanes to alcohols is catalyzed by a trimer alkane hydroxylase system, including AlkB (alkane hydroxylase), AlkG (rubredoxin), and AlkT (rubredoxin reductase), followed by dehydrogenation to corresponding aldehydes by the alcohol dehydrogenases AlkJ and AlcA, and the conversion to fatty acids catalyzed by the aldehyde dehydrogenases AlkH and AldA. The gene product of *alkS* is a positive regulator for the *alkBFGHJKL* operon in the presence of *n*-alkanes, while the *alkF* gene encodes a nonfunctional rubredoxine. The *alkL* gene product is postulated to play an important role in alkane uptake in *P. putida*. Several bacteria with the ability to degrade C5–C10 alkanes, such as *Acinetobacter*, also contain alkane hydroxylases that belong to a distinct family of soluble cytochrome P450 monooxygenases (Maier et al. 2001). Alkane hydroxylases involved in long-chain length alkane (>C18) degradation are distinct from the enzymes described above. One such long-chain alkane hydroxylase, AlmA, is an alkane monooxygenase present in many bacteria, such as *Acinetobacter* and *Alcanivorax*. A second hydroxylase is LadA, which is a thermophilic soluble alkane monooxygenase from *Geobacillus thermodenitrificans* and capable of C15–C36 alkane hydroxylation (Feng et al. 2007; Wentzel et al. 2007).

Cyclo-aliphatic compounds can be degraded by a large range of bacteria, including *Acinetobacter* and *Rhodococcus* (Kostichka et al. 2001; Cheng et al. 2002). These bacteria use a set of chromosomally encoded enzymes containing monooxygenases, dehydrogenases, and hydrolases. Even though the different bacteria use the similar enzymes, the corresponding gene orientation may differ significantly.

PAHs have been reported to be biodegraded by many bacteria and by several metabolic pathways (Mallick et al. 2011). Aerobic catabolism is typically initiated by a dioxygenase-mediated hydroxylation of the aromatic ring into a *cis*-dihydrodiol that is re-aromatized to a diol intermediate by a dehydrogenase reaction. Subsequent ring cleavage by a multicomponent dioxygenase, generally including reductase, ferredoxin, and oxygenase subunits, results in intermediates which finally are converted to TCA cycle intermediates (Mallick et al. 2011). Bacteria can also degrade PAHs via the cytochrome P450-mediated pathway, with the production of *trans*-dihydrodiols (Moody et al. 2004). Biochemical pathways of PAH degradation have mainly been investigated for naphthalene and phenanthrene (Lu et al. 2011). In *P. putida* G7, naphthalene degradation genes (*nah*) are located on the NAH7 plasmid, organized in two operons. The genes encoding the enzymes for the conversion of naphthalene to salicylate are located in the *nal* operon (*nahAaAbAcAdBFCED*), and enzymes for further conversion of salicylate to pyruvate and acetaldehyde are encoded by genes located in the *sal* operon (*nahGTHINLOMKJY*) (Simon et al. 1993; Peng et al. 2008). The operons are under the control of the positive NahR regulator, induced by salicylate. The *nah* genes have been shown to be more than 90% identical in different *Pseudomonas* strains (Peng et al. 2008), and similar naphthalene operons are reported for *Ralstonia* spp. (*nag*) and *Comamonas* ssp. (Goyal and Zylstra 1997; Zhou et al. 2002).

Hydrocarbon degradation pathway genes have been used to evaluate the degradation potential of isolates from contaminated sites and also to assess the catabolic capacity of microbial communities (Bordenave et al. 2008; Païssé et al. 2010, 2011; Zhang et al. 2010). AlkB alkane hydroxylase systems, first described in *P. putida* GPo1 (Kok et al. 1989; van Beilen et al. 1994), are highly widespread in nature with more than 250 AlkB homologues identified in at least 45 bacterial species (Hazen et al. 2016). The RHD (ring-hydroxylating dioxygenase) genes, involved in PAH catabolism (PAH-RHD), have also been successfully used as markers to describe petroleum degradation capacities in different environments (Bordenave et al. 2008; Païssé et al. 2012).

Hydrocarbon-catabolizing bacteria are ubiquitously abundant, also in cold and pristine environments such as the Arctic and the Antarctic. Most studies on hydrocarbon biodegradation in cold environments have been performed on oil-contaminated soil in Arctic, Alpine, and Antarctic areas (Margesin and Schinner 2001; Margesin and Miteva 2011). Two strains of *Pseudomonas* sp. isolated from Arctic soils degraded *n*C5 to *n*C12-alkanes, toluene, and naphthalene at both 5 °C and 25 °C and possessed both the *alk* and *nah* catabolic pathways (Whyte et al. 1997). The psychrotrophic *Rhodococcus* sp. strain Q15 metabolized a broad degree of C10 to C21 alkanes, branched alkanes, and a substituted cycloalkane present in diesel fuel at 5 °C, and the strain contained at least four alkane monooxygenase gene homologues (*albB1*, *alkB2*, *alkB3*, and alkB4) (Whyte et al. 1998). It is assumed that the explanation for the presence of the four monooxygenases in one strain is that each alkane monooxygenase is specific for a certain range of alkanes (Whyte et al. 2002a). Colony-hybridization methods showed that *alkB* genes were more abundant in cold-adapted cultures than in mesophilic cultures from the same origin, while *alkM* genes were less abundant (Whyte et al. 2002b).

A variety of genes involved in hydrocarbon biodegradation have been examined in hydrocarbon-polluted and pristine soils from the Arctic and the Antarctic soil environments. Genes for alkane monooxygenase (*alkB*), naphthalene dioxygenase (*ndoB*), toluene dioxygenase (*todC1*), catechol-2,3-dioxygenase (*xylE* and *cat23*), and biphenyl dioxygenase (*bphA*) were detected in Antarctic petroleum-contaminated and control soils and at higher frequencies than in polluted soils from Brazil (Luz et al. 2004). Genes encoding alkane monooxygenase (*alkB*) and aromatic ring cleavage (PAH-RHD) examined in Arctic Canadian diesel-contaminated soil during a bioremediation action showed decrease in *alkB*/PAH-RDH ratios over periods of 1–4 years indicating a shift from degradation of easily degradable alkanes to more complex PAH compounds (Yergeau et al. 2012). Similar observations were also described from bioremediation actions in hydrocarbon-contaminated Antarctic soil (Powell et al. 2006).

27.3.2.2 Anaerobic Degradation

For anaerobic sulfate- and nitrate-reducing bacteria, shorter chain length alkanes are more recalcitrant than mid- to long-chain length alkanes. Short-length hydrocarbons (up to C17) can accumulate in anaerobic environments due to lack of evaporation and can exert toxic effects on microorganisms resulting in impaired degradation. Furthermore, degradation of branched alkanes, such as pristane and phytane, has been shown to be more efficient than for normal alkanes in sulfate-reducing bacteria (Hasinger et al. 2012). Anaerobic degradation of petroleum hydrocarbons contributes to degradation in marine low-temperature anoxic environments, and low-temperature degradation of PAH-compounds have been reported with enrichment cultures from Arctic soils under anoxic and nitrate-reducing conditions at 7 °C (Eriksson et al. 2003).

Different catabolic pathways have been described for anaerobic hydrocarbon degradation. Carbonic group addition has been reported in sulfate-, nitrate-, iron-reducing bacteria; methanogenic consortia; and anaerobic phototrophs. This pathway type is characterized by the addition of fumarate (or other carbonic groups) to a carbon atom of hydrocarbons. Initially, fumarate is added to a hydrocarbon (e.g. toluene) by a benzylsuccinate synthase (encoded by *bbsABC*), resulting in an unstable benzyl radical intermediate which attacks the unsaturated bond of a fumarate molecule to form benzylsuccinate. Then CoA is added via a succinyl-CoA/benzylsuccinate CoA-transferase reaction (encoded by *bbsEF*), and additional oxidation reactions result in benzyl-CoA, which is further oxidized to aliphatic compounds that are attacked by hydrolytic and oxidative reactions. Similar degradation routes have been reported for xylene, o-toluidine, and cresol isomers in sulfate-reducing bacteria (Leuthner and Heider 2000; Rotaru et al. 2010). Recent habitat studies employing an alkane-activating (methylalkyl) succinate synthase gene marker (*masD*, coding for the catalytic subunit) demonstrated ubiquitous presence and diversity of anaerobic alkane degraders in cold marine sediments (Gittel et al. 2015). A fertilization study of a 20-year oil fuel spill in Antarctic soil showed increased biodegradation in anaerobic soils, corresponding with a shift in the denitrifier community composition and an increased abundance of denitrifiers and benzoyl-CoA reductase (Powell et al. 2006).

Other anaerobic hydrocarbon biodegradation pathways include oxygen-independent hydroxylation (Szaleniec et al. 2007), carboxylation of unsubstituted carbon atoms (Laban et al. 2009; Jobst et al. 2010), hydration of the double and triple bond of alkenes and alkynes (Brodkorb et al. 2010), and reverse methanogenesis (Rabus et al. 2016).

27.3.2.3 Microbial Metabolism in Sea Ice

Although sea ice may be regarded to be "frozen seawater," the conditions for the microbes differ significantly between seawater and marine ice, not only with respect to temperature. In first-year ice, brine inclusion networks are generated by salting-out processes. The brine channels represent saline liquid niches at subzero temperatures for microbial motility and respiration (Junge et al. 2003, 2004, 2006). At these extreme conditions of low temperature and high salinity, laboratory experiments have shown metabolism and growth of psychrophilic bacteria at temperatures below −10 °C (Breezee et al. 2004; Junge et al. 2006).

Microbes living in marine ice may adapt to their environments in several ways. Psychrophilic bacteria may produce extracellular polymeric substances (exopolysaccharides, EPS), which are associated with halotolerance and may have a cryoprotective role in the sea ice brine channels, as well as binding essential cationic trace metals (Nichols et al. 2005; Krembs et al. 2011). Enzymes in bacteria isolated from sea ice may be cold-active, with catalytic activities well below the freezing point of seawater (Groudieva et al. 2004). Microbes exposed to large drops in temperature may produce cold-shock, cold-acclimation, and anti-freeze proteins (Berger et al. 1996; Raymond et al. 2007). Cold-shock proteins are supposed to be involved in protein translation regulation, while cold-acclimation proteins show high catalytic activity at a low temperature and rapid inactivation at a moderate temperature (Fukunaga et al. 1999). Anti-freeze proteins will block the formation of additional ice crystals that may damage cellular membranes (Boetius et al. 2015). Psychrophilic and halophilic bacteria are also capable of adjusting the lipid composition of their membranes in such a way that the proton permeability at the respective growth temperature remains constant by altering fatty acid composition and protein content of the membrane (Chintalapati et al. 2004). Protein flexibility is important for adaptions to low temperatures to avoid reduced stability (Fields 2001). In a study of the psychrophilic bacterium *Colwellia psychrerythraea* 34H, comparative genomic analysis suggested that psychrophilic life is most likely conferred not by a unique set of genes, but by a collection of synergistic changes in overall genome content and amino acid composition (Methé et al. 2005).

Some studies from oil-polluted Arctic soil indicated biodegradation at subzero temperatures (Rike et al. 2003; Børresen and Rike 2007), and other studies have shown that oil in ice stimulates bacterial growth (Delille et al. 1997; Brakstad et al. 2008). However, biodegradation of hydrocarbons in ice still needs to be documented and is probably mainly confined to soluble compounds migrating in the brine channels (Brakstad et al. 2008; Faksness and Brandvik 2008b).

27.3.2.4 Biodegradation and Flocculation

Oil biodegradation in seawater may result in the generation of macroscopic aggregates. These aggregates, or "flocs," have been observed during experimental biodegradation studies with dispersed oil (Macnaughton et al. 2003; Bælum et al. 2012). We observed rapid flocculation in oil biodegradation studies performed in Arctic seawater (Svalbard) at 1–2 °C (Fig. 27.1). In our experience, these "flocs"

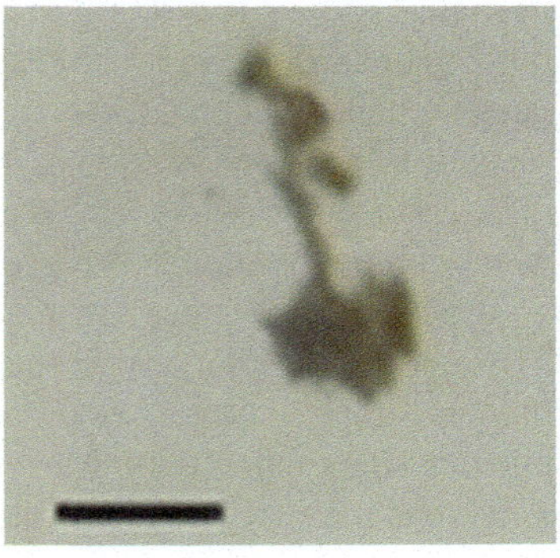

Fig. 27.1 A "floc" generated during biodegradation of a paraffinic oil at 1–2 °C in unpolluted Arctic (scale 0.5 cm). Photo: Roman Netzer, SINTEF

are fragile structures, and generation of turbulence like magnetic stirring rapidly disintegrates the structures. "Flocs" were also detected in field samples from DWH deep sea plume (Hazen et al. 2010).

Analyses of "flocs" from field studies or laboratory experiments showed distribution of microorganisms, diverse polysaccharides and EPS, oil, and oil degradation products in the typical "floc" structure (Hazen et al. 2010; Bælum et al. 2012). Time-related analyses during biodegradation studies suggested that the oil tended to initially concentrate in the "floc" material, but was subsequently degraded, while the biological material in the "flocs" increased (Bælum et al. 2012), and the "floc" is therefore considered to be an active site of oil biodegradation. Analyses of EPS material from "floc"-producing microorganisms isolated from crude petroleum oil showed that these polymers were glycolipids (Zaki et al. 2011). Specific bacterial groups are known to be producers of the EPS material associated with the "floc," and several typical EPS-producing bacteria were observed during the DWH spill, including *Halomonas*, *Alteromonas*, *Colwellia*, and *Pseudoalteromonas*. One purpose of this polymeric material is to increase the solubilization of aromatic hydrocarbons, so these may become more available for biodegradation (Gutierrez et al. 2013).

This "floc" material may be related to marine snow formed as part of the biological pump, bringing particulate organic material to deepwaters (Alldredge and Silver 1988; De La Rocha and Passow 2007). In association with diatoms and mineral particles, it has been suggested that these aggregates transported oil residues from the oil spill to the seabed of the Gulf of Mexico (Valentine et al. 2014; Passow 2016).

27.4 Modeling of Hydrocarbon Biodegradation

27.4.1 Biodegradation Rates

Biodegradation data determined as degradation rates have been included in some numerical models as part of the prediction of the fate of the oil after a spill. Due to its complexity, it is difficult to use one degradation rate for crude oils. Hydrocarbons also distribute between water and oil after discharge to seawater, resulting in phase-separated degradation rates. For simplicity, biodegradation rates of oil in the prediction model OSCAR has been separated in 25 compound groups based on different boiling points, and the degradation rates are further separated between water-soluble fractions, dispersed oil, and sedimented oil (Reed et al. 2000).

Biodegradation rates are determined by nonlinear regression analyses, using first- or second-order approaches. In environments with low and relative constant biomass, like seawater, rate coefficients may be determined by first-order rate kinetics (Battersby 1990). Most biodegradation studies of oil involving rate determinations actually use first-order rates, with or without lag phases, and with half-lives determined from the rate coefficients (ln2/*k1*) (Prince et al. 2007, 2013; Venosa and Holder 2007; Brakstad et al. 2015b; Wang et al. 2016).

More advanced biodegradation models have also been developed for dispersed oil, either as single- or multiple-substrate models. A single component model was developed in which each oil droplet was viewed as a porous non-biodegradable asphaltene matrix filled with biodegradable and non-biodegadable fractions of de-asphaltened oil. The asphaltene content controlled the diffusivity and thereby the bioavailability of oil compounds. The oil was considered as one substrate, using oxygen uptake as measure of substrate utilization (Uraizee et al. 1997). Several multi-substrate growth kinetic models have been published for pure and mixed cultures and for di- to multicomponent substrates. Most of these could be fitted to data of two or three substrates, but failed significantly the description of growth on more numerous mixed substrates (Kovárová-Kovar and Egli 1998). A physical-metabolic model was used to assess the mixing of deepwater bacterial communities and their capacity for hydrocarbon and oxygen consumption during the DWH oil spill (Valentine et al. 2012). Carbon from oil compounds was tracked during mineralization, as well as oxidized extracellular intermediate and biomass of the bacteria degrading the primary and secondary (extracellular intermediate) consumers. In this work, no less than 26 modeled primary and another 26 secondary bacterial operational metabolic types (OMT) were included. The authors assigned 22 substrate classes to the OMTs, which served as substrate for several OMTs, but did not include substrate competition effects on the growth kinetics.

27.4.2 Temperature

Arrhenius plots have often been used to predict the relations between temperature and biological processes (Arrhenius 1889). Ideally, these plots should be linear. The Q10 value is the ratio of the rate constant at a specific temperature to the rate constant at a temperature 10 °C lower. The Q10 values depend on the activation energy (E_a) of the specific chemical reaction, which is directly related to the activation enthalpy (ΔH^*); the lower the activation enthalpy, the lower the Q10 value. A Q10 of 2.0 describes a doubling of the rate coefficient ($k1$) for every 10 °C. Efforts have been made to suggest Q10 values for oil biodegradation. Comparison of data from marine oil biodegradation experiments performed at different temperatures indicated that the Q10 value of 2 was a fairly good approximation in a temperature range of 5–27 °C (Bagi et al. 2013).

However, in crude oil different compound groups may follow different temperature-related biodegradation profiles. This is exemplified in Fig. 27.2, showing biotransformation half-lives during biodegradation of dispersed crude oil at different temperatures (1 °C, 5 °C and 13 °C). Toluene and 3-ring PAHs, which are completely or partly water-soluble, showed different temperature-related biodegradation from C17–C18 saturates, which were associated with the oil phase. Thus, Q10 approaches for predictions of crude oil biodegradation at different

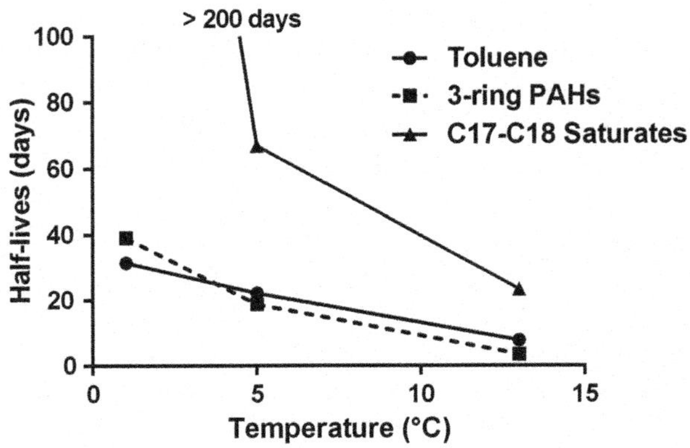

Fig. 27.2 Half-lives of toluene, 3-ring PAHs and C17–C18 saturates in natural unamended seawater at 1 °C, 5 °C, and 13 °C

temperatures should be used with caution, especially at low temperatures, where the oil properties are changed.

27.5 Oil Bioremediation in Cold Marine Environments

Most reported bioremediation attempts have focused on developing biostimulation strategies, typically by applying dispersants, fertilizers stimulating degradation, or the combination of these and other treatments, to accelerate the biodegradation processes by the indigenous microbes in the polluted environment. Bioremediation processes, if successful, are cost-effective and reduce the environmental impacts of marine oil spills (Swannell et al. 1996; Prince and Clark 2004; Prince 2008; Atlas and Hazen 2011). An alternative to biostimulation is bioaugmentation, which involves the inoculation of exogenous microbial cultures with high biodegradation potentials for contaminants. Bioaugmentation approaches have been reported to improve biodegradation of hydrocarbons from oil spills in cold soil or marine sediments and may be used in combination with fertilizers (Margesin and Schinner 1997; Ruberto et al. 2003; Tyagi et al. 2011). Bacterial mats from marine oil-contaminated sites have also been suggested for use in the degradation of coastal oil spills, although these are of greater relevance for spills in non-Arctic areas (Cohen 2002). Various methods and strategies for bioremediation have been reviewed (Lee and Merlin 1999; Prince 2010).

27.5.1 Biostimulation

Biostimulation is a strategy for secondary cleanup of stranded oil and includes addition of nutrients or other methods to enhance the capability of the indigenous microbial communities to degrade environmental pollutions. However, other methods may also be regarded as biostimulatory actions, for instance the use of chemical dispersants for oil spills in seawater, since this may aid in oil biodegradation by increasing the oil surface area accessible to the oil-degrading microbes.

27.5.1.1 Shoreline Sediments

Most biostimulation activities have focused on stranded oil, with application of fertilizers to increase natural degradation by the indigenous microorganisms. Biostimulation treatment is often combined with mechanical treatment to improve oxygen and nutrient availability.

In marine environments, some growth- and biomass-stimulating factors are essential for oil biodegradation, especially nitrogen and phosphorus, and the addition of these nutrients is common practice in bioremediation. Balanced nutrient availability is important for biodegradation and the composition of hydrocarbon-degrading communities, since nutrient amendments, in some instances, can inhibit microbial activity (Braddock et al. 1997). It is, therefore, important to avoid excess nutrients, which can cause detrimental effects, such as eutrophication. During biostimulation, molar carbon/nitrogen/phosphorus ratios of 100/10/1 have often been used (e.g., Obbard et al. 2004; Beolchini et al. 2010). However, results from laboratory studies have also shown that certain microbial populations may require more than one N/P ratio for optimal degradation of different hydrocarbons (Smith et al. 1998). Nutrient products are available as briquettes, granules, or liquid fertilizers. Liquid inorganic fertilizers have proven effective but require frequent application, and therefore oleophilic slow-release nutrient formulations have been developed, which promote hydrocarbon degraders at the oil–water interface.

For improved results, bioremediation may be combined with other cleanup procedures. Surf washing and the use of surfactants may increase the surface area of the oil and hence increase oil degradation. *Ex situ* technologies like land farming (spreading the polluted sediments over a larger area for better oxygenation), composting, and biopiling may be used for treating oily waste during spill treatment (Lynch and Moffat 2005) although these approaches have seen limited application in polar environments.

Arctic field biostimulation experiments were conducted on beaches at Svalbard in the 1980s with the slow-release oleophilic fertilizer Inipol EA22. The results of these studies indicated that application of the fertilizer to oil in beach sediments resulted in increased oil biodegradation in coarse sediments, but not in finer sediments (Sveum and Ladousse 1989). A full-scale trial of several remediation processes at Svalbard in 1997, with a fuel oil in mixed intertidal shorelines, included remediation methods like sediment relocation (surf washing), mixing (tilling), and bioremediation (Guénette et al. 2003; Sergy et al. 2003). The introduction of soluble or slow-release fertilizers resulted in approximately doubled biodegradation rates over a period of one year in the oiled sediments that received fertilizers when compared to non-treated oiled sediments, and no acute toxicity was associated with the bioremediation treatment (Prince et al. 2003). Also mixing/

tilling seemed to result in increased microbial activity for limited periods by increasing sediment permeability (Owens et al. 2003).

Biostimulation field experiments have also been conducted in Antarctic environments, in intertidal sandy beaches on the main island of the Kerguelen Archipelago. A crude weathered oil was added to several closures in the beaches, and different fertilizers were added to the top of the oil, including the slow-release Inipol EAP 22 and various experimental mixtures consisting of dry fish compost (with or without supplements of urea), phosphate, and charged or neutral surfactants. Oil was depleted after 300 days in both untreated and treated sediments at seawater temperatures of 3–4 °C, and the various fertilizers accelerated the biodegradation rates. It was also observed that a fertilizer with a lipidic surfactant reduced the toxicity of the oil during the last 3 months of the experiment (Pelletier et al. 2004).

Bioremediation was used as an oiled-beach cleaning technology on a full-scale oil spill in Arctic environments during the Exxon Valdez accident in March 1989. This spill in Prince William Sound, Alaska, resulted in the release of 42 million liters of Alaskan North Slope crude oil (Atlas and Hazen 2011). Bioremediation was used extensively, employing the fertilizers Inipol EAP 22 and Customblen (slow-release granulated fertilizer). Approximately, 50,000 kg nitrogen and 5000 kg phosphorus were applied to the shorelines over the summers of 1989–1992 (Bragg et al. 1994). For a low-energy beach containing both surface and subsurface oil and treated with both fertilizers, it was estimated that the fertilizers enhanced oil biodegradation 3–5 times compared to untreated beaches (Bragg et al. 1993).

27.5.1.2 Seawater and the Use of Chemical Dispersants

Most remediation actions in seawater have rather focused on mechanical removal, e.g., by use of oil booms and skimmers, than on accelerated biodegradation. As an alternative to mechanical treatments, chemical dispersants have been used. Dispersants are used primarily to remove oil from the water surface in order to reduce the impacts on seabird and mammal populations close to an oil spill, but this treatment has also been suggested to improve hydrocarbon biodegradation in the seawater column. Dispersants were also used in subsurface areas during the DWH spill to reduce surfacing of the oil, and subsequent impacts on the GoM shorelines (Atlas and Hazen 2011). Dispersants are mixtures of surface-active agents and reduce the surface tension of the oil, resulting in the formation of small oil droplet dispersions which will rapidly dilute in the water column. The dispersants reduce the average oil-droplet size, resulting in higher surface/volume ratios, and thereby increase the bioavailability of the dispersed oil. Studies of the correlation between the droplet area of dispersed oil and oil degradation indicated that both dispersed area and dispersant chemistry control the degradation and depend on the surfactant blend hydrophile/lipophile balance and treatment levels (Varadaraj et al. 1995).

The effect of dispersants on biodegradation has been debated. Several biodegradation studies have emphasized their positive effects, promoting oil biodegradation by reducing the oil droplet size, thereby generating larger oil surface areas for the oil-degrading bacteria to attach. It has been shown that the dispersant treatment increased oil biodegradation at low seawater temperatures ranging from −1 °C to 5 °C, using both seawater from temperate and Arctic sources (Siron et al. 1995; Venosa and Holder 2007; McFarlin et al. 2014b). However, other studies have reported no or even negative effects of the dispersants on oil biodegradation (Lindstrom and Braddock 2002; Macnaughton et al. 2003; Kleindienst et al. 2015). These studies have often been performed with concentrations of oil/dispersant higher than expected in the environment (Lee et al. 2013) or have investigated water-accommodated fractions rather than dispersed oil (Kleindienst et al. 2015).

Fertilizers have also stimulated biodegradation of crude oils in cold seawater under controlled experimental conditions. The slow-release fertilizer Inipol EAP 22 was added to Antarctic seawater contaminated with crude oil in a mesocosm study, resulting in enhanced concentrations of heterotrophic and oil-degrading bacteria, and in increased rates of biodegradation during a 40-day experiment both in ice-covered and ice-free seawater (Delille et al. 1998).

27.5.1.3 Sea Ice

Only a few studies have investigated the possibilities for accelerated biodegradation of petroleum in sea ice. In this environment, the microbes are primarily active in the high-salinity brine channels where fluidity is maintained at temperatures well below the freezing point. During freezing periods, the brine channels, which may constitute up to 30% of the ice volume, may reach salinities >80 psu, while ice melting results in ice surface salinities well below that of seawater in the upper parts of the ice (Krembs et al. 2001). Water-soluble oil components have been shown to migrate out through the ice into brine channels and the underlying water, at low concentrations but over long periods (Faksness and Brandvik 2008a, b). Stimulation of oil degradation in ice and brine channels could therefore limit possible toxic effects on sea ice algae growing in and underneath the sea ice.

A study on bioremediation of diesel fuel and "Arabian light" crude oil in land-fast Antarctic ice showed that the oils induced a negative effect on ice-microalgae growth, reducing the phytoplankton bloom, but addition of the fertilizer slow-release Inipol EAP 22 stimulated the microalgae development (Fiala and Delille 1999). During the winter of 2004, a field bioremediation experiment was set up by the Alfred Wegener Institute in Van Mijenfjord on Svalbard, in which slow-release fertilizers were applied to oil-polluted Arctic fjord ice. Biodegradation of oil frozen into ice was not stimulated at sub-zero temperatures. However, bacterial communities changed significantly in response to the addition of nutrients to released oil in sea ice melt pools (Gerdes and Dieckmann 2006).

27.5.2 Bioaugmentation

Bioaugmentation has been proposed as a bioremediation method for soil and sediments, sometimes in combination with biostimulation treatments (Tyagi et al. 2011). Introduction of exogenic hydrocarbonoclastic bacteria for detoxification of hydrocarbon-polluted cold environments has been reported with variable success. Several bioaugmentation studies have been performed in Arctic and Antarctic soil systems (Camenzuli and Freidman 2015).

Bioaugmentation studies have also been reported in marine environments, although none of these were from cold waters. A laboratory biodegradation and toxicity study of 12 commercially available bioaugmentation products applied to weathered oil (Alaska North Slope) in seawater at 20 °C showed that three of the products enhanced biodegradation better than nutrient-amended controls, but only one product resulted in reduced toxicity (Aldrett et al. 1997). In a marine sediment microcosm study, the aromatic hydrocarbon-degrading bacterial strain *Cycloclasticus* sp. E2 was identified to play an important role during degradation of naphthalene in combination with biostimulation treatment (Miyasaka et al. 2006). Also in seawater systems, bioaugmentation methods have been suggested in combination with biostimulation (Nikolopoulou et al. 2013). It should also be noted that typical oil-degrading bacteria like *Pseudoalteromonas*, *Marinobacter*, *Oleispira*, and *Alcanivorax* were isolated from pristine seawater at Svalbard (Crisafi et al. 2016).

27.6 Conclusions

The cold areas on Earth are exposed to petroleum exploration, production, and transport, increasing the risk of accidental oil discharges in these environments. Since hydrocarbon-degrading microorganisms are present in most of these environments, they may play a crucial role during decontamination of toxic oil compounds in the seawater column and in sediments. It has been shown that oil pollution in marine ice stimulated bacterial growth, indicating hydrocarbon biodegradation in the contaminated ice, probably biodegradation of water-soluble compounds present in the brine channels of the ice. As our knowledge of microbial ecology, metabolic processes, and interactions with their cold environments are constantly increasing, we will be able to design improved regimes for bioremediation in these vulnerable environments. In this respect, advances in chemical and molecular biology methods, and data processing of high-resolution data, are important. The recent focus on interactions between oil bacteria and other organisms during oil biodegradation will be important to predict the fate of oil spills in the environments including polar and deepwater.

Bioremediation is still a “promising” technology for treatment of polluted cold marine environments. This treatment is cheap, labor-effective, generates no harmful by-products, and may be suited for secondary clean-up actions in remote Arctic and Antarctic locations, in combination with other technologies. Novel tools for microbial and chemical analysis will also be important for determining restitution of oil-polluted environments during remediation actions.

References

Aldrett S, Bonner JS, Mills MA, Autenrieth RL, Stephens FL (1997) Microbial degradation of crude oil in marine environments tested in a flask experiment. Water Res 31(11):2840–2848

Alldredge AL, Silver MW (1988) Characteristics, dynamics and significance of marine snow. Prog Oceanogr 20(1):41–82

Arrhenius S (1889) Über die Reaktionsgeschwindigkeit bei der Inversion von Rohrzucker durch Säuren. Z Phys Chem 4:226–248

Atlas RM (1981) Microbial degradation of petroleum hydrocarbons: an environmental perspective. Microbiol Rev 45(1):180

Atlas RM (1995) Petroleum biodegradation and oil spill bioremediation. Mar Pollut Bull 31 (4):178–182

Atlas RM, Bartha R (1972) Degradation and mineralization of petroleum in sea water: limitation by nitrogen and phosphorous. Biotechnol Bioeng 14(3):309–318

Atlas RM, Hazen TC (2011) Oil biodegradation and bioremediation: a tale of the two worst spills in US history. Environ Sci Technol 45(16):6709–6715

Austin RN, Groves JT (2011) Alkane-oxidizing metalloenzymes in the carbon cycle. Metallomics 3(8):775–787

Bælum J, Borglin S, Chakraborty R, Fortney JL, Lamendella R, Mason OU, Auer M, Zemla M, Bill M, Conrad ME (2012) Deep-sea bacteria enriched by oil and dispersant from the Deepwater Horizon spill. Environ Microbiol 14(9):2405–2416

Bagi A, Pampanin DM, Brakstad OG, Kommedal R (2013) Estimation of hydrocarbon biodegradation rates in marine environments: a critical review of the Q10 approach. Mar Environ Res 89:83–90

Bagi A, Pampanin DM, Lanzén A, Bilstad T, Kommedal R (2014) Naphthalene biodegradation in temperate and arctic marine microcosms. Biodegradation 25(1):111–125

Bano N, Ruffin S, Ransom B, Hollibaugh JT (2004) Phylogenetic composition of Arctic Ocean archaeal assemblages and comparison with Antarctic assemblages. Appl Environ Microbiol 70 (2):781–789

Battersby N (1990) A review of biodegradation kinetics in the aquatic environment. Chemosphere 21(10-11):1243–1284

Beolchini F, Rocchetti L, Regoli F, Dell'Anno A (2010) Bioremediation of marine sediments contaminated by hydrocarbons: experimental analysis and kinetic modeling. J Hazard Mater 182(1):403–407

Berger F, Morellet N, Menu F, Potier P (1996) *Cold* shock and cold acclimation proteins in the psychrotrophic bacterium *Arthrobacter* globiformis SI55. J Bacteriol 178(11):2999–3007

Bienhold C, Boetius A, Ramette A (2012) The energy-diversity relationship of complex bacterial communities in Arctic deep-sea sediments. ISME J 6(4):724–732

Boetius A, Anesio AM, Deming JW, Mikucki JA, Rapp JZ (2015) Microbial ecology of the cryosphere: sea ice and glacial habitats. Nat Rev Microbiol 13(11):677–690. doi:10.1038/nrmicro3522

Bordenave S, Goñi-urriza M, Vilette C, Blanchard S, Caumette P, Duran R (2008) Diversity of ring-hydroxylating dioxygenases in pristine and oil contaminated microbial mats at genomic and transcriptomic levels. Environ Microbiol 10(12):3201–3211

Børresen M, Rike A (2007) Effects of nutrient content, moisture content and salinity on mineralization of hexadecane in an Arctic soil. Cold Reg Sci Technol 48(2):129–138

Bowman JP, McCuaig RD (2003) Biodiversity, community structural shifts, and biogeography of prokaryotes within Antarctic continental shelf sediment. Appl Environ Microbiol 69 (5):2463–2483

Braddock JF, Ruth ML, Catterall PH, Walworth JL, McCarthy KA (1997) Enhancement and inhibition of microbial activity in hydrocarbon-contaminated arctic soils: implications for nutrient-amended bioremediation. Environ Sci Technol 31(7):2078–2084

Bragg JR, Prince RC, Harner EJ, Atlas RM (1993) Bioremediation effectiveness following the Exxon Valdez spill. In: International oil spill conference, vol 1. American Petroleum Institute, Washington, DC, pp 435–447

Bragg JR, Prince RC, Harner EJ, Atlas RM (1994) Effectiveness of bioremediation for the Exxon Valdez oil spill. Nature 368:413–418

Brakstad OG (2008) Natural and stimulated biodegradation of petroleum in cold marine environments. In: Margesin R, Schinner F, Marx J-C, Gerday C (eds) Psychrophiles: from biodiversity to biotechnology. Springer, Berlin, pp 389–407

Brakstad OG, Bonaunet K (2006) Biodegradation of petroleum hydrocarbons in seawater at low temperatures (0–5°C) and bacterial communities associated with degradation. Biodegradation 17(1):71–82

Brakstad OG, Bonaunet K, Nordtug T, Johansen Ø (2004) Biotransformation and dissolution of petroleum hydrocarbons in natural flowing seawater at low temperature. Biodegradation 15 (5):337–346

Brakstad OG, Nonstad I, Faksness L-G, Brandvik PJ (2008) Responses of microbial communities in Arctic sea ice after contamination by crude petroleum oil. Microb Ecol 55(3):540–552

Brakstad OG, Nordtug T, Throne-Hoist M, Netzer R, Stoeckel DM, Atlas RM (2015a) Microbial communities related to biodegradation of dispersed Macondo oil at low seawater temperature with Norwegian coastal seawater. Microb Biotechnol 8(6):989–998

Brakstad OG, Nordtug T, Throne-Holst M (2015b) Biodegradation of dispersed Macondo oil in seawater at low temperature and different oil droplet sizes. Mar Pollut Bull 93(1):144–152

Brandvik PJ (1997) Optimisation of oil spill dispersants on weathered oils. A new approach using experimental design and multivariate data analyses. Norwegian University of Science and Technology, NTNU Trondheim, Norway

Breezee J, Cady N, Staley J (2004) Subfreezing growth of the sea ice bacterium "*Psychromonas ingrahamii*". Microb Ecol 47(3):300–304

Brinkmeyer R, Knittel K, Jurgens J, Weyland H, Amann R, Helmke E (2003) Diversity and structure of bacterial communities in Arctic versus Antarctic pack ice. Appl Environ Microbiol 69(11):6610–6619

Brodkorb D, Gottschall M, Marmulla R, Lüddeke F, Harder J (2010) Linalool dehydratase-isomerase, a bifunctional enzyme in the anaerobic degradation of monoterpenes. J Biol Chem 285(40):30436–30442

Camenzuli D, Freidman BL (2015) On-site and in situ remediation technologies applicable to petroleum hydrocarbon contaminated sites in the Antarctic and Arctic. Pol Res 34(24492). doi:10.3402/polar.v34.24492

Camilli R, Reddy CM, Yoerger DR, Van Mooy BA, Jakuba MV, Kinsey JC, McIntyre CP, Sylva SP, Maloney JV (2010) Tracking hydrocarbon plume transport and biodegradation at Deepwater Horizon. Science 330(6001):201–204

Chen E, Overall JC, Phillips C (1974) Spreading of crude oil on an ice surface. Can J Chem Eng 52 (1):71–74

Cheng Q, Thomas S, Rouviere P (2002) Biological conversion of cyclic alkanes and cyclic alcohols into dicarboxylic acids: biochemical and molecular basis. Appl Microbiol Biotechnol 58(6):704–711

Chintalapati S, Kiran M, Shivaji S (2004) Role of membrane lipid fatty acids in cold adaptation. Cell Mol Biol (Noisy-le-Grand) 50(5):631–642

De La Rocha DL, Passow U (2007) Factors influencing the sinking of POC and the efficiency of the biological carbon pump. Deep Sea Res II Top Stud Oceanogr 54(5):639–658

Cohen Y (2002) Bioremediation of oil by marine microbial mats. Int Microbiol 5(4):189–193
Comeau AM, Li WKW, Tremblay J-É, Carmack EC, Lovejoy C (2011) Arctic Ocean microbial community structure before and after the 2007 Record Sea Ice Minimum. PLoS One 6(11): e27492
Cowie ROM, Maas EW, Ryan KG (2011) Archaeal diversity revealed in Antarctic sea ice. Antarct Sci 23(6):531–536
Crisafi F, Giuliano L, Yakimov MM, Azzaro M, Denaro R (2016) Isolation and degradation potential of a cold-adapted oil/PAH-degrading marine bacterial consortium from Kongsfjorden (Arctic region). Rend Lincei Sci Fis Nat 27(1):261–270
Cui Z, Lai Q, Dong C, Shao Z (2008) Biodiversity of polycyclic aromatic hydrocarbon-degrading bacteria from deep sea sediments of the Middle Atlantic ridge. Environ Microbiol 10 (8):2138–2149
Daly KL, Passow U, Chanton J, Hollander D (2016) Assessing the impacts of oil-associated marine snow formation and sedimentation during and after the deepwater horizon oil spill. Anthropocene 13:18–33. doi:10.1016/j.ancene.2016.01.006
Delille D, Basseres A, Dessommes A (1997) Seasonal variation of bacteria in sea ice contaminated by diesel fuel and dispersed crude oil. Microb Ecol 33(2):97–105
Delille D, Basseres A, Dessommes A (1998) Effectiveness of bioremediation for oil-polluted Antarctic seawater. Polar Biol 19(4):237–241
Deppe U, Richnow HH, Michaelis W, Antranikian G (2005) Degradation of crude oil by an arctic microbial consortium. Extremophiles 9(6):461–470
Dong C, Bai X, Sheng H, Jiao L, Zhou H, Shao Z (2015) Distribution of PAHs and the PAH-degrading bacteria in the deep-sea sediments of the high-latitude Arctic Ocean. Biogeosciences 12(7):2163–2177. doi:10.5194/bg-12-2163-2015
Dubbels BL, Sayavedra-Soto LA, Arp DJ (2007) Butane monooxygenase of 'Pseudomonas butanovora': purification and biochemical characterization of a terminal-alkane hydroxylating diiron monooxygenase. Microbiology 153(6):1808–1816
Dubinsky EA, Conrad ME, Chakraborty R, Bill M, Borglin SE, Hollibaugh JT, Mason OU, Piceno YM, Reid FC, Stringfellow WT, Tom LM, Hazen TC, Andersen GL (2013) Succession of hydrocarbon-degrading bacteria in the aftermath of the deepwater horizon oil spill in the Gulf of Mexico. Environ Sci Technol 47(19):10860–10867. doi:10.1021/es401676y
Eriksson M, Sodersten E, Yu Z, Dalhammar G, Mohn WW (2003) Degradation of polycyclic aromatic hydrocarbons at low temperature under aerobic and nitrate-reducing conditions in enrichment cultures from northern soils. Appl Environ Microbiol 69(1):275–284
Faksness L-G, Brandvik PJ (2008a) Distribution of water soluble components from oil encapsulated in Arctic sea ice: Summary of three field seasons. Cold Reg Sci Technol 54(2):106–114
Faksness LG, Brandvik PJ (2008b) Distribution of water soluble components from Arctic marine oil spills - A combined laboratory and field study. Cold Reg Sci Technol 54(2):97–105
Feng L, Wang W, Cheng J, Ren Y, Zhao G, Gao C, Tang Y, Liu X, Han W, Peng X (2007) Genome and proteome of long-chain alkane degrading Geobacillus thermodenitrificans NG80-2 isolated from a deep-subsurface oil reservoir. Proc Natl Acad Sci U S A 104(13):5602–5607
Fiala M, Delille D (1999) Annual changes of microalgae biomass in Antarctic sea ice contaminated by crude oil and diesel fuel. Polar Biol 21(6):391–396
Fields PA (2001) Review: Protein function at thermal extremes: balancing stability and flexibility. Comp Biochem Physiol A Mol Integr Physiol 129(2):417–431
Fingas M, Hollebone B (2003) Review of behaviour of oil in freezing environments. Mar Pollut Bull 47(9):333–340
Fukunaga N, Sahara T, Takada Y (1999) Bacterial adaptation to low temperature: Implications for cold-inducible genes. J Plant Res 112(2):263–272
Galand PE, Casamayor EO, Kirchman DL, Lovejoy C (2009) Ecology of the rare microbial biosphere of the Arctic Ocean. Proc Natl Acad Sci U S A 106(52):22427–22432
Garneau M-È, Michel C, Meisterhans G, Fortin N, King TL, Greer CW, Lee K (2016) Hydrocarbon biodegradation by Arctic sea-ice and sub-ice microbial communities during microcosm

experiments, Northwest Passage (Nunavut, Canada). FEMS Microbiol Ecol 92(10). doi:10.1093/femsec/fiw130

Gerdes B, Brinkmeyer R, Dieckmann G, Helmke E (2005) Influence of crude oil on changes of bacterial communities in Arctic sea-ice. FEMS Microbiol Ecol 53(1):129–139

Gerdes B, Dieckmann G (2006) Biological degradation of crude oil in sea ice. In: Juurmaa K (ed) Abstract of the report "Growth Project GRD-2000-30112 ARCOP Technology and Environment, WP 6 Workshop Activities". Helsinki, Finland

Gittel A, Donhauser J, Røy H, Girguis PR, Jørgensen BB, Kjeldsen KU (2015) Ubiquitous presence and novel diversity of anaerobic alkane degraders in cold marine sediments. Front Microbiol 6:1414. doi:10.3389/fmicb.2015.01414

Gough M, Rowland S (1990) Characterization of unresolved complex mixtures of hydrocarbons in petroleum. Nature 344(6267):648–650

Goyal A, Zylstra G (1997) Genetics of naphthalene and phenanthrene degradation by *Comamonas testosteroni*. J Ind Microbiol Biotechnol 19(5-6):401–407

Groudieva T, Kambourova M, Yusef H, Royter M, Grote R, Trinks H, Antranikian G (2004) Diversity and cold-active hydrolytic enzymes of culturable bacteria associated with Arctic sea ice, Spitzbergen. Extremophiles 8(6):475–488

Grund A, Shapiro J, Fennewald M, Bacha P, Leahy J, Markbreiter K, Nieder M, Toepfer M (1975) Regulation of alkane oxidation in *Pseudomonas putida*. J Bacteriol 123(2):546–556

Guénette CC, Sergy GA, Owens EH, Prince RC, Lee K (2003) Experimental design of the Svalbard shoreline field trials. Spill Sci Technol Bull 8(3):245–256

Gutierrez T, Berry D, Yang T, Mishamandani S, McKay L, Teske A, Aitken MD (2013) Role of bacterial exopolysaccharides (EPS) in the fate of the oil released during the deepwater horizon oil spill. PLoS One 8(6):e67717

Harayama S, Kasai Y, Hara A (2004) Microbial communities in oil-contaminated seawater. Curr Opin Biotechnol 15(3):205–214

Harayama S, Kishira H, Kasai Y, Shutsubo K (1999) Petroleum biodegradation in marine environments. J Mol Microbiol Biotechnol 1(1):63–70

Hasinger M, Scherr KE, Lundaa T, Bräuer L, Zach C, Loibner AP (2012) Changes in iso-and n-alkane distribution during biodegradation of crude oil under nitrate and sulphate reducing conditions. J Biotechnol 157(4):490–498

Hatam I, Charchuk R, Lange B, Beckers J, Haas C, Lanoil B (2014) Distinct bacterial assemblages reside at different depths in Arctic multiyear sea ice. FEMS Microbiol Ecol 90(1):115–125. doi:10.1111/1574-6941.12377

Hatam I, Lange B, Beckers J, Haas C, Lanoil B (2016) Bacterial communities from Arctic seasonal sea ice are more compositionally variable than those from multi-year sea ice. ISME J 10 (10):2543–2552. doi:10.1038/ismej.2016.4

Hazen TC, Dubinsky EA, DeSantis TZ, Andersen GL, Piceno YM, Singh N, Jansson JK, Probst A, Borglin SE, Fortney JL (2010) Deep-sea oil plume enriches indigenous oil-degrading bacteria. Science 330(6001):204–208

Hazen TC, Prince RC, Mahmoudi N (2016) Marine oil biodegradation. Environ Sci Technol 50 (5):2121–2129

Head IM, Jones DM, Roling WFM (2006) Marine microorganisms make a meal of oil. Nat Rev Microbiol 4(3):173–182

Hearn EM, Patel DR, Lepore BW, Indic M, van den Berg B (2009) Transmembrane passage of hydrophobic compounds through a protein channel wall. Nature 458(7236):367–370

Hokstad JN, Daling PS, Buffagni M, Johnsen S (1999) Chemical and ecotoxicological characterisation of oil–water systems. Spill Sci Technol Bull 5(1):75–80

Inagaki F, Sakihama Y, Inoue A, Kato C, Horikoshi K (2002) Molecular phylogenetic analyses of reverse-transcribed bacterial rRNA obtained from deep-sea cold seep sediments. Environ Microbiol4 (5):277-286

Jobst B, Schuhle K, Linne U, Heider J (2010) ATP-dependent carboxylation of acetophenone by a novel type of carboxylase. J Bacteriol 192(5):1387–1394

Johansen Ø, Rye H, Cooper C (2003) DeepSpill-Field study of a simulated oil and gas blowout in deep water. Spill Sci Technol Bull 8:433–443

Johnson J, Hill R (2003) Sediment microbes of deep-sea bioherms on the Northwest Shelf of Australia. Microb Ecol 46(1):55–61

Junge K, Eicken H, Deming JW (2003) Motility of *Colwellia psychrerythraea* strain 34H at subzero temperatures. Appl Environ Microbiol 69(7):4282–4284

Junge K, Eicken H, Deming JW (2004) Bacterial activity at −2 to −20°C in Arctic wintertime sea ice. Appl Environ Microbiol 70(1):550–557

Junge K, Eicken H, Swanson BD, Deming JW (2006) Bacterial incorporation of leucine into protein down to −20°C with evidence for potential activity in sub-eutectic saline ice formations. Cryobiology 52(3):417–429

Kessler JD, Valentine DL, Redmond MC, Du M, Chan EW, Mendes SD, Quiroz EW, Villanueva CJ, Shusta SS, Werra LM (2011) A persistent oxygen anomaly reveals the fate of spilled methane in the deep Gulf of Mexico. Science 331(6015):312–315

Khelifi N, Grossi V, Hamdi M, Dolla A, Tholozan J-L, Ollivier B, Hirschler-Réa A (2010) Anaerobic oxidation of fatty acids and alkenes by the hyperthermophilic sulfate-reducing archaeon *Archaeoglobus fulgidus*. Appl Environ Microbiol 76(9):3057–3060

Killops S, Al-Juboori M (1990) Characterisation of the unresolved complex mixture (UCM) in the gas chromatograms of biodegraded petroleums. Org Geochem 15(2):147–160

King G, Kostka J, Hazen T, Sobecky P (2015) Microbial responses to the deepwater horizon oil spill: from coastal wetlands to the deep sea. Ann Rev Mar Sci 7:377–401

Kirchman DL, Cottrell MT, Lovejoy C (2010) The structure of bacterial communities in the western Arctic Ocean as revealed by pyrosequencing of 16S rRNA genes. Environ Microbiol 12(5):1132–1143

Kleindienst S, Seidel M, Ziervogel K, Grim S, Loftis K, Harrison S, Malkin SY, Perkins MJ, Field J, Sogin ML (2015) Chemical dispersants can suppress the activity of natural oil-degrading microorganisms. Proc Natl Acad Sci U S A 112(48):14900–14905

Kok M, Oldenhuis R, Van Der Linden M, Raatjes P, Kingma J, van Lelyveld PH, Witholt B (1989) The Pseudomonas oleovorans alkane hydroxylase gene. Sequence and expression. J Biol Chem 264(10):5435–5441

Kostichka K, Thomas SM, Gibson KJ, Nagarajan V, Cheng Q (2001) Cloning and characterization of a gene cluster for cyclododecanone oxidation in *Rhodococcus ruber* SC1. J Bacteriol 183 (21):6478–6486

Kotani T, Yurimoto H, Kato N, Sakai Y (2007) Novel acetone metabolism in a propane-utilizing bacterium, Gordonia sp. strain TY-5. J Bacteriol 189(3):886–893

Kovárová-Kovar K, Egli T (1998) Growth kinetics of suspended microbial cells: from single-substrate-controlled growth to mixed-substrate kinetics. Microbiol Mol Biol Rev 62 (3):646–666

Krembs C, Eicken H, Deming JW (2011) Exopolymer alteration of physical properties of sea ice and implications for ice habitability and biogeochemistry in a warmer Arctic. Proc Natl Acad Sci U S A 108(9):3653–3658

Krembs C, Mock T, Gradinger R (2001) A mesocosm study of physical-biological interactions in artificial sea ice: effects of brine channel surface evolution and brine movement on algal biomass. Polar Biol 24(5):356–364

Laban NA, Selesi D, Jobelius C, Meckenstock RU (2009) Anaerobic benzene degradation by Gram-positive sulfate-reducing bacteria. FEMS Microbiol Ecol 68(3):300–311

Le Borgne S, Paniagua D, Vazquez-Duhalt R (2008) Biodegradation of organic pollutants by halophilic bacteria and archaea. J Mol Microbiol Biotechnol 15(2-3):74–92

Learman DR, Henson MW, Thrash JC, Temperton B, Brannock PM, Santos SR, Mahon AR, Halanych KM (2016) Biogeochemical and microbial variation across 5500 km of Antarctic surface sediment implicates organic matter as a driver of benthic community structure. Front Microbiol 7:284. doi:10.3389/fmicb.2016.00284

Lee K, Merlin X (1999) Bioremediation of oil on shoreline environments: development of techniques and guidelines. Pure Appl Chem 71(1):161–171

Lee K, Nedwed T, Prince RC, Palandro D (2013) Lab tests on the biodegradation of chemically dispersed oil should consider the rapid dilution that occurs at sea. Mar Pollut Bull 73 (1):314–318

Lee K, Weise AM, St-Pierre S (1996) Enhanced oil biodegradation with mineral fine interaction. Spill Sci Technol Bull 3(4):263–267

Leuthner B, Heider J (2000) Anaerobic toluene catabolism of *Thauera aromatica*: the bbs operon codes for enzymes of beta oxidation of the intermediate benzylsuccinate. J Bacteriol 182 (2):272–277

Lindstrom JE, Braddock JF (2002) Biodegradation of petroleum hydrocarbons at low temperature in the presence of the dispersant Corexit 9500. Mar Pollut Bull 44(8):739–747

Lofthus S, Netzer R, Lewin A, Brakstad OG (2015) Successions of bacteria adhering to oil surfaces during biodegradation of crude oil in natural seawater at temperatures from 0 to 20°C. In: Abstracts of the 115th General Meeting of the American Society for Microbiology, New Orleans, Louisiana, May 30-June 2 2015

Lu Z, Deng Y, Van Nostrand JD, He Z, Voordeckers J, Zhou A, Lee Y-J, Mason OU, Dubinsky EA, Chavarria KL (2012) Microbial gene functions enriched in the Deepwater Horizon deep-sea oil plume. ISME J 6(2):451–460

Lu X, Zhang T, Fang HH-P, Leung KM, Zhang G (2011) Biodegradation of naphthalene by enriched marine denitrifying bacteria. Int Biodeter Biodegr 65(1):204–211

Luria CM, Ducklow HW, Amaral-Zettler LA (2014) Marine bacterial, archaeal and eukaryotic diversity and community structure on the continental shelf of the western Antarctic Peninsula. Aquat Microb Ecol 73(2):107–121

Luz A, Pellizari V, Whyte L, Greer C (2004) A survey of indigenous microbial hydrocarbon degradation genes in soils from Antarctica and Brazil. Can J Microbiol 50(5):323–333

Lynch JM, Moffat AJ (2005) Bioremediation–prospects for the future application of innovative applied biological research. Ann Appl Biol 146(2):217–221

Macnaughton SJ, Swannell R, Daniel F, Bristow L (2003) Biodegradation of dispersed Forties crude and Alaskan North Slope oils in microcosms under simulated marine conditions. Spill Sci Technol Bull 8(2):179–186

Maier T, Förster H-H, Asperger O, Hahn U (2001) Molecular characterization of the 56-kDa CYP153 from *Acinetobacter* sp. EB104. Biochem Biophys Res Commun 286(3):652–658

Mallick S, Chakraborty J, Dutta TK (2011) Role of oxygenases in guiding diverse metabolic pathways in the bacterial degradation of low-molecular-weight polycyclic aromatic hydrocarbons: a review. Crit Rev Microbiol 37(1):64–90

Margesin R, Miteva V (2011) Diversity and ecology of psychrophilic microorganisms. Res Microbiol 162(3):346–361

Margesin R, Schinner F (1997) Efficiency of indigenous and inoculated cold-adapted soil microorganisms for biodegradation of diesel oil in alpine soils. Appl Environ Microbiol 63 (7):2660–2664

Margesin R, Schinner F (1999) Biological decontamination of oil spills in cold environments. J Chem Technol Biotechnol 74(5):381–389

Margesin R, Schinner F (2001) Biodegradation and bioremediation of hydrocarbons in extreme environments. Appl Microbiol Biotechnol 56(5-6):650–663

Mason OU, Han J, Woyke T, Jansson JK (2014) Single-cell genomics reveals features of a Colwellia species that was dominant during the deepwater horizon oil spill. Front Microbiol 5:29–36

McFarlin K, Leigh MB, Perkins R (2014a) Biodegradation of oil and dispersed oil by Arctic marine microorganisms. In: International oil spill conference proceedings, May 2014, vol 1. American Petroleum Institute, pp 300317. doi:10.7901/2169-3358-2014-1-300317.1

McFarlin KM, Prince RC, Perkins R, Leigh MB (2014b) Biodegradation of dispersed oil in arctic seawater at -1°C. PLoS One 9(1):e84297

Methé BA, Nelson KE, Deming JW, Momen B, Melamud E, Zhang X, Moult J, Madupu R, Nelson WC, Dodson RJ (2005) The psychrophilic lifestyle as revealed by the genome sequence of *Colwellia psychrerythraea* 34H through genomic and proteomic analyses. Proc Natl Acad Sci U S A 102(31):10913–10918

Michaud L, Lo Giudice A, Saitta M, De Domenico M, Bruni V (2004) The biodegradation efficiency on diesel oil by two psychrotrophic Antarctic marine bacteria during a two-month-long experiment. Mar Pollut Bull 49(5–6):405–409

Miyasaka T, Asami H, Watanabe K (2006) Impacts of bioremediation schemes on bacterial population in naphthalene-contaminated marine sediments. Biodegradation 17(3):227–235

Mohanty G, Mukherji S (2008) Biodegradation rate of diesel range n-alkanes by bacterial cultures *Exiguobacterium aurantiacum* and *Burkholderia cepacia*. Int Biodeter Biodegr 61(3):240–250

Moody DE, Walsh SL, Rollins DE, Neff JA, Huang W (2004) Ketoconazole, a cytochrome P450 3A4 inhibitor, markedly increases concentrations of Levo-acetyl-α-methadol in ppioid-naive individuals. Clin Pharmacol Ther 76(2):154–166

Moriya K, Horikoshi K (1993) Isolation of a benzene-tolerant bacterium and its hydrocarbon degradation. J Ferment Bioeng 76(3):168–173

Nichols CM, Bowman JP, Guezennec J (2005) Effects of incubation temperature on growth and production of exopolysaccharides by an Antarctic sea ice bacterium grown in batch culture. Appl Environ Microbiol 71(7):3519–3523

Nikolopoulou M, Pasadakis N, Kalogerakis N (2013) Evaluation of autochthonous bioaugmentation and biostimulation during microcosm-simulated oil spills. Mar Pollut Bull 72(1):165–173

North EW, Adams E, Schlag ZZ, Sherwood CR, He RR, Hyun KHK, Socolofsky SA (2011) Simulating oil droplet dispersal from the deepwater horizon spill with a lagrangian approach. In: Liu Y, Macfadyen A, Ji Z-G, Weisber RH (eds) Monitoring and modeling the deepwater horizon oil spill: a record-breaking enterprise. American Geophysical Union, Washington, DC, pp 217–226. doi:10.1029/2011GM001102

NRC (1985) Oil in the sea Inputs, fates, and effects, vol 1. National Academy Press, Washington, DC

Obbard J, Ng K, Xu R (2004) Bioremediation of petroleum contaminated beach sediments: use of crude palm oil and fatty acids to enhance indigenous biodegradation. Water Air Soil Pollut 157 (1-4):149–161

Owens EH, Sergy GA, Guénette CC, Prince RC, Lee K (2003) The reduction of stranded oil by in situ shoreline treatment options. Spill Sci Technol Bull 8(3):257–272

Païssé S, Duran R, Coulon F, Goñi-Urriza M (2011) Are alkane hydroxylase genes (alkB) relevant to assess petroleum bioremediation processes in chronically polluted coastal sediments? Appl Microbiol Biotechnol 92(4):835–844

Païssé S, Goñi-Urriza M, Coulon F, Duran R (2010) How a bacterial community originating from a contaminated coastal sediment responds to an oil input. Microb Ecol 60(2):394–405

Païssé S, Goñi-Urriza M, Stadler T, Budzinski H, Duran R (2012) Ring-hydroxylating dioxygenase (RHD) expression in a microbial community during the early response to oil pollution. FEMS Microbiol Ecol 80(1):77–86

Passow U (2016) Formation of rapidly-sinking, oil-associated marine snow. Deep Sea Res II Top Stud Oceanogr 129:232–240. doi:10.1016/j.dsr2.2014.10.001

Passow U, Ziervogel K, Asper V, Diercks A (2012) Marine snow formation in the aftermath of the deepwater horizon oil spill in the Gulf of Mexico. Environ Res Lett 7(3):035301

Payne JR, Hachmeister LE, McNabb GD Jr, Sharpe HE, Smith GS, Menen CA (1991) Brine-induced advection of dissolved aromatic hydrocarbons to arctic bottom waters. Environ Sci Technol 25(5):940–951

Payne JR, McNabb GD (1991) Oil-weathering behavior in Arctic environments. Polar Res 10 (2):631–662

Pelletier E, Delille D, Delille B (2004) Crude oil bioremediation in sub-Antarctic intertidal sediments: chemistry and toxicity of oiled residues. Mar Environ Res 57(4):311–327

Peng R-H, Xiong A-S, Xue Y, Fu X-Y, Gao F, Zhao W, Tian Y-S, Yao Q-H (2008) Microbial biodegradation of polyaromatic hydrocarbons. FEMS Microbiol Rev 32(6):927–955

Powell SM, Ferguson SH, Bowman JP, Snape I (2006) Using real-time PCR to assess changes in the hydrocarbon-degrading microbial community in Antarctic soil during bioremediation. Microb Ecol 52(3):523–532

Prince RC (2005) The microbiology of marine oil spill bioremediation. In: Ollivier B, Magot M (eds) Petroleum microbiology. American Society of Microbiology, Washington, DC, pp 317–335

Prince RC (2008) Petroleum spill bioremediation in marine environments. Crit Rev Microbiol 19 (4):217–240

Prince R (2010) Bioremediation of marine oil spills. In: Timmis KN Timmis KN, McGenity TJ, van der Meer JR, de Lorenzo V (eds) Handbook of hydrocarbon and lipid microbiology. Springer, pp 2617-2630

Prince RC, Bare RE, Garrett RM, Grossman MJ, Haith CE, Keim LG, Lee K, Holtom GJ, Lambert P, Sergy GA (2003) Bioremediation of stranded oil on an Arctic shoreline. Spill Sci Technol Bull 8(3):303-312

Prince R, Clark J (2004) Bioremediation of marine oil spills. Stud Surf Sci Catal 151:495–512

Prince RC, McFarlin KM, Butler JD, Febbo EJ, Wang FC, Nedwed TJ (2013) The primary biodegradation of dispersed crude oil in the sea. Chemosphere 90(2):521–526

Prince RC, Parkerton TF, Lee C (2007) The primary aerobic biodegradation of gasoline hydrocarbons. Environ Sci Technol 41(9):3316–3321

Rabus R, Boll M, Golding B, Wilkes H (2016) Anaerobic degradation of p-alkylated benzoates and toluenes. J Mol Microbiol Biotechnol 26(1-3):63–75

Raymond JA, Fritsen C, Shen K (2007) An ice-binding protein from an Antarctic sea ice bacterium. FEMS Microbiol Ecol 61(2):214–221

Reddy CM, Arey JS, Seewald JS, Sylva SP, Lemkau KL, Nelson RK, Carmichael CA, McIntyre CP, Fenwick J, Ventura GT (2012) Composition and fate of gas and oil released to the water column during the deepwater horizon oil spill. Proc Natl Acad Sci U S A 109 (50):20229–20234

Redmond MC, Valentine DL (2012) Natural gas and temperature structured a microbial community response to the deepwater horizon oil spill. Proc Natl Acad Sci U S A 109 (50):20292–20297

Reed M, Daling PS, Brakstad OG, Singsaas I, Faksness L-G, Hetland B, Ekrol N (2000) OSCAR2000: a multi-component 3-dimensional oil spill contingency and response model. In: Proceedings of the 23 Arctic and Marine Oilspill Program (AMOP) Technical Seminar, Albeta, Canada, June 14-16, 2000

Ribicic D, Netzer R, Lewin A, Brakstad OG (2015) Biodegradation of dispersed oil at low seawater temperature (5°C). Comparison of chemistry and microbiology In: Abstracts of the 115th general meeting of the American Society for Microbiology, New Orleans, Louisiana, May 30-June 2 2015

Rike AG, Haugen KB, Børresen M, Engene B, Kolstad P (2003) In situ biodegradation of petroleum hydrocarbons in frozen arctic soils. Cold Reg Sci Technol 37(2):97–120

Rojo F (2009) Degradation of alkanes by bacteria: Minireview. Environ Microbiol 11 (10):2477–2490

Rotaru AE, Probian C, Wilkes H, Harder J (2010) Highly enriched Betaproteobacteria growing anaerobically with p-xylene and nitrate. FEMS Microbiol Ecol 71(3):460–468

Ruberto L, Vazquez SC, Mac Cormack WP (2003) Effectiveness of the natural bacterial flora, biostimulation and bioaugmentation on the bioremediation of a hydrocarbon contaminated Antarctic soil. Int Biodeter Biodegr 52(2):115–125

Ruff SE, Probandt D, Zinkann A-C, Iversen MH, Klaas C, Würzberg L, Krombholz N, Wolf-Gladrow D, Amann R, Knittel K (2014) Indications for algae-degrading benthic microbial communities in deep-sea sediments along the Antarctic Polar Front. Deep Sea Res II Top Stud Oceanogr 108:6–16. doi:10.1016/j.dsr2.2014.05.011

Sawamura S, Nagaoka K, Machikawa T (2001) Effects of pressure and temperature on the solubility of alkylbenzenes in water: volumetric properties and hydrophobic hydration. J Phys Chem B 105:2429–2436

Schedler M, Hiessl R, Valladares Juárez AG, Gust G, Müller R (2014) Effect of high pressure on hydrocarbon-degrading bacteria. AMB Express 4(1):7. doi:10.1186/s13568-014-0077-0
Schwarz J, Walker J, Colwell R (1974) Deep-sea bacteria: growth and utilization of hydrocarbons at ambient and in situ pressure. Appl Microbiol 28(6):982–986
Sergy GA, Guénette CC, Owens EH, Prince RC, Lee K (2003) In-situ treatment of oiled sediment shorelines. Spill Sci Technol Bull 8(3):237–244
Simon MJ, Osslund TD, Saunders R, Ensley BD, Suggs S, Harcourt A, Wen-chen S, Cruder DL, Gibson DT, Zylstra GJ (1993) Sequences of genes encoding naphthalene dioxygenase in *Pseudomonas putida* strains G7 and NCIB 9816-4. Gene 127(1):31–37
Siron R, Pelletier E, Brochu C (1995) Environmental factors influencing the biodegradation of petroleum hydrocarbons in cold seawater. Arch Environ Contam Toxicol 28(4):406–416
Si-Zhong Y, Hui-Jun J, Zhi W, Rui-Xia H, Yan-Jun J, Xiu-Mei L, Shao-Peng Y (2009) Bioremediation of oil spills in cold environments: a review. Pedosphere 19(3):371–381
Smith AJ, Flemings PB, Fulton PM (2014) Hydrocarbon flux from natural deepwater Gulf of Mexico vents. Earth Planet Sci Lett 395:241–253
Smith VH, Graham DW, Cleland DD (1998) Application of resource-ratio theory to hydrocarbon biodegradation. Environ Sci Technol 32(21):3386–3395
Stout SA, Payne JR (2016) Macondo oil in deep-sea sediments: Part 1–sub-sea weathering of oil deposited on the seafloor. Mar Pollut Bull 111(1):365-380
Sutton P, Lewis C, Rowland S (2005) Isolation of individual hydrocarbons from the unresolved complex hydrocarbon mixture of a biodegraded crude oil using preparative capillary gas chromatography. Org Geochem 36(6):963–970
Sveum P, Ladousse A (1989) Biodegradation of oil in the Arctic: enhancement by oil-soluble fertilizer application. In: International Oil Spill Conference 1989, vol 1. American Petroleum Institute, Washington, DC, pp 439–446
Swannell R, Lee K, McDonagh M (1996) Field evaluations of marine oil spill bioremediation. Microbiol Rev 60(2):342–365
Szaleniec M, Hagel C, Menke M, Nowak P, Witko M, Heider J (2007) Kinetics and mechanism of oxygen-independent hydrocarbon hydroxylation by ethylbenzene dehydrogenase. Biochemistry 46(25):7637–7646
Teske A, Durbin A, Ziervogel K, Cox C, Arnosti C (2011) Microbial community composition and function in permanently cold seawater and sediments from an Arctic fjord of Svalbard. Appl Environ Microbiol 77(6):2008–2018
Tyagi M, da Fonseca MMR, de Carvalho CCCR (2011) Bioaugmentation and biostimulation strategies to improve the effectiveness of bioremediation processes. Biodegradation 22 (2):231–241
Uraizee FA, Venosa AD, Suidan MT (1997) A Model for diffusion controlled bioavailability of crude oil components. Biodegradation 8(5):287–296
Valentine DL, Fisher GB, Bagby SC, Nelson RK, Reddy CM, Sylva SP, Woo MA (2014) Fallout plume of submerged oil from Deepwater Horizon. Proc Natl Acad Sci U S A 111 (45):15906–15911
Valentine DL, Kessler JD, Redmond MC, Mendes SD, Heintz MB, Farwell C, Hu L, Kinnaman FS, Yvon-Lewis S, Du M (2010) Propane respiration jump-starts microbial response to a deep oil spill. Science 330(6001):208–211
Valentine DL, Mezić I, Maćešić S, Črnjarić-Žic N, Ivić S, Hogan PJ, Fonoberov VA, Loire S (2012) Dynamic autoinoculation and the microbial ecology of a deep water hydrocarbon irruption. Proc Natl Acad Sci U S A 109(50):20286–20291
van Beilen JB, Wubbolts MG, Witholt B (1994) Genetics of alkane oxidation by *Pseudomonas oleovorans*. Biodegradation 5(3-4):161–174
van den Berg B (2005) The FadL family: unusual transporters for unusual substrates. Curr Opin Struct Biol 15(4):401–407
Van Hamme JD, Singh A, Ward OP (2003) Recent advances in petroleum microbiology. Microbiol Mol Biol Rev 67(4):503–549

Varadaraj R, Robbins ML, Bock J, Pace S, MacDonald D (1995) Dispersion and biodegradation of oil spills in water. In: Proceedings to the 1995 international oil spill conference. American Petroleum Institute, Washington, DC, pp 101–106

Venosa A, Holder E (2007) Biodegradability of dispersed crude oil at two different temperatures. Mar Pollut Bull 54(5):545–553

Venosa AD, Zhu X (2003) Biodegradation of crude oil contaminating marine shorelines and freshwater wetlands. Spill Sci Technol Bull 8(2):163–178

Wang B, Lai Q, Cui Z, Tan T, Shao Z (2008) A pyrene-degrading consortium from deep-sea sediment of the West Pacific and its key member *Cycloclasticus* sp. P1. Environ Microbiol 10 (8):1948–1963

Wang J, Sandoval K, Ding Y, Stoeckel D, Minard-Smith A, Andersen G, Dubinsky EA, Atlas R, Gardinali P (2016) Biodegradation of dispersed Macondo crude oil by indigenous Gulf of Mexico microbial communities. Sci Total Environ 557:453-468

Watson J, Jones D, Swannell R, Van Duin A (2002) Formation of carboxylic acids during aerobic biodegradation of crude oil and evidence of microbial oxidation of hopanes. Org Geochem 33 (10):1153–1169

Wentzel A, Ellingsen TE, Kotlar HK, Zotchev SB, Throne-Holst M (2007) Bacterial metabolism of long-chain n-alkanes. Appl Microbiol Biotechnol 76(6):1209–1221

Whyte LG, Bourbonniere L, Greer CW (1997) Biodegradation of petroleum hydrocarbons by psychrotrophic *Pseudomonas* strains possessing both alkane (alk) and naphthalene (nah) catabolic pathways. Appl Environ Microbiol 63(9):3719–3723

Whyte LG, Hawari J, Zhou E, Bourbonnière L, Inniss WE, Greer CW (1998) Biodegradation of variable-chain-length alkanes at low temperatures by a psychrotrophic *Rhodococcus* sp. Appl Environ Microbiol 64(7):2578–2584

Whyte LG, Schultz A, Van Beilen J, Luz A, Pellizari V, Labbé D, Greer C (2002b) Prevalence of alkane monooxygenase genes in Arctic and Antarctic hydrocarbon-contaminated and pristine soils. FEMS Microbiol Ecol 41(2):141–150

Whyte L, Smits T, Labbe D, Witholt B, Greer C, Van Beilen J (2002a) Gene cloning and characterization of multiple alkane hydroxylase systems in *Rhodococcus strains* Q15 and NRRL B-16531. Appl Environ Microbiol 68(12):5933–5942

Yakimov MM, Gentile G, Bruni V, Cappello S, D'Auria G, Golyshin PN, Giuliano L (2004) Crude oil-induced structural shift of coastal bacterial communities of rod bay (Terra Nova Bay, Ross Sea, Antarctica) and characterization of cultured cold-adapted hydrocarbonoclastic bacteria. FEMS Microbiol Ecol 49(3):419–432

Yakimov MM, Giuliano L, Gentile G, Crisafi E, Chernikova TN, Abraham W-R, Lünsdorf H, Timmis KN, Golyshin PN (2003) *Oleispira antarctica* gen. nov., sp. nov., a novel hydrocarbonoclastic marine bacterium isolated from Antarctic coastal sea water. Int J Syst Evol Microbiol 53(3):779–785

Yergeau E, Sanschagrin S, Beaumier D, Greer CW (2012) Metagenomic analysis of the bioremediation of diesel-contaminated Canadian high arctic soils. PLoS One 7(1):e30058

Zaki S, Farag S, Elreesh GA, Elkady M, Nosier M, El Abd D (2011) Characterization of bioflocculants produced by bacteria isolated from crude petroleum oil. Int J Environ Sci Technol 8(4):831–840

Zhang T, Gannon SM, Nevin KP, Franks AE, Lovley DR (2010) Stimulating the anaerobic degradation of aromatic hydrocarbons in contaminated sediments by providing an electrode as the electron acceptor. Environ Microbiol 12(4):1011–1020

Zhou N-Y, Al-Dulayymi J, Baird MS, Williams PA (2002) Salicylate 5-hydroxylase from *Ralstonia* sp. strain U2: a monooxygenase with close relationships to and shared electron transport proteins with naphthalene dioxygenase. J Bacteriol 184(6):1547–1555

Chapter 28
Bioremediation of Petroleum Hydrocarbon Spills in Cold Terrestrial Environments

Charles W. Greer and David F. Juck

Abstract Petroleum hydrocarbons, mostly fuels, are the principal source of energy in polar regions and as such their transport, storage, and use can lead to the contamination of soil and water. Hydrocarbons are natural, ubiquitous organic compounds that can serve as carbon and energy sources for many different organisms, the most abundant and diverse being the bacteria. Natural seeps in marine and freshwater (e.g., Athabasca watershed) environments, coupled with the presence of biologically produced hydrocarbons, mean that organisms able to metabolize hydrocarbons have access to these substrates almost all the time naturally and, as such, have evolved many different metabolic pathways to accommodate these substrates. Hydrocarbon-contaminated soils in cold environments present numerous challenges related to the physical and chemical properties of the soil, accessibility of the hydrocarbons and other essential nutrients, soil water content, temperature and other parameters inherent in these, as well as limitations associated with site remoteness and infrastructure availability. Although the literature on hydrocarbon biodegradation potential in cold environments is extensive, there remains a limited amount of information on case studies of large-scale bioremediation projects. This chapter presents an overview of the site-specific abiotic and biotic factors that have the most influence on the bioremediation of hydrocarbon-contaminated cold environment soils and presents a detailed case study of a successful bioremediation project in the high Arctic.

Contents

C.W. Greer (✉) • D.F. Juck
Energy, Mining and Environment, National Research Council Canada, 6100 Royalmount Ave., Montreal, QC, Canada, H4P 2R2
e-mail: Charles.greer@cnrc-nrc.gc.ca

R. Margesin (ed.), *Psychrophiles: From Biodiversity to Biotechnology*,
DOI 10.1007/978-3-319-57057-0_28

28.1 Introduction

Since the first review on the degradation of hydrocarbons by bacteria by Claude ZoBell (1946), numerous reviews and thousands of research papers have been published on this subject. One of the earlier and most-comprehensive reviews by Atlas (1981) laid out much of the groundwork for what was known at the time and fuelled many additional studies to fill in the gaps in our knowledge.

The natural attenuation of petroleum hydrocarbons in the marine environment is well known and has been extensively documented. Following the recent Deepwater Horizon accident in the Gulf of Mexico, rapid biodegradation was observed at depths of more than 1 km, where the water temperature was 4 °C (Hazen et al. 2010; Mason et al. 2012). Laboratory studies have also demonstrated that natural microbial assemblages from cold marine environments can rapidly degrade crude oil at sub-zero (McFarlin et al. 2014; Garneau et al. 2016) or near-zero temperatures (Brakstad and Bonaunet 2006).

The biodegradation potential of hydrocarbons in contaminated soils in cold regions has also been well documented. Several recent reviews have described many of the encountered limitations and challenges (Yang et al. 2009; Naseri et al. 2014; Camenzuli and Freidman 2015). Low temperature environments are usually remote, have limited infrastructure and numerous environmental constraints. In many cases, temperature is not the most important limiting factor for remediation: other factors such as soil moisture content/liquid water, nutrient availability, and the presence and activity of indigenous microbial hydrocarbon-degrading populations are important, and these factors are also intimately linked to temperature.

Bioremediation is by its nature the biological remediation or natural attenuation of contaminated sites. It is considered as a cost-effective and efficient method to remove many organic contaminants from soil systems. A wide range of prokaryotic and eukaryotic organisms are capable of metabolizing hydrocarbons, and these are dominated by eubacteria comprising 7 of the 24 current major phyla (Prince et al. 2010), and the microbial communities that inhabit various polar, alpine, and temperate hydrocarbon-contaminated soils have recently been reviewed (Greer et al. 2010). In polar and alpine areas, both abiotic and biotic factors are essential contributors to successful bioremediation. The low temperatures, which can fluctuate extensively over the course of a day causing freeze-thaw cycling, have a profound influence on biotic and abiotic factors. Site-specific conditions can be highly variable and can dictate the final outcome of any treatment strategy.

Abiotic and biotic factors contribute to the weathering of spilled hydrocarbons in cold environments and influence the observed rate of biodegradation as well as soil toxicity (Schafer et al. 2007, 2009; Naseri et al. 2014). Abiotic factors that are the most important are nutrients, temperature, and moisture. As hydrocarbon-degrading bacteria can be aerobic or anaerobic, both groups are usually involved and the involvement of each group is dictated by the local environmental conditions. Temperature is clearly one of the most important influential factors as the rates of chemical and enzymatic reactions are strongly affected by temperature. In addition, the bioavailability of the hydrocarbon substrates and other nutrients and electron acceptors such as oxygen are dictated in part by temperature.

Camenzuli and Freidman (2015) recently presented a review of remediation technologies to address hydrocarbon-contaminated sites in Arctic and Antarctic environments. The technologies reviewed included a number of operational strategies including landfarming, biopiles, and phytoremediation, as well as electrokinetic remediation and permeable reactive barriers. These approaches have been used in a variety of remediation projects in cold environments with varying degrees of success depending on the influences of the local physicochemical, geological, and biological parameters. The importance of site-specific conditions cannot be overstated as these will have profound effects on the likelihood of successful bioremediation.

28.2 Soil Moisture

In polar and alpine environments, moisture is critical since many of these environments experience desert-like conditions. The microbes that are able to survive and thrive in these environments are adapted to low-moisture conditions, something that is often not considered sufficiently during the design of bioremediation treatments.

Many cold soil environments have limited water availability, and the extent of precipitation can be quite low. Living cells require liquid water for survival and metabolism. Although determining the lowest temperature for life remains elusive, it is clear that bacteria are able to function at sub-zero temperatures providing there is liquid water available (Mykytczuk et al. 2013; Goordial et al. 2016). This is the result of the lowering of the freezing point of water in saline and hyper-saline environments. In cold environments that are subject to multiple freeze-thaw cycles over short periods of time, soil properties, nutrient availability, the activity of the indigenous microbial population, and the physical state of the hydrocarbon substrates are affected, which in turn will affect biodegradation activity.

Børresen and Rike (2007) examined the effects of salt, moisture, and nutrients on hexadecane mineralization in Arctic soil at 5 °C. They noted that although the acclimation period was extended at higher salt concentrations, the overall extent of mineralization was not affected, but that at higher soil moisture content, oxygen availability appeared to be decreased. In other studies on the effects of added nutrients in Arctic soils, there was a clear negative effect of higher nutrient concentrations on biodegradation activity (Walworth et al. 2001, 2007).

28.3 Nutrients

Cold environments typically don't support significant amounts of plant biomass and so the soil is limited in organic matter, which creates a less favorable growth environment for many microbes. There is an intimate relationship between nutrient concentrations and moisture content in polar and high elevation alpine soil systems. The addition of nutrients as an amendment to promote hydrocarbon degradation activity must be carefully controlled. Nutrients are critical since nitrogen and phosphorus are required elements during microbial metabolism of carbon compounds. The addition of high concentrations of nutrients in low moisture soils will result in increasing the ionic strength of the liquid water, which will inhibit microbial activity (Walworth et al. 2001, 2007; Børresen and Rike 2007). In addition, the freeze-thaw cycles that are frequent in cold environment soils will result in a dynamic redistribution of the added nutrients, affecting their bioavailability. Following a hydrocarbon spill, depending on the amount spilled, there is a large amount of carbon available for microbial growth, while other essential nutrients such as nitrogen and phosphorus may be in very limited supply (Margesin and Schinner 2001). As biodegradation proceeds, unless there is a continuous resupply of these essential nutrients, biodegradation activity will become limited. Although different C:N:P ratios have been suggested, a value of 100:15:1 has been reported for optimal microbial growth (Filler et al. 2006). Given the limited water availability in polar soils and the preceding discussion, it is prudent to add the necessary nutrients in smaller doses over a period of time and not all at once, with close monitoring of the soil moisture level and dissolved nutrient concentrations.

28.4 Temperature

ZoBell (1973) was one of the first to report on hydrocarbon biodegradation at below 0 °C, indicating that the bioremediation of polar and cold environment soils was a possibility. Colwell et al. (1978) found that Metula crude oil degradation in the Antarctic marine environment was considerably faster at 3 °C than at 22 °C, showing that temperature was not the most important factor in determining in situ biodegradation rates.

Although temperature may not be the limiting factor in cold environments, the physical state of the substrate and the availability of liquid water to support microbial functioning are limiting (Atlas 1981). Hydrocarbon biodegradation in soils at below freezing temperatures has been observed in the laboratory (Børresen and Rike 2007) and in the field (Rike et al. 2003). In addition, rapid biodegradation of crude oil in the Arctic marine environment at below freezing temperatures has been demonstrated (McFarlin et al. 2014; Garneau et al. 2016). These observations provide evidence for the biodegradation potential of hydrocarbons at below freezing temperatures and indicate that there are bacteria that are well-adapted to these temperatures. The injection of warmed and humidified air to enhance bioremediation in cold soils has also been performed to increase the effective yearly time period for biodegradation activity (Sanscartier et al. 2009; Couto et al. 2014; King et al. 2014). In some cases, this has been quite successful, but in others, the success has been limited due to the effects of too high temperatures on some microorganisms and the increased evaporation rate of soil water.

From a biochemical perspective, temperature is a major factor in reaction kinetics and growth. In recent years, the identification of bacteria that can metabolize at temperatures well below 0 °C has shed light on the adaptations that enhance membrane stability, ensure liquid water in the cytoplasm, and enzyme modifications that enhance low temperature activity. These studies have demonstrated that bacteria are capable of surviving and thriving in environments previously thought to have very limited biological activity. Akbari and Ghosal (2015) and Chang et al. (2011) showed how diurnal and seasonal temperature fluctuations could influence the rate of hydrocarbon biodegradation, as well as the evolution of different microbial communities. At a constant low incubation temperature of 5 °C, biodegradation rates were significantly reduced over those at 5–15 °C. The bacterial community that was rich in the alkane degradation gene *alkB* was dominated by Gammaproteobacteria at the constant lower temperature and has been seen elsewhere (Yergeau et al. 2012). Members of the Actinobacteria, another important hydrocarbon-degrading phylum, were present at all temperatures. The increased degradation rates observed were related to the increase in microbial activity as the temperature increased and show that the microbial populations appear well adapted to respond rapidly to changing temperatures. This bodes well for the bioremediation of hydrocarbon-contaminated soils in cold temperature environments.

28.5 Case Studies

There have been very few case studies of hydrocarbon biodegradation in cold environments: most published studies have been performed in temperate environments. One case study by Dibble and Bartha (1979) in a New Jersey wheat field contaminated by a kerosene spill used liming, fertilization, and frequent tilling and after 2 years, surface soil concentrations were insignificant. After 1 year, the soil was returned to a "near-normal" functional state. In strong contrast, Sexstone et al. (1978) found very low biodegradation rates in hydrocarbon-contaminated Alaskan soil, with hydrocarbons still present 28 years after the spill. This highlights the basic difference between performing bioremediation in temperate regions as compared to permanently cold environments.

Mohn et al. (2001) performed small-scale bioremediation field studies at an Arctic and a subarctic tundra site. They found enhanced biodegradation activity in soils that had been biostimulated, with and without peat as a bulking agent and a microbial inoculum from a hydrocarbon-degrading enrichment culture. In terms of larger-scale bioremediation projects, Juck et al. (2007) reported on a multi-year on site/in situ bioremediation project at Eureka in the Canadian High Arctic, Ellesmere Island, Nunavut. Soil contaminated by a surface diesel fuel spill affected an area of more than 3200 m^2, part of the area being adjacent to the station's potable water supply. The bioremediation approach included initial solid and liquid fertilizer application with tilling of the upper 10–20 cm of soil, followed by yearly application of liquid fertilizer and tilling. Within four summer seasons of treatment, the active layer soil PHC contamination had been reduced by 85% and the supra-permafrost soil (at a depth of ca. 1 m) concentrations had been reduced by 50%.

28.6 Bioremediation Program at CFS-Alert

To put many of these challenges into perspective, a case study is presented to illustrate a large-scale bioremediation project at Canadian Forces Station (CFS) Alert, the most northerly, permanently inhabited location in the world.

28.6.1 Introduction

CFS-Alert is located on the north-eastern tip of Ellesmere Island, Nunavut (82°30′ N, 62°20′ W), approximately 800 km south of the geographic North Pole and 40 km south of the northernmost point in Canada. Built in 1950 as a weather station, Alert became a Canadian Military Communications Research Facility in 1956. At its peak of activity in the 1970s, 200 military staff were stationed there throughout the year. CFS-Alert is now a remotely operated listening facility, and staffing has been reduced to approximately 50–100 operational personnel. Due to the remote nature of the site and logistical constraints, all electricity, heating, and vehicle fuelling requirements are derived from JP8 jet fuel (referred to as "diesel") stored in two separate tank farms: one adjacent to the airstrip apron (lower tank farm) and the second located adjacent to the main station (upper tank farm), approximately 1.3 km from the airstrip. The movement of fuel between tank farms, the transfer of fuel to vehicles, and stand-alone reservoirs and historic activities on site constitute the principal avenues by which petroleum hydrocarbon (PHC) contamination has occurred at CFS-Alert. The most important climatic conditions impacting biological activity at CFS-Alert include low temperatures (average summer air temperatures of −0.4 °C, 3.4 °C, and 0.8 °C in June, July, and August, respectively) and desert conditions (total average annual precipitation of 158.3 mm) (Government of Canada website: http://climate.weather.gc.ca/climate_normals/results_1981_2010_e.html?stnID=1731&lang=e&StationName=Alert&SearchType=Contains&stnNameSubmit=go&dCode=5&dispBack=1).

A preliminary evaluation of the potential for bioremediation of the PHC-contaminated soil at CFS-Alert was performed in a laboratory feasibility study (Greer et al. 2007). Based on the results of the feasibility study, which

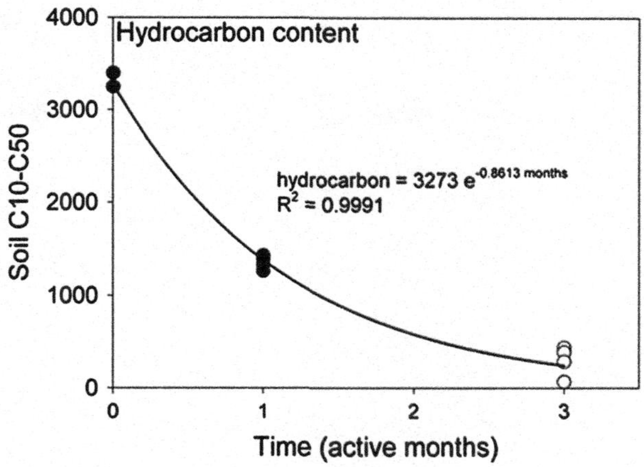

Fig. 28.1 Residual PHC in the pilot-scale biopile project initially, after 1 month and then in July the following year (effectively 3 months of above-zero temperatures). Each data point represents a composite sample

demonstrated effective biodegradation activity following nutrient addition, a small-scale pilot study (small biopile) was initiated on site in July 2006.

On site monitoring of the small-scale pilot study was performed after 1 month and yearly thereafter. The residual hydrocarbon concentration in the biopile soil was reduced by 60% after 1 month of incubation (July–August) and by July of the following year, the average residual PHC concentration was below 500 mg kg^{-1} (Fig. 28.1). A detailed analysis of the evolution of the indigenous bacterial community during this pilot study showed that the bacterial community structure changed over time, initially being dominated by Proteobacteria and later by Actinobacteria (Yergeau et al. 2012).

28.6.2 *Source of Diesel-Contaminated Soil and Biopile Design*

In September of 2006, the first of two large spills occurred, and the approach deployed in the small-scale demonstration project was rapidly scaled up into a large-scale biopile project. These "fresh spill" sites (2006 Diesel Pipeline Spill and 2007 Standby Power Plant Spill) are described below.

The diesel pipeline spill of September 2006 was due to the rupture of an expansion joint adjacent to the airstrip apron, along the diesel pipeline used to transfer fuel between the lower and upper tank farms (Fig. 28.2). Within hours of the spill event, the lower extent of the impacted area was excavated and a soil dam built to capture any impacted runoff created during the ubsequent spring freshet. There was minimal penetration of the diesel fuel at this time due to the frozen state of the surface soil. The excavated material was placed in the small biopile area ("small" in Fig. 28.2).

Fig. 28.2 The 2006 diesel pipeline spill site. The *arrow* indicates where the pipeline break occurred and the *white outline* indicates the extent of soil excavated. *Small* and *large* refer to the two biopiles (aerial photo NRC)

During the winter of 2006–2007, the large biopile ("large" in Fig. 28.2) was designed, materials purchased, and transported to CFS-Alert. Starting in June 2007, the large biopile area was constructed: it consisted of a 30 cm packed gravel base covered with an impermeable membrane, overlain with geotextile and 15 cm of packed gravel. The exterior berm surrounding the area was approximately 2 m in height and the final biopile area measured 40 × 100 m. The purpose of the construction was to have an on site platform for isolating and treating contaminated soils from current and future hydrocarbons spills.

In July 2007, the remaining diesel impacted area (outlined in white in Fig. 28.2) was excavated until clean soil was reached in all areas, to a depth of approximately 1 m. The impacted soil was placed into the large biopile in ca. 2 m high windrows. The total volume of excavated soil was estimated at 3000 m^3. A control pile, to monitor the effectiveness of the applied treatment system, was created during transfer of the contaminated soil into the biopile area. A single dump truck load of contaminated soil (ca. 8 m^3) was placed in the biopile area and separated from the windrows by ca. 10 m.

The second fuel spill occurred at the standby power plant in August, 2007, due to human error during the re-filling of a fuel tank located at the rear of the Standby Power Plant (Fig. 28.3). The area impacted by diesel adjacent to the splashdown area (outlined in white in Fig. 28.3) was immediately excavated (within 24–48 h of the event) until clean soil was reached in all areas, at a depth of approximately 1 m. The excavated soil was transported to the large biopile area where it was placed in separate windrows. The volume of soil excavated was estimated at 250 m^3.

Fig. 28.3 Standby Power Plant Spill of 2007. The area outlined in *white* indicates the extent of the soil excavated (aerial photo Google Earth)

28.6.3 Biopile Treatment Approach

Based on the results of the previous lab and pilot scale field work, as well as an understanding of the site-specific limitations on heavy equipment access and operator experience, and the difficulty of accessing CFS-Alert, the treatment plan established was based on the philosophy of keeping operations and soil manipulations simple and to a minimum. Treatment of the contaminated soil consisted of the annual manual application of small quantities of slow release mono-ammonium phosphate (MAP) (ca. 150 kg per application, which translates into a yearly fertilization rate of 21 mg N L^{-1} soil water) followed by turning of the soil using heavy equipment to mix in the MAP and aerate the soil. The fresh spill soils undergoing treatment in the biopile were initially divided into two areas, the soil from the 2006 Diesel Pipeline spill and the soil from the 2007 Standby Power Plant spill. A summary of the activities performed during the treatment from 2007 to 2015 (application of MAP and turning of soils) is presented in Table 28.1. After several years of soil turning, the original division between the two areas of the biopile had become obscured during the mixing, so from this point on the soils were treated as a single unit during treatment and monitoring.

The control biopile, as described in Sect. 28.6.2, was not treated with MAP nor was it turned at any point during the treatment process. The control biopile was sampled for chemical and microbiological analyses at the same time as the large biopile.

Table 28.1 Summary of large biopile (fresh spill soil) treatment system activities from 2007 to 2015

Soils	Application of MAP[a] (0.05 kg m^{-3} per application)	Turning event
Large biopile (fresh spill soil)	08/2007	
	07/2008	07/2008
	08/2009	08/2009
	06/2010, 08/2010	06/2010, 08/2010
	07/2011, 08/2011	07/2011
	07/2012, 08/2012	08/2012
	08/2013	08/2013
	07/2014	07/2014, 08/2014
	07/2015	07/2015
Control soil	None	None

[a]Mono-ammonium phosphate

28.6.4 Bioremediation Performance Monitoring

Samples for chemical analyses were collected annually. At the beginning of the treatment, samples were collected from the large biopile in mid-August before MAP was added or turning was performed. The samples destined for chemical analyses were transported to an accredited lab and PHC fraction 1–4 (F1–F4), and benzene, toluene, ethylbenzene, and total xylenes (BTEX) analyses were performed, as prescribed by the Canadian Council of Ministers of the Environment (CCME) (2008). During later field seasons (2013, 2014, 2015) samples for chemical analysis were collected in early July and in mid-August. A final sampling campaign was performed in August 2016, immediately preceding the movement of cleaned soils into the original excavation located at the 2006 Diesel Pipeline spill site (Fig. 28.2).

All samples collected from the treatment area were composites: 5–6 sub-samples of equal volume were collected from within the sample area at depths of at least 20 cm, from randomly selected areas (e.g., tops and bottoms) of the windrows from holes dug manually with a shovel. Each sample was then collected using fresh disposable gloves and taken from newly exposed soil (i.e., soil not contacted by the shovel). Sub-samples were mixed together, large stones removed, and then the samples were transferred into glass sample jars with Teflon coated caps, and the sample jars were completely filled with soil to avoid contaminant volatilization. Field and transport blanks prepared by the analytical labs, containing clean sand, were also prepared and shipped along with the test samples. The samples collected from the control biopile were single samples, taken from depths of at least 20 cm from the top of the pile using a clean shovel, new disposable gloves, and soil which was not contacted by the shovel. The samples were transferred into glass sample jars with Teflon coated caps, as described above.

Chemical analyses were performed at least yearly on all biopile soils. Comparison guidelines for the assessment of soil contamination at CFS-Alert were based on the CCME Canada-Wide standards (2008). PHC F1–F4 and BTEX results for the duration of the treatment campaign were compared to the more stringent CCME (2008) commercial/industrial generic levels for coarse soils. The CCME commercial/industrial land use category is assigned to a site where the primary activity involves the production, manufacture, or storage of materials and where public access is restricted. Children are not permitted continuous access or occupancy on commercial/industrial sites. This land use category corresponds to the activities occurring at CFS-Alert.

Total Kjeldahl nitrogen (TKN) analyses were also performed on the fresh spill soils from 2011 to 2015 to monitor for total nitrogen concentrations in the soils. TKN analyses were performed on four composite soils (five sub-samples per composite) collected prior to performing the annual addition of MAP.

Based on the observed biodegradation success (below) of the treatment approach on the "fresh spill" soils (i.e., those where treatment commenced within 1 year of the spill event), several "aged spill" sites (contamination occurred more than 12 years before treatment) were excavated in 2011 and subjected to the same treatment approach. Soils from three different areas were excavated and transported to a separate section of the large biopile treatment area and mixed together, and the previously established bioremediation methodology of annual application of MAP (25 kg per application, which translates into a yearly average of 150 mg N L^{-1} soil water) and turning was applied (summary of activities in Table 28.2). The aged spill soils, from two separate areas around the Main Power Plant and from another area along the Diesel Pipeline, were combined and treated in a single biopile. The total volume of soil excavated from the two Main Power Plant sites and the Diesel Pipeline site was estimated at 85 m^3.

Residual hydrocarbon analyses (PHC F1–F4 and BTEX) of the large biopile samples demonstrated that only PHC F2 was above the CCME guideline concentration of 260 mg kg^{-1} in all of the samples collected initially from both the fresh spill and aged spill sites. All other components (PHC-F1, -F3, -F4, and BTEX)

Table 28.2 Summary of aged spill biopile treatment system activities from 2011 to 2016

Soils	Application of MAP[a] (0.3 kg m^{-3} per application)	Turning event
Large biopile (aged spill soil)	07/2011	
	07/2012	08/2012
	08/2013 (×2)	08/2013
	07/2014 (×2)	08/2014
	07/2015	07/2015
	08/2016	08/2016

[a]Mono-ammonium phosphate

remained below CCME guidelines throughout the duration of the treatment (results not shown).

The evolution of PHC F2 concentrations over time in the large biopile treatment area for both fresh and aged spill soils is presented in Fig. 28.4. The average F2 concentration at the start of the treatment for the fresh spill soil (2007) and aged spill soil (2011) was approximately 2000 mg kg^{-1}, which dereased to below the CCME guideline by 2013 and 2014, respectively. Although the average PHC F2 concentrations were below CCME guidelines by 2013 and 2014, samples from some areas remained above the 260 mg kg^{-1} target. Treatment was continued until August 2015, at which time all samples from the fresh spill soils were below the CCME PHC F2 guideline with an average of 72 mg kg^{-1}. For the aged spill soils, no hotspots were detected during the 2016 sampling campaign and the average PHC F2 concentration was 106 mg kg^{-1}.

Nitrogen (TKN) analyses of the final five seasons of the fresh spill soils (Fig. 28.5) demonstrated that nitrogen levels were never significantly elevated in the soils and that they remained relatively stable. When compared to the significant decrease in average PHC concentrations over the course of the treatment, from 2000 mg kg^{-1} to 72 mg kg^{-1}, nitrogen neither appeared to be limiting with respect to biodegradation activity nor did it accumulate in the soils when the PHC carbon was consumed and removed from the system.

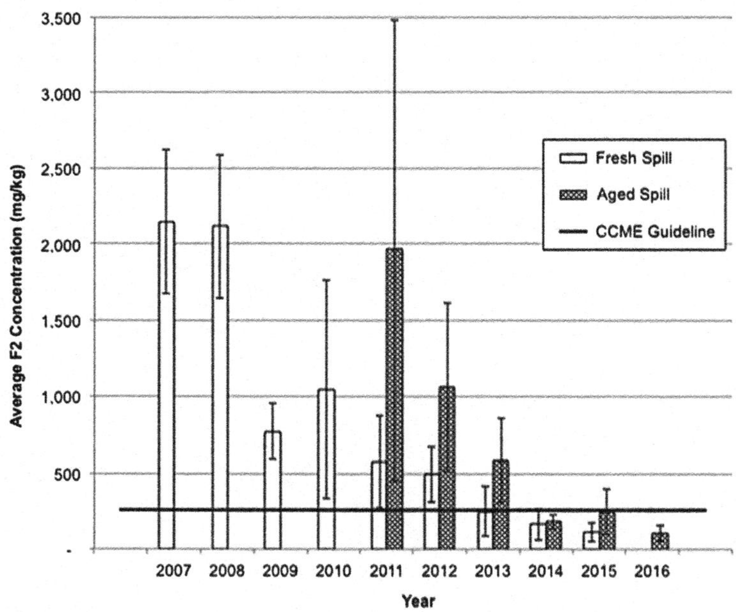

Fig. 28.4 Petroleum Hydrocarbon (PHC) Fraction 2 (F2) concentrations over time in the fresh spill (*open bars*) and aged spill (*hatched bars*) soils. The *black line* indicates the CCME PHC F2 guideline of 260 mg kg^{-1}. Error bars represent one standard deviation

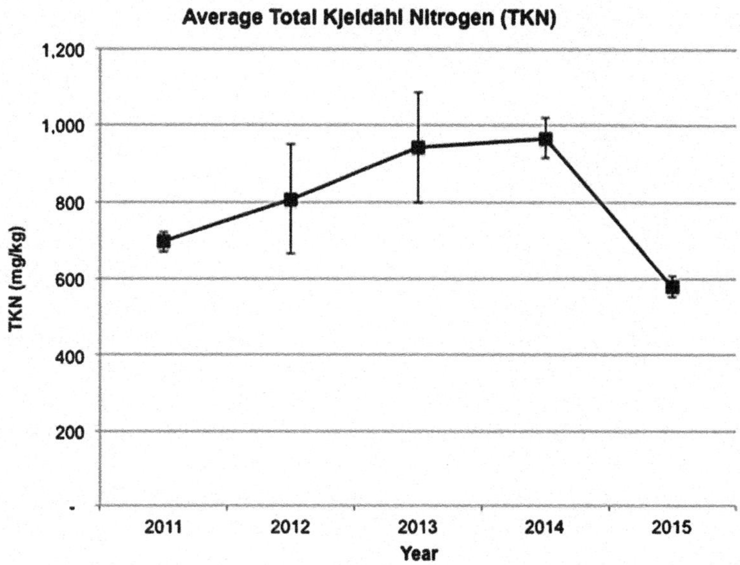

Fig. 28.5 Average TKN concentrations in fresh spill soils from 2011 to 2015. Error bars represent one standard deviation

The concentration of PHC F2 in the control biopile was quite variable over time (results not shown) with an overall average throughout the duration of the project of 1873 mg kg^{-1}. This indicates that the vast majority of the observed PHC degradation was due to the stimulated biological activity associated with the annual addition of MAP and turning of the biopiles. The recorded loss of PHC from the control biopile was most likely due to volatilization.

It is of particular interest that the time required to reach the CCME PHC F2 guidelines in the aged spill soils was less than that required in the new spill soils, even though they both started out at roughly the same concentration. This, at first glance may be counter intuitive, as aged contaminants are generally less bioavailable and longer treatment times are often required. However, fresh contamination, especially with lower molecular mass PHC, is known to increase soil toxicity (Cermak et al. 2010). The observed rapid rate of biodegradation was most likely due to the difference in total MAP added to the soils; over the course of the treatment, the total concentration of MAP added to the fresh spill and aged spill soils was 0.25 g and 1.2 g MAP kg^{-1} soil, respectively. While the concept of adding more MAP to speed up biodegradation appears to hold in this example, lab based experiments using fresh spill soils demonstrated that when elevated concentrations of MAP were used to stimulate the biodegradation of PHCs, an inhibitory effect on biodegradation activity was noted (results not shown). The periodic addition of low concentrations of MAP was the most appropriate method to stimulate the desired long-term biodegradation activities observed during the biopile treatment process.

Overall, the ability to biodegrade both fresh and aged PHC contamination in soil under the extreme climatic conditions encountered at CFS-Alert was clearly demonstrated and highlights the feasibility of using biologically based treatment processes at challenging sites across the Arctic.

28.7 Conclusions

The bioremediation of hydrocarbon-contaminated soils in cold environments is not only feasible, but has been successfully accomplished, in spite of the numerous challenges imposed by remoteness, limited infrastructure, and harsh environmental conditions, including both abiotic and biotic pressures. Site-specific conditions will impact the bioavailability of hydrocarbon substrates, other essential nutrients, the presence of liquid water, and the necessary microorganisms to perform the degradation, all of which must be considered in the design and implementation of bioremediation approaches. Although longer treatment times can be anticipated, bioremediation can be a cost-effective and efficient means to address hydrocarbon contamination in cold soil environments. The recent successful completion of a large-scale bioremediation project at CFS-Alert in the Canadian high Arctic provided considerable insight into how to address the various challenges that can be encountered in these extreme environments. It is clear that additional large-scale demonstration projects are required to provide a more thorough validation of application of bioremediation technologies in cold regions under the variety of site-specific and harsh environmental conditions that would be encountered for much of the year.

Acknowledgment The authors gratefully acknowledge the contributions of Drew Craig, Don Kovanen, Andrew Tam and Chris McRae of the Environment Office at 8 Wing Trenton, Department of National Defence Canada.

References

Akbari A, Ghoshal S (2015) Effects of diurnal temperature variation on microbial community and petroleum hydrocarbon biodegradation in contaminated soils from a sub-Arctic site. Environ Microbiol 17(12):4916–4928

Atlas RM (1981) Microbial degradation of petroleum hydrocarbons: an environmental perspective. Microbiol Rev 45:180–209

Børresen MN, Rike AG (2007) Effects of nutrient content, moisture content and salinity on mineralization of hexadecane in an Arctic soil. Cold Reg Sci Technol 48:129–138

Brakstad OG, Bonaunet K (2006) Biodegradation of petroleum hydrocarbons in seawater at low temperatures (0–5°C) and bacterial communities associated with degradation. Biodegradation 17:71–82

Camenzuli D, Freidman BL (2015) On-site and in situ remediation technologies applicable to petroleum hydrocarbon contaminated sites in the Antarctic and Arctic. Polar Res 34:24492. doi:10.3402/polar.v34.24492

Canadian Council of Ministers of the Environment (CCME) (2008) Canada-wide standards for petroleum hydrocarbons (PHC) in soil, Table 1 revised 2008. http://www.ccme.ca/files/Resources/csm/phc_cws/phc_standard_1.0_e.pdf

Cermak JH, Stephenson GL, Birkholtz D, Wang Z, Dixon DG (2010) Toxicity of petroleum distillates to soil organisms. Environ Toxicol Chem 29(12):2685–2694

Chang W, Whyte L, Ghoshal S (2011) Comparison of the effects of variable site temperatures and constant incubation temperatures on the biodegradation of petroleum hydrocarbons in pilot-scale experiments with field-age contaminated soils from a cold regions site. Chemosphere 82:872–878

Colwell RR, Mills AL, Walker JD, Garcia-Tello P, Campos-P V (1978) Microbial ecology studies of the Metula spill in the Straits of Magellan. J Fish Res Board Can 35:573–580

Couto N, Fritt-Rasmussen J, Jensen PE, Højrup M, Rodrigo AP, Ribeiro AB (2014) Suitability of oil bioremediation in an Artic soil using surplus heating from an incineration facility. Environ Sci Pollut Res 21:6221–6227

Dibble JT, Bartha R (1979) Rehabilitation of oil-inundated agricultural land: a case history. Soil Sci 128:56–60

Filler DM, Reynolds CM, Snape I, Daugulis AJ, Barnes DL, Williams PJ (2006) Advances in engineered remediation for us in the Arctic and Antartica. Polar Rec 42:111–120

Garneau M-E, Michel C, Meisterhans G, Fortin N, King TL, Greer CW, Lee K (2016) Hydrocarbon biodegradation by Arctic sea-ice and sub-ice microbial communities during microcosm experiments, Northwest Passage (Nunavut, Canada). FEMS Microbiol Ecol 92(10):fiw130. doi:10.1093/femsec/fiw130

Goordial J, Davila A, Lacelle D, Pollard W, Marinova MW, Greer CW, DiRuggiero J, McKay CP, Whyte LG (2016) Nearing the cold-arid limits of microbial life in permafrost of an upper dry valley, Antarctica. ISME J 10:1613–1624

Greer CW, Masson C, Beaumier D, Kovanen D, Craig D (2007) Biotreatability assessment and on site biotreatment monitoring of hydrocarbon contaminated soil at CFS-Alert, Nunavut. In: Biggar K, Cotta G, Nahir M, Mullick A, Buchko J, Ho A, Guigard S (eds) Assessment and remediation of contaminated sites in Arctic and cold climates (ARCSACC), Edmonton, AB, Canada, pp 3–11

Greer CW, Whyte LG, Niederberger TD (2010) Microbial communities in hydrocarbon contaminated temperate, tropical, alpine and polar soils. In: Timmis KN, McGenity TJ, van der Meer JR, de Lorenzo V (eds) Handbook of hydrocarbon and lipid microbiology. Springer, Berlin, pp 2313–2328

Hazen TC, Dubinsky EA, DeSantis TZ, Andersen GL, Piceno YM, Singh N, Jansson JK, Probst A, Borglin SE, Fortney JL, Stringfellow WT, Bill M, Conrad ME, Tom LM, Chavarria KL, Alusi TR, Lamendella R, Joyner DC, Spier C, Baelum J, Auer M, Zemla ML, Chakraborty R, Sonnenthal EL, D'Haeseleer P, Holman HYN, Osman S, ZM L, Van Nostrand J, Deng Y, Zhou JZ, Mason OU (2010) Deep-sea oil plume enriches indigenous oil-degrading bacteria. Science 330:204–208

Juck DF, Labbe D, Buchko J, Domijan E, Whyte LG Greer CW (2007) Monitoring of a high Arctic site impacted by petroleum hydrocarbons: lessons learned from microbial community characterization. In: Biggar K, Cotta G, Nahir M, Mullick A, Buchko J, Ho A, Guigard S (eds) Assessment and remediation of contaminated sites in Arctic and cold climates (ARCSACC), Edmonton, AB, Canada, pp 167–173

King MM, Kinner NE, Deming D, Simonton JA, Belden LM (2014) Bioventing of no. 2 fuel oil: effects of air flowrate, temperature, nutrient amendment, and acclimation. Remediat J 24:47–60

Margesin R, Schinner F (2001) Bioremediation (natural attenuation and biostimulation) of diesel-oil-contaminated soil in an alpine glacier skiing area. Appl Environ Microbiol 67:3127–3133

Mason OU, Hazen TC, Borglin S, Chain PS, Dubinsky EA, Fortney JL, Han J, Holman HY, Hultman J, Lamendella R (2012) Metagenome, metatranscriptome and single-cell sequencing reveal microbial response to Deepwater Horizon oil spill. ISME J 6:1715–1727

McFarlin KM, Prince RC, Perkins R, Leigh MB (2014) Biodegradation of dispersed oil in Arctic seawater at -1°C. PLoS ONE 9(1):e84297. doi:10.1371/journal.pone.0084297

Mohn WW, Radziminski CZ, Fortin M-C, Reimer KJ (2001) On site bioremediation of hydrocarbon-contaminated Arctic tundra soils in inoculated biopiles. Appl Microbiol Biotechnol 57:242–247

Mykytczuk NCS, Foote SJ, Omelon CR, Southam G, Greer CW, Whyte LG (2013) Bacterial growth at -15°C; molecular insights from the permafrost bacterium *Planococcus halocryophilus* Or1. ISME J 7:1211–1226

Naseri M, Barabadi A, Barabady J (2014) Bioremediation treatment of hydrocarbon-contaminated Arctic soils: influencing parameters. Environ Sci Pollut Res 21:11250–11265

Prince RC, Gramain A, McGenity TJ (2010) Prokaryotic hydrocarbon degraders. In: Timmis KN (ed) Handbook of hydrocarbon and lipid microbiology. Springer, Berlin, pp 1672–1692

Rike AG, Haugen KB, Borreson M, Engene B, Kolstad P (2003) In situ biodegradation of petroleum hydrocarbons in frozen Arctic soils. Cold Reg Sci Technol 37:97–120

Sanscartier D, Zeeb B, Kch I, Reimer KJ (2009) Bioremediation of diesel-contaminated soil by heated and humidified biopile system in cold climates. Cold Reg Sci Technol 55:167–173

Schafer AN, Snape I, Siciliano SD (2007) Soil biogeochemical toxicity end points for sub-Antarctic islands contaminated with petroleum hydrocarbons. Environ Toxicol Chem 26 (5):890–897

Schafer AN, Snape I, Siciliano SD (2009) Influence of liquid water and soil temperature on petroleum hydrocarbon toxicity in Antarctic soil. Environ Toxicol Chem 28(7):1409–1415

Sexstone A, Everett K, Jenkins T, Atlas RM (1978) Fate of crude and refined oils in north slope soils. Arctic 31:339–347

Walworth J, Braddock J, Woolard C (2001) Nutrient and temperature interactions in bioremediation of cryic soils. Cold Reg Sci Technol 32:85–91

Walworth J, Pond A, Snape I, Rayner J, Ferguson S, Harvey P (2007) Nitrogen requirements for maximizing petroleum bioremediation in a sub-Antarctic soil. Cold Reg Sci Technol 48:84–91

Yang S-Z, Jin H-J, Wei Z, He R-X, Ji Y-J, Li X-M, S-P Y (2009) Bioremediation of oil spills in cold environments: a review. Pedosphere 19(3):371–381

Yergeau E, Sanschagrin S, Beaumier D, Greer CW (2012) Metagenomic analysis of the bioremediation of diesel-contaminated Canadian high Arctic soils. PLoS ONE 7(1):e30058. doi:10.1371/journal.pone.0030058

ZoBell CE (1946) Action of microorganisms on hydrocarbons. Bacteriol Rev 10:1–49

ZoBell CE (1973) Bacterial degradation of mineral oils at low temperatures. In: Ahearn DG, Meyers SP (eds) The microbial degradation of oil pollutants. Publication no. LSU-SG-73-01, Center for Wetland Resources, Louisiana State University, Baton Rouge, p. 153–161

Chapter 29
Rhizoremediation in Cold Climates

Kim Yrjälä, Timo P. Sipilä, and Shinjini Mukherjee

Abstract Rhizoremediation has become increasingly interesting as it offers several solutions to environmental problems by making use of plants. The International Phytotechnology Society (http://phytosociety.org/) defines phytotechnology as the strategic use of plants to solve environmental problems by remediating the qualities and quantities of our soil, water, and air resources and by restoring ecosystem services in managed landscapes. Plants always interact with belowground microbes, bacteria, fungi, and archaea, and even aboveground epi/endophytic microbes. The recent adoption of omics techniques has led to much widened understanding of soil microbial communities, and conditions that promote predictable activity in contaminated soils with effects on plants. These methods have in microbial ecology brought out new concepts like plant microbiome describing the wide array of microorganisms living and interacting in different ways with the plant. The identification of increasing numbers of microbes associated with plants helps to notify new functional groups of microbes that are and become important for applications of phytotechnology. In the boreal cold climate freezing and thawing of soil occurs, which shapes the active microbial communities in a peculiar way. Also in the cold season soil, microbes perform tasks important for ecosystem functioning. The plants create a suitable environment for microbes especially in the rhizosphere where root exudates are excellent food for microbes in the vicinity of roots. Woody plants have received increased attention, especially poplars (*Populus* spp.), when it was recognized that they can reduce the level of trichloroethylenes in soil with the help of endophytes. Poplars have successfully been used for rhizoremediation of petroleum and other hydrocarbon compounds.

K. Yrjälä (✉) • S. Mukherjee
MEM-Group, Department of Biosciences, University of Helsinki, PO Box 56 (Viikki Biocenter 1), 00014 Helsinki, Finland
e-mail: kim.yrjala@helsinki.fi

T.P. Sipilä
MEM-Group, Department of Biosciences, University of Helsinki, PO Box 56 (Viikki Biocenter 1), 00014 Helsinki, Finland

Division of Plant Biology, Department of Biosciences, Viikki Plant Science Centre, University of Helsinki, Helsinki, Finland

R. Margesin (ed.), *Psychrophiles: From Biodiversity to Biotechnology*,
DOI 10.1007/978-3-319-57057-0_29

Contents

29.1 Introduction

Rhizoremediation has been coined as an alternative in situ method for cleanup of polluted soils. It relies on the use of plants and their associated microbes for biodegradation of organic compounds. The biodegradation of harmful compounds takes place mainly in the rhizosphere of plants which is a hotspot for microbial activity in soil (Gerhardt et al. 2009; Glick 2010). The method fulfills the criteria of sustainability in that it takes us of nature's own system developed along the evolution of different organisms on Earth. While the plant through photosynthesis is producing sugars with the energy from light, it releases secondary metabolites as exudates in the root zone that stimulates microbial activity in soil (Pilon-Smits 2005). It is considered that the abundance of microbes in the rhizosphere most often is elevated compared to the bulk soil due to the diverse growth substrates released by the plant. Manmade chemicals like herbicides and pesticides often containing aromatic and aliphatic ring structures are not completely new for plant-associated microbes since several plant secondary metabolites also contain similar compounds (Singer et al. 2003; Martin et al. 2014). Plants can take up pollutants from soil if the compounds are enough water-soluble. Very hydrophobic compounds cannot enter the plant tissue. The plant may produce catabolic enzymes for degradation of organics that have entered into the plant. Xenobiotics within plant tissues can also be transported between plant compartments and even evaporated through the leaves in the process of phytovolatilization.

The cold climate is a challenge for rhizoremediation, a biological process involving microbes and their catabolic enzymes, where the temperature is an important factor. The mean temperatures in Scandinavia belonging to the boreal climate region range from +21 °C in July to −10 °C in January (Climate-data.org; https://en.climate-data.org/location/134283/). For a considerable part of the year (5–6 months depending on location), the medium temperature is below 0 °C.

Rhizoremediation is ultimately dependent on the microbial diversity in soil, but the plant itself, its health and growth, is a crucial part of a successful bioremediation process. Pollutants in soil not only have strong effects on the soil microbes but also on the plant (Mackova et al. 2006; Gerhardt et al. 2009; Glick 2010). The plant experiences environmental stress that retards growth of roots, hampering the remediation process. The positive interaction between the plant and microbes can reduce plant stress.

A somewhat conceptual and even scientific misconception is that there could be phytoremediation without effects of microbes. Microbes have, however, established themselves in all possible ecosystems. The misconception is partly result of poor education about the facts of microbial life and the microbial contribution to life on our planet (Barberán et al. 2016). The science of ecology including biology has not incorporated the recent findings and knowledge of microbial ecology. From modern microbial ecology, we have learned that microbes inhabit also the plant (Hardoim et al. 2015), and they may influence most mechanisms we know about phytoremediation. The advances in omics for studying the great diversity of microorganisms living in association with plants are opening new horizons.

The awareness of polluted soils has very recently received increased attention (Finnish Environment 2015) when a strategy for polluted soil risk management was done and published by the Ministry of Environment. Sustainability has become a demanding factor in dealing with these environmental problems, and there is a growing need to find alternatives to the present widely adopted extensive excavation of polluted soils. In situ treatment methods are called upon as green technology for more effective use of available resources.

29.2 Rhizoremediation as Type of Bioremediation

Bioremediation is now a widely accepted solution for the treatment of petroleum hydrocarbon (PHC)-contaminated soils in temperate climates (Dobson et al. 2004) and is increasingly viewed as an appropriate clean-up option for this type of contamination in cold climates (Aislabie et al. 2006). Its practical application in cold regions has received far less attention than in temperate ones. Laboratory and field-scale studies have shown the potential of this technology for the treatment of PHC-contaminated soil in cold temperature (Filler et al. 2001; Aislabie et al. 2006; Schiewer and Niemeyer 2006).

Why use plants for bioremediation? The best answer to the question is maybe the so-called rhizosphere effect which refers to rhizosphere priming. It is the consequence of the energy fixed by photosynthesis, which is used for making sugars that are further allocated to roots. Plants can be seen as energy pumps translocating solar energy to the soil and allowing soil microorganisms to grow and degrade toxic compounds. Phytoremediation or the use of plants for cleanup of soils involves several processes like uptake and transformation of organic contaminants, biodegradation of contaminants in the rhizosphere, and also extraction of heavy metals to the roots and translocation to other plant compartments (Cunningham et al. 1995; Macek et al. 2000; Kuiper et al. 2004). Rhizoremediation is strongly relying on the rhizosphere effect when the abundance and activity of bacteria is elevated, and mycorrhizal symbionts are increasing the root mass and root surface area (Thomas and Cébron 2016).

Rhizoremediation has successfully been applied for several types of organic contaminants including pesticides, herbicides, explosives, and petroleum hydrocarbons (Kidd et al. 2008; Mackova et al. 2009; Gaskin and Bentham 2010; Ramos et al. 2010; Slater et al. 2011; Mukherjee et al. 2015). The rhizosphere has received increased attention in form of publications since the year 2000 (Tibbett et al. 2012). As a consequence, Rhizosphere 1 congress was first time held in Munich 2004, and Rhizosphere 4 was held 2015 in Maastricht. The rhizosphere has been defined as the area around the root extending 2 mm from the root surface (Lynch and de Leij 1990). This area is a hotspot for degradation of hydrocarbons due to elevated microbial biomass and activity. This rhizosphere effect has well been documented (Anderson et al. 1993; Nie et al. 2010).

The outcome of hitherto conducted rhizoremediation varies a lot. Much of variation is explained by different experimental designs. Soil type and choice of plants and type of contaminants evidently affect the results (Liste and Prutz 2006; Olson et al. 2007). Plant species selection is of fundamental importance and the best policy is to use native plants and if possible special genotypes or clones known to be best suitable for combating pollution stress.

29.3 Root Exudation Priming Biodegradation

The allocation of fixed photosynthetic energy to the root as biosynthesized exudates is the basis for the remedial process in the rhizosphere. The soluble fraction of root exudates consists of low molecular mass carbohydrates, amino acids, organic acid anions, and various secondary metabolites (Gerhardt et al. 2009; Martin et al. 2014). Low molecular mass secondary metabolites are excreted from the roots. The plant itself has the ability to produce laccases and peroxidases that can take part in catabolizing aromatic compounds and polyaromatic hydrocarbons (PAHs) (Fig. 29.1). Fungal root symbionts may also contribute with extracellular broad specificity enzymes to the catabolism of aromatics (Harms et al. 2011; Lenoir et al. 2016).

29.3.1 Cometabolism

Cometabolism is a fundamental mechanism in biodegradation, especially in rhizoremediation, where recalcitrant contaminants like four and five ring PAHs can partly be degraded by microbes that do not get enough energy for growth, but

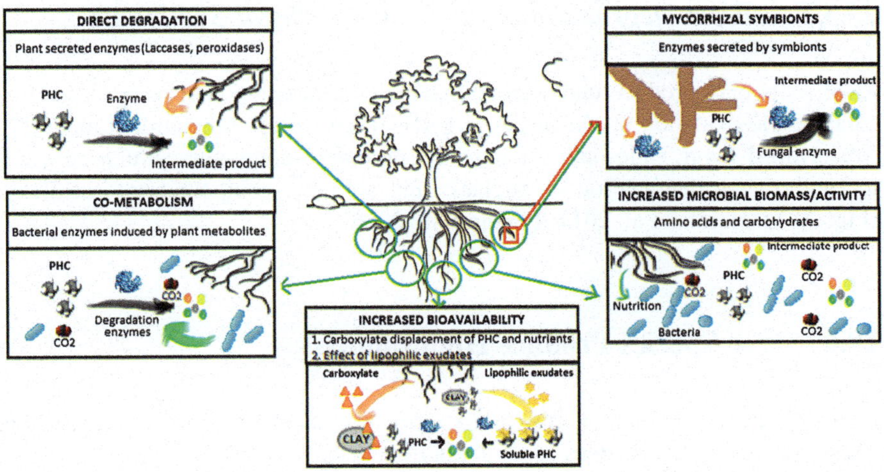

Fig. 29.1 Mechanisms how root exudates can enhance biodegradation of petroleum hydrocarbons (figure designed by Paola Diaz Londono)

can simultaneously use other carbon compounds for energy and propagation. Degradation of polychlorinated biphenyls (PCBs) has also been found to occur cometabolically (Furukawa et al. 1983; Kilpi et al. 1988).

Enzyme activities on several structural analogs are important for cometabolism in that natural compounds structurally resembling the contaminants induce degradation. Biphenyl is a non-halogenated parent compound of PCBs and can activate microbes in PCB degradation. Enzymes needed for breakdown of biphenyl are also needed in PCB degradation and are induced by this structural analog (Yrjälä 1997; Sipilä et al. 2010).

Flavonoids produced by the plant often contain aromatic structures that soil microbes have become familiar with. It is thought that through evolution bacteria have developed catabolic enzymes and pathways to degrade these compounds (Singh et al. 2004; Mackova et al. 2006).

29.3.2 Improved Bioavailability

The prerequisite for bioremediation of hydrocarbons via rhizoremediation is that the contaminant is directly available for the degrading microbes. Solubility of strongly hydrophobic compounds can be improved with detergents (Christofi and Ivshina 2002; Singh et al. 2008). The plant exudates can improve the bioavailability of pollutants since some of them are lipophilic and function as detergents (Read et al. 2003). Also microorganisms can secrete compounds that can function as surfactants, and their production can be induced upon environmental stress (Mulligan 2009). Even in polar regions, bacteria have been isolated that degrade hydrocarbons and produce biosurfactants (Malavenda et al. 2015).

29.3.3 Direct Degradation and via Mycorrhizal Symbionts

Plant roots may produce and secrete enzymes like laccases and peroxidases that catalyze the oxidation of, for instance, PAHs of petroleum (Gerhardt et al. 2009). Mycorrhizal fungi, endo- and ectomycorrhiza, may excrete extracellular enzymes to catalyze the degradation of aliphatic and aromatic compounds of petroleum (Fig. 29.1) (Harms et al. 2011; Lenoir et al. 2016).

29.4 Tree Species Produce High Biomass

Poplar (*Populus* spp.) and willow (*Salix* spp.) are often used for phytoremediation. They are fast growing trees with a high capacity of water uptake (Hoag and Hornsby 1992). Red mulberry (*Morus rubra*) and birches (*Betula* spp.) have also been used for phytoremediation (Vervaeke et al. 2003; de Carcer et al. 2007; Sipilä et al. 2008; Yrjälä et al. 2010a). Poplar and willows have commercial uses as firewood (e.g., *P. tremuloides* and *S. viminalis*). The major advantages of trees in phytoremediation are:

1. Perennial growth and accumulation with no need for annual harvest
2. Well-developed root system in mature trees
3. Effective water uptake, which can be used to stabilize the polluted site
4. Production of litter that provides nutrients to poor soils after decomposition
5. Some potential tree species for phytoremediation are also of economic value (Golan-Goldhirsh et al. 2004)

Extensive research efforts tackles use of woody species for phytoremediation of soils and groundwater from organic pollutants (Corseuil and Moreno 2001; Hong et al. 2001; Palmroth et al. 2002; Mukherjee et al. 2015), aromatic hydrocarbons (Thompson et al. 1998), and halogenated compounds (Doni et al. 2012). Mycorrhizal fungi infect trees, and the effect of the symbiotic association on uptake of pollutants was studied (Schnabel and White 2001). Some tree species can be seen as the first "immigrants" on such sites (Golan-Goldhirsh et al. 2004). In search for suitable plants for the phytoremediation of polluted sites, the use of trees is a good option also giving substantial biomass.

29.5 *Populus* Trees for Rhizoremediation

Poplar (*Populus* spp.), like willow, is a member of the *Salicaeae* family that have the ability to cross within the genus, giving rise to a number of potential genotypes (Monclus et al. 2009). They also readily form hybrids utilized by foresters to maximize growth rates and yield (Klopfenstein et al. 1997). Poplar can be

Table 29.1 Studies involving poplars for remediation of organic contaminants

Populus species	Organic contaminant	Theme of study	Reference
Populus deltoides × *P. nigra* DN34	Atrazine	Vegetative uptake and degradation of atrazine in the rhizosphere	Burken and Schnoor (1996)
Populus deltoides × *P. nigra* DN-34, Imperial Carolina	Atrazine, benzene, toluene, and xylenes (BTX)	First report of enumeration of specific microbial populations (total heterotrophs, denitrifiers, pseudomonads, BTX degraders, and atrazine degraders) in *Populus* rhizosphere	Jordahl et al. (1997)
Populus trichocarpa × *P. deltoides*	Trichloro-ethylene (TCE)	A comprehensive study demonstrating efficient metabolism of TCE by axenic cultures of *Populus* cells, poplar cuttings, and in a field trial with poplar trees	Gordon et al. (1998)
Populus deltoides × *P. nigra* DN-34	Benzene	Laboratory experiments investigating the toxicity response of poplar cuttings to benzene exposure, contaminant distribution in plant tissues, contaminant degradation in the soil profile, and contaminant volatilization from the soil and plant tissues	Burken et al. (2001)
Populus deltoides × *P. nigra* DN-34, Imperial Carolina	Methyl tertbutyl ether (MTBE)	A three phase study involving (A) a laboratory bioreactor study that examined the fate and transport of ^{14}C-radiolabeled MTBE in hybrid poplars; (B) mathematical modelling investigating the influence of deep-rooted trees on unsaturated and saturated groundwater flow; (C) a field study at a site with MTBE-contaminated groundwater where hybrid poplar trees were planted	Hong et al. (2001)
Populus nigra	Diesel	A report on effect of diesel fuel on plant growth and on the rhizosphere microflora. T-RFLP fingerprinting of rhizosphere bacterial communities and isolation and identification of hydrocarbon-degrading strains were performed, and these strains were checked for the presence of *alkB* gene	Tesar et al. (2002)
Populus deltoids × *P. trichocarpa*	Volatile organic compounds (VOCs)	Phytoremediation field experiment demonstrating that hybrid	Hirsh et al. (2003)

(continued)

Table 29.1 (continued)

Populus species	Organic contaminant	Theme of study	Reference
		poplar trees mitigate the migration of a groundwater plume of VOCs	
Populus deltoids × *P. nigra* DN-34	RDX explosives	Poplar tissue cultures and leaf crude extracts were shown to mineralize RDX upon exposure to light	Van Aken et al. (2004)
P. trichocarpa × *P. deltoides* cv. Hoogvorst	Toluene	First report of in planta horizontal gene transfer among plant-associated endophytic bacteria. Inoculation of *Burkholderia cepacia* VM1468 containing pTOM-Bu61 plasmid (carrying genes for toluene degradation) had a positive effect on plant growth in the presence of toluene	
Populus trichocarpa × *P. deltoides* cv. "Hazendans" and "Hoogvorst"	BTEX	Endophytic bacteria were isolated from the root, stem, and leaf of two cultivars of poplar tree growing on a site contaminated with BTEX compounds. They were further characterized genotypically (16S rRNA sequencing) and phenotypically by their tolerance to target pollutants. The endophytic bacteria exhibited marked spatial compartmentalization within the plant	Moore et al. (2006)
Populus deltoides × *P. nigra* DN-34	RDX explosives	First report showing that the exposure of poplar plants to RDX results in the induction of several genes that are involved in explosive detoxification such as glutathione S-transferases (GSTs), cytochrome P-450s (CYPs), NADPH-dependent reductases, and peroxidases	Tanaka et al. (2007)
Populus deltoides × *P. nigra* DN34	Polychlorinated biphenyls (PCB)	Report on the transport of PCBs through whole plants designed for use in treatment at disposal facilities	Liu and Schnoor (2008)
Transgenic *Populus tremula* × *tremuloides* var. Etropole	2,4,6-trinitrotoluene (TNT)	Transgenic hybrid poplars expressing bacterial nitroreductase gene, *pnrA*, were shown to tolerate and take up greater amounts of TNT	Van Dillewijn et al. (2008)
Populus deltoides × (*P. trichocarpa* ×	TCE	First in situ inoculation of poplar trees, growing on a TCE-contaminated site, with the	Weyens et al. (2009)

(continued)

Table 29.1 (continued)

Populus species	Organic contaminant	Theme of study	Reference
P. deltoides) cv. Grimminge		TCE-degrading strain *Pseudomonas putida* W619-TCE, which was established and enriched in planta as a poplar root endophyte and by further horizontal gene transfer of TCE metabolic activity to members of the poplar's endogenous endophytic population	
Populus tremula × *P. alba* and *P. trichocarpa*	Chlorpyrifos	Hydroponic study demonstrating the efficiency of poplars in uptake and translocation of Chloropyrifos	Lee et al. (2012)
Populus nigra (var. italica)	Petroleum hydrocarbons (PHC), PCBs, and metals	Phytoremediation field study at a site historically contaminated with PHCs, PCBs, and metals; poplars together with horse manure treatment were shown to be effective in remediation. Biogeochemical parameters were monitored by SDS PAGE	Doni et al. (2012)
Populus deltoides × *P. nigra*	Polyaromatic hydrocarbons (PAHs)	Efficacy of *Burkholderia fungorum* DBT1 (a strain isolated from oil refinery discharge and capable of degrading dibenzothiophene, phenanthrene, naphthalene, and fluorene) was demonstrated as an endophyte in poplars during 18-weeks greenhouse experiment	Andreolli et al. (2013)
Salix viminalis L.	TCE	TCE added to cylinders filled with compost above water-saturated quartz sand	Schöftner et al. (2016)

considered as an important model woody plant. Sequencing of the whole genome of *Populus trichocarpa* and the extensively expressed sequence tag (EST) database are part of research of great importance for *Populus* species helping to improve the practical use of this tree in several applications (Sterky et al. 2004). *Populus tremuloides* is the fastest growing tree species in Finland and in other parts of Scandinavia (Pulkkinen 2014), and it is suited for clonal propagation and routine transformation (Bradshaw et al. 2000; Taylor et al. 2002; Wullschleger et al. 2002). Due to the favorable properties, hybrid poplars have been used as cash crops for pulp and energy and in past decades hybrid and transgenic poplars have extensively been used in remediation of organic contaminants (Table 29.1).

29.6 Monitoring Rhizoremediation in Cold Climates

The relatively short growing season in the boreal climate zone slows down the rhizoremediation process, and psychrophilic microbes become important in the process, since they are adapted to propagate in low temperatures. The soil freezes for several months in winter (4–6 months) and is subjected to freeze and thaw cycles that shape the microbial community structures in rhizosphere-associated soil.

The low temperatures make biological processes above and below ground much slower. The belowground diversity of microbes in soil is still huge (Torsvik and Øvreås 2002), and the catabolic potential of the soil is an interplay of the existing different taxa contributing to remediation. Laboratory-degradation studies have only shown those taxa that are readily growing in the laboratory. One major challenge that plant biologists and microbiologists face when studying these interactions is that many groups of microbes that inhabit this zone are not cultivable in the laboratory (Singh et al. 2004).

29.6.1 Molecular Markers for Monitoring

Recent advances in molecular technologies like next generation sequencing (NGS), applied to genomics and handling of massive data in bioinformatics, offer exciting opportunities to link structural diversity with the processes occurring in the rhizosphere. The diversity of microbes in woody plant rhizosphere can be studied by means of marker genes. The by far most common marker gene is the 16S rRNA phylogenetic marker gene which is excellent for identification of bacteria and archaea (Juottonen et al. 2005) using the extensive reference database with 3.4 million sequences (September 2016 release https://rdp.cme.msu.edu/). Monitoring the structural microbial diversity gives the phylogenetic background of microbes in soil. For soil functioning, and in this case for remediation, the functional diversity becomes more important. Functional marker genes encode enzymes that can be amplified from environmental DNA by polymerase chain reaction (PCR) (Table 29.2). For the determination of the catabolic community diversity related to the biodegradation potential, a protocol for the assessment of catabolic marker genes in polluted soils was developed. The upper pathway extradiol dioxygenase gene was used to assess bulk and rhizosphere soil bacterial communities degrading aromatic compounds (Sipilä et al. 2006). This aromatic ring-cleavage gene has successfully been used to assess the catabolic potential in soil. Birch rhizosphere-associated soil contained a higher diversity of aromatic ring-degrading enzymes than bulk soil (Sipilä et al. 2008). Other functionally important enzymes in different ecosystems have been amplified (Galand et al. 2002; Yrjälä et al. 2004, 2010a, 2011).

Table 29.2 PCR primers are designed to target genes coding enzymes needed for bacterial aromatic compound degradation

Primer name	Primer sequences	Target enzyme	Metabolic-degradation pathway	Target taxa	Product size (bp)	Environmental community	Reference
fwdSP9	CAGTACAAYTCCTACACVACBG	6-Oxocyclohex-1-ene-1-carbonyl-coenzyme A hydrolases	Anaerobic benzoate degradation	Beta, Delta, Alpha	300	Not amplified	Kuntze et al. (2008)
revASP1	CMATGCCGATYTCCTGRC						
TBMD-F	GCCTGACCATGGATGCSTACTGG	a-subunits of multi-component mono-oxygenases (sub-family R.1)	Phenol and benzene oxygenase	Gamma, Beta	640	Yes (diversity, not estimated)	Hendrickx et al. (2006)
TBMD-R	CGCCAGAACCACTTGTCRRTCCA						
TMOA-F	CGAAACCGGCTTYACCAAYATG	a-subunits of multi-component mono-oxygenases (sub-family R.3)	Toluene monooxygenase	Gamma, Beta	505	Yes (DGGE)	Hendrickx et al. (2006)
TMOA-R	ACCGGGATATTTYTCTTCSAGCCA						
XYLA-F	CCAGGTGGAATTTTCAGTGGTTGG	Electron transfer component of mono-oxygenases	Xylene monoxygenase	Gamma	510	Yes (diversity, not estimated)	Hendrickx et al. (2006)
XYLA-R	AATTAACTCGAAGCGCCCACCCCA						
TODC1-F	CAGTGCCGCCAYCGTGGYATG	Multi-component aromatic dioxygenases	BTEX dioxygenation	Gamma, Beta, Actino	510	Yes (diversity, not estimated)	Hendrickx et al. (2006)
TODC1-R	GCCACTTCCATGYCCRCCCCA						
XYLE1-F	CCGCCGACCTGATCWSCATG	catechol 2,3-dioxygenase I.2.A subfamily	Lower meta-pathway	Gamma	242	Yes (diversity not estimated)	Hendrickx et al. (2006)
XYLE1-R	TCAGGTCAKCACGGTCAKGA						
XYLE2-F	GTAATTCGCCCTGGCTAYGTICA	catechol 2,3-dioxygenase I.2.B subfamily	Lower meta-pathway	Alpha	906	Yes (diversity, not estimated)	Hendrickx et al. (2006)
XYLE2-R	GGTGTTCACCGTCATGAAGCGBTC						
CDO-F	CATGTCAACATGCGCGTAATG	catechol 2,3-dioxygenase I.2.C subfamily	Phenol degration meta-pathway	Gamma	225	Yes (diversity, not estimated)	Hendrickx et al. (2006)
CDO-R	CATGTCTGTGTTGAAGCCGTA						

(continued)

Table 29.2 (continued)

Primer name	Primer sequences	Target enzyme	Metabolic-degradation pathway	Target taxa	Product size (bp)	Environmental community	Reference
TBUE-F	CTGGATCATGCCCTGTTGATG	catechol 2,3-dioxygenase I.2.C subfamily	Phenol degration meta-pathway	Beta	444	Yes (diversity, not estimated)	Hendrickx et al. (2006)
TBUE-R	CCACAGCTTGTCTTCACTCCA						
TODE-F	GGATTTCAAACTGGAGACCAG	catechol 2,3-dioxygenase I.3.B subfamily	Toluene meta pathway	Gamma	246	Yes (diversity, not estimated)	Hendrickx et al. (2006)
TODE-R	GCCATTAGCTTGCAGCATGAA						
BP-F	TCTAYCTVCGNATGGAYHDBTGGCA	Extradiol dioxygenase I.3.E subfamily	Upper meta-pathway ring-cleavage dioxygenase	Alpha, Gamma, Beta, Actino	467	Yes (RFLP)	Sipilä et al. (2006)
BP-R	TGVTSNCGNBCRTTGCARTGCATGAA						
Rieske_f	TGYMGICAYMGIGG	Dioxygenases targeting non polar substrates	Dioxygenation on several pathways	Alpha, Gamma, Beta, Actino	78	Yes (sequencing)	Ni Chadhain et al. (2006)
Rieske_r	CCANCCRTGRTANSWRCA						
FRT5A	TYRARGCYAACTGGAA	a-subunits of multi-component aromatic dioxygenases	Initial PAH dioxygenation	Beta, Gamma	491	Yes (RFLP)	Bordenave et al. (2008)
FRT6A	TACCACGTBGGTTGGAC						
FRT3B	CATGTCTTTTTCKACVATGGC	a-subunits of multi-component aromatic dioxygenases	Initial PAH dioxygenation	Beta, Gamma	437	Yes (RFLP)	Bordenave et al. (2008)
FRT4B	GWHDCYGTYTCCATRTTGTC						
NAPH-1F	TGGCTTTTCYTSACBCATG	a-subunits of multi-component aromatic dioxygenases	Initial dioxygenation of naphthalene	Gamma, Beta	896	Yes (DGGE)	Gomes et al. (2007)
NAPH-1R	DGRCATSTCTTTTTCBAC						
bzAQ4F	GTGGGCACCGGNTAYGGNMG	a-subunits of benzoyl-CoA reductase	Anaerobic benzoate degradation	Alpha, Beta, Gamma	484	Yes (sequencing)	Song and Ward (2005)
bzAQ4R	GGTTCTTGGCGAYNCCNCCNGT						

Primers designed after 2005 are listed
Alpha Alphaproteobacteria, *Beta* Betaproteobacteria, *Gamma* Gammaproteobacteria, *Actino* Actinobacteria

29.7 Rhizoremediation as Successional Dynamics of Microbes in Polluted Soil

Degradation of the different organic compounds of PHC is a stepwise catabolic process, and the addition or spread of contaminants causes severe changes of soil microbial populations and communities. The initial situation at the onset of PHC-spreading into soil (1% soil concentration) was studied in greenhouse after some initial dose tests to assess the tolerance of hybrid aspens to PHC (Mukherjee et al. 2013). A drastic change of the soil bacterial community structures was seen immediately after the PHC addition and most probable number (MPN) assay revealed an increase in the abundance of culturable PHC-degraders in the polluted rhizosphere during first 2 weeks. Ecological succession refers to a process of more or less deterministic developments in the composition or structure of an ecological community (Connell and Slatyer 1977; Walker and Del Moral 2003). Succession has so far been a well-known concept derived from plant ecology, however, microbial succession has received much less attention (Fierer et al. 2010). Recent methodological advancements have made it possible for microbial ecologists to overcome limitations in analysis of temporal changes of highly diverse and rapidly changing microbial communities in various ecosystems (Wertz et al. 2007; Kowalchuk et al. 2000; Banning et al. 2011; Podell et al. 2014). Studies on secondary succession can play an instrumental role in understanding the resistance and resilience of microbial communities in response to perturbations (Griffiths and Philippot 2013).

29.7.1 Secondary Succession in Rhizoremediation of Petroleum: A Case Study

Petroleum contains toxic chemicals that reduce the growth of many microbial taxa, but in nutrient-poor soils with low organic contents, the added hydrocarbons are a resource for other taxa to propagate. Soil quality becomes a major factor effecting the outcome of remediation. The level and duration of pollution influence the results of rhizoremediation when aged pollution decreases the bioavailability of contaminants. Fresh petroleum contamination was subjected to poplar rhizoremediation in Southern Finland where different clones of hybrid aspen (*Populus tremula*) were used. The seasonality of bacterial communities in rhizosphere-associated soil was monitored and the temporal development was followed (Mukherjee et al. 2015). The petroleum level was 0.79% (w/w) at the start of remediation, but turned out to be quite toxic to the hybrid aspen plants. Fertilization was not used for the low nutrient sandy soil at the beginning of remediation in order to be able to see the direct response of plants to the petroleum contamination.

The rhizosphere-associated and unvegetated soils were sampled at seven different time points during 2 years and the sampling points were chosen to take into account the seasonal variations (Mukherjee et al. 2015). DNA was isolated from the polluted soil and from controls and was PCR-amplified. To get detailed data about the bacterial succession, 454 pyrosequencing was adopted. Successional patterns of bacterial communities after PHC-pollution in aspen rhizosphere in the field were demonstrated. Seasonality had a significant impact on bacterial communities during winter months. November 2011 and February 2012 communities were different from the spring (May 2012) and summer (August 2012) communities, using 16S rRNA as marker gene in this long-term field experiment (Mukherjee et al. 2015).

Experimental details: The 16S rRNA gene was amplified by PCR. 454 pyrosequencing yielded a total of 3215 OTUs (97% sequence similarity) from 34 samples after removing singletons. The standard operating procedure (http://www.mothur.org/wiki/Schloss_SOP) was followed for the analysis (Schloss et al. 2009). In order to determine the co-occurrence patterns of OTUs across seven sampling time points, hierarchical clustering was performed separately. Top 50 most abundant OTUs were selected, and spearman's correlation coefficients based on the average linkage criteria were calculated in Cluster 3.0 software. The resulting heatmaps with hierarchical clustering were visualized in Java Tree view.

Non-metric multidimensional scaling (NM-MDS) and hierarchical clustering (Fig. 29.2) grouped the bacteria into distinct clusters. Bacterial communities in contaminated rhizosphere-associated soil and bulk soil from June 2010, September 2010 and May 2011 grouped in the "early phase." The samples from August 2011 to February 2012 grouped in the "late phase."

29.7.2 Microbial Functional Diversity in Secondary Succession

Two different marker genes, *edo* (extradiol dioxygenase gene) and *alkB* (alkane monooxygenase gene), were used for elucidating the functional diversity in PHC-polluted rhizosphere and bulk soil: *edo* as a marker for bacteria-degrading aromatic hydrocarbons and *alkB* as a marker for bacteria-degrading aliphatic hydrocarbons. Similar to the 16S rRNA communities, a clear succession was displayed by EDO (extradiol dioxygenase amino-acid sequences) communities. Differences in rhizosphere and bulk communities were more prominent in the late phase of succession. We also observed a steady rise in EDO diversity during the course of succession.

Experimental details: 454 pyrosequencing of extradiol dioxygenase amplicons was performed from 34 samples taken at seven time points from rhizosphere and unvegetated soil during the 2-year study period. A total of 22,391 good quality sequences were clustered into 153 EDO OTUs (90% amino-acid sequence similarity). Co-occurrence patterns of EDO OTUs were determined by hierarchical

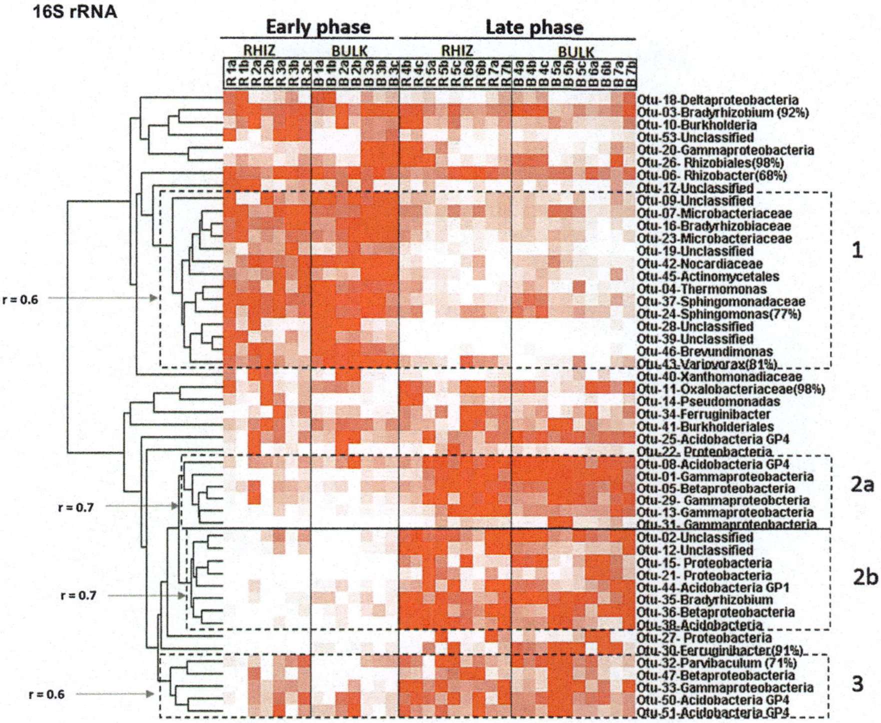

Fig. 29.2 Hierarchical clustering of 50 most abundant 16S rRNA OTUs. Closest phylogenetic identification is given for each OTU. Spearman's correlations of the co-occurring OTUs forming clusters are indicated. RHIZ, rhizosphere; BULK, unvegetated samples (Mukherjee 2014)

clustering to differentiate between the members of late phase and early phase communities.

In order to identify the putative hosts of the extradiol dioxygenase sequences detected in this study, we constructed a phylogenetic tree (Fig. 29.3) with the OTU representative sequences; closest hits were taken from NCBI database (-nr) and sequences from a study by Sipilä et al. (2008). Members of the previously mentioned early phase communities grouped in two different clusters of the bacterial EDO phylogenetic tree. In the late phase of succession, community composition of these bacterial EDOs was much more diverse and the sequence dissimilarity higher according to the EDO tree. OTUs dominating in the late phase showed sequence similarity to *Sphingomonas*, *Sphingopyxis*, *Novosphingobium*, *Burkholderia*, *Cycloclasticus*, *Polaromonas*, *Ralstonia*, *Pseudomonas*, and *Rhodococcus* type extradiol dioxygenases.

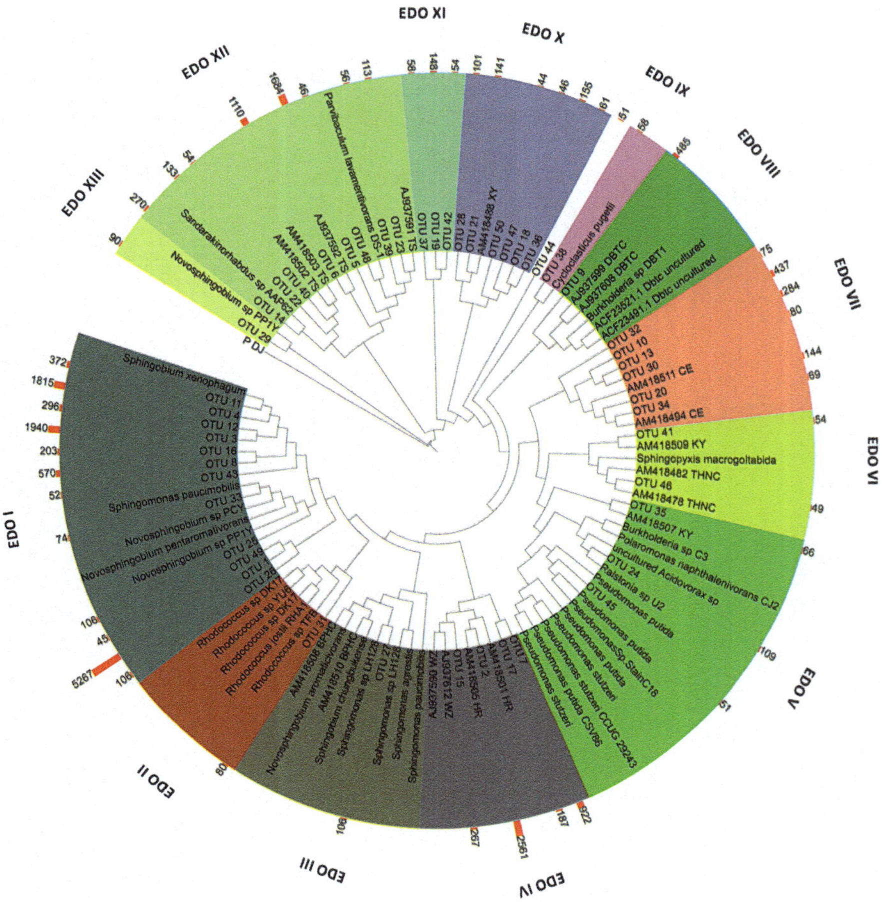

Fig. 29.3 Maximum-likelihood tree of extradiol dioxygenase (EDO) amino-acid sequences. The bootstrap consensus tree was inferred from 500 replicates. The 50 most abundant OTUs were included in the tree along with the closest hits from NCBI non-redundant protein database (Mukherjee 2014)

29.8 Woody Plant Stress in Polluted Soil

The addition of hydrocarbon pollutants, like PAHs, to planted soil confers stress to the plant. This was studied using heavy metal-tolerant birch clones (Tervahauta et al. 2009). Two birch clones, HA and WA, originating from metal-contaminated sites were exposed for 3 months to soils (sand–peat ratio 1:1 or 4:1) spiked with a mixture of PAHs: anthracene, fluoranthene, phenanthrene, pyrene. PAH degradation differed between the two birch clones and also by the soil type. The statistically most significant elimination ($p \leq 0.01$), i.e., 88% of total PAHs, was observed in the more sandy soil planted with birch, the clearest positive effect being found with *Betula pubescens* (HA) clone for phenanthrene. Birch clone HA did, however, not grow in the experimental conditions and displayed chlorosis and higher level of HsP70 protein than the WA clone. Proteomic profiling supported these results.

29.9 Rhizosphere and Endophytic Bacteria

Endophytic microbes are today defined as microbes that spend at least part of their life cycle within plant tissue in the different plant compartments, roots, stem, and leaves (Hardoim et al. 2015). An interesting finding was more than a decade ago, when it was found that certain bacterial catabolic genotypes (*alkB*, *ndoB*, *ntdAa*, and *ntnM*) were enriched in the interior of plant roots, in so-called endophytic bacteria in response to pollution in a contaminant-dependent manner (Siciliano et al. 2001). Plasmids may contain catabolic genes and even whole pathways for degradation of xenobiotics (Sipilä 2009; Sipilä et al. 2010). The pTOM toluene-degradation plasmid of *Burkholderia cepacia* G4 was introduced into *B. cepacia* L. S.2.4, a natural endophyte of yellow lupine (*Lupinus luteus*). After surface-sterilized lupine seeds were successfully inoculated with the recombinant strain, the engineered endophytic bacteria strongly degraded toluene, resulting in a marked decrease in its phytotoxicity (Barac et al. 2004). Epiphytic and endophytic bacteria associated with hybrid aspen were isolated in a rhizoremediation study with PAHs. The largest diversity of associated bacteria was found in the rhizosphere (Yrjälä et al. 2010b). Some of the isolated bacteria were able to grow on PAHs and identified as *Burkholderia* species. One of the isolates was further characterized, *Burkholderia fungorum* 95, and was shown to reduce selenite to elemental selenium detected as intracellular nanoparticles (Khoei et al. 2017). Other examples of capacities of endophytic bacteria are the Cr(VI)-reducing bacterial strains *Pseudomonas aeruginosa* Rb-1 and *Ochrobactrum intermedium* Rb-2 that were found to colonize wheat roots. While detoxifying Cr(VI), they were detected in root cortex cells simultaneously increasing wheat root growth (Batool et al. 2015). One important function of these plant-associated microbes is plant growth promotion and protection against biotic and abiotic stress (Hardoim et al. 2015). The origin of endophytic microbes found in the plant has not been known. The root microbiome of *Arabidopsis* was determined using 454-pyrosequencing of the 16SrRNA gene (Lundberg et al. 2012). Very recent findings suggest that soil type defines the composition of root-inhabiting bacterial communities and host genotype determines their ribotype profiles to a limited extent (Bulgarelli et al. 2012).

29.9.1 Protection Against Biotic and Abiotic Stress

Endophytes may induce so-called induced systemic resistance (ISR) giving higher tolerance to pathogens (Zamioudis and Pieterse 2012). Fungal endophytes produce alkaloids that can inhibit growth of pathogens and herbivores (Fletcher and Harvey

1981). Bacterial endophytes can produce antimicrobial compounds, such as 4-hydroxybenzoate (Taghavi et al. 2009). Endophytes produce siderophores for iron acquisition which can play a role in pathogen–host interactions. *Methylobacterium* strains have been shown to be involved in suppression of citrus chlorosis (Araujo et al. 2002). Fungal strains of *Neotyphdodium* spp. have been shown to increase tolerance to draught, nitrogen starvation, and water stress (Ravel and Courty 1997). ACC (1-aminocyclopropane-1-carboxylate) deaminase is a bacterial enzyme that lowers the levels of ethylene in the plant. Ethylene is a plant hormone and a key regulator of colonization of plant tissue. It inhibits the elongation of roots stems and leaves. *Burkholderia phytofirmans* PsJN has ACC deaminase activity and promotes root growth (Sun et al. 2009).

29.9.2 Plant Growth Promotion

Inoculation of hybrid poplar with *Enterobacter* sp. 638 increased the biomass of the plant. The results indicate that an increase in the environmental and economic viability of poplar as a biofuel feedstock is possible when inoculated with endophytic bacteria like *Enterobacter* sp. 638 (Rogers et al. 2012). Endophytes with appropriate characteristics can be inoculated to phytoremediation plants like poplar to help dealing with toxic chemicals and organic contaminants (Weyens et al. 2015). Phytohormone production by endophytes leads to changes in plant host; the production of auxin and gibberellins is found in root endophytes (Shi et al. 2009; Khan et al. 2012). Several root endophytes fix nitrogen. Most efficient N-fixer is *Gluconacetobacter diazotrophicus* living in symbiosis with sugarcane (Dong et al. 1994). The inoculation of hybrid poplar with *Burkholderia fungorum* DBT1, a dibenzothiophene degrader, resulted in elongated roots, although total growth was not stimulated compared to control plants in an 18-week greenhouse study with PAH-polluted soil. The inoculated plant treatment was most effective in removing PAHs (Andreolli et al. 2013).

29.10 Spatial Heterogeneity in Assessing Rhizoremediation Potential of Soil

For in situ treatments of polluted soils, site investigation is crucial and the extent and level of pollution has to be estimated. Spatial heterogeneity is typical for the spread of pollution around the polluted area; aging of PAHs is a common phenomenon. The spatial uncoupling between biodegradation and soil respiration seemed to be governed by the aging of PAHs in the soil (Bengtsson et al. 2010). This puts constraints on sampling, and not only the chemical property of the soil but also the biology of the soil needs attention. Extensive sampling of a 1.7-ha creosote-polluted site and analysis of soil respiration, enzyme activity, and bacterial diversity gave valuable site information for implementing rhizoremediation (Mukherjee et al. 2014). Taking use of geostatistical approach (GIS) revealed how the bacteria find their own niches in heterogenic pollution. Geostatistical methods are valuable for visualization of ecosystem data like of soil quality over different scales (Enwall et al. 2010). The geographical information system became a very useful tool in visualizing the results from soil analysis and can as well be used in rhizoremediation monitoring.

29.11 Conclusions

Soils are providing fundamental ecosystem services like production of biomass, storage, filtration and transformation of nutrients, substances, and water. They are best habitats for both species and genetic biodiversity. Unfortunately, soils are not very well protected from organic and inorganic pollutants as well as excess of nutrients, but the microbial communities adjust to these disturbances providing capacities, together with plants, for remediation and improvement of the quality of water. Rhizoremediation has in the new Millennium provided green solutions to environmental problems with the increased understanding of plant–microbe interactions also through omics approaches. The recognition of the huge functional diversity of microbes together with appreciation of the biomass produced through rhizoremediation with carbon fixing potential, will greatly enhance the practical use of this type of phytotechnology.

Acknowledgements I want to thank BSc Paola Diaz Londono for planning and designing the figure illustrating rhizoremediation.

References

Aislabie J, Saul DJ, Foght JM (2006) Bioremediation of hydrocarbon-contaminated polar soils. Extremophiles 10:171–179

Anderson TA, Guthrie EA, Walton BT (1993) Bioremediation in the rhizosphere. Environ Sci Technol 27:2630–2636

Andreolli M, Lampis S, Poli M, Gullner G, Biró B, Vallini G (2013) Endophytic *Burkholderia fungorum* DBT1 can improve phytoremediation efficiency of polycyclic aromatic hydrocarbons. Chemosphere 92:688–694

Araujo WL, Marcon J, Maccheroni W Jr, Van Elsas JD, Van Vuurde JWL, Azevedo JL (2002) Diversity of endophytic bacterial populations and their interaction with *Xylella fastidiosa* in citrus plants. Appl Environ Microbiol 68:4906–4914

Banning NC, Gleeson DB, Grigg AH, Grant CD, Andersen GL, Brodie EL, Murphy DV (2011) Soil microbial community successional patterns during forest ecosystem restoration. Appl Environ Microbiol 77:6158–6164

Barac T, Taghavi S, Borremans B, Provoost A, Oeyen L, Colpaert JV, Vangronsveld J, Van Der Lelie D (2004) Engineered endophytic bacteria improve phytoremediation of water-soluble, volatile, organic pollutants. Nat Biotechnol 22:583–588
Barberán A, Hammer TJ, Madden AA, Fierer N (2016) Microbes should be central to ecological education and outreach. J Microbiol Biol Educ 17:23–28
Batool R, Yrjälä K, Hasnain S (2015) Study on cellular changes and potential endotrophy of wheat roots due to colonization of chromium reducing bacteria. Int J Sci Technol 12:3263–3272
Bengtsson G, Törneman N, Yang X (2010) Spatial uncoupling of biodegradation, soil respiration, and PAH concentration in a creosote contaminated soil. Environ Poll 158:2865–2871
Bordenave S, Goni-urriza M, Vilette C, Blanchard S, Caumette P, Duran R (2008) Diversity of ringhydroxylating dioxygenases in pristine and oil contaminated microbial mats at genomic and transcriptomic levels. Environ Microbiol 10:3201–3211
Bradshaw H, Ceulemans R, Davis J, Stettler R (2000) Emerging model systems in plant biology: poplar (*Populus*) as a model forest tree. J Plant Growth Regul 19:306–313
Bulgarelli D, Rott M, Schlaeppi K, Themaat v, Loren EV, Ahmadinejad N, Assenza F, Rauf P, Huettel B, Reinhardt R, Schmelzer E (2012) Revealing structure and assembly cues for *Arabidopsis* root-inhabiting bacterial microbiota. Nature 488:91–95
Burken JG, Schnoor JL (1996) Phytoremediation: plant uptake of atrazine and role of root exudates. J Environ Eng 122:958–963
Burken JG, Ross C, Harrison LM, Marsh A, Zetterstrom L, Gibbons JS (2001) Benzene toxicity and removal in laboratory phytoremediation studies. Pract Periodical Hazard, Toxic, Radioact Waste Manag 5:161–171
Christofi N, Ivshina I (2002) Microbial surfactants and their use in field studies of soil remediation. J Appl Microbiol 93:915–929
Connell JH, Slatyer RO (1977) Mechanisms of succession in natural communities and their role in community stability and organization. Am Nat 111:1119–1144
Corseuil HX, Moreno FN (2001) Phytoremediation potential of willow trees for aquifers contaminated with ethanol-blended gasoline. Water Res 35:3013–3017
Cunningham SD, Berti WR, Huang JW (1995) Phytoremediation of contaminated soils. Trends Biotechnol 13:393–397
de Carcer DA, Martin M, Karlson U, Rivilla R (2007) Changes in bacterial populations and in biphenyl dioxygenase gene diversity in a polychlorinated biphenyl-polluted soil after introduction of willow trees for rhizoremediation. Appl Environ Microbiol 73:6224–6232
Dobson R, Schroth MH, Schuermann A, Zeyer J (2004) Methods to assess the amenability of petroleum hydrocarbons to bioremediation. Environ Toxicol Chem 23:929–937
Dong Z, Canny MJ, McCully ME, Roboredo MR, Cabadilla CF, Ortega E, Rodes R (1994) A nitrogen-fixing endophyte of sugarcane stems (a new role for the apoplast). Plant Physiol 105:1139–1147
Doni S, Macci C, Peruzzi E, Arenella M, Ceccanti B, Masciandaro G (2012) In situ phytoremediation of a soil historically contaminated by metals, hydrocarbons and polychlorobiphenyls. J Environ Monitoring 14:1383–1390
Enwall K, Throback IN, Stenberg M, Soderstrom M, Hallin S (2010) Soil resources influence spatial patterns of denitrifying communities at scales compatible with land management. Appl Environ Microbiol 76:2243–2250
Fierer N, Nemergut D, Knight R, Craine JM (2010) Changes through time: integrating microorganisms into the study of succession. Res Microbiol 161:635–642
Filler DM, Lindstrom JE, Braddock JF, Johnson RA, Nickalaski R (2001) Integral biopile components for successful bioremediation in the Arctic. Cold Reg Sci Technol 32:143–156
Finnish Environment (2015) Valtakunnallinen pilaantuneiden maa-alueiden riskienhallintastrategia. Suomen ympäristö 10/2015. Helsinki, Finland
Fletcher L, Harvey I (1981) An association of a *Lolium* endophyte with ryegrass staggers. N Z Vet J 29:185–186

Furukawa K, Tomizuka N, Kamibayashi A (1983) Metabolic breakdown of Kaneclors (polychlorobiphenyls) and their products by *Acinetobacter* sp. Appl Environ Microbiol 46:140–145

Galand PE, Saarnio S, Fritze H, Yrjälä K (2002) Depth related diversity of methanogen Archaea in Finnish oligotrophic fen. FEMS Microbiol Ecol 42:441–449

Gaskin SE, Bentham RH (2010) Rhizoremediation of hydrocarbon contaminated soil using Australian native grasses. Sci Total Environ 408:3683–3688

Gerhardt KE, Huang XD, Glick BR, Greenberg BM (2009) Phytoremediation and rhizoremediation of organic soil contaminants: potential and challenges. Plant Sci 176:20–30

Glick BR (2010) Using soil bacteria to facilitate phytoremediation. Biotechnol Adv 28:367–374

Golan-Goldhirsh A, Barazani O, Nepovim A, Soudek P, Smrcek S, Dufkova L, Krenkova S, Yrjala K, Schröder P, Vanek T (2004) Plant response to heavy metals and organic pollutants in cell culture and at whole plant level. J Soils Sediments 4:133–140

Gomes NCM, Borges LR, Paranhos R, Pinto FN, Krögerrecklenfort E, Mendonça-Hagler LCS, Smalla K (2007) Diversity of ndo genes in mangrove sediments exposed to different sources of polycyclic aromatic hydrocarbon pollution. Appl Environ Microbiol 73:7392–7399

Gordon M, Choe N, Duffy J, Ekuan G, Heilman P, Muiznieks I, Ruszaj M, Shurtleff BB, Strand S, Wilmoth J, Newman LA (1998) Phytoremediation of trichloroethylene with hybrid poplars. Environ Health Perspect 106(Suppl 4):1001–1004

Griffiths BS, Philippot L (2013) Insights into the resistance and resilience of the soil microbial community. FEMS Microbiol Rev 37:112–129

Hardoim PR, van Overbeek LS, Berg G, Pirttila AM, Compant S, Campisano A, Doring M, Sessitsch A (2015) The hidden world within plants: ecological and evolutionary considerations for defining functioning of microbial endophytes. Microbiol Mol Biol Rev 79:293–320

Harms H, Schlosser D, Wick LY (2011) Untapped potential: exploiting fungi in bioremediation of hazardous chemicals. Nat Rev Microbiol 9:177–192

Hendrickx B, Dejonghe W, Faber F, Boenne W, Bastiaens L, Verstraete W, Top EM, Springael D (2006) PCR-DGGE method to assess the diversity of BTEX mono-oxygenase genes at contaminated sites. FEMS Microbiol Ecol 55:262–273

Hirsh S, Compton H, Matey D, Wrobel J, Schneider W (2003) Five year pilot study: Aberdeen proving ground Maryland. In: McCutcheon SC, Schnoor JL (eds) Phytoremediation: transformation and control of contaminants. Wiley, Hoboken, NJ, pp 635–659

Hoag DL, Hornsby AG (1992) Coupling groundwater contamination with economic returns when applying farm pesticides. J Environ Qual 21:579–586

Hong Z, Zhang Z, Olson JM, Verma DP (2001) A novel UDP-glucose transferase is part of the callose synthase complex and interacts with phragmoplastin at the forming cell plate. Plant Cell 13:769–779

Jordahl JL, Foster L, Schnoor JL, Alvarez PJ (1997) Effect of hybrid poplar trees on microbial populations important to hazardous waste bioremediation. Environ Toxicol Chem 16:1318–1321

Juottonen H, Galand PE, Tuittila E, Laine J, Fritze H, Yrjälä K (2005) Methanogen communities and bacteria along an ecohydrological gradient in a northern raised bog complex. Environ Microbiol 7:1547–1557

Khan AL, Hamayun M, Kang S, Kim Y, Jung H, Lee J, Lee I (2012) Endophytic fungal association via gibberellins and indole acetic acid can improve plant growth under abiotic stress: an example of *Paecilomyces formosus* LHL10. BMC Microbiol 12:1–14

Khoei NS, Lampis S, Zonaroa E, Yrjälä K, Bernardic P, Vallini G (2017) Insights into selenite reduction and biogenesis of elemental selenium nanoparticles by two environmental isolates of *Burkholderia fungorum*. New Biotechnol 34:1–11

Kidd P, Prieto-Fernández A, Monterroso C, Acea M (2008) Rhizosphere microbial community and hexachlorocyclohexane degradative potential in contrasting plant species. Plant Soil 302:233–247

Kilpi S, Himberg K, Yrjälä K, Backström V (1988) The degradation of biphenyl and chlorobiphenyls by mixed bacterial cultures. FEMS Microbiol Ecol 53:19–26

Klopfenstein NB, YoungWoo C, MeeSook K, Ahuja MR, Dillon MC, Carman RC, Eskew LG (eds) (1997) Micropropagation, genetic engineering, and molecular biology of *Populus*. General Technical Report RM-GTR-297. US Department of Agriculture, Forest Service, Rocky Mountain Research Station, Colorado

Kowalchuk GA, Stienstra AW, Stephen JR, Woldendorp JW (2000) Changes in the community structure of ammonia-oxidizing bacteria during secondary succession of calcareous grasslands. Environ Microbiol 2:99–110

Kuiper I, Lagendijk EL, Bloemberg GV, Lugtenberg BJ (2004) Rhizoremediation: a beneficial plant-microbe interaction. Mol Plant Microbe Interact 17:6–15

Kuntze K, Shinoda Y, Moutakki H, McInerney MJ, Vogt C, Richnow HH, Boll M (2008) 6-Oxocyclohex-1-ene-1-carbonyl-coenzyme A hydrolases from obligately anaerobic bacteria: characterization and identification of its gene as a functional marker for aromatic compounds degrading anaerobes. Environ Microbiol 10:1547–1556

Lee KY, Strand SE, Doty SL (2012) Phytoremediation of chlorpyrifos by populus and salix. Int J Phytoremediat 14:48–61

Lenoir I, Lounes-Hadj Sahraoui A, Fontaine J (2016) Arbuscular mycorrhizal fungal-assisted phytoremediation of soil contaminated with persistent organic pollutants: a review. Eur J Soil Sci 67:624–640

Liste H, Prutz I (2006) Plant performance, dioxygenase-expressing rhizosphere bacteria, and biodegradation of weathered hydrocarbons in contaminated soil. Chemosphere 62:1411–1420

Liu J, Schnoor JL (2008) Uptake and translocation of lesser-chlorinated polychlorinated biphenyls (PCBs) in whole hybrid poplar plants after hydroponic exposure. Chemosphere 73:1608–1616

Lundberg DS, Lebeis SL, Paredes SH, Yourstone S, Gehring J, Malfatti S, Tremblay J, Engelbrektson A, Kunin V, Del Rio TG (2012) Defining the core *Arabidopsis thaliana* root microbiome. Nature 488:86–90

Lynch JM, de Leij F (1990) Rhizosphere. Wiley Online Library, Chichester. http://www.els.net

Macek T, Mackova M, Kas J (2000) Exploitation of plants for the removal of organics in environmental remediation. Biotechnol Adv 18:23–34

Mackova M, Dowling D, Macek T (eds) (2006) Phytoremediation and rhizoremediation. Springer, Dordrecht

Mackova M, Prouzova P, Stursa P, Ryslava E, Uhlik O, Beranova K, Rezek J, Kurzawova V, Demnerova K, Macek T (2009) Phyto/rhizoremediation studies using long-term PCB-contaminated soil. Environ Sci Poll Res 16:817–829

Malavenda R, Rizzo C, Michaud L, Gerçe B, Bruni V, Syldatk C, Hausmann R, Giudice AL (2015) Biosurfactant production by Arctic and Antarctic bacteria growing on hydrocarbons. Polar Biol 38:1565–1574

Martin BC, George SJ, Price CA, Ryan MH, Tibbett M (2014) The role of root exuded low molecular weight organic anions in facilitating petroleum hydrocarbon degradation: current knowledge and future directions. Sci Total Environ 472:642–653

Monclus R, Villar M, Barbaroux C, Bastien C, Fichot R, Delmotte FM, Delay D, Petit JM, Brechet C, Dreyer E, Brignolas F (2009) Productivity, water-use efficiency and tolerance to moderate water deficit correlate in 33 poplar genotypes from a *Populus deltoides* x *Populus trichocarpa* F1 progeny. Tree Physiol 29:1329–1339

Moore FP, Barac T, Borremans B, Oeyen L, Vangronsveld J, Van der Lelie D, Campbell CD, Moore ER (2006) Endophytic bacterial diversity in poplar trees growing on a BTEX-contaminated site: the characterisation of isolates with potential to enhance phytoremediation. Syst Appl Microbiol 29:539–556

Mukherjee S (2014) Successional and spatial patterns of bacterial communities in hydrocarbon-contaminated soils and *Populus* rhizosphere. Dissertationes Biocentri Viikki Universitatis Helsingiensis 27/2014, Helsinki, Finland

Mukherjee S, Heinonen M, Dequvire M, Sipilä T, Pulkkinen P, Yrjälä K (2013) Secondary succession of bacterial communities and co-occurrence of phylotypes in oil-polluted *Populus* rhizosphere. Soil Biol Biochem 58:188–197

Mukherjee S, Juottonen H, Siivonen P, Quesada CL, Tuomi P, Pulkkinen P, Yrjälä K (2014) Spatial patterns of microbial diversity and activity in an aged creosote-contaminated site. ISME J 8:2131–2142

Mukherjee S, Sipilä T, Pulkkinen P, Yrjälä K (2015) Secondary successional trajectories of structural and catabolic bacterial communities in oil-polluted soil planted with hybrid poplar. Mol Ecol 24:628–642

Mulligan CN (2009) Recent advances in the environmental applications of biosurfactants. Curr Opin Colloid Interphace Sci 14:372–378

Ni Chadhain SM, Norman RS, Pesce KV, Kukor JJ, Zylstra GJ (2006) Microbial dioxygenase gene population shifts during polycyclic aromatic hydrocarbon biodegradation. Appl Environ Microbiol 72:4078–4087

Nie M, Yang Q, Jiang LF, Fang CM, Chen JK, Li B (2010) Do plants modulate biomass allocation in response to petroleum pollution. Biol Lett 6:811–814

Olson PE, Castro A, Joern M, DuTeau NM, Pilon-Smits EA, Reardon KF (2007) Comparison of plant families in a greenhouse phytoremediation study on an aged polycyclic aromatic hydrocarbon–contaminated soil. J Environ Qual 36:1461–1469

Palmroth MR, Pichtel J, Puhakka JA (2002) Phytoremediation of subarctic soil contaminated with diesel fuel. Bioresour Technol 84:221–228

Pilon-Smits E (2005) Phytoremediation. Annu Rev Plant Biol 56:15–39

Podell S, Emerson JB, Jones CM, Ugalde JA, Welch S, Heidelberg KB, Banfield JF, Allen EE (2014) Seasonal fluctuations in ionic concentrations drive microbial succession in a hypersaline lake community. ISME J 8:979–990

Pulkkinen P (2014) Finnish Broadcasting Company (Yle), Science (Tiede), 30 January 2014. http://yle.fi/uutiset/3-7060319

Ramos J, Duque E, Van Dillewjin P, Daniels C, Krell T, Espinosa-Urgel M, Ramos-González M, Rodríguez S, Matilla M, Wittich R (2010) Removal of hydrocarbons and other related chemicals via the rhizosphere of plants. In: Timmis KN, McGenity TJ, van der Meer JR, de Lorenzo V (eds) Handbook of hydrocarbon and lipid microbiology. Springer, Berlin, pp 2575–2581

Ravel C, Courty C (1997) Beneficial effects of *Neotyphodium lolii* on the growth and the water status in perennial ryegrass cultivated under nitrogen deficiency or drought stress. Agronomie 17:173–181

Read D, Bengough A, Gregory P, Crawford JW, Robinson D, Scrimgeour C, Young IM, Zhang K, Zhang X (2003) Plant roots release phospholipid surfactants that modify the physical and chemical properties of soil. New Phytol 157:315–326

Rogers A, McDonald K, Muehlbauer MF, Hoffman A, Koenig K, Newman L, Taghavi S, Lelie D (2012) Inoculation of hybrid poplar with the endophytic bacterium *Enterobacter* sp. 638 increases biomass but does not impact leaf level physiology. GCB Bioenergy 4:364–370

Schiewer S, Niemeyer T (2006) Soil heating and optimized nutrient addition for accelerating bioremediation in cold climates. Polar Rec 42:23–31

Schloss PD, Westcott SL, Ryabin T, Hall JR, Hartmann M, Hollister EB, Lesniewski RA, Oakley BB, Parks DH, Robinson CJ, Sahl JW, Stres B, Thallinger GG, Van Horn DJ, Weber CF (2009) Introducing mothur: open-source, platform-independent, community-supported software for describing and comparing microbial communities. Appl Environ Microbiol 75:7537–7541

Schnabel WE, White DM (2001) The effect of mycorrhizal fungi on the fate of aldrin: phytoremediation potential. Int J Phytoremediat 3:221–241

Schöftner P, Watzinger A, Holzknecht P, Wimmer B, Reichenauer TG (2016) Transpiration and metabolisation of TCE by willow plants – a pot experiment. Int J Phytoremediat 18:686–692

Shi Y, Lou K, Li C (2009) Promotion of plant growth by phytohormone-producing endophytic microbes of sugar beet. Biol Fertil Soils 45:645–653

Siciliano SD, Fortin N, Mihoc A, Wisse G, Labelle S, Beaumier D, Ouellette D, Roy R, Whyte LG, Banks MK (2001) Selection of specific endophytic bacterial genotypes by plants in response to soil contamination. Appl Environ Microbiol 67:2469–2475

Singer AC, Crowley DE, Thompson IP (2003) Secondary plant metabolites in phytoremediation and biotransformation. Trends Biotechnol 21:123–130

Singh BK, Millard P, Whiteley AS, Murrell JC (2004) Unravelling rhizosphere–microbial interactions: opportunities and limitations. Trends Microbiol 12:386–393

Singh N, Sethunathan N, Megharaj M, Naidu R (2008) Bioavailability of sorbed pesticides to bacteria: an overview. Dev Soil Sci 32:73–82

Sipilä T (2009) Plasmids and aromatic degradation in *Sphingomonas* for bioremediation: aromatic ring cleavage genes in soil and rhizosphere. Dissertationes bioscientiarum molecularium Universitatis Helsingiensis in Viikki 31/2009, Helsinki, Finland

Sipilä TP, Riisiö H, Yrjälä K (2006) Novel upper meta-pathway extradiol dioxygenase gene diversity in polluted soil. FEMS Microbiol Ecol 58:134–144

Sipilä TP, Keskinen A, Åkerman M, Fortelius C, Haahtela K, Yrjälä K (2008) High aromatic ring-cleavage diversity in birch rhizosphere: PAH treatment-specific changes of IE 3 group extradiol dioxygenases and 16S rRNA bacterial communities in soil. ISME J 2:968–981

Sipilä TP, Väisänen P, Paulin L, Yrjälä K (2010) *Sphingobium* sp. HV3 degrades both herbicides and polyaromatic hydrocarbons using ortho-and meta-pathways with differential expression shown by RT-PCR. Biodegradation 21:771–784

Slater H, Gouin T, Leigh MB (2011) Assessing the potential for rhizoremediation of PCB contaminated soils in northern regions using native tree species. Chemosphere 84:199–206

Song B, Ward BB (2005) Genetic diversity of benzoyl coenzyme A reductase genes detected in denitrifying isolates and estuarine sediment communities. Appl Environ Microbiol 71:2036–2045

Sterky F, Bhalerao RR, Unneberg P, Segerman B, Nilsson P, Brunner AM, Charbonnel-Campaa L, Lindvall JJ, Tandre K, Strauss SH, Sundberg B, Gustafsson P, Uhlen M, Bhalerao RP, Nilsson O, Sandberg G, Karlsson J, Lundeberg J, Jansson S (2004) A *Populus* EST resource for plant functional genomics. Proc Natl Acad Sci U S A 101:13951–13956

Sun Y, Cheng Z, Glick BR (2009) The presence of a 1-aminocyclopropane-1-carboxylate (ACC) deaminase deletion mutation alters the physiology of the endophytic plant growth-promoting bacterium *Burkholderia phytofirmans* PsJN. FEMS Microbiol Lett 296:131–136

Taghavi S, Garafola C, Monchy S, Newman L, Hoffman A, Weyens N, Barac T, Vangronsveld J, van der Lelie D (2009) Genome survey and characterization of endophytic bacteria exhibiting a beneficial effect on growth and development of poplar trees. Appl Environ Microbiol 75:748–757

Tanaka S, Brentner LB, Merchie KM, Schnoor JL, Yoon JM, Aken BV (2007) Analysis of gene expression in poplar trees (*Populus deltoides* × *nigra*, DN34) exposed to the toxic explosive hexahydro-1, 3, 5-trinitro-1, 3, 5-triazine (RDX). Int J Phytoremediat 9:15–30

Taylor PM, Medd JM, Schoenborn L, Hodgson B, Janssen PH (2002) Detection of known and novel genes encoding aromatic ring-hydroxylating dioxygenases in soils and in aromatic hydrocarbon-degrading bacteria. FEMS Microbiol Lett 216:61–66

Tervahauta AI, Fortelius C, Tuomainen M, Åkerman ML, Rantalainen K, Sipilä T, Lehesranta SJ, Koistinen KM, Kärenlampi S, Yrjälä K (2009) Effect of birch (*Betula* spp.) and associated rhizoidal bacteria on the degradation of soil polyaromatic hydrocarbons, PAH-induced changes in birch proteome and bacterial community. Environ Poll 157:341–346

Tesar M, Reichenauer TG, Sessitsch A (2002) Bacterial rhizosphere populations of black poplar and herbal plants to be used for phytoremediation of diesel fuel. Soil Biol Biochem 34:1883–1892

Thomas F, Cébron A (2016) Short-term rhizosphere effect on available carbon sources, phenanthrene degradation, and active microbiome in an aged-contaminated industrial soil. Front Microbiol 7:1–15

Thompson PL, Ramer LA, Guffey AP, Schnoor JL (1998) Decreased transpiration in poplar trees exposed to 2,4-6-trinitrotoluene. Environ Sci Technol 17:902–906

Tibbett M, Ryan M, Kertesz MA (2012) Rhizosphere 3: where plants meet soils down-under. Plant Soil 358:1–5

Torsvik V, Øvreås L (2002) Microbial diversity and function in soil: from genes to ecosystems. Curr Opin Microbiol 5:240–245

Van Aken B, Yoon JM, Just CL, Schnoor JL (2004) Metabolism and mineralization of hexahydro-1, 3, 5-trinitro-1, 3, 5-triazine inside poplar tissues (*Populus deltoides* × *nigra* DN-34). Environ Sci Technol 38:4572–4579

Van Dillewijn P, Couselo JL, Corredoira E, Delgado A, Wittich R, Ballester A, Ramos JL (2008) Bioremediation of 2, 4, 6-trinitrotoluene by bacterial nitroreductase expressing transgenic aspen. Environ Sci Technol 42:7405–7410

Vervaeke P, Luyssaert S, Mertens J, Meers E, Tack F, Lust N (2003) Phytoremediation prospects of willow stands on contaminated sediment: a field trial. Environ Poll 126:275–282

Walker LR, Del Moral R (2003) Primary succession and ecosystem rehabilitation. University Press, Cambridge

Wertz S, Degrange V, Prosser JI, Poly F, Commeaux C, Guillaumaud N, Le Roux X (2007) Decline of soil microbial diversity does not influence the resistance and resilience of key soil microbial functional groups following a model disturbance. Environ Microbiol 9:2211–2219

Weyens N, Van Der Lelie D, Artois T, Smeets K, Taghavi S, Newman L, Carleer R, Vangronsveld J (2009) Bioaugmentation with engineered endophytic bacteria improves contaminant fate in phytoremediation. Environ Sci Technol 43:9413–9418

Weyens N, Beckers B, Schellingen K, Ceulemans R, Van der Lelie D, Newman L, Taghavi S, Carleer R, Vangronsveld J (2015) The potential of the Ni-resistant TCE-degrading *Pseudomonas putida* W619-TCE to reduce phytotoxicity and improve phytoremediation efficiency of poplar cuttings on a Ni-TCE co-contamination. Int J Phytoremediat 17:40–48

Wullschleger SD, Jansson S, Taylor G (2002) Genomics and forest biology: *Populus* emerges as the perennial favorite. Plant Cell 14:2651–2655

Yrjälä K (1997) Polyphasic taxonomy, ecology, and genetics of plasmid-encoded sromatic cmp-pathway of dtrain HV3: a *Sphingomonas* sp. Dissertationes Biocentri Viikki Universitatis Helsingiensis 14/1997, Helsinki, Finland

Yrjälä K, Katainen R, Jurgens G, Saarela U, Saano A, Romantschuk M, Fritze H (2004) Wood ash fertilization alters the forest humus Archaea community. Soil Biol Biochem 36:199–201

Yrjälä K, Keskinen A, Åkerman M, Fortelius C, Sipilä TP (2010a) The rhizosphere and PAH amendment mediate impacts on functional and structural bacterial diversity in sandy peat soil. Environ Poll 158:1680–1688

Yrjälä K, Mancano G, Fortelius C, Åkerman M-L, Sipilä TP (2010b) The incidence of Burkholderia in epiphytic and endophytic bacterial cenoses in hybrid aspen grown on sandy peat. Boreal Environ Res 15:81–96

Yrjälä K, Tuomivirta T, Juottonen H, Putkinen A, Lappi K, Tuittila E, Penttilä T, Minkkinen K, Laine J, Peltoniemi K (2011) CH4 production and oxidation processes in a boreal fen ecosystem after long-term water table drawdown. Global Change Biol 17:1311–1320

Zamioudis C, Pieterse CM (2012) Modulation of host immunity by beneficial microbes. Mol Plant-Microbe Interact 25:139–150